Biomechatronics

Biomechatronics

Marko B. Popovic

ELSEVIER

ACADEMIC PRESS

An imprint of Elsevier

Academic Press is an imprint of Elsevier
125 London Wall, London EC2Y 5AS, United Kingdom
525 B Street, Suite 1650, San Diego, CA 92101, United States
50 Hampshire Street, 5th Floor, Cambridge, MA 02139, United States
The Boulevard, Langford Lane, Kidlington, Oxford OX5 1GB, United Kingdom

Notices

Knowledge and best practice in this field are constantly changing. As new research and experience broaden our understanding, changes in research methods, professional practices, or medical treatment may become necessary.

Practitioners and researchers must always rely on their own experience and knowledge in evaluating and using any information, methods, compounds, or experiments described herein. In using such information or methods they should be mindful of their own safety and the safety of others, including parties for whom they have a professional responsibility.

To the fullest extent of the law, neither the Publisher nor the authors, contributors, or editors, assume any liability for any injury and/or damage to persons or property as a matter of products liability, negligence or otherwise, or from any use or operation of any methods, products, instructions, or ideas contained in the material herein.

Library of Congress Cataloging-in-Publication Data
A catalog record for this book is available from the Library of Congress

British Library Cataloguing-in-Publication Data
A catalogue record for this book is available from the British Library

ISBN 978-0-12-812939-5

For information on all Academic Press publications
visit our website at https://www.elsevier.com/books-and-journals

Publisher: Mara Conner
Acquisition Editor: Chris Katsaropoulos
Editorial Project Manager: Andrae Akeh
Production Project Manager: Surya Narayanan Jayachandran
Cover Designer: Miles Hitchen

Typeset by SPi Global, India

Contents

Contributors

Yunus Ziya Arslan Istanbul University, Istanbul, Turkey

Philipp Beckerle Technische Universität Dortmund, Dortmund; Technische Universität Darmstadt, Darmstadt, Germany

Matthew P. Bowers Worcester Polytechnic Institute, Worcester, MA, United States

Pinar Boyraz Chalmers University of Technology, Gothenburg, Sweden; Istanbul Technical University, Istanbul, Turkey

Ana Djuric Wayne State University, Detroit, MI, United States

Ivo Dobrev University Hospital Zurich, University Zurich, Zürich, Switzerland

Gregory Fischer Worcester Polytechnic Institute, Worcester, MA, United States

Adam D. Goodworth University of Hartford, Hartford, CT, United States

Keiji Iramina Kyushu University, Fukuoka, Japan

Songbai Ji Worcester Polytechnic Institute, Worcester, MA, United States

Michelle J. Johnson University of Pennsylvania, Philadelphia, PA, United States

Vukica Jovanovic Old Dominion University, Norfolk, VA, United States

Krishnanand N. Kaipa Old Dominion University, Norfolk, VA, United States

Derya Karabulut Istanbul University, Istanbul, Turkey

Kathleen A. Lamkin-Kennard Rochester Institute of Technology, Rochester, NY, United States

Minas Liarokapis The University of Auckland, Auckland, New Zealand

Ming Luo Stanford University, Stanford, CA, United States

Jun Ohta Nara Institute of Science and Technology (NAIST), Nara, Japan

Cagdas Onal Worcester Polytechnic Institute, Worcester, MA, United States

Faruk Ortes Istanbul University, Istanbul, Turkey

Taskin Padir Northeastern University, Boston, MA, United States

Marko B. Popovic Worcester Polytechnic Institute, Worcester, MA, United States

Dmitry A. Sinyukov Northeastern University, Boston, MA, United States

Hiroyuki Tashiro Kyushu University, Fukuoka, Japan

Yasuo Terasawa Nidek Co., Ltd., Aichi, Japan

Kimberly Tetreault Gaylord Hospital, Wallingford, CT, United States

Karen L. Troy Worcester Polytechnic Institute, Worcester, MA, United States

Steffen Willwacher German Sport University Cologne, Cologne, Germany

1 Introduction

Marko B. Popovic
WORCESTER POLYTECHNIC INSTITUTE, WORCESTER, MA, UNITED STATES

Born at the turn of the 21st century, the word biomechatronics refers to an interdisciplinary field that closely merges biological and mechatronics systems.

Here, mechatronics refers to technology combining electronics and mechanical engineering according to its dictionary definition.

The word biomechatronics, also initially written as bio-mechatronics, has been in use since the late 1990s. Since then, numerous authors have attempted to provide a concise and accurate definition that would most appropriately describe this popular field. Unfortunately, the majority of them failed to some extent either because they artificially limited its notion to only a very specialized subfield, typically, closely related to their own research theme or topic, or because they regrettably missed to fully recognize a fine ambiguity in the level of unification of biological and synthetic systems, here, specifically mechatronics systems.

For example, consider a human driving a car. Is that a biomechatronic system? Probably not, if the car is just an old-fashioned 20th-century car, a machine which is merely mechanically operated by human driver. And probably yes, if the car is an advanced 21st-century car which interfaces more closely with human driver and directly affects the process of driving through a controlled feedback loop. Imagine that this car can sense its environment, own state, and state of the human driver. It can understand user intent, has different levels of communication with a user, and can provide automatic breaking or steering when necessary. For example, interior car cameras can decipher user's attention and direction in which the driver is glimpsing at, based on the orientation of the driver's head and location of irises and eye pupils. Moreover, tremor in operation of steering wheel or gas pedal can be removed or even car seat and specifically neck support could change stiffness levels based on vibration levels introduced by different road conditions. Clearly, this is a prime example of biomechatronic system and human-machine merging.

Hence, biomechatronics is quite a wide field covering every system that closely relates biological and mechatronics system within a functionally unified entity.

First wave of usage of this word was rather naturally focused on advanced human prosthetic and brace technologies, specifically robots that are either attached to or wrapped around the human body as well as on advanced human rehabilitation technologies. However, biomechatronic practices have also been advanced, as it is probably less known, in the context of robotics and automation biotechnologies for bioproduction and control in agriculture, food processing, and pharmaceutical industries.

Biomechatronics. https://doi.org/10.1016/B978-0-12-812939-5.00001-X

Due to its interdisciplinary nature, wide range of applicability, and impact that it creates, biomechatronics holds a promise to be one of the most influential innovative research directions defining the 21st century.

Due to the expertise of its authors, this book is mainly focused on human-centered biomechatronics. All chapters addressing concrete examples of biomechatronic system have a human directly looped in the system. The one exception toward the end of this book is Chapter 17 on bioinspired robots that only indirectly deals with human beings and animals; either through biologically motivated design or through co-robot (collaborative robot) concept. There is also some mention in the concluding Chapter 18 of assistive systems for animals.

This choice of human-centered content certainly does not imply that other areas of biomechatronics are of any lesser value. It is just the available knowledge and bandwidth of this team of authors which lead to the choice of material covered in this book. For readers with more interest, for example, in biomechatronic design in biotechnology one may review the book by Carl-Fredrik Mandenius and Mats Bjorkman published by Willey in 2011 [1].

Kinematics and dynamics relevant for manipulation, locomotion, postural balance, and propulsion in fluids are addressed in Chapter 2. Standard terminology is introduced and several conventional computational and problem-solving techniques are reviewed. These theoretical "tools" and methods are frequently utilized in the Biomechatronics research. However, as a drawback, Chapter 2 can be easily expanded to cover the entire book due to the nature of its content, and hence material presented here is somewhat condensed and limited in scope. Readers who find this material too difficult to follow but are still motivated to learn more on these topics may want to review a well-written book by John J. Craig published by Pearson/Prentice-Hall in 2005 [2]. Actually, a whole variety of textbooks and online video lectures exist that readers might find handy [3–12]. Still further, readers may also opt to completely skip Chapter 2 and without much consequence and overdue focus their attention on other materials presented in this book. When needed, reader may clearly return to Chapter 2 for clarification or look into one of the many useful references on these topics.

Review and comparison of biological and synthetic organismal building blocks in terms of passive material properties, actuation, sensing, and control/intelligence is a grand theme of Chapters 3–5. Chapter 3 discusses biological and synthetic actuators. Chapter 4 addresses biological and synthetic sensors. Chapter 5 focuses on biological and synthetic control/intelligence. All these are quite relevant in the context of biomechatronics research. In order to decide how best to interface with biological system one should clearly know a lot on biological system including its building block elements. Moreover, that knowledge can also serve as a good inspiration to engineer high fidelity synthetic counterparts. Still further, it is always good to have a "rich menu," that is, multiple options on the table in terms of synthetic elements; different problems may require different approaches and corresponding components. Finally, it is a quite intellectually rewarding experience to compare living and engineered systems and their building block elements.

Majority of books focused on robotics follow this actuator-sensor-control organization with good reason. After all, robot is nothing else than synthetic organism. According to "Biomechanics and Robotics" book [13]: "Robotics is the branch of Science, Technology and Art that deals with robots, that is, artificial 'organisms' that convey lifelike appearance by physical actions (movements, etc.) and perform tasks autonomously by using an active sensory-control system, or with guidance, typically by teleoperation, in which case, some of the functions of an active sensory-control system are taken over by operator."

Ever since the end of the 19th century we have been engineering "things" composed of sensors, actuators, and control systems intertwined such that these "things" have a lifelike appearance and often useful practical applications. These "things" are synthetic organisms, also called robots.

Nikola Tesla, Serbian inventor who lived most of his life in the United States, is credited with the creation of the first robot at the end of the 19th century. He invented a remote control and patented a radio-controlled robot-boat (referred to as teleautomaton) on November 8, 1898. In September 1898, Tesla demonstrated his invention. He used radio waves to move a robot-boat in a small pool of water in Madison Square Garden in New York City during the Electrical Exhibition of 1898. According to his own words, inspiration for this invention was the human organism; instead of eyes, the boat used small antennas that would pick up radio ways. This sensory input was then processed by simple AND logical control gate (approximately half century before the invention of transistor) instead of by brain and appropriate propulsion would be accordingly created [13].

The word robot was eventually introduced in the play *R.U.R. (Rossum's Universal Robots)* by Czech author Karel Capek published in 1920. Karel Capek's robots were built in the human form. They were allowed to physically interact with humans and operate in the same workplaces as humans do. Karel Capek would probably be amused to learn that such robots are now also called cobots or co-robots (from collaborative robots).

Chapter 6 addresses neural interfaces, for example, high-resolution peripheral neuromuscular interfaces and cortical microelectrode technologies. These interfaces can be used among others for diagnosis and monitoring or they can be an integral part of the biomechatronics system. One of the critical challenges in the field of advanced prosthesis, braces, exoskeletons/exomusculatures, and even wheelchairs is communication of user intent, that is, biological control command followed by appropriate actuation of synthetic (or engineered) system resulting in corresponding movement of the entire system. Moreover, to close the feedback loop it is also often highly desirable to communicate synthetic sensory output back to human neural control system; problem which in practice proved even harder than communication of biological command. Finally, new directions of research also address neural prosthesis as well as assistive and augmentative brain add-ons, that is, technologies that could deal with various brain-related deficiencies and maybe also fundamentally influence ways how we interface with the digital world, store and process our memories, communicate with each other, how long do we live, and how we perceive ourselves, that is, our likely substantially evolved identities. Clearly, the development of these technologies is likely to create a strong influence on the entire human race. And hence one cannot but wonder about the ethical consequences. However, as usual when technology is concerned, it is often not the technology that is good or evil.

Chapter 7 is focused on synthetic organs, tissues, and support systems. Numerous systems that are already in wide use as well as those that are still being actively developed and researched are addressed in this chapter. Topics include pacemakers, implantable artificial hearts, heart valves, blood vessels/stents, hearing aids, middle ear and cochlear implants, brainstem implants, artificial cornea, intraocular lens, visual prosthesis including retinal, optic nerve, or cortical prosthesis, breast prosthesis, dental implants, artificial skin in the context of wound dressing and cultured skin, artificial dura mater, etc. The cardiopulmonary bypass (CPB), or heart-lung machine, artificial respirator, artificial pancreatic islet, and artificial dialyzer are addressed as examples of support systems. It appears that the accelerated advancement of Health Science and Technology are bringing us closer and closer to the day when the majority of human body parts from macro to micro, that is, cellular or molecular level, will be easily exchangeable with engineered systems made of synthetic or biological elements. To assist that process, concepts addressed in the following chapter might prove to be very critical.

Chapter 8 discusses biomechatronics and robotics at the molecular and cellular levels. Imagine, soon in the future, a new generation of microrobots, nanorobots, and even biohybrid robots that could effectively assist with diagnosis, health monitoring, therapy, surgery, gene therapy, cell manipulation, cancer detection and treatment, drug delivery, tissue engineering, detoxification, etc. They operate at tiny scales where, for example, adhesion, Van der Waals, and capillary forces dominate and where our intuitive understanding of physics at macroscale is not as applicable. Hence, one needs to consider that viscous fluid dynamics at low Reynolds numbers are also referred to as Stokes flows and contemplate on various ratios including advection to diffusion, momentum diffusivity to thermal diffusivity, viscous forces to surface tension, gravity to surface tension, the molecular mean free path to the characteristic system length scale, and so on. What is the propulsion mechanism of these robots? Where is their energy coming from? How are they controlled? What type of actions they might be able to perform? The interested reader may want to review this chapter in great detail to find answers to these and other important questions.

Advanced assistive prosthesis, braces, and exoskeletons/exomusculatures/exosuits are addressed in Chapters 9–11, respectively. The robots that intimately interact with the human body probably necessitate the least introduction as they represent the more traditional biomechatronics systems. In terms of research, profound progress has been made in the last two decades in terms of novel materials and device architectures, rapid prototyping and manufacturing techniques, biological inspiration, novel actuation technologies, sensing, adaptive control and learning as well as human-machine physical and control (invasive or noninvasive, unidirectional or bidirectional) interfaces.

Numerous research highlights in this area clearly exemplify that modern advanced assistive biomechatronics devices can be quite cost effective; no reason to price single actuated joint degree-of-freedom robot as three brand new cars (!). In terms of actual products on the market, progress has been a bit slower with a few great exceptions and most of these devices are still priced unreasonably high (from the perspective of the average actual user).

Apparently, big companies in this field have a lot of "inertia" and they tend to stick to their line of successful products including their high price tags while introducing only minimal changes to refresh their brand and keep up with other big companies that typically do the same. Also, instead of utilizing a substantial portion of their profit on research and development, they tend to just buy smaller companies and their research products and corresponding intellectual property. On the other hand, smaller companies often struggle to break even after several years of intense and expensive R&D phase, followed by efforts to get their product approved with the appropriate code, and subsequent, often unsuccessful, attempts to get onto a market dominated by the big companies as a never-heard-of-brand. Similarly, health insurance companies cannot keep up with such high price tags and they keep finding innovative ways to explain to health insurers why they do not need more expensive robotic assistive devices, often without a good connection with reality. The similar business-driven destiny is shared by many other fields and the solution, as usual, is a more competitive market at the level of big companies, hence no monopolies, which ultimately drives prices down as well as more disruptive technologies introduced by smaller players in this field, which should be probably incentivized by some government mechanisms, as a way to diversify and advance product quality.

Chapter 12 discusses biomechatronics within physical therapy and rehabilitation, for example, support systems for function restoration and short-term assistance. In addition to physical therapy professionals and rehabilitation scientists, engineers also play an important role in physical therapy and rehabilitation. Biomechatronics have been integrated into devices for truly innovative treatments and for highly specific assessments. This chapter describes some of the most important biomechatronic products in this field separately for upper limb therapy, lower limb therapy (primarily gait training), and balance therapy. Within each of these sections is a separate description of design strategy, treatment technologies, and assessment technologies. Finally, a central concept emphasized throughout the chapter is the need for biomechantronics technology to match resources available with individual patient's abilities and limitations.

Everyone, who has ever trained for a sport or has worked as a coach knows that the process of improving physical performance is rather challenging and that there is no unique and most optimal way that works equally well for everyone. It really depends on many factors. Highly skilled, very experienced coaches typically first get to know their trainees well and then carefully gauge when is the right moment to put more attention on a certain type of exercise and what is the right level of that activity. Even with best efforts from both the sides this endeavor may or may not be fruitful. Similarly, in the field of physical therapy and rehabilitation there is no unique and most optimal way that works equally well for all patients at any moment in time. Rather, there is a variety of treatments that can be more or less beneficial depending on patient's physical condition and motivation.

Advanced wheelchairs and other mobility assistance including walkers, crutches, and canes are addressed in Chapter 13. Although it is hard to visualize a wheelchair as super advanced robot, this view may change soon. Imagine a wheelchair that is self-propelled and can move over rough terrain, up and down the stairs, follow another wheelchair or person, plan its best route, navigate through narrow spaces, and find its way through a crowd of people. It could also assist a person in getting up from or lying down on a bed. It could help someone stand up, so they could press the top button in the elevator or get cereal from the top shelf in their kitchen. It could help a person use the restroom. It could also take voice commands, and possibly, it could talk and discuss weather conditions or discuss the best store to buy a new pair of gloves. All these elements have already been successfully tested in practice and it will only take a little while before they become integrated into a single wheelchair. Should not this wheelchair be considered a robot? Decisively, yes. It would be an amazing robot! Large portion of Chapter 13 is also dedicated to addressing the problems of communication of user intent, especially for paralyzed people. Clearly, developments in this field could completely revolutionize human-machine mergence. The wheelchair as a propulsion platform can also be added with robotic arm(s) that can still further assist with numerous tasks and activities of everyday life (ADL). Discussion of this topic is addressed in more detail in Chapter 14.

Chapter 14 discusses several topics. People with disabilities and the elderly often require assistance with feeding tasks and with personal hygiene tasks such as bathing, brushing teeth, using restroom, combing hair, shaving, etc. The first part of Chapter 14 is focused on feeding systems and it also discusses robotic arms attached to a more conventional wheelchair platform. Second part of Chapter 14 discusses robotic nurses and other systems. Finally, the third part of Chapter 14 addresses advanced robotic massage systems as prime examples of biomechatronics systems.

Chapter 15 is dedicated to robotic surgery which is one of the fastest growing areas within the biomechatronics systems, as it provides certain advantages to both surgeons and patients. From the perspective of surgeons the procedure may be improved both in safety and effectiveness, therefore, alleviating the tiredness. The robots can reduce the physical and cognitive effort paid by the surgeon, especially in suture-intensive operations. From the perspective of the patients, robotic surgery usually means less postoperative pain, shorter hospital times, and less damage to tissues. A general overview of robotic surgery is provided while focusing on specific developments on hyper-redundant, continuum, and soft-material robotic platforms. This chapter also provides a wide and comprehensive outlook on the implications of human-machine interaction and autonomy levels in robotic surgery. To explain better the new developments in robotic surgery front, two case studies are selected reporting on the state-of-the-art applications in robotic ear surgery and hyper-redundant semiautonomous robotic platforms.

Chapter 16 is focused on biomechanics and biomechatronics in sports, exercise, and entertainment. Biomechanics is the study of the structure and function of biological systems such as humans, animals, plants, organs, and cells by means of the methods of mechanics. This chapter provides a comprehensive overview of biomechanics experimental procedures, data analysis, modeling, and simulation. This chapter also overviews several examples of advanced biomechatronics systems for sports, exercise, and entertainment.

Biologically inspired robotics is a grand theme of Chapter 17. As previously mentioned, this chapter discusses biologically motivated design and the co-robot (collaborative robot) concept. This chapter is clearly very fundamental for robotics in general. Nature and especially the human beings were the driving inspiration for the development and advancement of robotics from early days. Even nowadays, biological inspiration is driving the progress in robotics away from the 20th century conventional high-gain position controlled rigid robots and decisively toward a new natural soft robots and more biomimetic systems.

Chapter 18 provides a brief summary of questions and answers presented in this book, discusses, and resolves some of the typical misconceptions, and finally, points toward the *new dawn of Biomechatronics*, i.e., future where engineered systems and elements from a micro to macrolevel will easily interface with or exchange the biological ones. This is that "magical" place in our collective evolved future where our defining principles as a human race will change; hopefully, in beneficial directions. This is a very exciting journey and probably the crown of all scientific, health related, and engineering efforts of our civilization.

This book is intended for students at the undergraduate and graduate levels, their course instructors, and researchers in biomechatronics, biomechanics, robotics, and biomedical engineering as well as for the advanced high school students, and the general public. The book is written such as to provide a "big picture" of the 21st-century state-of-the-art science and technology by offering a pedagogical introduction to a large variety of topics, a review of some of the historical developments, and the most up-to-date insights into the modern day biomechatronics as witnessed by a team of leaders in this interdisciplinary but unified field. It is quite worth noting that this team consists of experts from all around the globe and hence it transcends various boundaries and barriers with an overarching goal to better humanity as a whole (Fig. 1.1).

The biomechatronics book may serve as primary textbook in course on biomechatronics at the graduate and advanced undergraduate level by providing a complete account of the majority of popular topics in biomechatronics. The book may be also used in the courses on biologically inspired engineering, biomedical engineering, bioengineering, biotechnology, engineering in medicine, biomechanics, and robotics-related courses in general as part of programs in robotics engineering, computer science, mechanical and electrical engineering, biomedical engineering, health science and technology, bioengineering, biotechnology, etc.

To facilitate its acceptance as part of the standard university curricula Chapters 19 and 20 provide a number of homework problems and some solutions that course instructors and students may find useful.

FIG. 1.1 The Wall Mural (1989) depicting journey into the future and its author Marko B Popovic.

References

[1] C.-F. Mandenius, M. Björkman, Biomechatronic Design in Biotechnology: A Methodology for Development of Biotechnological Products, John Wiley & Sons, 2011.

[2] J.J. Craig, Introduction to Robotics: Mechanics and Control, vol. 3, Pearson/Prentice Hall, Upper Saddle River, NJ, 2005.

[3] L.D. Landau, E.M. Lifshitz, Mechanics: Course of Theoretical Physics, third ed., vol. 1, Butterworth-Heinemann, 1976 ISBN: 9780750628969.

[4] S.T. Thornton, J.B. Marion, Classical Dynamics of Particles and Systems, Cengage Learning India, 2012. ISBN: 9788131518472.

[5] H. Goldstein, J. Safko, C.P. Poole, Classical Mechanics, Pearson, 2013. ISBN: 9781292026558.

[6] D. Kleppner, R.J. Kolenkow, An Introduction to Mechanics, McGraw-Hill, New York, NY, 1973. ISBN: 9780070350489.

[7] H. Asada, J.-J.E. Slotine, Robot Analysis and Control, John Wiley & Sons, 1986.

[8] R.P. Paul, Robot Manipulators: Mathematics, Programming, and Control: The Computer Control of Robot Manipulators, The MIT Press, 1981.

[9] H. Choset, K.M. Lynch, S. Hutchinson, G. Kantor, W. Burgard, L. Kavraki, S. Thrun, Principles of Robot Motion: Theory, Algorithms, and Implementations, MIT Press, 2005.

[10] K. Oussama, "Introduction to Robotics" Video lectures, Stanford Center for Professional Development, Stanford University, 2008. https://www.youtube.com/view_play_list?p=65CC0384A1798ADF (retrieved/accessed 03.09.18).

[11] J. Vandiver, D. Gossard, 2.003SC Engineering Dynamics, Massachusetts Institute of Technology: MIT OpenCourseWare, 2011. https://ocw.mit.edu License: Creative Commons BY-NC-SA.

[12] D. Chakrabarty, P. Dourmashkin, M. Tomasik, A. Frebel, V. Vuletic, 8.01SC Classical Mechanics, Massachusetts Institute of Technology: MIT OpenCourseWare, https://ocw.mit.edu, 2016 License: Creative Commons BY-NC-SA.

[13] M.B. Popovic, "Biomechanics and Robotics", 351 pages, Copyright © 2014. Pan Stanford Publishing Pte. Ltd., Singapore, 2013. doi:10.4032/9789814411387 ISBN 978-981-4411-37-0 (Hardcover), 978-981-4411-38-7 (eBook).

2 Kinematics and Dynamics

Marko B. Popovic, Matthew P. Bowers
WORCESTER POLYTECHNIC INSTITUTE, WORCESTER, MA, UNITED STATES

CHAPTER OUTLINE

2.1 Introduction

Kinematics and dynamics relevant for manipulation, locomotion, postural balance, and propulsion in fluids are addressed in this chapter. Standard terminology is introduced and several conventional computational and problem-solving techniques are reviewed. These theoretical "tools" and methods are frequently utilized in biomechatronics research. However, as a drawback, Chapter 2 can be easily expanded to cover an entire book due to the nature of its content; hence, material presented here is somewhat condensed and limited in scope. Readers who find this material too difficult to follow but

Biomechatronics. https://doi.org/10.1016/B978-0-12-812939-5.00002-1

are still motivated to learn more on these topics may want to review a well-written book by John J. Craig published by Pearson/Prentice-Hall in 2005 [1]. Actually, a whole variety of textbooks and online video lectures exist that readers might find handy [2–11]. Still further, readers may also opt to completely skip this chapter and without much consequence and overdue, focus their attention on other material presented in this book. When needed, reader may clearly return to Chapter 2 for clarification, or look into one of the many useful references on these topics.

2.2 Kinematics

Probably, the simplest example of dynamic system that would suffice to illustrate majority of the concepts we intend to cover in this brief crash course on manipulator (and locomotor) kinematics and dynamics is a planar two-link manipulator in gravitational field depicted in Fig. 2.1. This system has two actuated joints: joint 1 in between ground and link 1, and joint 2 in between link 1 and link 2. Joint 1 is attached to stationary point in the lab inertial frame. We refer to the nonactuated end of manipulator as end effector.

One may draw a parallel between these two-link manipulator and human arm constrained to move in sagittal plane. Joints O1 and O2 relate to shoulder and elbow, and end effector relates to wrist and fist.

2.2.1 Forward Kinematics

Forward kinematics refers to process of obtaining position and velocity of end effector, given the known joint angles and angular velocities. For example, if shoulder and elbow

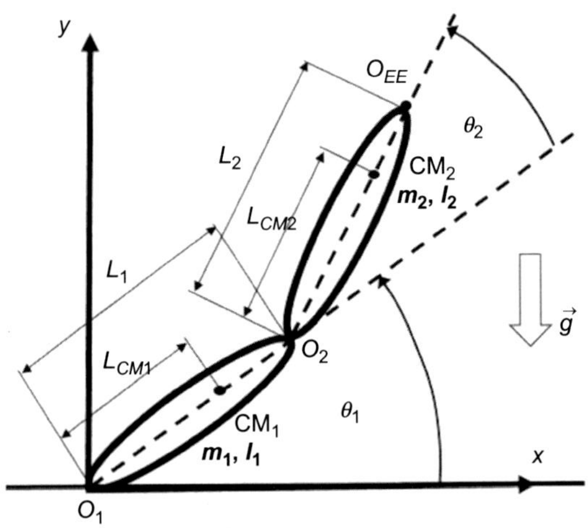

FIG. 2.1 Planar two-link manipulator with two actuated joints O_1, O_2 in gravitational field and with joint O_1 affixed to lab inertial frame. Joints angles, θ_1, θ_2, links dimensions L_1, L_2, center of masses, CM_1, CM_2, proximal distances to center of masses L_{CM1}, L_{CM2}, links masses m_1, m_2, moments of inertia about links' center of masses I_1, I_2, and end-effector O_{EE} are depicted.

joint angles are given for arm in sagittal plane, the goal is to find Cartesian coordinates of wrist/fist.

In our particular case we are looking for transformation

$$\theta_1, \theta_2 \rightarrow x_{EE}, y_{EE} \tag{2.1}$$

In this example, joint 1 is located at the origin. Hence, coordinates of joint 2 are

$$x_2 = L_1 \cos\theta_1 \tag{2.2}$$

$$y_2 = L_1 \sin\theta_1 \tag{2.3}$$

Therefore, coordinates of end effector are

$$x_{EE} = x_2 + L_2 \cos(\theta_1 + \theta_2) = L_1 \cos\theta_1 + L_2 \cos(\theta_1 + \theta_2) \tag{2.4}$$

$$y_{EE} = y_2 + L_2 \sin(\theta_1 + \theta_2) = L_1 \sin\theta_1 + L_2 \sin(\theta_1 + \theta_2) \tag{2.5}$$

By taking time derivative one could now obtain end-effector velocity

$$\dot{x}_{EE} = -\dot{\theta}_1 L_1 \sin\theta_1 - (\dot{\theta}_1 + \dot{\theta}_2) L_2 \sin(\theta_1 + \theta_2) \tag{2.6}$$

$$\dot{y}_{EE} = \dot{\theta}_1 L_1 \cos\theta_1 + (\dot{\theta}_1 + \dot{\theta}_2) L_2 \cos(\theta_1 + \theta_2) \tag{2.7}$$

If instead of planar system, we dealt with more complex three-dimensional (3D) system, this calculation would be clearly a bit more demanding. Still further, it may happen that the number of degrees of freedom of "input" variables and "output" variables do not match. For example, both joints could be two-degrees-of-freedom joints defining a three-degree-of-freedom trajectory of point-like end effector. Hence, based on joints' four-degrees-of-freedom "input", one would obtain only three-degrees-of-freedom "output." This redundancy would simply mean that there may be infinitely many joint angle solutions for single position of end effector. This can be visualized as infinitely many possible locations of joint 2 along the circle obtained as the intersection of two spheres centered at joint 1 and end effector with radii equal to appropriate links' lengths (Fig. 2.2).

The *kinematic state* or just *state* of manipulator is typically (like in physics) fully defined with joint angles and angular velocities. In certain cases, like here, state is also fully defined with end-effector position and velocity; however this is not true, as discussed above, for general case.

2.2.2 Inverse Kinematics

Inverse kinematics is just opposite to forward kinematics. It refers to process of obtaining joint angles from known coordinates of end effector. For example, if wrist/fist Cartesian coordinates are known, the goal is to decipher shoulder and elbow joint angles for arm in sagittal plane. Clearly if "input" and "output" degrees of freedom are not matched, the inverse kinematics may be futile as there may be either infinitely many or no solutions.

In our particular case we are looking for transformation

$$x_{EE}, y_{EE} \rightarrow \theta_1, \theta_2 \tag{2.8}$$

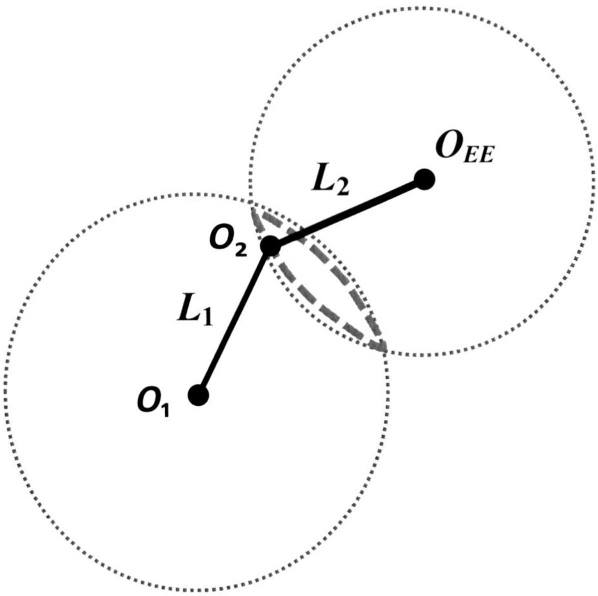

FIG. 2.2 Circle obtained as intersection of two spheres centered at joint 1 and end effector with radii equal to appropriate links' lengths.

Note that even in our case with well-matched degrees of freedom, there may be no solution, one single solution or two solutions for inverse kinematics corresponding, respectively, to no intersection between circles centered at joint 1 and end effector, and with radii equal to appropriate links' lengths, two circles touching each other defining singular position of joint 2, and two intersecting points defining two possible solutions for joint 2 (Fig. 2.3).

This can be quantified by setting equations for two circles as

$$(x_{EE} - x_2)^2 + (y_{EE} - y_2)^2 = L_2^2 \tag{2.9}$$

$$(x_1 - x_2)^2 + (y_1 - y_2)^2 = L_1^2 \tag{2.10}$$

and solving for x_2, y_2. If solution exists, one can then solve for θ_1, θ_2.

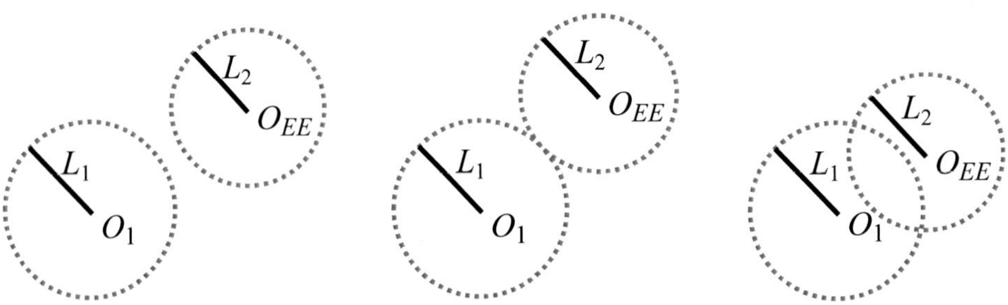

FIG. 2.3 No intersection between circles centered at joint 1 and end effector, two circles touching each other defining singular position of joint 2, and two intersecting points defining two possible solutions for joint 2.

Another approach would be to use law of cosine and find angle $\measuredangle O_1O_2O_{EE}$ as

$$\measuredangle O_1O_2O_{EE} = \cos^{-1}\frac{L_1^2 + L_2^2 - \left(x_{EE}^2 + y_{EE}^2\right)}{2L_1L_2} \tag{2.11}$$

and then obtain

$$\theta_2 = \pi - \measuredangle O_1O_2O_{EE} \tag{2.12}$$

Similarly, one could use law of cosine and obtain

$$\measuredangle O_{EE}O_1O_2 = \cos^{-1}\frac{L_1^2 + \left(x_{EE}^2 + y_{EE}^2\right) - L_2^2}{2L_1\sqrt{x_{EE}^2 + y_{EE}^2}} \tag{2.13}$$

and then obtain

$$\theta_1 = \tan^{-1}\frac{y_{EE}}{x_{EE}} - \measuredangle O_{EE}O_1O_2 \tag{2.14}$$

Finally, expressing everything in terms of known quantities gives

$$\theta_1 = \tan^{-1}\frac{y_{EE}}{x_{EE}} - \cos^{-1}\frac{L_1^2 + \left(x_{EE}^2 + y_{EE}^2\right) - L_2^2}{2L_1\sqrt{x_{EE}^2 + y_{EE}^2}} \tag{2.15}$$

$$\theta_2 = \pi - \cos^{-1}\frac{L_1^2 + L_2^2 - \left(x_{EE}^2 + y_{EE}^2\right)}{2L_1L_2} \tag{2.16}$$

2.2.3 Jacobian of Coordinate Transformation

Next, we want to introduce forward kinematics in infinitesimal form. In our particular case we are looking for

$$d\theta_1, d\theta_2 \rightarrow dx_{EE}, dy_{EE} \tag{2.17}$$

In the context of arm constrained to sagittal plane analogy, here we want to calculate by how much wrist/fist Cartesian coordinates change based on infinitesimal change of shoulder and elbow joint angles.

Following definition of total derivative

$$dx_{EE} = \frac{\partial x_{EE}}{\partial \theta_1}d\theta_1 + \frac{\partial x_{EE}}{\partial \theta_2}d\theta_2 = -[L_1\sin\theta_1 + L_2\sin(\theta_1+\theta_2)]d\theta_1 - L_2\sin(\theta_1+\theta_2)d\theta_2 \tag{2.18}$$

$$dy_{EE} = \frac{\partial y_{EE}}{\partial \theta_1}d\theta_1 + \frac{\partial y_{EE}}{\partial \theta_2}d\theta_2 = [L_1\cos\theta_1 + L_2\cos(\theta_1+\theta_2)]d\theta_1 + L_2\cos(\theta_1+\theta_2)d\theta_2 \tag{2.19}$$

Similarly, inverse kinematics version in infinitesimal form can be expressed as

$$d\theta_1 = \frac{\partial \theta_1}{\partial x_{EE}}dx_{EE} + \frac{\partial \theta_1}{\partial y_{EE}}dy_{EE} \tag{2.20}$$

$$d\theta_2 = \frac{\partial \theta_2}{\partial x_{EE}}dx_{EE} + \frac{\partial \theta_2}{\partial y_{EE}}dy_{EE} \tag{2.21}$$

By substituting $d\theta_1$ and $d\theta_2$ in the above equation for dx_{EE}

$$dx_{EE} = \frac{\partial x_{EE}}{\partial \theta_1}\left(\frac{\partial \theta_1}{\partial x_{EE}} dx_{EE} + \frac{\partial \theta_1}{\partial y_{EE}} dy_{EE}\right) + \frac{\partial x_{EE}}{\partial \theta_2}\left(\frac{\partial \theta_2}{\partial x_{EE}} dx_{EE} + \frac{\partial \theta_2}{\partial y_{EE}} dy_{EE}\right) \tag{2.22}$$

$$dx_{EE} = \left(\frac{\partial x_{EE}}{\partial \theta_1}\frac{\partial \theta_1}{\partial x_{EE}} + \frac{\partial x_{EE}}{\partial \theta_2}\frac{\partial \theta_2}{\partial x_{EE}}\right) dx_{EE} + \left(\frac{\partial x_{EE}}{\partial \theta_1}\frac{\partial \theta_1}{\partial y_{EE}} + \frac{\partial x_{EE}}{\partial \theta_2}\frac{\partial \theta_2}{\partial y_{EE}}\right) dy_{EE} \tag{2.23}$$

one obtains two equalities

$$\frac{\partial x_{EE}}{\partial \theta_1}\frac{\partial \theta_1}{\partial x_{EE}} + \frac{\partial x_{EE}}{\partial \theta_2}\frac{\partial \theta_2}{\partial x_{EE}} = 1 \tag{2.24}$$

$$\frac{\partial x_{EE}}{\partial \theta_1}\frac{\partial \theta_1}{\partial y_{EE}} + \frac{\partial x_{EE}}{\partial \theta_2}\frac{\partial \theta_2}{\partial y_{EE}} = 0 \tag{2.25}$$

Similarly, by substituting $d\theta_1$ and $d\theta_2$ in the above equation for dy_{EE}

$$dy_{EE} = \frac{\partial y_{EE}}{\partial \theta_1}\left(\frac{\partial \theta_1}{\partial x_{EE}} dx_{EE} + \frac{\partial \theta_1}{\partial y_{EE}} dy_{EE}\right) + \frac{\partial y_{EE}}{\partial \theta_2}\left(\frac{\partial \theta_2}{\partial x_{EE}} dx_{EE} + \frac{\partial \theta_2}{\partial y_{EE}} dy_{EE}\right) \tag{2.26}$$

$$dy_{EE} = \left(\frac{\partial y_{EE}}{\partial \theta_1}\frac{\partial \theta_1}{\partial x_{EE}} + \frac{\partial y_{EE}}{\partial \theta_2}\frac{\partial \theta_2}{\partial x_{EE}}\right) dx_{EE} + \left(\frac{\partial y_{EE}}{\partial \theta_1}\frac{\partial \theta_1}{\partial y_{EE}} + \frac{\partial y_{EE}}{\partial \theta_2}\frac{\partial \theta_2}{\partial y_{EE}}\right) dy_{EE} \tag{2.27}$$

one obtains other two equalities

$$\frac{\partial y_{EE}}{\partial \theta_1}\frac{\partial \theta_1}{\partial x_{EE}} + \frac{\partial y_{EE}}{\partial \theta_2}\frac{\partial \theta_2}{\partial x_{EE}} = 0 \tag{2.28}$$

$$\frac{\partial y_{EE}}{\partial \theta_1}\frac{\partial \theta_1}{\partial y_{EE}} + \frac{\partial y_{EE}}{\partial \theta_2}\frac{\partial \theta_2}{\partial y_{EE}} = 1 \tag{2.29}$$

Eqs. (2.18) and (2.19) can be expressed in matrix notation as

$$\begin{pmatrix} dx_{EE} \\ dy_{EE} \end{pmatrix} = \begin{pmatrix} \frac{\partial x_{EE}}{\partial \theta_1} & \frac{\partial x_{EE}}{\partial \theta_2} \\ \frac{\partial y_{EE}}{\partial \theta_1} & \frac{\partial y_{EE}}{\partial \theta_2} \end{pmatrix} \begin{pmatrix} d\theta_1 \\ d\theta_2 \end{pmatrix} \tag{2.30}$$

and we refer to the above 2×2 matrix as *Jacobian* of $\theta_1,\ \theta_2 \rightarrow x_{EE},\ y_{EE}$ coordinate transformation:

$$\boldsymbol{J}_{\theta_1,\theta_2\rightarrow x_{EE},y_{EE}} = \begin{pmatrix} \frac{\partial x_{EE}}{\partial \theta_1} & \frac{\partial x_{EE}}{\partial \theta_2} \\ \frac{\partial y_{EE}}{\partial \theta_1} & \frac{\partial y_{EE}}{\partial \theta_2} \end{pmatrix} \tag{2.31}$$

Similarly, the Jacobian of $x_{EE},\ y_{EE} \rightarrow \theta_1,\ \theta_2$ coordinate transformation is defined as

$$\boldsymbol{J}_{x_{EE},y_{EE}\rightarrow\theta_1,\theta_2} = \begin{pmatrix} \frac{\partial \theta_1}{\partial x_{EE}} & \frac{\partial \theta_1}{\partial y_{EE}} \\ \frac{\partial \theta_2}{\partial x_{EE}} & \frac{\partial \theta_2}{\partial y_{EE}} \end{pmatrix} \tag{2.32}$$

Equalities (2.24), (2.25), (2.28), and (2.29) can now prove useful to show that

$$\boldsymbol{J}_{\theta_1,\theta_2\to x_{EE},y_{EE}}\boldsymbol{J}_{x_{EE},y_{EE}\to\theta_1,\theta_2}=\boldsymbol{J}_{x_{EE},y_{EE}\to\theta_1,\theta_2}\boldsymbol{J}_{\theta_1,\theta_2\to x_{EE},y_{EE}}=\begin{pmatrix}1&0\\0&1\end{pmatrix}\equiv\mathbf{1} \tag{2.33}$$

$$\left(\boldsymbol{J}_{\theta_1,\theta_2\to x_{EE},y_{EE}}\right)^{-1}=\boldsymbol{J}_{x_{EE},y_{EE}\to\theta_1,\theta_2} \tag{2.34}$$

$$\left(\boldsymbol{J}_{x_{EE},y_{EE}\to\theta_1,\theta_2}\right)^{-1}=\boldsymbol{J}_{\theta_1,\theta_2\to x_{EE},y_{EE}} \tag{2.35}$$

Jacobian of coordinate transformation $q_1, \ldots, q_{N_q}\to f_1, \ldots, f_{N_f}$ with $N_f\le N_q$ is defined as

$$\begin{pmatrix}df_1\\\vdots\\df_{N_f}\end{pmatrix}=\begin{pmatrix}\frac{\partial f_1}{\partial q_1}&\cdots&\frac{\partial f_1}{\partial q_{N_q}}\\\vdots&\ddots&\vdots\\\frac{\partial f_{N_f}}{\partial q_1}&\cdots&\frac{\partial f_{N_f}}{\partial q_{N_q}}\end{pmatrix}\begin{pmatrix}dq_1\\\vdots\\dq_{N_q}\end{pmatrix}=\boldsymbol{J}_{q_1,\ldots,q_{N_q}\to f_1,\ldots,f_{N_f}}\begin{pmatrix}dq_1\\\vdots\\dq_{N_q}\end{pmatrix}$$

2.3 Dynamics

2.3.1 Virtual Work, "Inverse" and "Forward" Dynamics, Inverse and Forward Dynamics

Till now we have discussed only kinematics quantities. Next, we turn to dynamics of system in the context of joint torques and end-effector's forces. As a warm-up exercise let us relate joint torques and end-effector forces in massless limit and with no frictional or other energy losses.

Quite generally, the infinitesimal work by joints will be utilized to infinitesimally change kinetic and potential energy of the system, also some of that energy will be lost due to friction, and finally, some amount will be transferred to end effector such that end effector can perform infinitesimal work on the environment. This statement can be quantified as

$$dW_{joints}=dK+dU_{gravity}+dW_{friction}+dW_{end\ effector} \tag{2.36}$$

In massless limit and with no frictional losses this becomes

$$dW_{joints}=dW_{end\ effector} \tag{2.37}$$

Eq. (2.37) is basis of *virtual work* approach that often proves quite useful in practice. In our particular case, Eq. (2.37) can be expressed as

$$\tau_1 d\theta_1+\tau_2 d\theta_2=F_{x\,EE}dx_{EE}+F_{y\,EE}dy_{EE} \tag{2.38}$$

Or by help of Eqs. (2.18) and (2.19) or equivalently Eq. (2.30):

$$F_{x\,EE}dx_{EE}+F_{y\,EE}dy_{EE}=F_{x\,EE}\left(\frac{\partial x_{EE}}{\partial\theta_1}d\theta_1+\frac{\partial x_{EE}}{\partial\theta_2}d\theta_2\right)+F_{y\,EE}\left(\frac{\partial y_{EE}}{\partial\theta_1}d\theta_1+\frac{\partial y_{EE}}{\partial\theta_2}d\theta_2\right) \tag{2.39}$$

Therefore,

$$\tau_1 d\theta_1 + \tau_2 d\theta_2 = \left(F_{x\,EE}\frac{\partial x_{EE}}{\partial\theta_1} + F_{y\,EE}\frac{\partial y_{EE}}{\partial\theta_1}\right)d\theta_1 + \left(F_{x\,EE}\frac{\partial x_{EE}}{\partial\theta_2} + F_{y\,EE}\frac{\partial y_{EE}}{\partial\theta_2}\right)d\theta_2 \tag{2.40}$$

and thus,

$$\tau_1 = F_{x\,EE}\frac{\partial x_{EE}}{\partial\theta_1} + F_{y\,EE}\frac{\partial y_{EE}}{\partial\theta_1} \tag{2.41}$$

$$\tau_2 = F_{x\,EE}\frac{\partial x_{EE}}{\partial\theta_2} + F_{y\,EE}\frac{\partial y_{EE}}{\partial\theta_2} \tag{2.42}$$

For our particular case we already obtained partial derivatives so joint torques can be expressed as

$$\tau_1 = -[L_1\sin\theta_1 + L_2\sin(\theta_1+\theta_2)]F_{x\,EE} + [L_1\cos\theta_1 + L_2\cos(\theta_1+\theta_2)]F_{y\,EE} \tag{2.43}$$

$$\tau_2 = -L_2\sin(\theta_1+\theta_2)F_{x\,EE} + L_2\cos(\theta_1+\theta_2)F_{y\,EE} \tag{2.44}$$

This is "*inverse dynamics*" transformation

$$F_{x\,EE}, F_{y\,EE} \to \tau_1, \tau_2 \tag{2.45}$$

explicitly expressed here in the context of virtual work method in massless limit and with no frictional and other losses.

In the context of arm constrained in sagittal plane analogy, Eqs. (2.43) and (2.44) can be utilized to estimate shoulder and elbow joint torques based on x, y forces that wrist/fist applies to the rest of world.

The essence of virtual work method can be illustrated in pure matrix notation too. Consider scalar product

$$\vec{\tau}\cdot d\vec{\theta} = (\tau_1\ \ \tau_2)\begin{pmatrix} d\theta_1 \\ d\theta_2 \end{pmatrix} = (F_{x\,EE}\ \ F_{y\,EE})\begin{pmatrix} dx_{EE} \\ dy_{EE} \end{pmatrix} = \vec{F}_{EE}\cdot d\vec{r}_{EE} \tag{2.46}$$

By inserting unit matrix $\boldsymbol{I}$ one obtains

$$(\tau_1\ \ \tau_2)\begin{pmatrix} d\theta_1 \\ d\theta_2 \end{pmatrix} = (F_{x\,EE}\ \ F_{y\,EE})\boldsymbol{J}_{\theta_1,\theta_2\to x_{EE},y_{EE}}\boldsymbol{J}_{x_{EE},y_{EE}\to\theta_1,\theta_2}\begin{pmatrix} dx_{EE} \\ dy_{EE} \end{pmatrix} \tag{2.47}$$

and by help of Eq. (2.32) one obtains

$$(\tau_1\ \ \tau_2)\begin{pmatrix} d\theta_1 \\ d\theta_2 \end{pmatrix} = (F_{x\,EE}\ \ F_{y\,EE})\boldsymbol{J}_{\theta_1,\theta_2\to x_{EE},y_{EE}}\begin{pmatrix} d\theta_1 \\ d\theta_2 \end{pmatrix} \tag{2.48}$$

As this needs to hold for every $\begin{pmatrix} d\theta_1 \\ d\theta_2 \end{pmatrix}$ one may conclude that

$$(\tau_1\ \ \tau_2) = (F_{x\,EE}\ \ F_{y\,EE})\boldsymbol{J}_{\theta_1,\theta_2\to x_{EE},y_{EE}} = (F_{x\,EE}\ \ F_{y\,EE})\begin{pmatrix} \frac{\partial x_{EE}}{\partial\theta_1} & \frac{\partial x_{EE}}{\partial\theta_2} \\ \frac{\partial y_{EE}}{\partial\theta_1} & \frac{\partial y_{EE}}{\partial\theta_2} \end{pmatrix} \tag{2.49}$$

This expression may be now multiplied from right with inverse matrix such that

$$\left(F_{x\,EE}\;\;F_{y\,EE}\right)=\left(\tau_1\;\;\tau_2\right)\begin{pmatrix}\dfrac{\partial x_{EE}}{\partial\theta_1} & \dfrac{\partial x_{EE}}{\partial\theta_2}\\ \dfrac{\partial y_{EE}}{\partial\theta_1} & \dfrac{\partial y_{EE}}{\partial\theta_2}\end{pmatrix}^{-1}=\left(\tau_1\;\;\tau_2\right)\boldsymbol{J}_{x_{EE},y_{EE}\rightarrow\theta_1,\theta_2} \tag{2.50}$$

and finally we obtain "*forward dynamics.*" In the context of virtual work method in massless limit and with no frictional and other losses

$$\left(F_{x\,EE}\;\;F_{y\,EE}\right)=\left(\tau_1\;\;\tau_2\right)\begin{pmatrix}\dfrac{\partial\theta_1}{\partial x_{EE}} & \dfrac{\partial\theta_1}{\partial y_{EE}}\\ \dfrac{\partial\theta_2}{\partial x_{EE}} & \dfrac{\partial\theta_2}{\partial y_{EE}}\end{pmatrix} \tag{2.51}$$

$$F_{x\,EE}=\tau_1\frac{\partial\theta_1}{\partial x_{EE}}+\tau_2\frac{\partial\theta_2}{\partial x_{EE}} \tag{2.52}$$

$$F_{y\,EE}=\tau_1\frac{\partial\theta_1}{\partial y_{EE}}+\tau_2\frac{\partial\theta_2}{\partial y_{EE}} \tag{2.53}$$

that is, given the joint torques we obtained the end-effector's force

$$\tau_1,\tau_2\rightarrow F_{x\,EE},F_{y\,EE} \tag{2.54}$$

In the context of arm constrained in sagittal plane analogy, Eqs. (2.52) and (2.53) can be utilized to estimate *x*, *y* forces that wrist/fist applies to the rest of the world based on shoulder and elbow joint torques.

Reader may have noticed that both "inverse" and "forward" dynamics have been written within quotation marks. The reason is that many consider *forward dynamics* as a process of obtaining kinematics trajectories, given joint torques. On the similar lines, *inverse dynamics* is a process of obtaining joint torques from known kinematics trajectories.

Next, we show how to relate these quantities within several computational methods.

2.3.2 Newton's Equations

Next, we take into account masses and analyze dynamics of our two-link template system by utilizing *Newton-Euler* approach based on Newton's laws of motion.

Newton's second law states that force equals rate of change of linear momentum

$$\vec{F}=\frac{d\vec{p}}{dt} \tag{2.55}$$

Similarly, torque (or moment of force) equals rate of change of angular momentum

$$\vec{\tau}=\frac{d\vec{L}}{dt} \tag{2.56}$$

Rotational version of 2nd Newton law

$$\vec{\tau}_i + \left(\sum_j \vec{r}_{ij} \times \vec{F}_j\right)_{one\ side} = \frac{d\vec{L}_{i\ one\ side}}{dt}$$

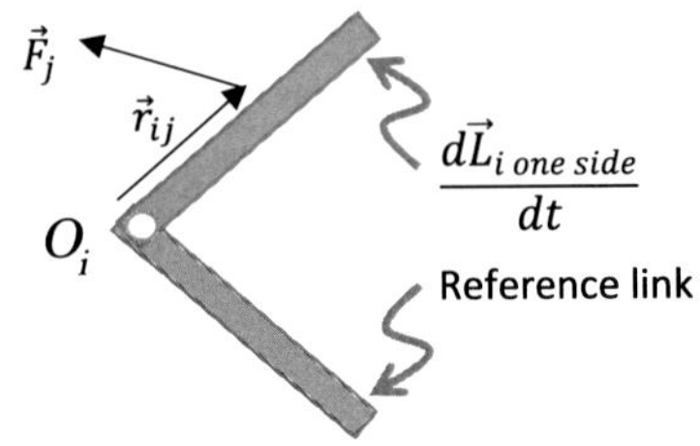

If there are no external torques the angular momentum of entire system is zero and this also means that torque that link 1 applies on link 2 is equal in magnitude and opposite in direction to the torque that link 2 applies on link 1. This suggests "no double counting criteria" or "one side" concept. "One side" means that one link out of two that are adjacent to joint is chosen as reference and rotational version of Newton's second law is expressed relative to that reference link. Hence, only moment of force (physical and inertial) with forces acting on one side of the joint matters and only rate of change of angular momentum on the same side of joint contributes.

Hence, dynamics of joint 2 can be expressed as

$$\vec{\tau}_2 + \left(\vec{r}_{CM2} - \vec{r}_{O2}\right) \times \left(m_2 \vec{g} - m_2 \vec{a}_{O2}\right) + \left(\vec{r}_{EE} - \vec{r}_{O2}\right) \times \vec{F}_{EE*} = I_{2\,(O2)} \left(\ddot{\theta}_1 + \ddot{\theta}_2\right) \hat{k} \tag{2.57}$$

The first term on the left-hand side of $\vec{\tau}_2$ is torque generated by joint 2, the second term is torque generated by gravity and inertial forces (as frame with coordinate origin attached to joint 2 is a noninertial frame), and the third term is torque generated by external force applied to end effector. Finally, according to Eq. (2.56), and thanks to fictitious torque generated by inertial force, sum of these torques is equal to the rate of change of angular momentum of link 2 about joint 2, as obtained in noninertial frame of joint 2, and which can be expressed as product of moment of inertia of link 2 about joint 2, and angular acceleration of link 2. Here, we assumed that there is no moment due to friction.

According to Steiner's parallel axis theorem, moment of Inertia of link 2 about joint 2 is

$$I_{2\,(O2)} = I_2 + m_2 L_{CM2}^2 \tag{2.58}$$

The acceleration of joint 2 in inertial lab frame is the sum of tangential and radial accelerations

$$\vec{a}_{O2} = L_1 \left(\ddot{\theta}_1 \frac{-y_{O2}\hat{\boldsymbol{i}} + x_{O2}\hat{\boldsymbol{j}}}{L_1} - \dot{\theta}_1^2 \frac{x_{O2}\hat{\boldsymbol{i}} + y_{O2}\hat{\boldsymbol{j}}}{L_1} \right) = -\left(\ddot{\theta}_1 y_{O2} + \dot{\theta}_1^2 x_{O2}\right)\hat{\boldsymbol{i}} + \left(\ddot{\theta}_1 x_{O2} - \dot{\theta}_1^2 y_{O2}\right)\hat{\boldsymbol{j}} \tag{2.59}$$

Note that if joint 2 motions were not constrained to plane, one would also need to consider other terms including Coriolis forces. We will return to this shortly.

And gravitation acceleration is

$$\vec{g} = -g\hat{\boldsymbol{j}} \tag{2.60}$$

Hence, second term forces are

$$m_2\vec{g} - m_2\vec{a}_{O2} = m_2\left[\left(\ddot{\theta}_1 y_{O2} + \dot{\theta}_1^2 x_{O2}\right)\hat{\boldsymbol{i}} + \left(-g - \ddot{\theta}_1 x_{O2} + \dot{\theta}_1^2 y_{O2}\right)\hat{\boldsymbol{j}}\right] \tag{2.61}$$

with corresponding moment arm

$$\vec{r}_{CM2} - \vec{r}_{O2} = L_{CM2}\left[\cos(\theta_1+\theta_2)\hat{\boldsymbol{i}} + \sin(\theta_1+\theta_2)\hat{\boldsymbol{j}}\right] \tag{2.62}$$

such that

$$\begin{aligned}&\left(\vec{r}_{CM2} - \vec{r}_{O2}\right)\times\left(m_2\vec{g} - m_2\vec{a}_{O2}\right) = m_2 L_{CM2}\left[\cos(\theta_1+\theta_2)\left(-g - \ddot{\theta}_1 x_{O2} + \dot{\theta}_1^2 y_{O2}\right)\right.\\ &\left.-\sin(\theta_1+\theta_2)\left(\ddot{\theta}_1 y_{O2} + \dot{\theta}_1^2 x_{O2}\right)\right]\hat{\boldsymbol{k}} = m_2 L_{CM2} L_1\left[\cos(\theta_1+\theta_2)\left(-\frac{g}{L_1} - \ddot{\theta}_1\cos\theta_1 + \dot{\theta}_1^2\sin\theta_1\right)\right.\\ &\left.-\sin(\theta_1+\theta_2)\left(\ddot{\theta}_1\sin\theta_1 + \dot{\theta}_1^2\cos\theta_1\right)\right]\hat{\boldsymbol{k}}\end{aligned} \tag{2.63}$$

Similarly, moment of end-effector forces can be obtained first by expressing moment arm

$$\vec{r}_{EE} - \vec{r}_{O2} = L_2\left[\cos(\theta_1+\theta_2)\hat{\boldsymbol{i}} + \sin(\theta_1+\theta_2)\hat{\boldsymbol{j}}\right] \tag{2.64}$$

and then by finding vector product

$$\left(\vec{r}_{EE} - \vec{r}_{O2}\right)\times\vec{F}_{EE*} = L_2\left[\cos(\theta_1+\theta_2)F_{y\,EE*} - \sin(\theta_1+\theta_2)F_{x\,EE*}\right]\hat{\boldsymbol{k}} \tag{2.65}$$

Finally, by adding all terms one obtains version of Newton's second law for joint 2

$$\begin{aligned}&\tau_2 + m_2 L_{CM2} L_1\left[\cos(\theta_1+\theta_2)\left(-\frac{g}{L_1} - \ddot{\theta}_1\cos\theta_1 + \dot{\theta}_1^2\sin\theta_1\right) - \sin(\theta_1+\theta_2)\left(\ddot{\theta}_1\sin\theta_1 + \dot{\theta}_1^2\cos\theta_1\right)\right]\\ &+ L_2\left[\cos(\theta_1+\theta_2)F_{y\,EE*} - \sin(\theta_1+\theta_2)F_{x\,EE*}\right] = \left(I_2 + m_2 L_{CM2}^2\right)\left(\ddot{\theta}_1 + \ddot{\theta}_2\right)\end{aligned} \tag{2.66}$$

Or by grouping joint torque end and end-effector torque

$$\tau_2 + \tau_{2FEE*} = m_2 L_1 L_{CM2}\cos\theta_2\ddot{\theta}_1 + \left(m_2 L_{CM2}^2 + I_2\right)\left(\ddot{\theta}_1 + \ddot{\theta}_2\right) + m_2 L_1 L_{CM2}\dot{\theta}_1^2\sin\theta_2 + g m_2 L_{CM2}\cos(\theta_1+\theta_2) \tag{2.67}$$

Similar version of Newton's second law can be also obtained for joint 1.
Moment of inertia of link 1 about joint 1 is

$$I_{1\,(O1)} = I_1 + m_1 L_{CM1}^2 \tag{2.68}$$

Rate of change of angular momentum of link 1 about joint 1 is

$$\frac{dL_{1\,(O1)}}{dt} = I_{1\,(O1)}\ddot{\theta}_1 \tag{2.69}$$

Angular momentum of link 2 about joint 1 is

$$\vec{L}_{2\,(O1)} = \left(\vec{r}_{CM2} - \vec{r}_{O1}\right) \times m_2 \dot{\vec{r}}_{CM2} + I_2(\dot{\theta}_1 + \dot{\theta}_2)\hat{\boldsymbol{k}} \tag{2.70}$$

Here,

$$\vec{r}_{CM2} - \vec{r}_{O1} = [L_1 \cos\theta_1 + L_{CM2}\cos(\theta_1+\theta_2)]\hat{\boldsymbol{i}} + [L_1 \sin\theta_1 + L_{CM2}\sin(\theta_1+\theta_2)]\hat{\boldsymbol{j}} \tag{2.71}$$

$$\dot{\vec{r}}_{CM2} = -\left[\dot{\theta}_1 L_1 \sin\theta_1 + (\dot{\theta}_1+\dot{\theta}_2)L_{CM2}\sin(\theta_1+\theta_2)\right]\hat{\boldsymbol{i}} + \left[\dot{\theta}_1 L_1\cos\theta_1 + (\dot{\theta}_1+\dot{\theta}_2)L_{CM2}\cos(\theta_1+\theta_2)\right]\hat{\boldsymbol{j}} \tag{2.72}$$

$$\left(\vec{r}_{CM2} - \vec{r}_{O1}\right) \times m_2 \dot{\vec{r}}_{CM2} = m_2\{[L_1\cos\theta_1 + L_{CM2}\cos(\theta_1+\theta_2)]\left[\dot{\theta}_1 L_1\cos\theta_1 + (\dot{\theta}_1+\dot{\theta}_2)L_{CM2}\cos(\theta_1+\theta_2)\right]$$

$$+ [L_1\sin\theta_1 + L_{CM2}\sin(\theta_1+\theta_2)]\left[\dot{\theta}_1 L_1\sin\theta_1 + (\dot{\theta}_1+\dot{\theta}_2)L_{CM2}\sin(\theta_1+\theta_2)\right]\}$$

$$\hat{\boldsymbol{k}} = m_2\left[\dot{\theta}_1 L_1^2 + (2\dot{\theta}_1+\dot{\theta}_2)L_1 L_{CM2}\cos\theta_2 + (\dot{\theta}_1+\dot{\theta}_2)L_{CM2}^2\right]\hat{\boldsymbol{k}} \tag{2.73}$$

Rate of change of angular momentum of link 2 about joint 1 is

$$\frac{d\vec{L}_{2\,(O1)}}{dt} = \{m_2\left[\ddot{\theta}_1 L_1^2 + (2\ddot{\theta}_1+\ddot{\theta}_2)L_1 L_{CM2}\cos\theta_2 - \dot{\theta}_2(2\dot{\theta}_1+\dot{\theta}_2)L_1 L_{CM2}\sin\theta_2 + (\ddot{\theta}_1+\ddot{\theta}_2)L_{CM2}^2\right]$$

$$+ I_2(\ddot{\theta}_1+\ddot{\theta}_2)\}\hat{\boldsymbol{k}} \tag{2.74}$$

By adding Eqs. (2.69) and (2.74), one obtains rate of change of angular momentum of entire planar two-link manipulator about joint 1

$$\frac{d\left(\vec{L}_{1\,(O1)} + \vec{L}_{2\,(O1)}\right)}{dt} = \{(I_1 + m_1 L_{CM1}^2)\ddot{\theta}_1 + I_2(\ddot{\theta}_1+\ddot{\theta}_2) + m_2\left[\ddot{\theta}_1 L_1^2 + (2\ddot{\theta}_1+\ddot{\theta}_2)L_1 L_{CM2}\cos\theta_2\right.$$

$$\left. -\dot{\theta}_2(2\dot{\theta}_1+\dot{\theta}_2)L_1 L_{CM2}\sin\theta_2 + (\ddot{\theta}_1+\ddot{\theta}_2)L_{CM2}^2\right]\}\hat{\boldsymbol{k}} \tag{2.75}$$

Hence, the version of Newton's second law for joint 1 is

$$\vec{\tau}_1 + \left(\vec{r}_{CM} - \vec{r}_{O1}\right) \times m\vec{g} + \left(\vec{r}_{EE} - \vec{r}_{O1}\right) \times \vec{F}_{EE*} = \frac{d\left(\vec{L}_{1\,(O1)} + \vec{L}_{2\,(O1)}\right)}{dt} \tag{2.76}$$

Here,

$$\left(\vec{r}_{CM} - \vec{r}_{O1}\right) \times m\vec{g} = -\{m_1 L_{CM1}\cos\theta_1 + m_2[L_1\cos\theta_1 + L_{CM2}\cos(\theta_1+\theta_2)]\}g\hat{\boldsymbol{k}} \tag{2.77}$$

$$\left(\vec{r}_{EE} - \vec{r}_{O1}\right) \times \vec{F}_{EE*} = \{[L_1\cos\theta_1 + L_2\cos(\theta_1+\theta_2)]F_{y\,EE*} - [L_1\sin\theta_1 + L_2\sin(\theta_1+\theta_2)]F_{x\,EE*}\}\hat{\boldsymbol{k}} \tag{2.78}$$

Finally,

$$\tau_1 - \{m_1 L_{CM1}\cos\theta_1 + m_2[L_1\cos\theta_1 + L_{CM2}\cos(\theta_1+\theta_2)]\}g + [L_1\cos\theta_1 + L_2\cos(\theta_1+\theta_2)]F_{y\,EE*}$$

$$-[L_1\sin\theta_1 + L_2\sin(\theta_1+\theta_2)]F_{x\,EE*} = (I_1 + m_1 L_{CM1}^2)\ddot{\theta}_1 + I_2(\ddot{\theta}_1+\ddot{\theta}_2)$$

$$+ m_2\left[\ddot{\theta}_1 L_1^2 + (2\ddot{\theta}_1+\ddot{\theta}_2)L_1 L_{CM2}\cos\theta_2 - \dot{\theta}_2(2\dot{\theta}_1+\dot{\theta}_2)L_1 L_{CM2}\sin\theta_2 + (\ddot{\theta}_1+\ddot{\theta}_2)L_{CM2}^2\right] \tag{2.79}$$

Or by grouping joint torque end and end-effector torque

$$\tau_1 + \tau_{1FEE*} = \left(m_1 L_{CM1}^2 + I_1 + m_2 L_1^2\right)\ddot{\theta}_1 + m_2 L_1 L_{CM2}\left[\cos\theta_2\left(2\ddot{\theta}_1 + \ddot{\theta}_2\right) - \sin\theta_2\left(2\dot{\theta}_1 + \dot{\theta}_2\right)\dot{\theta}_2\right] + \left(m_2 L_{CM2}^2 + I_2\right)\left(\ddot{\theta}_1 + \ddot{\theta}_2\right) + g[(m_1 L_{CM1} + m_2 L_1)\cos\theta_1 + m_2 L_{CM2}\cos(\theta_1 + \theta_2)] \quad (2.80)$$

2.3.3 Connection Between Virtual Work and Newton Euler Approach

For static case Eqs. (2.66) and (2.79) become

$$\tau_2 - m_2 L_{CM2} g\cos(\theta_1 + \theta_2) + L_2\left[\cos(\theta_1 + \theta_2)F_{y\,EE*} - \sin(\theta_1 + \theta_2)F_{x\,EE*}\right] = 0 \quad (2.81)$$

$$\tau_1 - \{m_1 L_{CM1}\cos\theta_1 + m_2[L_1\cos\theta_1 + L_{CM2}\cos(\theta_1 + \theta_2)]\}g + [L_1\cos\theta_1 + L_2\cos(\theta_1 + \theta_2)]F_{y\,EE*} - [L_1\sin\theta_1 + L_2\sin(\theta_1 + \theta_2)]F_{x\,EE*} = 0 \quad (2.82)$$

And in the massless limit they further simplify to

$$\tau_1 = -\tau_{1FEE*} = [L_1\sin\theta_1 + L_2\sin(\theta_1 + \theta_2)]F_{x\,EE*} - [L_1\cos\theta_1 + L_2\cos(\theta_1 + \theta_2)]F_{y\,EE*} \quad (2.83)$$

$$\tau_2 = -\tau_{2FEE*} = L_2\sin(\theta_1 + \theta_2)F_{x\,EE*} - L_2\cos(\theta_1 + \theta_2)F_{y\,EE*} \quad (2.84)$$

By comparison with virtual work result, that is, Eqs. (2.43) and (2.44), we confirm Newton's third law of action reaction, that is,

$$\vec{F}_{EE*} = -\vec{F}_{EE} \quad (2.85)$$

Because forces applied from the rest of the world on end effector are equal in magnitude and opposite in direction to forces that end effector applies onto the rest of the world.

2.3.4 Euler-Lagrange Method

Another method, instead of Newton-Euler method, that we could use to study dynamics of our template example is *Euler-Lagrange method*. It is based on function L called *Lagrange function* or *Lagrangian*. Lagrangian is defined as the difference between kinetic and potential energies of the system, that is,

$$L = K - U \quad (2.86)$$

In our particular case potential energy is

$$U = (m_1 y_{CM1} + m_2 y_{CM2})g = \{m_1 L_{CM1}\sin\theta_1 + m_2[L_1\sin\theta_1 + L_{CM2}\sin(\theta_1 + \theta_2)]\}g \quad (2.87)$$

and kinetic energy is

$$K = \frac{1}{2}m_1\dot{\vec{r}}_{CM1}\cdot\dot{\vec{r}}_{CM1} + \frac{1}{2}I_1\dot{\theta}_1^2 + \frac{1}{2}m_2\dot{\vec{r}}_{CM2}\cdot\dot{\vec{r}}_{CM2} + \frac{1}{2}I_2\left(\dot{\theta}_1 + \dot{\theta}_2\right)^2 \quad (2.88)$$

Note that kinetic energy of link is expressed as kinetic energy of its center of mass (CM) and kinetic energy of link spinning about its CM.

Velocity squared of link 1 CM is

$$\dot{\vec{r}}_{CM1}\cdot\dot{\vec{r}}_{CM1}=L_{CM1}^2\dot{\theta}_1^2 \tag{2.89}$$

Velocity squared of link 2 CM is

$$\dot{\vec{r}}_{CM2}\cdot\dot{\vec{r}}_{CM2}=\left[\dot{\theta}_1L_1\sin\theta_1+\left(\dot{\theta}_1+\dot{\theta}_2\right)L_{CM2}\sin\left(\theta_1+\theta_2\right)\right]^2+\left[\dot{\theta}_1L_1\cos\theta_1+\left(\dot{\theta}_1+\dot{\theta}_2\right)L_{CM2}\cos\left(\theta_1+\theta_2\right)\right]^2 \tag{2.90}$$

as obtained by application of Eq. (2.72). Hence

$$\dot{\vec{r}}_{CM2}\cdot\dot{\vec{r}}_{CM2}=L_1^2\dot{\theta}_1^2+2L_1L_{CM2}\dot{\theta}_1\left(\dot{\theta}_1+\dot{\theta}_2\right)\cos\theta_2+L_{CM2}^2\left(\dot{\theta}_1+\dot{\theta}_2\right)^2 \tag{2.91}$$

Therefore, kinetic energy of our system is

$$K=\frac{1}{2}m_1L_{CM1}^2\dot{\theta}_1^2+\frac{1}{2}I_1\dot{\theta}_1^2+\frac{1}{2}m_2\left[L_1^2\dot{\theta}_1^2+2L_1L_{CM2}\dot{\theta}_1\left(\dot{\theta}_1+\dot{\theta}_2\right)\cos\theta_2+L_{CM2}^2\left(\dot{\theta}_1+\dot{\theta}_2\right)^2\right]+\frac{1}{2}I_2\left(\dot{\theta}_1+\dot{\theta}_2\right)^2 \tag{2.92}$$

The *Euler-Lagrange equations* defining Euler-Lagrange method is a set of second-order differential equation, one per degree of freedom, in the form

$$\frac{d}{dt}\left(\frac{\partial L}{\partial \dot{q}}\right)-\frac{\partial L}{\partial q}=\tau_q \tag{2.93}$$

Here, q is either θ_1 or θ_2 and corresponding τ_q can be both joint torque and torque contribution from external force, applied to end effector in our case.

Next, we find all derivatives needed to write down Euler-Lagrange Eq. (2.93).

From Eq. (2.87) it follows that

$$\frac{\partial U}{\partial \theta_1}=g\left[\left(m_1L_{CM1}+m_2L_1\right)\cos\theta_1+m_2L_{CM2}\cos\left(\theta_1+\theta_2\right)\right] \tag{2.94}$$

$$\frac{\partial U}{\partial \theta_2}=gm_2L_{CM2}\cos\left(\theta_1+\theta_2\right) \tag{2.95}$$

$$\frac{\partial U}{\partial \dot{\theta}_1}=0 \tag{2.96}$$

$$\frac{\partial U}{\partial \dot{\theta}_2}=0 \tag{2.97}$$

From Eq. (2.92) it follows that

$$\frac{\partial K}{\partial \theta_1}=0 \tag{2.98}$$

$$\frac{\partial K}{\partial \theta_2}=-m_2L_1L_{CM2}\dot{\theta}_1\left(\dot{\theta}_1+\dot{\theta}_2\right)\sin\theta_2 \tag{2.99}$$

$$\frac{\partial K}{\partial \dot{\theta}_1}=\left[m_1L_{CM1}^2+I_1+m_2\left(L_1^2+2L_1L_{CM2}\cos\theta_2\right)\right]\dot{\theta}_1+m_2L_1L_{CM2}\cos\theta_2\dot{\theta}_2+\left(m_2L_{CM2}^2+I_2\right)\left(\dot{\theta}_1+\dot{\theta}_2\right) \tag{2.100}$$

$$\frac{\partial K}{\partial \dot{\theta}_2} = m_2 L_1 L_{CM2} \cos\theta_2 \dot{\theta}_1 + \left(m_2 L_{CM2}^2 + I_2\right)\left(\dot{\theta}_1 + \dot{\theta}_2\right) \tag{2.101}$$

$$\frac{d}{dt}\left(\frac{\partial K}{\partial \dot{\theta}_1}\right) = \left(m_1 L_{CM1}^2 + I_1 + m_2 L_1^2\right)\ddot{\theta}_1 + m_2 2 L_1 L_{CM2}\left(\cos\theta_2 \ddot{\theta}_1 - \sin\theta_2 \dot{\theta}_1 \dot{\theta}_2\right)$$

$$+ m_2 L_1 L_{CM2}\left(\cos\theta_2 \ddot{\theta}_2 - \sin\theta_2 \dot{\theta}_2^2\right) + \left(m_2 L_{CM2}^2 + I_2\right)\left(\ddot{\theta}_1 + \ddot{\theta}_2\right) = \left(m_1 L_{CM1}^2 + I_1 + m_2 L_1^2\right)\ddot{\theta}_1$$

$$+ m_2 L_1 L_{CM2} \cos\theta_2 \left(2\ddot{\theta}_1 + \ddot{\theta}_2\right) - m_2 L_1 L_{CM2} \sin\theta_2 \left(2\dot{\theta}_1 + \dot{\theta}_2\right)\dot{\theta}_2 + \left(m_2 L_{CM2}^2 + I_2\right)\left(\ddot{\theta}_1 + \ddot{\theta}_2\right) \tag{2.102}$$

$$\frac{d}{dt}\left(\frac{\partial K}{\partial \dot{\theta}_2}\right) = m_2 L_1 L_{CM2}\left(\cos\theta_2 \ddot{\theta}_1 - \sin\theta_2 \dot{\theta}_1 \dot{\theta}_2\right) + \left(m_2 L_{CM2}^2 + I_2\right)\left(\ddot{\theta}_1 + \ddot{\theta}_2\right) \tag{2.103}$$

By grouping Eqs. (2.94), (2.96), (2.98), and (2.102) we obtain Euler-Lagrange equation (2.93) for $q = \theta_1$

$$\tau_1 = \left(m_1 L_{CM1}^2 + I_1 + m_2 L_1^2\right)\ddot{\theta}_1 + m_2 L_1 L_{CM2} \cos\theta_2 \left(2\ddot{\theta}_1 + \ddot{\theta}_2\right) - m_2 L_1 L_{CM2} \sin\theta_2 \left(2\dot{\theta}_1 + \dot{\theta}_2\right)\dot{\theta}_2$$
$$+ \left(m_2 L_{CM2}^2 + I_2\right)\left(\ddot{\theta}_1 + \ddot{\theta}_2\right) + g\left[\left(m_1 L_{CM1} + m_2 L_1\right)\cos\theta_1 + m_2 L_{CM2} \cos\left(\theta_1 + \theta_2\right)\right] \tag{2.104}$$

By grouping Eq. (2.95), (2.97), (2.99), and (2.103) we obtain Euler–Lagrange equation (2.93) for $q = \theta_2$

$$\tau_2 = m_2 L_1 L_{CM2} \cos\theta_2 \ddot{\theta}_1 + \left(m_2 L_{CM2}^2 + I_2\right)\left(\ddot{\theta}_1 + \ddot{\theta}_2\right) + m_2 L_1 L_{CM2} \dot{\theta}_1^2 \sin\theta_2 + g m_2 L_{CM2} \cos\left(\theta_1 + \theta_2\right) \tag{2.105}$$

2.3.5 Euler-Lagrange Method Versus Newton-Euler Method

Interested reader should compare this result with Eqs. (2.80) and (2.67). By utilizing Euler-Lagrange method, we obtained Eqs. (2.104) and (2.105), which are identical to Eqs. (2.80) and (2.67) obtained by utilizing Newton-Euler method! The torques in Eqs. (2.104) and (2.105) are sums of joint torques and torques due to external forces as in Eqs. (2.80) and (2.67).

This should not come as a surprise because Newton-Euler method is equivalent to Euler-Lagrange method.

In practice, often it is easier to use Euler-Lagrange method as all that is needed is Lagrangian, Eq. (2.86), and the rest comes automatically, for example, by automatic computations of derivatives. In difference, Newton-Euler method requires directly obtaining equations of motion, task, whose complexity (read difficulty on both physics conceptual and computational level) typically grows exponentially with number of degrees of freedom. Hence Euler-Lagrange method provides decomposition of that complexity and better organization of computational and conceptual tasks that also provides a better way of error tracking.

To illustrate increasing complexities of Newton-Euler method let us imagine, for example, that instead of having single-degree-of-freedom joint 1, we have two-degrees-of-freedom joint 1. The joint 2 then accelerates along the three-dimensional path. How this affects the equation of motion for single-degree-of-freedom joint 2?

Well, first we need to realize that noninertial frame attached to joint 2 is both accelerating in translational sense as well as changing orientation of its axes with respect to axes of lab inertial frame.

It is easy to deal with translational accelerations; one just needs to make sure that appropriate three-dimensional inertial force proportional to three-dimensional accelerations of joint 2 is considered in Eq. (2.57). But how to deal with the rotating axis?

2.3.6 Dynamics in Noninertial Accelerating and Rotating Coordinate Frame

Consider inertial reference frame O and noninertial reference frame O′ depicted in Fig. 2.4. Quite generally, the origin of O′ can be accelerating and O′ axes can be rotating with instantaneous angular velocity $\vec{\omega}$.

The radius vector of any point can be expressed in both frames and related to each other as

$$\vec{r}=\vec{r}_{oo'}+\vec{r}\,' \tag{2.106}$$

The effect of angular velocity can be understood in the context of time evolution of radius vector in the lab frame. To simplify the problem, imagine that both $\vec{r}_{oo'}$ and $\vec{r}\,'$ are constant vectors. Furthermore, imagine that $\left|\vec{r}_{oo'}\right|=0$. Then for infinitesimally small period of time $\triangle t$,

$$\vec{r}\,(t+\Delta t)=\vec{r}\,(t)+\left[\vec{\omega}\times\vec{r}\,(t)\right]\Delta t \tag{2.107}$$

It is easy to verify this vector relationship. For example, if $\vec{\omega}=\omega_z\hat{k}$, that is, rotations are about the z axis, the z component of vector will not change, that is,

$$z(t+\Delta t)=z(t)=z(t)+\left[\omega_z\hat{k}\times z(t)\hat{k}\right]\cdot\hat{k}\Delta t \tag{2.108}$$

The right-hand side is just the z component of Eq. (2.107) and the last term is obviously zero.

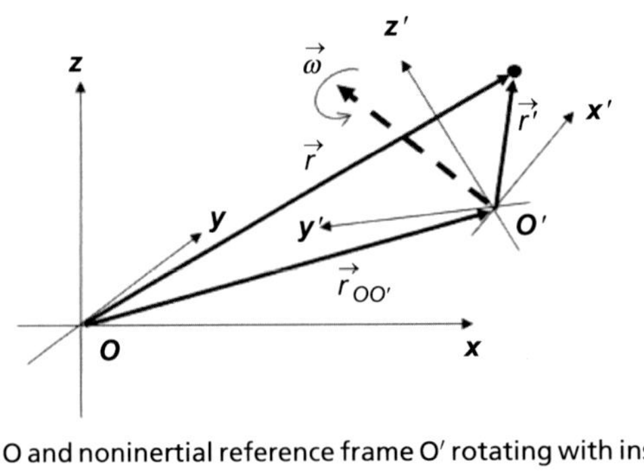

FIG. 2.4 Inertial reference frame O and noninertial reference frame O′ rotating with instantaneous angular velocity $\vec{\omega}$.

The x and y components will clearly change after being rotated by an angle $\omega_z \triangle t$

$$x(t+\Delta t) = x(t)\cos\omega_z\Delta t - y(t)\sin\omega_z\Delta t \tag{2.109}$$

$$y(t+\Delta t) = x(t)\sin\omega_z\Delta t + y(t)\cos\omega_z\Delta t \tag{2.110}$$

For infinitesimally small angle this becomes

$$x(t+\Delta t) = x(t) - y(t)\omega_z\Delta t = x(t) + \left[\omega_z\hat{k}\times y(t)\hat{j}\right]\cdot\hat{i}\Delta t \tag{2.111}$$

$$y(t+\Delta t) = x(t)\omega_z\Delta t + y(t) = y(t) + \left[\omega_z\hat{k}\times x(t)\hat{i}\right]\cdot\hat{j}\Delta t \tag{2.112}$$

with right-hand sides being x and y components of Eq. (2.107).

Eq. (2.107) holds for any vector, that is,

$$\frac{d\vec{A}}{dt} = \vec{\omega}\times\vec{A}(t) \tag{2.113}$$

Now, armed with this knowledge we can address the most general cases and relate variables in inertial and noninertial frame.

For example, taking the first derivative of Eq. (2.106) gives

$$\frac{d\vec{r}}{dt} = \frac{d\vec{r}_{oo'}}{dt} + \frac{d\vec{r}\,'}{dt} + \vec{\omega}\times\vec{r}\,' \tag{2.114}$$

And taking second derivative gives

$$\frac{d^2\vec{r}}{dt^2} = \frac{d^2\vec{r}_{oo'}}{dt^2} + \frac{d^2\vec{r}\,'}{dt^2} + \vec{\omega}\times\frac{d\vec{r}\,'}{dt} + \frac{d\left(\vec{\omega}\times\vec{r}\,'\right)}{dt} + \vec{\omega}\times\left(\vec{\omega}\times\vec{r}\,'\right) \tag{2.115}$$

Here

$$\frac{d\left(\vec{\omega}\times\vec{r}\,'\right)}{dt} = \frac{d\vec{\omega}}{dt}\times\vec{r}\,' + \vec{\omega}\times\frac{d\vec{r}\,'}{dt} \tag{2.116}$$

Hence

$$\frac{d^2\vec{r}}{dt^2} = \frac{d^2\vec{r}_{oo'}}{dt^2} + \frac{d^2\vec{r}\,'}{dt^2} + 2\,\vec{\omega}\times\frac{d\vec{r}\,'}{dt} + \frac{d\vec{\omega}}{dt}\times\vec{r}\,' + \vec{\omega}\times\left(\vec{\omega}\times\vec{r}\,'\right) \tag{2.117}$$

Now we are ready to address the version of Newton's law in noninertial frame.

The correct Newton's law as expressed by Eq. (2.56) holds only in inertial frame.

For example, for single material point with mass m the torque about origin O′ is

$$\vec{\tau} = \vec{r}\,'\times\vec{F} = \vec{r}\,'\times m\frac{d^2\vec{r}}{dt^2} = \vec{r}\,'\times m\left[\frac{d^2\vec{r}_{oo'}}{dt^2} + \frac{d^2\vec{r}\,'}{dt^2} + 2\,\vec{\omega}\times\frac{d\vec{r}\,'}{dt} + \frac{d\vec{\omega}}{dt}\times\vec{r}\,' + \vec{\omega}\times\left(\vec{\omega}\times\vec{r}\,'\right)\right] \tag{2.118}$$

If $\left|\vec{\omega}\right| = 0$ then

$$\vec{\tau} - \vec{r}\,'\times m\frac{d^2\vec{r}_{oo'}}{dt^2} = \vec{r}\,'\times m\frac{d^2\vec{r}\,'}{dt^2} = \frac{d}{dt}\left(\vec{r}\,'\times m\frac{d\vec{r}\,'}{dt}\right) - \frac{d\vec{r}\,'}{dt}\times m\frac{d\vec{r}\,'}{dt} = \frac{d}{dt}\left(\vec{r}\,'\times m\frac{d\vec{r}\,'}{dt}\right) = \frac{d}{dt}\left(\vec{r}\,'\times\vec{p}\,'\right) = \frac{d\vec{L}\,'}{dt} \tag{2.119}$$

Hence, for observer in primed system that trusts in calculation of angular momentum and its rate using only primed variables without any reference to acceleration of the origin O′, there exists a fictitious torque

$$\vec{\tau}_{fictitious} = -\vec{r}\,' \times m\frac{d^2\vec{r}_{oo'}}{dt^2} = \vec{r}\,' \times \left(-m\frac{d^2\vec{r}_{oo'}}{dt^2}\right) = \vec{r}\,' \times \vec{F}_{inertialtranslations} \tag{2.120}$$

One could integrate this expression over all masses and obtain integral version

$$\vec{\tau}_{fictitious} = -\int\left(\vec{r}\,' \times \frac{d^2\vec{r}_{oo'}}{dt^2}\right)dm = -\left(\int\vec{r}\,'dm\right) \times \frac{d^2\vec{r}_{oo'}}{dt^2} = -m\vec{r}\,'_{CM} \times \frac{d^2\vec{r}_{oo'}}{dt^2} \tag{2.121}$$

With $\vec{r}\,'_{CM}$ being the CM in primed, that is, noninertial frame.

We already used this fictitious torque, without a proof, hence trusting in reader's faith in Eq. (2.57), for the purpose of deriving equations of motion for joint 2 using the Newton-Euler method.

Now we turn to case with $|\vec{\omega}| \neq 0$. Here

$$\vec{\tau} + \vec{r}\,' \times \left[-m\frac{d^2\vec{r}_{oo'}}{dt^2} - 2m\,\vec{\omega} \times \frac{d\vec{r}\,'}{dt} - m\frac{d\vec{\omega}}{dt} \times \vec{r}\,' - m\vec{\omega} \times \left(\vec{\omega} \times \vec{r}\,'\right)\right] = \frac{d\vec{L}\,'}{dt} \tag{2.122}$$

or in terms of familiar inertial forces

$$\vec{\tau} + \vec{r}\,' \times \left[\vec{F}_{acc.translations} + \vec{F}_{Coriolis} + \vec{F}_{acc.rotations} + \vec{F}_{Centrifugal}\right] = \frac{d\vec{L}\,'}{dt} \tag{2.123}$$

including Coriolis fictitious force

$$\vec{F}_{Coriolis} = -2m\,\vec{\omega} \times \frac{d\vec{r}\,'}{dt} \tag{2.124}$$

as well as inertial fictitious force due to angular acceleration of rotating axes

$$\vec{F}_{acc.rotations} = -m\frac{d\vec{\omega}}{dt} \times \vec{r}\,' \tag{2.125}$$

and finally, with fictitious centrifugal force

$$\vec{F}_{Centrifugal} = -m\vec{\omega} \times \left(\vec{\omega} \times \vec{r}\,'\right) \tag{2.126}$$

Integration of masses gives integral version of fictitious torque, for example for Coriolis force

$$\vec{\tau}_{Coriolis} = -2\int\vec{r}\,' \times \left(\vec{\omega} \times \frac{d\vec{r}\,'}{dt}\right)dm = -2\omega_i\hat{e}_p\int\varepsilon_{lkp}\varepsilon_{ijk}r'_l\frac{dr'_j}{dt}dm \tag{2.127}$$

or by using standard formula for triple vector product

$$\vec{\tau}_{Coriolis} = -2\int\vec{r}\,' \times \left(\vec{\omega} \times \frac{d\vec{r}\,'}{dt}\right)dm = -2\int\left[\vec{\omega}\left(\vec{r}\,' \cdot \frac{d\vec{r}\,'}{dt}\right) - \frac{d\vec{r}\,'}{dt}\left(\vec{r}\,' \cdot \vec{\omega}\right)\right]dm \tag{2.128}$$

Second term could be rewritten as

$$\frac{d\vec{r}\,'}{dt}\left(\vec{r}\,'\cdot\vec{\omega}\right)=\frac{d}{dt}\left[\vec{r}\,'\left(\vec{r}\,'\cdot\vec{\omega}\right)\right]-\vec{r}\,'\left(\frac{d\vec{r}\,'}{dt}\cdot\vec{\omega}\right)-\vec{r}\,'\left(\vec{r}\,'\cdot\frac{d\vec{\omega}}{dt}\right) \tag{2.129}$$

Hence,

$$\vec{\tau}_{Coriolis}=-2\int\left[\vec{\omega}\frac{1}{2}\frac{d}{dt}\left(\vec{r}\,'\cdot\vec{r}\,'\right)-\frac{d}{dt}\left[\vec{r}\,'\left(\vec{r}\,'\cdot\vec{\omega}\right)\right]+\vec{r}\,'\left(\frac{d\vec{r}\,'}{dt}\cdot\vec{\omega}\right)+\vec{r}\,'\left(\vec{r}\,'\cdot\frac{d\vec{\omega}}{dt}\right)\right]dm \tag{2.130}$$

Now it is handy to introduce dyadic product

$$\int|\vec{\boldsymbol{r}}\,'\vec{\boldsymbol{r}}\,'|dm=-\tilde{\boldsymbol{I}}+\frac{1}{2}Tr\left(\tilde{\boldsymbol{I}}\right)\tilde{\mathbf{1}} \tag{2.131}$$

$$\int|\vec{\boldsymbol{r}}\,'\vec{\boldsymbol{r}}\,'|dm=\int\begin{pmatrix}x'^2 & x'y' & x'z'\\ x'y' & y'^2 & y'z'\\ x'z' & y'z' & z'^2\end{pmatrix}dm \tag{2.132}$$

And moment of inertia tensor

$$\tilde{\boldsymbol{I}}=\int\begin{pmatrix}y'^2+z'^2 & -x'y' & -x'z'\\ -x'y' & x'^2+z'^2 & -y'z'\\ -x'z' & -y'z' & x'^2+y'^2\end{pmatrix}dm \tag{2.133}$$

with trace, that is, sum of diagonal elements

$$Tr\left(\tilde{\boldsymbol{I}}\right)=\int 2\left(x'^2+y'^2+z'^2\right)dm \tag{2.134}$$

The first term in Eq. (2.130) can be expressed in terms of trace,

$$-\vec{\omega}\int\frac{d}{dt}\left(\vec{r}\,'\cdot\vec{r}\,'\right)dm=-\vec{\omega}\frac{1}{2}\frac{d}{dt}Tr\left(\tilde{\boldsymbol{I}}\right) \tag{2.135}$$

The second term can be also conveniently expressed in terms of moment of inertia tensor

$$2\int\frac{d}{dt}\left[\vec{r}\,'\left(\vec{r}\,'\cdot\vec{\omega}\right)\right]dm=2\int\frac{d}{dt}\left[|\vec{\boldsymbol{r}}\,'\rangle\langle\vec{\boldsymbol{r}}\,'|\vec{\omega}\right]dm=2\frac{d}{dt}\left\{\left[-\tilde{\boldsymbol{I}}+\frac{1}{2}Tr\left(\tilde{\boldsymbol{I}}\right)\tilde{\mathbf{1}}\right]_{ij}\hat{e}_i\omega_j\right\} \tag{2.136}$$

And similarly, the fourth term

$$-2\int\vec{r}\,'\left(\vec{r}\,'\cdot\frac{d\vec{\omega}}{dt}\right)dm=-2\int|\vec{\boldsymbol{r}}\,'\rangle\langle\vec{\boldsymbol{r}}\,'|\frac{d\vec{\omega}}{dt}>dm=-2\left[-\tilde{\boldsymbol{I}}+\frac{1}{2}Tr\left(\tilde{\boldsymbol{I}}\right)\tilde{\mathbf{1}}\right]_{ij}\hat{e}_i\frac{d\omega_j}{dt} \tag{2.137}$$

The third term in Eq. (2.130) depends somewhat on details of motion, for example, if

$$\frac{d\vec{r}\,'}{dt}=\vec{\omega}_*\times\vec{r}\,' \tag{2.138}$$

then

$$-2\int \vec{r}'\left(\frac{d\vec{r}'}{dt}\cdot\vec{\omega}\right)dm = -2\int \vec{r}'\left[\left(\vec{\omega}_*\times\vec{r}'\right)\cdot\vec{\omega}\right]dm = -2\int \vec{r}'\left[\left(\vec{\omega}\times\vec{\omega}_*\right)\cdot\vec{r}'\right]dm$$
$$= -2\left[-\tilde{\boldsymbol{I}}+\frac{1}{2}Tr\left(\tilde{\boldsymbol{I}}\right)\tilde{\mathbf{1}}\right]_{ij}\hat{e}_i\varepsilon_{klj}\omega_k\omega_{*l} \quad (2.139)$$

Therefore,

$$\vec{\tau}_{Coriolis} = -\vec{\omega}\frac{1}{2}\frac{d}{dt}Tr\left(\tilde{\boldsymbol{I}}\right)+2\frac{d}{dt}\left\{\left[-\tilde{\boldsymbol{I}}+\frac{1}{2}Tr\left(\tilde{\boldsymbol{I}}\right)\tilde{\mathbf{1}}\right]_{ij}\hat{e}_i\omega_j\right\}-2\left[-\tilde{\boldsymbol{I}}+\frac{1}{2}Tr\left(\tilde{\boldsymbol{I}}\right)\tilde{\mathbf{1}}\right]_{ij}\hat{e}_i\varepsilon_{klj}\omega_k\omega_{*l}$$
$$-2\left[-\tilde{\boldsymbol{I}}+\frac{1}{2}Tr\left(\tilde{\boldsymbol{I}}\right)\tilde{\mathbf{1}}\right]_{ij}\hat{e}_i\frac{d\omega_j}{dt} \quad (2.140)$$

Similar integral expressions can be easily obtained for fictitious torques due to fictitious forces $\vec{F}_{acc.rotations}$ and $\vec{F}_{Centrifugal}$.

By this point, reader should have already got a good grasp of increasing complexities of directly obtaining equations of motion within Newton-Euler method. These complexities quickly become unsurmountable and untraceable with increasing number of degrees of freedom. And hence, in practice, many stick with Euler-Lagrange method requiring only proper expression for single Lagrange function or Lagrangian.

For example, following Euler-Lagrange method one does not need to know the exact expressions for torques due to fictitious forces in non*inertial* system! A big advantage.

2.3.7 Hamilton Method

There exists yet another method to address dynamics of the system which still further decompose these computational complexities. It is called *Hamilton method.*

Consider generalized momenta defined as

$$p_q = \frac{\partial L}{\partial \dot{q}} \quad (2.141)$$

With L being Lagrange function we introduced in Eq. (2.86) and q is generalized coordinate, that is, degree of freedom. It is either θ_1 or θ_2 in case of our favorite planar 2 link template. Hence,

$$p_{\theta 1} = \frac{\partial L}{\partial \dot{\theta}_1} = \left[m_1L_{CM1}^2 + I_1 + m_2\left(L_1^2 + 2L_1L_{CM2}\cos\theta_2\right)\right]\dot{\theta}_1 + m_2L_1L_{CM2}\cos\theta_2\dot{\theta}_2 + \left(m_2L_{CM2}^2 + I_2\right)\left(\dot{\theta}_1 + \dot{\theta}_2\right)$$
$$= A\dot{\theta}_1 + B\dot{\theta}_2 \quad (2.142)$$

$$p_{\theta 2} = \frac{\partial L}{\partial \dot{\theta}_2} = m_2L_1L_{CM2}\cos\theta_2\dot{\theta}_1 + \left(m_2L_{CM2}^2 + I_2\right)\left(\dot{\theta}_1 + \dot{\theta}_2\right) = C\dot{\theta}_1 + D\dot{\theta}_2 \quad (2.143)$$

This can be also rewritten in matrix notation,

$$\begin{pmatrix} p_{\theta 1} \\ p_{\theta 2} \end{pmatrix} = \begin{pmatrix} A & B \\ C & D \end{pmatrix} \begin{pmatrix} \dot{\theta}_1 \\ \dot{\theta}_2 \end{pmatrix} \tag{2.144}$$

One could now solve for angular velocities,

$$\dot{\theta}_1 = \frac{Dp_{\theta 1} - Bp_{\theta 2}}{AD - BC} \tag{2.145}$$

$$\dot{\theta}_2 = \frac{Ap_{\theta 2} - Cp_{\theta 1}}{AD - BC} \tag{2.146}$$

or in matrix notation,

$$\begin{pmatrix} \dot{\theta}_1 \\ \dot{\theta}_2 \end{pmatrix} = \begin{pmatrix} A & B \\ C & D \end{pmatrix}^{-1} \begin{pmatrix} p_{\theta 1} \\ p_{\theta 2} \end{pmatrix} = \frac{1}{AD - BC} \begin{pmatrix} D & -B \\ A & -C \end{pmatrix} \begin{pmatrix} p_{\theta 1} \\ p_{\theta 2} \end{pmatrix} \tag{2.147}$$

The kinetic energy, see Eq. (2.92), we obtained previously within the scope of Euler-Lagrange method can be expressed as quadratic function in angular velocities and also written in matrix form

$$\begin{aligned} 2K &= \left(m_1 L_{CM1}^2 + I_1\right)\dot{\theta}_1^2 + m_2\left[L_1^2\dot{\theta}_1^2 + 2L_1 L_{CM2}\dot{\theta}_1\left(\dot{\theta}_1 + \dot{\theta}_2\right)\cos\theta_2 + L_{CM2}^2\left(\dot{\theta}_1 + \dot{\theta}_2\right)^2\right] + I_2\left(\dot{\theta}_1 + \dot{\theta}_2\right)^2 \\ &= \left(\dot{\theta}_1 \ \ \dot{\theta}_2\right) \begin{pmatrix} m_1 L_{CM1}^2 + I_1 + m_2\left(L_1^2 + 2L_1 L_{CM2}\cos\theta_2 + L_{CM2}^2\right) + I_2 & m_2 L_{CM2}\left(L_1\cos\theta_2 + L_{CM2}\right) \\ m_2 L_{CM2}\left(L_1\cos\theta_2 + L_{CM2}\right) & m_2 L_{CM2}^2 + I_2 \end{pmatrix} \begin{pmatrix} \dot{\theta}_1 \\ \dot{\theta}_2 \end{pmatrix} \end{aligned} \tag{2.148}$$

By using transformation equation (2.147), one could now express kinetic energy as quadratic function in generalized momenta:

$$\begin{aligned} K &= \left(\dot{\theta}_1 \ \ \dot{\theta}_2\right) \begin{pmatrix} K_{11} & K_{12} \\ K_{21} & K_{22} \end{pmatrix} \begin{pmatrix} \dot{\theta}_1 \\ \dot{\theta}_2 \end{pmatrix} = \frac{1}{(AD - BC)^2}\left(p_{\theta 1} \ \ p_{\theta 2}\right) \begin{pmatrix} D & A \\ -B & -C \end{pmatrix} \begin{pmatrix} K_{11} & K_{12} \\ K_{21} & K_{22} \end{pmatrix} \begin{pmatrix} D & -B \\ A & -C \end{pmatrix} \begin{pmatrix} p_{\theta 1} \\ p_{\theta 2} \end{pmatrix} \\ &= \frac{1}{(AD - BC)^2}\left(p_{\theta 1} \ \ p_{\theta 2}\right) \begin{pmatrix} D^2 K_{11} + DA(K_{12} + K_{21}) + A^2 K_{22} & -DBK_{11} - DCK_{12} - ABK_{21} - ACK_{22} \\ -BDK_{11} - BAK_{12} - CDK_{21} - CAK_{22} & B^2 K_{11} + BC(K_{12} + K_{21}) + C^2 K_{22} \end{pmatrix} \begin{pmatrix} p_{\theta 1} \\ p_{\theta 2} \end{pmatrix} \end{aligned} \tag{2.149}$$

Hamilton function or *Hamiltonian* can be now constructed as

$$H = \left(\sum_{i=1}^{2} \dot{\theta}_i p_{\theta i}\right) - L \tag{2.150}$$

And it is straightforward to show that

$$H(q, p) = K(q, p) + U(q) \tag{2.151}$$

To simplify the problem in hand, first we will consider the case without forces, that is, end-effector's $F_{xEE*} = F_{yEE*} = 0$, and also zero joint torques, that is, $\tau_1 = \tau_2 = 0$. In this case, Hamiltonian representing total energy of the system is a constant of motion.

Hamilton's equations are defined as

$$\frac{dp}{dt} = -\frac{\partial H}{\partial q}, \frac{dq}{dt} = \frac{\partial H}{\partial p} \tag{2.152}$$

For each generalized coordinate there are two first-order differential equations. Reader may recall that there was only one second-order differential equation per degree of freedom in Euler-Lagrange method.

For our specific system:

$$\frac{dp_{\theta 1}}{dt} = -\frac{\partial H}{\partial \theta_1}, \frac{d\theta_1}{dt} = \frac{\partial H}{\partial p_{\theta 1}} \tag{2.153}$$

$$\frac{dp_{\theta 2}}{dt} = -\frac{\partial H}{\partial \theta_2}, \frac{d\theta_2}{dt} = \frac{\partial H}{\partial p_{\theta 2}} \tag{2.154}$$

that is, four first-order differential equations instead of two second-order differential equations as obtained within Euler–Lagrange method.

In case of nonzero joint torques these equation are formulated as

$$\frac{dp_{\theta 1}}{dt} = \tau_1 - \frac{\partial H}{\partial \theta_1}, \frac{d\theta_1}{dt} = \frac{\partial H}{\partial p_{\theta 1}} \tag{2.155}$$

$$\frac{dp_{\theta 2}}{dt} = \tau_2 - \frac{\partial H}{\partial \theta_2}, \frac{d\theta_2}{dt} = \frac{\partial H}{\partial p_{\theta 2}} \tag{2.156}$$

And in case of end-effector's forces these equations become

$$\frac{dp_{\theta 1}}{dt} = \tau_1 + \tau_{1FEE*} - \frac{\partial H}{\partial \theta_1}, \frac{d\theta_1}{dt} = \frac{\partial H}{\partial p_{\theta 1}} \tag{2.157}$$

$$\frac{dp_{\theta 2}}{dt} = \tau_2 + \tau_{2FEE*} - \frac{\partial H}{\partial \theta_2}, \frac{d\theta_2}{dt} = \frac{\partial H}{\partial p_{\theta 2}} \tag{2.158}$$

Euler-Lagrange and Hamilton method are equivalent and they give the same result. The only difference is in computational method. Euler-Lagrange method requires solving second-order differential equation per degree of freedom and Hamilton method requires solving two first-order differential equations per degree of freedom. Both methods require knowledge of two times number of degree of freedom boundary conditions.

2.3.8 Lagrange Multipliers, Forces of Constraints, and Ground (or Base) Reaction Forces

Till now we addressed joint torques and end-effector forces. However, we still have not discussed ground (or base) reaction forces. Traditionally, the word "reaction" symbolizes the dependence of this force on kinematics and other external forces.

Ground (or manipulator base) reaction force can be simply obtained from Newton's second law

$$F_{grX} = (m_1 + m_2)\ddot{x}_{CM} - F_{xEE*} \tag{2.159}$$

$$F_{grY} = (m_1 + m_2)(\ddot{y}_{CM} + g) - F_{yEE*} \tag{2.160}$$

Another way to obtain ground reaction force would be to use Euler-Lagrange equations with two extra constrained degrees of freedom and two equations of constraint. This way of computation of ground reaction forces is much more complex than provided by Eqs. (2.159) and (2.160), but we nonetheless present it to illustrate concept of *Lagrange multipliers.*

The two extra "degrees of freedom" could be coordinates of joint 1, that is, (x_1, y_1) and two extra equations of constraints could be

$$g_1(\theta_1, \theta_2, x_1, y_1) = x_1 = 0 \tag{2.161}$$

$$g_2(\theta_1, \theta_2, x_1, y_1) = y_1 = 0 \tag{2.162}$$

The Euler-Lagrange equations with *undetermined multipliers* are

$$\left\{\frac{d}{dt}\left(\frac{\partial L}{\partial \dot{q}}\right) - \frac{\partial L}{\partial q}\right\}_{constrained} = \tau_q + \sum_{j=1}^{2} \lambda_j(t)\frac{\partial g_j}{\partial q} \tag{2.163}$$

With two extra degrees of freedom "new" Lagrangian can be obtained from "old" one

$$\begin{aligned} L &= L_{old} - \frac{1}{2}(m_1 + m_2)\left[\left(\dot{x}_{CM}^2 + \dot{y}_{CM}^2\right) - (\dot{x}_{CM} + \dot{x}_1)^2 - \left(\dot{y}_{CM} + \dot{y}_1\right)^2\right] - (m_1 + m_2)gy_1 \\ &= L_{old} + \frac{1}{2}(m_1 + m_2)\left[2\dot{x}_{CM}\dot{x}_1 + {\dot{x}_1}^2 + 2\dot{y}_{CM}\dot{y}_1 + {\dot{y}_1}^2\right] - (m_1 + m_2)gy_1 \end{aligned} \tag{2.164}$$

The two "old" Euler-Lagrange equations will not provide anything new. However, the two new Euler-Lagrange equations, that is, one for x_1

$$\left\{\frac{d}{dt}\left(\frac{\partial L}{\partial \dot{x}_1}\right) - \frac{\partial L}{\partial x_1}\right\}_{constrained} = (m_1 + m_2)\ddot{x}_{CM} \tag{2.165}$$

$$F_{xEE*} + \sum_{j=1}^{2} \lambda_j(t)\frac{\partial g_j}{\partial x_1} = F_{xEE*} + \lambda_1(t) \tag{2.166}$$

$$(m_1 + m_2)\ddot{x}_{CM} = F_{xEE*} + \lambda_1(t) \tag{2.167}$$

and one for x_2

$$\left\{\frac{d}{dt}\left(\frac{\partial L}{\partial \dot{y}_1}\right) - \frac{\partial L}{\partial y_1}\right\}_{constrained} = (m_1 + m_2)\ddot{y}_{CM} + (m_1 + m_2)g \tag{2.168}$$

$$F_{yEE*} + \sum_{j=1}^{2} \lambda_j(t)\frac{\partial g_j}{\partial y_1} = F_{yEE*} + \lambda_2(t) \tag{2.169}$$

$$(m_1 + m_2)\ddot{y}_{CM} + (m_1 + m_2)g = F_{yEE*} + \lambda_2(t) \tag{2.170}$$

provide nice interpretation for Lagrange multipliers. Namely, the forces of constraints are exactly ground reaction forces we obtained earlier

$$\lambda_1(t) = F_{grX} \tag{2.171}$$

$$\lambda_2(t) = F_{grY} \tag{2.172}$$

2.3.9 Locomotor Dynamics, Ground Reaction Forces, and ZMP

Imagine adding additional segment that we could interpret as "foot" attached to O_1 joint of our template system. The foot is assumed flat on the ground and subject to ground reaction forces (Fig. 2.5).

Eq. (2.76) for joint 1, formulated within the Newton-Euler method, is now modified to

$$\vec{\tau}_1 + \left(\vec{r}_{CM\,w/foot} - \vec{r}_{O1}\right) \times (m - m_{foot})\,\vec{g} + \left(\vec{r}_{EE} - \vec{r}_{O1}\right) \times \vec{F}_{EE*} = \frac{d\left(\vec{L}_{1\,(O1)} + \vec{L}_{2\,(O1)}\right)}{dt} \tag{2.173}$$

with $\vec{r}_{CM\,w/foot}$ being CM of links 1 and 2 and m_{foot} being the foot mass.

Note that there is no moment due to ground reaction forces in Eq. (2.173). If we added the term that would be an erroneous example of "double counting," as torques applied to adjacent links are equal in magnitude and opposite in direction.

One may now express,

$$\vec{r}_{CMw/foot} = \frac{m\vec{r}_{CM} - m_{foot}\vec{r}_{CMfoot}}{m - m_{foot}} = \vec{r}_{CM} - \frac{m_{foot}\left(\vec{r}_{CMfoot} - \vec{r}_{CM}\right)}{m - m_{foot}} \tag{2.174}$$

and rewrite Eq. (2.173) as

$$\vec{\tau}_1 + \left(\vec{r}_{CM} - \vec{r}_{O1}\right) \times (m - m_{foot})\,\vec{g} - \left(\vec{r}_{CMfoot} - \vec{r}_{CM}\right) \times m_{foot}\,\vec{g} + \left(\vec{r}_{EE} - \vec{r}_{O1}\right) \times \vec{F}_{EE*} = \frac{d\left(\vec{L}_{1\,(O1)} + \vec{L}_{2\,(O1)}\right)}{dt} \tag{2.175}$$

Because foot is not moving, $\vec{L}_{foot\,(O1)} = 0$ and

$$\frac{d\left(\vec{L}_{1\,(O1)} + \vec{L}_{2\,(O1)}\right)}{dt} = \frac{d\left(\vec{L}_{1\,(O1)} + \vec{L}_{2\,(O1)} + \vec{L}_{foot\,(O1)}\right)}{dt} = \frac{d\vec{L}_{CM\,(O1)}}{dt} + \frac{d\left(\vec{L}_{1\,(CM)} + \vec{L}_{2\,(CM)} + \vec{L}_{foot\,(CM)}\right)}{dt} \tag{2.176}$$

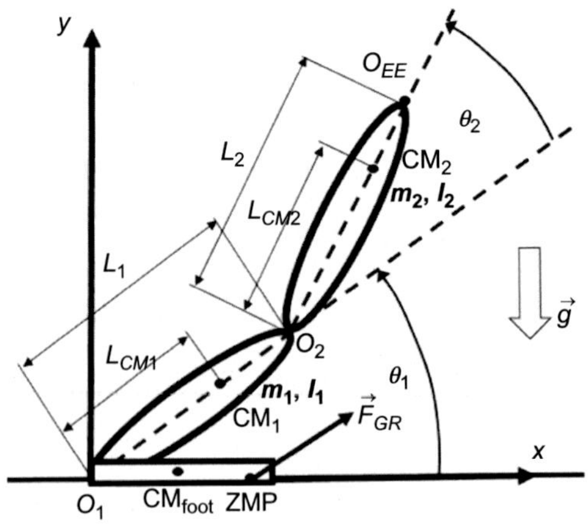

FIG. 2.5 Template with added foot segment.

Here, we utilized the fact that the angular momentum is sum of the CM orbital angular momentum and spin angular momentum.

Rate of change of orbital angular momentum is by definition,

$$\frac{d\vec{L}_{CM\,(O1)}}{dt}=\frac{d\left[\left(\vec{r}_{CM}-\vec{r}_{O1}\right)\times m\vec{v}_{CM}\right]}{dt}=\left(\vec{r}_{CM}-\vec{r}_{O1}\right)\times m\vec{a}_{CM} \tag{2.177}$$

To address the rate of change of spin angular momentum, it is prudent to introduce the *zero moment point (ZMP)* concept [12, 13].

The most important notion of the ZMP concept is that it resolves the ground reaction force distribution to a single point such that horizontal component of ground moment about the CM can be expressed as

$$\vec{\tau}_{ground}\Big|_{horizontal}=\left[\left(\vec{r}_{ZMP}-\vec{r}_{CM}\right)\times\vec{F}_{GR}\right]_{horizontal} \tag{2.178}$$

with $\vec{F}_{GR}$ being the net ground reaction force.

Although for flat horizontal ground surfaces, the ZMP is equal to the center of pressure (CP), the points are distinct for irregular ground surfaces [13]. However, quite generally, ZMP or/and CP must be located within ground support polygon or minimal convex hull encompassing all foot-ground contact surfaces.

The rate of change of spin angular momentum is

$$\frac{d\left(\vec{L}_{1\,(CM)}+\vec{L}_{2\,(CM)}+\vec{L}_{foot\,(CM)}\right)}{dt}=\left(\vec{r}_{EE}-\vec{r}_{CM}\right)\times\vec{F}_{EE*}+\left(\vec{r}_{ZMP}-\vec{r}_{CM}\right)\times\vec{F}_{GR} \tag{2.179}$$

One could now combine derivative terms, expressed by Eqs. (2.177) and (2.179) and solve for

$$\vec{\tau}_{1}=\left(\vec{r}_{ZMP}-\vec{r}_{O1}\right)\times\vec{F}_{GR}+\left(\vec{r}_{CMfoot}-\vec{r}_{O1}\right)\times m_{foot}\,\vec{g} \tag{2.180}$$

As expected Eq. (2.180) is nothing else than rotational version of Newton's second law for the foot.

One should notice that torque $\vec{\tau}_1$ in Eq. (2.173) or (2.180) is numerically same as the one obtained by Eq. (2.76) for same motion of links 1 and 2 and same end-effector force. Novelty here is only Eq. (2.180) relating joint torque and ground reaction force.

2.3.10 Static Versus Dynamic Balance

The notion of *static balance* is most closely related to the notion of equilibrium in physics. The necessary condition is that both net external forces and net external torques must be zero. Thus, ground reaction force must exactly balance the body weight. Still further, ZMP from which ground reaction force originates must coincide with the vertical projection of CM onto ground (see Fig. 2.6A).

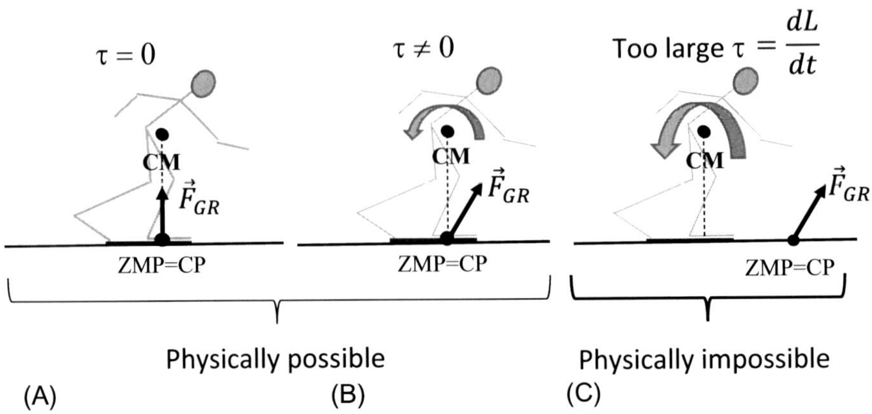

FIG. 2.6 Relationship between ZMP and moment about CM.

The most important notion of the ZMP concept [12, 13], applicable to both single and multi-leg ground support phases, is that it resolves the ground reaction force distribution to a single point such that horizontal component of moment about the CM can be expressed as in Eq. (2.178).

This moment is also equal to the rate of change of spin angular momentum [14], that is, the rate of change of angular momentum about CM,

$$\vec{\tau} = \frac{d\vec{L}_{CM}}{dt} \tag{2.181}$$

Although for flat horizontal ground surfaces, the ZMP is equal to the CP, the points are distinct for irregular ground surfaces [13]. However, quite generally, ZMP or/and CP must be located within the ground support polygon or minimal convex hull encompassing all foot-ground contact surfaces. This in turn constrain the moment about CM, as emphasized in Fig. 2.6B and C.

Therefore, the necessary condition for static balance is that the vertical projection of the CM onto the ground must be within the foot ground support.

Static balance can be stable (if passive response provides restoring force for all disturbance directions), unstable (if passive response provides destabilizing force for all disturbance directions), or saddle (if system response forces are restoring for some disturbance directions, and destabilizing for others).

The desired posture is defined both by joint angles and by global orientation of body (here the CM height is considered a dependent variable providing that feet are in contact with the ground without penetrating it). Clearly, not all desired postures can be statically balanced. In Fig. 2.7 global orientation is varied while joint angles are kept the same. Posture (C) is not statically balanced, postures (A) and (E) exhibit unstable static balance, and postures (B) and (D) possess stable static balance.

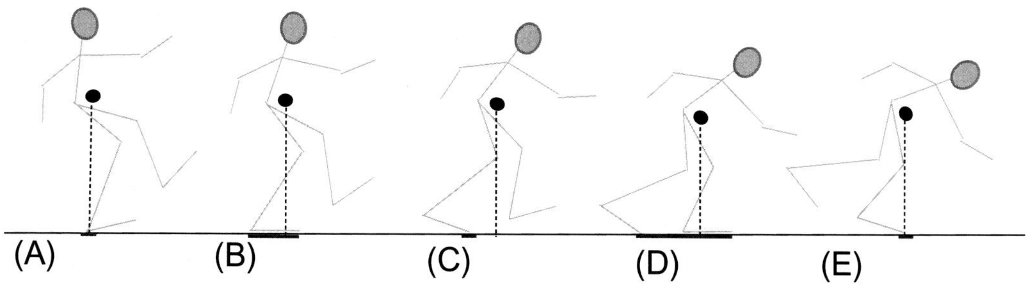

FIG. 2.7 Affecting static balance by varying global orientation for identical joint angles.

If goal is keeping the body "balanced" during specific dynamic movement, then successful realization of such activity is called *dynamic balance*. Clearly, a desired trajectory should be physically feasible (e.g., ZMP or CP should be inside physical ground support; limbs should not penetrate the ground or each other, etc.) and body should be capable of performing the desired task, that is, musculoskeletal system should be able to achieve desired joint angles, velocities, and accelerations.

Human walking is a well-known example of dynamically balanced activity [13,14]. Interestingly, the vertical projection of the CM is inside the support base only during the double-support phase (approximately 20% of the total walking action for younger population and about 30% for older population). Clearly, condition for dynamic balance is very different from condition for static balance. And response to errors and disturbances is not only passive but also a result of actively sensed, controlled, and actuated performance of human sensory-motor-control system.

2.3.11 Zero-Moment Versus Moment Balance Strategy

To illustrate the concept of two important balance strategies [13, 14] used for dynamic balancing, consider a simple system, Fig. 2.8 on the right, with massless foot (L_1), ankle joint (θ), and leg (L_2). A massive link is depicted rotating about its CM with the rate of change of spin angular momentum, denoted by the moment $\vec{\tau}$.

Quite generally, for any given system supported by a horizontal contact surface, one can express horizontal force as

$$F_{GRx} = M\left(a_{CMy} + g\right)\frac{x_{CM} - x_{ZMP}}{y_{CM}} - \frac{\tau_z}{y_{CM}} \tag{2.182}$$

which allows interpretation of the zero-moment balance term:

$$F_{GRx\,ZERO-MOMENT} = M\left(a_{CMy} + g\right)\frac{x_{CM} - x_{ZMP}}{y_{CM}} \tag{2.183}$$

and moment balance term:

$$F_{GRx\,MOMENT} = -\frac{\tau_z}{y_{CM}} \tag{2.184}$$

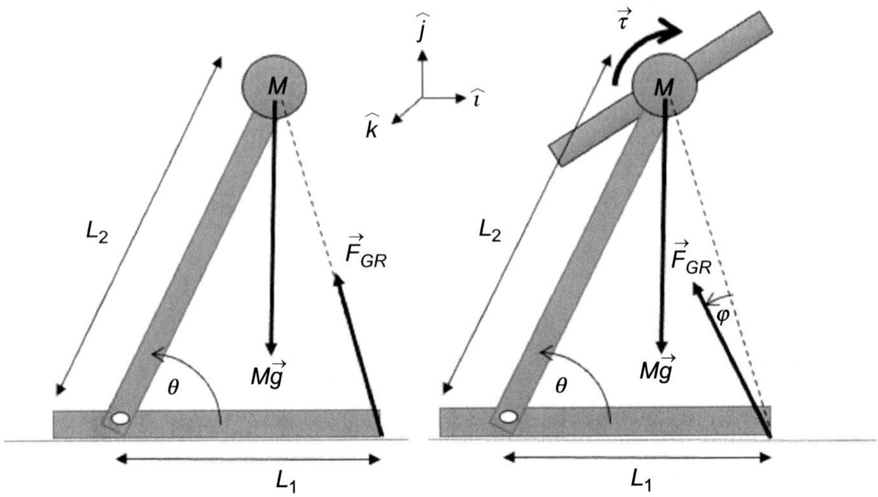

FIG. 2.8 Simplified system that can exhibit only zero-moment balance strategy (left) and system that can utilize both zero-moment and moment balance strategies (right).

For any given state and CM vertical acceleration, one can manipulate horizontal force by:

(1) repositioning of ZMP or CP, that is, *zero-moment balance strategy,* Eq. (2.183),
(2) modulation of moment, that is, *moment balance strategy,* Eq. (2.184),

while potentially changing support base (e.g., by stepping to new location).

Numerous simplified model of walking utilize zero-moment balance strategy either by addressing inverted pendulum dynamics or some variant of it, for example, with constant CM height. Experimental gait studies also indicate that both spin angular momentum and moment about CM are very small [14].

Clearly, the zero-moment balance strategy is more energy cost effective and it can be continuously used for a long time. However, the moment balance strategy is often necessary for dynamic movements and to correct sudden dynamic perturbations. Thus, it can easily dominate the system for at least short period of time. "Flywheel" strategy or moment balance strategy is often utilized to reject large perturbations and it is directly "wired" as one of human reflexes. For example, if a stance foot slips, the rest of body produces short but very fast movement, which resembles motion of the heavy link in the figure above. Note that the vertical projection of CM with zero speed of CM can be well outside the support base and while keeping the support base unchanged the "flywheel" moment strategy can restore CM back to statically balanced state above the support base. For example, consider the act of tight rope walking and the lateral stability that is enforced by the moment balance strategy.

2.3.12 Example of Reliable Balance Metric for Dynamic Balance

If desired task is validated by physics and the system is capable to perform it, it is now prudent to introduce some metric to judge how successful or unsuccessful "balancing" is. Is there a clear boundary between successful and unsuccessful "balancing"? Clearly, if a human or a robot falls we know the "balancing" was unsuccessful.

Can we tell that a fall is imminent even before it happens? A reliable metric of "balancing" must be able to distinctively capture that fine line between successful and unsuccessful "balancing."

Quite generally, a fall is imminent if there is no action within system capabilities to prevent it.

Going back to the example in Fig. 2.8 on the right imagine that for a given ankle joint angle θ system gets a sudden push to the right resulting in an angular speed $\dot{\theta}$ in the clockwise direction. The goal now is to find the maximal or critical clockwise angular speed $\dot{\theta}_{critical}$ of the ankle joint for a given θ such that angular speed of M can be brought to zero before CM reaches the ground.

The resulting critical ankle angular speed for a given θ then provides the metrics that can capture a fine line between balanced and imbalanced dynamic states. In other words, for a given ankle joint θ we can judge whether an imbalanced state can be stabilized or not. If the ankle angular speed is larger than the critical ankle angular speed, the system is out of balance. If the ankle angular speed is smaller than the critical ankle angular speed, the system is still balanced or can be stabilized.

Our reliable metric of "balancing" can be now expressed by using different:<AQ: >

$$\dot{\theta}_{critical}(\theta) - \dot{\theta} > 0 \ \textit{balance successful} \tag{2.185}$$

$$\dot{\theta}_{critical}(\theta) - \dot{\theta} < 0 \ \textit{fall imminent} \tag{2.186}$$

To simplify the problem, we may only require that the CM ground projection converges to the edge of the foot support polygon (like in the case of static balance) without ever crossing it. Although, in reality, the CM ground projection can even leave the support base and due to nonzero moments about CM eventually return to the support base, hence restoring the static balance condition.

Classical mechanics' equations are reversible in time. Hence, we could assume that the initial condition is asymptotically close to the CM ground projection coinciding with the front edge of foot support polygon and with zero velocity of the CM. Subsequently, we will anticipate the largest possible force generating the counterclockwise motion of leg.

Consider Newton's second law:

$$M\vec{g} + \vec{F}_{GR} = ML_2\dot{\theta}^2\left(-\cos\theta\hat{i} - \sin\theta\hat{j}\right) + ML_2\ddot{\theta}\left(-\sin\theta\hat{i} + \cos\theta\hat{j}\right) \tag{2.187}$$

The first and second terms on the right-hand side is proportional to radial and tangential accelerations, respectively.

In the special case of zero moment about the CM, the largest possible ground reaction force originates at the front edge of the support base and points in direction of the CM; see the model on the left in Fig. 2.8. Hence,

$$\vec{F}_{GR}(\tau=0)=\left|\vec{F}_{GR}(\tau=0)\right|\frac{(L_2\cos\theta-L_1)\hat{i}+L_2\sin\theta\hat{j}}{\sqrt{(L_2\cos\theta-L_1)^2+(L_2\sin\theta)^2}} \tag{2.188}$$

In the case of nonzero moment, the direction of ground reaction force is rotated by an angle which depends on angle θ and moment τ about the CM:

$$\varphi=\sin^{-1}\frac{\tau}{F_{GR}\sqrt{(L_2\cos\theta-L_1)^2+(L_2\sin\theta)^2}} \tag{2.189}$$

And ground reaction force can be expressed as

$$\vec{F}_{GR}(\tau\neq0)=\left|\vec{F}_{GR}(\tau\neq0)\right|\frac{[(L_2\cos\theta-L_1)\cos\varphi-L_2\sin\theta\sin\varphi]\hat{i}+[(L_2\cos\theta-L_1)\sin\varphi+L_2\sin\theta\cos\varphi]\hat{j}}{\sqrt{(L_2\cos\theta-L_1)^2+(L_2\sin\theta)^2}} \tag{2.190}$$

We could substitute this expression into force equation (2.187) and analyze each component separately.

$$F_{GR}\frac{(L_2\cos\theta-L_1)\cos\varphi-L_2\sin\theta\sin\varphi}{\sqrt{(L_2\cos\theta-L_1)^2+(L_2\sin\theta)^2}}=-ML_2(\dot{\theta}^2\cos\theta+\ddot{\theta}\sin\theta) \tag{2.191}$$

$$-Mg+F_{GR}\frac{(L_2\cos\theta-L_1)\sin\varphi+L_2\sin\theta\cos\varphi}{\sqrt{(L_2\cos\theta-L_1)^2+(L_2\sin\theta)^2}}=-ML_2(\dot{\theta}^2\sin\theta-\ddot{\theta}\cos\theta) \tag{2.192}$$

For example, we could solve for magnitude of ground reaction force ($F_{GR}=\left|\vec{F}_{GR}(\tau\neq0)\right|$) using Eq. (2.191) and then substitute it to Eq. (2.192) to obtain

$$-Mg-ML_2(\dot{\theta}^2\cos\theta+\ddot{\theta}\sin\theta)\frac{(L_2\cos\theta-L_1)\sin\varphi+L_2\sin\theta\cos\varphi}{(L_2\cos\theta-L_1)\cos\varphi-L_2\sin\theta\sin\varphi}=-ML_2(\dot{\theta}^2\sin\theta-\ddot{\theta}\cos\theta) \tag{2.193}$$

And this allows us to solve for the critical ankle angular acceleration:

$$\ddot{\theta}(\tau\neq0)=\frac{L_2\dot{\theta}^2(\sin\theta-\gamma\cos\theta)-g}{L_2(\cos\theta+\gamma\sin\theta)}>\ddot{\theta}(\varphi=0)=\frac{L_2\dot{\theta}^2\left(\sin\theta-\dfrac{L_2\sin\theta}{L_2\cos\theta-L_1}\cos\theta\right)-g}{L_2\left(\cos\theta+\dfrac{L_2\sin\theta}{L_2\cos\theta-L_1}\sin\theta\right)} \tag{2.194}$$

with

$$\gamma=\frac{(L_2\cos\theta-L_1)\sin\varphi+L_2\sin\theta\cos\varphi}{(L_2\cos\theta-L_1)\cos\varphi-L_2\sin\theta\sin\varphi} \tag{2.195}$$

This second-order differential equation can be solved by integration based on two initial conditions and applied moment τ about the CM. For example, $\tau=const.$ or it can be specified as $\tau=f(\theta)$.

It is not difficult to show that angular acceleration with a nonzero moment is always larger than angular acceleration with a zero moment, that is,

$$\ddot{\theta}(\tau \neq 0) > \ddot{\theta}(\tau = 0) = \frac{g(L_1 - L_2\cos\theta) - \dot{\theta}^2 L_2 L_1 \sin\theta}{L_2(L_2 - L_1\cos\theta)} \tag{2.196}$$

This also means that critical speed with a nonzero moment is always larger than the critical speed with a zero moment. Hence, this critical speed defining "reservoir of stability" is larger for systems that can utilize the moment balance strategy.

2.4 Propulsion in Fluids

The drag force proportional to the square of velocity is mainly responsible for the biological and biologically inspired propulsion in fluids characterized with high value of Reynolds number, as it is in the case for all birds, flying insects, and fish. The main contribution comes from a pressure drag force that is only concerned with the component of relative velocity that is perpendicular to the animal's body surface and acts in the same direction.

This topic will be revisited in Chapter 17 in the section on locomotion in fluids. Here, the goal is to introduce the basic concept and illustrate it with a simple example.

To obtain the net pressure drag force, one may want to integrate infinitesimal contributions across the entire surface area. The magnitude of this infinitesimal force is

$$dF_{\perp} = \frac{1}{2} C_D \rho v_{\perp}^2 dA \tag{2.197}$$

with symbol $\perp$ specifying the perpendicular direction, coefficient of drag in this microscopic description being idealized $C_D \cong 2$, and ρ being the fluid mass density.

Let us illustrate the effect of this force on plants. Specifically, imagine a flower petal "free falling."

To simplify this problem, we will artificially constrain the petals motion to two dimensions, described by horizontal x and vertical y coordinates. Also, assume that the petal of length L has a uniform mass distribution with the CM exactly in the middle of the petal. Its motion is completely described by the trajectory of the CM location $x_{CM}(t)$, $y_{CM}(t)$ and trajectory of the angle with respect to the horizontal axis $\theta(t)$.

The drag force will affect both the CM trajectory and orientation of the petal.

The relative velocity with respect to the still air of the small infinitesimal element of the petal at location r in the petal's local frame, that is, with $-L/2 < r < L/2$, is

$$\dot{x}(r) = \dot{x}_{CM} - r\dot{\theta}\sin\theta \tag{2.198}$$

$$\dot{y}(r) = \dot{y}_{CM} + r\dot{\theta}\cos\theta \tag{2.199}$$

that is,

$$\vec{v}(r) = (\dot{x}_{CM} - r\dot{\theta}\sin\theta)\hat{i} + (\dot{y}_{CM} + r\dot{\theta}\cos\theta)\hat{j} \tag{2.200}$$

Here we will assume that $|\dot{x}_{CM}|, |\dot{y}_{CM}| \gg |r\dot{\theta}|$ for $-L/2 < r < L/2$.

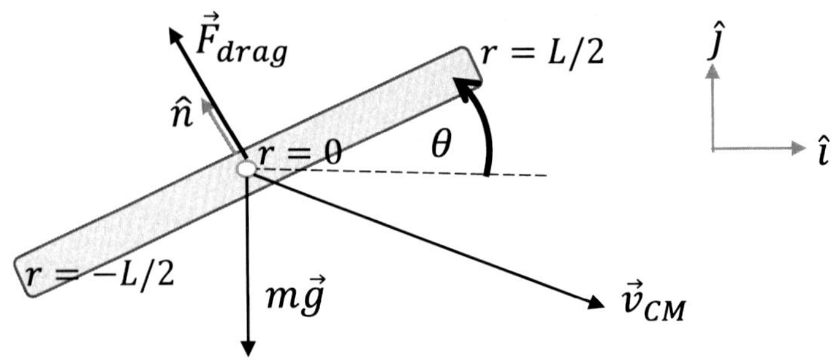

FIG. 2.9 For depicted center of mass velocity and positive (negative) angular velocity drag torque or moment of drag force will act into (out of) the page and decrease the angular speed and hence also decrease the spin angular momentum.

To simplify this even further let us imagine that the petal CM velocity and orientation is as shown in Fig. 2.9.

The perpendicular direction to the petal's surface is

$$\hat{n} = -\sin\theta\hat{i} + \cos\theta\hat{j} \tag{2.201}$$

Hence,

$$v_{\perp}^2 = \left[-\sin\theta(\dot{x}_{CM} - r\dot{\theta}\sin\theta) + \cos\theta(\dot{y}_{CM} + r\dot{\theta}\cos\theta)\right]^2 \tag{2.202}$$

and the infinitesimal drag force is

$$dF_{\perp}(r)\hat{n} = \frac{1}{2}C_D\rho\left[-\sin\theta(\dot{x}_{CM} - r\dot{\theta}\sin\theta) + \cos\theta(\dot{y}_{CM} + r\dot{\theta}\cos\theta)\right]^2 Wdr\left(-\sin\theta\hat{i} + \cos\theta\hat{j}\right) \tag{2.203}$$

with W being the width along the constrain, that is, the ignored z direction.

The net drag force can now be calculated as

$$\vec{F}_{drag} = \int_{-L/2}^{L/2} dF_{\perp}(r)\hat{n} \tag{2.204}$$

and the net moment generated by the drag force is

$$\vec{\tau}_{drag} = \int_{-L/2}^{L/2} rdF_{\perp}(r)\hat{z} \tag{2.205}$$

One could now figure out the acceleration of the petal using Newton's second law as

$$m\vec{a} = \vec{F}_{drag} + m\vec{g} \tag{2.206}$$

and similarly the angular acceleration as

$$I\alpha = \tau_{drag} \tag{2.207}$$

with m, I being the petal's mass and moment of inertia about petal's CM.

Given the initial state as described with $x, y, \theta, \dot{x}, \dot{y}$, and $\dot{\theta}$, one could now generate the entire future evolution of this system.

References

[1] J.J. Craig, Introduction to Robotics: Mechanics and Control, vol. 3, Pearson/Prentice Hall, Upper Saddle River, NJ, 2005.

[2] L.D. Landau, E.M. Lifshitz, Mechanics: Course of Theoretical Physics, third ed., vol. 1, Butterworth-Heinemann, 1976. ISBN: 9780750628969.

[3] S.T. Thornton, J.B. Marion, Classical Dynamics of Particles and Systems, Cengage Learning India, 2012. ISBN 9788131518472.

[4] H. Goldstein, J. Safko, C.P. Poole, Classical Mechanics, in: Pearson, 2013 ISBN: 9781292026558.

[5] D. Kleppner, R.J. Kolenkow, An Introduction to Mechanics, McGraw-Hill, New York, NY, 1973. ISBN: 9780070350489.

[6] H. Asada, J.-J.E. Slotine, Robot Analysis and Control, John Wiley & Sons, 1986.

[7] R.P. Paul, Robot Manipulators: Mathematics, Programming, and Control: The Computer Control of Robot Manipulators, The MIT Press, 1981.

[8] H. Choset, K.M. Lynch, S. Hutchinson, G. Kantor, W. Burgard, L. Kavraki, S. Thrun, Principles of Robot Motion: Theory, Algorithms, and Implementations, MIT Press, 2005.

[9] O. Khatib, "Introduction to Robotics" Video lectures, Stanford Center for Professional Development, Stanford University, 2008. https://www.youtube.com/view_play_list?p=65CC0384A1798ADF (retrieved/accessed 03.09.18).

[10] J. Vandiver, D. Gossard, 2.003SC Engineering Dynamics, Massachusetts Institute of Technology: MIT OpenCourseWare, 2011. https://ocw.mit.edu License: Creative Commons BY-NC-SA.

[11] D. Chakrabarty, P. Dourmashkin, M. Tomasik, A. Frebel, V. Vuletic, 8.01SC Classical Mechanics, Massachusetts Institute of Technology: MIT OpenCourseWare, 2016. https://ocw.mit.edu. License: Creative Commons BY-NC-SA.

[12] M. Vukobratovic, A.A. Frank, D. Juricic, On the stability of biped locomotion, IEEE Trans. Biomed. Eng. 17 (1) (1970) 25–36.

[13] M.B. Popovic, A. Goswami, H. Herr, Ground reference points in legged locomotion: definitions, biological trajectories and control implications, Int. J. Robotics Res. 24 (12) (2005) 1013–1032.

[14] H. Herr, M.B. Popovic, Angular momentum in human walking, J. Exp. Biol. 211 (2008) 467–481.

3

Actuators

Marko B. Popovic*, Kathleen A. Lamkin-Kennard†, Philipp Beckerle‡,§, Matthew P. Bowers*

*WORCESTER POLYTECHNIC INSTITUTE, WORCESTER, MA, UNITED STATES †ROCHESTER INSTITUTE OF TECHNOLOGY, ROCHESTER, NY, UNITED STATES ‡TECHNISCHE UNIVERSITÄT DORTMUND, DORTMUND, GERMANY §TECHNISCHE UNIVERSITÄT DARMSTADT, DARMSTADT, GERMANY

CHAPTER OUTLINE

3.1 Introduction

Synthetic and natural actuators can actively generate force and motion. They are also characterized by passive mechanical properties. Here, synthetic actuators are addressed, starting with synthetic muscles that more closely resemble biological ones. Then, other types of artificial actuators are briefly reviewed. The second part of this chapter focuses on biological muscles.

Biomechatronics. https://doi.org/10.1016/B978-0-12-812939-5.00003-3

3.2 Synthetic Muscles

3.2.1 Pneumatic Artificial Muscle (PAM); McKibben Muscle

The biologically inspired air-based pneumatic artificial muscles (PAMs) were invented in 1950s by the physician Joseph L. McKibben for use in prosthetic limbs. The Bridgestone rubber company (Japan) commercialized the idea in the 1980s under the name of rubbertuators [1].

As shown in Fig. 3.1, McKibben Artificial Muscle consists of an inner rubber member, typically silicone or latex tube, containing air and an outer member in the form of woven shell with nonstretchable fibers. The muscle is in a relaxed state when air pressure inside inner member is equal to the surrounding atmospheric pressure. The muscle is in a contracted state when air pressure is larger than the atmospheric pressure.

The physical principle of PAM operation is rather simple. Pressurization and addition of air causes PAM to expands radially thus contracting axially (i.e., longitudinally). Conversely, the depressurization and the removal of air cause PAM to shrink radially and elongate axially. The role of the woven shell and elastic tubing combined is to guide shape change as function of air pressure and external loading combined.

Due to pressurization and the addition of a small amount of air, the muscle undergoes nonconformal shape change. Specifically, the muscle expands in radial or transversal (perpendicular to the muscle's axis) directions while shortening in longitudinal or axial (along the muscle's axis) direction. This process also introduces larger side surface area subject to the air pressure force in axial direction. The process of elongation stops when the muscle reaches equilibrium state. In the equilibrium state the retraction force due to woven shell and elastic tubing combined is balanced with pneumatic force and possibly external loading force along the axial direction.

This is a highly nonlinear system and, hence, a bit cumbersome for fine control.

The exertion distance is limited by the tightness of the weave and by the helical angle of inelastic threads forming the muscle webbing (see Appendix A); a very loose weave and small helical angle allows greater bulging, which further twists individual fibers in the weave.

Imagine an extreme, rather unrealistic case with PAM undergoing a nonconformal shape change from cylinder to sphere (Fig. 3.2). The outer contour of a cylinder, with fibers

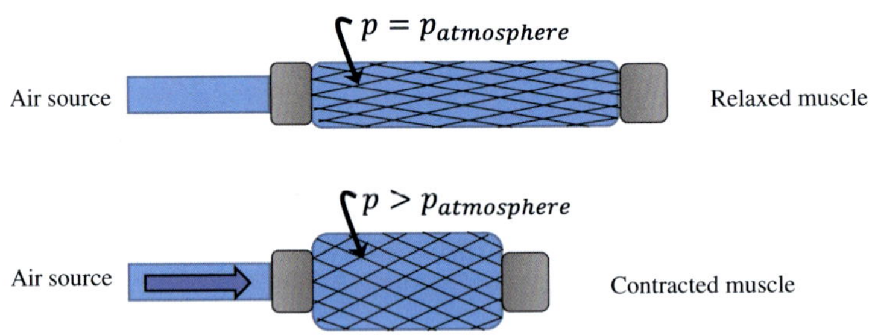

FIG. 3.1 McKibben artificial muscle.

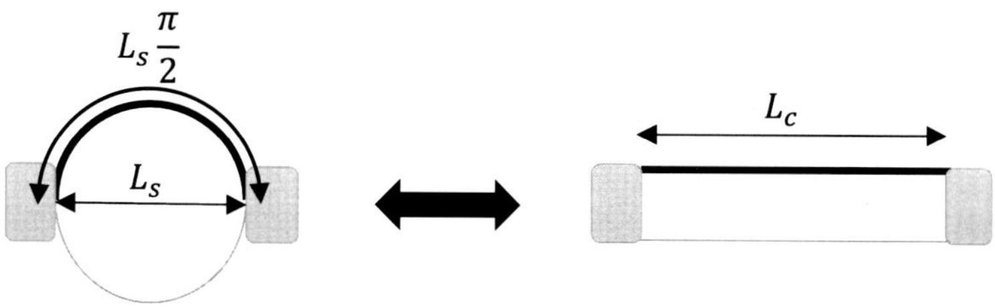

FIG. 3.2 Estimation of maximal strain for McKibben muscle.

oriented more axially, must be equal or larger than a sphere's outer contour, with fibers oriented more radially. The length change is $L_c \to L_s \le 2L_c/\pi$ with L_c being cylinder length and L_s being sphere length. Hence strain is

$$\text{strain} = \frac{L_C - L_S}{L_s} = \frac{L_C}{L_s} - 1 \ge \frac{\pi}{2} - 1 = 0.57 \tag{3.1}$$

In practice McKibben muscles typically have a maximum strain of about 0.4.

Biological muscles have an average maximum strain of 1.0.

In the context of biological realism both biological muscle and McKibben muscle soften radially while elongating axially and they bulge and stiffen radially while contracting axially. However, McKibben muscle has a substantially smaller strain than a biological muscle.

McKibben style muscles can be modeled from an analysis of the system using a virtual work balance. The energy associated with the input pressure can be transformed into a length contraction and force production such that

$$F = \frac{P}{4n^2\pi}\left(3L^2 - b^2\right) \tag{3.2}$$

where F is force, P the gauge pressure, L the muscle length, and n and b refer to the number of wraps and braid strand length, respectively [2].

The critical angle of McKibben muscle's braided helically wound webbing is addressed in Appendix.

McKibben style muscles can be utilized with water or other liquids as the working fluid. The use of water as an actuating fluid will increase the stiffness of the actuator, improve the response time, and increase the pressure and force bandwidth of the actuator [3]. The utilization of the muscles with water as the working fluid also enables the muscles to be utilized underwater at depths without concerns for the effects of the compressibility of air.

The power output of McKibben muscle is directly related to air source powering mechanism, for example, pump. This limitation is a disadvantage of McKibben muscle. Other synthetic muscles like Hydro Muscle [4–6] exploit the concept of power augmentation.

Substantial radial expansion and deflation of McKibben's elastomer introduces substantial energy that is not transferred from source to actuated load [6].

Advantages of McKibben muscle are biological realism in the sense of muscle bulging when contracting, ease of manufacturing, and cost-effectiveness.

The disadvantages are nonbiologically realistic strains, muscle power output directly proportional to power output of air or liquid source mechanism (e.g., pump), with only smaller amount of source energy actually transferred onto load (as it is instead used for radial expansion of the muscle).

Silicon tube-based designs have additional losses of the order of 30%. As verified experimentally [6] the silicone tubing has large amounts of hysteresis seen during the extension and retraction stress-strain curve, with about 30% of the input energy being lost. Latex tube-based designs perform much better in that respect with only a few percent being lost each cycle [6].

In the case of air being the working fluid there are also large energy losses due to air heating; air is a compressible fluid that releases heat while going through cyclic density change.

In principle, as already addressed, one could also use other types of fluids, for example, water. In the case of water which is an incompressible fluid, the energy losses due to heating are minimized.

Efficiency is one of the most critical performance metrics for an actuator. Previous studies of pneumatic McKibben muscles obtained an efficiency of 0.22 under quasi-stationary isotonic (constant muscle tension) shortening conditions [7]. This result is obtained without considering return fluid work.

3.2.2 Hydro Muscle

Hydro Muscles, a new type of hydraulically actuated muscles, originally developed at Worcester Polytechnic Institute (WPI) Popovic Labs, and its pneumatic counterpart the Air Muscles (also called air actuated Hydro Muscles), are linear actuators resembling ordinary biological muscles in terms of active dynamic output, passive material properties, and appearance (except pronounced bulging as McKibben muscle) [4–6].

The Hydro Muscle has a single water flow opening and comprises an inner and an outer structural element. The inner element is a smooth tubing made of an elastic material such as latex (performs better than silicon in terms of energy efficiency). Original, first version of Hydro Muscle, has an added non-stretchable sleeve, outer element, made of a soft, inelastic material, such as polyester, to limit radial expansion while promoting lengthwise expansion when pressurized (Fig. 3.3).

When the elastic contractile force due to the elongation of inner tubing is larger than the force produced by the pressure of the actuating fluid, the Hydro Muscle will exert a pulling force. When the force produced by the pressure of the actuating fluid is larger than the elastic contractile force due to elongation of inner tubing, the Hydro Muscle will produce a pushing force.

The Hydro Muscle is inexpensive and convenient to manufacture and operate. In comparison with conventional actuators with similar dynamic properties, Hydro Muscles are drastically (roughly two orders of magnitude) more cost-effective.

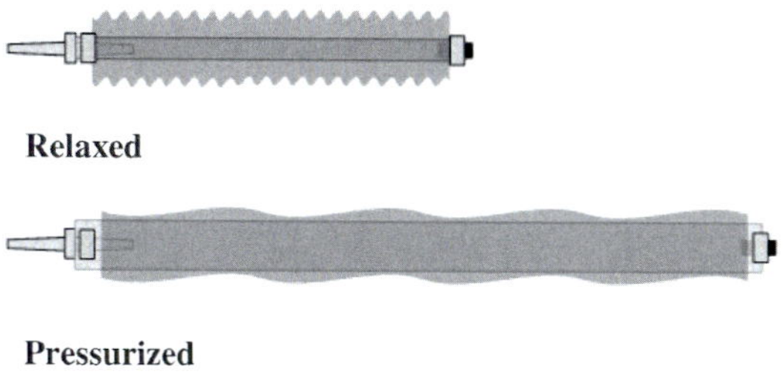

FIG. 3.3 Hydro Muscle.

The muscles can be constructed by inserting a section of latex tubing into nylon tubular sleeve, constraining the radius of the latex during expansion. Lengthwise, the otherwise wrinkled inelastic tubular sleeve, if completely stretched, defines the maximal muscle elongation (Fig. 3.4). The original Hydro Muscle's outer layer in the form of polyester fabric or cloth is allowed to naturally fold, possibly multiple times over itself; this feature introduces potentially extremely large strain, easily in excess of 300%.

Another option is tubular sleeve that is nonstretchable in radial direction and easily stretchable (without wrinkles) in axial, or lengthwise direction (Fig. 3.5).

The muscle architecture with the maximum similarities with the Hydro Muscle is the pneumatic rubber artificial muscle (PRAM) [8, 9], originally developed at Okayama University, which is a silicon-based pneumatic actuator with polyester bellows system (somewhat similar to accordion), with inner and outer bellows' radius tightly limiting radial expansion of the silicon tube and possibly reinforced with tape on one side such that actuator naturally bends in that direction when pressurized. The PRAM has been used in the

FIG. 3.4 Hydro Muscle with nonstretchable nylon tubular outer restraining member during control tests.

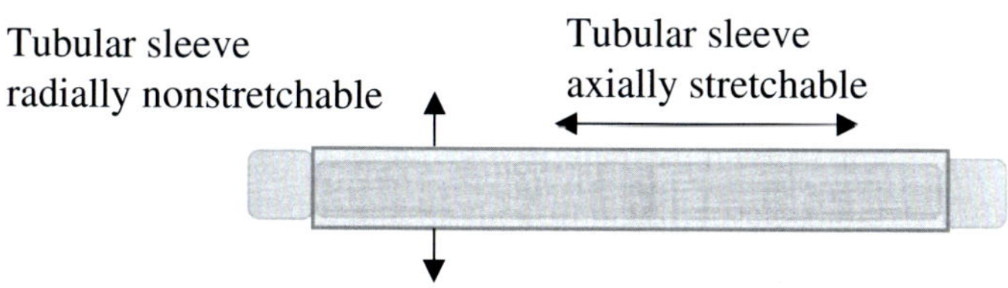

FIG. 3.5 Tubular outer restraining member radially nonstretchable and axially stretchable.

context of upper as well as lower extremities power assist devices. Hydro Muscle and PRAM differ in terms of actuating fluid (water vs air), materials used for elastic tubing (latex vs silicon), and different outer elements (sheathing vs bellows).

More distant designs still similar to the Hydro Muscle are PneuFlex [10], originally developed at TU Berlin, and fiber-reinforced actuator (FRA) [11], originally developed at the Harvard Wyss Institute. Both PneuFlex and FRA utilize air, are silicon based, and have typically helix-like fibers either bonded to the surface (PneuFlex) or embedded deeper within silicon (FRA). The concept of corrugated inner elastic member, additional embedded or external sleeve helix-like fibers, and additional rigid rings surrounding inner elastic member and helix-like fibers, date back to patent by Ove Larsson in 1980s [12].

The Hydro Muscle's outer layer is connected to elastic noncorrugated, flat, cylindrical tubing only at the muscle ends, it does not involve helix-like fibers or rigid rings, and it does not apply substantial forces along the axial direction as muscle axially expands or contracts.

The helix-like nonstretchable fiber structure, as in Ref. [12], may introduce substantial forces along the axial direction. Depending on the helical angle the introduction of fluid into the muscle, that is, pressurization and increase in muscle volume may result either in axial contraction and radial expansion, as with McKibben muscle, for smaller helical angles in respect to axial axis, or may result in axial elongation, similar to Hydro Muscle, but in difference to Hydro Muscle also radial contraction, for larger helical angles in respect to axial axis. For certain critical helical angle, the muscle will not change length independent of amount of pressurization [12]. For more details see the Appendix.

Efficiency [6] of the Hydro Muscle is determined next and compared to a hydraulically actuated McKibben muscle as well as to conventional hydraulic and pneumatic actuators (Fig. 3.6).

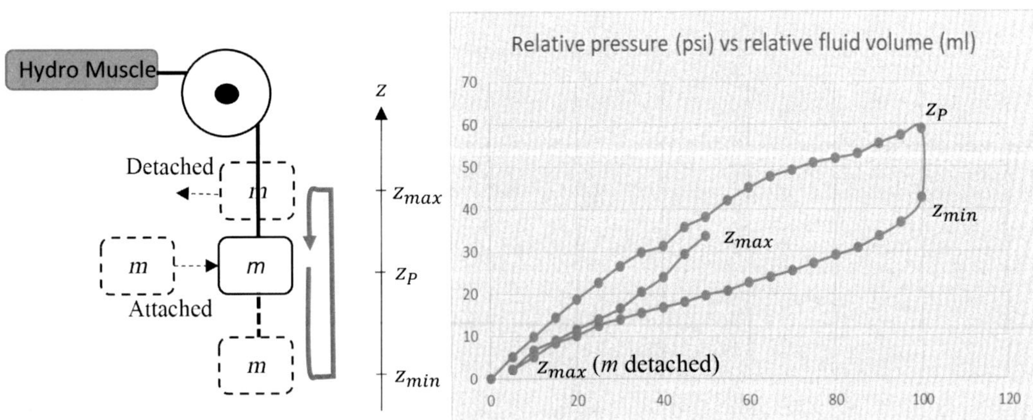

FIG. 3.6 Work cycle for Hydro Muscle efficiency measurement [6].

A hanging mass m was connected to the muscle by a cord that passes without sliding over a small pulley. Hydraulic work was obtained by integrating the product of the measured pressure p and the incremental changes in fluid volume, dV.

Efficiency of the Hydro Muscle *without* considering return fluid work is obtained as positive work performed by the muscle on mass, $mg(z_{max}-z_{min})$, over the sum of the positive work performed by gravity, $mg(z_p-z_{min})$, and the positive work done by the hydraulic system during pressurization of the muscle

$$W_+(z_{max}\rightarrow z_p)=W_+^{F\neq 0}(z=z_{max})+W_+^{F=0}(z_{max}\rightarrow z_p). \tag{3.3}$$

Here z_p is the height of the load when it first engages the muscle during the work cycle (and then drops to z_{min} while stretching the muscle) and z_{max} is the maximal height of the load when the fluid pressure is lowered to atmospheric pressure. The weight is then removed and the muscle length is kept constant while the hydraulic system performs positive work, $W_+{}^{F\neq 0}(z=z_{max})$, by adding fluid until the muscle force, F, becomes zero. Subsequently the hydraulic system performs positive work, $W_+{}^{F=0}(z_{max}\rightarrow z_p)$ with $F=0$ in quasi-static regime until z_p is again reached.

For reference this is compared with the efficiency of the McKibben muscle *without* considering return fluid work, obtained as $mg(z_{max}-z_{min})$ over positive work performed by the hydraulic system, $W_+(z_{min}\rightarrow z_{max})$.

Finally, efficiency of the Hydro Muscle considering return fluid work is obtained as

$$\varepsilon=\frac{mg(z_{max}-z_{min})+|W_-(z_{min}\rightarrow z_{max})|}{mg(z_p-z_{min})+W_+(z_{max}\rightarrow z_p)}. \tag{3.4}$$

$W_-(z_{min}\rightarrow z_{max})$ being the negative work performed by the hydraulic system, that is, positive work performed by the muscle on the hydraulic system. Again, the denominator is the total energy added to the muscle over the work cycle.

Fluidic muscles, that is, Hydro Muscle and Air Muscle, have several practical advantages over standard hydraulic and pneumatic actuators. These muscles are soft, lightweight, inexpensive, easy to manufacture from commercially available materials, easy to modify both in terms of mechanical properties and in the way they interface with supporting “skeletal” structure. Moreover, they have inherent elasticity which in turn makes them suitable for power augmentation paradigm; release of muscle's stored energy onto load can be much faster than transferring a pump's energy into the muscle [6].

Hydro Muscles are highly efficient actuators; there are almost no losses due to heating of working fluid, and latex within 100% strain range performs almost as an ideal spring. The Hydro Muscle has a greater efficiency than the McKibben muscle. According to Fig. 3.7 the Hydro Muscle maximal efficiency is 33% greater than that of the McKibben muscle maximal efficiency despite the latter having larger fluid input volume. This is as expected due to fact that a large portion of the McKibben's hydraulic energy is not transferred onto the load as it is instead used for radial expansion of the muscle. This loss is significantly less in the case of the Hydro Muscle due to the nonstretchable sheathing

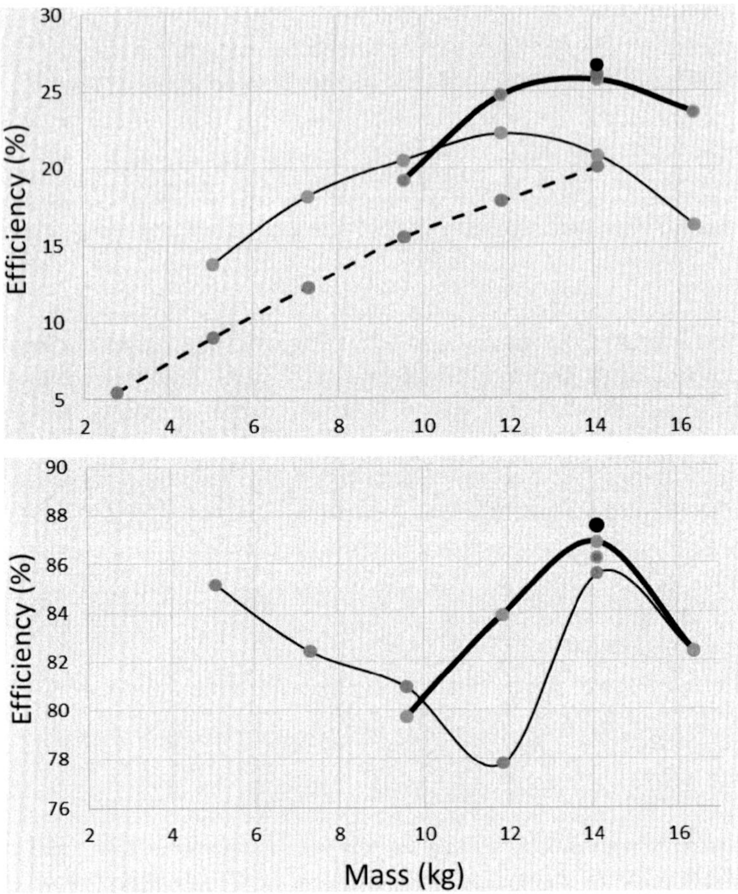

FIG. 3.7 Efficiency without (top) and with (bottom) return flow work: Hydro Muscle without lubricant stretched with 100 mL (thin line) and 150 mL (thick line) fluid input. Hydro Muscle with lubricant (gray and black dots). McKibben muscle with 200 mL fluid input without return flow work consideration (dash line). Both muscles utilized 1.91 cm inner diameter, 2.86 cm outer diameter, and 18.3 cm long latex tube in a relaxed state. For more details see Ref. [6].

preventing elastic tubing from substantial radial expansion. The efficiency of the pneumatic McKibben muscle, that is, PAM, is even smaller due to the air being compressible and thermodynamic heating losses. As discussed silicon-based designs have additional losses of the order of 30%. In comparison dielectric electroactive polymer (DEAP)-based muscles have energy conversion efficiency of about 18%.

Maximal efficiency plotted versus actuation strain for various synthetic muscles is shown in Fig. 3.8. Authors of the study [13] on three-dimensional (3D) printed silicone rubber matrix with ethanol distributed throughout in microbubbles utilizing liquid-vapor transition refer to their muscle as "soft actuator." Hydro Muscle is referred to as "fluidic elastomer actuator" (FEA). McKibben muscle is referred to as PAM.

Actually, Hydro Muscle could easily strain even >300%.

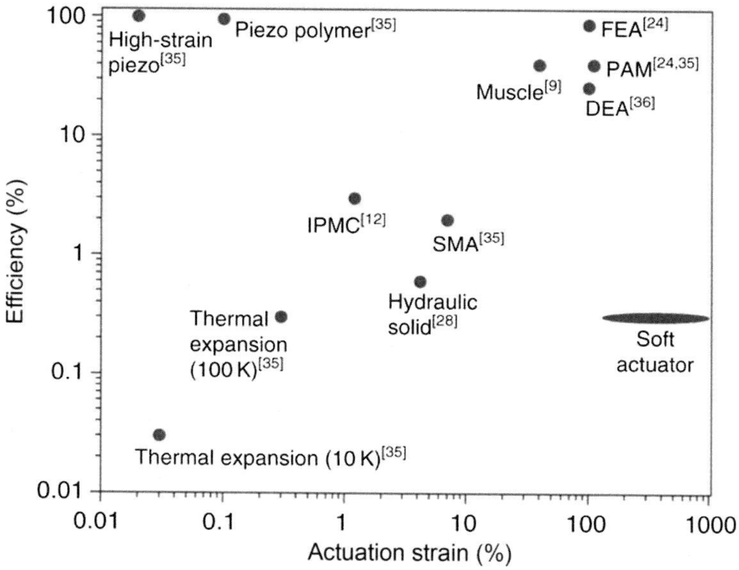

FIG. 3.8 Maximal efficiency plotted versus actuation strain for various actuating methods. Hydro Muscle referred to as "fluidic elastomer actuator" (FEA) is ranked as top synthetic muscle in terms of combined efficiency and strain. *Adapted from A. Miriyev, K. Stack, H. Lipson, Soft material for soft actuators, Nat. Commun. 8(1) (2017) 596.*

According to Fig. 3.8, Hydro Muscle is ranked as top synthetic muscle in terms of combined efficiency and strain. It should come as no surprise that this invention generated the largest intellectual property (IP) fee in WPI history.

A notable feature that the Hydro Muscles have in common with conventional hydraulic and pneumatic actuators are that a single motor unit (i.e., a pump) can power many actuated degrees of freedom while electrical motors with nonfluidic actuation can typically actuate only a single degree of freedom. Moreover, in the case of a wearable system, a conventional pump can be easily positioned more proximally to the center of mass without adding distal masses close to the joints. Finally, hydraulic-based transmissions are often more easily implemented than gearing mechanisms for pure mechanical transmissions.

The presented type of fluidic muscle has not been pressurized in excess of 1 MPa (145 psi), hence limiting the maximum loading force to approximately 100 N per cm^2. If more force is needed several muscles can be aligned in parallel or one could operate a single muscle with very large cross-sectional area.

An experimental platform [14] for the study of a biologically inspired muscular approach to joint actuation by synthetic gastrocnemius, soleus, and tibialis anterior muscles has been designed and the prototype tested, see Fig. 3.9. Three major lower leg muscle groups, that actuate a one-degree-of-freedom ankle, have a multi-fiber composition allowing for active control of stiffness and force similar to recruitment in biological muscles.

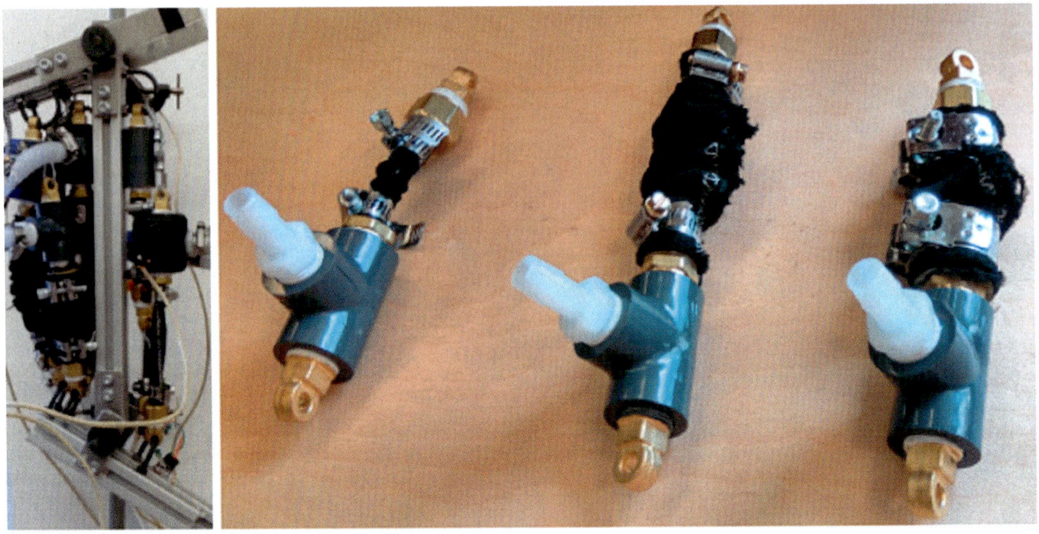

FIG. 3.9 Multi-fiber muscles for ankle and knee joints (left). Hydro Muscle-based fibers of artificial muscles, from left to right: the tibialis anterior, the gastrocnemius, and the soleus (right). For more details see Ref. [14].

Possible drawbacks of Hydro Muscles compared to conventional rigid hydraulic and pneumatic systems are related to their inherent mechanical compliance and elasticity. They are more ideal in performing pulling versus pushing tasks; a pressurized Hydro Muscle could bend when subject to a constant distance between attachment points (e.g., due to heavy load). McKibben muscle is a bit less prone to bending.

Bowing can be remedied with a rigid telescoping element or nonstretchable sheathing, that is, "skin" enclosing the "skeletal" structure and muscle. The later approach can be used to improve the muscle force-joint torque profile by increasing the angle between the muscle and the supporting structure, Fig. 3.10A.

If some bowing is desirable the inner member wall can have variable elastic properties and a different cross section along the muscle length. Furthermore, the inner member wall can have variable elastic properties and thickness as a function of the azimuthal angle; for example, one side of the inner member wall can be reinforced such that pressurizing actuating fluid will cause the Hydro Muscle to curve toward opposite side, Fig. 3.10B.

Another way [6, 15, 16] to address the bowing problem is to modify the muscle by adding granular media (elastic or rigid) between the latex tube and the muscle's outer layer, Fig. 3.10C. When the muscle is pressurized, the granular media is compressed uniformly, thus stiffening the interior support of the muscle, which decreases or completely eliminates bowing.

3.2.3 Pure Fluidic Elastomer Actuators (FEAs)

Pure FEAs is another class of actuators. Sometimes even Hydro Muscle and/or McKibben muscle are categorized as class of FEA. Pure FEAs also utilize fluidic actuation but typically

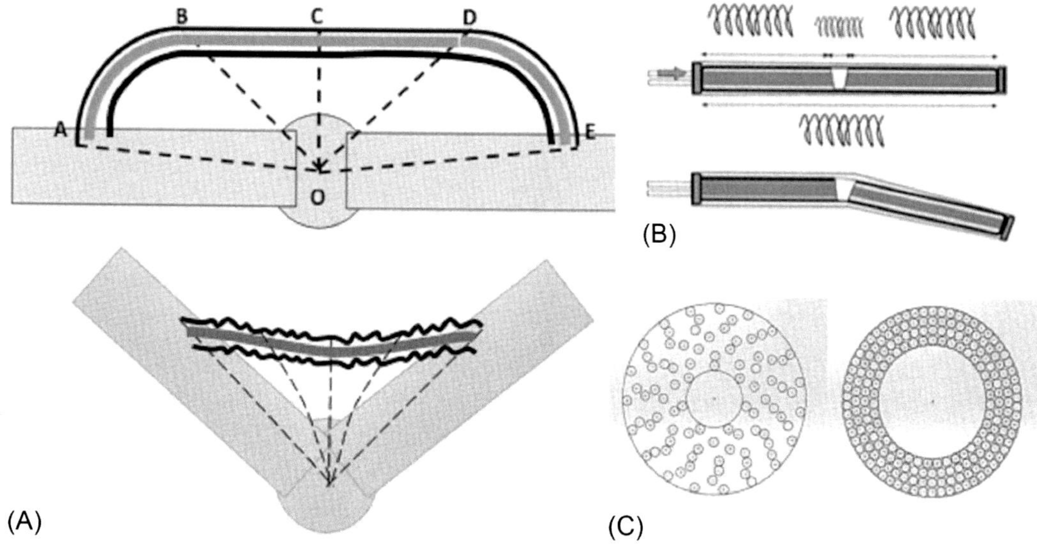

FIG. 3.10 (A) Sketch of guiding "skin" (curve ABCDE) improving force versus joint (O) torque profile for bowing Hydro Muscle. (B) Exploiting different stiffness to achieve desired bending. (C) Granular media "sandwiched" between elastomer tube and tubular outer restraining member [6, 15, 16], affecting overall tendency for bowing.

applied through easily moldable silicon-based materials. But in difference to Hydro Muscle (or McKibben muscle) pure FEA by definition does not utilize characteristic tubular webbing (or woven shell) or some other nonelastomeric element surrounding and guiding elastomer expansion-contraction. Instead, the shape change and direction of force is directly obtained by elastomer architecture alone. Different geometries and silicon-based material thicknesses introduce a whole variety of movement patterns and shape changes. In terms of aesthetic and appearance true FEAs provide true soft robotic "magic" (Fig. 3.11).

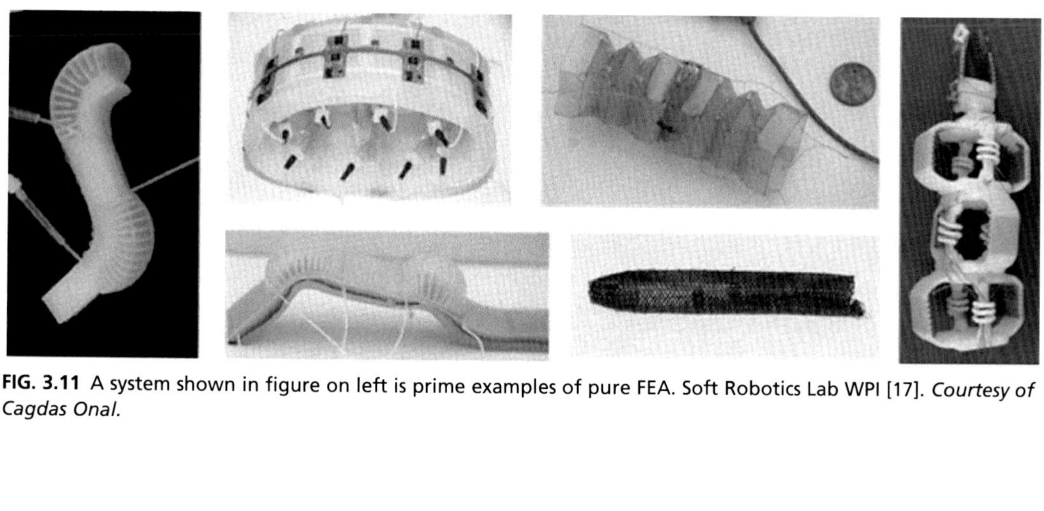

FIG. 3.11 A system shown in figure on left is prime examples of pure FEA. Soft Robotics Lab WPI [17]. *Courtesy of Cagdas Onal.*

The disadvantage of pure FEA is that these systems are typically too soft. Only small fluid pressures are allowed as application of higher pressure would likely damage unconstrained expansion of silicon-based material. Hence, forces are typically small. Still further stiffness is very low as governed by low-pressure fluid (typically air, which is easily compressible fluid) and by low value of silicon Young's modulus. Finally, due to use of silicon these actuators typically have a very low efficiency.

The FEAs as well as Hydro Muscle and McKibben Muscle are further discussed in more details in Chapter 17 on biologically inspired robotics.

3.2.4 Cable-Driven "Muscles"

Typical cable-driven muscle consists of an actuated pulley and a cable in series with an elastic element. An example of cable-driven prosthetic hand [18, 19] is shown in Fig. 3.12.

The standard system architectures with cable actuation having a role of muscle-like actuation are shown in Fig. 3.13. Each configuration has an elastic element in series with cable to mimic muscle-like stiffness. Essentially cable has the role of tendon, elastic element provides passive characteristic of the real muscle (e.g., Hill-like muscle model, discussed later, has one elastic element in series and one elastic element in parallel), and active muscle force is introduced via motorized spool.

For wearable application relatively heavy elements including motorized spool are typically positioned closer to the body's center of mass to minimize user's metabolic cost; limbs move much more than the back or waist areas and thus adding more mass to the limbs is typically avoided. The elastic element can be positioned distal, Fig. 3.13A,

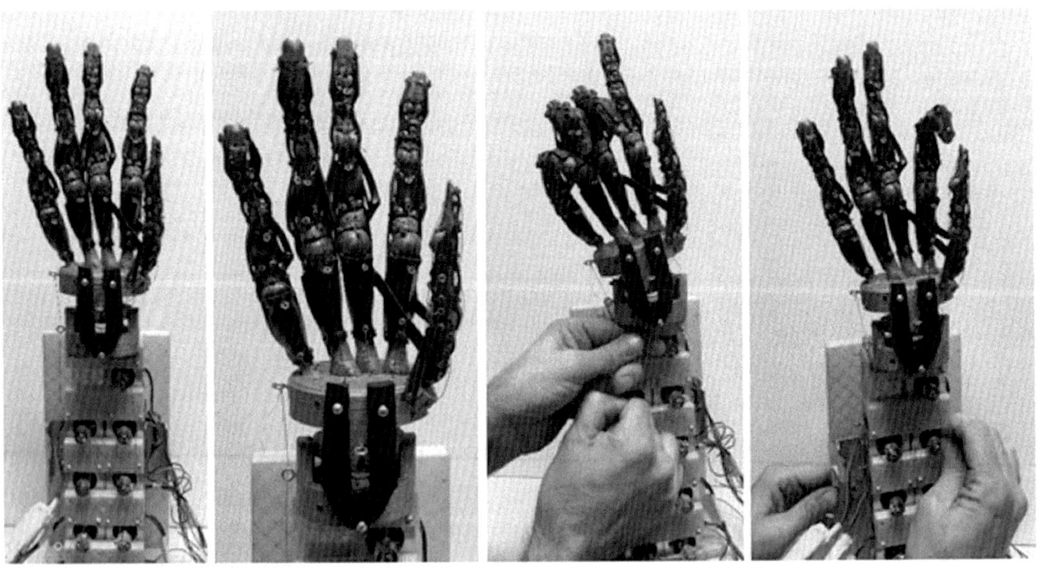

FIG. 3.12 Cable-driven muscle-tendon system of the WPI Accurate Prosthetic Hand [18, 19].

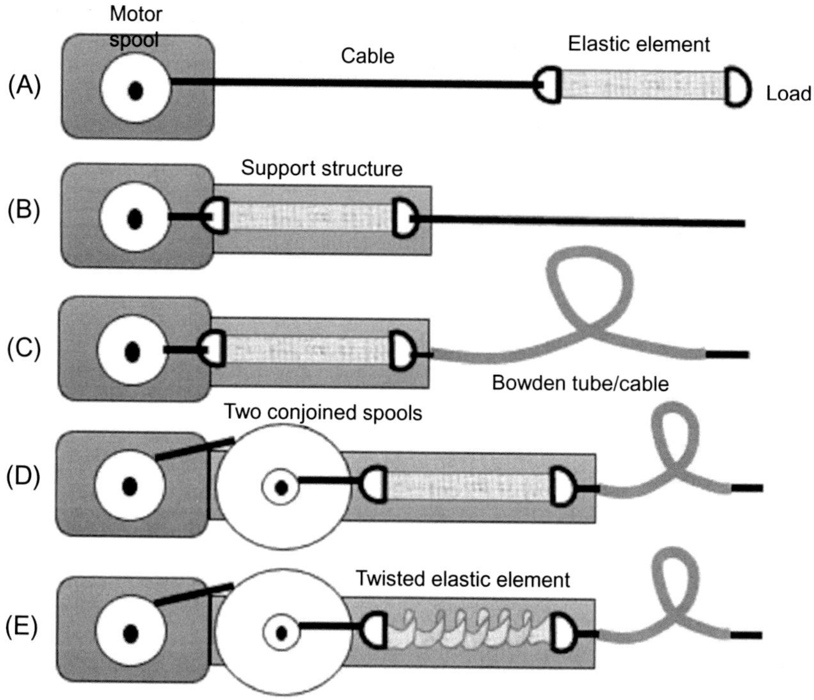

FIG. 3.13 Cable in series with elastic element, various system architectures.

or more proximal, Fig. 3.13B, and potentially added with support structure such that elastic element always stretches in desired, controlled way.

Bowden cable is depicted in Fig. 3.13C. The role of the Bowden cable is to set roughly the length of cable between conduit tube's attachments points (very useful for position control), possibly allow for both push-and-pull actuation, and to regulate cable bending and potentially undesirable interfering with other elements of the system and the user. The push actuation is possible only if very tiny length of cable is allowed to leave Bowden tube, and hence this sets limitation on travel distance of the cable. The push actuation also requires elastic element that can apply forces in both directions.

Contrary to some beliefs the role of Bowden cable/tube is not to "magically" redirect input force. As illustrated in Fig. 3.14 the magnitude of output cable force is smaller than the magnitude of input cable force (though if friction is small they can be almost equal). Because of the different direction of the two, there are forces introduced at the attachment points of Bowden tube that clearly affect the overall dynamics.

Two additional conjoined spools are depicted in Fig. 3.13D. They augment force and decrease travel distance of the cable affecting load. To change stiffness of elastic element one can also consider twisting it as depicted in Fig. 3.13E. The twisting can be performed by another actuator and hence one can simultaneously control both position and

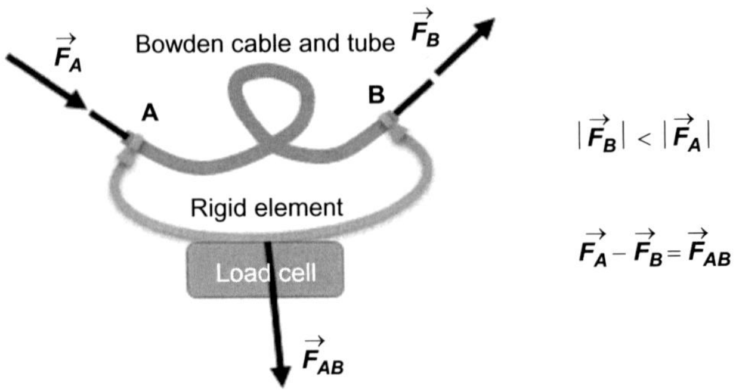

FIG. 3.14 Bowden cable force balance in the static case.

stiffness. Clearly there are also other ways that one can modulate the stiffness. Real biological muscles modulate simultaneously both joint angle position and stiffness.

One could also consider adding another elastic element in parallel with motorized spool, Fig. 3.15; the role of that element is to generate assistive force for motor and also store elastic energy. This architecture is very similar to the Hill model discussed in this chapter.

Advantages of cable actuation is that heavy elements (motor, spool, etc.) can be positioned proximal to body's center of mass while force is transmitted via light cable to distal end effector. Also, if elastic element in series can apply forces in both direction, the cable added with Bowden tube can both pull and push though with limited distance traveled.

The disadvantage compared to fluid systems is that each cable's actuated degree of freedom requires a dedicated motor unit; for hydraulic or pneumatic system typically only one strong motor (e.g., pump) is needed. Furthermore, fluid-based transmissions are often more easily implemented than systems with pulleys or gearing mechanisms for pure mechanical transmissions.

Hence, a large number of powerful motorized units required for soft-robotics Exomusculature [1, 4, 5, 20, 21], that is, network of a large number of cable-driven artificial muscle fibers embedded in a suit (or brace), might be a problem due to motorized units' mass, size, and power requirements.

Researchers are addressing ways to keep cable-driven actuation within system architectures that attempt to keep some of the good features characteristic of fluid-based system. The one-to-many (OTM) [22–24] assisted Exomusculature is one such attempt. All cable-driven artificial muscles are actuated only by a single (or several) motor. The

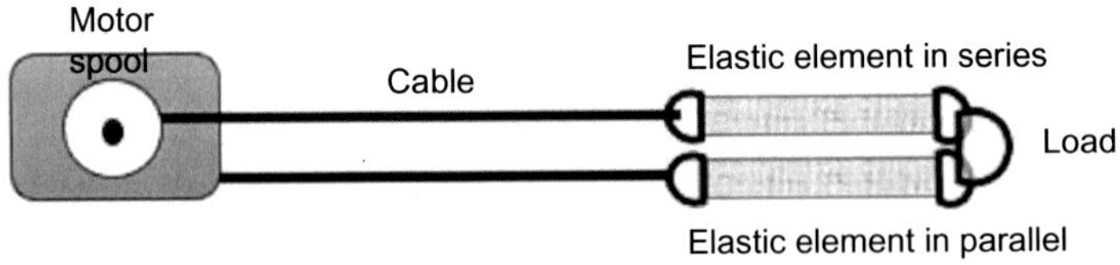

FIG. 3.15 Hill-type muscle model with elastic element in series and another elastic element in parallel.

OTM system utilizes active clutches to store and release energy through passive springs. The same muscle consists of many fibers where each fiber is an independently actuated clutch-spring system. Hence, the muscle's stiffness is varied by recruiting different fibers, exactly as in the human body. Although there are several working OTM prototypes, this approach is still under development. In the same way that fluid-based systems depend on advanced valves, the OTM systems depend on advanced clutches.

3.2.4.1 Cable-Driven Artificial Muscle Attachment: From Mechanical Engineering to Anatomy

During the period 2010–12, group at WPI advised by one of the authors (MP) and Prof. Robert Howe's group at Harvard University collaborated on the soft robotic brace for stroke rehabilitation [20, 21] (Fig. 3.16), as an example of possibly practical

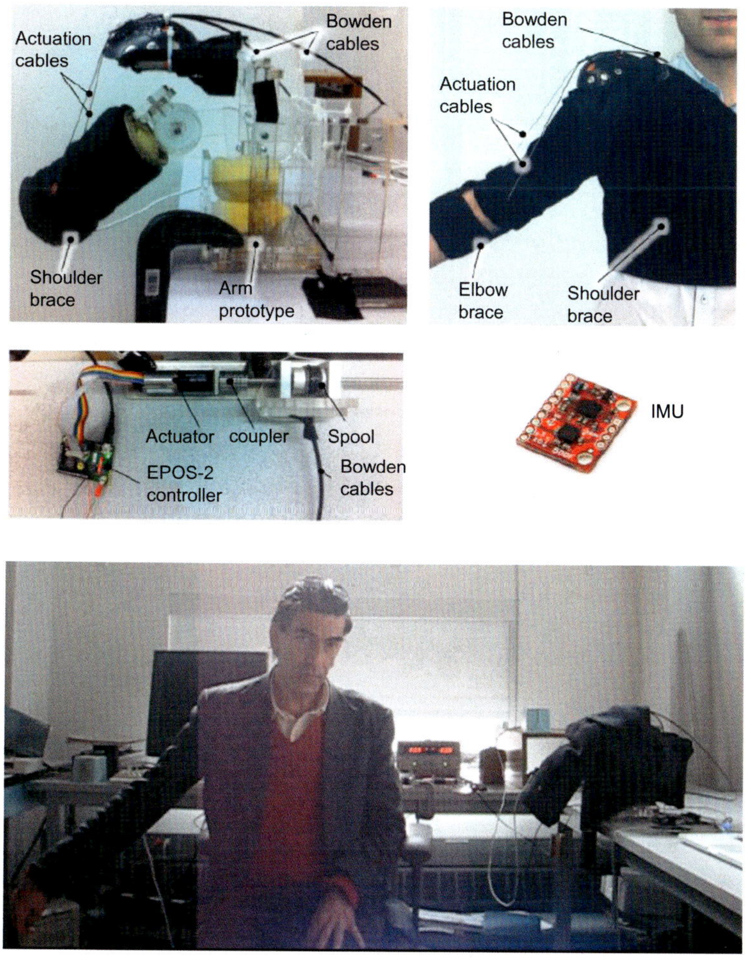

FIG. 3.16 Example of the soft robotics Exomusculature (top), the Exomusculature shoulder brace, on the right, attached to a mannequin arm with added weights is teleoperated by Professor Popovic (bottom).

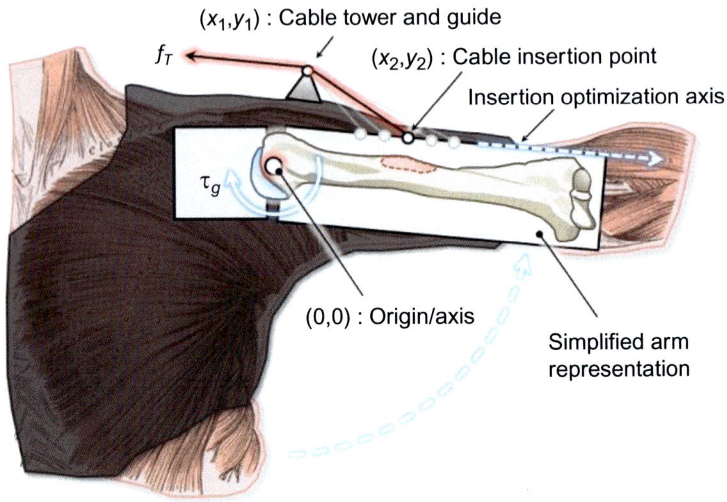

FIG. 3.17 Planar single muscle model.

Exomusculature. Ignacio Galiana from Centre for Automation and Robotics UPM-CSIC, Madrid, Spain joined these efforts in summer 2011.

The author and collaborators studied different criteria defining locations of artificial muscle insertions (i.e., attachments) into upper-arm brace.

The human shoulder has a very complex anatomical structure with approximately five degrees of freedom. Luckily, even a very simplified planar model can be sufficient to provide interesting insights into muscle-attachment architecture. For example, by using a single muscle theoretical model (Fig. 3.17), the minimization of *force into joint* and the minimization of *cable tension* suggest insertions (from the greater tubercle) at 0.104 and 0.136 m, respectively. The experimentally measured mean distance from the greater tubercle to the proximal and distal insertion of the deltoid muscle in humans is 0.061 and 0.158 m, respectively. Hence the theoretical prediction is exactly in the central area of experimentally measured range—from Robotics to Biomechanics and Human Anatomy {!}

3.2.5 Liquid-Vapor Transition and Chemical Reaction-Based "Muscles"

A team of engineers at Columbia University recently created a soft composite material claimed to be able to achieve very large strains (even up to 900%) and that can lift a thousand times its own mass [13]. Cost-effective muscle (Fig. 3.18) combines high stress (up to 1.3 MPa) with low density ($0.84\,g\,cm^{-3}$).

Team used a 3D printed silicone rubber matrix with ethanol distributed throughout in 1 mm microbubbles. The system in the shape of muscle was then actuated by application of current through low power, thin wire. The wire would heat up and lift the temperature of the muscle to liquid-vapor transition temperature (78.4°C). Just formed high vapor pressure would then press inner walls and expand microbubbles.

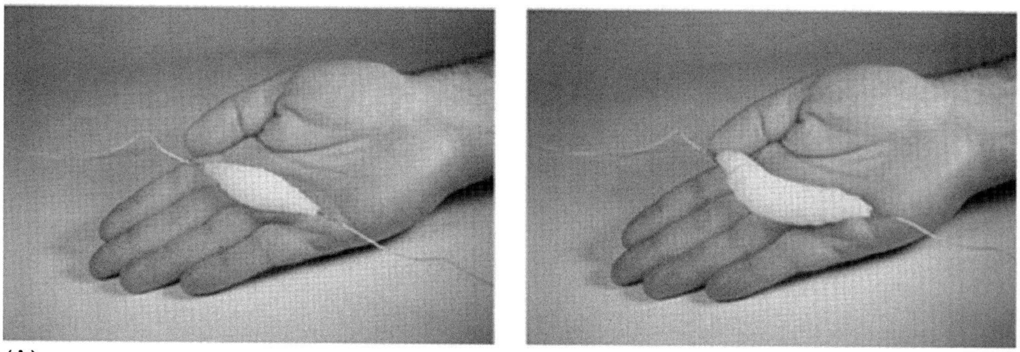

(A) (B)

FIG. 3.18 Soft artificial muscle based on liquid-vapor transition. The muscle is composed of ethanol distributed throughout the solid silicone elastomer matrix. (A) Electrically actuated muscle including thin resistive wire in a rest position on a human hand. (B) Expanded muscle actuated (8V, 1A). *Adapted from A. Miriyev, K. Stack, H. Lipson, Soft material for soft actuators, Nat. Commun. 8(1) (2017) 596.*

This is a promising research direction. Unfortunately for a repetitive work cycle, one would also need a vapor-liquid transition, which takes time. As reported it takes about 100s to complete the work cycle compared to a part of the second for biological muscles. The other disadvantage is the relatively low efficiency of this actuator, see Fig. 3.8.

Another approach to soft robotics muscle is even attempting to bypass the electronic components and battery [25], Fig. 3.19. A team of roboticists from WPI, MIT, and Harvard considers instances of mechanisms made from distributed elastomer actuators to

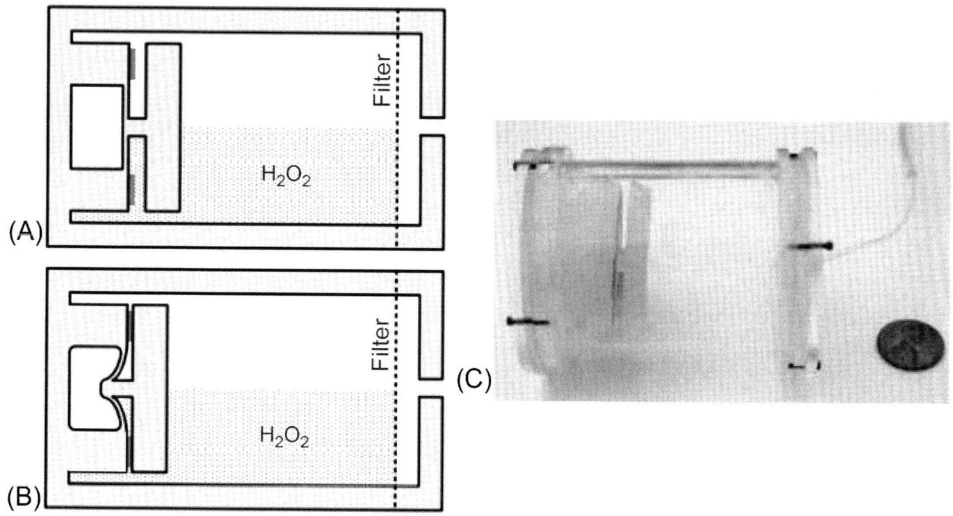

FIG. 3.19 Decomposition of hydrogen peroxide (H_2O_2) solution into oxygen (O_2) gas. Sketch of operation of self-regulated seal mechanism: catalyst unsealed (A) and catalyst sealed (B). Actual device (C) [25]. *Courtesy of Cagdas Onal.*

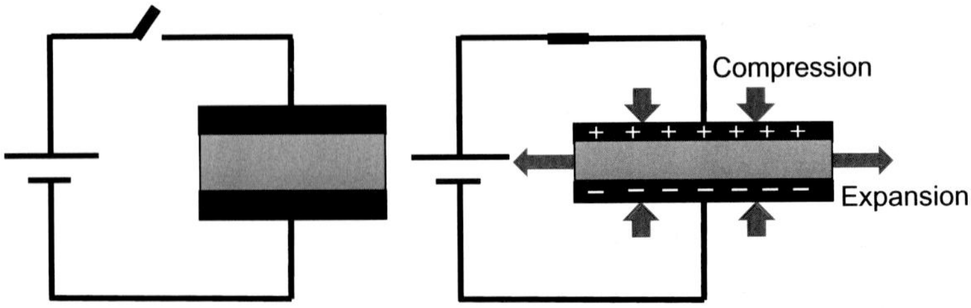

FIG. 3.20 Schematic (side view) showing the structure and the actuation mechanism of a simple dielectric elastomer actuator (black layer represents the conductive material, gray represents the dielectric material). When a voltage threshold is reached, the attraction force between the conductive layers induces a Maxwell stress on the dielectric layer and deforms it.

generate motion using a chemical means of pressure generation. A mechanical feedback loop controls the chemical decomposition of hydrogen peroxide into oxygen gas in a closed container to self-regulate the actuation pressure. This on-demand pressure generator, called the pneumatic battery, bypasses the need for electrical energy by the direct conversion of chemical to mechanical energy.

These portable and silent devices are equivalent to an electric battery and pump or compressor and regulator as they offload desired pressure via a controlled gas generating chemical process. An aqueous solution of hydrogen peroxide (H_2O_2) is the fuel in these devices. It provides high power and is easily replaced with a fresh solution when depleted. Pure H_2O_2 has a theoretical energy density of 2.7kJ/g, one of the highest in common monopropellant fuels. While this optimal value cannot be fully utilized at room temperature operation, it clearly indicates the potential of H_2O_2 as a practical power source. The small catalyst pack ignites decomposition of H_2O_2 into oxygen (O_2) gas in a closed container, Fig. 3.20A. When the desired O_2 pressure is reached a mechanical self-regulator seals off the catalyst pack, Fig. 3.20B on the left, chemical reaction stops, and high pressure gas is released through hydrophobic filter and subsequent passive check valve to, for example, McKibben muscle, Hydro Muscle, or some of wide variety of FEAs previously discussed. This portable pump design can operate in any orientation.

3.3 Electroactive Polymers

The electroactive polymers (EAPs) are plastics that move in response to electricity or an electric field. A typical characteristic of an EAP is that it can undergo a large amount of deformation while sustaining large forces. In the late 1990s, it was demonstrated that some EAPs can exhibit up to 380% strain, which is much more than that any ceramic actuator can achieve (piezo ceramics can strain only up to part of 1%) [1].

There are two main types of EAPs: ionic and electronic.

Ionic EAPs (which include ionic polymer gels, ionomeric polymer metal composites, conductive polymers, and carbon nanotubes) work on the basis of electrochemistry—the mobility or diffusion of charged ions. Even very low voltages will cause substantial actuation. They generally need to be wet and so they must be sealed within flexible coatings. A major shortcoming of many ionic EAPs is that as long as the electricity is on, the material will keep moving and if the voltage is above a certain level, electrolysis takes place, which causes irreversible damage to the material.

In contrast, electronic EAPs (such as ferroelectric polymers, electrets, dielectric elastomers (DEs), and electrostrictive graft elastomers) are driven by electric fields. They require relatively high voltages (>1 KV) which may be problematic for commercial applications but in return, electronic EAPs can react quickly and deliver strong mechanical forces. They do not need a protective coating and require almost no current to hold a position.

Actuation principle is typically based on electrostatic attraction between coating surfaces (capacitor plates) which in turn squeeze coated dielectric material (between capacitor plates) causing dielectric to strain, that is, actuate along direction parallel to coating surfaces.

The EAPs have garnered attention from scientists and engineers in many different fields over the past two decades due to their unique properties. The EAPs are electrical transducers that respond to electrical signals with significant changes in size or shape or generate electrical signals in response to changes in size or shape. Since 2000, when Pelrine et al. [26] reported that approximately 215% actuation could be achieved with acrylic elastomers, EAPs have become of particular interest to researchers working in the field of robotics, particularly soft robotics. The working principle of EAPs resembles, in a sense, the behavior of biological muscles. Therefore, since their first development, EAPs have been studied for their potential applications as actuators or artificial muscles.

The DEAPs, also referred to as 'dielectric elastomers' or DEs, are a class of electronic EAPs. The physical assembly of a DEAP resembles a capacitor, with a highly elastic dielectric polymer medium sandwiched between two electrodes. When an electric field is introduced, the conductive parallel plate electrodes attract each other due Coulombic forces. The attraction force translates to the elastic dielectric as a Maxwell stress and deforms the dielectric polymer. Therefore actuation (compression) occurs in a direction normal to the parallel plates as seen in Fig. 3.20. In order to conserve the total volume, the actuator expands in the lateral direction. The DEAPs have many advantages over other actuators, including a rapid response rate, high actuation ratios, flexibility, lightness, and stretchability. However, a significant drawback of DEAPs is that they require high-voltage input, of the order of 3–6 kV, in order to provide reasonable displacement and output force [27–29].

The magnitude of the reversible deformation of a DEA device is approximated by

$$\frac{\Delta z}{z_0} = \epsilon_0 \epsilon_r \frac{U^2}{d^2 Y} \tag{3.5}$$

where U is the magnitude of the applied voltage, Y the Young's modulus of the dielectric layer, d the separation distance between the parallel plates, and ϵ_0 and ϵ_r are permittivity of free space and relative permittivity of the dielectric layer, respectively. Thus, the actuation ratio of a DEAP can be increased by increasing the input voltage, decreasing the elastic modulus of the dielectric polymer, or decreasing the distance between the parallel plates.

For robotic applications, it is generally more practical to work with low-operating voltages, thus decreasing the elastic modulus or the thickness of the dielectric polymer to increase the actuation ratios are preferred over increasing the input voltage. Elastic modulus is an intrinsic material property; therefore, it depends on the material being used in the device. Decrease in the separation distance, on the other hand, can be achieved independent of the material choice. While this would increase the actuation ratio, it will diminish the total actuation distance.

Ionic polymer metal composites (IPMCs) are ionic EAPs that use the movement of ions to induce actuation or generate voltages for sensing applications. Similar to DEAPs, IPMCs are typically made with a thin flexible membrane, usually Nafion or Flemion, with noble metal electrodes plated on two sides. Cations are implanted inside the membrane to neutralize the system. The ions carry both positive and negative charges and, when a low voltage is applied, ionic movement causes actuation of the polymers. Each layer in the polymer contributes to the overall mechanical and electromechanical properties of the materials.

Advantages of IPMCs are that large actuation displacements can be achieved under relatively low-induced forces due to the relatively low stiffness of the materials. The IPMCs can also be easily manufactured and have low energy consumption. The primary disadvantage of IPMCs is that the materials are very sensitive to dehydration.

3.4 Shape-Memory Alloys and Shape-Memory Polymers

A shape-memory alloy (SMA, smart metal, memory metal, memory alloy, muscle wire, and smart alloy) is an alloy that "remembers" its original, cold-forged shape by returning to that pre-deformed shape when heated. This material is an extremely lightweight, solid-state alternative to conventional actuators such as hydraulic, pneumatic, and motor-based systems. SMAs have applications in industries, including medical and aerospace [1].

In most SMAs, a temperature change of only about 10°C is necessary to initiate this phase change. The two phases, which occur in SMAs, are martensite and austenite. Martensite is a relatively soft and easily deformed phase of SMAs, which exist at lower temperatures. Austenite, the stronger phase of SMAs, occurs at higher temperatures (Fig. 3.21).

The three main types of SMAs are the copper-zinc-aluminum-nickel, copper-aluminum-nickel, and nickel-titanium (NiTi) alloys, but SMAs can also be created by alloying zinc, copper, gold, and iron. NiTi alloys are generally more expensive and change from austenite to martensite upon cooling.

The SMA can provide large displacements and forces. The main problems with SMAs are relatively slow system response times due to issues with temperature control, in particular cooling and, hence, one often need to make trade-offs and minimize the mass.

The SMA typically uses a large amount of current, but has high power density, able to lift a thousand times its own weight though with low strain (<10%).

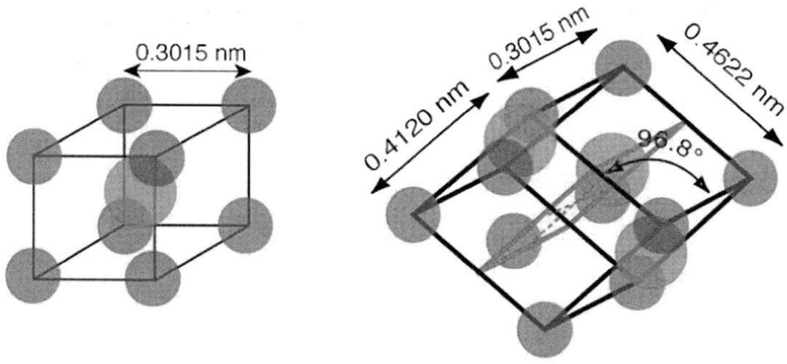

FIG. 3.21 Austenite (hot) and martensite (cool) crystalline alloy phases.

3.5 Variable Stiffness/Impedance Actuators

Variable stiffness/impedance actuators are an emerging class of actuators in applications such as assistive robotics. Similar to human muscles these actuators can control joint stiffness in addition to motion. The working principle of these actuators stems from the series elastic actuator [30] and the mechanical impedance adjuster [31], both of which introduced an elastic element between the motor and the load as shown in Fig. 3.22.

Besides introducing elastic elements, that is, coil and leaf springs, these early concepts were capable of modulating joint stiffness either by force control of a fixed-stiffness

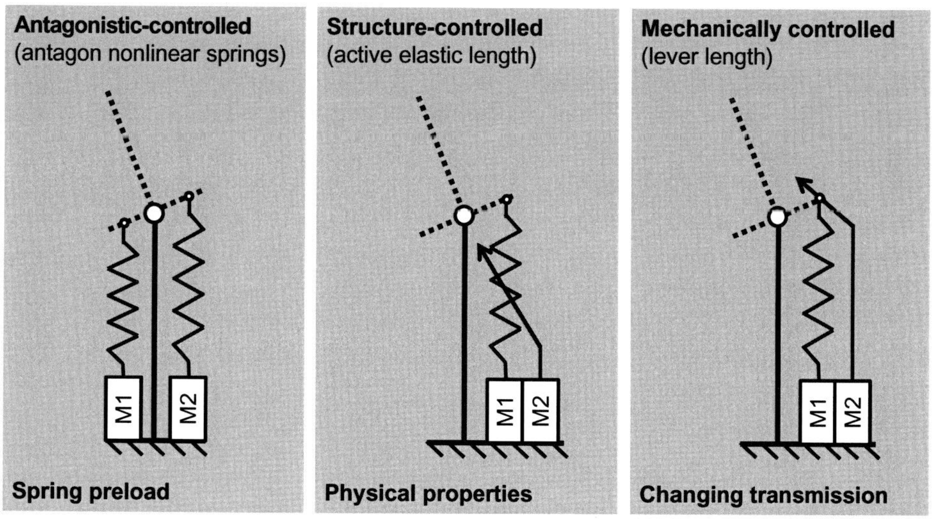

FIG. 3.22 Functional principles and implementations of variable stiffness actuators. Categorized according to Ref. [32] (bold face) and related to Ref. [33] (bold italic).

mechanism or by modifying the active spring length. The potential advantages of these actuators in the context of force control, safe human-robot interaction, and energy efficiency led to the development of various actuator concepts with variable stiffness.

In the milieu of robotic and assistive applications, for example, walking robots or prosthetic limbs, this allows one to partially rely on feedforward control and modulate the mechanical behavior by setting appropriate stiffness. For instance, during running only limited feedback control can be performed during ground contact, but joint or leg stiffness could be adapted in advance.

Going beyond altering stiffness, recent actuator design and control algorithms also enable changing the damping and inertial characteristics and are, thus, designated as variable impedance actuators [33]. While both these properties can simply be modified by control, approaches to vary damping physically exist [34].

Focusing on variable stiffness, different fundamental working principles are found. The categorization by Van Ham et al. [32] distinguishes equilibrium-controlled, structure-controlled, mechanically controlled, and antagonistic-controlled stiffness actuators. While equilibrium-controlled actuators implement a physical spring, stiffness is solely modulated by means of control. In contrast, actuators based on the latter three principles, which are sketched in Fig. 3.22, adapt their stiffness properties physically. Fig. 3.22 also relates the exemplary mechanisms to the similar, but not completely coherent, categorization by Vanderborght et al. [33]. All sketches show a link (dotted) moving around a rotational joint. The two motors M1 and M2 induce the motion and/or to adjust joint stiffness. The antagonistic-controlled principle (left in Fig. 3.22) relies on two or more actuators that are working against each other through fixed-stiffness, nonlinear springs. Structure-controlled elastic actuators (middle of Fig. 3.22) vary in the physical structure of the elastic element, for example, its moment of inertia or the effective elastic length. In the mechanically controlled principle (right in Fig. 3.22), apparent stiffness is set by modifying the connection between the springs and structure, that is, a change of pretension or preload.

3.5.1 Elastic Element Configuration and Design

Besides the stiffness variation principle, the dynamic and energetic properties of elastic actuators depend on the implementation of the elastic elements and, even more important, their configuration in relation to the motors [35].

Contemporary variable stiffness actuators mainly implement an elastic element in series with or in parallel to the actuator as shown in Fig. 3.23. While series elastic actuators are especially good for interfacing with the human body and beneficial for safe human-robot interaction [36, 37], both series and parallel elastic actuators can improve energy efficiency by exploiting passive dynamics [35] and stability by avoiding jerk peaks. Actuator designs that combine serial and parallel elastic elements are investigated [38, 39]. All these design configurations are able to decrease peak power and energy in comparison to direct drive, and selecting an appropriate configuration is always a case-specific solution that can be determined from inverse dynamics simulations and needs to consider effects such as

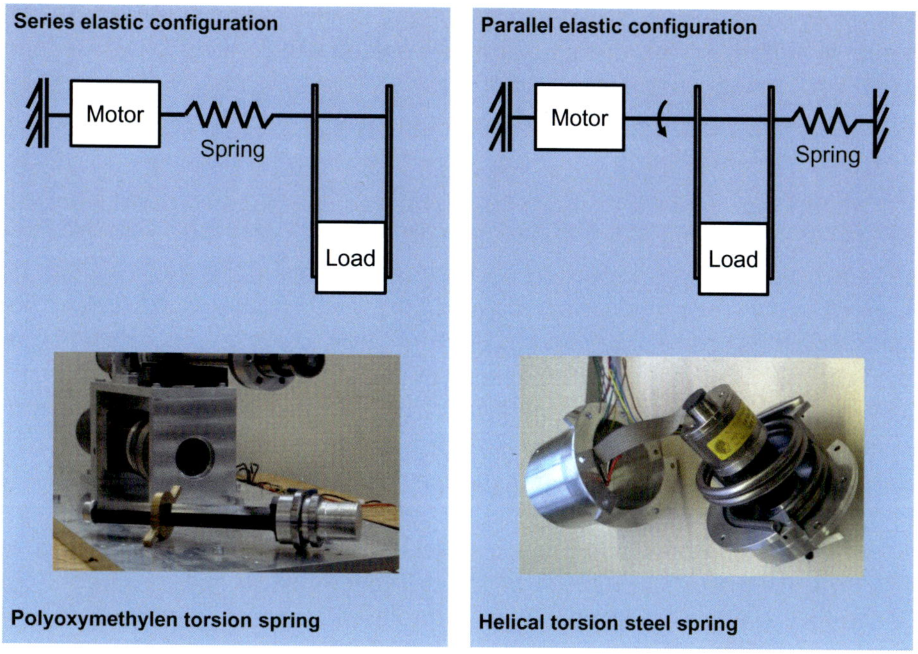

FIG. 3.23 Sketches of series (upper left) and parallel (upper right) actuator-elasticity configurations with exemplary implementations (bottom) and description of the spring type used in the example (bold italic).

gravitation, inertia, and friction [35, 38]. Generally, operation is energy efficient at antiresonance for series elastic configurations, while the stiffness of parallel elastic ones should be set to match their natural frequency to the trajectory frequency [35, 38].

3.6 A Brief Review of Nonbiologically (or Less Biologically) Inspired Conventional Actuators

Here nonbiologically (or less biologically) inspired conventional actuators are briefly reviewed [1, 40–43].

Nonbiological actuators are mechanical, electrical, or electromechanical systems that produce linear or rotary motions. Actuators convert some form of energy, electrical or mechanical, to mechanical work. Nonbiological actuators are classified primarily by the type of motion produced. Linear actuators convert energy into translational motion while rotary actuators convert energy to rotational motion. The outputs from a linear actuator are force and displacement while the outputs from a rotational actuator are torque and angular displacement.

In addition to the classification by the type of motion produced, actuators can also be classified by the type of medium used to supply energy to the system or the basic design configuration of the actuator. Both linear and rotary actuators can utilize pneumatic, hydraulic, or electrical energy to generate actuation. Electrical energy is the most common energy source although hydraulic oil or compressed air can be used to transfer energy to hydraulic or pneumatic actuators, respectively.

3.6.1 Linear Actuators

Linear actuators can be used to move loads, such as components or an assembly, in a straight line. A typical pneumatic linear actuator consists of a piston inside a cylinder. The pressure from an external source is used to move the piston inside the cylinder. Linear force is generated as the pressure increases and the piston moves to the desired position. Pneumatic cylinders can be single acting, double acting, or more complex configurations. For single-acting cylinders, force is developed in only one direction and an internal spring or gravity is typically used to return the piston to a base position upon the removal of the air pressure. Double-acting cylinders use air pressure to generate force in both extending and retracting directions. Single-acting cylinders consume less air than equivalently sized double-acting cylinders, are a simple design configuration, and require less valves and piping. However, the spring operation in a single-acting cylinder can become inconsistent over time and the return spring side of the cylinder vents to atmosphere. In contrast, double-acting cylinders are available in a broader range of bore and stroke sizes. Double-acting cylinders, however, cannot be held in mid-position and long-stroke cylinders may require additional guiding for the piston rod.

The basic operating principle of a linear hydraulic actuator is similar to a linear pneumatic actuator. The primary difference is the operating fluid. A typical hydraulic cylinder utilizes an incompressible fluid rather than compressed air to move the piston. Hydraulic cylinders can be single acting, double acting, ram type, or consist of telescopic rod connections.

An electric linear actuator utilizes electrical energy to generate torque. A typical electric linear actuator consists of a lead screw connected to an electric motor. As the motor turns, the lead screw is turned. A threaded lead or ball nut is prevented from rotating with the screw. The nut is then driven along the threads as the screw rotates. The direction of rotation of the screw determines the direction in which the nut moves.

Pneumatic linear actuators offer several advantages over hydraulic or electric actuators. Pneumatic actuators utilize a simple design and principle of operation, are low cost, can be utilized at a range of temperatures, do not utilize hazardous materials, and can generate precise linear motion. The primary disadvantages of linear actuators include the compressibility of the air used which leads to low forces and slower speeds at low pressures, the need for additional components such as valves and regulators, and the inherent need for a compressor or compressed air.

Hydraulic actuators offer other advantages over pneumatic and electric actuators. Hydraulic actuators are better suited than pneumatic actuators for high force applications. Hydraulic actuators also have a high horsepower-to-weight ratio, minimal loss of pressure through the supply lines over long distances, and can hold force constant for an extended period of time due to the incompressibility of the supply fluid. Primary disadvantages include the potential to leak fluids and the need for companion parts, such as motors, pumps, or noise reduction components.

Electric linear actuators provide the most accurate option for precise positioning applications. Electric actuators can also be integrated with other components, such as encoders, to allow for better control of the actuator. Electric actuators are also quieter than pneumatic or hydraulic actuators and do not have the potential to the release environmental hazards, such as that might occur with hydraulic actuators. However, electric actuators are costlier than hydraulic or pneumatic actuators, cannot be used in hazardous environments, may overheat with extended use, and may have limited force output. Examples of electric linear actuators include solenoids and relays.

3.6.2 Rotary Actuators

Rotary actuators transform pneumatic, hydraulic, or electric energy to mechanical rotation. Pneumatic rotary actuators utilize the pressure of compressed air to generate oscillatory rotary motions. The two most common configurations of pneumatic rotary actuators include rack and pinion and vane configurations. Rack-and-pinion actuators consist of a piston and rack that move linearly and cause a pinion gear and output shaft to rotate. Rack-and-pinion actuators can consist of single, double, or multiple racks. A vane actuator consists of a vane mounted on a central shaft enclosed in a cylindrical chamber. The vane rotates upon pressurization and continues to rotate until it reaches the end of the stroke. Air pressure applied to the other side of the vane causes it to rotate the shaft in the opposite direction. Key advantages of pneumatic rotary actuators are the simplicity of use, durability, high force output, and ability to be used in hazardous environments.

Hydraulic rotary actuators are typically utilized for applications requiring high torques. Common design configurations for hydraulic rotary actuators include piston type, vane type, or gear type. In piston-type actuators, pressurized hydraulic fluid is used to displace a piston and generate rotational motion. An example of a piston-type actuator is a swash plate motor. In a swash plate motor, pressurized oil is used to push a piston back and forth and generate rotation of a swash plate. The swash plate is connected to a shaft which rotates as the plate rotates. In vane-type actuators, the operating principle is similar to that of a pneumatic rotary actuator. Pressurized fluid is used to produce rotation of the vane which produces rotational motion of the central shaft. In a gear-type configuration, pressurized fluid moves meshed gears to generate rotary motion of the shaft connected to the gears. Key advantages of hydraulic rotary actuators include simplicity of design and the ability to generate high torques and low speeds.

Electric rotary actuators are commonly used in mechatronic systems. Examples of rotary electrical actuators include direct current (DC) motors, synchronous and asynchronous motors, alternating current (AC) motors, stepper motors, and servomotors. A current-carrying conductor placed in a magnetic field will experience a force proportional to the flux density of the magnetic field, the current flowing in the conductor, and the length of the conductor. The force and back electromotive force (EMF) will generate rotation and associated torque.

Reader is probably already familiar with basic operating principle involving rotor, stator, commutator, and brushes. Rotor is rotational part, stator is stationary part, commutator is a rotary electrical switch that periodically reverses the current direction between the rotor and the external circuit, and finally brush transfer current between the stationary and rotational parts.

The motor principle: When a current-carrying conductor is located in an external magnetic field perpendicular to the conductor, the conductor experiences a force perpendicular to itself and to the external magnetic field. The right-hand rule can be used to determine the direction of the force experienced by the conductor. If the right thumb points in the direction of the current in the conductor and the fingers of the right hand point in the direction of the external magnetic field, then the force on the conductor is directed outward from the palm of the right hand.

An electric motor consists of a permanent external field magnet (stator) and a coiled conducting ammeter (rotor) which is free to rotate within the field magnet. Brushes and a commutator (designed differently if AC or DC current is supplied to the armature) connect the armature to an external voltage source. The speed of rotation of a motor depends on the amount of current flowing through it, the number of coils on the armature, the strength of the field magnet, the permeability of the armature, and the mechanical load connected to the shaft (Fig. 3.24).

The stator is in the stationary electrical component. It consists of a group of individual electromagnets arranged in such a way that they form a hollow cylinder, with one pole of each magnet facing toward the center of the group. The rotor is the rotating electrical component. It also consists of a group of electromagnets arranged around a cylinder, with the poles facing toward the stator poles. The rotor, obviously, is located inside the stator and is mounted on the motor's shaft. The rotor will follow and rotate with the magnetic field of the stator as unlike magnetic poles attract each other and like poles repel (Fig. 3.25).

Brush motors utilize the wound armature and brushless motors utilize permanent-magnet rotor. Conventional brush-commutated DC motors use brushes commonly made

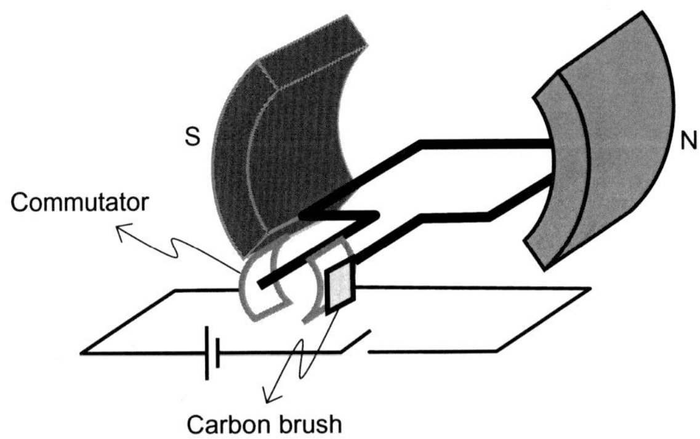

FIG. 3.24 Motor principle: Brushed DC motor *Adapted from M.B. Popovic, Biomechanics and robotics, 351 p., Copyright © 2014, Pan Stanford Publishing Pte. Ltd., Singapore, ISBN 978-981-4411-37-0 (Hardcover), 978-981-4411-38-7 (eBook). November 2013. doi:10.4032/9789814411387.*

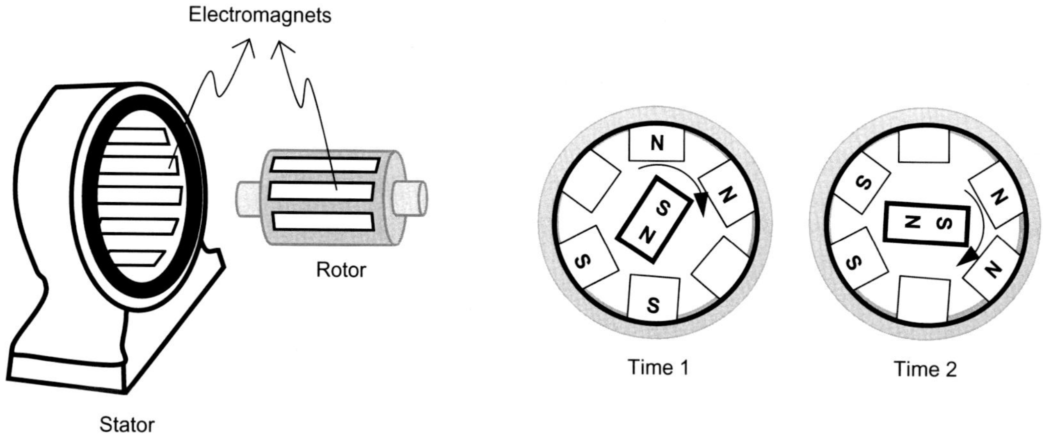

FIG. 3.25 AC motor with a permanent-magnet rotor, wound stator, and rotor-position sensor. *Adapted from M.B. Popovic, Biomechanics and robotics, 351 p., Copyright © 2014, Pan Stanford Publishing Pte. Ltd., Singapore, ISBN 978-981-4411-37-0 (Hardcover), 978-981-4411-38-7 (eBook). November 2013. doi:10.4032/9789814411387.*

of graphite with some metal content. Brushless motors commutate electronically using a permanent-magnet rotor, wound stator, and rotor-position sensor.

In synchronous motors, the speed of the rotor and frequency of the current flowing through motor are synchronous. In asynchronous motors, the speed and frequency are not the same.

Servomotor typically operates at low power and can be used to position mechanical elements with precision. It consists of small DC motor, potentiometer, and a control circuit. As the motor rotates, the resistance of the potentiometer changes. When the motor shaft reaches the target position, power to the motor is removed. It utilizes proportional control where speed is proportional to the difference between the target position versus the actual position. It is high-efficiency motor since it only runs as much as needed. It is controlled by pulse width modulation. Servos can be AC or DC. DC motor is most common and less expensive.

Stepper motors are brushless, synchronous, DC motors that divide the rotational motion into an equal number of steps. Stepper motors are capable of continuous motion, unlike servomotors, ideal for quick positioning in short distance, capable of high torque output, high accuracy, and low-cost option.

3.7 Biological Actuators: Muscles

Here the biological muscle classification, structure, and function are briefly reviewed [1].

Muscles are biological actuators that contract under stimuli and provide linear actuation.

They exhibit the property of scale invariance—their mechanism works equally efficiently at all sizes, which is why fundamentally the same muscle tissue powers both small insects and big animals.

Muscle is able to operate for billions of cycles over a period of a hundred years or more. This exceptional performance is made possible by the ability to regenerate proteins in situ.

Both muscle's force and stiffness can be graded by controlling the number of fibers that are activated in parallel, a process known as recruitment. Moreover, as not all muscle fibers within the same muscle are attached to the tendon-bone at the same points, muscle torque can also be graded by recruitment; for example, for the same joint state (joint angle and angular speed), different recruitment may produce identical force and stiffness but different torque. Regarding recruitment, it should be noted that each of the human skeletal muscles has >100 individually controlled motor units. Recruitment allows muscle performance to be optimized over a wide range of body states and loads.

There are three types of muscles: *striated, cardiac, and smooth.*

Striated muscle has a characteristically cross-striped appearance. The most striated muscle is skeletal muscle, involved in the rotation of bones around joints. Other striated muscles move the eyes and serve as valves to check the flow of blood and other fluids. Aside from creating motion across joints throughout the body, skeletal muscles help to protect the skeletal system and the organs contained within them.

Cardiac muscles cells are also striped, much like skeletal muscle cells. Cardiac muscles contract automatically to squeeze the walls of the heart inward. They produce about 100,000 heartbeats per day.

Smooth muscles lack the characteristic cross-striations, but contain the same contractile proteins. They are important as linings of the gastrointestinal tract that churn and propel food through the tract, as linings of blood vessels that control their diameters and, thus, flow through them, as valves that control the passage of gases and fluids in the body, and as controllers at many other places in the body.

The most abundant tissue in the body is skeletal muscle, accounting for approximately 40% of the body weight. There are between 320 and 425 pairs of identical bilateral skeletal muscles; this number varies depending on whether or not one defines several muscles as different parts of a single muscle.

Most skeletal muscles are attached to bones by bundles of collagen fibers known as tendons.

Passive and active dynamic properties of muscles are illustrated in Fig. 3.26.

Hill's muscle model provides a popular-state equation applicable to skeletal muscle that has been stimulated to show tetanic contraction. It relates tension to velocity with regard to the internal thermodynamics,

$$(v+b)(F+a)=b(F_0+a), \tag{3.6}$$

where F is the tension (or load) in the muscle, v is the velocity of contraction, F_0 is the maximum isometric tension (or load) generated in the muscle, a is the coefficient of shortening heat, where v_0 is the maximum velocity, when $F=0$, and finally $b=a\frac{v_0}{F_0}$.

A tetanic contraction (or tetanized state) occurs when a motor unit has been maximally stimulated by its motor neuron. This occurs when a muscle's motor unit is stimulated at a sufficiently high frequency of multiple impulses. Each stimulus causes a twitch. If stimuli

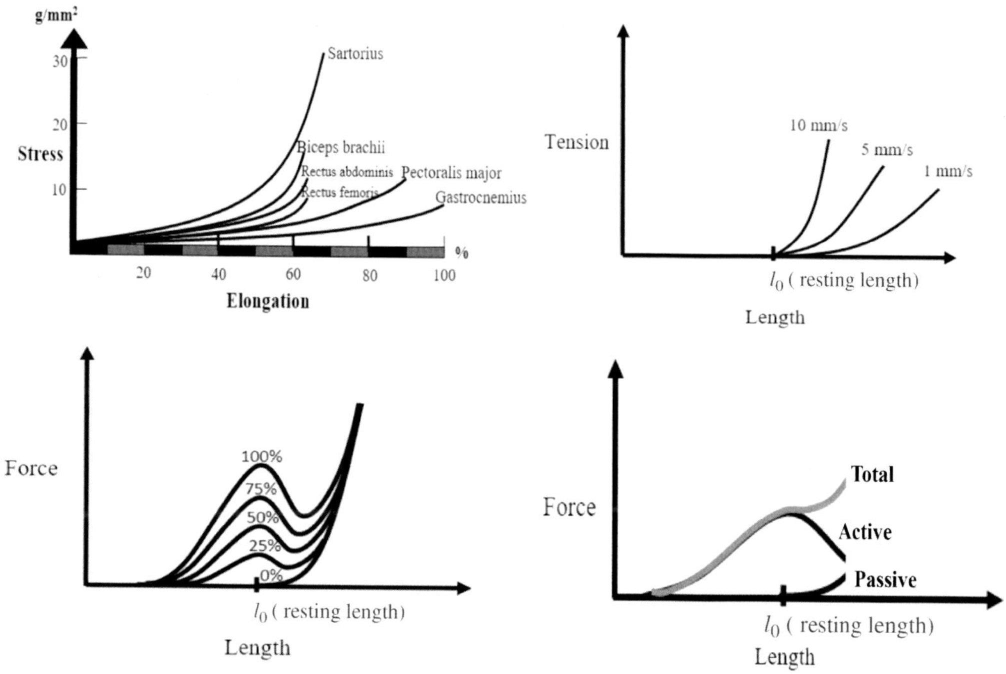

FIG. 3.26 Muscle passive and active forces. Top left, muscle passive force for zero velocity: The stress-elongation graph (based on Ref. [44]). Top right, muscle passive force at various speeds: the tension-length graph (based on Ref. [45]). Bottom left, muscle force at various percentages of activation. Bottom right, muscle's total force: sum of active and passive forces. *Adapted from M.B. Popovic, Biomechanics and robotics, 351 p., Copyright © 2014, Pan Stanford Publishing Pte. Ltd., Singapore, ISBN 978-981-4411-37-0 (Hardcover), 978-981-4411-38-7 (eBook). November 2013. doi:10.4032/9789814411387.*

are delivered slowly enough, the tension in the muscle will relax between successive twitches, but if stimuli are delivered at a high frequency, then the twitches will add up, resulting in tetanic contraction. When tetanized, the contracting tension in the muscle remains constant in a steady state. This tetanized state represents the maximal contraction.

Hill's equation demonstrates that the relationship between F and v is hyperbolic. Therefore, the higher the load applied to the muscle, the lower the contraction velocity. Similarly, if the contraction velocity is higher, then the muscle tension will be lower. However, this hyperbolic form has been found to fit the empirical data only during isotonic contractions near resting length. (*Note*: isotonic = constant force, Isometric = constant length.)

Skeletal muscle has a rather complex structure. For example, consider human biceps brachii. Total number of muscle fibers in human biceps brachii is between 150,000 and 450,000 (structures in parallel). Each muscle fiber contains 1000–2000 myofibrils (again structures in parallel). There is about 2000 myofilaments per myofibril (once again structures in parallel). Finally, there is about 100,000 sarcomers per myofilament (building

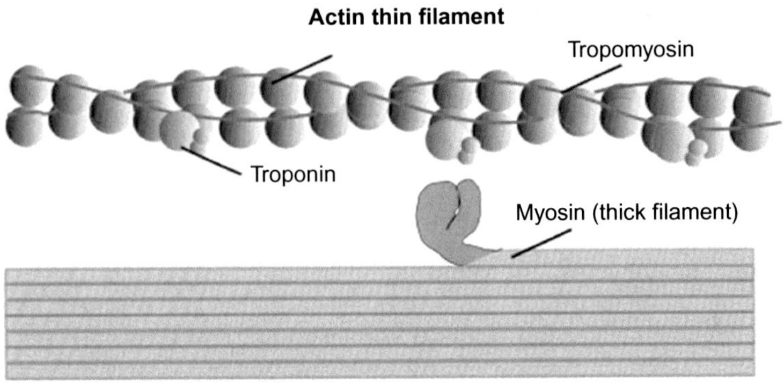

FIG. 3.27 Thin filament with two strands of actin alpha helix, top, and myosin (tick filament), bottom.

blocks in series). Therefore there is roughly between 3.0×10^{16} and 1.8×10^{17} sarcomers, which represent the muscles' elementary actuation building blocks.

The motor unit, which comprises a single motor nerve and all of the muscle fibers innervated by it, controls the graded response. In the biceps brachii, for example, a single motor unit may innervate on average 150 fibers. This means that human biceps brachii has about 1000 motor units that allow for different muscle recruitment strategies (!)

Each muscle fiber can be between 10 and 100 μm in diameter. Each myofibril is roughly 1 μm in diameter and run the length of the fiber. The individual myofibrils are made up of myofilaments, which are small protein fibers. Two of the important proteins are myosin, which forms the thick filament (about 16 nm in diameter), and actin, which forms the thin filament (about 6 nm in diameter) (Fig. 3.27).

The thin filaments are actually composed of three different types of protein: actin, tropomyosin, and troponin. Together, these are termed the regulatory protein complex.

Actin is a globular protein arranged as a chain of repeating units, forming two strands of an alpha helix. Interdigitated between the actin strands are rod-shaped proteins termed tropomyosin. There are six to seven actin molecules per tropomyosin. Attached to the tropomyosin at regular intervals is the troponin complex.

The troponin complex is made up of three subunits: troponin-T (TN-T), which attaches to the tropomyosin; troponin-C (TN-C), which serves as a binding site for Ca^{++} during excitation-contraction coupling (four Ca^{++} can bind per TN-C); and troponin-I (TN-I), which inhibits the myosin-binding site on the actin.

Myosin, forming the thick filaments, has a long, fibrous tail and a globular head, which binds to actin. The myosin head also binds to adenosine triphosphate (ATP), which is the source of energy for muscle movement. Myosin can only bind to actin when the binding sites on actin are exposed by calcium ions. There are about 300 molecules of myosin (in series) per thick filament. Each myosin contains two heads that are the site of the myosin ATPase, an enzyme that hydrolyzes ATP required for actin and myosin cross-bridge formation. These heads interact with a binding site on actin.

The structure and the function of myosin are strongly conserved across species, to the extent that rabbit muscle myosin will bind to actin from an amoeba.

While muscles can vary both in overall architecture and dimension (from micrometers to meters in length) the muscles' fundamental building blocks, sarcomeres, are identical across all known species. (!)

In 1954, two ground-breaking papers were published back-to-back in *Nature*, a well-known scientific magazine. They independently showed that muscle shortens as a result of the sliding between two sets of filaments containing the proteins myosin and actin.

The paper, "Structural changes in muscle during contraction: Interference microscopy of living muscle fibers" was authored by A. F. Huxley and R. Niedergerke. The paper "Changes in the cross-striations of muscle during contraction and stretch and their structural interpretation" was authored by H. E. Huxley and J. Hanson.

These two papers uncovered the fascinating story of one of the most intriguing of all biological problems—that of the conversion of chemical energy to mechanical work. Today, this theory of muscle contraction is known as sliding filament theory. Surprisingly, no Nobel Prize has ever been awarded to the creators of the sliding filament theory (!).

According to the sliding filament theory of muscle contraction, individually linked sarcomeres shorten (contract) when the filaments slide past each other. The movement of the actin filaments toward the middle of the sarcomere occurs as the myosin cross-bridges attach themselves to the actin sites.

3.7.1 Actuation of Biological Muscles

To actuate a muscle, the musculocutaneous nerve of peripheral nervous system transfers the wave of axon depolarization to individual muscle fibers via motor units. Each motor unit has some number of motor nerves that extend to individual muscle fibers by way of a neuromuscular junction (also called synapse). The electric events at the synapse initiate the stepwise actions that result in muscle contraction (otherwise known as excitation-contraction coupling).

When the motor nerve is depolarized, acetylcholine is released from the axon terminals at the neuromuscular junction. The acetylcholine binds to the receptor sites on the motor end-plate membrane, that is, a specialized connecting point of the muscle's sarcolemma (the fine transparent tubular sheath which envelops the fibers of skeletal muscles) with the axon terminal. The neurotransmitter, acetylcholine, increases the permeability of the motor end plate to sodium and potassium ions, which produces an end-plate potential. This potential depolarizes the sarcolemma that creates a muscle action potential that is propagated throughout the muscle membrane, causing depolarization of the transverse tubules, that is, T-tubules which are extensions of the cell membrane that penetrate into the center of skeletal and cardiac muscle cells. This leads to the release of calcium from the terminal cisternae of the sarcoplasmic reticulum (SR), a tubule system within individual fibers that run lengthwise with the fibers.

The release of calcium from the SR increases the concentration of calcium within the fiber. This calcium is then used to initiate contraction, given the affinity of troponin to calcium. As troponin attaches to calcium, it produces a movement of the tropomyosin molecule that frees up the actin site so that the charged cross-bridge can contact the site resulting in the liberation of energy from the ATP molecule. This released energy is used to drive the cross-bridge inwardly toward the mid-line of the sarcomere.

Actin activates the myosin ATPase found on the myosin cross-bridge. ATP (the energy currency of the muscle fiber) is hydrolyzed into adenosine diphosphate (ADP) and phosphate. The splitting of ATP releases stored energy to produce the movement of the myosin cross-bridges, which in turn produces sliding of the thick and thin filaments past each other. Then, the breaking of the actin-myosin connection (to allow for another actin combination with myosin) requires another ATP on the myosin cross-bridge. The cycle of binding and unbinding of actin with myosin cross-bridges along the actin filament is continued with a high sarcoplasmic concentration of calcium.

When Ca^{++} binds to TN-C, there is a conformational change in the troponin complex such that TN-I moves away from the myosin binding site on the actin, thereby, making it accessible to the myosin head. When Ca^{++} is removed from the TN-C, the troponin complex resumes its inactivated position, thereby, inhibiting the myosin-actin binding.

A single cycle of attachment, swivel, and detachment of the myosin head will produce a linear translation of the myofilaments of about 10 nm. Once all of the cross-bridges in a myofibril cycle synchronized, a relative movement equal to about 1% of the muscle length will occur. Muscle, of course, shortens by >1%. The total shortening of a sarcomere during contraction may exceed 1000 nm; therefore, the relative movement of a thin and thick filament would be half this amount or 500 nm. To achieve this magnitude of change in total length when each cross-bridge cycle produces a 10-nm shortening, a minimum of 50 cycles must occur.

The flexor muscles of the human upper arm can contract at the rate of 8 m/s, during which they can shorten by as much as 10 cm. This gives a contraction rate of 160 nm/ms for a sarcomere. If a stroke of the cross-bridge is taken to be 10 nm, then at this rate, there will be a minimum of 16 strokes/ms. Thus, the swivel time for the cross-bridge must be of the order of 60 ns. Calculations for the frog's sartorius muscle, which can shorten up to 4 cm/s, indicate a swivel time of about 1 ms, but this contraction occurs at a lower temperature than those in mammals.

A Appendix: Braided, Helically Wound Mesh for McKibben Like Artificial Muscle

Consider cylindrical webbing consisting of inelastic threads meshed in the form of intersecting clockwise and counterclockwise helixes. Imagine single helix unwrapped into plane; single thread is now represented by constant length diagonal b of a rectangle with one side equal to the actuator length L and other side equal to $R\theta$ where R is radius of the actuator and θ is the wrapping angle (i.e., the number of thread windings is equal to $n = \theta/2\pi$).

The change in the volume of the actuator in zeroth-order approximation can be expressed as

$$\Delta V = \Delta\left(LR^2\pi\right) = \Delta\left(LR^2\theta^2\pi/\theta^2\right) = \Delta\left[L\left(b^2 - L^2\right)\frac{1}{4\pi n^2}\right] = \left(b^2 - 3L^2\right)\frac{1}{4\pi n^2}\triangle L \quad (3.7)$$

Adding fluid into actuator, that is, $\triangle V > 0$, may result in the shortening of the actuator, that is, $\Delta L < 0$, as is the case for McKibben muscle, if $b < \sqrt{3}L$ which corresponds to the angle of helix (angle formed by diagonal b and side of length L) smaller than $\varphi_{critical} = \cos^{-1}\frac{1}{\sqrt{3}} = 54.7^o$. Note that shortening of the muscle will increase angle of helix.

Vice versa, adding fluid into actuator, that is, $\triangle V > 0$, may result in the elongation of the actuator, that is, $\triangle L > 0$, if angle of helix is larger than $\varphi_{critical} = 54.\,7^o$. Note that elongation of the muscle will decrease angle of helix.

Hence in both cases continued addition of the fluid will eventually result in helix angle asymptotically approaching critical angle $\varphi_{critical}$. In the close vicinity of the critical angle, it is close to impossible to add any more fluid.

Experimentally obtained critical angle $\beta = 54^o$ has been reported by Larsson [12].

Note that Hydro Muscle is not characterized by helically wound webbing or restraining mechanisms associated with it. For that and other reasons, Hydro Muscle has much better strain and energy efficiency as outer restraining member is not applying substantial restraining forces along axial direction (Fig. 3.28).

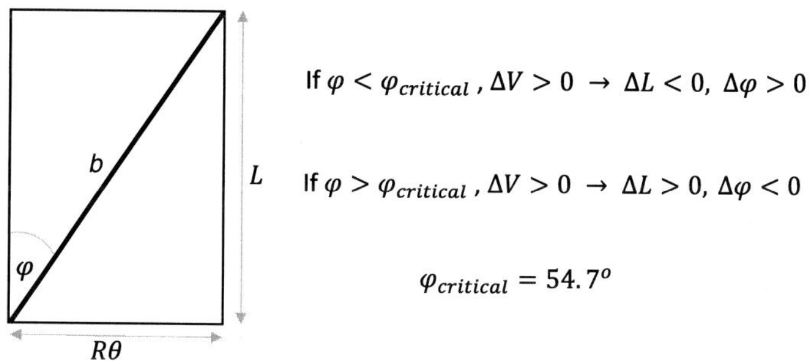

FIG. 3.28 Single helical thread unwrapped into plane.

References

[1] M.B. Popovic, Biomechanics and Robotics, 351 pages, Copyright © 2014. Pan Stanford Publishing Pte. Ltd., Singapore, 2013. doi:10.4032/9789814411387 ISBN 978-981-4411-37-0 (Hardcover), 978-981-4411-38-7 (eBook).

[2] S. Davis, N. Tsagarakis, J. Canderle, D.G. Caldwell, Enhanced modelling and performance in braided pneumatic muscle actuators, Int. J. Robot. Res. 22 (2003) 213–227.

[3] J.B. Chipka, M.A. Meller, E. Garcia, Efficiency testing of hydraulic artificial muscles with variable recruitment using a linear dynamometer, in: Bioinspiration, Biomimetics, and Bioreplication, vol. 9429, International Society for Optics and Photonics, 2015.

[4] M.B. Popovic, C. Onal, G. McCarthy, N. Corso, D. Effraimidis, and B. Jennings (2015) Actuators and Methods of Use" (Hydro Artificial Muscles). United States Patent and Trademark Office, Assignee Worcester Polytechnic Institute. Serial No.: 62/011,830. Filed: June 13, 2014. US Patent 20,150,359,698, 2015

[5] G. McCarthy, D. Effraimidis, B. Jennings, N. Corso, C. Onal and M. B. Popovic (2014) Hydraulically actuated muscle (HAM) exo-musculature, in Robot Makers: The Future of Digital Rapid Design

and Fabrication of Robots (RoMa) Workshop, the 2014 Robotics: Science and Systems Conference, Berkeley, CA, July 12, 2014.

[6] S. Sridar, C.J. Majeika, P. Schaffer, M. Bowers, S. Ueda, A.J. Barth, J.L. Sorrells, J.T. Wu, T. R. Hunt, M. Popovic, Hydro Muscle—a novel soft fluidic actuator, in: 2016 IEEE International Conference on Robotics and Automation (ICRA), 2016, , pp. 4014–4021.

[7] C.P. Chou, B. Hannaford, Measurement and modeling of McKibben pneumatic artificial muscles, IEEE Trans. Robot. Automat. 12 (1996) 90–102.

[8] T. Noritsugu, H. Yamamoto, D. Sasaki, M. Takaiwa, Wearable power assist device for hand grasping using pneumatic artificial rubber muscle, in: SICE Annual Conference in Sapporo, August 4–6, 2004, p. 420.

[9] T. Noritsugu, et al., Wearable power assist device for standing up motion using pneumatic rubber artificial muscles, J. Robot. Mechatron. 19 (6) (2007) 619–628.

[10] R. Deimel, O. Brock, A compliant hand based on a novel pneumatic actuator, in: 2013 IEEE International Conference on Robotics and Automation, ICRA, 2013, , pp. 2047–2053.

[11] K.C. Galloway, P. Polygerinos, C.J. Walsh, R.J. Wood, Mechanically programmable bend radius for fiber-reinforced soft actuators, in: 16th International Conference on Advanced Robotics (ICAR), 2013 Montevideo, Uruguay, 2013, , pp. 1–6.

[12] O. Larsson, "Flexible actuator", Assignee Komatsu Ltd, Tokyo, Japan, US Patent Number 4,777,868, Date of Patent Oct. 18, 1988.

[13] A. Miriyev, K. Stack, H. Lipson, Soft material for soft actuators, Nat. Commun. 8 (1) (2017) 596.

[14] M. Bowers, C. Harmalkar, A. Agrawal, A. Kashyap, J. Tai, M. Popovic, Design and test of biologically inspired multi-fiber Hydro Muscle actuated ankle, in: Proceedings of 2017 IEEE International Workshop on Advanced Robotics and its Social Impacts, March 8–10, 2017, University of Texas at Austin, Austin, TX, USA, 2017.

[15] M. Popovic, S. Sridar, C. Majeika, N. Deisadze, and E. Giancarlo (2016) "Variable Stiffness Devices and Methods of Use". United States Patent and Trademark Office, Assignee Worcester Polytechnic Institute. US Patent App. 15/154,443, 2016.

[16] M. Bowers, C. Harmalkar, S. Sridar, C. Majeika, C. Kaan, G. Iannacchione, M. Popovic, An approach to HydroBone and other variable stiffness structures, in: Proceedings of The 20th International Conference on Composite Structures (ICCS20) in Paris, France, 4–7 September, 2017.

[17] Soft Robotics Lab, Worcester Polytechnic Institute, http://softrobotics.wpi.edu/ (retrieved 15.03.18).

[18] M.A. Larrier J.r, E. Saint-Elme, C. Kracinovich, D. Renshaw, Accurate Prosthetic Hand, in: WPI Major Qualifying Project, Advisor Marko Popovic, Co-advisor Karen Troy, Worcester, Polytechnic Institute, Worcester, MA, 2017.

[19] E. Saint-Elme, M.A. Larrier Jr., C. Kracinovich, D. Renshaw, K. Troy, M. Popovic, Design of a biologically accurate prosthetic hand, in: IEEE RAS International Symposium on Wearable & Rehabilitation Robotics Houston, TX November 5–8, 2017.

[20] S.B. Kesner, L. Jentoft, F.L. Hammond, R.D. Howe, M.B. Popovic, Design considerations for an active soft orthotic system for shoulder rehabilitation, in: 33rd Annual International IEEE EMBS Conference, August 30–September 02, 2011, Boston, USA, 2011.

[21] I. Galiana, F.L. Hammond, R.D. Howe, M.B. Popovic, Wearable soft robotic device for post-stroke shoulder rehabilitation: identifying misalignments, in: 2012 IEEE/RSJ International Conference on Intelligent Robots and Systems, October 7–12, 2012, Portugal, 2012.

[22] T. Hunt, C. Berthelette, G.S. Iannacchione, S. Koehler, M.B. Popovic, Soft robotics variable stiffness exo-musculature, one-to-many concept, and advanced clutches, in: IEEE ICRA 2012 Workshop: Variable Stiffness Actuators Moving the Robots of Tomorrow, St. Paul, Minnesota, May, vol. 14. 2012, 2012.

[23] T. Hunt, C. Berthelette, M.B. Popovic, Linear one-to-many (OTM) system, in: IEEE International Conference on Technologies for Practical Robot Applications (TePRA), 2013, IEEE, 2013, , pp. 1–6.

[24] M. B. Popovic, C. J. Berthelette, M. DiPinto, J. D. Sareault, and T. R. Hunt "Actuation Systems and Methods" U.S. Patent Application No. 14/903,126 Filed January 6, 2016. U.S. Patent No. 9,879,767 Issued January 30, 2018.

[25] C.D. Onal, X. Chen, G.M. Whitesides, D. Rus, Soft mobile robots with on-board chemical pressure generation, in: Robotics Research, Springer, Cham, 2017, , pp. 525–540.

[26] [3]R. Pelrine, High-speed electrically actuated elastomers with strain greater than 100%, Science 287 (5454) (2000) 836–839.

[27] Y. Bar-Cohen, Electroactive polymers as artificial muscles: a review, J. Spacecraft Rockets 39 (6) (2002) 822–827.

[28] Bar-Cohen, Electroactive polymers as artificial muscles—reality and challenges, Adv. Piezoelectr. Mater. (2001) 1–13.

[29] G. Kovacs, L. During, S. Michel, G. Terrasi, Stacked dielectric elastomer actuator for tensile force transmission, Sensors Actuators A: Phys. 155 (2009) 299–307.

[30] G.A. Pratt, M.M. Williamson, Series elastic actuators, in: Intelligent Robots and Systems 95. IEEE/RSJ International Conference on "Human Robot Interaction and Cooperative Robots", Proceedings, 1995, Vol. 1, 1995, , pp. 399–406.

[31] T. Morita, S. Sugano, Design and development of a new robot joint using a mechanical impedance adjuster, ICRA (1995) 2469–2475.

[32] R. Van Ham, T.G. Sugar, B. Vanderborght, K.W. Hollander, D. Lefeber, Compliant actuator designs, IEEE Robot. Autom. Mag. 16 (3) (2009).

[33] B. Vanderborght, A. Albu-Schäffer, A. Bicchi, E. Burdet, D.G. Caldwell, R. Carloni, M. G. Catalano, O. Eiberger, W. Friedl, G. Ganesh, M. Garabini, Variable impedance actuators: a review, Robot. Autonom. Syst. 61 (12) (2013) 1601–1614.

[34] M. Laffranchi, N. Tsagarakis, D.G. Caldwell, A compact compliant actuator (CompAct™) with variable physical damping, in: IEEE International Conference on Robotics and Automation (ICRA), 2011, 2011, May, , pp. 4644–4650.

[35] T. Verstraten, P. Beckerle, R. Furnémont, G. Mathijssen, B. Vanderborght, D. Lefeber, Series and parallel elastic actuation: impact of natural dynamics on power and energy consumption, Mech. Mach. Theory 102 (2016) 232–246.

[36] M. Zinn, B. Roth, O. Khatib, J.K. Salisbury, A new actuation approach for human friendly robot design, Int. J. Robot. Res. 23 (4–5) (2004) 379–398.

[37] M. Zinn, O. Khatib, B. Roth, J.K. Salisbury, Playing it safe [human-friendly robots], IEEE Robot. Autom. Mag. 11 (2) (2004) 12–21.

[38] P. Beckerle, J. Wojtusch, S. Rinderknecht, O. von Stryk, Analysis of system dynamic influences in robotic actuators with variable stiffness, Smart Struct. Syst. 13 (4) (2014) 711–730.

[39] G. Mathijssen, D. Lefeber, B. Vanderborght, Variable recruitment of parallel elastic elements: series–parallel elastic actuators (SPEA) with dephased mutilated gears, IEEE/ASME Trans. Mechatron. 20 (2) (2015) 594–602.

[40] G.S. Hegde, Mechatronics, Jones and Bartlett Publishers, 2010.

[41] http://blog.parker.com/know-your-pneumatics-single-or-double-acting-choosing-the-right-cylinder (retrieved 15.03.18).

[42] http://www.machinedesign.com/linear-motion/what-s-difference-between-pneumatic-hydraulic-and-electrical-actuators (retrieved 15.03.18).

[43] http://blog.parker.com/know-your-pneumatics-rotary-actuator-basics (retrieved 15.03.18).

[44] D.R. Soderquist, Sensory Processes, SAGE, California, 2002.

[45] D. Winter, Biomechanics and Motor Control of Human Movement, John Wiley, New York, 1990.

4

Sensors: Natural and Synthetic Sensors

Kathleen A. Lamkin-Kennard*, Marko B. Popovic†
*ROCHESTER INSTITUTE OF TECHNOLOGY, ROCHESTER, NY, UNITED STATES †WORCESTER POLYTECHNIC INSTITUTE, WORCESTER, MA, UNITED STATES

CHAPTER OUTLINE

Biomechatronics. https://doi.org/10.1016/B978-0-12-812939-5.00004-5

4.1 Introduction

The nervous system is responsible for coordination of physiological processes. Sensory receptors provide information to the nervous system in response to a broad range of signal types. Biomechatronic systems seek to replicate these structure function relationships. However, the complexities of the human nervous system are difficult to replicate synthetically. Synthesizing information from multiple sensors is also challenging. In this chapter, a discussion of natural sensors and their synthetic counterparts is provided.

4.2 Natural Sensors

4.2.1 The Nervous System

The ultimate function of the nervous system is to control bodily activities. The nervous system controls the motor functions of the nervous system which include contraction of skeletal muscles, contraction of smooth muscles, and secretion by both endocrine and exocrine glands throughout the body. The muscles and glands are called *effectors* since they perform the functions prescribed by the signals sent from the nervous system. The nervous system consists of the autonomic nervous system (ANS) and the peripheral nervous system (PNS). The ANS controls smooth muscles, glands, and internal bodily systems. The function of the PNS is to relay signals between the brain and spinal cord and the rest of the body.

The PNS is comprised of the components of the nervous system outside the brain and spinal cord including the spinal nerves and their roots and branches, the cranial nerves, the peripheral nerves, and neuromuscular junctions. The cranial nerves are nerves that terminate in or originate from the brain. The spinal nerves terminate at or originate from the spinal cord. Electrochemical signals are sent between the central nervous system and the rest of the body via the nerves of the PNS. The PNS is subdivided into the ANS and the somatic nervous system. Fig. 4.1 shows a comparison of the functions associated with the ANS versus the somatic nervous system.

The ANS is responsible for involuntary responses, such as pupil constriction or dilation and salivation, to regulate physiological functions. The ANS is always active, but can exist in either a sympathetic or parasympathetic state. The state the system is a consequence of the surrounding environment and activities and will impact the neurotransmitters released by the system. In a sympathetic state, neurotransmitters, such as epinephrine and norepinephrine, are released to increase heart rate and blood flow to certain muscles. The sympathetic system is activated in situations of danger or stress and generates what is known as the "fight or flight" response. As heart rate and blood flow are increased, the activity associated with other noncritical physiological functions, such as digestion, is decreased. In the parasympathetic state, acetylcholine is the primary neurotransmitter released. The parasympathetic system is sometimes referred to as the "rest and digest" system and is responsible for activities that occur when the body is at rest, such as increasing intestinal and gland activity, slowing of the heart rate, salivation, or digestion.

Comparison of the somatic and the autonomic divisions of the nervous system.

Somatic	Autonomic
Involves both afferent and efferent pathways	Involves only the efferent pathway
Voluntary activities, including locomotion (effectors are skeletal muscles)	Involuntary activities (effectors are cardiac muscle, smooth muscles, fat cells, and glands)
Efferent signals originate at the cerebral cortex as a conscious decision and activate neurons in the brainstem or spinal cord	Unconscious signals originate in hypothalamus, brain stem, and spinal cord and activate target neurons that lie in the peripheral nervous system
Brainstem and spinal cord neuron exerts direct control over skeletal muscle (one neuron system, where there are no intermediate synapses between the CNS and the target organs)	Target neurons in the peripheral nervous system are grouped in ganglia and their axons exert direct control over smooth muscle, cardiac muscle, glands, and fat cells. (Two neuron system, where the efferent neurons synapse once outside the CNS before the signals reach the target organs.)
Efferent axons are myelinated (fast conduction)	Postsynaptic axons are nonmyelinated (slow conduction); presynaptic axons are myelinated
Neuromuscular junctions are specific and localized	Synapses at the target organs for the axons of the ANS may be diffuse (varicosities)
Target organs (skeletal muscles) are always stimulated into action	ANS used sereral different neurotransmitter molecules (predominantly ACh and norepinephrine, NE)

FIG. 4.1 Functions of the somatic and autonomic nervous systems [1].

The somatic nervous system is known as the voluntary nervous system and is responsible for transmission of sensory and motor information to and from the central nervous system. The transmission of information from the central nervous system to the body occurs via efferent nerves. Efferent nerves are also known as motor nerves. Afferent, or sensory, nerves relay sensory information from sensory receptors throughout the body to the central nervous system.

4.3 Sensory Receptors

Sensory receptors are specialized neurons that sense various forms of environmental energy. Information from the sensory receptors is transmitted as a train of nerve impulses to the central nervous system via afferent nerves. The central nervous system then analyzes the sensory information and utilizes it to generate motor commands.

The basic mechanism of transduction is that local currents are produced at nerve endings. All receptors have one factor in common. Regardless of type of stimulus, the net effect is to change the receptor potential. The change in potential alters membrane permeability which allows ions to diffuse more or less easily through the membrane and impacts the overall transmembrane potential.

Different receptors can be excited in several different ways. For example, mechanical deformation stretches the receptor membrane and opens ion channels. Chemical stimuli directly open ion channels. Thermal changes alter the permeability of the membrane. Electromagnetic radiation changes the membrane permeability or characteristics of the membrane and allows ions to flow more easily through membrane channels.

4.4 Sensory Receptors Classified by Stimulus Type Detected

Sensory receptors can be classified into multiple categories based on the type of stimulus detected including chemoreceptors, mechanoreceptors, thermoreceptors, nociceptors, and electromagnetic receptors.

4.4.1 Chemoreceptors

Chemoreceptors are sensory cells in the body capable of responding to a broad range of chemical signals, including taste, smell, arterial oxygen, blood carbon dioxide levels, and osmolality. Chemoreceptors constantly monitor chemical changes in the body and are essential to the maintenance of physiological homeostasis.

The *gustatory system* is responsible for *taste* and contains taste cells, afferent gustatory nerves, and brain structures responsible for processing the sense of taste [3]. There are about 10,000 taste buds on the human tongue. They are mainly contained in raised structures called papillae (about 200 on the tongue). Papillae are classified into four groups based on shape: fungiform, foliate, circumvallate, and filiform. Each taste bud contains 50–100 chemoreceptors, or taste receptor cells (TRCs), with taste receptors that sense different tastes or mediate biological processes based on taste detection. Four subtypes of TRCs have been identified to date. Type I are glial-like cells and detect salty taste. Type II cells express G-protein-coupled receptors (GPCRs) and are associated with the detection of sweet, umami, and bitter tastes. Sour stimuli are detected by Type III cells. Type IV cells are likely associated with stem or progenitor taste cells [4]. The GPCRs can be further subdivided into the taste 1 receptor (T1R) family and the taste 2 receptor (T2R) family. The T1R family contains multiple subtypes, including T1R2 (T1R member 2) and T1R3 (T1R member 3) that act as sweet taste receptors. T1R1-T1R2 receptors are responsible for detection of umami while T2 receptors detect bitter tastes [3, 4]. Taste receptors are also found in the palate and early parts of the digestive system such as the larynx and upper esophagus. There are three cranial nerves that innervate the tongue; the vagus nerve, the glossopharyngeal nerve, and the facial nerve. The glossopharyngeal nerve and the chorda tympani branch of the facial nerve innervate the taste receptors (Fig. 4.2).

Smell or olfactory chemoreceptors for vertebrates are olfactory sensory neurons in the olfactory epithelium. The mucus overlying the epithelium acts as a solvent for odor molecules. This mucus flows constantly and is replaced approximately every 10 min. Odor receptor neurons in the olfactory epithelium detect the odors then send signals to the olfactory bulbs. From there, the signals are sent along olfactory tracts to the olfactory cortex of the brain. Humans have about $10\,cm^2$ ($1.6\,in.^2$) of olfactory epithelium, whereas some dogs have $170\,cm^2$ ($26\,in.^2$). A dog's olfactory epithelium is also considerably more densely innervated, with 100 times more receptors per square centimeter (Fig. 4.3).

Peripheral and *central* chemoreceptors regulate respiratory activity to maintain arterial blood PO_2, PCO_2, and pH. The human respiratory system tries to maintain constant levels of oxygen (O_2), carbon dioxide (CO_2), and hydrogen ions (H^+) in the arterial blood to ensure these substances are available to meet cellular metabolic requirements and to maintain the acid-base balance in the body.

The peripheral chemoreceptors are located in the carotid and aortic bodies. The carotid bodies are connected to the respiratory centers in the brainstem and monitor

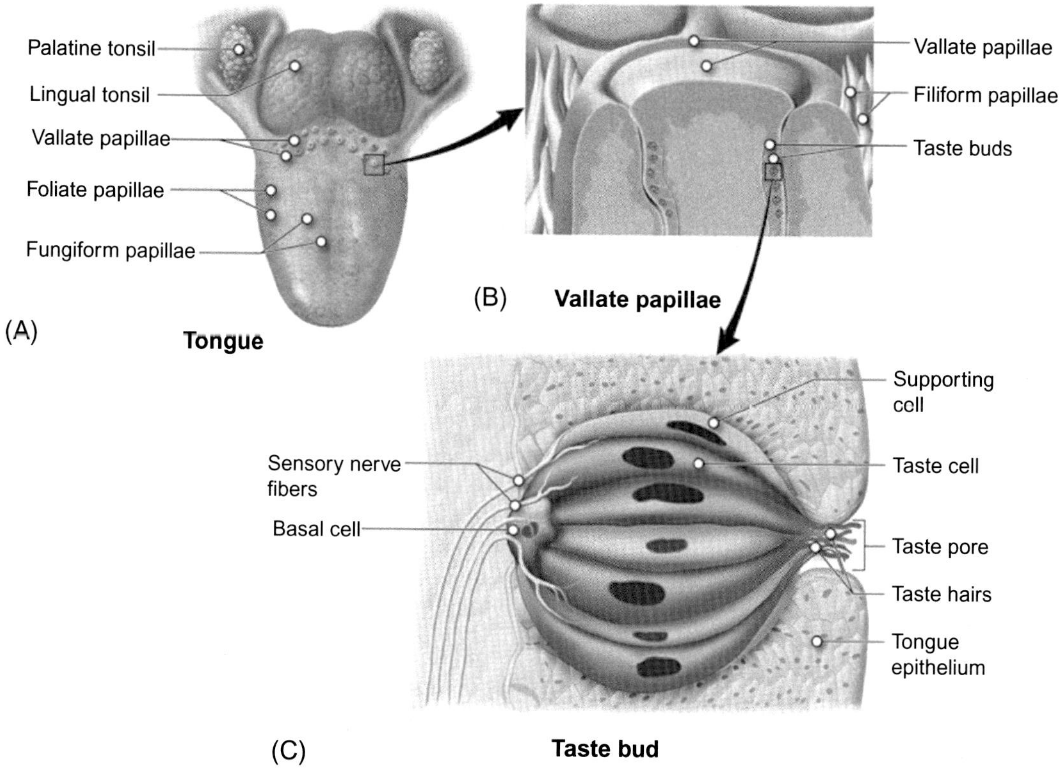

FIG. 4.2 Overview of the gustatory system [2].

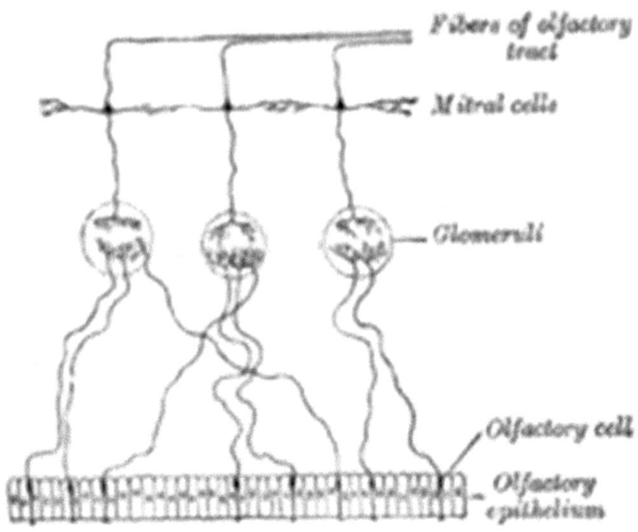

FIG. 4.3 The olfactory sensory neurons.

PO_2, PCO_2, and pH. As the arterial PO_2 decreases, the peripheral chemoreceptors in the carotid body increase the firing rate from the carotid sinus nerve to modulate respiration [5]. The aortic bodies are connected to the cardiovascular centers in the brainstem and are responsible for the cardiovascular response to chemical factors in the blood linked to respiration due to changes in PO_2 and PCO_2 [5].

The central chemoreceptors are located beneath the ventrolateral surface of the medulla and are sensitive to changes in arterial CO_2 and pH. As arterial PCO_2 increases, the CO_2 content of the cerebrospinal fluid (CSF) increases. The blood-brain-barrier is permeable to CO_2, but not hydrogen ions, so CO_2 will diffuse into the CSF. As CO_2 diffuses into the CSF, it reacts with water from the surrounding tissues to produce carbonic acid. The carbonic acid dissociates and causes the spontaneous release of free hydrogen ions. The increase in the hydrogen ion concentration in the CSF reduces the pH and stimulates the chemosensitive neurons. As the CO_2 decreases, the pH increases, and the central chemoreceptors reduce their activity below their basal rate.

Osmoreceptors are sensory chemoreceptors located in the anterior hypothalamus that detect change in osmotic pressure and plasma osmolarity help to regulate thirst. Osmoreceptors also regulate osmolality in the kidneys. As the osmolarity of the blood changes, the diffusion of water into or out of the osmoreceptors changes. For example, if the osmolarity increases due to an increase in plasma sodium concentration, the osmoreceptors will shrink. The contraction of the osmoreceptors causes them to send signals that stimulate the posterior pituitary to release antidiuretic hormone (ADH) into the blood [6]. ADH subsequently acts on the kidneys to reduce urine excretion and promote water retention.

4.4.2 Mechanoreceptors

Mechanoreceptors are sensory receptors that respond to mechanical deformation of the receptor or surrounding tissue. Mechanoreceptors are involved in hearing, detection of equilibrium, skin tactile sensing, deep tissue sensing, and sensing of arterial pressure.

Hearing or audition involves the transduction of sound waves into neural signals via mechanoreceptors in the inner ear. As shown in Fig. 4.4, sound vibrations from the outer ear are mechanically conducted from the eardrum through a series of tiny bones to the inner ear. Sound vibrations from the middle ear enter the cochlea at the basal end of the cochlea. The cochlea in the inner ear has over 30,000 hair cells that amplify the incoming acoustic signal and encode the frequency content. Outer hair cells primarily amplify sound vibrations, while inner hair cells detect those vibrations and excite the nerve fibers of the cochlear or auditory nerve. The basal end of the cochlea encodes the higher end of

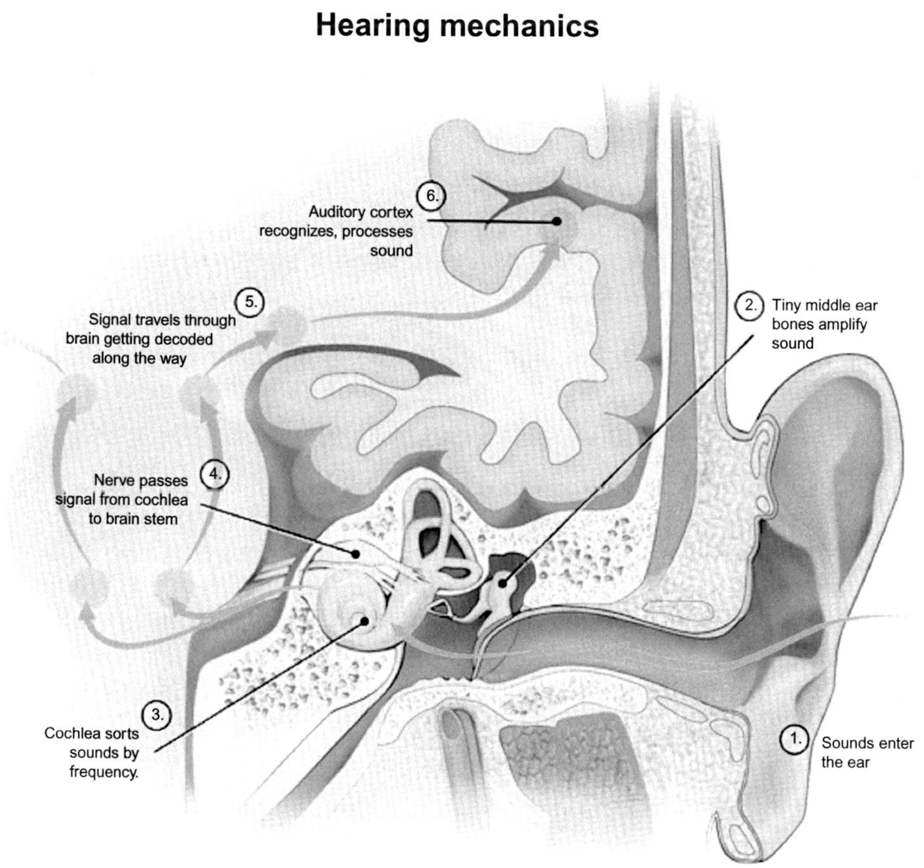

FIG. 4.4 The transmission of sound waves through the ear results in audition. Sounds travel from the outer ear to mechanoreceptors in the inner ear that excite the auditory nerve [7].

the audible frequency range while the apical end of the cochlea encodes the lower end of the frequency range.

The inner ear also hosts *vestibular* mechanoreceptors that excite the vestibular neurons of the auditory nerve. The vestibular mechanoreceptors communicate a sense of balance and spatial orientation. The vestibular system contains semicircular canals that provide information about rotational motions. Angular accelerations produce relative movements of endolymph in the semicircular canals. The movement of fluid pushes on the cupula in the ear which contains mechanosensitive hairs cells that transduce the mechanical signal into an electrical signal.

Touch or somatic sense receptors are located in the dermis, the bottom layer of skin. There are about 20 different types of nerve endings in the dermis. They can be activated by movement (mechanoreceptor), pressure (mechanoreceptor), chemical (chemoreceptor), and/or temperature (thermoreceptors). Another activation method is from the vibrations generated as a finger moves across a surface and feels these vibrations. The vibration sensing is directly related to function of lamellar (or Pacinian) corpuscles, which are one of the four major types of mechanoreceptors.

Baroreceptors are mechanoreceptors located in the carotid sinus and in the aortic arch that help to regulate arterial blood pressure. Baroreceptors are stimulated by changes in arterial pressure. Increases in pressure stretch or strain the baroreceptors and transmit signals to the central nervous system. The parasympathetic nervous system is activated while the sympathetic nervous system is inhibited in response to baroreceptor activation. Inhibition of the sympathetic system leads to a drop in peripheral resistance while the parasympathetic activation increases the vagal tone on the sinoatrial (SA) node and decreases the heart rate and contractility. The net effect is that blood pressure is decreased. Conversely, blood pressure can be increased with sympathetic activation and parasympathetic inhibition.

4.4.3 Thermoreceptors

Thermoreceptors receptors are specialized nerve cells that detect changes in temperature. Both *cold* and *warm* receptors are present in the skin. Warming results in an increase in the firing rate of the warm receptor while cooling results in an increase in the cold receptor discharge rate. At moderate skin temperatures, both receptors may be active. However, as the skin is warmed, the cold receptors become inactive and as the skin is cooled, the warm receptors inactivate. Temperatures that are likely to damage an organism are sensed by a subcategory of nociceptors.

4.4.4 Nociceptors

A *nociceptor* is a sensory receptor that responds to a potentially damaging stimulus by sending nerve signals to the spinal cord and brain. This process, called nociception, usually causes the perception of pain. Nociceptors were discovered by Charles Scott Sherrington in 1906 and respond to either physical or chemical damage in the tissue [6].

4.4.5 Electromagnetic Receptors

Photoreceptors are specialized neurons found in the retina that convert light into electrical signals that stimulate physiological processes. Signals from the photoreceptors are sent through the optic nerve to the brain for processing. The retina, which is the size of thumbnail, is filled with approximately 150 million light-sensitive cells. Two different types of receptors in the retina of each eye are responsible for the perception of fine detail and color (cones) and brightness (rods). Rods are located in the outer region of the retina while cones are located primarily in the central portion.

4.5 Sensory Receptors Classified by Stimulus Location

Another method for classifying sensory receptors is by the location of the stimulus. *Charles Scott Sherrington (1857–1952)* was an English neurophysiologist, histologist, bacteriologist, and pathologist. He received the Nobel Prize in Physiology or Medicine in 1932. In 1906, Sir Sherrington introduced the terms "*exteroception*," "*interoception*," and "*proprioception*." These terms are used to classify sensory receptors by stimulus location.

4.5.1 Proprioceptors

Proprioception from Latin *proprius* meaning "one's own," "individual," and "perception" is the sense of the relative position of neighboring parts of the body and the strength being employed in movement. Kinesthesia is the sense of body motion. Although proprioception and kinesthesia may appear to be directly related, there exists an ongoing debate among experts if there are receptors that directly communicate speed information.

The proprioceptive sense is believed to be composed of information from sensory neurons located in the inner ear (motion and orientation) and in the stretch receptors located in the muscles and the ligaments. The latter have been found in fruit flies, nematode worms, African clawed frogs, and zebrafish. However, the corresponding human proprioceptor has yet to be discovered. Examples of proprioceptors include the golgi tendon organ and muscle spindles.

4.5.2 Interoceptors

Interoception is the sense by which one perceives the internal organs (organ movements, pain, hunger, thirst, etc.). Internal nociceptors, chemoreceptors, and stretch receptors are types of interoceptors. Specific examples includes pulmonary stretch receptors in the lungs, which are part of the respiratory rate control system, chemoreceptors in the circulatory system measuring salt and prompting the thirst sensation when needed, and stretch receptors in the gastrointestinal tract, measuring gas distension that may result in colic pain.

4.5.3 Extroreceptors

Exteroception is the sense by which one perceives the outside world. The "exteroceptors" are the sensors for gathering information from outside the body and are embedded within organs such as the eyes, ears, mouth, nose, and skin. Exteroreceptors can sense acoustic, chemical, electric, optical, magnetic, mechanical, radiation, or thermal energy.

4.6 Synthetic Biological Sensors

4.6.1 Overview and Challenges

Biomechatronic devices require sensing capabilities for localization, obstacle detection, sensing for specific tasks, haptic feedback, interaction with the environment, or for protection. Humans have an exceptionally complex system for sensing many types of external stimuli and transmitting the response to the central nervous system for processing. However, replication of physiological capabilities using synthetic sensors can be challenging. The human nervous system continuously and seamlessly integrates feedback from multiple sensory receptors and processes the information to develop a coordinated motor response. The integration and processing of information from multiple sensors in real time using synthetic sensors is not as easily accomplished.

Synthesizing and understanding the relevance of the information to perform a coordinated response is also a challenge. In the nervous system, extraneous information is disregarded and only sensory information needed at that point in time is utilized to generate an appropriate motor response. For example, although the PNS continuously senses the feel of clothing on the skin, under most conditions, the information is disregarded and humans are not consciously aware of the feel of the clothing on their skin unless there is a change in the sensory input. Humans are also capable of compensating for a lack of information in cases of sensor deprivation by integrating data from other sensors with an overlapping scope. The following sections describe some of the most common types of synthetic sensors used for biomechatronic systems.

4.7 Synthetic Sensors

Sensors are devices that map an environmental attribute to a physical measurement. Synthetic sensors operate as transducers, similar to natural biological sensors, and convert energy from one form to another. Synthetic sensors can be classified by a number of means. For example, sensors can be classified as proprioceptive or exteroreceptive, similar to biological sensors. Proprioceptive sensors refer to sensing of values measured internal to the system while exteroreceptive sensors measure quantities external to the system. Examples of proprioceptive sensors include motor encoders, potentiometers, or gyroscopes. Examples of exteroreceptive sensors include touch sensors, humidity sensors, or force sensors.

Another method for classifying sensors relies on whether the sensors are active or passive. Active sensors emit energy into the environment and measure a response to the emitted energy. Passive sensors only receive energy inputs as a data source. Sensors can also be classified as contact or noncontact or visual or nonvisual. The most common method of classification is based on the type of energy transduced. The general classifications are similar to those described previously for biological systems and include chemical, electric, optical, mechanical, and thermal sensors. Table 4.1 provides the commonly measured quantities for each type of stimulus.

4.7.1 Chemical Sensors

A chemical sensor is a device or instrument that determines the detectable presence, concentration, or quantity of an analyte [9]. Chemical sensors are analogous to the chemoreceptors in the human body. A chemical sensor typically contains two main components: a receptor and a transducer. The receptor transforms chemical information about a sample into a form of energy. The transducer then transforms the energy into some form of analytical signal [8].

Two of the most important features of any chemical sensor include its selectivity and sensitivity. Analytes can be present in liquid, gas, or solid phases and may only be transiently present in the sample. Analytes may also be present over a range of dimensional scales from liters to picoliters. Selectivity involves the differential response to a particular molecular structure or its associated reactivity [9]. Over 10 billion known molecular substances exist, so selectively identifying substances is a critical, but challenging sensor function. Sensitivity relates to how well the quantity or concentration of the molecule to be detected, the analyte, can be measured. Chemical sensors frequently can detect 10^{-9} M concentrations or less [8, 9]. Achieving a high level of sensitivity is also a challenging, but essential aspect of the sensor function. Sensor design often requires a trade-off between selectivity and sensitivity. Other factors of importance for sensor design include response time, packaging size, and limit of detection [9].

The methods used for detection of an analyte can vary greatly between sensors. Some sensors directly bind the analyte while others consume the analyte via a chemical reaction and use the product from the reaction to generate a signal. Another method involves passing the sample through a thin film that only allows the target molecule to pass through. A direct-read electrochemical sensor uses the diffusion of charged molecules to detect

Table 4.1 Commonly Measured Sensor Quantities Based on Stimulus Type

Type of Stimulus	Measured Quantities
Chemical	Concentration
Electrical	Voltage, current, charge, electric field, conductivity, resistance
Optical	Absorption, reflection, refractive index
Mechanical	Displacement, velocity, acceleration, torque, stress, strain, pressure
Thermal	Temperature, specific heat, heat flux, thermal conductivity

changes in conductivity, current, or potential due to the presence of the analyte [8, 9]. A surface acoustic wave sensor sends acoustic waves across a surface between electrodes. The sensor detects changes in wave velocity or losses of intensity if the target molecule binds to the surface [8, 9].

Some of the most common applications for chemical sensors include detection of toxic chemicals, gases, or biomolecules. Detection of toxic chemicals can range from detection of ground-based chemicals, such as petroleum or mercury in soil, to the detection of deadly chemical warfare agents, such as nerve agents. Gas monitoring devices can be used to detect common gases like CO_2, O_2, or CO as well as toxic industrial compounds and chemical warfare agents. Chemical sensors can also be used for clinical and medical applications including detection of glucose, pregnancy and ovulation tests, enzyme-linked immunosorbent assay (ELISA), or western blot analyses.

Chemical sensing is a sensing modality that has traditionally been neglected in robotics since the technical challenges associated with implementing chemical sensing are substantial. However, integration of chemical sensing capabilities into biomechatronic systems could add a new dimension of functional capabilities to robotic systems. Just as animals rely on their sense of smell to detect food, other animals, or nests, robotic systems equipped with olfactory capabilities could offer added environmental detection capabilities, methods for control, or the capability to detect physiological quantities of interest. Some recent examples of biorobotic systems in which chemical sensing has been utilized include plume tracking robots, a robotic fish, and electronic nose (e-nose) systems [9–11].

Plume tracking is a strategy utilized by animals, such as moths, for odor tracking. A plume results from the interaction of a flowing fluid with a source of volatile chemicals [11]. The distribution of chemicals in the plume is nonuniform so animals use the flow direction to steer toward the source. The same strategy has been implemented into flying and underwater biomimetic robotic devices using single volatile organic compound (VOC) gas sensors or sensor arrays and sophisticated two-dimensional (2D) or three-dimensional (3D) plume tracking algorithms [12]. Robo-moth was designed to track aerial plumes by mimicking the flight of moths in 3D [11] while RoboLobster was designed to track fluorescent dye plumes underwater [12].

Chemical sensors were used by Ravalli et al. [13] to control the swimming patterns of a bio-inspired artificial fish for monitoring water quality. Hydrogen ions in the environment were detected using an electrochemical multi-sensor platform. The pH measurements were used to control the tail movement of the fish robot.

A more novel approach to integrating chemical sensing into robotics is to use the neural activity from locusts to monitor odorants [14]. The antennae of locusts are capable of sophisticated chemical sensing. Sensors were placed on the locust's back to read neural activity and nanoparticles tattoos that generate heat were used to steer the locusts toward a target. The long-term goal was to better understand the mechanisms underlying animal olfaction to aid in the design of biomimetic sensing systems [14].

e-nose devices have been widely investigated over the past 25 years with potential applications in the areas of environmental monitoring, quality control systems, explosive detection, or detection of illicit drugs [15]. A typical e-nose device, shown in Fig. 4.5,

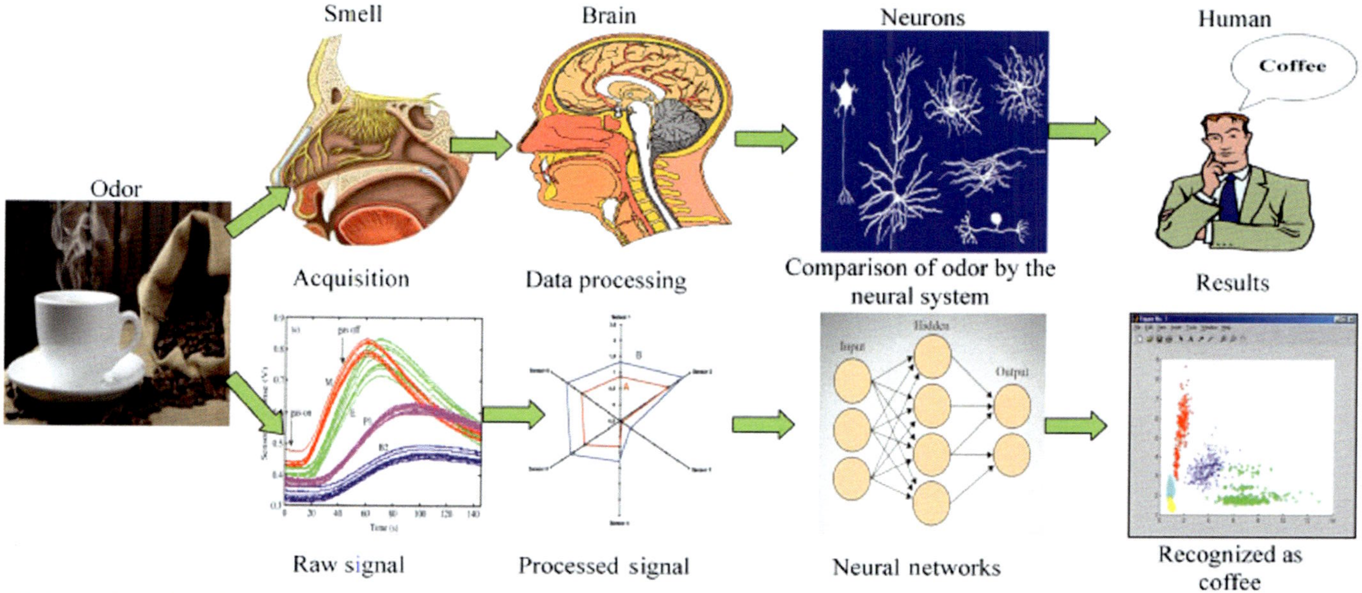

FIG. 4.5 Comparison of the functions between a biological nose and an electronic nose. *Reproduced from M. Ghasemi-Varnamkhastia, J. Lozanob, Electronic nose as an innovative measurement system for the quality assurance and control of bakery products: a review Eng. Agric. Environ. 9(4) (2016) 365–374.*

consists of an array of chemical sensors and algorithms that process the sensor responses and interpret the findings [15]. A strong interest has emerged among the robotics community to integrate e-nose devices into robotic systems. If integrated into an autonomous robot, an e-nose could be used to aid in controlling the device, provide essential information about its environment, and potentially eliminate the need to place humans or animals in hazardous environments.

4.7.2 Electric Sensors

Electric sensors can be used to measure voltage, current, charge, electric field strength, resistance, or capacitance and have a broad range of configurations. Electric sensors are commonly used to provide feedback for control of robotic devices. Some examples of electric sensors are described below.

A *limit switch* is an electromechanical device commonly used to detect the single position of a moving part. A typical limit switch contains an actuator linked to a set of contacts. As an object contacts the actuator, it will either make or break an electrical connection. A limit switch is configured to be either normally open or normally closed. Limit switches are commonly used in robotics to determine if components are touching or to ensure that movements do not exceed some limit.

Current sensors detect electric current in a wire and generate a signal proportional to the current. Current sensors can be either open or closed loop. Open-loop current sensors measure AC and DC currents. Open-loop sensors have a lower cost, lower operating power requirement, and smaller footprint than closed-loop sensors. Closed-loop sensors offer fast responses, high linearity, and low temperature drift. Current sensing is not as commonly used in robotics as other sensing modalities, but can be used to know the amount of power being used for a particular robot component.

Resistive sensors act as variable resistors. The resistance of a resistive sensor is changed in response to some type of input. Resistance can vary with temperature, light, pressure, moisture, chemical composition, or sound.

Capacitors consist of two electrodes separated by a dielectric material. Capacitance is a measure of the amount of charge that a capacitor can hold for a particular voltage. If a voltage is applied to the electrodes, an electric field is created between them. A *capacitive sensor* converts changes in properties of the dielectric material into an electric signal. Capacitive sensors can be used for proximity sensing, position sensing, measurement, or humidity sensing [16]. Capacitive sensors are widely used for tactile or touch sensing applications in robotics.

4.7.3 Optical Sensors

Optical sensors can be used for a broad range of applications ranging from simple detection of distance to providing artificial vision to a system for object recognition. Infrared sensors can also be used to detect distances based on the detection of modulated light.

Photoresistors, or light-dependent resistors, are light-sensitive variable resistors. The resistance depends on the intensity of light. The resistance decreases with increasing light intensity and increases with decreasing intensity, thus photoresistors can be used in light-sensitive detector circuits.

Photodiodes are devices that consume light energy to produce an electric current. Photodiodes are also referred to as photodetectors or photo-sensors. Common applications of photodiodes for robotics include the detection of the presence of light, color, position, or intensity. Combinations of light-emitting diodes (LEDs) and photodiodes are commonly used together in many optical sensor systems.

Infrared sensors are commonly used for sensing of objects and detection of distances. An infrared sensor emits a pulse of infrared light from an emitter. As shown in Fig. 4.6, the light will be reflected and hit a detector if an object is present. Based on the angle of the returning infrared light, the distance to the object can be calculated. Infrared sensors are used in robotics for distance measurement, object detection and avoidance, and terrain sensing.

Encoders provide information about the relative position and motion of an object. *Optical encoders* can be used to convert the motion of a motor shaft to an analog or digital signal for controls. Optical encoders use LED light shining through an encoding disk with lines which interrupt the beam of light to a photosensor. The output is a series of square waves that can be related to the relative position. Encoders can be linear or rotary. Fig. 4.7 depicts the operation of a linear encoder. Encoders can also be absolute or incremental.

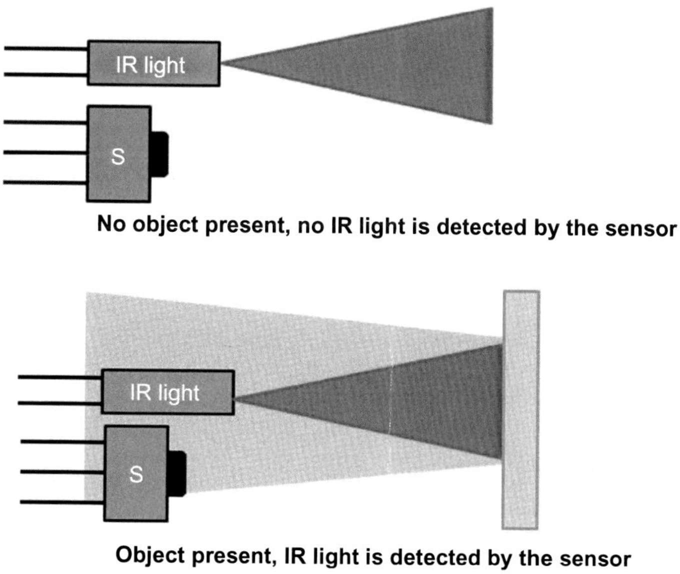

FIG. 4.6 Principle of operation of an infrared (IR) sensor. If an object is present, the IR light will be reflected and detected by the sensor (*S*). *Reproduced from Z. Kappassov, J. Corrales, V. Perdereau, Tactile sensing in dexterous robot hands—review, Robot. Autonom. Syst. 74(A) (2015) 195–220.*

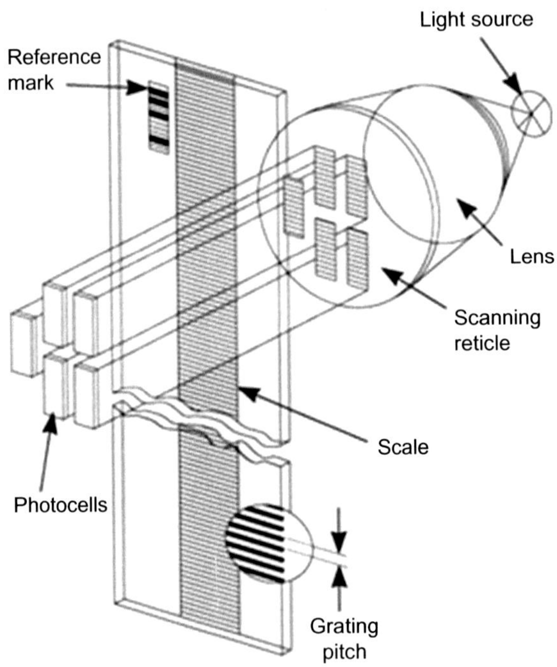

FIG. 4.7 Components of a linear encoder. *Reproduced from http://www.anaheimautomation.com/manuals/forms/encoder-guide.php#sthash.eTWRyY2U.dpbs (accessed 16.03.18).*

Absolute encoders indicate the current position of a rotating shaft while incremental encoders provide information about the relative motion of the shaft.

The most complicated application of optical sensors is for providing artificial vision. Vision refers the process of capturing an image in real time and converting it to a form that can be analyzed. Vision can be used for detecting the presence of an object, position, and movement, recognition of patterns or features, or complex scene analysis.

4.7.4 Mechanical Sensors

Mechanical sensors detect some form of mechanical deformation and translate the deformation into an electrical signal. The mechanical deformation can occur due to a variety of stimuli. Mechanical sensors are one of the most commonly used classes of sensors for biomechatronic applications.

Force sensors can be used to detect tensile, compressive, shear, torsional, bending, or frictional forces.

Touch sensors are devices which measure the parameters of contact between the sensor and an object. Touch sensors measure a contact force at a defined point. Tactile sensors detect and measure the spatial distribution of forces perpendicular to a predetermined sensory area and interpret the spatial information [17]. A simple tactile sensor consists

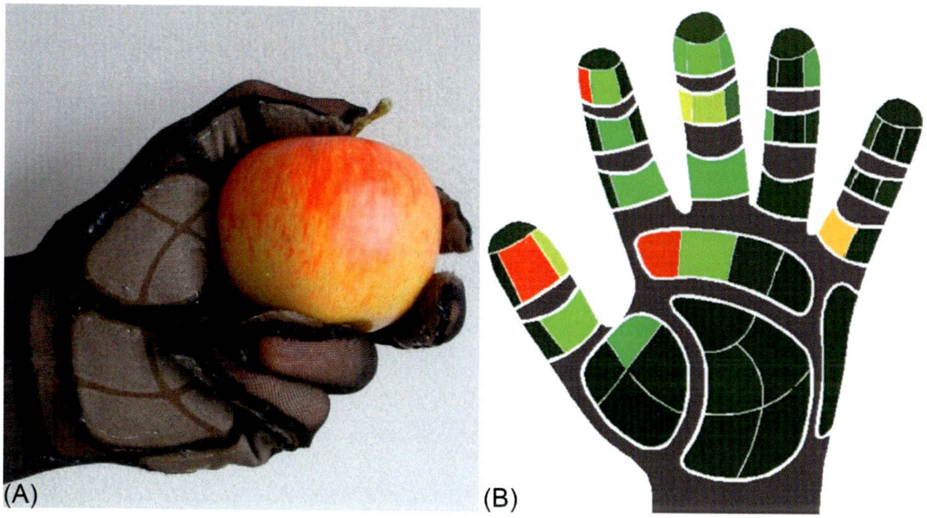

FIG. 4.8 (A) Tactile data glove and (B) tactile information from the data glove. *Reproduced from http://bdml.stanford.edu/Main/TactileSensing (accessed 16.03.18).*

of an array of simple touch sensors arranged in a matrix configuration. As individual sensors become active, an imprint of the component being sensed can be constructed.

Tactile sensing is seemingly trivial for humans, but complex to implement for a robotic system. The sense of touch allows for detection of texture, shape, hardness, or material type and helps with the detection of slipping objects [19]. Fig. 4.8A depicts a tactile data glove that uses conductive rubber to obtain tactile information during grasping [18]. The tactile information obtained from the glove is shown in Fig. 4.8B [18].

Acoustic wave sensors rely on the detection of mechanical, or acoustic, waves. As an acoustic wave propagates through or across the surface of a material, the velocity and/or amplitude of the wave changes. Acoustic proximity sensors can be used to detect the presence of and determine the distance to an object at close range. An acoustic proximity sensor analyzes the phase relationship between a transmitted wave and its corresponding reflected wave. Ultrasonic distance sensors are some of the most common acoustic sensors utilized for robotic applications. Ultrasonic sensors are good for distance measurement or mapping the proximity of objects surrounding a robot. However, acoustic sensors can be adversely impacted by the presence of external noise or cloth.

Acoustic sensors can also be used to detect or potentially distinguish between different sounds. Acoustic sensors offer potential for speech recognition or for recognizing abnormal sounds. A common type of acoustic sensor for sound recognition is a microphone. Acoustic sensors can be tuned to respond only to particular frequency ranges.

Inertial sensors, such as accelerometers or gyroscopes, sense static, and dynamic acceleration or angular acceleration, respectively. Accelerometers measure the vibration or acceleration of motion of an object. An accelerometer contains a piezoelectric material which, when compressed, produces an electrical charge that is proportional to the applied force. For a constant mass, Newton's second law states that charge is proportional to the force. When selecting an accelerometer, the mass of the accelerometer should be significantly smaller than the mass of the system being measured.

Gyroscopes help to monitor the orientation of an object in motion. The method of detection relies on the Coriolis effect and detection of vertical vibrations. An inertial measurement unit (IMU) combines an accelerometer and gyroscope to sense up to six degrees of freedom using a single device. IMUs are widely used for the measurement of precise positions or to study the body movements.

Piezoresistive sensors respond to changes in pressure, bending, or force. Piezoresistive sensors rely on the changes in resistance that occurs when a mechanical load is imposed on a conductor or semiconductor. Mathematically, the relationship can be expressed as

$$\Delta R = R_0 \alpha P \tag{4.1}$$

where ΔR is the change in resistance, R_0 is the unstressed resistance, and α is the piezoresistive coefficient (m^2/N). Piezoresistive sensors are commonly used for pressure detection or as strain sensors. Strain sensors detect how much a component changes in length under a load.

4.7.5 Thermal Sensors

Thermal sensors measure changes in temperature. Temperature is the measure of the average kinetic energy of the molecules of a gas, liquid, or solid. Some of the most common methods for sensing thermal stimuli include bimetallic strips, thermocouples, thermometers, or thermistors. Infrared cameras can also be used for the detection of low-level heat sources.

4.8 Sensor Fusion and Integration

The concept of sensor fusion attempts to replicate the capability of the central nervous system to process sensory inputs from multiple sensors simultaneously. For robotic devices, feedback from one sensor is typically not enough, particularly for implementation of control algorithms. Sensor fusion can be used to compensate for deficiencies in information by utilizing feedback from multiple sensors. The deficiencies associated with individual sensors to calculate particular types of information can be compensated for by combining the data from multiple sensors. The net effect of sensor fusion is that the resulting information should have less uncertainty than would have occurred if the sensors were used individually. Sensor fusion can also help to compensate for sensor noise, limited accuracy, failure, or a lack of information about a particular aspect of the environment. In addition, sensor fusion can be used when the ideal sensor of choice is cost prohibitive. Sensor fusion can utilize information directly from sensors or historic sensor data or can use indirect information from prior knowledge about the system inputs [20].

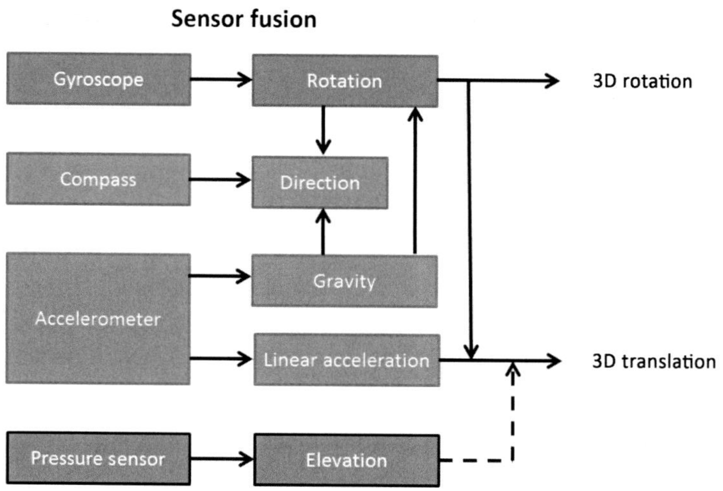

FIG. 4.9 Fusion of data from multiple sensors to get information about 3D rotation and translation.

Fig. 4.9 illustrates the concept of sensor fusion. Information is acquired from multiple sensor types and combined to get information about 3D rotations and translations.

Sensor integration refers to intelligently combining data from multiple sources to derive synthesized information for subsequent processing. A variety of algorithms have been developed for integrating data from multiple sources and accounting for uncertainties associated with the data sources. Examples include discrete Bayesian methods, Kalman filtering, or neural networks.

4.9 Integrated Systems for Obtaining Sensory Feedback

The previous section provided information about individual sensor types. By integrating sensor types, more sophisticated sensor systems can be created.

4.9.1 Force Plates

Force plates are mechanical sensing systems designed to measure the ground reaction forces and moments involved in human movements. A force plate relies on the use of load cells to determine forces. The load cells may contain piezoelectric elements, strain gauges, or beam load cells [21, 22]. As force is applied to the plate, the sensors distort thereby causing measurable voltage changes that are proportional to the applied force. Placing the sensors in different orientations enables the direction and magnitude of forces in 3D to be obtained. Other information that can be obtained includes the center of pressure, the center of force, and the moment around each of the axes [26]. Fig. 4.10 shows the forces and moments that are measured when a subject is in contact with a 3D force platform.

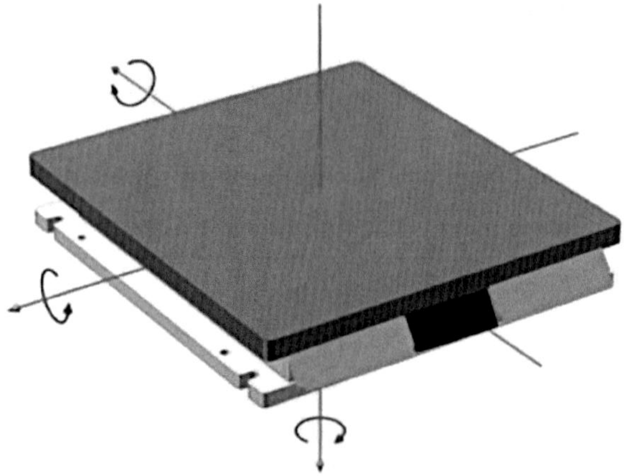

FIG. 4.10 Representative force plate system. The lines depict the forces and moments that can be measured.

In addition to data that can be directly measured from the force plate, other information can be derived. Newton's second law states that the sum of the forces acting on a body is proportional to the mass (m) times the acceleration (dv/dt) such that

$$\sum F = m\frac{dv}{dt} \tag{4.2}$$

For calculation of vertical jump height, only the forces acting on the body in the z-direction need to be considered. The two forces are the ground reaction force ($F_z(t)$) and the force due to the mass of the body (mg). Substituting into Eq. (4.2) yields

$$\sum F_z = m\frac{dv_z}{dt} = F_z(t) - mg \tag{4.3}$$

If Eq. (4.3) is solved for dv_z/dt, the expression can be integrated to get an expression for the velocity of the center of mass of the whole body as a function of time, such that

$$v_z(t) = \frac{1}{m}\int (F_z(t) - mg)dt \tag{4.4}$$

An impulse is a time-dependent force that causes a change in momentum. The vertical impulse (I_z) can be calculated by integrating the measured vertical ground reaction force until the point of takeoff, when the force goes to zero, such that

$$I_z = \int_{t_1}^{t_2} F_z(t)dt = m\int_{t_1}^{t_2} v_z(t)dt \tag{4.5}$$

In Eq. (4.5), t_1 is the time before takeoff when standing on the plate and t_2 is the takeoff time. The center of mass velocity at takeoff, v_{to}, is given by

$$v_{to} = \frac{1}{m}\int_{t_1}^{t_2} (F_z(t) - mg)dt \tag{4.6}$$

The jump height, which is a function of the total impulse, can be calculated using energy methods or time of flight calculations. Using the energy method, the kinetic energy of the whole body center of mass at takeoff (*to*) can be equated to the potential energy at peak jump height (*ph*). Mathematically,

$$KE_{to} + PE_{to} = KE_{ph} + PE_{ph} \tag{4.7}$$

Assuming that the potential energy is zero at takeoff and the kinetic energy goes to zero when the vertical velocity falls to zero at peak jump height

$$KE_{to} = PE_{ph} \tag{4.8}$$

If $KE_{to} = \frac{1}{2}mv_{to}^2$ and $PE_{ph} = mgh_{EM}$, the peak jump height can be calculated using

$$h_{EM} = \frac{v_{to}^2}{2g} \tag{4.9}$$

where m is mass and h_{EM} denotes the peak height calculated using energy methods.

4.9.2 Motion Capture Systems

Motion capture systems utilize a combination of optical, mechanical, and/or electrical sensors to obtain kinematic data. There are several different types of motion capture systems: optical, inertial, electromagnetic, and electromechanical. The most common type of system is the optical system. There are two primary types of optical systems, marker based and markerless. Marker-based optical systems use cameras to track the movements of each of several markers which are attached to the subject [24]. Marker-based optical systems are the "gold standard" for motion capture since they have a higher accuracy compared to markerless systems. Fig. 4.11 depicts a typical marker-based optical capture system.

Markerless systems use cameras to identify the subject in space, then use a software program to fit a skeleton onto the subject. The primary advantages of a markerless system are that no markers are required. Markerless systems are cheaper and more portable although many systems are still in the research phase since the system relies of software computations to track the subject's movements. Markerless systems are nonintrusive, but require tight-fitting clothing. Computationally, real-time and final data errors tend to be higher than with marker-based systems and more cleanup is required during postprocessing.

Overall optical systems are adaptable, have high accuracy, high temporal resolution (120+ frames/s), and minimal interference with the subject's normal movements. However, optical systems can be expensive, particularly for systems with large numbers of cameras.

Inertial motion capture systems are the second most commonly used type of motion capture system. An inertial system consists of sensors with gyroscopes and accelerometers that are attached to the subject. Angular velocity and linear acceleration are captured in six axes. The data from the system are transferred wirelessly. Body position and

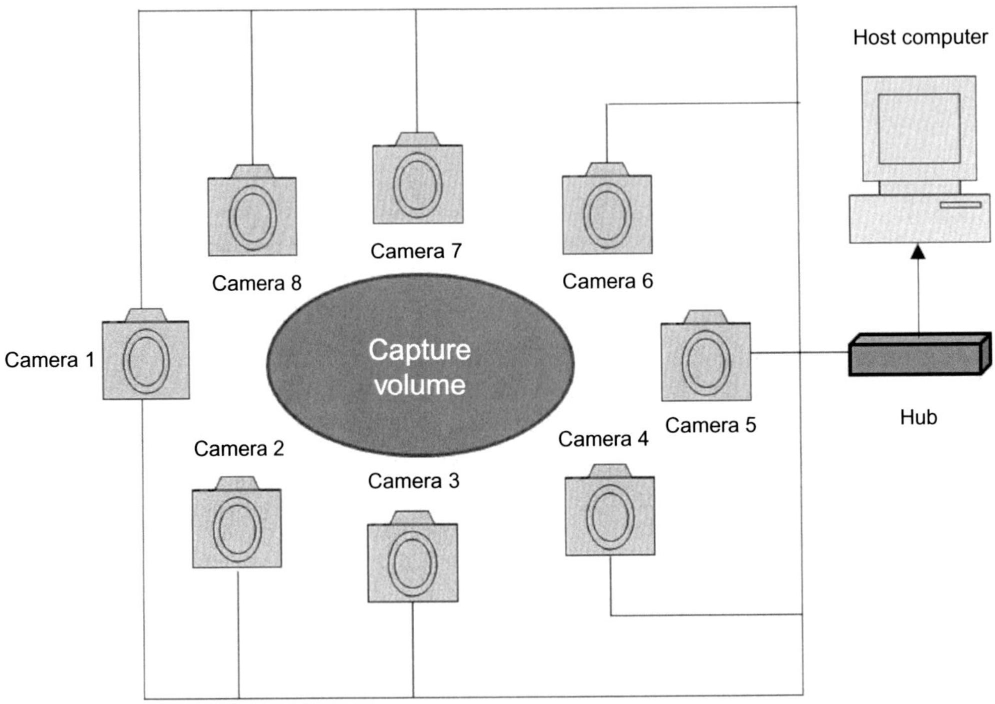

FIG. 4.11 Representative optical motion capture system.

movement, relative to a known starting position, can be determined from the kinematic data captured. Inertial systems are less expensive than optical systems, are self-contained, and can be used in cases when optical markers would be obscured from view. However, drift and low sampling rates can be problematic [25, 26]. Fig. 4.12 depicts an inertial motion capture system in use for tracking of hockey goalie motions. The sensors are placed below the goalie pads to accurately track the joint movements rather than the movements of the pads. An optical system would require placement on the pads to ensure a proper line of sight.

Electromagnetic motion capture systems measure the magnetic flux from three orthogonal coils on both the transmitter and each receiver. Data captured with electromagnetic systems can be very noisy since signals are prone to interference from any magnetic and electrical components in the environment. The capture volumes for electromagnetic systems are also dramatically smaller than those of optical systems.

Electromechanical systems combine potentiometers and exoskeletons to capture kinematic data. The exoskeletons are worn by subject and utilize joint sensors and gyros to detect motions. Rods are also connected to potentiometers for detection of motion. Electromechanical systems provide accurate real-time information, but can be restrictive to the subject and need to be scaled to the body size. Additionally, most electromechanical systems are tethered.

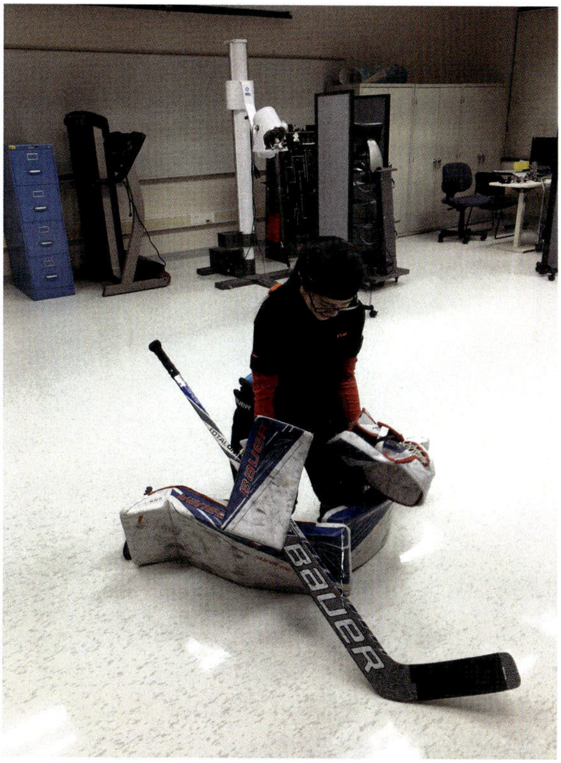

FIG. 4.12 Inertial motion capture system in use for tracking of hockey goalie motions.

Motion capture is widely used for gait studies. Motion capture has also been used as a training device for task-oriented training in neurological rehabilitation and integrated into design processes to improve the design of devices for physically disabled users.

4.9.3 VO_2/CO_2 Sensors

$VO_{2\ max}$ is the measurement of the maximum amount of oxygen that an individual can utilize during intense, or maximal exercise. $VO_{2\ max}$ is the measurement of the number of milliliters of oxygen used in 1 min/kg of body weight. VO_2 is one factor that may help to predict an athlete's capacity to perform sustained exercise. VO_2 is measured using oxygen gas sensors and is temperature compensated for use in different environments. VO_2 sensors are becoming widely used in fitness trackers or other wearables.

VCO_2 is the amount of carbon dioxide exhaled after transporting oxygen through the body. CO_2 sensors are used for measuring VCO_2. The most common methods for sensing use CO_2 infrared or chemical sensing methods. Chemical CO_2 gas sensors with sensitive layers based on polymer- or heteropolysiloxane have the principal advantage of having low energy consumption, and can be reduced in size to fit into microelectronic-based systems.

4.9.4 Electrocardiography

Electrocardiography utilizes electrical sensors to record the electrical activity of the heart. Electrocardiograms (ECGs or EKGs) are used to monitor heart status. An ECG gives two major kinds of information. The time intervals on an ECG can be used to determine the length of time that it takes for an electrical wave to pass through the heart. The length of time to travel through the heart is an indicator if the electrical activity of the heart is normal, slow, fast, or irregular. Measurement of the amount of electrical activity passing through the heart can provide information about the relative size of the heart to determine if it is distended or overworked.

The different peaks in an ECG correspond to how the electrical wave travels through the heart. Fig. 4.13 depicts a standard ECG waveform.

The right and left atria or upper chambers contribute to the first wave called a *P* wave. As the electrical impulse moves to the bottom chambers, the signal follows a flat line. The QRS complex occurs when the electrical wave moves through the right and left ventricles. The *T* wave corresponds to electrical recovery and a return to the resting state for the ventricles.

4.9.5 Electromyography

Electromyography (EMG) also relies on electrical sensors to generate a signal. EMG is used to assess the health of muscles and motor neurons. The motor neurons transmit the electrical signals that cause the muscles to contract. An EMG uses electrodes to transmit or detect electrical signals. During a needle EMG, a needle electrode is inserted directly into

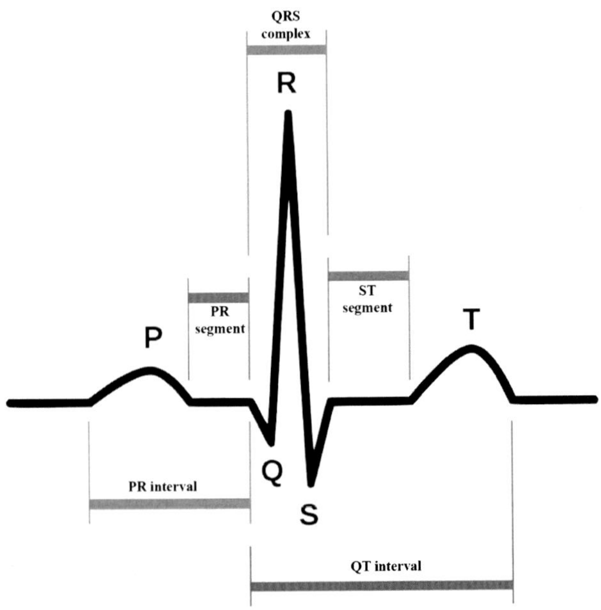

FIG. 4.13 Typical ECG waveform.

a muscle and the electrical activity of the muscle is recorded. A nerve conduction study uses surface electrodes to measure the strength and velocity of signals traveling between two or more points. EMGs can be used to identify nerve dysfunction, muscle dysfunction, or problems with nerve-to-muscle signal transmission. Further analysis of the signals can be used to detect medical abnormalities, activation level, or recruitment order, or to analyze the biomechanics of human or animal movement.

There are two kinds of EMG: surface EMG and intramuscular EMG. Surface EMG assesses muscle function by recording muscle activity from the surface above the muscle on the skin. Surface electrodes are able to provide only a limited assessment of the muscle activity. Surface EMG can be recorded by a pair of electrodes or by a more complex array of multiple electrodes. More than one electrode is needed because the EMG recordings display the potential difference (voltage difference) between two separate electrodes. The limitations of this approach are that surface electrode recordings are restricted to superficial muscles, are influenced by the depth of the subcutaneous tissue at the site of the recording which can be highly variable depending on the weight of a patient, and cannot reliably discriminate between the discharges of adjacent muscles.

Intramuscular EMG can be performed using a variety of different types of recording electrodes. The simplest approach is a monopolar needle electrode. This can be a fine wire inserted into a muscle with a surface electrode as a reference or two fine wires inserted into muscle referenced to each other. Fine wire recordings are most commonly used for research or kinesiology studies.

Robotic therapy utilizing EMG signals provides an opportunity to aid in early active rehabilitation. Surface electromyography (s EMG) can be detected during voluntary contractions in severely affected patients with very little to no function in their fingers and hand. The robotic-aided EMG interface increases the active self-participation for patients without voluntary finger force or movement. EMG signals can also be used for movement control for exoskeletal robots.

4.10 Conclusions and Future Perspective

The human body is equipped with a vast range of sensing capabilities. Feedback from the sensors is seamlessly integrated and used to control both voluntary and involuntary processes. Synthetic sensors successfully mimic many of the individual sensing capabilities of the human body, but fusion of signals from multiple sensors requires complex algorithms for processing. For biomechatronic systems to truly have the sensing capabilities of their physiological counterparts, much work in the area of sensing still remains.

References

[1] https://courses.lumenlearning.com/epcc-austincc-ap1-2/chapter/divisions-of-the-autonomic-nervous-system/ (accessed 14.08.18).

[2] https://courses.lumenlearning.com/austincc-ap1/chapter/special-senses-taste-gustation/ (accessed 29.08.18).

[3] A.A. Bachmanov, N.P. Bosak, C. Lin, I. Matsumoto, M. Ohmoto, D.R. Reed, T.M. Nelson, Genetics of taste receptors, Curr. Pharmaceut. Design 20 (16) (2014) 2669–2683.

[4] A.A. Lee, C. Owyang, Sugars, sweet taste receptors, and brain responses, Nutrients 9 (7) (2017) 653, doi:10.3390/nu9070653.

[5] Pittman RN. (2011) Regulation of Tissue Oxygenation. San Rafael (CA): Morgan & Claypool Life Sciences; Chapter 5, Chemical Regulation of Respiration.

[6] J.E. Hall, Guyton and Hall Textbook of Medical Physiology, 13th ed., Elsevier, Philadelphia, PA, 2016.

[7] https://courses.lumenlearning.com/publicspeakingprinciples/chapter/chapter-4-hearing-versus-listening/ (accessed 29.08.18).

[8] Z.F. Hugh, Chemical sensors and microfluidics. J. Biosens. Bioelectron. 4 (2013), e117doi:10.4172/2155-6210.1000e117.

[9] National Academies Press, Expanding the Vision of Sensor Materials, Danvers, MA, (1995).

[10] H. Ishida, Y. Wada, H. Matsukura, Chemical sensing in robotic applications, IEEE Sens. J. 12 (11) (2012) 3163–3173.

[11] M.A. Willis, Chemical plume tracking behavior in animals and mobile robots, J. Inst. Navigat. 55 (2) (2008).

[12] L. Osório, G. Cabrita, L. Marques, Mobile robot odor plume tracking using three dimensional information, in: Proc. of European Conference on Mobile Robots (ECMR2011), 2011.

[13] A. Ravalli, C. Rossi, G. Marrazza, Bio-inspired fish robot based on chemical sensors, Sens. Actuators B: Chem. 239 (2016) 325–329.

[14] https://cacm.acm.org/news/204390-engineers-to-use-cyborg-insects-as-biorobotic-sensing-machines/fulltext (accessed 16.03.18).

[15] R. Trivino, D. Gaibor, J. Mediavilla, A. Guarnan, Challenges to embed an electronic nose on a mobile robot, IEEE Explore (2017), doi:10.1109/ANDESCON.2016.7836251.

[16] E. Terzic, J. Terzic, R. Nagarajah, M. Alamgir, Capacitive sensing technology, in: A Neural Network Approach to Fluid Quantity Measurement in Dynamic Environments, Springer, London, 2012.

[17] M.I. Tiwana, S.J. Redmond, N.H. Lovell, A review of tactile sensing technologies with applications in biomedical engineering, Sens. Actuators A: Phys. 179 (2012).

[18] http://bdml.stanford.edu/Main/TactileSensing (accessed 16.03.18).

[19] D. Ravinder, V. Maurizio, Robotic Tactile Sensing Technologies and System, Springer, Netherlands, 2013.

[20] http://www.kionix.com/sensor-fusion (accessed 16.03.18).

[21] G. Beckham, T. Suchomel, S. Mizuguchi, Force plate use in performance monitoring and sport science testing, New Stud. Athlet. 29 (2014) 25–37.

[22] R. Cross, Standing, walking, running, and jumping on a force plate, Am. J. Phys. 67 (4) (1999) 304–309.

[23] http://www.amti.biz/fps-guide.aspx (accessed 16.03.18).

[24] N.P. Linthorne, Analysis of standing vertical jumps using a force platform, Am. J. Phys. 69 (11) (2001) 1198–1204.

[25] http://psy.mq.edu.au/me2/mocap_v1/introduction.html (accessed 16.03.18).

[26] https://www.xsens.com/products/mti-100-series/ (accessed 16.03.18).

Further Reading

[27] M. Ghasemi-Varnamkhastia, J. Lozanob, Electronic nose as an innovative measurement system for the quality assurance and control of bakery products: a review, Eng. Agric. Environ. 9 (4) (2016) 365–374.

[28] http://www.anaheimautomation.com/manuals/forms/encoder-guide.php#sthash.eTWRyY2U.dpbs (accessed 16.03.18).

[29] Z. Kappassov, J. Corrales, V. Perdereau, Tactile sensing in dexterous robot hands—review, Robot. Autonom. Syst. 74 (A) (2015) 195–220.

5

Control and Physical Intelligence

Marko B. Popovic*, Kathleen A. Lamkin-Kennard†, Matthew P. Bowers*

*WORCESTER POLYTECHNIC INSTITUTE, WORCESTER, MA, UNITED STATES †ROCHESTER INSTITUTE OF TECHNOLOGY, ROCHESTER, NY, UNITED STATES

CHAPTER OUTLINE

5.1 Introduction: General Control Problem Revised

Sensors and actuators in both biological and artificial systems are connected to each other through a feedback-control system [1]. As shown in Fig. 5.1, the control system processes sensory data and, subsequently, sends commands to the motor system. Finally, the feedback loop is closed by the effects caused by the actuators and "everything else," defining new state of measured quantities.

Biomechatronics. https://doi.org/10.1016/B978-0-12-812939-5.00005-7

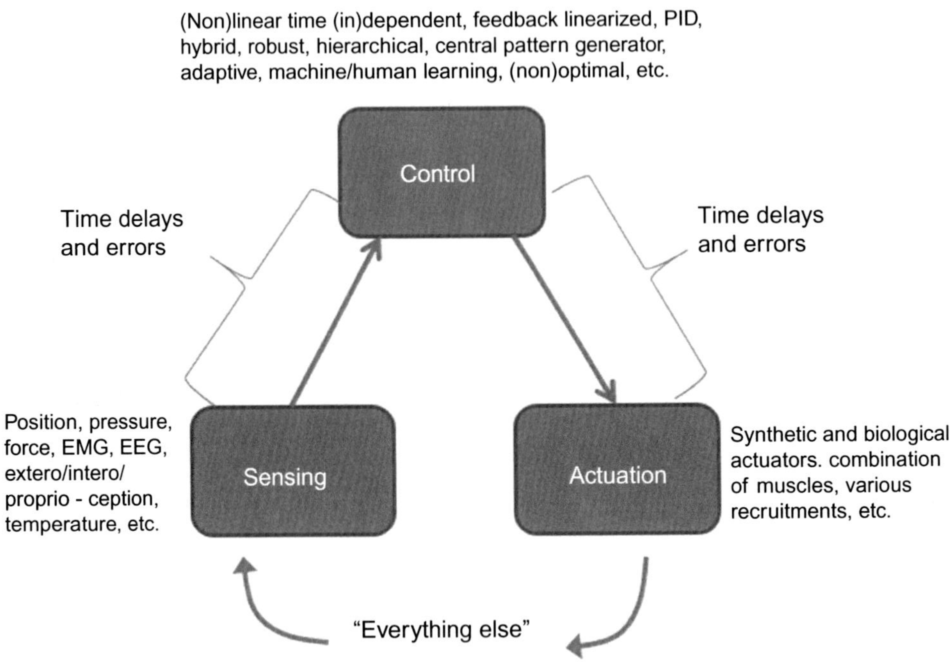

FIG. 5.1 Control feedback loop. *Adapted from M. Popovic, "Biomechanics and Robotics", 364 pages, book, Copyright © 2014 Pan Stanford Publishing Pte. Ltd., Singapore, ISBN: 978-981-4411-37-0 (Hardcover), 978-981-4411-38-7 (eBook), 2014.*

The control problem is defined only if the control goal and measure of error are specified.

The control goal is typically defined with desired value(s) of some function of measured quantities and/or desired value(s) of its higher derivatives.

Measure of error is typically defined with a metric articulating the level of deviation of actual measurement, that is, the measured value from the desired value.

Please note that the value of measurement is most certainly different from the actual value for continuous variables.

For example, for a one-dimensional (1D) system, the control goal may be defined with some particular value, for example, $x_{\text{desired}} = 1$, and the measure of error may be defined with some particular function, for example, $(x-1)^2$, where x is the value of actual sensory measurement [which may be different from the actual value (!)].

The control system is meaningless if over a substantial time period, the control goal cannot be expressed in relation to directly or indirectly measured quantities or if the measure of error is not defined in some way.

The measure of error can be also called an objective function, cost function (minimization), utility function [maximization if, e.g., $1/(x-1)^2$], or, in certain fields, energy function, or energy functional. A feasible solution that minimizes the measure of error is called an optimal solution.

The above control loop abstraction can be visualized as either a continuous or discrete *finite-speed* flow of information and actions. This flow depends on sensor performance, control, actuators, and "everything else." Hence, the complexities of these control-loop elements added with the complexities of goal and measure of error define the complexity of the control problem.

These complexities may be caused by inaccurate and imprecise sensors (inaccurate refers to absolute error and imprecise refers to inability to measure with smaller error), sensor to control transfer errors and latencies, control errors, control to actuator transfer errors and latencies, actuator errors, and finally by the dynamics of "everything else."

Analysis that goes beyond the feedback loop may find that these factors may depend on past events and history (e.g., causing battery to discharge and change voltage) or on a quantity that is not directly expressible in terms of measured quantities (e.g., surrounding temperature affecting specific resistivity of wires).

By no means is one limited to a 1D control problem. The control problem may clearly be multidimensional. For example, one may address control of x,y,z position of end-effector and/or one can simultaneously address control of the end-effector's force output in terms of F_x,F_y,F_z. If sensors allow, the problem can even be stated in terms of continuous functions (goal would be to track desired function as opposed to tracking a single desired value) or scalar, vector, or higher rank tensor field quantities paired with an appropriate measure of error.

Moreover, by no means is one limited to a single-time measure of error. The error can also be defined over a finite period of time and it can relate same or different quantities considered at different times.

5.2 PID Control Approach

To introduce and interpret the popular proportional-integral-derivative (PID) control approach in a simplified manner, let us first consider a 1D case and ignore time delays.

P stands for the term **proportional** to error (Fig. 5.2), defined as the difference between current and desired position (also called set point or target position)

$$\varepsilon = x_{current} - x_{desired} \tag{5.1}$$

where desired position may be time dependent and may not be known in advance (i.e., one should not assume knowledge of future desired trajectory). Here, we used position, but in principle one could use any other physical or abstract quantity.

I stands for **integral** term, that is, term proportional to integral of error

$$\int \varepsilon dt = \int (x_{current} - x_{desired}) dt \tag{5.2}$$

And, **D** stands for **derivative** term, that is, term proportional to derivative of error.

$$\dot{\varepsilon} = \frac{d\varepsilon}{dt} = \frac{d(x_{current} - x_{desired})}{dt} = \dot{x}_{current} - \dot{x}_{desired} \tag{5.3}$$

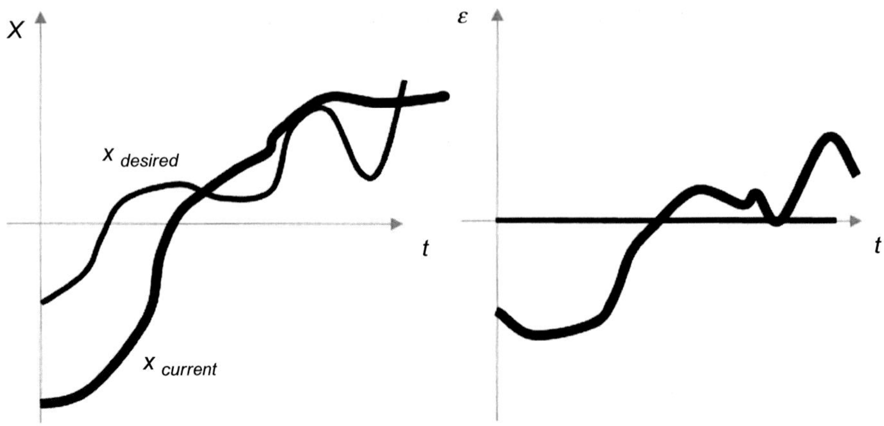

FIG. 5.2 Actual (or current) and desired trajectory (left). Error is difference between the two (right).

The weighted sum of these three terms constitute the **PID** control law created by control designer used to define, for example, commanded force $F_{command}$ sent to actuators

$$F_{command} = k_P \varepsilon + k_I \int \varepsilon dt + k_D \dot{\varepsilon} \tag{5.4}$$

Note that commanded force is specified here by only three parameters k_P, k_I, k_D, and error state.

The commanded force then affects real motive force F_{real} in some nontrivial way

$$F_{real} = f(F_{command}, \text{many other things}) = m\frac{d^2 x_{current}}{dt} = m\ddot{x}_{current} \tag{5.5}$$

and in accordance with second Newton's law.

The functional f depends on "many other things" which may include state quantities $x_{current}$ and $\dot{x}_{current}$ (less often $x_{desired}$ and $\dot{x}_{desired}$, unless there is physical object situated at $x_{desired}$ and some sort of physical interaction between the two objects) as well as other possibly time-dependent quantities like temperature (affecting resistivity and hence all resistances in the electronic circuit), nonideal electric wiring with slightly loose connections possibly affected by state of the motion (most likely by $\ddot{x}_{current}$, through inertial forces), the state of the battery, friction between mechanical elements, drag forces due to wind, and any other time-dependent external forcing. Hence, bottom line, in reality

$$F_{real} \neq F_{command} \tag{5.6}$$

The real physical force is generally very different from the commanded force. The function of functions, that is, functional f typically requires very large (if not an infinite) number of parameters to be fully specified. Hence, it is quite ambitious to expect to control force output with only three ordinary numbers, k_P, k_I, k_D. But good news is that it often works very well.

Quite generally the role of the three terms **P**, **I**, and **D** depends on the details of the actual physical system. Control designer fine tunes these parameters by trial and error method. Experienced control designer typically quickly recognizes the type of physical dynamics and relative importance of terms.

5.2.1 Trivial Case $F_{real} = F_{command}, x_{desired}(t) = 0 \quad \forall t$ and Effect of Individual *P*, *I*, and *D* terms

To get some "feel" for these terms individually, let us first consider the *trivial case* with

$$F_{real} = F_{command} \text{ and } x_{desired}(t) = 0 \text{ for all times} \tag{5.7}$$

that is, $x_{current} = \varepsilon$, and initial conditions

$$x_{current}(t=0) = X_0 \neq 0 \text{ and } \dot{x}_{current}(t=0) = 0 \tag{5.8}$$

The **P term** alone is "spring" term for $k_P < 0$ as

$$-|k_P|\varepsilon = m\ddot{\varepsilon} \tag{5.9}$$

And for given initial conditions, one obtains oscillatory solution

$$\varepsilon = X_0 \cos\sqrt{\frac{|k_P|}{m}}t \tag{5.10}$$

The **D term** alone is "damping" for $k_D < 0$ as

$$-|k_D|\dot{\varepsilon} = m\ddot{\varepsilon} \tag{5.11}$$

And for given initial conditions, one obtains constant solution

$$\varepsilon = X_0 \tag{5.12}$$

Finally, for the **I term** alone for $k_D < 0$, one could first address the following differential equation:

$$|k_I|\varepsilon = m\dddot{\varepsilon} \tag{5.13}$$

This type of equation has solution of the form $e^{\gamma t}$ such that

$$-|k_I| = m\gamma^3 \text{ or } \gamma^3 + \frac{|k_I|}{m} = 0 \tag{5.14}$$

with three possible solutions for γ

$$\gamma = -\sqrt[3]{\frac{|k_I|}{m}}, \quad \gamma = \sqrt[3]{\frac{|k_I|}{m}}\left(\frac{1}{2} \pm i\frac{\sqrt{3}}{2}\right) \tag{5.15}$$

Hence, solution for $x_{current}$ can be expressed as

$$\varepsilon(t) = c_1 e^{-\sqrt[3]{\frac{|k_I|}{m}}t} + e^{\sqrt[3]{\frac{|k_I|}{m}}\frac{t}{2}}\left[c_2 \cos\left(\frac{\sqrt{3}t}{2}\right) + c_3 \sin\left(\frac{\sqrt{3}t}{2}\right)\right] \tag{5.16}$$

For given initial conditions, Eq. (5.8), one obtains

$$X_0 = c_1 + c_2 \text{ and } 0 = \sqrt[3]{\frac{|k_I|}{m}}\left(-c_1 + \frac{c_2}{2}\right) + \frac{\sqrt{3}}{2}c_3 \tag{5.17}$$

One also requires $\ddot{\varepsilon}(t=0)=0$ because $\int \varepsilon dt|_{t=0}=0$; this condition gives

$$0 = \left(\frac{|k_I|}{m}\right)^{\frac{2}{3}}\left(c_1 + \frac{c_2}{4}\right) + \sqrt[3]{\frac{|k_I|}{m}}\frac{\sqrt{3}}{2}c_3 - \frac{3}{4}c_2 \tag{5.18}$$

The c_1 term in Eq. (5.16) decays exponentially. Unfortunately, c_2 and c_3 terms grow exponentially and to have a non-divergent solution for $t \to \infty$ one needs $c_2 = c_3 = 0$ which is not in agreement with our boundary conditions, Eq. (5.8). Hence, the **I** term alone gives an unstable solution.

Therefore, interestingly, even for the most trivial case, Eq. (5.7), with boundary conditions, Eq. (5.8), none of the terms individually cannot provide asymptotic convergence to desired position.

However, each pair could provide a convergent solution.

5.2.2 Trivial Case $F_{real} = F_{command}, x_{desired}(t) = 0 \quad \forall t$ and Effect of Paired *P* and *D* Terms

For example, consider **P** and **D** terms

$$F_{real} = -|k_P|\varepsilon - |k_D|\dot{\varepsilon} = m\ddot{\varepsilon} \tag{5.19}$$

This is standard damped oscillator with three types of solutions.

5.2.2.1 Underdamped or Slightly Damped

$$\frac{{k_D}^2}{m^2} < 4\omega_0^2 = 4\frac{|k_P|}{m} \tag{5.20}$$

with solution for $x_{current}$, Fig. 5.3 on the top left,

$$\varepsilon(t) = \mathrm{A}e^{-\frac{|k_D|}{2m}t}\cos(\omega_1 t + \varphi) \text{ and } \omega_1 = \frac{1}{2}\sqrt{4\omega_0^2 - \frac{{k_D}^2}{m^2}} \tag{5.21}$$

with A and φ defined by the initial conditions.

5.2.2.2 Critically Damped

$$\frac{{k_D}^2}{m^2} = 4\omega_0^2 = 4\frac{|k_P|}{m} \tag{5.22}$$

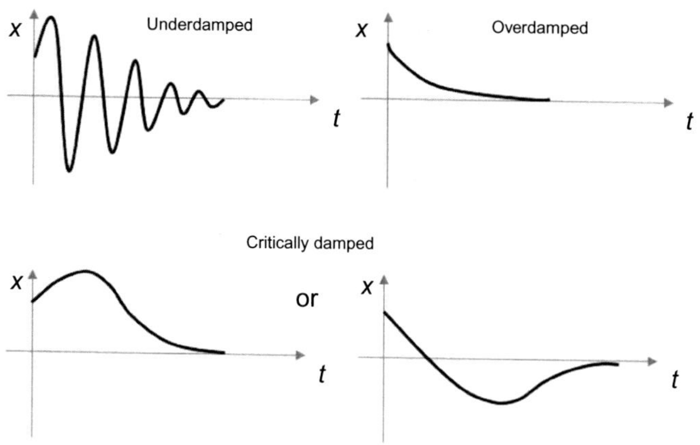

FIG. 5.3 Three standard solutions for damped oscillator.

with solution for $x_{current}$, Fig. 5.3 at the bottom,

$$\varepsilon(t) = (A + Bt)e^{-\frac{|k_D|}{2m}t} \tag{5.23}$$

with A and B defined by the initial conditions.

5.2.2.3 Overdamped

$$\frac{{k_D}^2}{m^2} > 4\omega_0^2 = 4\frac{|k_P|}{m} \tag{5.24}$$

with solution for $x_{current}$, Fig. 5.3 on the top right,

$$\varepsilon(t) = C_1 e^{\frac{-\frac{|k_D|}{m} + \sqrt{\frac{{k_D}^2}{m^2} - 4\omega_0^2}}{2}t} + C_2 e^{\frac{-\frac{|k_D|}{m} - \sqrt{\frac{{k_D}^2}{m^2} - 4\omega_0^2}}{2}t} \tag{5.25}$$

with C_1 and C_2 defined by the initial conditions.

5.2.3 Forced Oscillator $F_{real} = F_{command} + F_0 \cos(\omega t + \varphi_0), x_{desired}(t) = 0 \quad \forall t$ and Effect of Paired *P* and *D* terms

To expand "feel" for **P** and **D** terms beyond *trivial case*, let us consider slightly more complex dynamics of **forced oscillator**

$$F_{real} = F_{command} + F_0 \cos(\omega t + \varphi_0) = -|k_P|\varepsilon - |k_D|\dot{\varepsilon} + F_0 \cos(\omega t + \varphi_0) \quad \text{and} \quad x_{desired}(t) = 0 \tag{5.26}$$

with same initial conditions as specified by Eq. (5.8).

Particular solution is in the form

$$\varepsilon_{particular}(t) = A \cos(\omega t + \varphi) \tag{5.27}$$

with A and φ expressed as

$$A = \frac{F_0}{m} \frac{1}{\sqrt{\left(\omega_0^2 - \omega^2\right)^2 + \left(\omega \frac{|k_D|}{m}\right)^2}} \quad \text{and} \quad \varphi = \tan^{-1} \frac{\omega \frac{|k_D|}{m}}{\omega_0^2 - \omega^2} \tag{5.28}$$

and with $\omega_0^2 = \frac{|k_P|}{m}$ as before.

The solution to differential Eq. (5.26) is equal to the general solution, that is, one of the solutions, Eq. (5.21), (5.23), or (5.25), for damped oscillator plus particular solution expressed by Eq. (5.27).

Hence, the dynamics of the system are clearly very different due to the additional forcing term.

The amplitude, Eq. (5.28), has a maximum, also called a resonance, for ω a bit smaller than ω_0. The control designer may want to choose k_P that will give ω_0 with corresponding resonant angular frequency far away from actual driving angular frequency ω.

Let us reinstate what we already claimed before: there is no recipe to obtain the optimal values of k_P, k_I, k_D which are independent of actual physical dynamics. One set of parameters may be optimal for one physical problem and not optimal for another.

5.2.4 Extra Constant Force $F_{real} = F_{command} + F, x_{desired}(t) = 0 \quad \forall t$ and Effect of I Term

In practice, the control designer is typically not aware of all the details of physical system. For example, one may assume a damped oscillator case, Eq. (5.19), but "true" dynamics for example may happen to have an extra constant force term

$$F_{real} = -|k_P|\varepsilon - |k_D|\dot{\varepsilon} + F = m\ddot{\varepsilon} \tag{5.29}$$

This can be rewritten as

$$F_{real} = -|k_P|\left(\varepsilon - \frac{F}{|k_P|}\right) - |k_D|\dot{\varepsilon} = m\ddot{\varepsilon} \tag{5.30}$$

with steady state obtained for $\varepsilon = \frac{F}{|k_P|}$. The solution will be one of the three types of solutions for a damped oscillator depending on the parameters. Let us suppose it is an overdamped case as illustrated in Fig. 5.4 on the left.

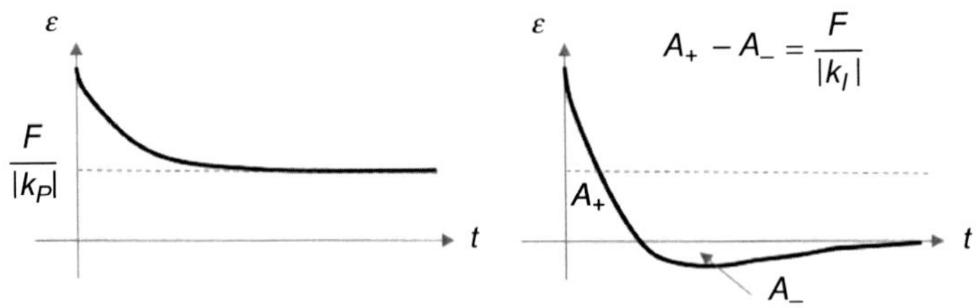

FIG. 5.4 Damped oscillator, overdamped case with nonzero steady state.

This unwanted behavior, that is, nonzero equilibrium or steady state can be regulated with integral **I** term

$$F_{real} = -|k_P|\varepsilon - |k_I|\int \varepsilon dt - |k_D|\dot{\varepsilon} + F = m\ddot{\varepsilon} \tag{5.31}$$

as illustrated in Fig. 5.4 on the right.

5.2.5 PID "Kitchen" and Heuristic Tuning Methods

In the **PID** "kitchen," the **P** term typically plays the role of a spring that drives the current position toward an equilibrium position related to the desired position and causes oscillations about that equilibrium point. The **D** term typically plays the role of damping that may decrease the amplitude or completely remove these oscillations. The **I** term plays a role like **P**, but has a "memory" of previous errors though integration, and it is typically used to regulate the steady-state error; one just needs to make sure that there is not much overshoot and that the solution is stable. In most practical cases, only two terms (**PD**, **PI**, or **ID**) are sufficient to provide working though nonoptimal solution.

There are several recipes [2], that is, more or less heuristic tuning methods for "rigid PID controllers" that interested reader could explore in more details including (in chronological order): Ziegler-Nichols rule (1942) [3], Cohen-Coon Parameters method (1953) [4], direct pole placement within internal model control (1982) [5], no overshoot rule (1989) [6], some overshoot rule (1989) [6], refined Ziegler-Nichols rule (1991) [7], Pessen integral rule (1994) [8], optimization based method by Argelaguet (1996) [9], etc. A number of methods which are able to provide automatic and adaptive tuning to changing process characteristics have been also developed. These include pattern recognition (1977) [10], relay auto-tuning by Åström and Hägglund (1984) [11], and stepwise parameter optimization (1987) [12].

5.2.6 PID in Laplace Domain and in Series RLC Electric Circuit

For a dynamical system with natural oscillatory behavior, or for those systems for which one may exploit superposition principle of oscillatory solutions, it may be beneficial to address problem in Fourier or Laplace domain. The transfer function in the Laplace domain of the PID controller, Eq. (5.4), is

$$F_{command}(s) = K_P + \frac{K_I}{s} + K_D s \tag{5.32}$$

With $s = \sigma + i\omega$ being a complex frequency such that

$$F_{command}(s) = \int_0^\infty F_{command}(t)e^{-st}dt \tag{5.33}$$

as is often done in analysis of electric circuits.

A lot of inspiration for **PID** control approach, for case with $F_{real}(t) = F_{command}(t) + F_{ext}(t)$, comes from, probably familiar to reader, in series RLC electric circuit consisting of resistor, solenoid, and capacitor connected in series.

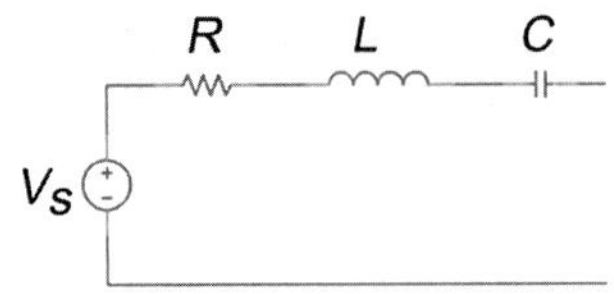

FIG. 5.5 Voltage source in series with in series RLC part of the electric circuit.

Consider part of the circuit shown in Fig. 5.5. Voltage on the right is equal to

$$u = V_s - iR - L\frac{di}{dt} - \frac{q}{C} = V_s - iR - L\frac{di}{dt} - \frac{\int idt}{C} \tag{5.34}$$

If the goal is to have current $i \to 0$ then this becomes a similar problem as the one expressed by Eq. (5.31); although instead of constant driving "force" as in Eq. (5.31) here one addresses alternating voltage source $V_s = V_0\, cos\,(\omega t + \varphi_0)$. The **PID** parameters are now expressed by electric resistance, inverse capacitance, and inductance

$$|k_P| \leftrightarrow R, |k_I| \leftrightarrow \frac{1}{C}, |k_D| \leftrightarrow L \tag{5.35}$$

For specific "driving" frequency ω, voltage on inductive element leads by 90°

$$v_L = L\frac{di}{dt} = L\frac{dI\cos\omega t}{dt} = -L\omega I\sin\omega t = L\omega I\cos\left(\omega t + \frac{\pi}{2}\right) = X_L I\cos\left(\omega t + \frac{\pi}{2}\right) \tag{5.36}$$

and quantity $X_L = L\omega$ is called inductive reactance.

Similarly, for specific "driving" frequency ω, voltage on capacitive element lags by 90° such that

$$i = \frac{dq}{dt} = \frac{dCv_C}{dt} = C\frac{dv_C}{dt} = C\frac{dV\cos\left(\omega t - \frac{\pi}{2}\right)}{dt} = -C\omega V\sin\left(\omega t - \frac{\pi}{2}\right) = I\cos\omega t \tag{5.37}$$

and quantity $X_C = \frac{1}{C\omega}$ is called capacitive reactance.

Finally, one may introduce impedance $Z = \sqrt{R^2 + \left(L\omega - \frac{1}{C\omega}\right)^2} = \frac{V}{I}$ and utilize phasor diagram to solve for the relative phase of in series RLC part of the circuit in respect to phase of alternating voltage source.

5.3 Error and Time Delays in Time Domain

Let us now briefly address time delays in time domain.

The previously presented formalism with continuous variables without time delays may be affected by the fact that both sensory information and actuation command may have substantial time delays.

Lets consider a simplified discrete situation with equal time sequences Δt such that the force command received by actuators at time instance t_n is expressed in terms of error at time instance $t_n - \Delta t = t_{n-1}$. However, if desired position is not known in advance, but rather obtained through sensory input, the error at time instance t_{n-1} may be expressed for example as

$$\varepsilon_I(t_{n-1}) = x_{current}(t_{n-1}) - x_{desired}(t_{n-2}) \tag{5.38}$$

or as

$$\varepsilon_{II}(t_{n-1}) = \varepsilon(t_{n-2}) = x_{current}(t_{n-2}) - x_{desired}(t_{n-2}) \tag{5.39}$$

On those lines, derivative of error can be expressed as

$$\dot{\varepsilon}_I(t_{n-1}) = \frac{x_{current}(t_{n-1}) - x_{current}(t_{n-2})}{\Delta t} - \frac{x_{desired}(t_{n-2}) - x_{desired}(t_{n-3})}{\Delta t} \tag{5.40}$$

or as

$$\dot{\varepsilon}_{II}(t_{n-1}) = \dot{\varepsilon}(t_{n-2}) = \frac{x_{current}(t_{n-2}) - x_{current}(t_{n-3})}{\Delta t} - \frac{x_{desired}(t_{n-2}) - x_{desired}(t_{n-3})}{\Delta t} \tag{5.41}$$

And similarly, integral of error can be expressed as

$$[(\cdots + x_{current}(t_{n-1})) - (\cdots + x_{desired}(t_{n-2}))]\Delta t \tag{5.42}$$

or as

$$[(\cdots + x_{current}(t_{n-2})) - (\cdots + x_{desired}(t_{n-2}))]\Delta t \tag{5.43}$$

Another option could be to guess desired position at time instance t_{n-1} using only first derivative as

$$x_{desiredguessed}(t_{n-1}) = x_{desired}(t_{n-2}) + \frac{x_{desired}(t_{n-2}) - x_{desired}(t_{n-3})}{\Delta t}\Delta t = 2x_{desired}(t_{n-2}) - x_{desired}(t_{n-3}) \tag{5.44}$$

And, then construct error at time instance t_{n-1} as

$$\varepsilon_{III}(t_{n-1}) = x_{current}(t_{n-1}) - x_{desiredguessed}(t_{n-1}) = x_{current}(t_{n-1}) - 2x_{desired}(t_{n-2}) + x_{desired}(t_{n-3}) \tag{5.45}$$

and subsequently calculate corresponding derivative and integral terms.

Still further one may consider also second derivative (acceleration) for guessing desired position at time instance t_{n-1}.

5.4 Stability

Stability is a very important factor for control systems. Having a system that goes unstable is clearly not desirable.

The stability of a system is measured by its ability to stay near and return to a specific desired state. A system is considered unstable if it continues to move farther away from the particular desired state without returning. A marginal system is between these two types of stability. It neither moves farther away from a common steady state, nor does is return

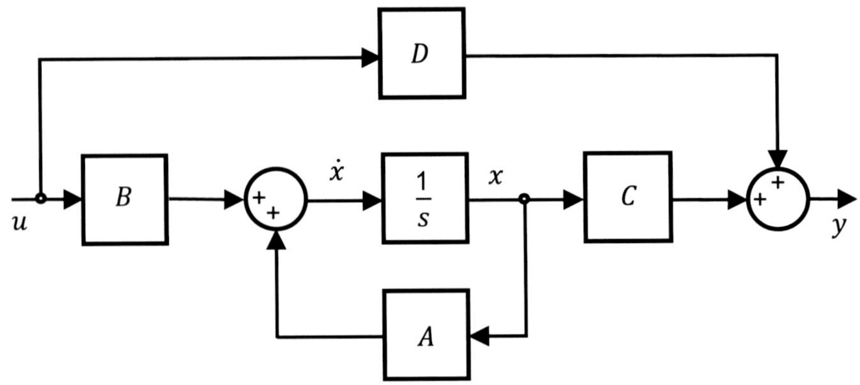

FIG. 5.6 Block diagram representation of the typical linear state-space equations.

to the desired state, and like instability, it is clearly not desirable as it results in system trapped where it should not be.

Different control systems allow different stability analysis approaches. Let us first consider linear systems.

The state-space representation of a linear system with p inputs, q outputs, and n state variables (Fig. 5.6) can be written in the following form [13]:

$$\dot{\boldsymbol{x}}(t) = A(t)\boldsymbol{x}(t) + B(t)\boldsymbol{u}(t) \tag{5.46}$$

$$\boldsymbol{y}(t) = C(t)\boldsymbol{x}(t) + D(t)\boldsymbol{u}(t) \tag{5.47}$$

where $\boldsymbol{x}(t) \in R^n$ is state vector, $\boldsymbol{y}(t) \in R^q$ is output vector, $\boldsymbol{u}(t) \in R^p$ is input (or control) vector, and $\dot{\boldsymbol{x}} = \frac{dx}{dt}$.

5.4.1 Stability and Eigenvalues of Linear Time-Invariant System

If matrices A, B, C, and D are time independent, then this system is said to be continuous-time linear time-invariant (LTI) system.

In the Laplace domain of continuous-time LTI system, one can express Eqs. (5.46), (5.47) as

$$s\boldsymbol{X}(s) = A\boldsymbol{X}(s) + B\boldsymbol{U}(s) \rightarrow (sI - A)\boldsymbol{X}(s) = B\boldsymbol{U}(s) \tag{5.48}$$

$$\boldsymbol{Y}(s) = C\boldsymbol{X}(s) + D\boldsymbol{U}(s) \tag{5.49}$$

This now allows us to study stability and natural response characteristics of a continuous-time LTI system by analyzing the eigenvalues of the matrix A.

The stability of a time-invariant state-space model can be determined by expressing the system's transfer function $\boldsymbol{X}(s)/\boldsymbol{U}(s)$ in factored form. Based on Eq. (5.48), the denominator of the transfer function is equal to the characteristic polynomial found by taking the determinant of $sI - A$, that is, $\lambda(s) = |sI - A|$. The roots of this polynomial (i.e., the eigenvalues) are the system transfer function's poles, that is, the singularities where the transfer function's magnitude is unbounded. These poles can be used to analyze whether the system is asymptotically stable or marginally stable.

In control theory, a continuous LTI is exponentially stable if and only if the system has eigenvalues (i.e., the poles of input-to-output systems) with strictly negative real parts (i.e., in the left half of the complex plane) [14]. It is easy to interpret this statement when going back to time domain; the largest contribution to exponentially stable system dynamics is coming from exponentially decaying functions.

Exponential stability is a form of asymptotic stability. Systems that are not LTI are exponentially stable if their convergence is bounded by an exponential decay [14].

An exponentially stable LTI system is one that will not "blow up" (i.e., give an unbounded output) when given a finite input or nonzero initial condition. Moreover, if the system is given a fixed, finite input (i.e., a step), then any resulting oscillations in the output will decay at an exponential rate, and the output will tend asymptotically to a new final, steady-state value. If the system is instead given a Dirac delta impulse as input, then induced oscillations will die away and the system will return to its previous value. If oscillations do not die away, or the system does not return to its original output when an impulse is applied, the system is instead marginally stable. For some systems, it is also proper to state that a system is exponentially stable over a certain range of inputs.

In the theory of dynamical systems and control theory, a LTI system is marginally stable if it is neither asymptotically stable nor unstable [14].

A linear system is called bounded-input bounded-output (BIBO) stable if its output will stay bounded for any bounded input [13].

An alternative approach to determining stability, which does not involve calculating eigenvalues, is to analyze the system's Lyapunov stability [15].

5.4.2 Lyapunov Stability

Consider an autonomous nonlinear dynamical system.

$$\dot{x} = f(x(t)) \quad \text{and} \quad x(0) = x_0 \tag{5.50}$$

and suppose that f has an equilibrium at x_e such that $f(x_e) = 0$.

Lyapunov stability [15] of an equilibrium means that solutions starting in vicinity to the equilibrium (within a distance δ from it) remain in vicinity forever (within a distance ϵ from it). Note that this must be true for any. Hence, the equilibrium is Lyapunov stable if

$$\forall \epsilon > 0, \exists \delta > 0, \|x(0) - x_e\| < \delta \Longrightarrow \|x(t) - x_e\| < \epsilon \quad \forall t \geq 0 \tag{5.51}$$

Asymptotic stability means that solutions that start close enough not only remain close enough but also eventually converge to the equilibrium. Hence, the equilibrium is asymptotically stable if it is Lyapunov stable and if

$$\exists \delta > 0, \|x(0) - x_e\| < \delta \Longrightarrow \lim_{t \to \infty} \|x(t) - x_e\| = 0 \tag{5.52}$$

Exponential stability means that solutions not only converge, but in fact converge faster than or at least as fast as a specified rate $\alpha\|x(0)-x_e\|^{-\beta t}$. In symbolic language, the equilibrium is exponentially stable if it is asymptotically stable and if

$$\exists\alpha>0,\exists\beta>0,\exists\delta>0,\|x(0)-x_e\|<\delta\Longrightarrow\|x(t)-x_e\|\leq\alpha\|x(0)-x_e\|^{-\beta t}\quad\forall t\geq 0 \tag{5.53}$$

5.4.3 Lyapunov Stability Criterion

Lyapunov's second method for stability also referred to as the Lyapunov stability criterion [15], makes use of a Lyapunov function $V(x)$ which has an analogy to the potential function of classical dynamics. Consider a function $V(x):R^n\rightarrow R$ such that

$$V(x=0)=0,\ V(x\neq 0)>0,\ \frac{dV}{dt}\leq 0 \tag{5.54}$$

Then V(x) is called a Lyapunov-candidate-function and the system is stable in the sense of Lyapunov. An additional condition called "properness" or "radial unboundedness" is required to conclude global stability and finally global asymptotic stability (GAS) follows similarly as above.

It is easy to visualize this approach in an analogy with a physical system (e.g., vibrating spring and mass) and considering the energy of such a system. If the system loses energy over time and the energy is never restored, then eventually the system must slow down and reach some final resting state.

This final state is called the attractor. However, finding a function that gives the precise energy of a physical system can be difficult, and for abstract mathematical systems, economic systems, or biological systems, the concept of energy may not be always applicable.

Lyapunov's realization was that stability can be proven without requiring knowledge of the true physical energy, provided a Lyapunov function can be found that satisfies the above constraints.

The Lyapunov's second method or the method of Lyapunov functions is the only universal technique for the investigation of the stability of the dynamical systems of the most general configurations and properties including all possible types of motions (nonstationary, stationary, periodic), nonlinearities, dimensionalities, parametric dependence, etc. The big advantage of the method is that it does not require the knowledge of the solutions of the ordinary differential equations (ODEs) governing the systems.

Readers interested to learn more on Lyapunov stability and application may consider studying the following references [15–22].

5.5 Feedback Linearization

Here, we introduce concept of feedback realization following presentation in Ref. [20]. The main idea is to introduce linear feedback law involving tracking error and higher derivatives, for example, in the form

$$c_0\varepsilon + c_1\dot{\varepsilon} + \ldots c_N\varepsilon^{(N)} = 0 \tag{5.55}$$

and ensure, through choice of coefficients, that the error is converging to zero for $t\rightarrow\infty$. The choice of feedback law also in turn affects the control law, which may be nonlinear, and which in the idealization assumed here satisfies $F_{command}=F_{real}$.

For example, consider dynamical equations for two-link manipulator that we already addressed numerous times in Chapter 2 using several different methods. Without going into details, the equations of motions are in the form

$$\boldsymbol{H}(\boldsymbol{q})\ddot{\boldsymbol{q}} + \boldsymbol{C}(\boldsymbol{q},\dot{\boldsymbol{q}})\dot{\boldsymbol{q}} + \boldsymbol{g}(\boldsymbol{q}) = \boldsymbol{\tau} \tag{5.56}$$

Now by rewriting

$$\ddot{\boldsymbol{q}} = \ddot{\boldsymbol{q}}_{\boldsymbol{current}} = \ddot{\boldsymbol{q}}_{\boldsymbol{desired}} + \ddot{\boldsymbol{\varepsilon}} = \ddot{\boldsymbol{q}}_{\boldsymbol{desired}} - \boldsymbol{2}\lambda\dot{\boldsymbol{\varepsilon}} - \lambda^2\boldsymbol{\varepsilon} + \ddot{\boldsymbol{\varepsilon}} + \boldsymbol{2}\lambda\dot{\boldsymbol{\varepsilon}} + \lambda^2\boldsymbol{\varepsilon} \tag{5.57}$$

with λ being a strictly positive number, one obtains both the control ("computed torque") law

$$\boldsymbol{H}(\boldsymbol{q})\left(\ddot{\boldsymbol{q}}_{\boldsymbol{desired}} - 2\lambda\dot{\boldsymbol{\varepsilon}} - \lambda^2\boldsymbol{\varepsilon}\right) + \boldsymbol{C}(\boldsymbol{q},\dot{\boldsymbol{q}})\dot{\boldsymbol{q}} + \boldsymbol{g}(\boldsymbol{q}) = \boldsymbol{\tau} \tag{5.58}$$

and linear feedback law

$$\ddot{\boldsymbol{\varepsilon}} + \boldsymbol{2}\lambda\dot{\boldsymbol{\varepsilon}} + \lambda^2\boldsymbol{\varepsilon} = \boldsymbol{0} \tag{5.59}$$

which ensures that the tracking error converges to zero.

The idea of feedback linearization is simply applied to a class of nonlinear systems, like our two-link manipulator, described by the so-called companion form, or controllability canonical form represented by

$$x^{(n)} = f(\boldsymbol{x}) + b(\boldsymbol{x})u \tag{5.60}$$

where u is the scalar control input, x is the scalar output of interest, $\boldsymbol{x}=\left[x, \dot{x}, \ldots, x^{(n-1)}\right]^T$ is the state vector, and $f(\boldsymbol{x})$ and $b(\boldsymbol{x})$ are nonlinear functions of the state.

5.6 Sliding Control

Sliding control is often referred to in the context of the robust, nonlinear control that deals with modeling, and maybe even unstructured (unmodeled) uncertainties. In practice, sliding control successfully deals with parametric imprecisions thanks to two typically high-gain controllers that "push" state vector in the direction of the so-called sliding (or switching) surface defined in the state space by control designer.

Two controllers push from two opposite sides of possibly time-dependent surface (Fig. 5.7). There is a trade-off of how close the state vector is to the surface and how large the gains of these controllers are which in turn also affect how often the state vector crosses this surface. Typically, high-gain translates to extremely high control activity and lots of switching between the two control regimes. Some authors even compare this behavior to a marble going through a narrow channel due to crack in concrete.

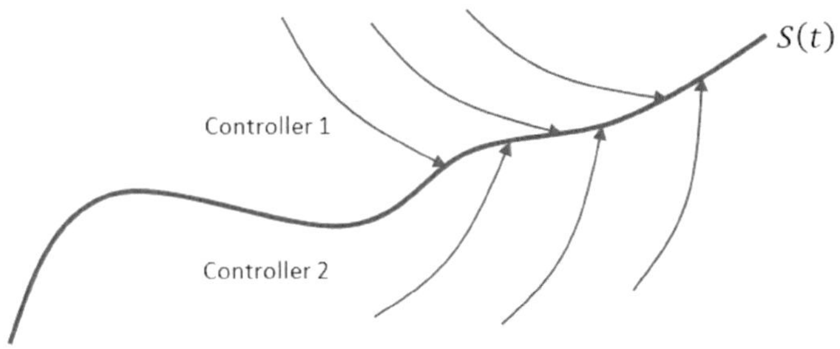

FIG. 5.7 Sliding surface and two discrete state control regimes.

The sliding control is also referred to in the context of hybrid control as a continuous-state vector variable that is controlled within two discrete state control regimes.

It is also possible to replace a switching, chattering control law by its smooth approximation, given the available bandwidth, for more details see Slotine and Li [20].

In essence, the sliding control method splits an N-dimensional control problem into an N − 1-dimensional sliding surface control problem (nominal part, similar to feedback linearizing or inverse control law) and 1D distance from sliding surface stabilization problem.

For example, following the methodology presented in Slotine and Li [20], the time-varying surface $S(t)$ can be defined in the state-space $\boldsymbol{R}^{(n)}$ by the scalar equation

$$s(x;t) = \left(\frac{d}{dt} + \lambda\right)^{n-1} \varepsilon = \left(\frac{d}{dt} + \lambda\right)^{n-1} (x_{current} - x_{desired}) = 0 \tag{5.61}$$

with a strictly positive constant λ.

For a typical state space with $n=2$, imagine that the goal is to track two-dimensional (2D) $\boldsymbol{x}_{\boldsymbol{desired}} = [x_{desired} \quad \dot{x}_{desired}]^T$. The scalar function is

$$s = \dot{\varepsilon} + \lambda\varepsilon = 0 \tag{5.62}$$

And one may note that for $\varepsilon > 0$ the $\dot{\varepsilon}$ is negative and vice versa for negative ε the $\dot{\varepsilon} > 0$. Still further, if $\varepsilon = 0$ then $\dot{\varepsilon} = 0$ and hence $x_{current} = x_{desired}$ for all times. The surface dynamics are automatically under control for $s=0$ and hence, the main challenge is to keep the scalar quantity s at zero.

Or consider a more general situation with $n=3$, such that the goal is to track three-dimensional (3D) $\boldsymbol{x}_{\boldsymbol{desired}} = [x_{desired} \quad \dot{x}_{desired} \quad \ddot{x}_{desired}]^T$. The scalar function is

$$s = \ddot{\varepsilon} + 2\lambda\dot{\varepsilon} + \lambda^2\varepsilon = 0 \tag{5.63}$$

And again, the main challenge is to keep the scalar quantity s at zero.

The first-order stabilization problem is typically expressed in terms of sliding condition

$$\frac{1}{2}\frac{d}{dt}(s^2) \leq -\eta|s| \tag{5.64}$$

with η being a strictly positive constant. This is just another way of saying that "distance" to the surface decreases along all system trajectories [20].

5.7 Adaptive Control

Adaptive control has in-advance introduced set of parameters of controller being adjusted in some automatic fashion.

Historically, there are two main types of adaptive control approaches: model-reference adaptive control (MRAC), Fig. 5.8 top, and self-tuning control (STC), Fig. 5.8 bottom.

While these two approaches share some similarities, the main difference between them comes from the fact that MRAC's process of parameter adjustment is directly influenced by both controller and plant performance and STC's process of parameter adjustment is only directly influenced by plant performance.

Specifically, in the MRAC approach, based on the controller output, i.e., plant input u, the plant performs and subsequently, based on difference between the plant output y and reference model output y_m, adjusted parameters $\hat{a}$ are estimated and sent to controller. Hence, the adaptation law block is affected by performance of both the controller and the plant. On the same lines, the reference model is a model of both the controller and plant.

This approach is also called direct adaptive control.

The success of this approach depends on the accuracy of the reference model (and reference set of parameters) and how well these parameters can be interpolated (typically in linear approximation) beyond close vicinity of reference model values.

In the case of a STC approach, the estimator adjusts parameters $\hat{a}$ that are sent to controller based only on plant input u and output y. Hence, this procedure is not directly influenced by controller; it is primarily focused on parametrizing model of plant.

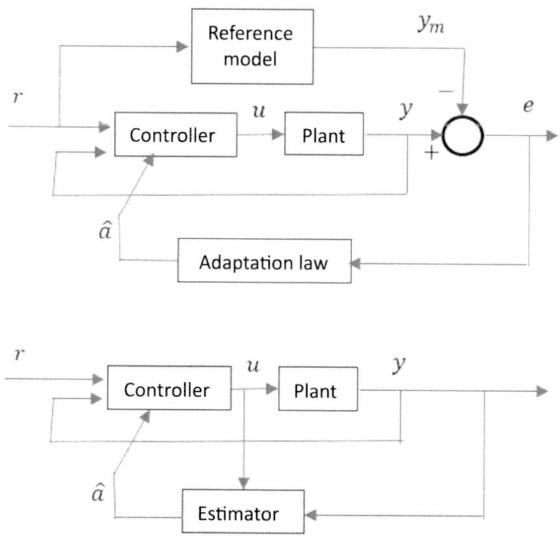

FIG. 5.8 Two main types of adaptive control approaches: model-reference adaptive control (MRAC), top, and self-tuning control (STC), bottom. *Based on J.-J.E. Slotine, W. Li, Applied Nonlinear Control, Prentice Hall, NJ, 1991.*

Because one needs to translate plant parameters into new set of control parameters, this approach is also called indirect adaptive control.

In the case of a linear plant, there are various techniques that can be employed by the estimator. The most popular method is the least square method and its extensions. Other techniques include pole placement, PID, linear quadratic control (LQR), minimum variance control, or H^∞ designs [20].

In case of substantial modeling inaccuracies, less elegant high-gain robust control methods may still prove more successful than adaptive control methods.

However, adaptive control might be very useful if a good model is available. The system performs experimentations in real time and can deduce its own structural parameters, battery state, external load characteristics, or even effect of temperature change (affecting both mechanical and electrical properties of the system), etc. and performs accordingly.

In a sense, some variant of adaptive control is most likely utilized in biological systems. We are continuously performing experimentations and refining our understanding of object we hold or grounds on which we walk.

5.8 Linearity and Predictability; Multidimensionality and Associated Problems

The control problem is affected by time delays. At the time instance when the controller analyzes sensory data, this sensory data are already outdated. Furthermore, at the time instance when the actuators perform action based on controller commands, this sensory data are even further outdated. Hence, controller needs to deal with delays in some way and predict behavior of tracked system. The successful tracking of desired trajectory is often closely related to the fact that the tracked system behaves linearly and hence it is quite predictable at small time scales as illustrated in Fig. 5.9. The function at time A can be closely estimated from the value of the function and its first derivative at time B. However, that is not the case with C and D.

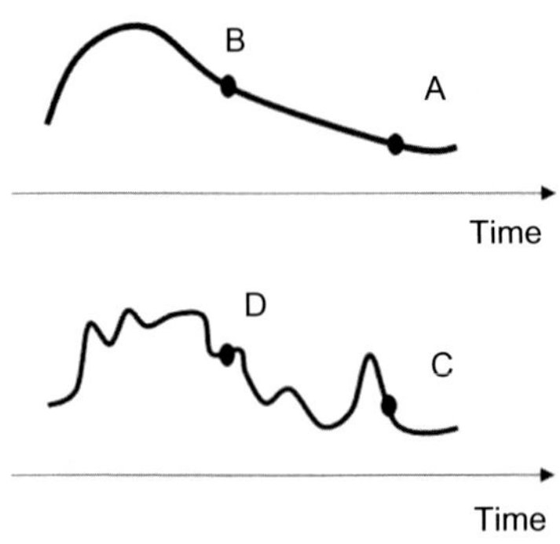

FIG. 5.9 A predictable versus unpredictable behavior.

The example in the figure may correspond to a simple experiment involving two students standing on opposite sides of a glass door, facing each other, and moving their pointer fingers along the transparent surface. One student may move the "target" finger randomly at will and the other student may try to align his/her finger with the one of his/her colleague [1].

The exact desired value at each instance is defined by the target position. The error at each instance is a 2D vector.

If the target is moving slow and along a smooth trajectory, the task to align fingers can be accomplished quickly. However, if the target is moving very fast in an unpredictable manner (e.g., frequent jerky movements), it could be very difficult to accomplish the task.

Sometimes, the higher derivative of desired quantity may be known, and the exact desired value is not known. For example, the control goal can be stated in terms of the global minimum of some measured quantity.

A controller typically slides down the slope and searches for a minimum. The problem, however, is that the system could get stuck at the local minimum, hence, missing the global minimum as illustrated in Fig. 5.10. Hence, the controller must explore a larger region of work space and these experimentations may be time consuming.

This is even more evident for multidimensional systems. Consider a problem with D dimensions. If the controller chooses N values along each dimension, then there is N to the power of D control points that the controller should explore. For example, consider $N=20$ control values. In D = 2 dimensions this corresponds to 400 control points, in D = 3 to 8000, in D = 10 to 10,240,000,000,000 control points, etc. In other words, the number of control points grows exponentially [1].

This is the famous Bellman's curse of dimensionality. The term was coined by Richard E. Bellman, celebrated for his invention of dynamic programming in 1953, when considering problems in dynamic optimization [23, 24]. The term "Bellman equation" is a type of problem named after its discoverer, in which a problem that would otherwise not be possible to solve is broken into a solution, based on the intuitive nature of the solver. Applied

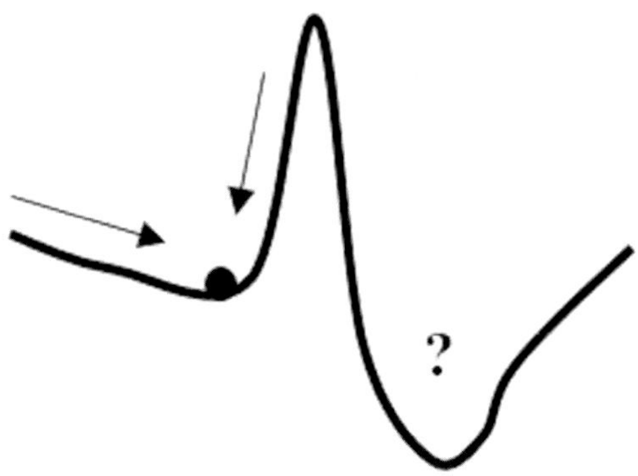

FIG. 5.10 The local versus the global minimum.

in control theory, economics, and medicine, it has become an important tool in using math to solve difficult problems.

Consider now a humanoid robot with say 50 degrees of freedom, and imagine that each "experiment" takes 1 ms and try to discover a global minimum of some performance metric by the above-proposed "brute-force" method. Time required is beyond imagination!

However, the problem may be solved by various methods. Among many approaches, one approach may be to utilize the fast, hierarchical-prioritized control method [25–27]. This method is extremely fast as the control of multi-degree robot is simplified to control applied to the product of a low-dimensional spaces where optimization can be quickly and reliably performed.

Another approach is the fast, robust policy optimization [28] which also works extremely well when the control laws are made up of collections of simple policies where only one component policy is active at a time. The robustness is achieved by simultaneously designing one control law for multiple models with potentially different model structures, which represent model uncertainty and unmodeled dynamics (i.e., the complexities mentioned before).

If an exemplary system exists, *learning by demonstration* method may also prove useful. Here, the artificial system attempts to "copy" successful realization of control task by an exemplary system, for example, biological organism [1]. This method typically utilizes data-driven simplification of the control problem.

Successfully demonstrated kinematics (or dynamics) are statistically analyzed, for example, by principal component analysis (PCA) method which decomposes data space into basis vectors or principle components (PCs) listed in order of their importance, that is in the order of amount of "data explained" as quantified by their variances.

Cost function, if known, may be optimized in a lower-dimensional subspace spanned by only a few top directions. The advantage is that optimization in a smaller subspace can be quickly computed in real time, that is, "on the fly." However, this approach can be problematic as perturbation residing outside predetermined subspace may quickly destabilize the system [1].

Another approach would be to address all control dimensions one by one in order of their importance, while performing 1D optimization along each direction. Clearly, the result of optimization on the entire space is different from that obtained by successive optimizations along each direction separately. However, in practice, this method has proven successful; it was effectively used to address arms', legs', and whole body *movement primitives* [29] expressed either in joint angular space, angular momentum space [29], etc. Many authors relate these movement primitives with central pattern generator (CPG) outputs in biological organisms.

If an exemplary system does not exist, one may perform first slow optimization, for example, by means of simulation, and then for purpose of fast real-time control, one could create a control library linking state (positions and velocities) and control goal with (near) optimal actuations. Hence, one may split large control space into enumerated bins, in other words, smaller volumes that may allow for further simplifying linearization of control problem. Here, the control system may first quickly determine rough position in nonlinear "library space" and then finely tune control output through the linear assumption [1].

5.9 Physical Intelligence

What is physical intelligence?

Physical intelligence is a control paradigm rooted in the estimation of the physical model and the state of the system (e.g., the robot, in the context of robotics, and relevant surrounding, i.e., "world") and a decision-making process in the context of the actions needed to achieve a desired physical performance [30].

While a fair comparison among two control approaches can be only made for the same robot (in the context of robotics) and "world," state, task, and success metric, practical physical intelligence seeks "more universal laws" that transcend these very specific situations and that can be applied across different robots and "worlds," states, tasks, and success metrics [30].

5.10 Control and Artificial Intelligence, Machine Learning, Data Mining

As already discussed at the beginning of this chapter, control is defined as an attempt to successfully achieve some goal in the context of the metric that quantifies distance to same goal. This definition encompasses rudimentary tasks like those described earlier with state position tracking of an in advance known desired trajectory. Very cleanly defined problem in mathematical language. However, the same definition can also encompass more complex problems with still very simple goals but addressed in the context of often more realistic situations with very limited sensory information, unknown complex conditions, and unknown complex dynamics of the system.

For example, high level simple goal might be to minimize distance to specific bottle at your friend's party. Because the bottle is out of sight, the exact distance to bottle is not known. And, you don't have superhuman abilities to see through walls. However, you were provided with a hint to look for the bottle in the fridge. Based on your commonsense knowledge, you assume that fridge must be in the kitchen and that your friend's place has one. But because this is the first time that you are visiting your friend you do not know where the kitchen is and hence you need a mean to explore that unknown condition. Because there are many very dynamic, say dancing people at the party and you do not want to bump to any of them or walls or furniture you also need a mean to deal with that unknown complex dynamics. To make a long story short after a bit of walking around, observing and maybe even making new friends you discovered the kitchen, fridge, and bottle and returned to your seat. And, you learned a great deal about your friend's flat, dynamics of the party, and you even located person that you would really like to get acquainted with. Clearly, some of this newly acquired knowledge might help you to even faster or more safely accomplish a similar goal next time (unless the content of the bottle already affected your cognitive and physical performance).

If one could decompose your physical and mental actions, one may realize that you effectively performed as if you were using many (probably majority) of the techniques associated with modern-day artificial intelligence (AI)-based intelligent control, machine learning and data mining, as well as techniques addressed earlier in the chapter. One could probably write an entire book on ***How you discovered the bottle and returned to your seat at your friend's party*** and describe everything step by step in clear mathematical language. And, if a computer embedded in ideal biomimetic humanoid body is supposed to process all these actions, it would probably take 10^{15} bits or even more.

Here, we will briefly review some of the modern-day AI-based intelligent control, machine learning and data mining concepts, and terminology. We also invite reader, if interested, to explore these topics elsewhere [31–36] in more details.

AI, also called machine intelligence (MI) is intelligence exhibited by machines, in contrast with the natural intelligence (NI). AI is concerned with "intelligent agents," that is, any device that perceives its environment and takes actions that maximize its chance of success at some goal. AI attempts to mimic "cognitive" functions that humans associate with human minds, such as "learning" and "problem solving." The traditional problems of AI research include reasoning, knowledge, planning, learning, natural language processing, perception, and the ability to move and manipulate objects.

Intelligent control is a class of control techniques that use various AI computing approaches like expert systems, machine learning and data mining, neural networks, Bayesian probability, (neuro-)fuzzy logic, evolutionary computation and genetic algorithms, as well as many other.

An AI *expert system* is a computer system that emulates the decision-making ability of a human expert by reasoning through bodies of knowledge, represented mainly as if-then rules. An expert system is divided into two subsystems: the inference engine and the knowledge base. The knowledge base represents facts and rules. The inference engine applies the rules to the known facts to deduce new facts. The first expert systems were created in the 1970s and then proliferated in the 1980s. Expert systems were among the first truly successful forms of AI.

Machine learning, a fundamental concept of AI research, is the study of computer algorithms that improve automatically through experience: "A computer program is said to learn from experience E with respect to some class of tasks T and performance measure P if its performance at tasks in T, as measured by P, improves with experience E" [36]. The interest in machine learning proliferated after the 1980s.

Data mining is the process of discovering patterns in large data sets. Data mining is the analysis step of the "knowledge discovery in databases" process, or KDD. The actual data mining task is the semiautomatic or automatic analysis of large quantities of data to extract previously unknown, interesting patterns such as groups of data records (cluster analysis), unusual records (anomaly detection), and dependencies (association rule mining and sequential pattern mining).

Machine learning and data mining often employ the same methods and overlap significantly, but while machine learning focuses on prediction, based on known properties learned from the training data, data mining focuses on the discovery of unknown properties in the data. Data mining uses many machine learning methods, but with different goals; on the other hand, machine learning also employs data mining methods as "unsupervised learning" or as a preprocessing step to improve learner accuracy.

There are several different types of learning approaches including: *(semi)-supervised learning* with in advance provided (in)complete training set of input/output data pairs, *active learning* with optimization of limited choice of objects for which training label can be requested, *reinforcement learning* with in real time provided award or penalty-based feedback, *unsupervised learning* with no labels given to the learning algorithm, leaving it on its own to find structure in its input.

Typical machine learning and data mining tasks are: *classification* (or *pattern recognition*) requiring assignment of inputs to in-advance set classes, *clustering* requiring dividing inputs into groups not known beforehand, *regression* attempting to find a function which models the data with the least error, *dimensional reduction* simplifying problem by discovering lower-dimensional space containing most of input, *anomaly detection* identifying unusual data records, *association rule learning* (dependency modeling) searching for relationships between variables, *sequential pattern mining* searching for relationships between (time or other) sequences, etc.

Neural network is a learning algorithm that is inspired by the structure and functional aspects of biological neural networks. Computations within this nonlinear statistical data modeling tool are structured in terms of an interconnected group of artificial neurons. They are used to model complex relationships between inputs and outputs, to find patterns in data, or to capture the statistical structure in an unknown joint probability distribution between observed variables. The main categories of networks are acyclic or feedforward neural networks (where the signal passes in only one direction) and recurrent neural networks (which allow feedback and short-term memories of previous input events). Among the most popular feedforward networks are perceptrons, multilayer perceptrons, and radial basis networks (Fig. 5.11).

Deep neural networks (DNNs) consist of multiple hidden layers. This approach tries to model the way the human brain processes light and sound into vision and hearing. Some successful applications of deep learning took place within computer vision and speech recognition. Deep learning is part of a broader family of machine learning methods based on learning data representations. DNN architectures generate compositional models in which the object is expressed as a layered composition of primitives. The extra layers enable composition of features from lower layers, potentially modeling complex data with fewer units than a similarly performing shallow network. DNNs are typically feedforward networks in which data flow from the input layer to the output layer without looping back.

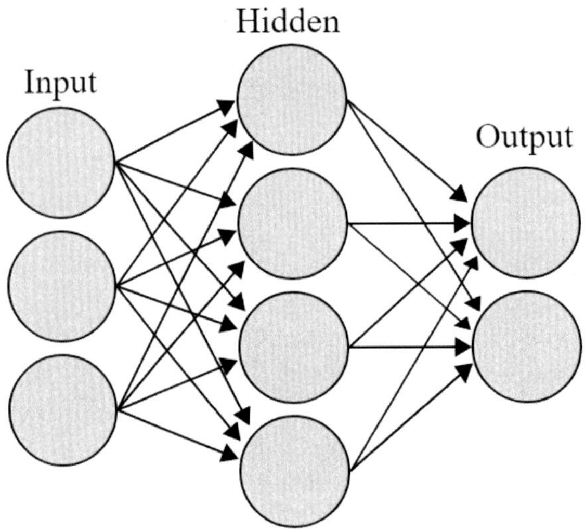

FIG. 5.11 Example of multilayered feedforward neural network.

One of the authors (MP) was recently involved with research on self-evolving neural networks where even the structure of neural network is not set in-advance but rather deals with nodes and links which can die or be formed. This area of research seems promising and can potentially lead to new approach to neural networks and AI.

5.11 Biological Neural Networks

Here, the basic aspects of biological control system, that is, central and peripheral nervous system (PNS) and its constituents are briefly reviewed following [1].

Neuroscience is the scientific study of the nervous system. It involves an interdisciplinary approach including biology, chemistry, pharmacology, computer science, engineering, AI, linguistics, mathematics, medicine and allied disciplines, philosophy, physics, and psychology.

More sophisticated studies of the brain followed the invention of the microscope and the development of a staining procedure by Camillo Golgi during the late 1890s. The procedure used a silver chromate salt to reveal the intricate structures of individual neurons. The technique was used by Santiago Ramón y Cajal and led to the formation of *the neuron doctrine*, the hypothesis that the functional unit of the brain is the neuron. Golgi and Ramón y Cajal shared the Nobel Prize for their extensive observations, descriptions, and categorizations of neurons throughout the brain.

The nervous system is an organ system containing a network of specialized cells called neurons that coordinate the actions of an animal and transmit signals between different parts of its body. In most animals, the nervous system consists of two parts, central and peripheral. The central nervous system (CNS) of vertebrates (such as humans) contains the brain, spinal cord, and retina. The PNS consists of sensory neurons, clusters of neurons called ganglia, and nerves connecting them to each other and to the CNS.

A nerve is an enclosed, cable-like bundle of axons (the long, slender projections of neurons) in the PNS. A nerve provides a common pathway for the electrochemical nerve impulses that are transmitted along each of the axons. In the CNS, the analogous structures are known as tracts.

The vertebrate nervous system can be divided into areas called gray matter and white matter. *Gray matter* (is better described as pink or light brown in living tissue) contains a high proportion of cell bodies of neurons (also called soma). Gray matter is found in clusters of neurons in the brain and spinal cord, and in cortical layers that line their surfaces. *White matter* is composed mainly of myelinated axons (see definition below), and takes its color from the myelin. White matter includes all of the nerves, and much of the interior of the brain and spinal cord.

There is an anatomical convention that a cluster of neurons in the brain or spinal cord is called a *nucleus,* whereas a cluster of neurons in the periphery is called a *ganglion.*

Nerves are categorized into three groups based on the direction that signals are conducted: *afferent nerves* conduct signals from sensory neurons to the CNS, for example, from the mechanoreceptors in skin. *Efferent nerves* conduct signals from the CNS along motor neurons to their target muscles and glands. *Mixed nerves* contain both afferent and efferent axons, and thus, conduct both incoming sensory information and outgoing muscle commands in the same bundle.

Nerves can be categorized into two groups based on where they connect to the CNS: *Spinal nerves* innervate much of the body, and connect through the spinal column to the spinal cord. *Cranial nerves* innervate parts of the head, and connect directly to the brain (especially to the brainstem).

5.11.1 Neuron

The CNS is composed entirely of two kinds of specialized cells: neurons (Fig. 5.12) and glia. The only notable differences between neurons and glial cells are neurons' possession of axons and dendrites, and capacity to generate action potentials.

There are two different kinds of processes or extensions extending away from the soma, i.e., from the neuron's main body containing the cell nucleus. The numerous relatively short processes or extensions are called dendrites and the one relatively lengthy process or extension is the axon.

The word "dendrite" comes from the ancient Greek word *dendron,* meaning *tree.*

The word "axon" is related to the Greek word *axis.*

In essence, the dendrites collect information that is processed by soma and forwarded along axon. The axon terminals are the short segments that branch away from the axon. At the very end of each axon terminal, there is a small enlargement called a terminal button.

In vertebrates, the axons of many neurons are sheathed in myelin, which is formed by either of two types of glial cells: *Schwann cells* ensheathing peripheral neurons and

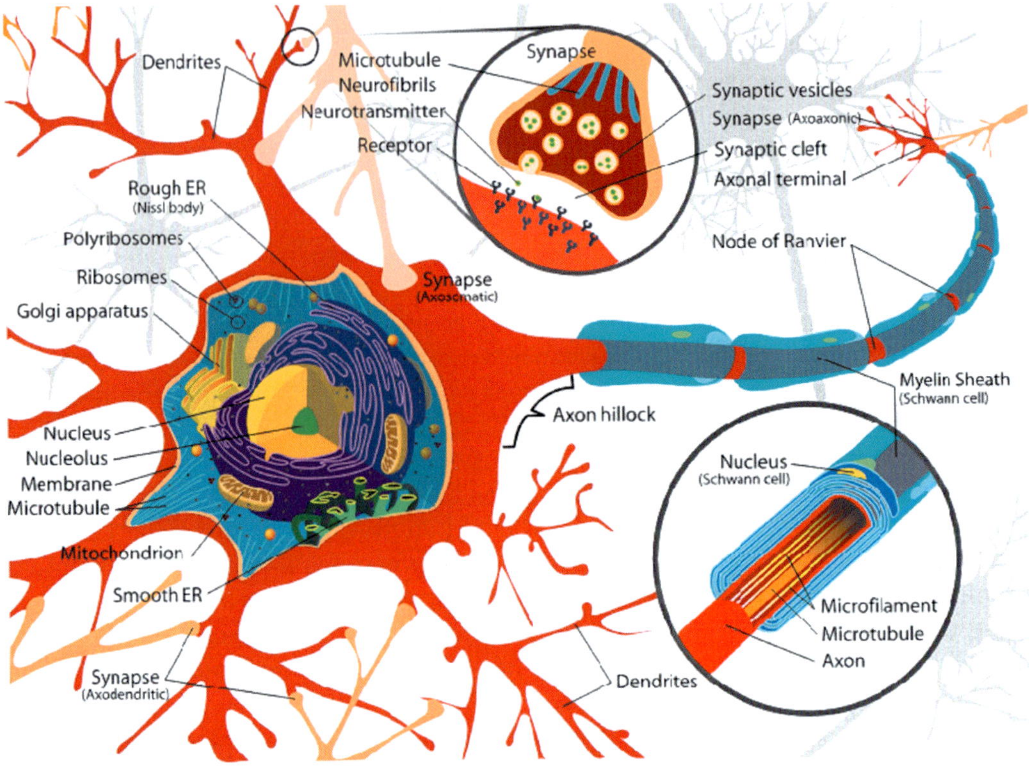

FIG. 5.12 Neuron.

oligodendrocytes insulating those of the CNS. Along myelinated nerve fibers, gaps in the sheath known as *nodes of Ranvier* occur at evenly spaced intervals. The myelination enables an especially rapid mode of electrical impulse propagation called saltation. The demyelination of axons is what causes the multitude of neurological symptoms found in the disease multiple sclerosis.

Neurotransmission (or synaptic transmission) is communication between neurons as accomplished by the movement of chemicals or electrical signals across a synapse.

An action potential is an electrical signal very much like the electrical signals in electronic devices. But whereas an electrical signal in an electronic device occurs because electrons move along a wire, an electrical signal in a neuron occurs because ions (sodium, mostly outside, and potassium, mostly inside) move across the neuronal membrane.

The protein membrane of a neuron acts as a barrier to ions. Ions move across the membrane through ion channels that open and close due to the presence of neurotransmitter.

When the concentration of ions on the inside of the neuron changes, the electrical property of the membrane itself changes. Normally, the membrane potential of a neuron rests as $-70\,\text{mV}$ (and the membrane is said to be polarized). The influx and outflux of ions (through ion channels during neurotransmission) will make the inside of the target neuron more positive (hence, depolarized). When this depolarization reaches a point of no return called a *threshold,* a large electrical signal is generated. A higher stimulus intensity is reflected in increased frequency of impulses, not in higher voltages.

The speed of propagation for mammalian motor neurons is 10–120 m/s, while for nonmyelinated sensory neurons it is about 5–25 m/s (nonmyelinated neurons fire in a continuous fashion, without the jumps; ion leakage allows effectively complete circuits, but slows the rate of propagation). Typically, linear connection between speed and diameter for myelinated axons is 6–9 m/s per micron diameter.

There may be as many as 10,000 different types of neurons!

There are three kinds of neurons: motor neurons (for conveying motor information), sensory neurons (for conveying sensory information), and interneurons (for conveying information between different types of neurons).

Alpha motor neurons (α-MNs) are large lower motor neurons of the brainstem and spinal cord. They innervate extrafusal muscle fibers of skeletal muscle and are directly responsible for initiating their contraction, see Chapter 3. α-MNs are distinct from gamma motor neurons, which innervate intrafusal muscle fibers of muscle spindles.

While their cell bodies are found in the CNS, α-MNs are also considered part of the somatic nervous system—a branch of the PNS—because their axons extend into the periphery to innervate skeletal muscles.

An α-MN and the muscle fibers it innervates is a *motor unit.*

Gamma motoneurons (γ-motoneurons), also called *gamma motor neurons,* are the efferent component of the fusimotor system, the system by which the CNS controls and modifies muscle spindle sensitivity. The fusimotor system refers to the combination of muscle spindles and γ-motoneurons. The function of the muscle spindle is to provide proprioceptive feedback for the movement, position, and extension of muscles.

γ-motoneurons are located in the brainstem and spinal cord and are smaller than their α-motoneuron counterparts, which are responsible for controlling skeletal muscle.

5.11.2 Central Pattern Generators

Central pattern generators (CPGs) are neural networks that produce rhythmic patterned outputs without sensory feedback (Fig. 5.13). CPGs have been shown to produce rhythmic outputs resembling normal "rhythmic motor pattern production" even in isolation from motor and sensory feedback from limbs and other muscle targets.

To be classified as a rhythmic generator, a CPG requires (1) two or more processes that interact such that each process sequentially increases and decreases, and (2) that, as a result of this interaction, the system repeatedly returns to its starting condition.

Motor systems are areas of the brain that are directly or indirectly involved in producing body movements, that is, in activating muscles.

Except for the muscles that control the eye, which are driven by nuclei in the midbrain, all the voluntary muscles in the body are directly innervated by motor neurons in the spinal cord and hindbrain.

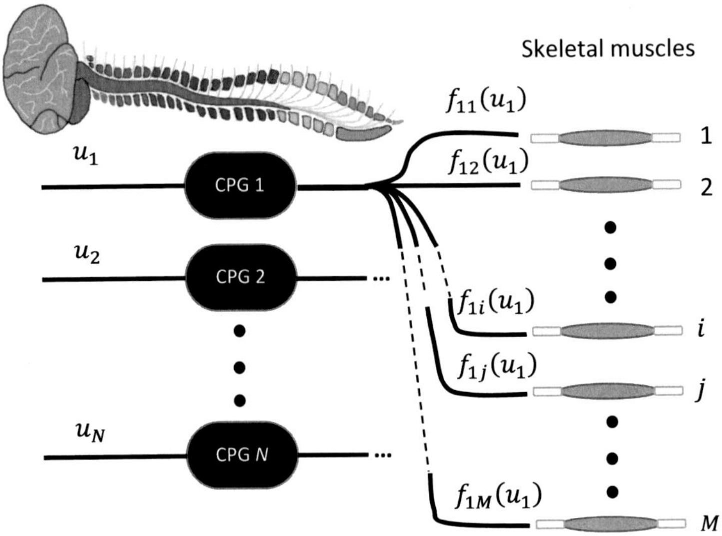

FIG. 5.13 Central pattern generator.

Spinal motor neurons are controlled both by neural circuits intrinsic to the spinal cord, and by inputs that descend from the brain. The intrinsic spinal circuits implement many reflex responses, and contain pattern generators for rhythmic movements such as walking or swimming. The descending connections from the brain allow for more sophisticated control.

References

[1] M. Popovic, "Biomechanics and Robotics", 364 pages, book, Copyright © 2014 Pan Stanford Publishing Pte. Ltd., Singapore, ISBN 978-981-4411-37-0 (Hardcover), 978-981-4411-38-7 (eBook), (2014).

[2] S. Hornsey "A review of relay auto-tuning methods for the tuning of PID-type controllers" Reinvent.: Int. J. Undergr. Res., 5, 2, www.warwick.ac.uk/reinventionjournal/issues/volume5issue2/hornsey (accessed 16.03.18).

[3] N.B. Nichols, J.G. Zeigler, Optimum settings for automatic controllers, Trans. ASME 64 (1942) 759–768.

[4] G.H. Cohen, G.A. Coon, Theoretical consideration of retarded control, Trans. ASME 75 (1953) 827–834.

[5] C.E. Garcia, M. Morari, Internal model control: a unifying review and some new results, Ind. Eng. Chem. Process Design Dev. 1 (1982) 308–323.

[6] D.E. Seborg, T.F. Edgar, D.A. Mellichamp, Process Dynamics and Control, Wiley, New York, 1989.

[7] C.C. Hang, K.J. Åström, W.K. Ho, Refinements of the Ziegler-Nichols tuning formula, Control Theory Appl. IEE Proc. D 21 (1991) 111–118.

[8] D.W. Pessen, A new look at PID-controller tuning, Trans. Am. Soc. Mech. Eng. J. Dyn. Syst. Measur. Control 116 (1994) 553–557.

[9] R. Argelaguet, A new tuning of PID controllers based on LQR optimisation, in: Preprints of Proceedings of PID '00: IFAC Workshop on Digital Control, 1996, pp. 303–308.

[10] E.H. Bristol, Pattern recognition: an alternative to parameter identification in adaptive control, Automatica 13 (1977) 179–202.

[11] K.J. Åström, T. Hägglund, Automatic tuning of simple regulators, in: Proceedings of IFAC 9th World Congress, Budapest, 1984, , pp. 1867–1872.

[12] F. Radke, R. Isermann, A parameter-adaptive PID controller with stepwise parameter optimisation, Automatica 23 (1987) 449–457.

[13] W.L. Brogan, Modern Control Theory, first ed., Quantum Publishers, Inc., 1974, p. 172

[14] D.N. Cheban, Global Attractors of Non-Autonomous Dissipative Dynamical Systems, 47, (2004).

[15] A.M. Lyapunov, The General Problem of the Stability of Motion (In Russian), (Doctoral dissertation) Univ. Kharkov, 1892 (English translations: (1) Stability of Motion, Academic Press, New-York/London, 1966; (2) The General Problem of the Stability of Motion (A. T. Fuller trans.), Taylor & Francis, London, 1992. Included is a biography by Smirnov and an extensive bibliography of Lyapunov's work).

[16] N.G. Chetaev, The Stability of Motion - Translated by Morton Nadler, Oxford, 1961 (200 p., Originally published in Russian in 1946 by ОГИЗ. Гос. изд-во технико-теорет. лит., Москва-Ленинград).

[17] N.P. Bhatia, G.P. Szegő, Stability Theory of Dynamical Systems, Springer, 2002. ISBN: 978-3-540-42748-3.

[18] G. Gandolfo, Economic Dynamics, third ed., Springer, Berlin, 1996, . ISBN: 3-540-60988-1, pp. 407–428.

[19] P.C. Parks, A. M. Lyapunov's stability theory—100 years on. IMA J, Math. Control Inf. 9 (4) (1992) 275–303, doi:10.1093/imamci/9.4.275.

[20] J.-J.E. Slotine, W. Li, Applied Nonlinear Control, Prentice Hall, NJ, 1991.

[21] G. Teschl, Ordinary Differential Equations and Dynamical Systems. Providence, American Mathematical Society, 2012. ISBN: 978-0-8218-8328-0.

[22] S. Wiggins, Introduction to Applied Nonlinear Dynamical Systems and Chaos, second ed., Springer Verlag, New York, 2003. ISBN: 0-387-00177-8.

[23] R.E. Bellman, Dynamic Programming (Princeton Landmarks in Mathematics), Princeton University Press, England, 2010.

[24] Richard Bellman—GHN: IEEE Global History Network, http://www.ieeeghn.org/wiki/index.php/Richard_Bellman (retrieved July 2012).

[25] O. Khatib, K. Yokoi, K.-S. Chang, D. Ruspini, T. Holmberg, A. Casal, Coordination and decentralized cooperation of multiple mobile manipulators, J. Robot. Syst. 13 (11) (1996).

[26] O. Khatib, L. Sentis, J.H. Park, J. Warren, Whole body dynamic behavior and control of human-like robots, Int. J. Human. Robot. 1 (1) (2004).

[27] L. Sentis, O. Khatib, Task-oriented control of humanoid robots through prioritization, in: IEEE RAS/RSJ International Conference on Humanoid Robots, Santa Monica, USA, 2004.

[28] C.G. Atkeson, Efficient robust policy optimization, in: American Control Conference, 2012.

[29] M. Popovic, A. Englehart, H. Herr, Angular momentum primitives for human walking: biomechanics and control, in: Proceedings. 2004 IEEE/RSJ International Conference on Intelligent Robots and Systems, 2004 (IROS 2004), vol. 2, IEEE, 2004, , pp. 1685–1691.

[30] Physical Intelligence and Legged Robotic Systems, Popovic Labs http://users.wpi.edu/~mpopovic/pages/physicalintelligence.html (retrieved 17.03.18).

[31] P.C. Jackson Jr., Introduction to Artificial Intelligence, second enlarged ed., Dover Publications, Inc., New York, 1985

[32] T. Rashid, Make Your Own Neural Network, first ed., CreateSpace Independent Publishing Platform, 2016 ISBN-10: 1530826608, ISBN-13: 978-1530826605.

[33] I. Goodfellow, Y. Bengio, A. Courville, Deep learning, Adaptive Computation and Machine Learning Series, The MIT Press, 2016.

[34] T. Hastie, R. Tibshirani, J. Friedman, The elements of statistical learning: Data mining, inference, and prediction, in: Springer Series in Statistics, 2009.

[35] A. Konar, Artificial Intelligence and Soft Computing: Behavioral and Cognitive Modeling of the Human Brain, CRC Press, 2000.

[36] T. Mitchell, Machine Learning, McGraw Hill, 1997, p. 2. ISBN: 0-07-042807-72.

6 Direct Neural Interface

Hiroyuki Tashiro*, Marko B. Popovic†, Keiji Iramina*, Yasuo Terasawa‡, Jun Ohta§

*KYUSHU UNIVERSITY, FUKUOKA, JAPAN †WORCESTER POLYTECHNIC INSTITUTE, WORCESTER, MA, UNITED STATES ‡NIDEK CO., LTD., AICHI, JAPAN §NARA INSTITUTE OF SCIENCE AND TECHNOLOGY (NAIST), NARA, JAPAN

CHAPTER OUTLINE

6.1 Introduction

A direct neural interface (DNI) is a direct communication pathway between the nervous system and an external device. DNI is also known by other terms like brain-machine interface (BMI), brain-computer interface (BCI), mind-machine interface (MMI), human-computer interface (HCI), and so forth. Even now, the definition of the term is ambiguous. Some researchers refuse to use the term BMI to represent connections to peripheral nerves while others would not use the term BCI to describe invasive interfaces. Still others think that neuromodulation devices such as deep brain stimulators (DBSs) are distinguished from BMI devices. In fact, it seems that the terms are used without distinction between invasive/noninvasive and central/peripheral interfaces [1]. However, the terms BMI and BCI are preferred to DNI [1]. A BMI is also defined as a device that translates neuronal information into instructions for external software or hardware such as a computer

Biomechatronics. https://doi.org/10.1016/B978-0-12-812939-5.00006-9

or robotic arm. In this chapter, we focus on examining the interface itself (Sections 6.1–6.4) and we briefly review various BMI applications (Section 6.5). Therefore, DNI is sparingly used as a term to describe the interface itself.

In 1929, Hans Berger recorded the human electroencephalogram (EEG) for the first time in history [2]. In the 1950s, Jose Delgado implanted electrodes into the brains of animals and humans and stimulated them [3, 4]. In the 1970s, Jacques J. Vidal succeeded in demonstrating cursor control with visual evoked potential [5]. This represents the first evidence that BCI can be used as an interface with an external device via the brain signal. In the same period, hearing regeneration using cochlear implant and visual regeneration through artificial vision system using electrodes implanted on the cerebrum, was demonstrated [6, 7]. These instances show that research and development into the practical uses of BMI blossomed in the 1970s.

Information can be transmitted in two directions: inward (input) and outward (output). Inward BMIs are developed to replace lost sensation using nervous stimulation. Electrical stimulation is mostly used for nerve stimulation, but magnetic stimulation [8], optical stimulation [9], ultrasound stimulation [10], and chemical stimulation have also been developed [11]. Outward BMIs or BCIs decode a behavioral decision from brain activity and feed the decoded information into external devices to control them [12]. Both inward and outward BMI use invasive and noninvasive approaches.

In BCIs (outward BMIs), electrophysiological and hemodynamic measurements are used to acquire the activity of the brain. The brain activity is measured using EEG, electrocorticography (ECoG), and magnetoencephalography (MEG). For BCIs using intracortical electrodes, action potential voltage waveforms are often sorted to separate individual neurons [13]. Fig. 6.1 shows recording sites for obtaining electrophysiological signals. EEG is recorded using surface electrodes on the scalp. ECoG is recorded by planner electrodes on the cortical surface. Action potentials or local-field potentials (LFPs) are recorder by microelectrodes penetrating into the cortex. The hemodynamic response is caused by the higher rate release of glucose from the blood to active neurons than to inactive

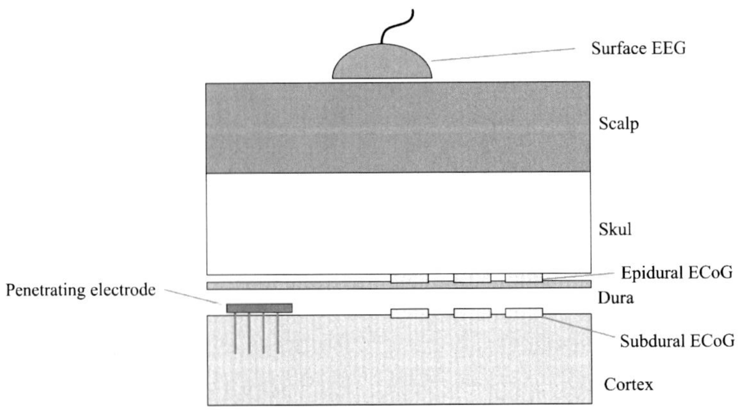

FIG. 6.1 Recording sites for obtaining electrophysiological signals.

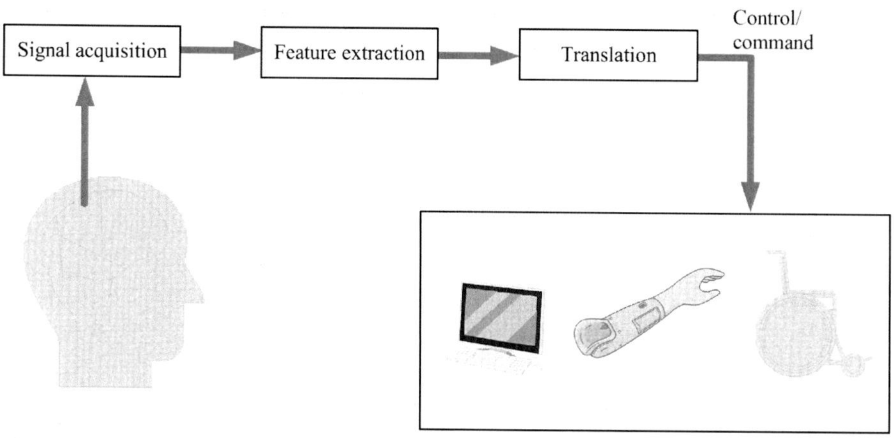

FIG. 6.2 The basic design of a BCI system.

neurons. These changes can be quantified by neuroimaging methods such as functional magnetic resonance imaging (fMRI) and functional near-infrared spectroscopy (fNIRS). These methods relate to neuronal activity more indirectly than electrophysiological measurements. The BCI records these brain signal, extracts specific features from them, and translates these features into control of the devices, machines, or computers (Fig. 6.2).

6.2 Theory of Electrical Recording

There are invasive and noninvasive approaches to record the electrical activity of the brain. Invasive approaches toward acquiring the activity of central nervous system (CNS) involve either implanting the electrode within the cerebrum or placing it on the surface. Obtaining EEG recording using an electrode array placed on the surface of the cerebrum is known as ECoG. A more invasive method can obtain better quality signals, but electrodes implanted in the cerebral tissue have problems of durability. Recording the EEG from the scalp is an effective and safe method but the spatial resolution and the frequency band are limited. The signals that are recorded range from tens to hundreds of microvolts when recording a single spike in brain activity with a needle electrode inserted into the cerebrum [14, 15]. The frequency of the slow LFP ranges from zero up to several hundreds of Hertz. Obtaining single-spike activities requires frequency bands from hundreds of Hertz up to several kilohertz. Similarly, there are noninvasive and invasive approaches toward recording the activities of the peripheral nervous system (PNS). The amplitude and constituent frequency components of a bioelectric signal that can be recorded on the body surface are as shown in Fig. 6.3. The bioelectric signal level from CNS and PNS is extremely low, making it very difficult to measure in the presence of noise.

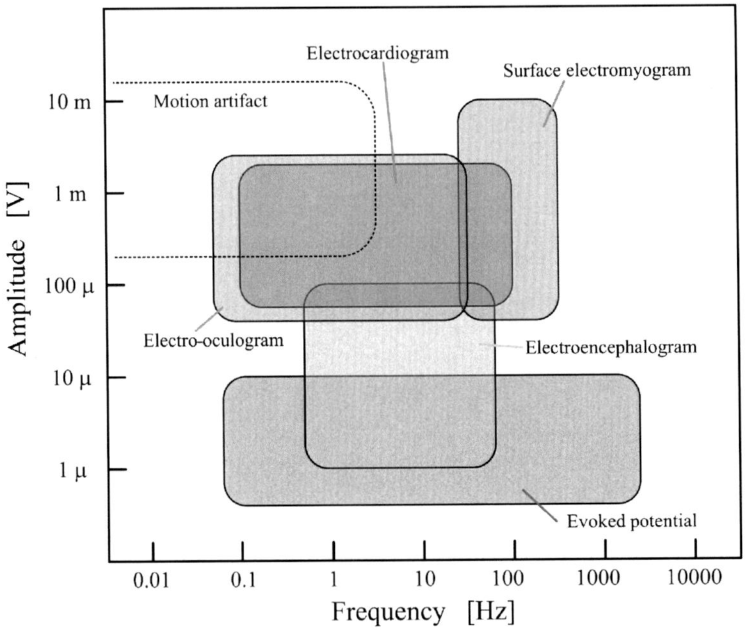

FIG. 6.3 Amplitude and frequency range of a bioelectric signal.

6.2.1 Recording Electrode

6.2.1.1 Material of Recording Electrode

The recording of electrical nerve activity is greatly affected by the characteristics of the recording electrode [16]. In the frequency range of the bioelectric signal, the human body is almost equivalent to saline. Electrical conduction in metals is due to free electrons. On the other hand, electrical conduction in the electrolytic solution is caused by ionic diffusion, or electron transfer between the electrode and electrolyte when an electrochemical reaction occurs. Therefore, when a metal electrode contacts an electrolytic solution, a potential difference caused by the energy gap is developed at the electrode-electrolyte interface. The potential difference is called a half-cell potential or polarization potential. In the case that charge transfer does not occur (no electrochemical reaction occurs on the electrode), the charge in the metal is attracted to the ions in the electrolytic solution. However, in order to cause a electrochemical reaction, it is necessary to adapt the potential difference above this energy gap. When the electrochemical reaction does not occur, ions of the same charge amount as the free electrons gathered on the electrode surface are attracted in order to maintain the electrical neutrality. This is called an electric double layer, and it can be electrically regarded as a capacitor (Fig. 6.4A). Refer to Ref. 17 for details of the principle that such electrochemical phenomenon occurs. Metals that can be used in the measurement of bioelectric signals, and their half-cell potentials and toxicity are shown in Table 6.1 [18–20].

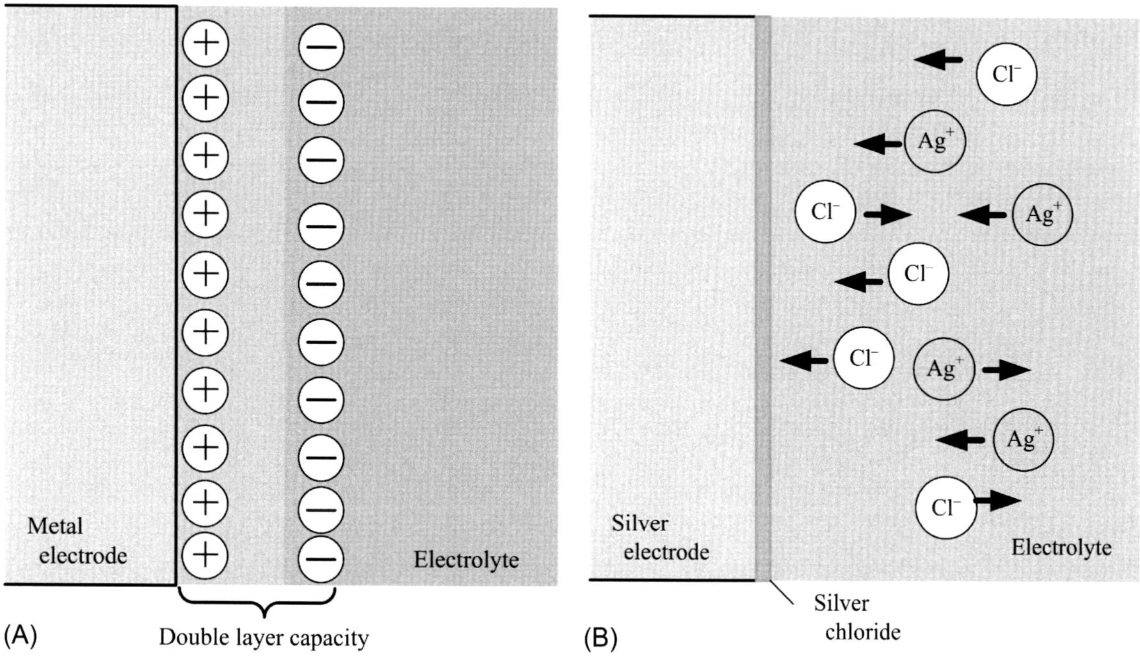

6.4 Boundary between electrode and electrolyte. (A) Electric double layer is formed on the surface of the irreversible electrode, and half-cell ential is generated. (B) Reversible reaction of silver chloride inhibits formation of electric double layer.

Table 6.1 Half-Cell Potential for Metal-Ion Interfaces and Toxicity in Biopotential Measurement

Material, Reaction	Potential (V)	Tissue Response
Silver (Ag)		Toxic
$Ag^+ + e \rightleftharpoons Ag$	+0.7996	
Silver-Silver Chloride (AgCl)		Toxic
$AgCl + e \rightleftharpoons Ag + Cl^-$	+0.2223	
Aluminum (Al)		Nontoxic
$Al^{3+} + 3e \rightleftharpoons Al$	−1.662	
Gold (Au)		Nontoxic
$Au^+ + e \rightleftharpoons Au$	+1.692	
$Au^{3+} + 2e \rightleftharpoons Au^+$	+1.401	
$Au^{3+} + 3e \rightleftharpoons Au$	+1.498	
$Au^{2+} + e \rightleftharpoons Au^+$	+1.8	
Copper (Cu)		Toxic
$Cu^+ + e \rightleftharpoons Cu$	+0.521	
$Cu^{2+} + e \rightleftharpoons Cu^+$	+0.153	
$Cu^{2+} + 2e \rightleftharpoons Cu$	+0.3419	
$Cu^{3+} + e \rightleftharpoons Cu^{2+}$	+2.4	

Continued

Table 6.1 Half-Cell Potential for Metal-Ion Interfaces and Toxicity in Biopotential Measurement—cont'd

Material, Reaction	Potential (V)	Tissue Response
Iron (Fe)		Toxic
$Fe^{2+} + 2e \rightleftharpoons Fe$	−0.447	
$Fe^{3+} + 3e \rightleftharpoons Fe$	−0.037	
$Fe^{3+} + e \rightleftharpoons Fe^{2+}$	+0.771	
Iridium (Ir)		Nontoxic
$Ir^{3+} + 3e \rightleftharpoons Ir$	+1.156	
Platinum (Pt)		Nontoxic
$Pt^{2+} + 2e \rightleftharpoons Pt$	+1.18	
Tantalum (Ta)		Nontoxic
$Ta^{3+} + 3e \rightleftharpoons Ta$	−0.6	
Titanium (Ti)		Nontoxic
$Ti^{2+} + 2e \rightleftharpoons Ti$	−1.630	
$Ti^{3+} + e \rightleftharpoons Ti^{2+}$	−0.9	
$Ti^{3+} + 3e \rightleftharpoons Ti$	−1.37	
Tungsten (W)		Nontoxic
$W^{3+} + 3e \rightleftharpoons W$	+0.1	

The potential is a value with reference to standard hydrogen electrode (SHE) (25°C) [18–20].

In an ideal recording electrode, the polarization potential generated at the boundary between the metal electrode and the electrolyte is constant. If the polarization potential fluctuates during the measurement, it is mixed into the signal as noise. Electrodes having such characteristics are called nonpolarizable or reversible electrodes. In the nonpolarizable electrode, since the electrochemical reaction always occurs on the electrode surface, the polarization potential is low and stable. The most common nonpolarizable electrode for biopotential measurement is the silver-silver chloride (Ag/AgCl) electrode. As shown in Fig. 6.4B, the potential of a Ag/AgCl electrode is determined by two equilibration reactions:

$$\left.\begin{array}{r} Ag \rightleftharpoons Ag^{+} + e^{-} \\ Ag^{+} + Cl^{-} \rightleftharpoons AgCl\downarrow \end{array}\right\} \tag{6.1}$$

These reactions inhibit the formation of an electric double layer, making the Ag/AgCl electrode realize a low impedance and low polarization potential. Unfortunately, Ag/AgCl electrodes are unsuitable for chronic implantation because of the toxicity of silver chloride [20]. If Ag/AgCl electrodes are used as skin surface electrodes for EEG or EMG recording, it is necessary to wet them with a paste or gel containing chloride ions in order to realize the reversible reaction of Eq. (6.1). Therefore, they are also called wet electrodes. However, many feels uncomfortable wearing BCI wet electrodes as they stick to scalp and hair by application of jelly-like/gelatin-like/gel-like gluing substance. Consequently, a spike-array dry electrode, also referred to as comb-shaped dry electrode, coated with Ag/AgCl is developed. It is composed of microscale pillars that contact the scalp directly or slightly penetrate the epidermoid scalp [20]. In order to improve the signal-to-noise (S/N) ratio of the recording by skin surface electrodes such as Ag/AgCl disc electrodes, it is generally important to reduce the contact impedance.

Noble metals that are stable in vivo like platinum are used as electrode materials for implantation. They belong to the class of polarizable or irreversible electrodes. The polarizable electrode forms a double-layer capacity at the boundary between the electrode surface and the electrolyte (Fig. 6.4A). If the actual surface area is enlarged by roughening the electrode surface, the electric double-layer capacity increases. Increase of the electric double-layer capacity decreases the electrode impedance. It especially affects the electrode impedance characteristics at low frequencies [21]. To improve the low-frequency characteristics of polarizable electrodes, it is effective to increase the actual surface area versus the geometric surface area. A platinum black electrode is a well-known polarizable electrode whose characteristics are improved by increasing the actual surface area. However, the toxicity of platinum black electrodes by residual lead has been reported [22].

6.2.1.2 Electrochemical Property of Recording Electrode

Due to the electric double layer, the electrode impedance has frequency dependence. Fig. 6.5A illustrates the simplified equivalent circuit model showing the electrode-electrolyte phase boundary for a recording electrode. The battery E_{hc} represents the half-cell potential. Because the direct current potential is rejected by a frequency filter, generally, the half-cell potential is not included in the equivalent circuit in many cases. C_P symbolizes the capacity of the electric double layer. The charge transfer resistance, which is related to the electrochemical reaction, is given by the resistance, R_{ct}. The lead resistance R_L is associated with the resistance of the lead, the electrode materials themselves, and volume resistance of solution. The impedance of the recording electrode shows frequency dependency as illustrated in Fig. 6.5B. Variations in the bias current of the amplifier cause fluctuations in the quantity of electric charge stored in the electric double layer. As a result, a change in the electrode potential generates electrode noise. This phenomenon is remarkable especially in implanted microelectrodes with small capacitance of electric double layer. Similarly, when the skin electrode moves with respect to the body, the charge distribution of the electric double layer is changed, and a fluctuation of the half-cell potential, called motion artifact, occurs temporarily. These artifacts are minimal for nonpolarizable electrodes such as Ag/AgCl electrodes. If it is not necessary to acquire a low-frequency signal, a capacitive electrode using a dielectric thin film such as silicon nitride can be used [23–25].

6.2.1.3 Type of Recording Electrode

Electrodes used for CNS recording are either implanted in the brain, or placed on the surface of the cerebral cortex or dura. The penetrating cortical electrode is needle shaped, and is defined as a single microwire, microwire bundles, or a multielectrode array (MEA) (Fig. 6.6A), used for measuring the action potential or LFP. Generally, the penetrating electrodes provide a better quality recording compared with the ECoG electrode.

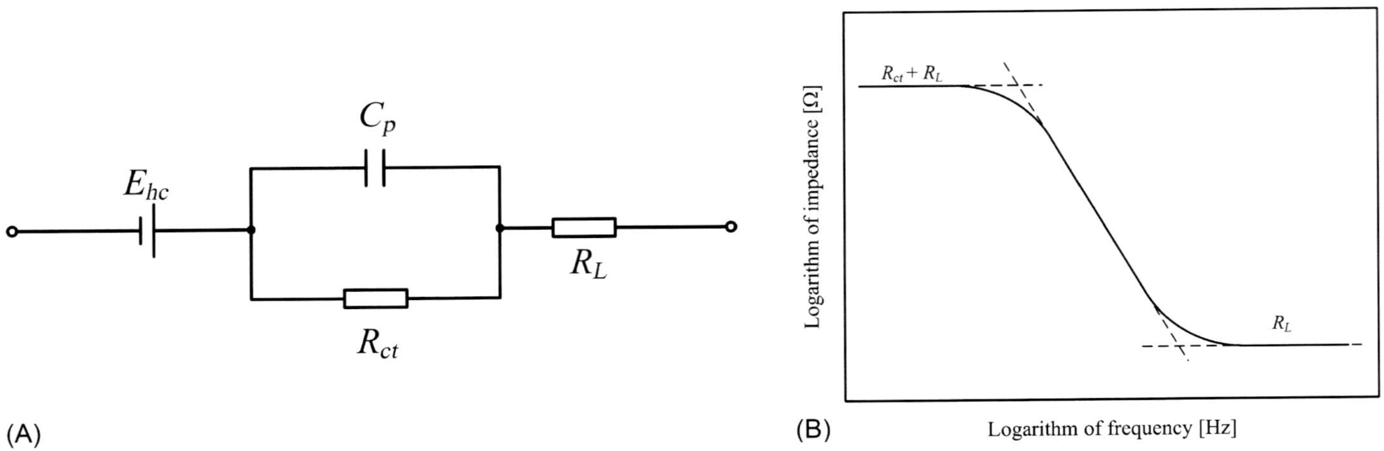

FIG. 6.5 Simplified equivalent circuit model of electrode-electrolyte boundary.

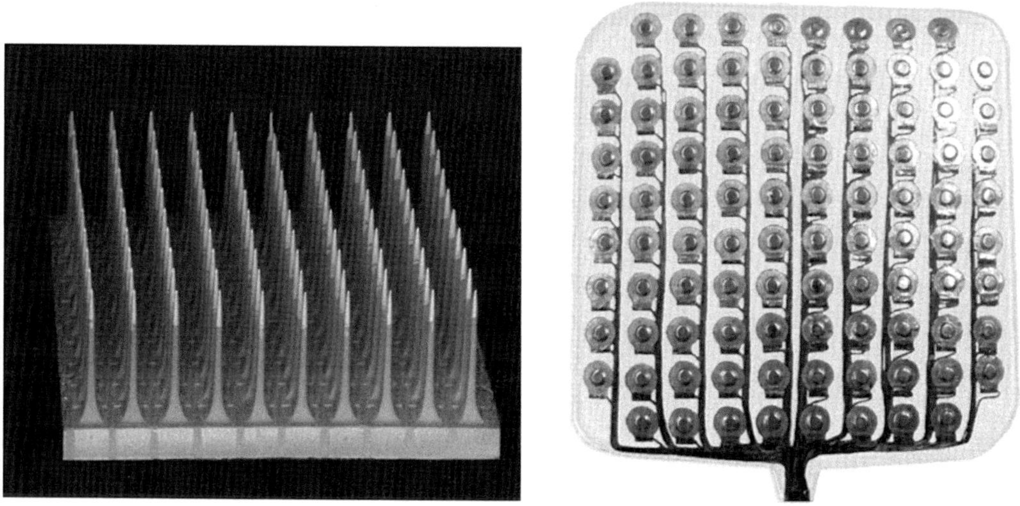

(A) Intracortical electrode array (B) ECoG electrode array

FIG. 6.6 Invasive EEG recording electrode used for BMI. *Photographs provided courtesy of Unique Medical Co., Ltd, Japan, and Blackrock Microsystems LLC, the USA.*

However, the brain floats in spinal fluid, which causes gliosis and encapsulation by fibroblastic tissue through mechanical stress with the rigid electrode [26]. Gliosis is a fibrous proliferation of glial cells caused in injured areas of the CNS. The glial cells are a generic term for cells that are not neurons constituting the nervous system. These are foreign body reactions in the CNS. When the glial cells and the fibroblastic tissue gather around the recording electrode, the distance to the nerve cell as a signal source becomes far. Moreover, R_L increases in the equivalent circuit shown in Fig. 6.5A. As a result, such penetrating electrodes have limited lifespan. For such occasions, soft neural probes constructed from polymer shanks have been developed using microfabrication technology [27]. Besides that, there are approaches with fine electrodes (<50 μm), a stent-based electrode, and so forth [28–31] (see Section 6.5.2 for details). Implanting non-penetrating flat-shaped electrodes on the surface of the dura or cortex is a less invasive approach (Fig. 6.6B). Depending on the place in which the electrode is implanted, the EEG recordings are classified into epicranial EEG, epidural ECoG, or epicortical ECoG. A better EEG signal can be acquired when the recording electrode is nearer to the cerebral cortex, although deterioration of the recorded signal due to scratching is liable to occur, similar to the insertion electrode approach. Therefore, more flexible ECoG electrodes are being developed [32]. Moreover, the ECoG electrode that realize signal recording with quality equivalent to that of penetrating electrodes was developed [33].

When electrodes are implanted in a living body, protein adsorption takes place, leading to the deterioration of the electrode characteristics [34, 35]. In particular, electrodes with porous surfaces are greatly affected by adsorbed proteins [36]. There is a report that the adsorbed proteins desorb through the adaptation of weak electric potential pulse, leading to the recovery of the electrode performance [37].

For skin surface electrodes used in PNS recording, the same electrodes employed in EEG recording can be used. Needle electrodes are applied for diagnostic EMG examinations [38]. For the BCI recording electrodes, a planar electrode, wire electrode, or needle MEA can be used [39]. Cuff electrode is commonly used as the electrode of choice for extracellular-potential recording of a myelinated nerve fiber [40]. The lead wires of electrodes chronically implanted into violently moving body parts such as upper and lower limbs might break. For this reason, EMG sensors have been developed in which both ends of a capsule-shaped device are capped with recording electrodes [41].

6.2.2 Electrode Configuration

Biopotential can be recorded by several electrode configurations, including monopolar and bipolar. The monopolar electrode setup is the simplest configuration for recording the potential difference between two points (Fig. 6.7A). In this configuration, the ground electrode must be placed at a no-signal location. If the noise level near the ground electrode and the recording electrode (different electrode) are the same, it is subtracted and canceled. When the noise or a signal generated from the vicinity thereof is added only to the ground electrode, it cannot be distinguished from the signal recorded by the different electrode. In a bipolar electrode setup, an indifferent electrode matched to the different electrode is placed near the different electrode as a second active electrode (Fig. 6.7B). Such an electrode configuration is very advantageous for the rejection of common-mode noise such as electric hum, variation of power supply voltage, myoelectric potential overlapped on nervous action potential, and so on. The electric hum is 50/60 Hz noise caused by the alternating current of the electrical mains. For the ground potential, it is also possible to use a common average reference obtained by average potential of multiple electrodes [42].

6.3 Electrical Stimulation

Electrical voltage is used to stimulate the nerve tissue of the subject and induce a neural response. Two electrodes are normally used, and stimulation is achieved by applying a constant voltage or current between the electrodes. One of the electrodes, called a stimulating or stimulation electrode, is placed in the vicinity of the target neural cell to be stimulated (e.g., a spinal ganglion cell for a cochlear implant). The other electrode is referred to as a return electrode because the stimulation current ejected from the stimulating electrode returns to the stimulator through it. The function of the return electrode is not to stimulate neural tissue but to allow the passage of electric current. For this reason, the return electrode often has a much larger surface area than the stimulating electrode. The return electrode is sometimes called the indifferent electrode or counter electrode due to historical practices. There are also cases, referred to as bipolar configuration, in

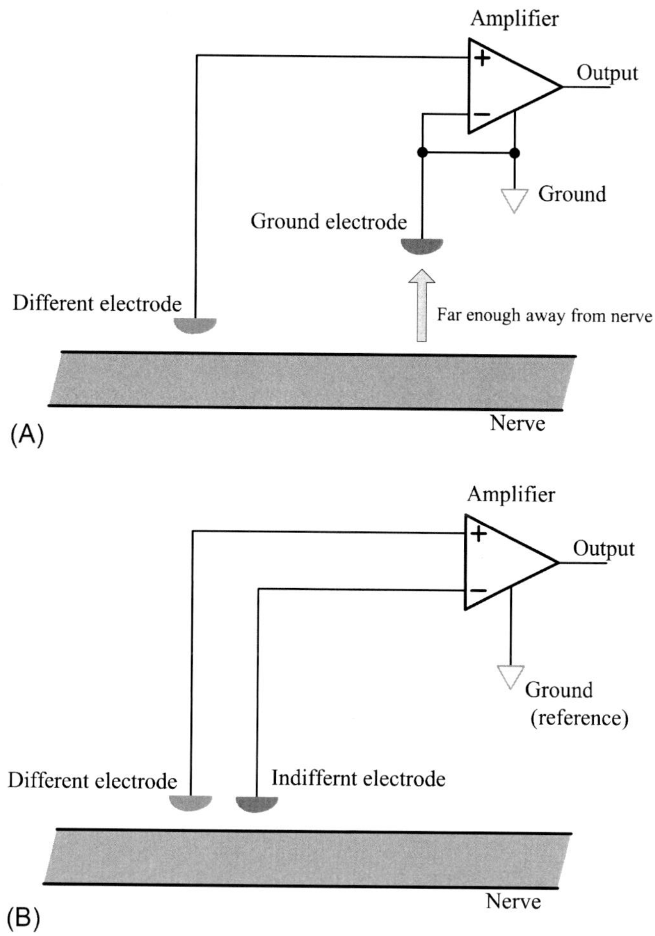

FIG. 6.7 Configuration of the two types of nervous activity recording.

which the return electrode is placed in the vicinity of the stimulating electrode. In some cases, referred to as monopolar configuration, return electrodes are located far from the stimulating electrodes. A great feature of bipolar stimulation is that the charge injected by the stimulating electrode is recovered by the neighboring return electrode, so the spatial spread of electrical current in the biological tissue is small. In other words, localized stimulation is possible. This is useful for applications that emphasize the need for high spatial resolution, such as visual prosthesis. Bipolar stimulation has the disadvantage of having a high threshold value, that is, greater electrical charge injection is required to induce a neural response compared to monopolar stimulation. On the other hand, the threshold for the monopolar configuration is lower than that of bipolar stimulation. However, the stimulation current is spatially spread in biological tissues, making stimulation less localized. Although using only one return electrode is typical for most applications, a plurality of independent return electrodes is employed to steer current path. Recently, a quasi-monopolar configuration that simultaneously uses both return electrodes of monopolar and bipolar configurations, has been proposed [43]. In this configuration, the return electrode in the vicinity of the stimulating electrode and the return electrode far from the stimulating electrode are used at the same time.

6.3.1 Theory

A stimulator voltage or current source is connected between the stimulating electrode and the return electrode. During electrical stimulation, a voltage is applied between the two electrodes to form a spatial distribution of potential gradients. Current flows according to the evoked potential gradient. A portion of this current reaches the target neural tissue to be stimulated and induces a neural response.

The potential within neurons is generally 60–70 mV lower than the potential outside the cell. This is called the resting membrane potential and represents the intracellular potential with reference to the extracellular potential. In normal neurotransmission, a number of synaptic inputs arrive at the same time to the dendrite of a neuron, making the membrane potential slightly high. If the membrane potential reaches the threshold value due to these synaptic inputs, the voltage-gated sodium channel opens, causing sodium ions to enter the cell, leading to a sharp rise in the membrane potential. This is called the action potential.

In electrical stimulation, the neuron is excited by forcibly changing the membrane potential instead of the above-mentioned synaptic inputs. Let us consider a case where cathodal stimulation is performed by the stimulating electrode, that is, a case where electrical current flows into the stimulating electrode. In this case, since the stimulating electrode has a negative potential with respect to the return electrode, the extracellular potential of the neuron near it is lowered. However, the intracellular potential of the neuron cannot immediately follow this potential fluctuation, so the membrane potential (the intracellular potential relative to the extracellular, as mentioned above) rises temporarily. In other words, neurons are depolarized by electrical stimulation. This is considered to open the voltage-gated sodium channel and evokes an action potential. On the contrary, when anodal stimulation is performed by the stimulating electrode (electrical current flows out of the stimulating electrode), the membrane potential of the neuron in the vicinity of the stimulating electrode is lowered ("hyperpolarized"). The mechanism of neural excitation by anodal stimulation is somewhat complicated. For more information about this mechanism (refer to Merrill et al. [44]).

It is an important to estimate the path and the intensity of the electrical current in a biological tissue during electrical stimulation. The current distribution is determined by the impedance of the living body between the stimulating and return electrodes. Since the impedance of the tissue is generally spatially nonuniform and nonlinear, the current distribution in the tissue is quite complex. Sometimes numerical simulation is employed to estimate such current distribution.

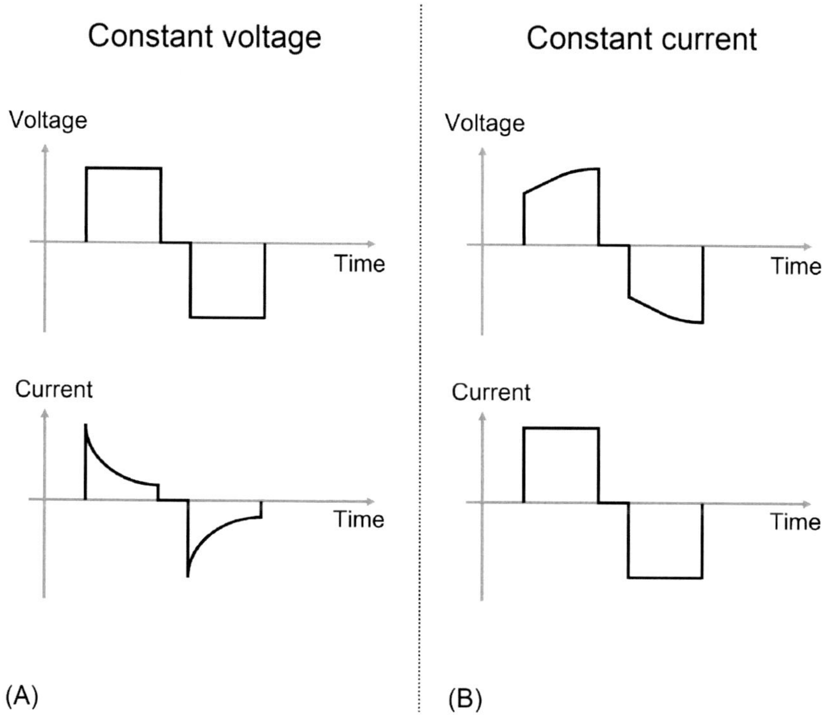

FIG. 6.8 Schematic description of a biphasic square wave pulse. (A) Constant voltage pulse. (B) Constant current pulse.

Constant voltage stimulation or constant current stimulation is generally used for electrical stimulation. In the constant voltage stimulation, a constant voltage is applied between the stimulating electrode and the return electrode during stimulation. In the case of constant voltage stimulation, the largest current flows immediately after the onset of stimulation. Thereafter, the current amplitude decreases exponentially (Fig. 6.8A). This is because a charged capacitor called an electric double layer exists between the electrode and the living tissue. On the other hand, in the case of constant current stimulation, the stimulator controls the voltage between the stimulating and return electrodes so that the current amplitude is constant during stimulation. In this case, the voltage between the stimulating and return electrodes immediately after the onset of stimulation is low but rises gradually (Fig. 6.8B). This rise corresponds to the above-mentioned double-layer charging.

In practical applications, it is extremely important to perform safe stimulation without damaging the biological tissue. Charge-balanced biphasic current pulses are widely used for safety, but charge balance alone is not enough. If the charge density (the value obtained by dividing the electrical charge of one phase of stimulation pulse by the electrode surface area) is excessive, or the charge amount (not charge density) is excessive, tissue damage could be induced. Both charge density and electrical charge affects the occurrence of tissue damage, and the relationship is the well-known Shannon's Law [45, 46].

Furthermore, it is important to understand the concept of charge injection capacity (CIC), also called charge limit or charge delivery capacity (CDC). The CIC is the maximum electrical charge density that can be injected without causing an irreversible electrochemical reaction. For example, consider applying an anodic-first biphasic current pulse (Fig. 6.8B) to a platinum stimulating electrode. In this case, the double layer is charged at the beginning of the first anodal phase. Continued flow of current further causes an oxidation reaction at the platinum surface. Finally, oxygen evolved at the electrode surface through the electrolysis of water. When the anodal phase ends and the cathodal phase begins, discharge of the double layer first takes place followed by the monatomic adsorption of hydrogen atoms on the electrode surface. When further electric current is applied, hydrogen gas is generated through the electrolysis of water. Among the above-mentioned reactions, the generation of oxygen and hydrogen is an irreversible reaction. Irreversible reactions should be avoided for safe electrical stimulation. This is because a reversible electrochemical reaction (e.g., adsorption of hydrogen atoms to the platinum surface) returns to the original state when the direction of current flow is reversed. This does not happen when the reaction is irreversible. Considering the case whereby oxygen is generated by the anodal phase of electrical stimulation, the generated oxygen diffuses into the body fluid, so that even if the direction of current is reversed thereafter, it does not return to water but rather leads to some other electrochemical reaction. When an irreversible reaction occurs, the pH near the electrode fluctuates and raises the risk of damage to the surrounding tissue. For example, in the oxygen evolution reaction, four hydrogen ions are generated per oxygen molecule due to the reaction

$$2H_2O \rightarrow O_2 + 4H^+ + 4e^- \tag{6.2}$$

This makes the body fluid in the vicinity of the electrode acidic. Other irreversible reactions such as corrosion of the stimulating electrode could also occur. In order to avoid such problematic reactions, it is important to perform electrical stimulations within the range of the CIC. It should be noted that the electrochemical reactions that may occur during electrical stimulation in real situations are extremely complicated and the above explanation is just a simplified version. For more detail on the CIC with platinum electrode interested reader may find reference [47] useful. It should equally be noted that the CIC is not a fixed constant determined by the electrode material, but depends on the measurement environment such as in vitro/in vivo [48] and the stimulation pulse parameter used for the CIC measurement [49].

6.3.2 Stimulation Electrode

Platinum is the most widely used electrode material. Platinum has higher charge injection capability than gold and is chemically stable. It is also advantageous that platinum electrodes can be formed by various methods such as sputtering, electroplating, and mechanical machining. Platinum black formed by electroplating is widely used as an electrode for experiments. Platinum black achieves extremely large surface area by precipitating granular platinum on the surface of the metal electrode, and its surface area is typically three orders of magnitude larger than the apparent surface area. If the surface area is large, the above-described CIC will also be high. However, platinum black is not used for long-term implantable devices because it is mechanically fragile. In addition, the toxicity of lead contained in the plating solution is a cause for concern [22]. Recently, platinum gray that uses electroplating similar to platinum black and can be formed without using lead has been proposed, and has been applied to a retinal prosthesis [50].

Titanium nitride (TiN) is also a commonly used electrode material. With TiN, charge injection is performed only by the charging/discharging of the double layer. For this reason, TiN on a smooth surface has extremely low CIC and is not practical as an electrode material. However, by choosing appropriate conditions for depositing a TiN film with sputtering, it is possible to create a very porous surface that has a high surface area. It is reported that porous TiN has a higher CIC than platinum [51].

Iridium oxide is one of the best electrode materials that have the highest CIC. Iridium oxide can inject electrical charge through reversible oxidation and reduction reactions between the Ir^{3+} and Ir^{4+} states of the oxide [51]. Several methods for fabricating iridium oxide are known. One of the most commonly used methods is called activation. To activate iridium, repeated voltage sweeps are applied in saline to pure iridium electrodes. This is called AIROF (activated iridium oxide film). In addition, other methods such as reactive sputtering (SIROF: sputtered iridium oxide film), electrochemical deposition method (EIROF: electrochemically formed IROF), and thermal method (TIROF: thermally formed IROF), are known. Although iridium oxide has a high CIC, it is known that delamination or deterioration of the iridium oxide film occurs when excessive voltage is applied. Therefore, it is important to control the electrode potential during current application.

In recent years, PEDOT [poly(ethylenedioxythiophene)], a conductive polymer has attracted attention as an electrode material. There are also reports that it has higher CIC than iridium oxide. In order to introduce electrical conductivity to PEDOT, it is necessary to mix it with an electron donor, and in many cases PSS [poly(styrene sulfonate)] is used. One of the advantages of PEDOT is that it can electropolymerize and electrodeposit on the electrode surface like the EIROF. Although PEDOT is not new as an industrial material, its history as an electrode material for neural interfaces is not long known. Its charge injection mechanism and its long-term stability as an electrode for neural stimulation are not sufficiently investigated and further research is expected in the future.

6.4 Optical Recording and Stimulation

Electrical recording and electrical stimulation of nerves have many advantages because of their high temporal resolution and ease of control. MEA applying micro electro mechanical systems (MEMS) technology made it possible to stimulate and record with such high spatial resolutions that would be unthinkable in the past. However, the structure of living tissues is finer beyond the current electrode fabrication techniques. The wiring for establishing an electrical connection between the electrode and the electronic circuit becomes more difficult to manufacture as the number of electrodes increases. In addition, miniaturization of stimulation electrodes has a physical limitation due to CIC. Optical recording and stimulation offers the possibility of realizing high spatial resolutions that cannot be achieved by the electrical approach [9, 52].

6.4.1 Optical Recording

Oxygen consumption is higher in regions where nerve activity is higher. By utilizing the difference in absorbance of oxygenated-hemoglobin (oxy-Hb) and deoxygenated-hemoglobin (deoxy-Hb) in near-infrared (NIR) light, it is possible to measure nerve activity. fNIRS is a noninvasive measurement method applying this principle. Laser Doppler flowmetry is another method of measuring nerve activity by the oxygen demand of the tissue, through the measurement of blood flow. When the nerve is activated, there will be a change in the reflection index of neuronal membranes and the geometry of neuronal components. By acquiring these signals by functional optical coherence tomography (fOCT), nerve activity can be analyzed. A surface plasmon resonance (SPR) sensor has been developed as a recording device. It detects the change in the refractive index of a thin metal layer. There are also more direct methods for recording of nerve activity using optical probes such as calcium indicators and voltage-sensitive probes (VSFPs). Such optical measurement techniques can be applied to both CNS and PNS [53, 54]. In addition, a method of implanting an integrated circuit (IC) in which light emitting and receiving elements are designed into a single chip for obtaining higher definition information has also been developed (Fig. 6.9) [54, 55].

6.4.2 Optical Stimulation

The study of optical neural stimulation began with the genetic modification of target nerves to introduce photosensitive receptors. By expressing channelrhodopsin-2 (ChR2), a rapidly gated blue (470 nm) light-sensitive cation channel single spike can be evoked by light stimulation [56]. Volvox-derived channelrhodopsin-1 (VChR1) (589 nm) and red-activatable channelrhodopsin (ReaChR) (610 nm) responsive to longer wavelength light were also actualized [57, 58].

Photosensitivity can be provided to nerves using neurotransmitters held in a photosensitive cage such as caged glutamine [59]. Approaches that provide photoresponsiveness by bonding optical isomers such as benzene to ion channels have also been developed [60].

It was reported that nerves can be stimulated by a mid-infrared light pulse whose wavelength is around 2000 nm [61]. This technique, called infrared neural stimulation (INS), is advantageous in invasive procedures because there is no modification of the target nerve. However, since the wavelength of the light used for stimulation is in the absorption wavelength band of water, there are problems of penetration and heating of the tissue.

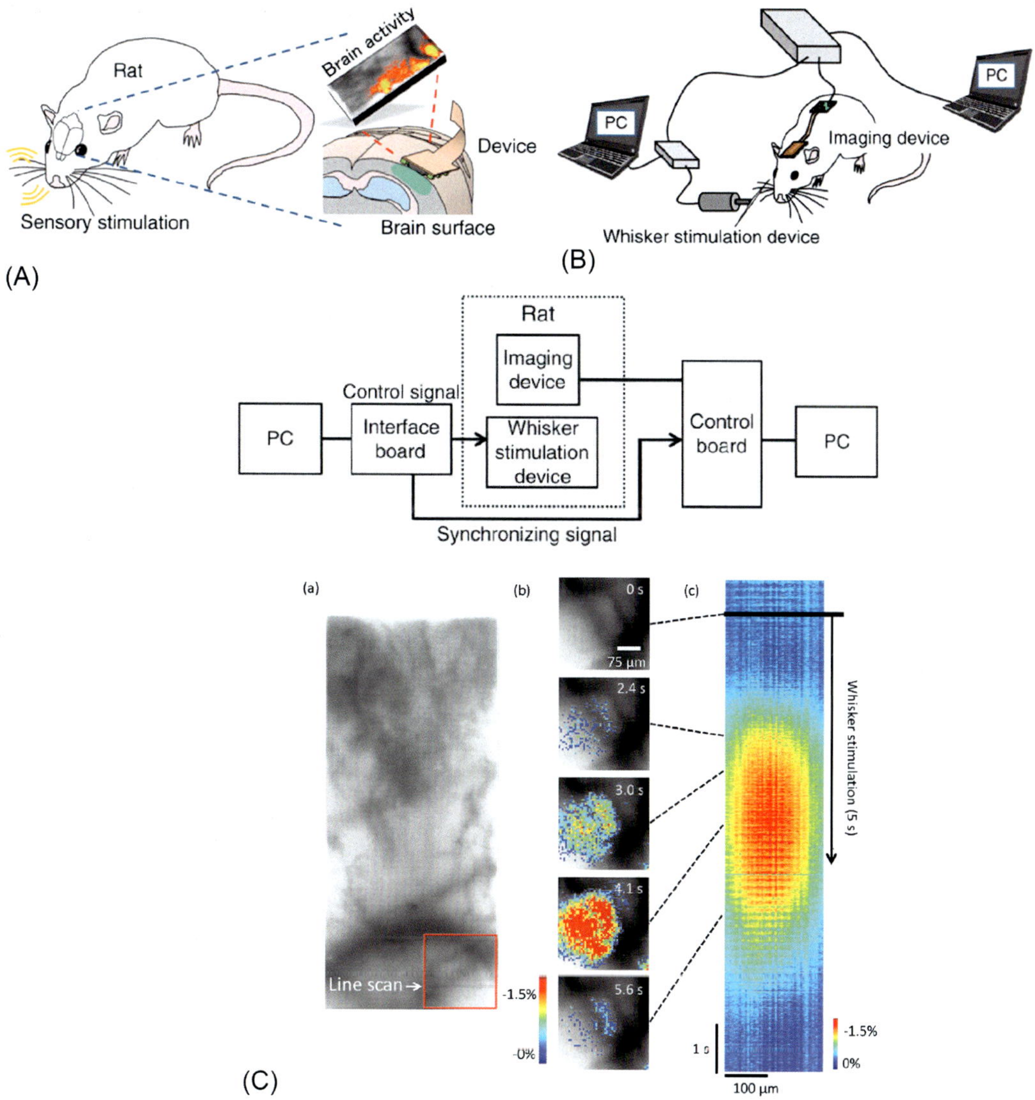

FIG. 6.9 Intrinsic signal imaging of brain function using a small implantable CMOS imaging device. (A) Small implantable CMOS imaging device. (B) A small implantable CMOS imaging system. (C) Flow diagram of the imaging system. (D) The intrinsic signal imaging with whisker stimulation. (A) The brain surface image was captured by the device. A square frame indicates a measurement area. (B) Images of the intrinsic signal responses of the right-whisker region of the left primary somatosensory cortex. 0 s indicates the stimulation onset. (C) Line-scan image of (B). The arrow indicates the sensory stimulation. The circular area indicates the activated brain surface [54]. *Copyright 2015, The Japan Society of Applied Physics.*

6.5 Applications of BMI

6.5.1 Overview of BMI Application

Unfortunately, we still know a very little about human brain's "hardware" and "software". Neuroscience and engineering are yet in early rudimentary stage. For example, neuron doctrine, that is, the basic concept that the nervous system is made up of discrete individual cells—neurons, was introduced based on scientific observations only 130 years ago through the work of Santiago Ramón y Cajal in 1888. This concept was finally confirmed in the 1950s through advent of electronic microscope. Today, it is rather common knowledge that nerve cells are indeed individual cells physically interconnected through synapses to form a tangible nervous system "bathed" within dynamic and reach-in-detail local electromagnetic field.

The current estimate is that there are around 100 billion ($\sim 10^{11}$) neurons in the human brain. Each neuron may be directly connected to up to 10,000 other neurons through electrical and chemical synapses. Each neuron can be also indirectly connected to several more thousands of neurons through ephaptic coupling via extracellular electric fields. A typical neuron fires 5–50 times every second. Hence, the number of these more discrete spiking processes within the brain is rather outstanding!

Regrettably, we still do not know what the overwhelming majority of these signals really represent. We have some understanding of neuro-geography, that is, in which areas of the brain some processes take a place. In addition, we have rudimentary preconcepts and a number of yet-to-be-tested hypotheses on how various neural networks function or how different memories are formed, stored, and further processed. However, we are still a long way from truly understanding the brain.

BMI is defined here as a communication pathway between a physically enhanced or wired brain and a device. BMI field obviously has a lot of overlap with neuroprosthetics, which is an area of neuroscience concerned with neural prostheses, that is, devices to replace the function of impaired nervous systems and brain related problems, or of sensory organs.

The most widely used neuroprosthetic device is the cochlear implant which, as of December 2012, had been implanted in more than 320,000 people worldwide [62]. There are also several neuroprosthetic devices that aim to restore vision, including retinal implants.

Not surprisingly, advancement of BMI field is not only helping accelerate acquiring fundamental knowledge on human brain's "hardware" and "software" but it is also opening a door to a wide variety of potentially very humane applications. These applications include introducing the possibility to abridge one's spinal cord injury (SCI) so that one can operate own limbs via braces or functional electrical stimulation (FES) of own muscles, or empowering amputee with opportunity to fluidly operate prosthetic limbs while potentially feeling what external sensors are "feeling" or BMI can be also used to drive wheelchair. Moreover, BMI and neuroprosthetics applications include giving someone the ability to hear (cochlear implants), see (retinal prosthesis), detect onset and possibly inhibit epileptic seizure, or potentially treat depression or address diseases like Parkinson, Alzheimer, Multiple Sclerosis, etc.

Advancements of BMI are also bringing us closer to the imaginative worlds of 20th century cyber pop culture including the concept of cyberspace as described by William Gibson's "Neuromancer" or by the "Matrix" movie trilogy, concept of telepresence as addressed by the movie "Avatar", or concepts of newly defined sense of identity in relation to afresh synthetic bodies and minds as illustrated by Isaac Asimov's "I, Robot" and Masamune Shirow's "Ghost in the Shell".

It seems very natural in the context of evolution (at least to one of the authors—MP) that the human race will eventually augment its abilities affecting quality of life and lifespan, and redesign, exchange and add new brain features while redefining its own identity.

6.5.2 Invasive BMI

There are more and less invasive BMI. A good example of more invasive BMI is a cortical implant which is in direct connection with the cerebral cortex of the brain. Among those, intracortical implants have electrodes that penetrate into the brain, while epicortical implants have electrodes that record fields along the surface. Since the intracortical implants go deeper into the brain, they typically require a stiffer electrode. However, due to micromotion in the brain, some flexibility is necessary in order to prevent injury to the brain tissue.

After several decades of experimenting mainly with primates, the intracortical BMI for restoring motor functions has been applied to human brain at the turn of century. One of the intracortical BMI-related research highlights which received a lot of media attention is the BrainGate project led by Brown University professor John Donoghue who also cofounded an early neurotechnology startup company, Cyberkinetics. In collaboration with Massachusetts General Hospital and Department of Veterans Affairs (VA), BrainGate project involved four tetraplegic study subjects, three of which suffered advanced amyotrophic lateral sclerosis (ASL) and one of which was a young man with SCI, C3 tetraplegic Matthew Nagle from Boston paralyzed from the neck down after being stabbed in 2001 at age 22 (Fig. 6.10) [14].

The implanted interface was a rather simple array of 96 hair-thin electrodes ("Utah Array") that sense the electromagnetic signature of nearby neurons firing. The MEA was surgically placed on the surface of his brain over the region of the motor cortex that controlled his dominant left hand and arm by neurosurgeon Gerhard Friehs in June 2004. The implant was connected to a computer by wires going through a small opening drilled in his skull. The computer was then trained to recognize Nagle's thought patterns and associate them with movements he was trying to achieve. This is an example of unidirectional devices as signals were not fed back to the motor cortex. Per FDA regulations and study protocol, the device was removed approximately a year later.

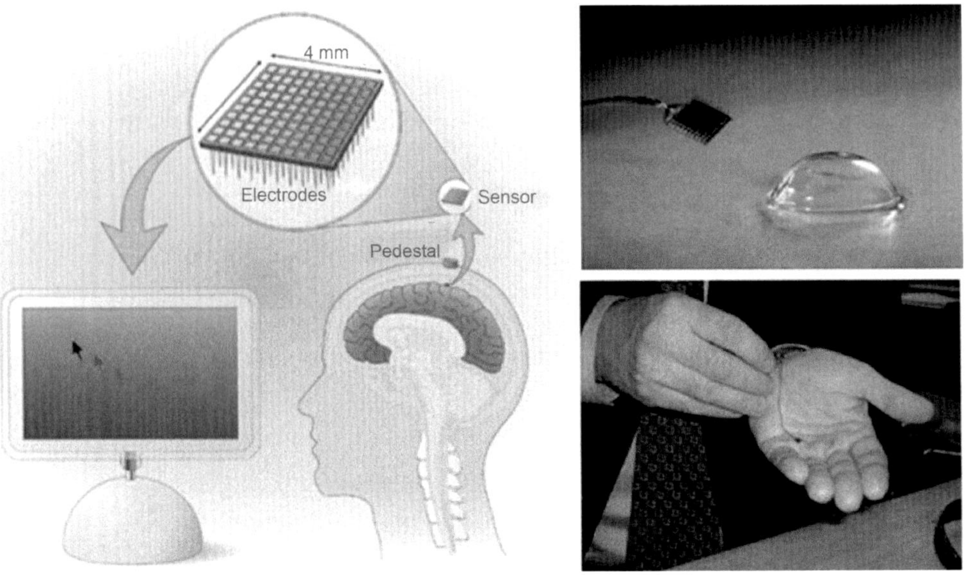

FIG. 6.10 BrainGate system and multielectrode array as example of intracortical implant [14, 63]. *Credit BrainGate, www.braingate.org. Courtesy of Leigh R. Hochberg, M.D., Ph.D.*

The results were rather amazing and they were probably equally a consequence of advanced computer technology as they were due to the ability of Matthew's brain to figure out how to use this simple interface. BrainGate implant allowed Nagle to control a robotic arm by thinking about moving his hand as well as to operate a computer cursor, lights, email, and TV. Matthew was the first person to use a BMI to restore functionality lost due to paralysis [64]. The results of this study were published in journal Nature in 2006 [14]. Nagle, shown in Fig. 6.11 on the left, died on July 24, 2007 in Stoughton, Massachusetts from sepsis. The BrainGate technology and related Cyberkinetic's assets are now owned by BrainGate, Co.

In May 2012, BrainGate researchers published another study in Nature demonstrating that two people paralyzed by brainstem stroke could control robotic arms for reaching and grasping [65]. One participant, Cathy Hutchinson, who has been unable to move or speak for 15 years, had the 96-channel sensor implanted in her brain's motor cortex in 2005. Hutchinson, shown in Fig. 6.11 on the right demonstrated ability to use the robotic arm to drink coffee from a bottle at her Boston home in Dorchester, Massachusetts [65, 66].

In December 2014, a group from the University of Pittsburgh reported on their experiment involving Jan Scheuermann, who has longstanding quadriplegia and had been taking part in the study for over 2 years. She had gone from giving "high fives" to the "thumbs-up" after increasing the maneuverability of the robotic arm from seven dimensions (7D) to 10 dimensions (10D) [67].

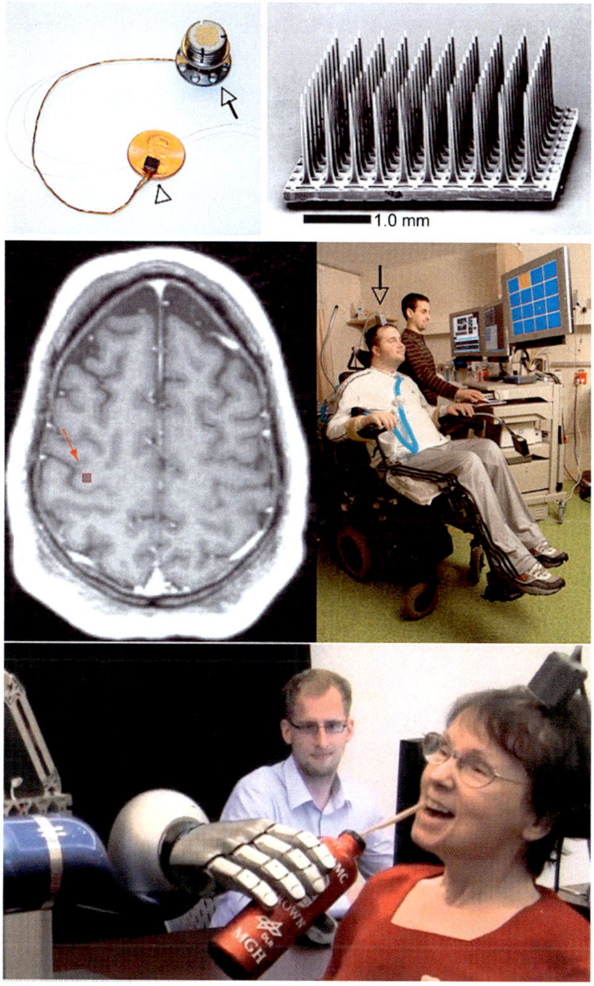

FIG. 6.11 BrainGate study experiments reported in 2006 (top) and 2012 (bottom) [14, 65, 66]. *Credit BrainGate, www.braingate.org. Courtesy of Leigh R. Hochberg, M.D., Ph.D.*

In addition to addressing motor functions, Intracortical implants also include a wide variety of potential other uses, for example, restoring vision to blind patients (by interfacing with visual cortex) or helping patients suffering from dementia (by interfacing with hippocampus believed to have important role for brain's full long-term memory capabilities).

Still further, in regard to motor function, there can be potentially better areas to place electrodes. In May 2015, the University of Southern California reported: "Through a clinical collaboration between Caltech, Keck Medicine of USC and Rancho Los Amigos National Rehabilitation Center, a 34-year-old paralyzed man is the first person in the world to have a neural prosthetic device implanted in a region of the brain where intentions are made, giving him the ability to perform a fluid hand-shaking gesture, drink a beverage, and even play 'rock, paper, scissors,' using a robotic arm." A pair of small electrode arrays were implanted in two parts of the posterior parietal cortex (PPC), one that controls reach and another that controls grasp [68].

Research on visual cortex-based BMI already started in early 1970, however many researchers believe that interfacing with retina or optic nerve might be easier to deal with. Here, optical sensory information is externally obtained and then imprinted to brain.

Researchers are also trying to determine the neural basis for memory by finding out how the brain encodes different memories in the hippocampus. In people whose brains have suffered damage from Alzheimer's, stroke, or injury, disrupted neuronal networks often prevent long-term memories from forming. Theodore Berger from the University of Southern California in Los Angeles designs chips that mimic the signal processing that he believes allows us to recall experiences and knowledge for more than a minute. Experiments with rats and monkeys brains brought some optimism. But Berger and his team have yet to conduct human tests. Here, the idea is to implant a bidirectional device that both records and plays processed signals back to brain [69]. In 2016 Berger also served as science advisor and Chief Technology Officer with $100 million startup Kernel founded by entrepreneur Bryan Johnson to enhance human intelligence by developing brain implants capable of linking people's thoughts to computers. Kernel recently acquired the MIT spin-off firm KRS, which is promising to make a data-driven revolution in understanding neurodegenerative diseases. However, it is unclear if Kernel's focus is still in direction of development of memory neuroprostheses, potentially due to difficulties related to invasive technologies [69].

Another famous entrepreneur also joined this neurotechnology startup wave. Elon Musk, Tesla and SpaceX founder, launched the company Neuralink in June 2016, registered in California as a "medical research" company, and focused on developing ultrahigh bandwidth BMIs, or "neural-laces" according to Wall Street journal, to connect humans and computers. Some speculate that this may have something to do with his beliefs that a "merger of biological intelligence and machine intelligence" would be necessary for humans to compete with machines [70–72].

One of the top challenges with developing stable intracortical BMI is related to biocompatibility or how the body will respond to a foreign object. And brain submerged into cerebrospinal fluid moves a lot such that too stiff electrodes may cause development of scar tissue around the electrodes. Most MEAs require neuronal cell bodies to be within 50 μm of the electrodes to provide the best function, and studies have shown that implanted animals have significantly reduced cell density within this range [28].

In February 2017, researchers at the University of Texas at Austin reported on new ultra-flexible, nanoelectronic thread (NET) brain probes that can achieve more reliable long-term neural recording than existing probes and do not elicit scar formation when implanted [29].

Another approach as reported by the University of Melbourne researchers in February 2016 might be the BMI consisting of a stent-based electrode (stentrode), which is implanted within a blood vessel in the brain, and records the type of neural activity that has been shown in preclinical trials to move limbs through an exoskeleton or to control bionic limbs [30].

In October 2015 researchers at Lund University also reported on new tailored electrodes, which they call three-dimensional (3D) electrodes, that are unique in that they are extremely soft and flexible in all three dimensions, in a way that enables stable recordings from the neurons over a long time [31].

In difference to the intracortical implants, the ECoG, or intracranial electroencephalography (iEEG) uses electrodes placed directly on the exposed surface of the brain to record electrical activity from the cerebral cortex but without penetrating the blood-brain barrier like invasive recording devices. It is said that ECoG provides a partially invasive compromise between the already discussed invasive techniques and a noninvasive EEG approach. Advantage over invasive methods is that it could cover larger areas and it can be moved from one area of brain to another and the advantage over EEG methods is that it features a higher spatial resolution and higher S/N ratio (as being positioned much closer to brain and without being obstructed by the presence of signal strength dampening skull and possibly dura matter, that is, thick membrane that is the outermost of the three layers of the meninges that surround the brain and spinal cord, and being responsible for keeping in the cerebrospinal fluid).

ECoG measures the electrical activity of the brain taken from beneath the skull in a similar way to noninvasive electroencephalography, but the electrodes are embedded in a thin plastic pad that is placed above the cortex, beneath the dura mater [73]. ECoG technologies functioning as BMI were first trialed in humans in 2004 by Eric Leuthardt and Daniel Moran from Washington University in St Louis. In a later trial, the researchers enabled a teenage boy to play Space Invaders using his ECoG implant [74].

The ECoG recording is performed from electrodes placed on the exposed cortex. In order to access the cortex, a part of the skull needs to be removed to expose the brain surface. Electrodes are then surgically implanted on the surface of the cortex; electrodes may either be placed outside the dura mater (epidural) or under the dura mater (subdural).

ECoG electrode arrays are typically arranged in a “crown” geometry consisting of 16 electrodes, each mounted on a ball and socket joint for ease in positioning. Transparent and flexible subdural strips and grids with electrodes are also widely used in various dimensions, having anywhere from 4 to 256 electrode contacts. Standard spacing between grid electrodes is 1 cm; individual electrodes are typically 5 mm in diameter. The electrodes sit lightly on the cortical surface to ensure that normal movements of the brain do not cause injury. A key advantage of strip and grid electrode arrays is that they may be slide underneath the dura mater into cortical regions not exposed by the craniotomy [75, 76].

Direct cortical electrical stimulation (DCES) is often used with ECoG; these currents are between 2 and 4 mA for somatosensory stimulation, and near 15 mA for cognitive stimulation. It is believed that the presence of these currents improves signal strength for at least several hours [77].

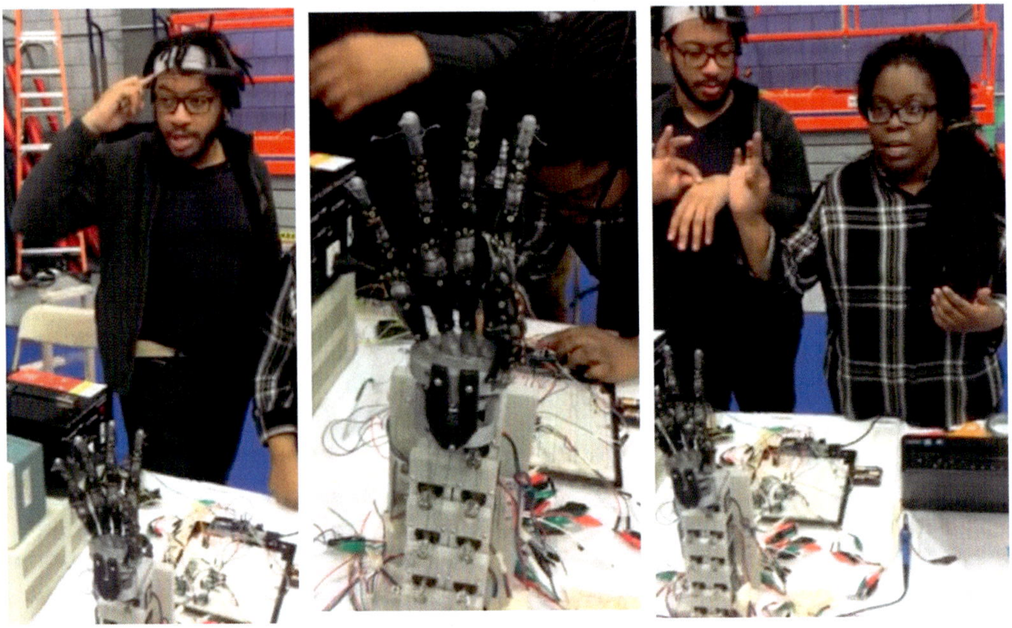

FIG. 6.12 WPI Popovic Labs EEG controlled accurate prosthetic hand demonstration at 2017 Cambridge Science Festival [82].

It seems that the same is true for noninvasive methods. A group of researchers from Singapore and Australia reported in December 2015 that EEG recordings following transcranial direct current stimulation (tDCS)—the application of an external electric current to the skull, provided much stronger deciphering of stroke patient's intent than without it in a study on 19 stroke patients with impaired arm movements [78].

While the ECoG is still considered to be the "gold standard" for defining epileptogenic zones, new research shows that the integration of structural MRI and scalp EEG might provide a noninvasive alternative to ECoG [79].

In October 2016, the University of Washington researchers at the National Science Foundation Center for Sensorimotor Neural Engineering (CSNE) reported on, for the first time in humans, use of brain surface stimulation (utilizing ECoG) to provide 'touch' feedback to direct movement [80].

Pure noninvasive EEG system can be also powerful BMI. In September 2015 researchers from the University of California, Irvine reported on the participant, who had been paralyzed for 5 years, and who walked along a 3.66-m-long course using an EEG-based system. The EEG-based BMI would send FES signal to stimulate muscles in his legs [81].

One of the authors also worked on noninvasive BMI. Author's group at WPI designed a highly biomimetic prosthetic hand. Using the commercially available Neurosky sensor it was shown that three distinct EEG patterns could successfully trigger execution of three

Table 6.2 Qualitative Approximate Evaluation on Characteristics of Noninvasive Functional Brain Imaging Methods

Characteristics	Low ◀			▶ High
Temporal resolution	PET	fMRI NIRS	EEG	MEG
Spatial resolution	NIRS	EEG	PET MEG	fMRI
Ease of use	fMRI PET	MEG	NIRS	EEG
Portability	fMRI MEG PET		EEG NIRS	
Cost of method	fMRI MEG PET		NIRS	EEG

distinct hand actions: pinch, hook, and point with 68% success rate. The system had been showcased at the 2017 Cambridge Science Festival using EEG signals alone. Out of about 30 visitors who tested thought control capabilities of the accurate prosthetic hand, the majority was able to generate a peace sign after trying the system for only a minute or two (Fig. 6.12) [82].

6.5.3 Noninvasive BMI

For a noninvasive BMI, acquiring brain signals from most individuals with minimal health risks is important. Therefore, noninvasive BMI techniques are based on noninvasive neuroimaging methods such as EEG, MEG, fNIRS, and fMRI. EEG and MEG can detect electrical activity generated by cortical neurons. By contrast, fNIRS, PET, and fMRI can detect hemodynamic responses generated by blood activity. Each measurement technique has different features related to spatial and temporal resolution. Table 6.2 lists a summary of the features of each neuroimaging method.

Among these imaging methods, owing to its high temporal resolution, low inter- and intra-subject variability, ease of use, low cost, and user safety, EEG is the most widely used method to detect the cortical activity using noninvasive BMI systems [83]. EEG signals have a very high temporal resolution alongside a minimally invasive acquisition protocol.

There are two types of EEG signals: evoked signals and spontaneous signals. Evoked signals are generated unconsciously as a response to external stimuli. Spontaneous EEG signals are generated by voluntary cortical activity without any external stimulation. Evoked EEG signals can be applied to exogenous BCIs. Exogenous BCIs have the advantages of requiring minimal training, having an easy and quick set up for signal measurement, and featuring a high bit rate of information transfer. However, they have the disadvantage of requiring attention to the external stimuli causing the signal. By contrast, endogenous BCIs used for measuring spontaneous EEG signals have the advantages of being independent to any stimulus, and operated at free will. However, they have the disadvantages of requiring training, not all users being able to control them, and having a lower bit rate of information transfer.

Among several types of evoked potentials, steady-state visual evoked potential (SSVEP) is commonly used in many BCI applications. SSVEPs are EEG signals that are generated when subjects perceive continuous periodic visual, auditory, or somatosensory stimuli. EEG signals are modulated by a flickering light, sound, or vibration. The EEG signal frequency is increased in order to reach the stimulus frequency [84]. Thus, SSVEPs are elicited by the repetitive visual stimuli. If the stimulus is a flickering image, the SSVEP at the visual cortex will have the same frequency of the flickering image. The flickering frequency is usually between 6 and 30 Hz [85].

BCIs based on SSVEP are applied to select a target by means of an eye gaze. The user visually fixes attention on a target and the BCI identifies the target through SSVEP features analysis [12]. SSVEP BCIs are classified into three categories based on the types of modulation: time modulated VEP (t-VEP) BCI, frequency modulated VEP (f-VEP) BCI, and pseudorandom code modulated VEP (c-VEP) [85].

In a t-VEP BCI, the flash sequences of the different targets related to the different stimulus are orthogonal or near orthogonal to each other. However, there is no user training in this system. In a f-VEP BCI, each target is flashed at a unique frequency. It generates a periodic sequence of evoked responses with the same frequency of stimuli. In a c-VEP BCI, pseudo-random sequences that determine the duration of ON and OFF are used. Signal modulations can optimize the information transfer rate of the BCI. t-VEP BCI has a low information transfer rate, which is almost less than 30 bits/min; f-VEP has a high information transfer rate: 30–60 bits/min; and c-VEP BCI has a very high information transfer rate of more than 100 bits/min [12].

P300 is an evoked potential that has a positive peak after 300 ms when the subject is exposed to infrequent auditory, visual, or somatosensory stimuli. The endogenic P300 response is generated by "odd-ball" stimulus among several frequent stimuli [86, 87]. The less probable the stimulus is, the larger the amplitude of the response peak it generates [88]. The use of P300-based BCI does not require any subject training; however, it requires repetitive stimuli. A typical application of P300-based BCI is an input of letters, numbers, symbols, or commands [89, 90]. The user of this BCI gazes at the desired letter while the rows and columns of this matrix are flashing randomly. The P300 response appears when the row or column which includes the target letter is flashed. As the S/N ratio of this response is low, the responded signals must be averaged over the trial in order to improve accuracy of choice.

Sensory motor rhythms are oscillations that have mu band frequency (7–13 Hz) and beta band (13–30 Hz) in the brain activity. Amplitude of sensory motor rhythms is related to not only any real motor task, but also imagery of motor activity in the brain. Thus, motor intention can modulate mu or beta signals without any motor output. In BCI methods, two kinds of amplitude modulations are used: event-related desynchronization (ERD) and event-related synchronization (ERS), which are caused by real motor activity or mental imagery [91]. ERD is the suppression of the amplitude of mu rhythm and ERS is an enhancement of amplitude. It is possible to use ERD and ERS for the design of endogenous BCIs, which are more useful than exogenous BCIs such as P300 VEP-based BCIs. Although the user might not need any training, it is recommended to take training before using ERD and ERS-based BCI systems in order to increase accuracy and speed of command control. Motor imagery which causes ERD or ERS is considered to be a useful cognitive tool for improving motor skills as well as for rehabilitation therapy of movement disorders [92, 93]. BCI systems were developed as an assistive technology for people with severe paralysis or suffering from lock-in syndrome. However, these systems are proposed as tools for diagnosing or treating many neurophysiological and neuropsychiatric disorders such as consciousness disorder (vegetative state, minimally conscious state, locked in syndrome) [94], ADHD [95], schizophrenia [96], and autistic spectrum disorder (ASD) [97].

NIRS is an optical spectroscopy that employs NIR light to measure the changes of the localized blood volume and oxygenation during neural activity [98]. NIRS can measure hemodynamic variations, namely variations in oxygenated hemoglobin (HbO) and deoxygenated hemoglobin (de-HbO) concentrations, in local cortical regions in real time. This occurs because the levels of absorption and reflection of NIR light by hemoglobin vary depending on the amount of hemoglobin combined with oxygen that exists in those regions [99]. NIR light penetrates the skull to a depth of approximately 30 mm below the surface of the scalp. The absorption of the infrared light photons is modulated by the fluctuations in tissue oxygenation [99]. Due to the limited depth of NIR light penetration, fNIRS can detect the activity of the outer cortical layer. Although the spatial resolution is quite low compared to fMRI, NIRS has a potential modality for BMI methods because of low cost, high portability, and an acceptable temporal resolution [12]. In addition, the fNIRS operation is easier than the EEG, which uses multiple electrodes. A limitation of using fNIRS for BCI is that the information transfer speed is limited by the inherent delay in the hemodynamic response. A High-density multi-distance source-detector NIRS BCI has been proposed in order to enhance performance [100]. Since the speed of the EEG can be utilized, the use of hybrid EEG-fNIRS systems has the potential as a noninvasive, portable, and wearable BCI system superior to EEG-BCI or fNIRS-BCI. The hybrid fNIRS-EEG system improves the classification accuracy and increases the number of control commands for BCI. Significant enhanced performance of classification accuracy [101] and optimal time slot for command generation [102] has been demonstrated by combining both fNIRS and EEG.

Although it differs from BMI, which operates computers or machines using brain signals, neurofeedback technologies that use brain signals as feedback signals to teach self-regulation of brain functions has progressed rapidly in recent years.

The neurofeedback-based fMRI is a type of biofeedback that self-regulates brain functions using real-time fMRI signals. The first fMRI neurofeedback system was developed in 2003 [103]. Using implicit protocols, external rewards, multivariate analysis, and connectivity analysis, it has been possible to explore causal involvement of modified brain activity and modified behavior. These techniques are integrated into innovative neurofeedback technology, decoded neural feedback (DecNef), and functional connection-based neural feedback (FCNef) [104].

DecNef is a method for reading brain information from brain signals using machine learning based on the specific brain activity pattern and performing feedback of the information to the subject in real time. Changes in the subject's perception and behavior can be found by inducing the activity of a specific brain region (target region) to a specific pattern (target pattern) by the DecNef method. The activities of the brain region used in the DecNef method have shown to cause the corresponding perception, cognition, and behavior changes [105].

FCNef was developed in 2015. FCNef changes the temporal correlation of fMRI signals between two brain regions in a network without the participants being aware of the meaning of neurofeedback provided to them [106]. Functional connectivity in the brain can be changed in both directions using FCNef training.

DecNef and FCNef have the unique ability to explore causal relationships between targeted brain functions and resultant behavioral changes, particularly via implicit protocols. DecNef and FCNef have great potential for interventions in psychiatric disorders which may be caused by abnormal brain functions and connectivity. The goal of DecNef is for subjects to learn to induce a specific activity pattern within a target brain region. By contrast, the goal of FCNef is to change specific functional connectivity between two brain regions.

For mental disorders such as schizophrenic disorder, bipolar disorder, depression, post-traumatic stress disorder (PTSD), and ASD, there seems to be abnormal connectivity in the brain functions. DecNef and FCNef are expected to contribute to the treatment of mental disorders and the development of new rehabilitation/learning support methods making full use of brain plasticity. FCNef has produced preliminary results in patients with depression as well as in adults with high-functioning ASD [107–109].

One of the most important applications of BMI is poststroke rehabilitation, which consists of repetitive motor training. The role of BMI is the enhancement of brain plasticity caused by motor training. Other kinds of techniques that enhance brain plasticity are noninvasive brain stimulations such as tDCS or repetitive transcranial magnetic stimulation (rTMS). tDCS stimulates the brain by a weak direct current through two electrodes, anode, and cathode. Anodal tDCS induces an increase of the neuronal excitability. By contrast, a cathodal tDCS causes neuronal inhibition. The rTMS can also stimulate the brain nerve and can cause the enhancement or inhibition of neuronal excitability by changing the stimulus frequency. These stimulus methods, tDCS and TMS, have the potential for improving recovery of motor functions using the brain's plasticity.

Combining tDCS and BMI or rTMS and BMI may improve neuroplasticity and functional recovery in poststroke patients with motor impairments. Hon et al. reported that combining tDCS and BMI may facilitate neuroplasticity, and suggested the potential

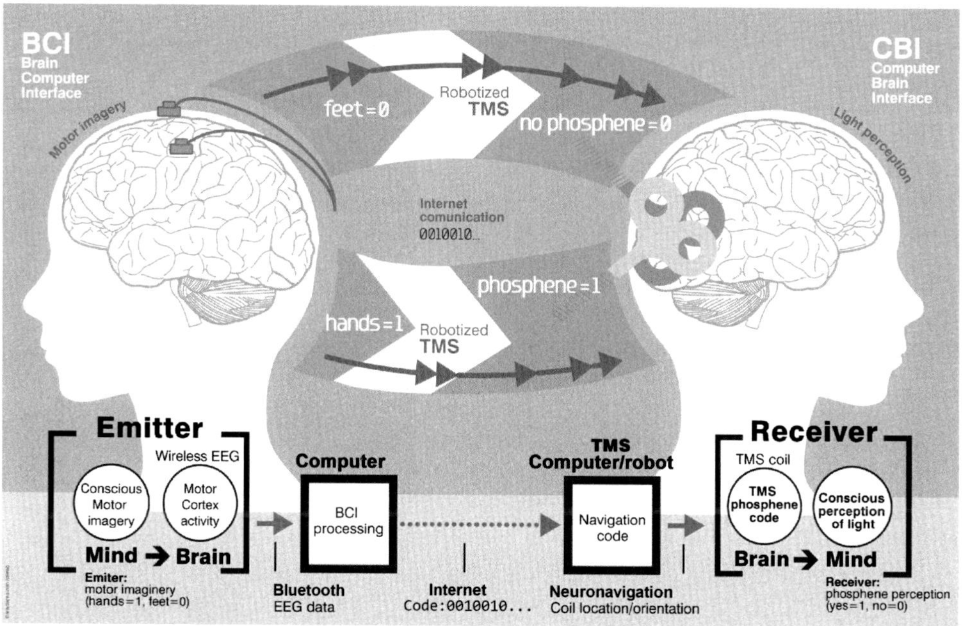

FIG. 6.13 Brain-to-brain (B2B) communication system overview. *From C. Grau, R. Ginhoux, A. Riera, T.L. Nguyen, H. Chauvat, M. Berg, J.L. Amengual, A. Pascual-Leone, G. Ruffini, Conscious brain-to-brain communication in humans using non-invasive technologies, PLoS ONE 9 (8) (2014) e105225, doi:10.1371/journal.pone.0105225.*

for refining rehabilitation strategies for stroke patients [110]. Johnson et al. investigated the effect of combining repetitive rTMS with BMI training on motor impairment after stroke, and showed the potential of both rTMS and virtual reality BCI training for motor recovery after stroke [111].

The recent development of BCI is also enabling a communication system between two brains (B to B). Combining BCI based on voluntary motor imagery-controlled EEG and CBI inducing conscious perception of phosphenes by TMS can establish a B to B communication system as follows: the BCI and CBI systems are connected through the internet; the signal detected by the BCI system is transferred through binary code to the CBI system; the CBI system transmits the signal to the brain as a sequence of magnetic phosphenes (Fig. 6.13) [112].

References

[1] F. Nijboer, J. Clausen, B.Z. Allison, P. Haselager, The Asilomar survey: stakeholders' opinions on ethical issues related to brain-computer interfacing. Neuroethics 6 (3) (2013) 541–578, doi:10.1007/s12152-011-9132-6.

[2] H. Berger, Über das elektrenkephalogramm des menschen. Arch. Psychiatr. Nervenkr. 87 (1) (1929) 527–570, doi:10.1007/BF01797193.

[3] J.M.R. Delgado, Permanent implantation of multilead electrodes in the brain, Yale J. Biol. Med. 24 (5) (1952) 351–358.

[4] J.M.R. Delgado, H. Hamlin, Surface and depth electrography of the frontal lobes in conscious patients. Electroencephalogr. Clin. Neurophysiol. 8 (3) (1956) 371–384, doi:10.1016/0013-4694(56) 90003-7.

[5] J.J. Vidal, Real-Time detection of brain events in EEG. Proc. IEEE 65 (5) (1977) 633–641, doi:10.1109/ PROC.1977.10542.

[6] M.M. Merzenich, R.P. Michelson, C.R. Pettit, R.A. Schindler, M. Reid, Neural encoding of sound sensation evoked by electrical stimulation of the acoustic nerve. Ann. Otol. Rhinol. Laryngol. 82 (4) (1973) 486–503, doi:10.1177/000348947308200407.

[7] W.H. Dobelle, M.G. Mladejovsky, J.P. Girvin, Artificial vision for the blind: electrical stimulation of visual cortex offers hope for a functional prosthesis. Science 183 (4123) (1974) 440–444, doi:10.1126/science.183.4123.440.

[8] S.W. Lee, F. Fallegger, B.D.F. Casse, S.I. Fried, Implantable microcoils for intracortical magnetic stimulation. Sci. Adv. 2 (12) (2016)e1600889doi:10.1126/sciadv.1600889.

[9] A.C. Thompson, P.R. Stoddart, E.D. Jansen, Optical Stimulation of Neurons. Curr. Mol. Imaging 3 (2) (2014) 162–177, doi:10.2174/2211555203666141117220611.

[10] E. Rezayat, I.G. Toostani, A review on brain stimulation using low intensity focused ultrasound. Basic Clin. Neurosci. 7 (3) (2016) 187–194, doi:10.15412/J.BCN.03070303.

[11] J. Rivnay, H. Wang, L. Fenno, K. Deisseroth, G.G. Malliaras, Next-generation probes, particles, and proteins for neural interfacing. Sci. Adv. 3 (6) (2017)e1601649doi:10.1126/sciadv.1601649.

[12] L.F. Nicolas-Alonso, J. Gomez-Gil, Brain computer interfaces, a review. Sensors (Basel) 12 (2) (2012) 1222–1279, doi:10.3390/s120201211.

[13] B.P. Christie, D.M. Tat, Z.T. Irwin, V. Gilja, P. Nuyujukian, J.D. Foster, S.I. Ryu, K.V. Shenoy, D. E. Thompson, C.A. Chestek, Comparison of spike sorting and thresholding of voltage waveforms for intracortical brain-machine interface performance. J. Neural. Eng. 12 (1) (2015) 016009, doi:10.1088/1741-2560/12/1/016009.

[14] L.R. Hochberg, M.D. Serruya, G.M. Friehs, J.A. Mukand, M. Saleh, A.H. Caplan, A. Branner, D. Chen, R.D. Penn, J.P. Donoghue, Neuronal ensemble control of prosthetic devices by a human with tetraplegia. Nature 442 (7099) (2006) 164–171, doi:10.1038/nature04970.

[15] J.D. Simeral, S.-P. Kim, M.J. Black, J.P. Donoghue, L.R. Hochberg, Neural control of cursor trajectory and click by a human with tetraplegia 1000 days after implant of an intracortical microelectrode array. J. Neural Eng. 8 (2) (2011)025027doi:10.1088/1741-2560/8/2/025027.

[16] M.J. Nelson, P. Pouget, E.A. Nilsen, C.D. Patten, J.D. Schall, Review of signal distortion through metal microelectrode recording circuits and filters. J. Neurosci. Methods 169 (1) (2008) 141–157, doi:10.1016/j.jneumeth.2007.12.010.

[17] A.J. Bard, L.R. Faulkner, Electrochemical Methods: Fundamentals and Application, John Wiley & Sons, Inc., Danvers, 2001

[18] P. Vanýsek, Electrochemical series, in: W.M. Haynes (Ed.), CRC Handbook of Chemistry and Physics, 92nd ed., CRC Press, Boca Raton, 2011–2012, , pp. 5-80–5-89.

[19] L.A. Geddes, R. Roeder, Criteria for the selection of materials for implanted electrodes. Ann. Biomed. Eng. 31 (7) (2003) 879–890, doi:10.1114/1.1581292.

[20] C. Im, J.-M. Seo, A review of electrodes for the electrical brain signal recording. Biomed. Eng. Lett. 6 (3) (2016) 104–112, doi:10.1007/s13534-016-0235-1.

[21] M.R. Neuman, Biopotential electrodes, in: J.D. Bronzine (Ed.), Medical Devices and Systems, The Biomedical Engineering Handbook, third ed., CRC Press, Boca Raton, 2006, , pp. 47-1–47-13.

[22] M. Schuettler, T. Doerge, S.L. Wien, S. Becker, A. Staiger, M. Hanauer, S. Kammer, T. Stieglitz, Cytotoxicity of platinum black, in: Proc. 10th Ann. Conf. Int. FES Soc., 5–8 July 2005, Montreal, Canada, 2005, , pp. S.343–S.345.

[23] B. Alizadeh-Taheri, R.L. Smith, R.T. Knight, An active, microfabricated, scalp electrode array for EEG recording. Sens. Actuators A 54 (1–3) (1996) 606–611, doi:10.1016/S0924-4247(97)80023-4.

[24] Y.M. Chi, T.-P. Jung, G. Cauwenberghs, Dry-contact and noncontact biopotential electrodes: methodological review. IEEE Rev. Biomed. Eng. 3 (2010) 106–119, doi:10.1109/RBME.2010.2084078.

[25] M.A. Lopez-Gordo, D. Sanchez-Morillo, F. Pelayo Valle, Dry EEG electrodes. Sensors (Basel) 14 (7) (2014) 12847–12870, doi:10.3390/s140712847.

[26] J.W. Salatino, K.A. Ludwig, T.D.Y. Kozai, E.K. Purcell, Glial responses to implanted electrodes in the brain. Nature Biomed. Eng. 1 (2017) 862–877, doi:10.1038/s41551-017-0177-7.

[27] A. Weltman, J. Yoo, E. Meng, Flexible, penetrating brain probes enabled by advances in polymer microfabrication. Micromachines 7 (10) (2016) 180, doi:10.3390/mi7100180.

[28] K.A. Potter, A.C. Buck, W.K. Self, J.R. Capadona, Stab injury and device implantation within the brain results in inversely multiphasic neuroinflammatory and neurodegenerative responses. J. Neural Eng. 9 (4) (2012) 046020, doi:10.1088/1741-2560/9/4/046020.

[29] L. Luan, X. Wei, Z. Zhao, J.J. Siegel, O. Potnis, C.A. Tuppen, S. Lin, S. Kazmi, R.A. Fowler, S. Holloway, A.K. Dunn, R.A. Chitwood, C. Xie, Ultraflexible nanoelectronic probes form reliable, glial scar-free neural integration. Sci. Adv. 3 (2) (2017)e1601966doi:10.1126/sciadv.1601966.

[30] T.J. Oxley, N.L. Opie, S.E. John, G.S. Rind, S.M. Ronayne, T.L. Wheeler, J.W. Judy, A.J. McDonald, A. Dornom, T.J.H. Lovell, C. Steward, D.J. Garrett, B.A. Moffat, E.H. Lui, N. Yassi, B.C.V. Campbell, Y. T. Wong, K.E. Fox, E.S. Nurse, I.E. Bennett, S.H. Bauquier, K.A. Liyanage, N.R. van der Nagel, P. Perucca, A. Ahnood, K.P. Gill, B. Yan, L. Churilov, C.R. French, P.M. Desmond, M. K. Horne, L. Kiers, S. Prawer, S.M. Davis, A.N. Burkitt, P.J. Mitchell, D.B. Grayden, C. N. May, T.J. O'Brien, Minimally invasive endovascular stent-electrode array for high-fidelity, chronic recordings of cortical neural activity. Nat. Biotechnol. 34 (3) (2016) 320–327, doi:10.1038/nbt.3428.

[31] J. Agorelius, F. Tsanakalis, A. Friberg, P.T. Thorbergsson, L.M.E. Pettersson, J. Schouenborg, An array of highly flexible electrodes with a tailored configuration locked by gelatin during implantation—initial evaluation in cortex cerebri of awake rats. Front. Neurosci. 9 (2015) 331, doi:10.3389/fnins.2015.00331.

[32] J. Kim, M. Lee, J.S. Rhim, P. Wang, N. Lu, D.-H. Kim, Next-generation flexible neural and cardiac electrode arrays. Biomed. Eng. Lett. 4 (2) (2014) 95–108, doi:10.1007/s13534-014-0132-4.

[33] W.S. Konerding, U.P. Froriep, A. Kral, P. Baumhoff, New thin-film surface electrode array enables brain mapping with high spatial acuity in rodents. Sci. Rep. 8 (2018) 3825, doi:10.1038/s41598-018-22051-z.

[34] A. Mercanzini, P. Colin, J.C. Bensadoun, A. Bertsch, P. Renaud, *In vivo* electrical impedance spectroscopy of tissue reaction to microelectrode arrays. IEEE Trans. Biomed. Eng. 56 (7) (2009) 1909–1918, doi:10.1109/TBME.2009.2018457.

[35] S.F. Lempka, S. Miocinovic, M.D. Johnson, J.L. Vitek, C.C. McIntyre, *In vivo* impedance spectroscopy of deep brain stimulation electrodes. J. Neural. Eng. 6 (4) (2009) 046001, doi:10.1088/1741-2560/6/4/046001.

[36] J. Selvakumaran, J.L. Keddie, D.J. Ewins, M.P. Hughes, Protein adsorption on materials for recording sites on implantable microelectrodes. J. Mater. Sci. Mater. Med. 19 (1) (2008) 143–151, doi:10.1007/s10856-007-3110-x.

[37] K.J. Otto, M.D. Johnson, D.R. Kipke, Voltage pulses change neural interface properties and improve unit recordings with chronically implanted microelectrodes. IEEE Trans. Biomed. Eng. 53 (2) (2006) 333–340, doi:10.1109/TBME.2005.862530.

[38] S.D. Nandedkar, D.B. Sanders, Recording characteristics of monopolar EMG electrodes. Muscle Nerve 14 (2) (1991) 108–112, doi:10.1002/mus.880140204.

[39] M. Ortiz-Catalan, R. Brånemark, B. Håkansson, J. Delbeke, On the viability of implantable electrodes for the natural control of artificial limbs: review and discussion. Biomed. Eng. Online 11 (2012) 33, doi:10.1186/1475-925X-11-33.

[40] G.E. Loeb, R.A. Peck, Cuff electrodes for chronic stimulation and recording of peripheral nerve activity. J. Neurosci. Methods 64 (1) (1996) 95–103, doi:10.1016/0165-0270(95)00123-9.

[41] D.R. Merrill, J. Lockhart, P.R. Troyk, R.F. Weir, D.L. Hankin, Development of an implantable myoelectric sensor for advanced prosthesis control. Artif. Organs 35 (3) (2011) 249–252, doi:10.1111/j.1525-1594.2011.01219.x.

[42] K.A. Ludwig, R.M. Miriani, N.B. Langhals, M.D. Joseph, D.J. Anderson, D.R. Kipke, Using a common average reference to improve cortical neuron recordings from microelectrode arrays. J. Neurophysiol. 101 (3) (2009) 1679–1689, doi:10.1152/jn.90989.2008.

[43] R. Cicione, M.N. Shivdasani, J.B. Fallon, C.D. Luu, P.J. Allen, G.D. Rathbone, R.K. Shepherd, C.E. Williams, Visual cortex responses to suprachoroidal electrical stimulation of the retina: effects of electrode return configuration, J. Neural Eng. 9 (3) 036009, doi:https://doi.org/10.1088/1741-2560/9/3/036009.

[44] D.R. Merrill, M. Bikson, J.G.R. Jefferys, Electrical stimulation of excitable tissue: design of efficacious and safe protocols. J. Neurosci. Methods 141 (2) (2005) 171–198, doi:10.1016/j.jneumeth.2004.10.020.

[45] R.V. Shannon, A model of safe levels for electrical stimulation. IEEE Trans. Biomed. Eng. 39 (4) (1992) 424–426, doi:10.1109/10.126616.

[46] S.F. Cogan, K.A. Ludwig, C.G. Welle, P. Takmakov, Tissue damage thresholds during therapeutic electrical stimulation, J. Neural Eng. 13 (2) 021001, doi:https://doi.org/10.1088/1741-2560/13/2/021001.

[47] S.B. Brummer, M.J. Turner, Electrical stimulation with Pt electrodes: II—estimation of maximum surface redox (theoretical non-gassing) limits, IEEE Trans. Biomed. Eng. 24 (5) (1977) 440–443.

[48] S.F. Cogan, In vivo and in vitro differences in the charge-injection and electrochemical properties of iridium oxide electrodes. in: Proceedings of 2006 International Conference of the IEEE Engineering in Medicine and Biology Society, New York, NY, 30 August–3 September 2006, Conf. Proc. IEEE Eng. Med. Biol. Soc.2006, , pp. 882–885, doi:10.1109/IEMBS.2006.259654.

[49] T.L. Rose, L.S. Robblee, Electrical stimulation with Pt electrodes. VIII. Electrochemically safe charge injection limits with 0.2 ms pulses (neuronal application). IEEE Trans. Biomed. Eng. 37 (11) (1990) 1118–1120, doi:10.1109/10.61038.

[50] D.D. Zhou, R.J. Greenberg, Microelectronic visual prosthesis. in: E. Greenbaum, D. Zhou (Eds.), Implantable neural prostheses 1, Springer, New York, 2009doi:10.1007/978-0-387-77261-5_1.

[51] S.F. Cogan, Neural stimulation and recording electrodes. Annu. Rev. Biomed. Eng. 10 (2008) 275–309, doi:10.1146/annurev.bioeng.10.061807.160518.

[52] S.A. Kim, S.B. Jun, In-vivo optical measurement of neural activity in the brain. Exp. Neurobiol. 22 (3) (2013) 158–166, doi:10.5607/en.2013.22.3.158.

[53] S. Sasaki, I. Yazawa, N. Miyakawa, H. Mochida, K. Shinomiya, K. Kamino, Y. Momose-Sato, K. Sato, Optical imaging of intrinsic signals induced by peripheral nerve stimulation in the *in vivo* rat spinal cord. Neuroimage 17 (3) (2002) 1240–1255, doi:10.1006/nimg.2002.1286.

[54] M. Haruta, Y. Sunaga, T. Yamaguchi, H. Takehara, T. Noda, K. Sasagawa, T. Tokuda, J. Ohta, Intrinsic signal imaging of brain function using a small implantable CMOS imaging device. Jpn. J. Appl. Phys. 54 (4S) (2015)04DL10doi:10.7567/JJAP.54.04DL10.

[55] T. Kobayashi, M. Haruta, K. Sasagawa, M. Matsumata, K. Eizumi, C. Kitsumoto, M. Motoyama, Y. Maezawa, Y. Ohta, T. Noda, T. Tokuda, Y. Ishikawa, J. Ohta, Optical communication with brain cells by means of an implanted duplex micro-device with optogenetics and Ca^{2+} fluoroimaging. Sci. Rep. 6 (2016) 21247, doi:10.1038/srep21247.

[56] E.S. Boyden, F. Zhang, E. Bamberg, G. Nagel, K. Deisseroth, Millisecond-timescale, genetically targeted optical control of neural activity. Nat. Neurosci. 8 (9) (2005) 1263–1268, doi:10.1038/nn1525.

[57] F. Zhang, M. Prigge, F. Beyrière, S.P. Tsunoda, J. Mattis, O. Yizhar, P. Hegemann, K. Deisseroth, Red-shifted optogenetic excitation: a tool for fast neural control derived from *Volvox* carteri. Nat. Neurosci. 11 (6) (2008) 631–633, doi:10.1038/nn.2120.

[58] J.Y. Lin, P.M. Knutsen, A. Muller, D. Kleinfeld, R.Y. Tsien, ReaChR: a red-shifted variant of channelrhodopsin enables deep transcranial optogenetic excitation. Nat. Neurosci. 16 (10) (2013) 1499–1508, doi:10.1038/nn.3502.

[59] R. Wieboldt, K.R. Gee, L. Niu, D. Ramesh, B.K. Carpenter, G.P. Hess, Photolabile precursors of glutamate: synthesis, photochemical properties, and activation of glutamate receptors on a microsecond time scale. Proc. Natl. Acad. Sci. U. S. A. 91 (19) (1994) 8752–8756, doi:10.1073/pnas.91.19.8752.

[60] I. Tochitsky, J. Trautman, N. Gallerani, J.G. Malis, R.H. Kramer, Restoring visual function to the blind retina with a potent, safe and long-lasting photoswitch. Sci. Rep. 7 (2017) 45487, doi:10.1038/srep45487.

[61] J. Wells, C. Kao, K. Mariappan, J. Albea, E.D. Jansen, P. Konrad, A. Mahadevan-Jansen, Optical stimulation of neural tissue *in vivo*. Opt. Lett. 30 (5) (2005) 504–506, doi:10.1364/OL.30.000504.

[62] NIH Publication No. 11-4798, Cochlear Implants, National Institute on Deafness and Other Communication Disorders, 2013 (retrieved 18.02.16).

[63] "BrainGate" Brain-Machine-Interface takes shape, New Atlas December 7th, 2004, https://newatlas.com/go/3503/, 2004 (retrieved 18.02.16).

[64] Richard Martin, Mind Control, Wired magazine March 5th, 2005, https://www.wired.com/2005/03/brain-3/, 2005 retrieved 18.02.16.

[65] L.R. Hochberg, D. Bacher, B. Jarosiewicz, N.Y. Masse, J.D. Simeral, J. Vogel, S. Haddadin, J. Liu, S.S. Cash, P. van der Smagt, J.P. Donoghue, Reach and grasp by people with tetraplegia using a neurally controlled robotic arm, Nature. 485 (7398) 372–375, doi:https://doi.org/10.1038/nature11076.

[66] Katie Moisse, Paralyzed Woman Moves Robotic Arm With Her Mind, ABC News May 16th, 2012, https://abcnews.go.com/Health/w_MindBodyNews/paralyzed-woman-moves-robotic-arm-mind/story?id=16353993, 2012 retrieved 18.02.16.

[67] B. Wodlinger, J.E. Downey, E.C. Tyler-Kabara, A.B. Schwartz, M.L. Boninger, J.L. Collinger, Ten-dimensional anthropomorphic arm control in a human brain–machine interface: difficulties, solutions, and limitations. J. Neural Eng. 12 (1) (2015)016011doi:10.1088/1741-2560/12/1/016011.

[68] T. Aflalo, S. Kellis, C. Klaes, B. Lee, Y. Shi, K. Pejsa, K. Shanfield, S. Hayes-Jackson, M. Aisen, C. Heck, C. Liu, R.A. Andersen, Decoding motor imagery from the posterior parietal cortex of a tetraplegic human. Science 348 (6237) (2015) 906–910, doi:10.1126/science.aaa5417.

[69] J. Cohen, Memory Implants: a maverick neuroscientist believes he has deciphered the code by which the brain forms long-term memories, MIT Technology Review <https://www.technologyreview.com/s/513681/memory-implants/>, 2012 (retrieved 18.02.16).

[70] A. Regalado, The Entrepreneur with the $100 Million Plan to Link Brains to Computers, MIT Technology Review March 16th, 2017, https://www.technologyreview.com/s/603771/the-entrepreneur-with-the-100-million-plan-to-link-brains-to-computers/, 2017 retrieved 18.02.16.

[71] AAAAA, Elon Musk wants to connect brains to computers with new company, The Guardian March 28th, 2017, https://www.theguardian.com/technology/2017/mar/28/elon-musk-merge-brains-computers-neuralink, 2017 (retrieved 18.02.16).

[72] J. Wu, R.P.N. Rao, How close are we to Elon Musk's brain-computer interface?, The Conversation, CNN April 12, 2017, https://edition.cnn.com/2017/04/12/health/brain-computer-interface-partner/index.html, 2017 retrieved 18.02.16.

[73] M.D. Serruya, J.P. Donoghue, Design principles of a neuromotor prosthetic device. in: K.W. Horch, G.S. Dhillon (Eds.), Neuroprosthetics: Theory and Practice, World Scientific Publishing, Singapore, 2004, , pp. 1158–1196, doi:10.1142/9789812561763_0040.

[74] T. Fitzpatrick, Teenager moves video icons just by imagination, press release October 9th, 2006, Washington University in St. Louis, https://source.wustl.edu/2006/10/teenager-moves-video-icons-just-by-imagination/, 2006 (retrieved 18.02.16).

[75] L. Schuh, I. Drury, Intraoperative electrocorticography and direct cortical electrical stimulation. Semin. Anesth. 16 (1) (1996) 46–55, doi:10.1016/s0277-0326(97)80007-4.

[76] N. Mesgarani, E.F. Chang, Selective cortical representation of attended speaker in multi-talker speech perception. Nature 485 (7397) (2012) 233–236, doi:10.1038/nature11020.

[77] P. Shenoy, K.J. Miller, J.G. Ojemann, R.P.N. Rao, Generalized features for electrocorticographic BCIs, IEEE Trans. Biomed. Eng. 55 (1) 273–280, doi:https://doi.org/10.1109/TBME.2007.903528.

[78] K.K. Ang, C. Guan, K.S. Phua, C. Wang, L. Zhao, W.P. Teo, C. Chen, Y.S. Ng, E. Chew, Facilitating effects of transcranial direct current stimulation on motor imagery brain-computer interface with robotic feedback for stroke rehabilitation, Arch. Phys. Med. Rehabil. 96 (3 Suppl). S79–S87, doi:https://doi.org/10.1016/j.apmr.2014.08.008.

[79] L. Ding, C. Wilke, B. Xu, X. Xu, W. van Drongelene, M. Kohrman, B. He, EEG source imaging: correlate source locations and extents with ECoG and surgical resections in epilepsy patients, J. Clin. Neurophysiol. 24 (2) 130–136, doi:https://doi.org/10.1097/WNP.0b013e318038fd52.

[80] J.A. Cronin, J. Wu, K.L. Collins, D. Sarma, R.P.N. Rao, J.G. Ojemann, J.D. Olson, Task-specific somatosensory feedback via cortical stimulation in humans. IEEE Trans. Haptics 9 (4) (2016) 515–522, doi:10.1109/TOH.2016.2591952.

[81] C.E. King, P.T. Wang, C.M. McCrimmon, C.C.Y. Chou, A.H. Do, Z. Nenadic, The feasibility of a brain-computer interface functional electrical stimulation system for the restoration of overground walking after paraplegia. J. NeuroEng. Rehabil 12 (2015) 80, doi:10.1186/s12984-015-0068-7.

[82] E. Saint-Elme, M.A. Larrier Jr., C. Kracinovich, D. Renshaw, K. Troy, M. Popovic, Design of a biologically accurate prosthetic hand, in: Proceedings of the International Symposium on Wearable Robotics (WeRob 2017). Houston, TX 5–8 (November, 2017) 97–98, doi:10.1109/WEROB.2017.8383866.

[83] A. Chancellor, Electroencephalography: maturing gracefully. Pract. Neurol. 9 (3) (2009) 130–132, doi:10.1136/jnnp.2009.176586.

[84] R.A. Ramadana, A.V. Vasilakosc, Brain computer interface: control signals review. Neuro computing 223 (2017) 26–44, doi:10.1016/j.neucom.2016.10.024.

[85] G. Bin, X. Gao, Y. Wang, B. Hong, S. Gao, VEP-based brain-computer interfaces: time, frequency, and code modulations. IEEE Comput. Intell. Mag. 4 (4) (2009) 22–26, doi:10.1109/MCI.2009.934562.

[86] L.A. Farwell, E. Donchin, Talking off the top of your head: toward a mental prosthesis utilizing event-related brain potentials, Electroencephalogr. Clin. Neurophysiol. 70 (6) (1988) 510–523.

[87] E. Donchin, D.B.D. Smith, The contingent negative variation and the late positive wave of the average evoked potential. Electroencephalogr. Clin. Neurophysiol. 29 (2) (1970) 201–203, doi:10.1016/0013-4694(70)90124-0.

[88] J. Polich, P.C. Ellerson, J. Cohen, P300, stimulus intensity, modality, and probability. Int. J. Psychophysiol. 23 (1–2) (1996) 55–62, doi:10.1016/0167-8760(96)00028-1.

[89] K. Tanaka, K. Matsunaga, H.O. Wang, Electroencephalogram-based control of an electric wheelchair. IEEE Trans. Robot. 21 (4) (2005) 762–766, doi:10.1109/TRO.2004.842350.

[90] E.M. Mugler, C.A. Ruf, S. Halder, M. Bensch, A. Kubler, Design and implementation of a P300-based brain-computer interface for controlling an internet browser. IEEE Trans. Neural Syst. Rehabil. Eng. 18 (6) (2010) 599–609, doi:10.1109/TNSRE.2010.2068059.

[91] G. Pfurtscheller, C. Neuper, Motor imagery and direct brain-computer communication. Proc. IEEE 89 (7) (2001) 1123–1134, doi:10.1109/5.939829.

[92] A. Vasilyev, S. Liburkina, L. Yakovlev, O. Perepelkina, A. Kaplan, Assessing motor imagery in brain-computer interface training: psychological and neurophysiological correlates. Neuropsychologia 97 (2017) 56–65, doi:10.1016/j.neuropsychologia.2017.02.005.

[93] S.R. Soekadar, S. Silvoni, L.G. Cohen, N. Birbaumer, Brain–machine interfaces in stroke neurorehabilitation. in: K. Kansaku, L.G. Cohen, N. Birbaumer (Eds.), Clinical Systems Neurosciences, Springer, Tokyo, 2015, , pp. 3–14, doi:10.1007/978-4-431-55037-2_1.

[94] L. Naci, M.M. Monti, D. Cruse, A. Kübler, B. Sorger, R. Goebel, B. Kotchoubey, A.M. Owen, Brain–computer interfaces for communication with nonresponsive patients. Ann. Neurol. 72 (3) (2012) 312–323, doi:10.1002/ana.23656.

[95] C.G. Lim, T.S. Lee, C. Guan, D.S.S. Fung, Y. Zhao, S.S.W. Teng, H. Zhang, K.R.R. Krishnane, A brain–computer interface based attention training program for treating attention deficit hyperactivity disorder. PLoS ONE 7 (10) (2012) e46692, doi:10.1371/journal.pone.0046692.

[96] T.P. Cothran, J.E. Larson, Brain-computer interface technology for schizophrenia. J. Dual Diagnosis 8 (4) (2012) 337–340, doi:10.1080/15504263.2012.723313.

[97] E.V.C. Friedrich, N. Suttie, A. Sivanathan, T. Lim, S. Louchart, J.A. Pineda, Brain-computer interface game applications for combined neurofeedback and biofeedback treatment for children on the autism spectrum. Front. Neuroeng. 7 (2014) 21, doi:10.3389/fneng.2014.00021.

[98] M. Ferrari, V. Quaresima, A brief review on the history of human functional near-infrared spectroscopy (fNIRS) development and fields of application. Neuroimage 63 (2) (2012) 921–935, doi:10.1016/j.neuroimage.2012.03.049.

[99] S. Coyle, T. Ward, C. Markham, G. McDarby, On the suitability of near-infrared (NIR) systems for next-generation brain–computer interfaces. Physiol. Meas. 25 (4) (2004) 815–822, doi:10.1088/0967-3334/25/4/003.

[100] J. Shin, J. Kwon, J. Choi, C.-H. Im, Performance enhancement of a brain-computer interface using high-density multi-distance NIRS. Sci. Rep. 7 (2017) 16545, doi:10.1038/s41598-017-16639-0.

[101] S. Fazli, J. Mehnert, J. Steinbrink, G. Curio, A. Villringer, K.R. Muller, B. Blankertz, Enhanced performance by a hybrid NIRS-EEG brain computer interface. Neuroimage 59 (1) (2012) 519–529, doi:10.1016/j.neuroimage.2011.07.084.

[102] Y. Tomita, F.B. Vialatte, G. Dreyfus, Y. Mitsukura, H. Bakardjian, A. Cichocki, Bimodal BCI using simultaneously NIRS and EEG. IEEE Trans. Biomed. Eng. 61 (4) (2014) 1274–1284, doi:10.1109/TBME.2014.2300492.

[103] N. Weiskopf, R. Veit, M. Erb, K. Mathiak, W. Grodd, R. Goebel, N. Birbaumer, Physiological self-regulation of regional brain activity using real-time functional magnetic resonance imaging (fMRI): methodology and exemplary data. Neuroimage 19 (3) (2003) 577–586, doi:10.1016/S1053-8119(03)00145-9.

[104] T. Watanabe, Y. Sasaki, K. Shibata, M. Kawato, Advances in fMRI real-time neurofeedback. Trends. Cogn. Sci. 21 (12) (2017) 997–1010, doi:10.1016/j.tics.2017.09.010.

[105] K. Shibata, T. Watanabe, Y. Sasaki, M. Kawato, Perceptual learning incepted by decoded fMRI neurofeedback without stimulus presentation. Science 334 (6061) (2011) 1413–1415, doi:10.1126/science.1212003.

[106] F. Megumi, A. Yamashita, M. Kawato, H. Imamizu, Functional MRI neurofeedback training on connectivity between two regions induces long-lasting changes in intrinsic functional network. Front. Hum. Neurosci. 9 (2015) 160, doi:10.3389/fnhum.2015.00160.

[107] N. Yahata, J. Morimoto, R. Hashimoto, G. Lisi, K. Shibata, Y. Kawakubo, H. Kuwabara, M. Kuroda, T. Yamada, F. Megumi, H. Imamizu, J.E. Náñez Sr, H. Takahashi, Y. Okamoto, K. Kasai, N. Kato, Y. Sasaki, T. Watanabe, M. Kawato, A small number of abnormal brain connections predicts adult autism spectrum disorder. Nat. Commun. 7 (2016) 11254, doi:10.1038/ncomms11254.

[108] T. Yamada, R. Hashimoto, N. Yahata, N. Ichikawa, Y. Yoshihara, Y. Okamoto, N. Kato, H. Takahashi, M. Kawato, Resting-state functional connectivity-based biomarkers and functional

MRI-based neurofeedback for psychiatric disorders: a challenge for developing theranostic biomarkers. Int. J. Neuropsychopharmacol. 20 (10) (2017) 769–781, doi:10.1093/ijnp/pyx059.

[109] N. Yahata, K. Kasai, M. Kawato, Computational neuroscience approach to biomarkers and treatments for mental disorders. Psychiatry Clin. Neurosci. 71 (4) (2017) 215–237, doi:10.1111/pcn.12502.

[110] X. Hong, Z.K. Lu, I. Teh, F.A. Nasrallah, W.P. Teo, K.K. Ang, K.S. Phua, C. Guan, E. Chew, K.-H. Chuang, Brain plasticity following MI-BCI training combined with tDCS in a randomized trial in chronic subcortical stroke subjects: a preliminary study. Sci. Rep. 7 (2017) 9222, doi:10.1038/s41598-017-08928-5.

[111] N.N. Johnson, J. Carey, B.J. Edelman, A. Doud, A. Grande, K. Lakshminarayan, B. He, Combined rTMS and virtual reality brain-computer interface training for motor recovery after stroke. J. Neural Eng. 15 (1) (2018)016009doi:10.1088/1741-2552/aa8ce3.

[112] C. Grau, R. Ginhoux, A. Riera, T.L. Nguyen, H. Chauvat, M. Berg, J.L. Amengual, A. Pascual-Leone, G. Ruffini, Conscious brain-to-brain communication in humans using non-invasive technologies. PLoS ONE 9 (8) (2014)e105225doi:10.1371/journal.pone.0105225.

7 Artificial Organs, Tissues, and Support Systems

Hiroyuki Tashiro*, Marko B. Popović[†], Ivo Dobrev[‡], Yasuo Terasawa[§]

*KYUSHU UNIVERSITY, FUKUOKA, JAPAN [†]WORCESTER POLYTECHNIC INSTITUTE, WORCESTER, MA, UNITED STATES [‡]UNIVERSITY HOSPITAL ZURICH, UNIVERSITY ZURICH, ZÜRICH, SWITZERLAND [§]NIDEK CO., LTD., AICHI, JAPAN

CHAPTER OUTLINE

7.1 Introduction

Disease or injuries may cause serious organ dysfunction potentially resulting in a life-threatening crisis. Even nonlife-threatening disabilities such as losing one's eyesight or hearing can cause the quality of life (QOL) to decrease drastically. In many of these situations, treatment with an artificial organ can be performed. The artificial organs are anticipated to perform functionally as substitutes for impaired or missing organs.

Biomechatronics. https://doi.org/10.1016/B978-0-12-812939-5.00007-0

According to some definitions, the artificial organ refers only to the device that is to be implanted or integrated into a human. Hence, a support system that requires a stationary power supply such as a heart-lung machine or a dialysis machine should not be classified as an artificial organ [1]. The artificial organs are expected to replace a wide spectrum of bodily functions, and extensively different technologies are used to achieve those goals. Therefore, it may be difficult to succinctly define artificial organs [2]. Galletti proposed the following comprehensive definition of an artificial organ, *"An artificial organ may be defined as a human-made device designed to replace, duplicate or augment, functionally or cosmetically, a missing, diseased, or otherwise incompetent part of the body, either temporarily or permanently, and which requires a non-biologic material interface with living tissue"* [3].

Attempts to replace dysfunctional or missing body parts with engineered devices have a history that is at least 3000–4000 years old. The oldest artifacts in the form of dentures and artificial toes (Fig. 7.1) which have been excavated date back to 2500 BC and 1500 BC, respectively [4].

FIG. 7.1 Wooden prosthesis attached to the forefoot by a textile lace. *Reprinted with permission from reference A.G. Nerlich, A. Zink, U. Szeimies, H.G. Hagedorn, Lancet 356 (9248) (2000) 2176–2179. Copyright 2000, Elsevier.*

Currently, not only material science, electrical and electronic engineering, mechanical engineering but also nanotechnology and tissue engineering are applied to the design of artificial organs [5–7].

7.2 Cardiovascular and Respiratory Devices

The circulatory system perfuses blood throughout the entire life. Circulating blood carries nutrients, waste products, and hormones. In the lungs, blood exchanges carbon dioxide and oxygen and has the important function of transporting oxygen and maintaining acid-base balance. Because circulation and respiratory failure cause life-threatening problems, the development of various artificial organs has been addressed from early times. Although the history of this development is quite long, a complete artificial heart and artificial lung have not been realized to the extent that is difficult to entirely solve the issue of hemocoagulative response to foreign matter.

7.2.1 Artificial Heart Lung—Circulation-Assisting Device—Artificial Heart

During some cardiac surgeries, the heartbeat must be temporarily suspended while an artificial heart-lung machine takes over the patient's heart and lung functions [8]. The artificial heart-lung machine is composed of a venous line, suction line, vent line, venous reservoir, oxygenator, blood infusion pump, arterial line, blood delivery line, and so forth (Fig. 7.2). There are two types of blood pumps: a roller pump (peristaltic tube pump) that sends blood by squeezing the tube (Fig. 7.3A), and a centrifugal pump (Fig. 7.3B). The centrifugal pump offers advantages such as less hemolysis, although the blood flow rate varies depending on the afterload. The peristaltic tube pump sends a blood volume corresponding to the rotation speed of the roller. Blood is oxygenated in the artificial lung (oxygenator). Carbon dioxide and oxygen are exchanged through the gas-permeable membrane. The gas exchange membrane is generally made of hollow fiber having a porous membrane in a straw-like shape. External perfusion, in which oxygen gas flows inside the hollow fiber, is generally employed. During extracorporeal circulation, the blood directly contacts with artificial material; therefore, anticoagulation by heparin is imperative.

The extracorporeal auxiliary circulation device assists the patient's heart for a long time. Venoarterial (VA) extracorporeal membrane oxygenation (ECMO) has almost the same configuration as the artificial heart-lung machine, which is a closed system without blood reservoir [9]. The centrifugal pump is used for the blood pump. Percutaneous cardiopulmonary support device (PCPS) is synonymous with VA ECMO. Veno-venous (VV) ECMO is used when the main purpose is to assist pulmonary function [9]. Intraaortic balloon pumping (IABP) boosts the diastolic blood pressure by inflating the balloon in the aorta to support peripheral circulation, and afterload is reduced by deflation [10]. The inflation/deflation needs to be synchronized with an electrocardiogram (ECG).

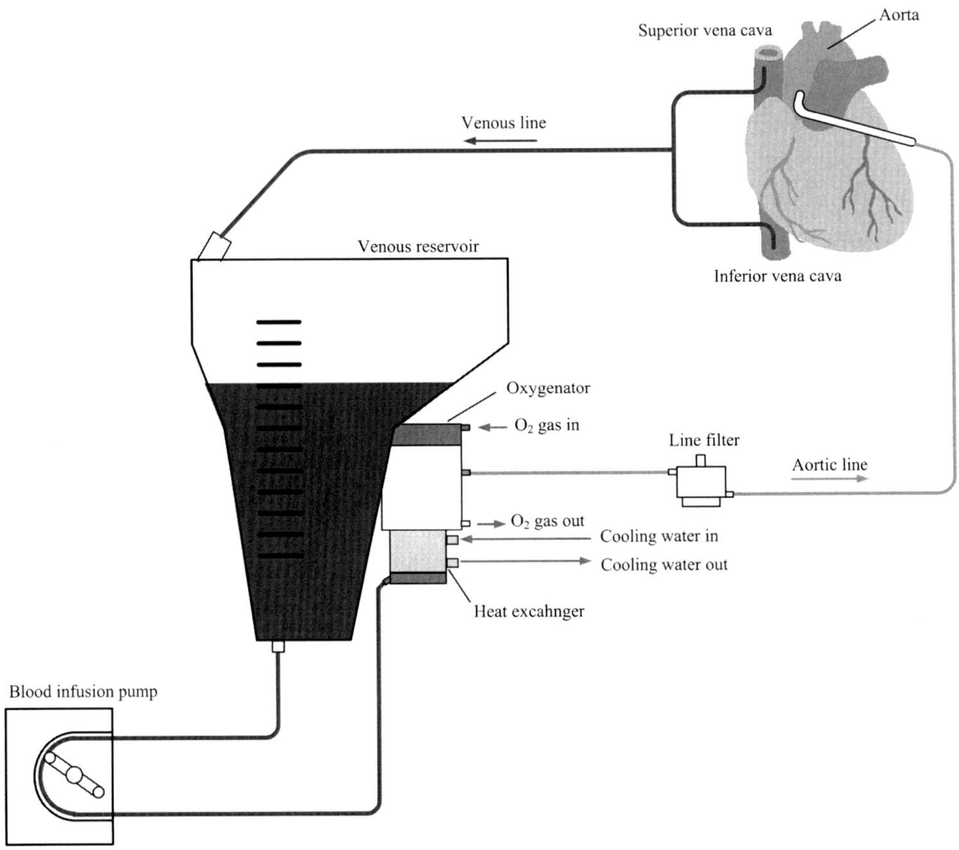

FIG. 7.2 Extracorporeal circulation circuit.

Attempts to completely replace the functions of the heart have been tried for quite a long time now. The ventricular assist device (VAD), using a pulsatile flow pump, was widely used as a bridge until heart transplantation [11]. At the present time, an implantable artificial heart utilizing a magnetic levitation pump for realizing miniaturization and high durability has been clinically used (Fig. 7.4). The AbioCor implantable replacement heart is shown in Fig. 7.5 [12]. The AbioCor implantable heart is the first completely self-contained artificial heart [13]. A hydraulic pump built in the AbioCor rotates at 10,000 revolutions per minute (rpm), creating a pressure differential in an incompressible fluid. As the fluid moves between the left and right side, blood is flowed out one ventricle at the same instant. As a result, it alternately sends blood to the lung and then to the body.

7.2.2 Artificial Heart Valve

The cardiac valve makes the blood flow unidirectional and prevents reverse flow. Assuming the heart rate of a person is 60 beat/min, then the heart valves repeat opening/closing

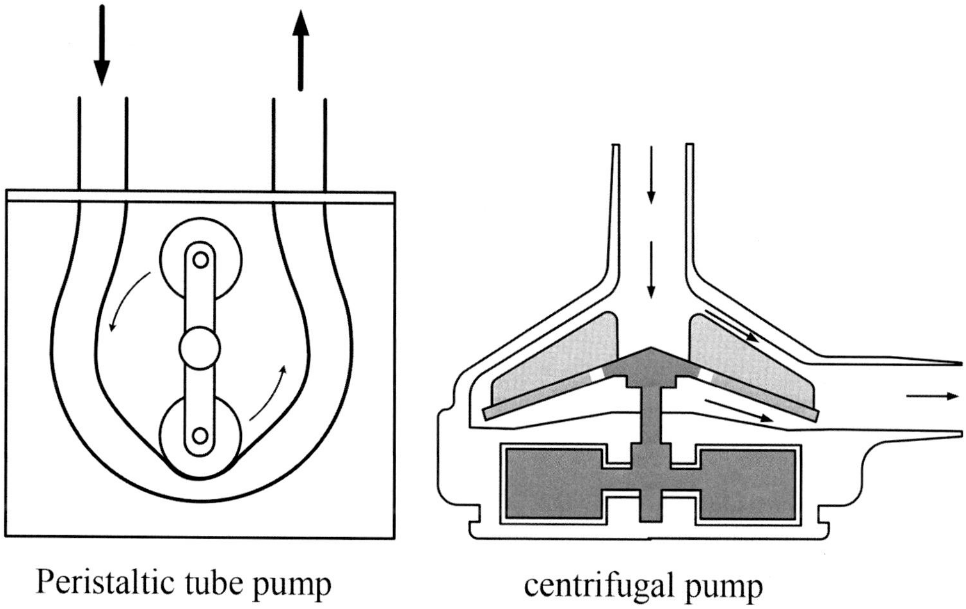

FIG. 7.3 Blood pump used for extracorporeal circulation.

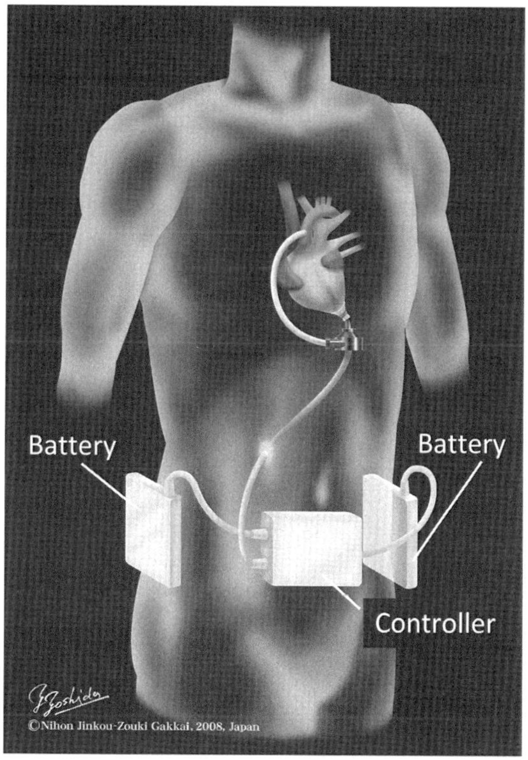

FIG. 7.4 Implantable artificial heart. *Copyright 2008, Japanese Society for Artificial Organs.*

FIG. 7.5 The AbioCor implantable replacement heart [12].

31,536,000 times in 1 year; therefore, high durability is required for artificial heart valves. There are two types of artificial heart valves: the biological valve made from living tissue extracted from cattle or pigs, and mechanical valve made of artificial material [14]. A biological valve resembles a patient's own valve and has better hydrodynamic performance than a mechanical valve; however, there is a problem of durability due to immunoreaction. A common type of mechanical valve is a bi-leaf valve (Fig. 7.6). A thrombus that occurs as a foreign body response to a prosthetic valve causes severe embolism. Therefore, patients implanted with prosthetic valves need to continue dosages of anticoagulants (warfarin potassium).

Mechanical heart valves (available since 1950s) all require lifelong treatment with anticoagulants (blood thinners), making the patient more prone to bleeding. These valves may last indefinitely. Mechanical valves are more commonly used in Asia and Latin America.

There are three major types of mechanical valves: caged-ball, tilting-disc, and bileaflet.

Caged-ball: Pressure difference between heart and aorta push the ball back and forth. In a natural heart valve, blood flows directly through the center of the valve (central flow). With a ball heart valve, there is no central flow. Ball valves also are known to damage or kill blood cells due to colliding with the ball.

The tilting disc valve (available since mid-1970s): The purpose in creating the tilting-disc valve was to restore the central blood flow. The disc opens and closes based on the same principles used in the ball valve design, except that the angular opening of this valve reduces damage to blood cells.

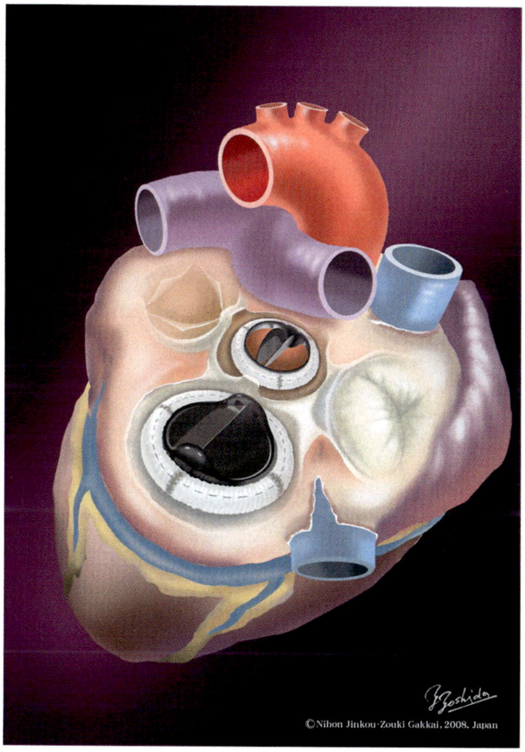

FIG. 7.6 Artificial heart valve. *Copyright 2008, Japanese Society for Artificial Organs.*

Bileaflet heart valve: The majority of mechanical valves today are bileaflet. They were introduced in the late 1970s. The bileaflet design consists of two semicircular leaflets that pivot on hinges. Bileaflet valves have the best central flow—the leaflets open completely, allowing very little resistance to blood flow. These valves correct the problem of central flow and blood cell damage; however, they allow some backflow.

All three types of valves are vulnerable to thrombus formation due to high shear stress, stagnation, and flow separation.

Tissue heart valves (available since 1960s) do not require the extensive use of anticoagulant drugs. They have better blood flow dynamics resulting in less red cell damage and, hence, less clot formation. However, traditional tissue valves, made of pig heart valves, will last on average 15 years before they require replacement. Tissue heart valves can come from a variety of sources: porcine (pig), bovine (cow), horse and homografts or allografts (human). The most commonly used heart valves in the United States and European Union are those that utilize tissue leaflets.

7.2.3 Artificial Blood Vessel

A blood vessel is a conduit for circulating blood all over the body. Artificial blood vessel is used for the replacement of blood vessels affected by aneurysms, and to serve as a bypass of stenosed vessels. The development of an artificial blood vessel with a diameter of <4 mm has been advanced [15]. Generally, it is made of polyethylene terephthalate fiber (Dacron) or polytetrafluoroethylene fiber (Teflon). The neointima is induced on the inner surface of the artificial blood vessel to realize an antithrombotic property. However, because it narrows the lumen of the artificial blood vessel, the development of a fine artificial blood vessel is difficult to achieve. A polyurethane artificial blood vessel with a porous structure inducing endocapillary cell has been developed to realize a finer diameter artificial blood vessel [16].

Stent is a device that maintains patency of blood vessel. Here, patency refers to the degree of openness, i.e., the relative absence of blockage of blood vessel. Aside from coronary stents, there are stents for large blood vessels such as the aorta, and nonvascular stents for biliary tract, ureter, urethra, airway, esophagus, and so forth. For coronary stents, a drug-eluting stent was developed, thus the problem of restenosis was greatly improved.

7.2.4 Pacemaker

Cardiac muscle contracts by the electrical stimulation from the sinus node to the myocardium transmitted through the stimulation conduction system. Pacemakers are used to treat dysfunctions of heart conduction system such as sick sinus syndrome and atrioventricular block (Fig. 7.7) [17].

An implanted pacemaker typically consists of two parts: (1) the pulse generator, i.e., a small metal container housing a battery and the electrical circuitry that regulates the rate of electrical pulses and (2) leads (electrodes), i.e., one to three flexible, insulated wires placed in a chamber, or chambers, of heart.

The stimulation electrode is placed in the right atrium and/or right ventricle or both. Left ventricular pacing, in which pacing electrodes are inserted into the left ventricle wall via the coronary sinus, is also used. It has the function of synchronizing the stimulation with the spontaneous ECG, or inhibiting it when spontaneous contraction is detected. Most pacemakers have sensors that detect body motion or breathing rate, which signals the pacemaker to increase the heart rate during exercise to meet the body's increased demand for blood and oxygen. Pacemaker is driven by a primary battery that occupies most of the volume of the generator. Replacement is necessary before the end of the battery life.

7.2.5 Artificial Respirator

The current artificial respirator realizes only the function of ventilation. A positive pressure ventilation that pushes air from a ventilator is used. In the past, the iron lung, which covers the patient's chest wall with a sealed container and performs negative pressure ventilation, was also used [18]. Control of the ventilator is via a pressure-limited type controlled by

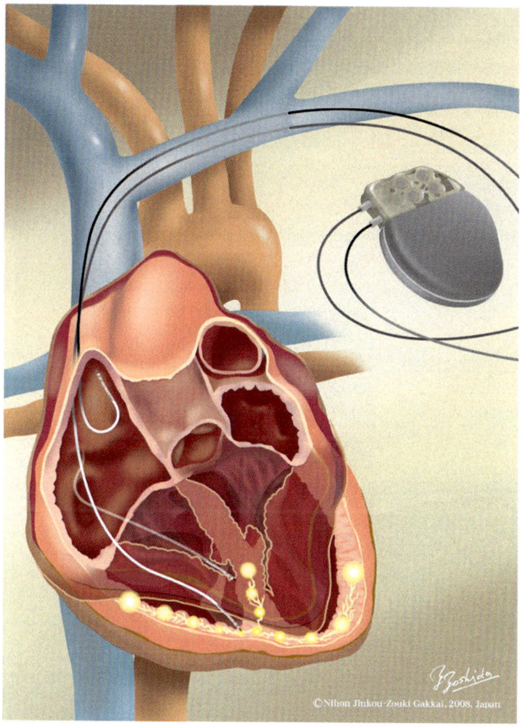

FIG. 7.7 Pacemaker. *Copyright 2008, Japanese Society for Artificial Organs.*

airway pressure and a volume-limited type controlled by ventilatory volume. In the case of pressure-limited control, monitoring the ventilation volume is necessary, whereas monitoring of airway pressure is essential to volume-limited control. A continuous positive airway pressure (CPAP) machine is used to treat sleep apnea syndrome as an aid to ventilation [19]. Although it does not have the function to directly replace the oxygenation of the blood, inhalation of high concentration oxygen supplements the oxygenation. To humidify the inspiratory air, a humidifier inserted in the respiratory circuit is often used. The artificial nose designed to protect respiratory tracts, to respond effectively and permanently to the consequences of laryngectomy, have heat and moisture exchanger. The humidification is introduced by trapping water vapor in the polyurethane filter from the expiratory air.

7.3 Metabolic and Digestive Devices

7.3.1 Artificial Dialyzer

The kidneys excrete unnecessary metabolic products, generated by cell activity, from the blood as urine. In the absence of a functioning kidney, artificial dialysis is used to condition the electrolyte concentration and the circulating blood volume, and excrete uremic toxin (Fig. 7.8). Artificial dialysis is performed by a dialyzer, which bundles approximately

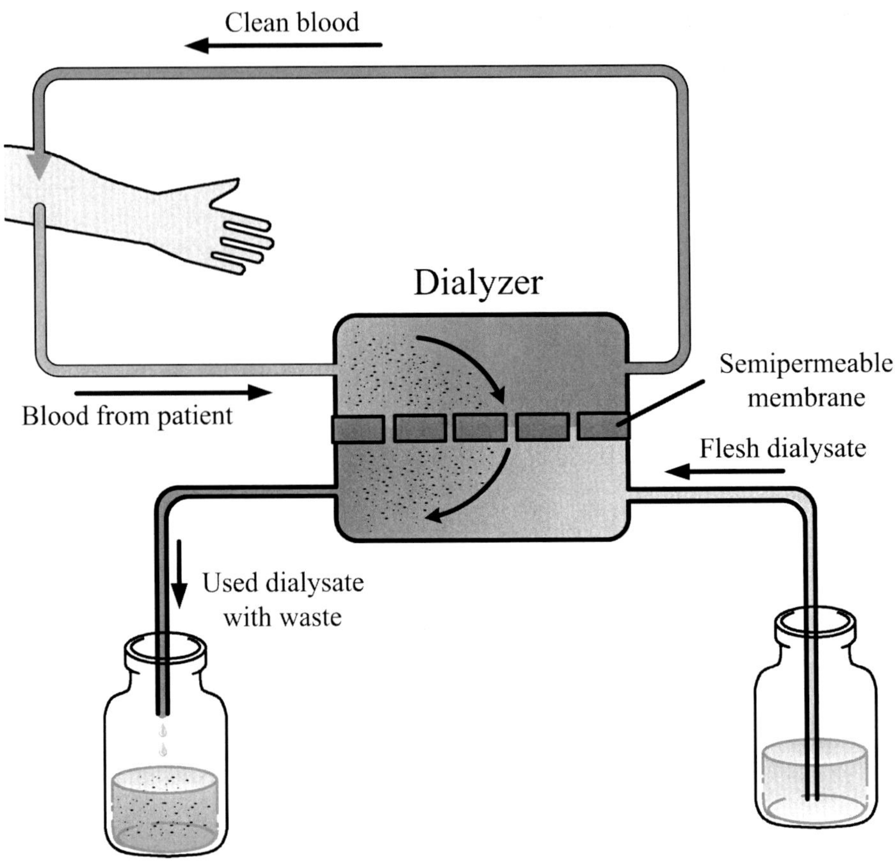

FIG. 7.8 Principle of hemodialysis.

10,000 hollow fibers of approximately 0.2 mm in diameter [20]. The blood flows inside the hollow fiber, and the dialysate conditioned to the blood component flows outside it. For the dialysis membrane, regenerated cellulose, polymethyl methacrylate (PMMA), polysulfone (PS), or the like, are used. Vascular access by arteriovenous fistula is created to ensure sufficient amount of blood removal. Generally, 4 h of hemodialysis (HD) is required three times per week. In hemofiltration (HF), blood is filtered in large quantities and the same amount of replacement fluid is replenished [20]. Hemodiafiltration (HDF) performs HD and HF at the same time. In continuous ambulatory peritoneal dialysis (CAPD), the dialysate is repeatedly stored in the peritoneal cavity and exchanged at regular intervals [21]. Although physical freedom is high, careful attention to infectious disease control is necessary in CAPD. Owing to the restriction of dialysis therapy, a complete implantable artificial kidney is desired; however, it has not yet been realized. Blood purification has also been applied to diseases other than renal failure, and has come to be called apheresis therapy. It includes therapeutic plasma exchange, which separates and discards plasma by ultrafiltration with fresh-frozen plasma (FPP) and extracorporeal adsorption therapy [22].

7.3.2 Artificial Pancreatic Islet

Insulin secreted from pancreatic islets is a hormone that controls the intake of glucose into somatic cells. Insulin injection therapy is employed for the treatment of diabetes. However, the method of administering a fixed amount of insulin at a fixed time interval is not appropriate due to fluctuations in blood glucose due to different food intake, i.e., meal content and amount. Artificial pancreatic islet, which physiologically controls the blood glucose, replaces the function of pancreatic islets α and β cells. It consists of a blood-glucose measuring part and insulin/glucose injector. Extracorporeal circulation equipment has been used clinically in Japan [23] and, furthermore, a portable device with an insulin pump attached to the patient's skin was also approved by the Food and Drug Administration (FDA) in 2016 [24]. The development of fully implantable devices is being addressed, although long-term implantable devices have not been realized because of the deterioration of blood glucose sensors in vivo.

7.4 Sensory Devices

Loss of sensation markedly decreases the QOL. Long before more modern advancement of cochlear implants and artificial retinas, a lost sensory function has been compensated by remaining-sensation substitutes such as sign language and the use of the tried-and-tested Braille. The success of the cochlear implant has shown that functional regeneration of lost body function can be realized. Artificial retina has been intensely developed as well. These are some of the successful examples of medical devices utilizing the brain-machine interface (BMI) technology, also addressed in Chapter 6.

7.4.1 Ear

7.4.1.1 Hearing Process

IMPORTANCE OF THE HEARING RELATIVE TO OTHER SENSES

Hearing is one out of a multitude of senses that enable humans to perceive their surroundings. Hearing plays a crucial role for communication, speech and language development, spatial orientation, environmental awareness and alertness, and overall QOL.

7.4.1.2 Definition of Sound and the Hearing Process in the Human Ear

The hearing process allows the perception of sound, which is defined as a time-varying physical disturbance in pressure, velocity, density, and temperature within a medium that propagates in space [25]. The human ear has evolved to perceive sounds typically propagated through air, also called air-conducted sound, where environmental sounds enter the ear directly via the pinna and the ear canal of the outer ear, as shown in Fig. 7.9. The sound waves then excite the tympanic membrane (eardrum), which in turn sets in motion the ossicles (malleus, incus, and stapes) in the middle ear. The motion of the stapes creates sound waves in the fluid of the cochlear, in the inner ear, which in turn sets in motion the organ of Corti and the hair cells within it. Once in motion, the hair cells produce a pulse

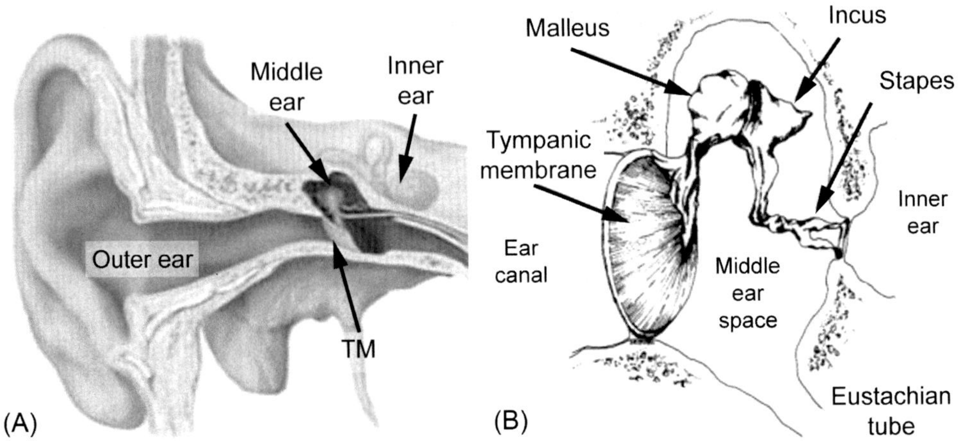

FIG. 7.9 Schematic of the ear with its primary components [25]: (A) the human ear consists of an external, middle, and inner ear with the tympanic membrane (TM) at the boundary of the external and middle ears; and (B) the middle-ear ossicles (i.e., malleus, incus, and stapes) mechanically couple the TM with the inner ear.

train of neural stimuli, which are then transmitted to cortical auditory centers of the brain, where the final psychophysiological perception of sound takes place. The ear is also sensitive, to a lesser degree, to sounds received via the body, also called body- or bone-conducted sound, where environmental sounds excite the human body (skull, skin, bones, etc.), which in turn stimulates cochlea fluid and is registered by the hair cells in a similar way to air-conducted sound. It is generally accepted that the primary purpose of the ear, specifically the middle ear, is to amplify the incident sound waves before transmitting them to the cochlea, to match the acoustic impedances difference between the outside air and cochlea fluid. Without this function, >99.9% of the sound energy would be reflected, instead of entering the cochlea [26].

7.4.1.3 Hearing Aids

A hearing loss (impairment) is a symptom of a variety of diseases, divided into three main categories: (1) conductive hearing loss, affecting the mechano-transduction processes in the outer and middle ear; (2) sensory-neural hearing loss, affecting the inner ear or the auditory nerve, connecting the cochlea with the brain; and (3) central hearing loss, affecting the high-level brain functions associated with hearing perception. It is estimated that >15% of the world population are affected by hearing loss. For nearly every type of hearing loss, there is a type of rehabilitative treatment [27]. Aside from surgical procedures on the outer or middle ear, or direct electrical stimulation of the auditory centers of the brain [28], treatment of hearing loss is mainly done through hearing aids in a wide range of transduction principles depending on the application location. One of the main challenges that these devices need to address is to match the sound perception capabilities of the human ear, specifically in terms of sensitivity, frequency bandwidth, and dynamic range. A generally accepted frequency bandwidth of a healthy human ear is 20–20,000 Hz, with best sensitivity at 1–4 kHz of 20 μPa (0 dB SPL, standard human hearing threshold), and a dynamic range of 120 dB, corresponding to 1,000,000:1 ratio between the quietest and the loudest sound that could be registered. Even with sounds at acoustical pressures typical for speech (0.01–0.1 Pa), the middle and inner ear structures undergo extremely small displacements, in the range of a nanometer or less, reaching picometer range at the limits of the hearing perception (hearing threshold). On the other hand, the ear has protective mechanisms allowing it to survive exposure to random quasi-static pressure variation (i.e., due to weather, elevator rides, airplane travel, etc.) of >20,000 Pa, or approximately 10^3–10^9 stronger pressure variations than its sensitivity threshold.

Current state-of-the-art acoustic hearing aids are technological wonders with respect to their size, computing power, and number of features, and as a result, they are effective for many situations. The design of most hearing aids can be divided into three main parts: acoustical receiver, audio processor, and actuator.

The acoustical receiver transduces the incoming pressure waves into electrical signals, digital or analog. Often multiple acoustical receivers are used simultaneously to allow for enhancing of signals from a chosen direction (typically in front of the user) and suppressing background sound from other directions. Most acoustical receivers are based on electret condenser microphones [29] or microelectromechanical systems (MEMS)-based capacitive microphones [30]. Some of the main design requirements for acoustical receivers for hearing aids is their size (a few cubic millimeters), power consumption (typically in the sub-milliwatt range), sensitivity (10–30 dB SPL), as well as dynamic range and frequency range (30–70 dB SNR, 0.1–8 kHz).

The audio processor performs real-time computational operations on the input signal(s), such that sounds at predefined frequency bandwidths are augmented in order to match with the available frequency and dynamic range of the specific user's ear [31]. In the simplest case, input sound is simply amplified, with nearly no processing, aside from simple high- or low-pass filtering. However, modern audio processors allow patients to perceive sounds with high dynamic and frequency range, even though the remaining limited hearing capacity of the user (both as a frequency and dynamic range) is less than that of the input sound. This is achieved through multiband compression algorithms, which are tuned (fitted) to the user's specific hearing loss characteristics. In addition, the audio processor aids in the differentiation of speech from other sounds in the vicinity of the user as well as improving spatial orientation through directional filtration of the input sounds, which could be further improved by wirelessly connecting two sounds processors, one on each ear, if available. Some devices allow for wireless connection with the user's cell phone (smartphone), enabling myriad of features such as direct phone call handling as well as dynamic adjustment of the characteristics of the sound output (modes of operation) by the user via a software application. Some of the main restrictions for audio processors are size (a few cubic centimeters, <50 g with battery) and power consumption (<10 mW on average).

The actuator in a hearing aid system is tasked with transducing the electrical signals from the audio processor into acoustic pressure, mechanical vibration, or a different type of electrical stimulation. There is a wide range of actuators being used in commercial or research systems, ranging in transduction principle and application. Below are a few examples of hearing aid systems, grouped by their application location in the ear.

7.4.1.4 Outer Ear Hearing Aids

One of the most common types of hearing aid designs capture sounds from the user's environment, process and amplify them, and then send them as sound waves into the ear canal. The amplified sound is then captured by the ear in an equivalent way to typical air-conducted sounds. Such devices typically have the acoustic receiver (e.g., microphone), sound processor, and actuator (e.g., acoustic speaker) into the same housing, suspended behind the pinna of the outer ear. Sound from the actuator is guided into the ear canal through a small tube such that it does not block the natural way for sounds to reach the eardrum. The device is worn by the user, but it is not attached in any surgical way to the body. Such open-canal configuration allows for better use of the remaining hearing capacity of the user, however, it provides a potential feedback pathway, where sound from the actuator affects significantly the receiver, which effectively limits the maximum usable dynamic range (effective gain) of the device. This leads to reduced effectiveness for understanding speech in noisy situations and their less-than-ideal perceived sound quality [32]. The feedback problem is partially alleviated by using designs that block the ear canal, however, this negates the remaining hearing capacity of the user. A potential solution to both problems is a new design that stimulates the ear via direct mechanical contact between the eardrum and a miniature vibratory actuator suspended on frame (custom made to each user) in the ear canal. The design mechanically decouples the actuator from the sound processor and receiver enclosure by utilizing light emitted by the sound processor to drive and control the actuator [33]. The approach of direct mechanical actuation of the eardrum has additional advantages over conventional acoustic stimulation hearing aids such as wider bandwidth and better sound quality due to reduced output signal distortion.

7.4.1.5 Middle Ear Implantable Hearing Aids

While acoustic hearing aids in the outer ear provide convenience of use, their transduction capabilities limit their use to treatment of only mild to medium hearing loss. However, treatment of more severe forms of hearing loss is addressed by implantable hearing aids, which can stimulate directly various structures of the middle ear, via a variety of actuation methods, not limited by the transducer technology of acoustic speakers. The stimulation is typically provided via miniature electromagnetic or piezoelectric actuator implanted in the middle ear, where it is in direct contact or even completely suspended on any of the ossicles, the eardrum or the round window [34]. In such a way, mechanical vibrations from the implanted actuator are coupled directly to the middle ear structures, without a significant transduction loss. The sound processor and the acoustic receiver (e.g., microphone) are typically not implanted, but attached to the head via an implanted magnet and radio frequency (RF) coil (receiver) for driving the actuator wirelessly through the intact skin. One major disadvantage of implantable hearing aids is the relatively complex surgical procedure as well as problems with long-term stability of the contact between the actuator and the middle ear.

7.4.1.6 Bone Conduction Hearing Aids

In the case of severe or complete conduction hearing loss, where the function of the middle ear is nearly or completely lost, bone conduction hearing aids allow for stimulation of the cochlea via vibration of the human skull and its contents [35]. In this case, the actuator, typically an electromagnetic transducer, is coupled to the skull bone via a range of possible options: elastic headband or sticky pads for skin contact, percutaneous screw implant into the skull, subcutaneous magnet attached to the skull for transcutaneous attachment (receiver), or direct subcutaneous implantation of the actuator. The actuator excites the surrounding bone and soft tissue, which in turn stimulate the whole head including the cochlea. Similar to the middle ear implants, the sound capture and processing electronics is attached externally to the patients head. In the case of nonimplanted actuator, all device components are combined into a single housing. Major disadvantages of this type of devices are potential cross-stimulation to the other (contralateral) ear, thus reducing sound directionality perception, as well as relatively high variability in actuator performance due to large variability in the sound transmission properties of patients' head.

7.4.1.7 Cochlear Implants

In the case of senso-neural hearing loss, resulting in reduced or complete lack of hair cells function, none of the abovementioned hearing aids are suitable. In such cases, the complete ear is bypassed, and a cochlea implant is surgically placed in the cochlea, where it provides direct stimulation of the intact auditory nerve [36]. The sound capture and processing electronics is similar to other hearing aids types and is attached externally to the patient's head. However, the actuator is an electrode array, consisting of 10–20 individual electrodes, arranged on a string-like structure approximately 10–20 mm in length and 0.3–0.5 mm in diameter. The electrode array is surgically implanted in the cochlea such that it is in close proximity to the auditory nerve endings (Fig. 7.10). The electrode arrangement is optimized relative to the natural tonotopic arrangement of the hair cells in the cochlea, where cells with sensitivity to lower frequency are located deeper into the cochlea

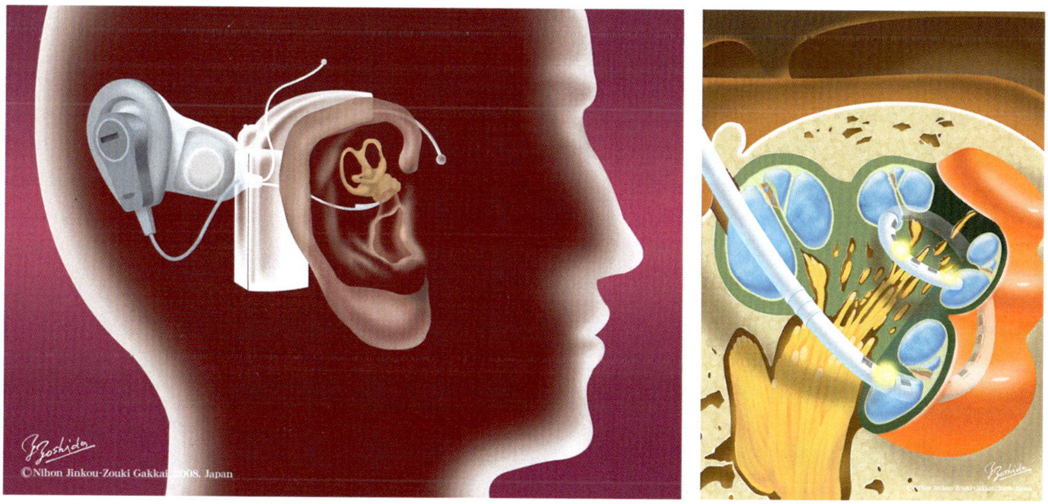

FIG. 7.10 Cochlear implant. *Copyright 2008, Japanese Society for Artificial Organs.*

canals (vestibuli), and vice versa. In this case, the sound processor decomposes the incoming sound into individual frequency bands (channels), related to the location of each electrode in the cochlea, and encodes the waveform of each band into pulse trains of electrical impulses for the direct stimulation of the intact auditory nerve. Major disadvantage of this type of hearing aid is cross talk between electrodes, limiting the number of usable frequency bands, which limits the overall fidelity and intelligibility of the perceived sound.

7.4.1.8 Brainstem Implants

In the case of retrocochlear hearing impairment, there is damage to the cochlea or auditory nerves, which precludes the use of cochlea implants or other types of hearing aids. In this case, an auditory brainstem implant is introduced surgically within the brainstem, bypassing the ear and the auditory nerve. The actuator in this case is an electrode array, similar in function to the one in the cochlea implant devices, but electrodes are arranged in a rectangular matrix, with a total of 10–15 electrodes [37]. Major disadvantages are poor to moderate hearing outcome, need for complex neurosurgical procedure for implantation and long training process for the user.

7.4.1.9 Balance

Directly connected to the cochlea in the inner ear is the vestibular system, and it provides crucial information for balancing. However, the process of balancing relies not only on the vestibular system, but also on other motion-sensitive inputs to the human nervous system provided by the somatosensory and the visual systems. The vestibular system acts as an inertia measurement unit (IMU) composed of three-axis gyroscopes and accelerometers. The visual system (eyes) provides estimation of the location and motion relative to a fixed or moving world frame (i.e., ground, sky, walls, etc.). The somatosensory is a complex network of various sensors, including the sense of touch, sense of position and movement (proprioception), and haptic perception. Each of the three sensory systems provides parts of the full information needed for balancing. Various procedures for treatment and testing of balance have been developed, which are based on stimulation and measuring the response of individual or all of the three sensory systems. One such commonly used test is the EquiTest, during which the test subject is supported on mobile platform and surroundings, the position of which can be dynamically controlled independently of each other [38]. In addition, the test subject's eyes are closed (blocked) for parts of the test procedure. This test elicits sensation of motion to the vestibular, visual, and somatosensory systems, independently of one another, enabling their individual quantitative evaluation. In terms of treatment, the vestibular system can be stimulated via cochlea implants, due to cross-sensitivity caused by the relative vicinity of the neural receptors of the vestibular system to the location of the cochlea implant. Such stimulation could be used for treatment of vertigo, for example. Similarly, kinematic mechanical setups or virtual reality systems (VR headset) could be used for providing controlled stimulation to the visual system.

7.4.2 Eye

7.4.2.1 Artificial Cornea

The artificial cornea is an artifact that can replace a cornea damaged by trauma or disease. Corneal transplantation surgery is the first choice to treat severe corneal diseases, but substitution by an artificial cornea is an option for patients who cannot tolerate a human donor cornea. Various artificial corneas have been studied, and Boston Keratoprosthesis, AlphaCor, and the procedure for osteo-odonto-keratoprosthesis (OOKP) are widely used.

Boston Keratoprosthesis, which is one of the most famous and validated artificial corneas, consists of a front part and a back plate made of PMMA, a corneal graft, and a locking ring made of titanium. The donor cornea is sandwiched between the front part and the back plate, and then sutured to the eye of the patient in the same manner as in corneal transplantation surgery. The main advantages of Boston Keratoprosthesis include its high retention rate, stability, and rapid establishment of good vision. However, from a cosmetic viewpoint, its appearance is not the same as that of a natural cornea.

AlphaCor is a single-piece implant made of poly(2-hydroxyethyl methacrylate) (PHEMA), and consists of a transparent center part and a peripheral part ("skirt") with a porous structure. It is implanted within the cornea through corneal incision. The host tissue and microvessel grow into the peripheral part of the implant, whereby the implant is fixated. The appearance of an implanted AlphaCor is very similar to that of normal cornea, which is a major advantage for the patients. The challenges for AlphaCor include a low survival rate after surgery and difficulty in achieving good vision after implantation surgery.

OOKP is a method that involves using a cylinder made of PMMA as the optical path by using a canine tooth and surrounding tissues of the patient. The PMMA cylinder penetrates a hole in the patient's canine teeth, and is placed subdermally in the body of the patient for a certain period of time, following which it is removed and implanted in the cornea. The advantage of the cornea in OOKP is its excellent durability and retention rate; however, its appearance is significantly different than that of a normal cornea. For more detail, please see the review [39].

7.4.2.2 Intraocular Lens

An intraocular lens (IOL) is a device that can replace a crystalline lens whose transparency has been lost owing to cataracts (Fig. 7.11). In the past, it was common to remove the crystalline lens through a large incision of 6 mm or more, and to implant a large and hard IOL. However, owing to technological advances, it has become possible to perform implantation surgery using a small incision of 3 mm or less. In modern surgery, the lens is emulsified with ultrasonic waves and aspirated from the eye, and then, the IOL is inserted into the lens capsule.

In general, IOLs are composed of a central optical section and a supporting section, which physically supports the optical section. Two types of IOLs are commonly used: a three-piece IOL in which the optical and supporting sections are separately formed, and a one-piece IOL in which the two sections are molded together. PMMA is commonly used as the material for IOL; however, polydimethylsiloxane is also used. Although the optical part has a diameter of approximately 6 mm, it can be folded because most modern IOLs are made of flexible materials, thus enabling insertion from small corneal incisions.

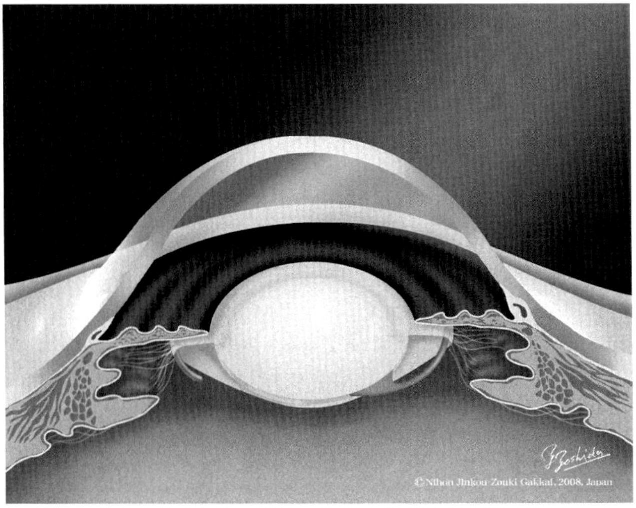

FIG. 7.11 Intraocular lens. *Copyright 2008, Japanese Society for Artificial Organs.*

The most common IOL is the monofocal IOL, which, as its name suggests, has only a single focus. When focus adjustment is necessary, the patient is required to use eyeglasses. Recently, multifocal IOLs having two or more focal points have been developed. Although multifocal IOLs have the advantage of being able to focus on a wide range, there are also inherent disadvantages such as perception of glare and halo, and higher difficulty in focusing on objects compared with single-focus lenses. In recent years, toric IOLs that can correct astigmatism have been commercialized. In the past, spherical lenses were common; however, most modern IOLs have a spherical lenses that correct spherical aberration.

Actual crystal lenses are not completely transparent, and short wavelengths such as of blue light tend to be absorbed by the lenses. Replacing such a lens with a transparent IOL can sometimes lead to an emphasized perception of blue. To avoid such a problem, colored IOLs having a transmittance spectrum similar to that of actual crystal lenses are widely used. Recently, accommodating IOLs, which change power in response to ciliary muscle contraction, have been developed.

In another application, phakic IOLs (PIOLs), which are IOLs that do not aim to cure cataract but to treat myopic refractive errors have been developed. A PIOL is an "implantable contact lens." PIOLs are implanted into the anterior of the posterior chamber of the cornea. PIOLs can treat a larger range of myopic refractive errors, and thus, they are recommended for patients who are suffering from severe myopia and encounter difficulties in laser-assisted in situ keratomileusis (LASIK). However, if a patient with PIOLs develops cataract, the PIOL needs to be removed.

7.4.2.3 Visual Prosthesis

A visual prosthesis is an artificial organ that transmits visual information by artificially stimulating a part of the visual nervous system of patients who have acquired blindness. In most cases, electrical stimulation is employed as the stimulation method. It is well known that a pseudo-light perception called "phosphene" can be elicited by applying electrical stimulation to the visual nervous system [40]. A visual prosthesis provides visual information to the patients by using multiple phosphene, such as in an electric scoreboard. Depending on the region to be stimulated, the visual prosthesis is classified into retinal prosthesis, optic nerve prosthesis, or cortical prosthesis. The retinal prosthesis is further classified into epiretinal prosthesis [41, 42], subretinal prosthesis [43, 44], and suprachoroidal prosthesis [45, 46] according to the difference in the locations of the stimulating electrodes (Fig. 7.12). The epiretinal prosthesis is advantageous in that the stimulation electrode and the retinal ganglion cell, which is one of the main targets of electrical stimulation, can be arranged close to each other. The challenge in the epiretinal approach is the fixation of the electrode array. To fix the electrode array onto the retina, a pin-shaped part known as the retinal tack is necessary. The invasiveness of the retinal tack is an issue because it penetrates the retina, the choroid, and the sclera. The subretinal prosthesis provides a better way to fix the electrode array because the array is sandwiched between the retina and the choroid. Another advantage is that the bipolar cell (another retinal cell that is a target for electrical stimulation) and the stimulation electrode are arranged close to each other, and therefore, the retina can be stimulated with a relatively small amount of current. On the other hand, implantation surgery requires a skilled surgeon. The required thick electrode array is difficult to implant owing to limited subretinal space.

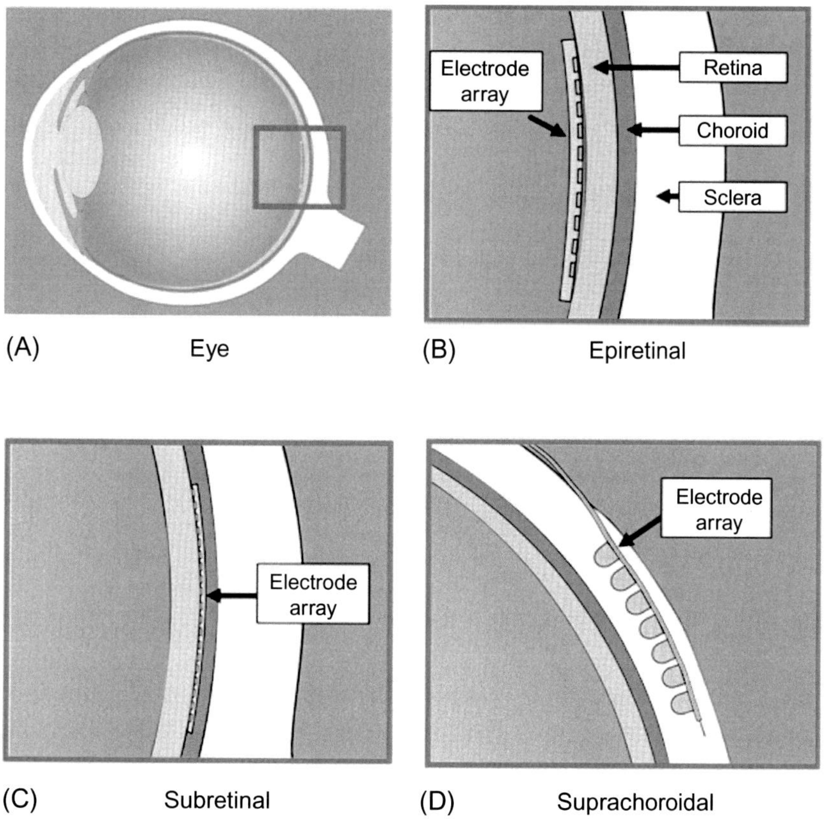

FIG. 7.12 Classification of retinal prosthesis.

In the suprachoroidal prosthesis, the stimulation electrode array is implanted inside the sclera, or between the choroid and the sclera. This approach has the advantage of being relatively safe because the stimulating electrode array is not placed inside the eye. For the same reason, suprachoroidal prosthesis allows implantation of a large electrode array. The challenge of this approach is that a large charge injection is required to excite the retina as compared with other methods because of the large distance between the stimulating electrode and the retina. Epiretinal prosthesis and subretinal prosthesis are currently approved in the United States and Europe.

Because retinal prosthesis involves the placement of the stimulation electrode array only on a part of the retina, the field-of-view that can be realized is limited (typically 10–15°). Moreover, research has been conducted on the electrical stimulation of the optic nerve, which connects the retina and the brain [47]. This method has the advantage that it induces phosphene throughout the field-of-view in principle because all axons from the retinal ganglion cell concentrate in the optic nerve. However, it is difficult to realize a high spatial resolution for optic nerve prosthesis because a large number of optic nerves (estimated at 1 million) is stimulated with only a few stimulating electrodes.

The retinal prosthesis is a device that stimulates the retina; therefore, the function of the optic nerve connecting the retina and the brain must be maintained. However, in diseases such as glaucoma, the optic nerve is damaged and retinal prosthesis cannot be applied. Meanwhile, visual prosthesis that stimulates the central part of the visual nervous system, such as the lateral geniculate nucleus (a relay center in the thalamus) [48] and visual cortex [49, 50], is being explored. These methods have the advantage that they can be applied to patients whose optic nerve is not functional. However, it is suggested that signal processing, which imitates the processing in the retina, and stimulation based on this processing is necessary because it bypasses information processing in the retina.

7.5 Orthopedic, Dentistry, Plastic, and Reconstructive Devices

7.5.1 Breast Prosthesis

Breast prostheses are usually developed as soft tissue prosthetic materials. They are implanted as a reconstruction after breast cancer surgery and there are implantations for cosmetic purposes as well. A breast implant consists of a silicone rubber bag with saline or silicone bag with silicone gel. Breast implantation has a long history, and sales of breast implants began in 1963; however, a product liability (PL) lawsuit occurred related to leakage of uncross-linked silicone oil. This trial triggered the establishment of standards concerning the biocompatibility of implant materials and the registration of the Biomaterials Access Assurance Act of 1998 (BAAA), which protects material suppliers from PL litigation [51]. It is an important case study that offers deep insights into the importance of biocompatibility of artificial materials used in implants, and the regulatory science for the approval of new medical devices.

7.5.2 Dental Implant

Various dental materials are used to restore oral function and form, such as chewing and pronunciation, caused by tooth loss, abnormality of dentition, and so forth. Not only metal materials such as gold alloy, silver alloy, gold-silver-palladium alloy, silver-tin amalgam but also ceramic and composite materials are used. Treatment with artificial dental roots using titanium or titanium alloy has also become common (Fig. 7.13).

7.5.3 Artificial Skin

Main applications of artificial skin are for wound dressing and cultured skin. An animal-derived collagen sheet and a synthetic polymer membrane are clinically used as the wound dressing material. Artificial dermis consisting of two layers of atelocollagen sponge and silicone film to prevent infection from the outside has also been developed [52]. Cultured skin is one of the most successful applications of tissue engineering. Cultured skin includes autologous cultured epidermis using the patient's own cells and allogeneic cultured epidermis using the cells from a donor [52].

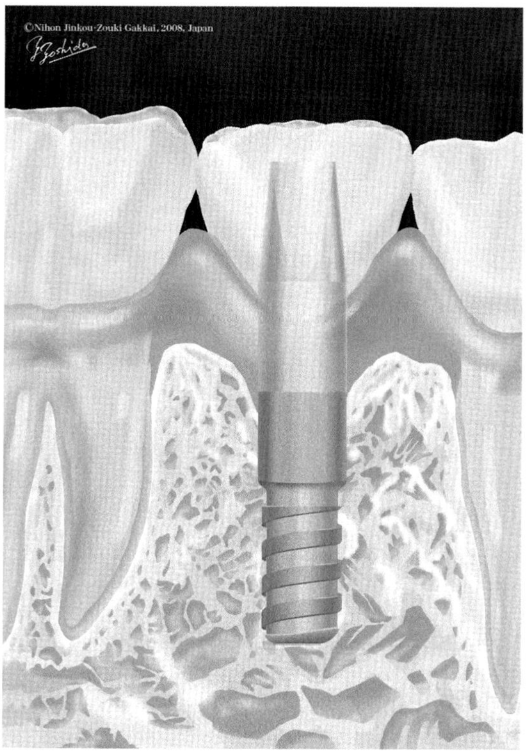

FIG. 7.13 Artificial dental root. *Copyright 2008, Japanese Society for Artificial Organs.*

7.5.4 Artificial Dura Mater

In the past, a section of frozen and dried tissue has been used as an effective patch material for brain surgery. It could be made ready for use simply by being soaked in water. However, infectious cases of Creutzfeldt-Jakob disease (CJD) caused by lyophilized dura mater have been reported, and the sale of it is prohibited now [53]. It is replaced with artificial dura mater using nonwoven fabric made of synthetic polymer such as expanded polytetrafluoroethylene (ePTFE) [54].

7.5.5 Artificial Bone and Artificial Joint

Orthopedic implant materials include artificial joints of the hip and knee joints employed for the treatment of rheumatism and osteoarthritis, for the bone cement used for fixation of fractures of limbs and spines, and as an artificial bone filling the defective part caused by bone tumors and autogenous bone graft. The development of biomaterials in this area has advanced dramatically. New alloys and bioinert ceramics with excellent biocompatibility and excellent abrasion properties are applied to artificial joints. Bioactive ceramics promoting bone formation overtime is applied for bone defects. Moreover, absorbable polymeric materials are applied to bone junctions. Hydroxyapatite is used as the bone filler, while stainless steel, cobalt-chromium alloy, titanium alloy, and so forth are used for the largely defective parts to maintain mechanical strength. There are artificial skulls made of alumina, and artificial bones using material coated with hydroxyapatite. Zirconia or alumina excellent in abrasion quality is used for the artificial femoral head, and titanium alloy is used for the stem inserted into the bone marrow cavity [55].

7.6 Neuromodulation

Neuromodulation is the alteration or modulation of nerve activity by delivering electrical or other physical stimuli directly to a target area. Clinically used representatives are deep brain stimulation (DBS), which is employed for the treatment of essential tremor and Parkinson‘s disease, and spinal cord stimulator (SCS), which is used to relieve chronic pain. Noninvasive methods include transcranial direct current stimulation (tDCS), transcutaneous electrical nerve stimulation (TENS or TNS), and repetitive transcranial magnetic stimulation (rTMS). Neuromodulation therapy has been studied as a treatment for chronic conditions such as Alzheimer’s disease and depression [56].

References

[1] R. Tang, Artificial organs, Bios 69 (3) (1998) 119–122.

[2] G. Catapano, G.J. Verkerke, Artificial organs. in: Z.O. Abu-Faraj (Ed.), Handbook of Research on Biomedical Engineering Education and Advanced Bioengineering Learning: Interdisciplinary Concepts, IGI Global, Hershey, 2012, p. 63, doi:10.4018/978-1-4666-0122-2.ch002.

[3] P.M. Galletti, Prostheses and artificial organs, in: J.D. Bronzino (Ed.), The Biomedical Engineering Handbook, CRC Press, Boca Raton, 1995, p. 1836.

[4] A.G. Nerlich, A. Zink, U. Szeimies, H.G. Hagedorn, Ancient Egyptian prosthesis of the big toe. Lancet 356 (9248) (2000) 2176–2179, doi:10.1016/S0140-6736(00)03507-8.

[5] G. Capatano, Artificial organs design: towards the integration of disciplines. in: Proceeding of 2011 1st Middle East Conference on Biomedical Engineering (MECBME), Sharjah, United Arab Emirates, 21–24 Feburuary, 2011, pp. 185–187, doi:10.1109/MECBME.2011.5752096.

[6] Y. Man, X. li, S. Wang, Recent advances in biohybrid materials for tissue engineering and regenerativemedicine. J. Mol. Eng. Mater 4 (1) (2016) 164001, doi:10.1142/S2251237316400013.

[7] P.S. Malchesky, Artificial organs and vanishing boundaries. Artif. Organs 25 (2) (2001) 75–88, doi:10.1046/j.1525-1594.2001.025002075.x.

[8] A.C. Passaroni, M.A. de Moraes Silva, W.B. Yoshida, Cardiopulmonary bypass: development of John Gibbon's heart-lung machine. Rev. Bras. Cir. Cardiovasc. 30 (2) (2015) 235–245, doi:10.5935/1678-9741.20150021.

[9] E. Tatsumi, Artificial lungs: current state and trends of clinical use and research and development. J. Artif. Organs 10 (1) (2007) 1–5, doi:10.1007/s10047-006-0356-x.

[10] D. Hou, F. Yang, X. Hou, Clinical application of intra-aortic balloon counterpulsation in high-risk patients undergoing cardiac surgery. Perfusion 33 (3) (2017) 178–184, doi:10.1177/0267659117734630.

[11] C. Feldmann, A. Chatterjee, A. Haverich, J.D. Schmitto, Left ventricular assist devices—a state of the art review. in: S. Islam (Ed.), Heart Failure: From Research to Clinical Practice, in: Adv. Exp. Med. Biol. Series, vol. 1067, Springer, Cham, 2013, pp. 287–294, doi:10.1007/5584_2018_145.

[12] M.B. Popović, Biomechanics and Robotics, Pan Stanford, Singapore, 2013.

[13] R.D. Dowling, L.A. Gray Jr., S.W. Etoch, H. Laks, D. Marelli, L. Samuels, J. Entwistle, G. Couper, G. J. Vlahakes, O.H. Frazier, Initial experience with the AbioCor implantable replacement heart system. J. Thorac. Cardiovasc. Surg. 127 (1) (2004) 131–141, doi:10.1016/j.jtcvs.2003.07.023.

[14] G.D. Dangas, J.I. Weitz, G. Giustino, R. Makkar, R. Mehran, Prosthetic heart valve thrombosis. J. Am. Coll. Cardiol. 68 (24) (2016) 2670–2689, doi:10.1016/j.jacc.2016.09.958.

[15] S. Pashneh-Tala, S. MacNeil, F. Claeyssens, The tissue-engineered vascular graft—past, present, and future, Tissue Eng. Part B Rev. 22 (1) (2016) 68–100, doi:10.1089/ten.teb.2015.0100.

[16] H. Bangerter, S. Boemke, R. Röthlisberger, V. Schwartz, M. Bergmann, M.D. Müller, V. Djonov, Combined maceration procedure permits advanced microsurgical dissection of Thiel-embalmed specimens. Ann. Anat. 210 (2017) 9 17, doi:10.1016/j.aanat.2016.10.008.

[17] O. Aquilina, A brief history of cardiac pacing, Images Paediatr. Cardiol. 8 (2) (2006) 17–81.

[18] A.S. Slutsky, History of mechanical ventilation. From vesalius to ventilator-induced lung injury. Am. J. Respir. Crit. Care Med. 191 (10) (2015) 1106–1115, doi:10.1164/rccm.201503-0421PP.

[19] V.L. Pinto, S. Sharma, Continuous Positive Airway Pressure (CPAP), StatPearls [Internet], StatPearls Publishing, Treasure Island, 2018. https://www.ncbi.nlm.nih.gov/books/NBK482178/ (Accessed 18.03.16).

[20] Z.J. Twardowski, History of hemodialyzers' designs. Hemodial. Int. 12 (2) (2008) 173–210, doi:10.1111/j.1542-4758.2008.00253.x.

[21] G. Panarello, H. de Baz, E. Cecchin, F. Tesio, Dialysis for the elderly: survival and risk factors, Adv. Perit. Dial. 5 (1989) 49–51.

[22] R. Hirano, K. Namazuda, J. Suemitsu, T. Harashima, N. Hirata, Plasma separation using a membrane. Transfus. Apher. Sci. 56 (5) (2017) 649–653, doi:10.1016/j.transci.2017.08.008.

[23] N. Kambe, S. Kawahito, N. Mita, K. Takaishi, T. Katayama, Y. Sakai, T. Soga, H. Kawano, M. Matsuhisa, M. Shimada, T. Kitagawa, H. Kitahata, Impact of newly developed, next-generation artificial endocrine pancreas. J. Med. Investig. 62 (1–2) (2015) 41–43, doi:10.2152/jmi.62.41.

[24] J.R. Castle, J.H. DeVries, B. Kovatchev, Future of automated insulin delivery systems. Diabetes Technol. Ther 19 (Suppl. 3) (2017) S67–S72, doi:10.1089/dia.2017.0012.

[25] L. Chittka, A. Brockmann, Perception space-the final frontier, PLoS Biol 3 (4) (2005) e137, doi:10.1371/journal.pbio.0030137.

[26] L.L. Beranek, Acoustics, Acoustical Society of America, New York, 1993.

[27] C. Mathers, A. Smith, M. Concha, Global Burden of Hearing Loss in the Year 2000. Global Burden of Disease, World Health Organization, Geneva, 2003. http://www.who.int/healthinfo/statistics/bod_hearingloss.pdf. Accessed 18.03.16.

[28] D.J. Hoare, P. Adjamian, M. Sereda, Electrical stimulation of the ear, head, cranial nerve, or cortex for the treatment of tinnitus: a scoping review. Neural Plast 2016 (2016) 5130503, doi:10.1155/2016/5130503.

[29] G.M. Sessler, J. Hillenbrand, Hearing aid microphones: from electret to piezoelectret transducers, in: Proceedings of SENSOR+TEST Conferences 2011, Nürnberg, Germany, 7–9 June 2011, Proc. Sensor 2011 (2011) 463–467, doi:10.5162/sensor11/c6.1.

[30] D.J. Young, M.A. Zurcher, M. Semaan, C.A. Megerian, W.H. Ko, MEMS capacitive accelerometer-based middle ear microphone. IEEE Trans. Biomed. Eng. 59 (12) (2012) 3283–3292, doi:10.1109/TBME.2012.2195782.

[31] H. Luts, K. Eneman, J. Wouters, M. Schulte, M. Vormann, M. Buechler, N. Dillier, R. Houben, W. A. Dreschler, M. Froehlich, H. Puder, Multicenter evaluation of signal enhancement algorithms for hearing aids. J. Acoust. Soc. Am. 127 (3) (2010) 1491–1505, doi:10.1121/1.3299168.

[32] S. Kochkin, MarkeTrak VIII: 25-year trends in the hearing health market, Hear. Rev. 16 (11) (2009) 12–31.

[33] J.P. Fay, R. Perkins, S.C. Levy, M. Nilsson, S. Puria, Preliminary evaluation of a light based contact hearing device for the hearing impaired. Otol. Neurotol 34 (5) (2013) 912–921, doi:10.1097/MAO.0b013e31827de4b1.

[34] G.R. Ball, Implantable Electromagnetic Hearing Transducer, U.S. Patent 5,554,096, Issued September 10, 1996.

[35] S. Stenfelt, R.L. Goode, Bone conducted sound: physiological and clinical aspects. Otol. Neurotol. 26 (6) (2005) 1245–1261, doi:10.1097/01.mao.0000187236.10842.d5.

[36] N. Dillier, J. Güntensperger, T. Spillmann, A computer-controlled test system for electrical stimulation of the auditory nerve of deaf patients with implanted multi-electrodes, in: M. Hoke, G. Kauffmann, E. Bappert (Eds.), Scand. Audiol (Suppl. 11) (1980) 163–170.

[37] M.S. Schwartz, S.R. Otto, R.V. Shannon, W.E. Hitselberger, D.E. Brackmann, Auditory brainstem implants. Neurotherapeutics 5 (1) (2008) 128–136, doi:10.1016/j.nurt.2007.10.068.

[38] B. Schwab, M. Durisin, G. Kontorinis, Investigation of balance function using dynamic posturography under electrical-acoustic stimulation in cochlear implant recipients. Int. J. Otolaryngol (2010) (2010) 978594, doi:10.1155/2010/978594.

[39] A. Gomaa, O. Comyn, C. Liu, Keratoprostheses in clinical practice – a review. Clin. Exp. Ophthalmol. 38 (2) (2010) 211–224, doi:10.1111/j.1442-9071.2010.02231.x.

[40] A.M. Potts, J. Inoue, D. Buffum, The electrically evoked response of the visual system (EER), Invest. Ophthalmol. Vis. Sci. 7 (3) (1968) 269–278.

[41] Y.H.-L. Luo, L. da Cruz, The ArgusR II retinal prosthesis system. Prog. Retin. Eye Res. 50 (2016) 89–107, doi:10.1016/j.preteyeres.2015.09.003.

[42] R. Hornig, M. Dapper, E. Le Joliff, R. Hill, K. Ishaque, C. Posch, R. Benosman, Y. LeMer, J.-A. Sahel, S. Picau, Pixium vision: first clinical results and innovative developments. in: V.P. Gabel (Ed.), Artificial Vision, Springer, Cham, 2017, pp. 99–113, doi:10.1007/978-3-319-41876-6_8.

[43] E. Zrenner, K.U. Bartz-Schmidt, D. Besch, F. Gekeler, A. Koitschev, H.G. Sachs, K. Stingl, The subretinal implant ALPHA: implantation and functional results. in: V.P. Gabel (Ed.), Artificial Vision, Springer, Cham, 2017, pp. 65–83, doi:10.1007/978-3-319-41876-6_6.

[44] J.F. Rizzo III, D.B. Shire, S.K. Kelly, P. Troyk, M. Gingerich, B. McKee, A. Priplata, J. Chen, W. Drohan, P. Doyle, O. Mendoza, L. Theogarajan, S. Cogan, J.L. Wyatt, Overview of the boston retinal prosthesis: challenges and opportunities to restore useful vision to the blind. in: Proceedings of 2011 Annual International Conference of the IEEE Engineering in Medicine and Biology Society, Boston, MA, 30 August–3 September, 2011, pp. 7492–7495, doi:10.1109/IEMBS.2011.6093610.

[45] L.N. Ayton, P.J. Blamey, R.H. Guymer, C.D. Luu, D.A.X. Nayagam, N.C. Sinclair, M. N. Shivdasani, J. Yeoh, M.F. McCombe, R.J. Briggs, N.L. Opie, J. Villalobos, P. N. Dimitrov, M. Varsamidis, M.A. Petoe, C.D. McCarthy, J.G. Walker, N. Barnes, A.N. Burkitt, C. E. Williams, R.K. Shepherd, P.J. Allen, Bionic Vision Australia Research Consortium, First-in-human trial of a novel suprachoroidal retinal prosthesis. PLoS ONE 9 (12) (2014) e115239, doi:10.1371/journal.pone.0115239.

[46] T. Fujikado, M. Kamei, H. Sakaguchi, H. Kanda, T. Endo, M. Hirota, T. Morimoto, K. Nishida, H. Kishima, Y. Terasawa, K. Oosawa, M. Ozawa, K. Nishida, One-year outcome of 49-channel suprachoroidal–transretinal stimulation prosthesis in patients with advanced retinitis pigmentosa. Invest. Ophthalmol. Vis. Sci 57 (14) (2016) 6147–6157, doi:10.1167/iovs.16-20367.

[47] C. Veraart, C. Raftopoulos, J.T. Mortimer, J. Delbeke, D. Pins, G. Michaux, A. Vanlierde, S. Parrini, M.-C. Wanet-Defalque, Visual sensations produced by optic nerve stimulation using an implanted self-sizing spiral cuff electrode. Brain Res 813 (1) (1998) 181–186, doi:10.1016/S0006-8993(98)00977-9.

[48] M.J. Kyada, N.J. Killian, J.S. Pezaris, Thalamic visual prosthesis project, in: V.P. Gabel (Ed.), Artificial Vision, Springer, Cham, 2017, pp. 177–189, doi:10.1007/978-3-319-41876-6_14.

[49] W.H. Dobelle, Artificial vision for the blind by connecting a television camera to the visual cortex, ASAIO J. 46 (1) (2000) 3–9.

[50] G.P. Kaskhedikar, Z. Hu, G. Dagnelie, P.R. Troyk, Proposed intracortical vision prosthesis system for phosphene mapping and psychophysical studies, in: Proceedings of 2013 6th International IEEE/EMBS Conference on Neural Engineering (NER), San Diego, CA, 6–8 November, 2013, pp. 880–882, doi:10.1109/NER.2013.6696075.

[51] C.F. Walter, E.P. Richards III, The biomaterials access assurance act of 1998, IEEE Eng. Med. Biol. Mag. 18 (2) (1999) 125–127.

[52] R.V. Shevchenko, S.L. James, S.E. James, A review of tissue-engineered skin bioconstructs available for skin reconstruction. J. R. Soc. Interface 7 (43) (2010) 229–258, doi:10.1098/rsif.2009.0403.

[53] T. Esmonde, C.J. Lueck, L. Symon, L.W. Duchen, R.G. Will, Creutzfeldt-Jakob disease and lyophilised dura mater grafts: report of two cases. J. Neurol. Neurosurg. Psychiatry 56 (9) (1993) 999–1000, doi:10.1136/jnnp.56.9.999.

[54] K. Yamada, S. Miyamoto, M. Takayama, I. Nagata, N. Hashimoto, Y. Ikada, H. Kikuchi, Clinical application of a new bioabsorbable artificial dura mater. J. Neurosurg. 96 (4) (2002) 731–735, doi:10.3171/jns.2002.96.4.0731.

[55] M. Navarro, A. Michiardi, O. Castaño, J.A. Planell, Biomaterials in orthopaedics. J. R. Soc. Interface 5 (27) (2008) 1137–1158, doi:10.1098/rsif.2008.0151.

[56] E.B. Plow, A. Pascual-Leone, A. Machado, Brain stimulation in the treatment of chronic neuropathic and non-cancerous pain. J. Pain 13 (5) (2012) 411–424, doi:10.1016/j.jpain.2012.02.001.

8

Molecular and Cellular Level—Applications in Biotechnology and Medicine Addressing Molecular and Cellular Level

Kathleen A. Lamkin-Kennard*, Marko B. Popovic†

*ROCHESTER INSTITUTE OF TECHNOLOGY, ROCHESTER, NY, UNITED STATES †WORCESTER POLYTECHNIC INSTITUTE, WORCESTER, MA, UNITED STATES

CHAPTER OUTLINE

Biomechatronics. https://doi.org/10.1016/B978-0-12-812939-5.00008-2

8.1 Introduction and Overview

Advances in microelectromechanical systems (MEMS) have led to the development of microscale and nanoscale robotic devices. Microrobots can be used for applications in health and medicine, such as minimally invasive diagnosis or micromanipulation of cells. Other applications of microrobotic devices include benchtop micromanufacturing or development of sensor networks for environmental monitoring [1]. Nanorobots are finding applications in drug delivery, identification of biohazards, in vivo chemical sensing, and transporting oxygen to regions of the body following major trauma [2].

Microrobots can either consist of synthetic materials, such as polymers, magnetic materials, elastomers, or composites [1]. In contrast, biohybrid robots are integrated with living components, such as single or multiple cells or microorganisms, and are powered by the chemical energy associated with the living components or surrounding environment [1].

Many challenges emerge when working with robotic devices at the micro- or nanoscales. At the microscale, physical principles begin to dominate the robot behavior and the fundamental physics associated with the length scale of the device must be considered. Miniature robots are frequently classified based on the size of the device. Robotic devices 1 mm to 10 cm in size are typically classified as millirobots. At this scale, macroscale volume-related forces, such as gravity, inertia, and buoyancy are the dominant forces acting on the robot. The Reynolds numbers of the fluids acting on the devices are typically greater than 1 [1]. The devices 1 μm to 1 mm in size are typically classified as microrobots. Reynolds numbers at this scale are typically close to or much less than one and surface- or perimeter-based forces, such as drag forces, viscous forces, friction, surface tension, and adhesion, dominate. The devices less than 1 μm in size are considered nanorobots. At this scale, nanoscale physical and chemical forces must be considered and effects due to stochastic Brownian motion are no longer negligible [1]. The Navier-Stokes equations do not accurately describe the fluid dynamics associated with the nanoscale so the Reynolds number is no longer relevant [1].

Other challenges that must be addressed at the micro- and nanoscales include a lack of options for actuation, sensing, and power. Traditional methods for actuation, such as motors or linear actuators, are too large and options for scaled versions are very limited. For sensing, the magnitudes of the physical phenomena to be sensed are often in the range of the noise or uncertainties associated with large-scale sensor networks. Finding lightweight options for power for macroscale robots can also be challenging and the problem is amplified at the nano- and microscales since few nanosized or microsized batteries currently exist. As a result, an increasing number of alternate strategies or novel methods for actuation, sensing, and power are being investigated along with fabrication methods for these necessary components.

Microscale or nanoscale biorobotic devices operating in the human body or interacting with human cells present both opportunities and challenges. The devices at these scales are at the same or smaller fundamental length scales than molecules or cells in the body. As a result, micro- or nanorobots can directly interact with or potentially penetrate cells to deliver a payload. However, cells and molecules can also exert unexpected forces on the robots or induce changes on the local microenvironment that can adversely impact robot behavior. Thus, when designing for interactions in the human body, there is a fundamental need to understand physiological processes at the micron or submicron scales that could have a significant impact on robot behavior. This chapter addresses many of the novel aspects that must be considered when designing robots to interact at the molecular and cellular levels.

8.2 Scaling Laws

Miniaturization requires a fundamental understanding of how physical phenomena scale. All quantities of physical interest have dimensions that can be expressed in terms of mass (m), length (l), or time (t) [3]. Scaling laws define the proportional relationships associated with a given parameter relative to one of these scales. For example, the volume of an object is proportional to l^3 while surface area scales as l^2. Thus, the ratio of surface area to volume is larger for a smaller object than for a larger object with a similar shape. Galileo Galilei was the first to describe the scaling laws in 1638 and noted that nature is not scale invariant [4].

Scaling laws can provide information about the structural and functional consequences of changing size or scale for similar structures or organisms. Two types of scaling laws exist. The first is related to the scaling of the physical size of the object. The second is related to the scaling of the physical behavior of an object or a system. Scaling can be represented mathematically using a power law relationship between two variables of interest, x and y, such that

$$y = ax^b \tag{8.1}$$

where a represents the y-intercept and b represent the slope graphically. If the slope b is equal to 1, the relationship is linear and all variables change proportionally with scaling. The relationship is said to undergo isometric, or isomorphic, scaling, and geometric integrity is maintained with scaling [3]. Allometric scaling occurs when different variables do not scale proportionally. Most living systems scale allometrically.

Allometric scaling can have profound effects at various scales. Consider an insect versus a human falling down a large distance. The insect will likely land uninjured while the human will likely be severely injured or killed. To further understand the effects of scaling, consider a cube of side 1 m. The volume of the cube is 1 m^3. However, if the length of the side is doubled, the volume of the cube becomes 8 m^3. Consider the effects on the mass of the cube if the density is 2.7 g/cm^3. Density (ρ) can be represented as

$$\rho = \frac{m}{V} \tag{8.2}$$

where m is the mass of an object and V is the volume of the object. For a cube with a side length of 1 m, the mass of the object is 2,700,000 g or 2700 kg while a cube with a side length 2 m will have a mass of 21,600,000 g or 21,600 kg. This is an eight times increase in mass.

To solve this, calculate the volume of the original cylinder, such that

$$V_1 = \pi {r_1}^2 h_1 = \pi (11.5^2)(175)\text{cm}^3 = 72{,}708.23\ \text{cm}^3$$

If the height and radius are doubled, the volume of the new cylinder becomes

$$V_2 = \pi {r_2}^2 h_2 = \pi (23^2)(350)\text{cm}^3 = 581{,}665.9\ \text{cm}^3$$

Using the density to calculate mass yields

$$m_1 = \rho V_1 = (1\ \text{g/cm}^3)(72{,}708.23\ \text{cm}^3) = 72{,}708.23\ \text{g} = 72.71\ \text{kg}$$

and

$$m_2 = \rho V_2 = (1\ \text{g/cm}^3)(581{,}665.9\ \text{cm}^3) = 581{,}665.9\ \text{g} = 581.67\ \text{kg}$$

To account for this increase in mass, a chain of physiological changes would need to occur. For example, the bones would need to increase significantly in diameter to support the mass. The muscle strength needed to move the mass would become much greater and the metabolic rate needed to support the increased muscle strength and mass of the human would also increase. At some point, an ultimate limit to the size of humans would be reached.

Physical parameters can scale in many ways and the combined effects of scaling must be considered. As shown in Example 8.1, an increase in size will lead to a much greater increase in mass. An increase in mass would likely require larger bones. However, the strength of bone is proportional to its cross-sectional area and scales as l^2. If the size of an animal is scaled up and the bones are proportional in size, at twice the size, the strength relative to weight is half as much as that for a smaller animal. Larger animals could thus be more prone to bone breakage at low force impacts. This is why the skeletal structure of large animals, such as an elephant or rhinoceros, has much thicker bones than those found in small animals [3].

EXAMPLE 8.1

If a human is approximated as a cylinder with a radius of 11.5 cm, a height of 175 cm, and a density of 1 g/cm^3, how much would the mass of the human change if the height and radius are doubled? What physiological changes would need to occur to account for this change in mass?

The scaling is of particular importance when designing micro- or nanoscale robotic systems. The physical effects that dominate at the macroscale may be neglected at the microscale while other factors become predominant. For example, at the macroscale, the weight of a system cannot be neglected. However, at small scales, weight, which scales as l^3, becomes relatively insignificant in comparison to air resistance, which scales as l^2. Many microscale robotic designs take advantage of scaling principles to mimic nature and impart unique capabilities, such as walking on water, which could not be achieved at the macroscale.

8.3 Physical Considerations at the Microscale

As previously described, surface and perimeter-based forces dominate at the microscale and can impact microscale robotic devices very differently than the volume-related forces that dominate at larger scales. The physical factors that must be considered at the microscale include the effects of microscale fluid dynamics, adhesion, Van der Waals forces, capillary forces, electrostatic forces, and frictional forces [1].

8.3.1 Microscale Fluid Dynamics

Microfluidics deals with the behavior of fluids at micron or submicron scales. The behavior of fluid flows at the microscale can have significant effects on microscale robotic devices. The Navier-Stokes equations utilize the principles of conservation of momentum to describe the relationships between velocity, pressure, temperature, and density of a continuous, incompressible moving fluid. The Navier-Stokes equation states that

$$\frac{\partial \boldsymbol{u}}{\partial t} + (\boldsymbol{u} \cdot \nabla)\boldsymbol{u} = -\frac{\nabla P}{\rho} + \nu \nabla^2 \boldsymbol{u} \tag{8.3}$$

where ρ represents the density of the fluid (kg/m^3), $\boldsymbol{u}$ the velocity of the fluid (m/s), and ν is the kinematic viscosity of the fluid (m^2/s). The Reynolds number associated with the flow of a fluid is an important dimensionless parameter that can aid in the prediction of flow patterns. The Reynolds number can be calculated using

$$\mathrm{Re} = \frac{\rho u L}{\mu} = \frac{u L}{\nu} \tag{8.4}$$

where ρ represents the density of the fluid (kg/m^3), u the velocity of the fluid (m/s), L the characteristic linear dimension (L), μ the dynamic viscosity of the fluid (kg/m s), and ν is the kinematic viscosity of the fluid (m^2/s). The Reynolds number describes the ratio of inertial forces to viscous forces in the fluid. At low Reynolds numbers, the inertial forces can be neglected and Eq. (8.3) reduces to

$$\nabla P = \mu \nabla^2 \boldsymbol{u} \tag{8.5}$$

using the relationship between the dynamic and the kinematic viscosity of the fluid such that

$$\nu = \frac{\mu}{\rho} \tag{8.6}$$

These low Reynolds number flows are also referred to as Stokes flows. For Stokes flows, the role of time becomes negligible and flows are reversible in nature. Laminar flow occurs at low Reynolds numbers where viscous forces dominate. Turbulent flow can occur at high Reynolds numbers where inertial forces dominate. At microscopic length scales, viscous forces dominate and most liquid microfluidic flows have Reynolds numbers less than 10. Momentum transport at microscopic scales is also diffusion dominated due to the effects of viscous forces.

Other dimensionless parameters of importance in microfluidics include the Peclet number (*Pe*), which provides information about the ratio of advection to diffusion. The Prandtl number (*Pr*) gives the ratio of momentum diffusivity to thermal diffusivity. The capillary number (*Ca*) provides information about the ratio of viscous forces to surface tension. The Bond number (*Bo*) is the ratio of gravity to surface tension. The Knudsen number (*Kn*) refers to the ratio of the molecular mean free path to the characteristic system length scale.

8.3.2 Microscale Adhesion

Adhesion is defined as the force required to separate two surfaces in contact in a given environment, in the normal direction of the contact surface [5, 6]. The maximum force needed for separation is referred to as the adhesion force, and the energy required for separation is regarded as the work of adhesion [6]. Understanding the adhesion between solid-solid surfaces at the microscale is important for the development of microrobotic devices. Adhesion can offer advantages for locomotion and transmission of forces, such as that can occur in nature. For example, insects are capable of exploiting adhesion forces at the microscale to carry large loads and move quickly despite their small size [5]. The Tabor parameter can be used to understand and predict adhesive interactions at the microscale.

The Johnson-Kendall-Roberts (JKR), Derjaguin-Muller-Toporov (DMT), and Dugdale (D) models are three different contact mechanics models that can be used to predict adhesive interactions at the micro- and nanoscales. The Tabor number is a dimensionless parameter that can be used to understand which model is most applicable. The Tabor parameter, λ, also known as the elasticity parameter, can be calculated using

$$\lambda = \left(\frac{8R_e W_{12}^2}{\pi h^3 K^2}\right)^{\frac{1}{3}} \tag{8.7}$$

where W_{12} is the work of adhesion and is a function of the adhesive force between the two materials. W_{12} depends on the surface energies of the materials, γ_1 and γ_2, and can be calculated using

$$W_{12} = \gamma_1 + \gamma_2 - \gamma_{12} \approx 2\sqrt{\gamma_1 \gamma_2} \tag{8.8}$$

In Eq. (8.7), R_e is the effective radius of contact between two spherical objects of radii R_1 and R_2 and can be calculated using

$$R_e = R_1^{-1} + R_2^{-1} \tag{8.9}$$

The parameter h is the typical separation distance between the points of contact for the two materials' atoms. A typical value for h is 0.165 nm [5]. K is the equivalent modulus of contact based on the Young's modulus of each material (E) and Poisson's ratio (ν). K can be calculated using

$$K = \frac{4}{3}\left(\frac{1-\nu_1^2}{E_1} + \frac{1-\nu_2^2}{E_2}\right)^{-1} \tag{8.10}$$

For λ values less than 0.6, where long-range surface forces act outside the contact area and elastic deformation is negligible, the DMT model is most applicable. If the long-range surface forces act inside the contact area, corresponding to λ values greater than 5, the JKR model is most appropriate [5]. For values between 0.6 and 5, the interface can be modeled as a crack and the D model should be used [5]. The adhesive, or pull-off force, P, can be calculated using.

$P = -2\pi RW_{12}$ for the DMT model or $P = -\frac{3}{2}\pi RW_{12}$ for the JKR model. For the D model, the pull-off force, P, is given by

$$P = -\left(\frac{7}{4} - \frac{1}{4}\frac{4.04\lambda^{\frac{1}{4}} - 1}{4.04\lambda^{\frac{1}{4}} + 1}\right)\pi RW_{12} \tag{8.11}$$

A microrobotic device can thus interact differently with different substrates due to differences in pull-off forces. If the device is immersed in fluid, forces in the liquid must be included and will likely dominate [5]. Fluid forces can lead to repulsion between surfaces which could aid in object motion [5]. Van der Waals forces, capillary, and electrostatic forces directly impact adhesion.

8.3.3 Van der Waals Forces

Van der Waals forces are distance-dependent interactions between molecules. The forces arise due to the movement of electrons in molecules and atoms as shown in Fig. 8.1. The inherent energy associated with molecules leads to the constant motion of electrons in a molecule and can lead to transient concentrations of electrons in a region of the molecule. The transient concentrations can lead to the formation of transient or permanent electric dipoles in a neighboring molecule. The net effect is attraction or repulsion between the molecules.

The Van der Waals forces play an important role in adhesion. The Van der Waals forces act at ranges of 0.2–20 nm and are always attractive for identical materials, but can be repulsive for different materials [5]. For microscale robotic devices, these forces may impose a limit on how the device acts on different surfaces or can be used to enhance interaction.

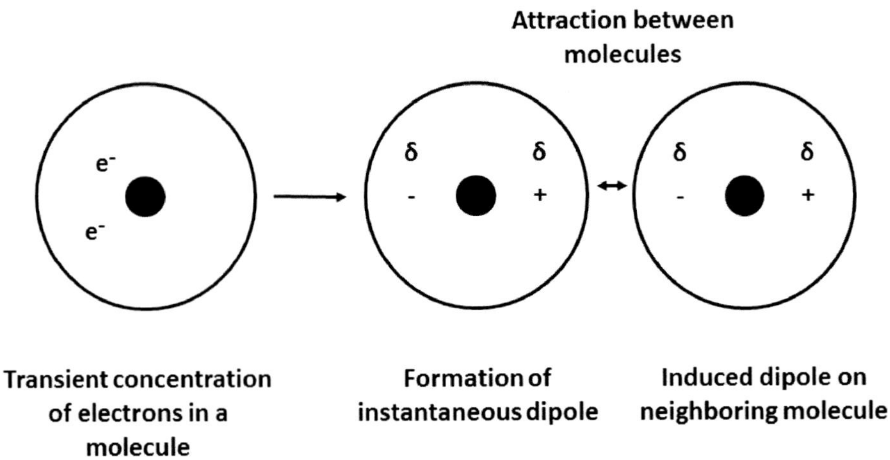

FIG. 8.1 Van der Waals attraction between molecules. The movement of electrons in a molecule (left) can lead to the formation of an instantaneous dipole (middle). The transient dipole induces a dipole on a neighboring molecule (right) and attraction occurs.

Two approaches can be used to calculate Van der Waals interactions as a function of separation distance. The Hamaker approach assumes that the force between two bodies is the sum of the interactions between all pairs of molecules [7, 8]. The second approach is based on Lifshitz theory and utilizes the Van der Waals interaction energy calculated using the electrodynamic properties of the interacting media [8]. The primary difference between the approaches involves the way in which the Hamaker constant is calculated. The Hamaker constant varies between materials and surface configurations. The simplest form for calculating the force of interaction between two semi-infinite surfaces is

$$F = -\frac{A_H}{12\pi l^2} \tag{8.12}$$

where F is the energy of interaction, A_H is the Hamaker constant, and l is the distance between surfaces.

Geckos have a unique ability to scale walls and adhere to ceilings. In 2000, Robert Full discovered that Van der Waals forces created between hairs on a gecko's foot and a surface enable adhesion [9]. Gecko feet have toepads with a large number microscopic hairs called setae. Each of the setae contains even smaller nanoscale projections called spatula as shown in Fig. 8.2. Van der Waals interactions between the spatulae and a surface generate adhesive forces. A single spatula can generate a force of approximately 0.4 μN and the combined effect from millions of spatulae generate an adhesive force close to 10 N or approximately 2.25 pounds [9]. On hydrophobic surfaces, Van der Waals forces have the primary impact on adhesion while on hydrophilic surfaces, capillary forces combine with Van der Waals forces to generate adhesive forces [9, 10].

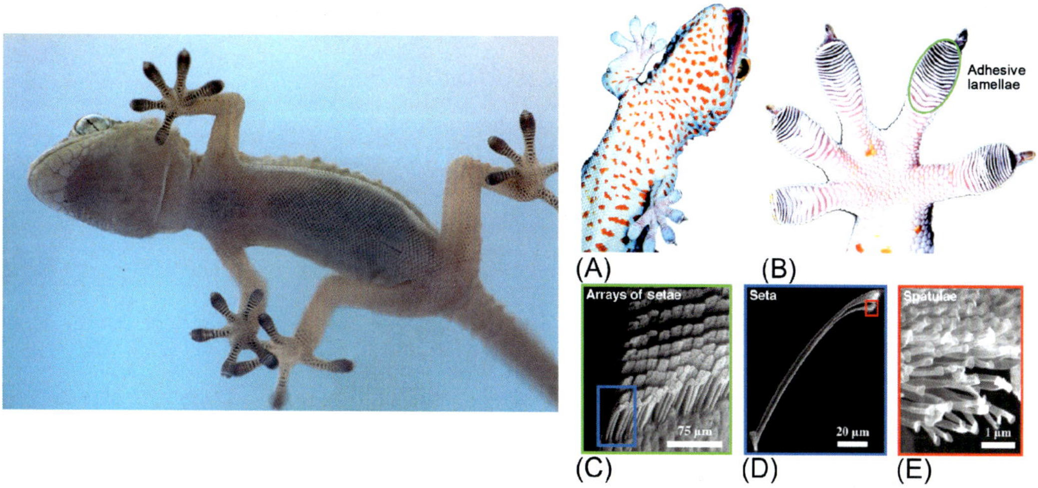

FIG. 8.2 Overview of a gecko adhesive system. (A) Macroscopic view of a tokay gecko climbing a vertical structure. (B) Ventral view of the gecko foot with adhesive lamellae. (C) Single lamella with arrays of setae. (D) Single seta. (E) Spatulae tips. *Reproduced from W.R. Hansen, K. Autumn, Evidence for self-cleaning in gecko setae, Proc. Natl. Acad. Sci. 102(2) (2005) 385–389, Copyright (2005) National Academy of Sciences, USA.*

8.3.4 Capillary Forces

Capillary forces can have a critical impact at the microscale. Capillary forces act at fluid-air-solid interfaces to minimize the surface energy of the interface. Surface tension or capillary forces scale with the perimeter of the wetted area $\sim[s^1]$. In the bulk region of a liquid, cohesive forces between the liquid molecules are balanced and shared with all the neighboring molecules. However, at the surface of the liquid, the molecules are not surrounded by liquid molecules on all sides and exert stronger attractive forces with neighboring molecules on and below the surface as shown in Fig. 8.3.

Surface tension is a consequence of the imbalances in attractive forces between molecules and is related to the decrease in surface energy that results from decreasing surface area. At the surface, a layer forms in which the molecules are strongly bonded, thereby making it harder to penetrate the surface of the liquid than to move submersed objects. Stronger interactions between molecules result in higher values of surface tension while lower values indicate that the molecules are not interacting as strongly. The dimensions for surface tension are expressed as force per unit length or energy per unit area. The surface tension of water is 72 dynes/cm at 25°C indicating that it would take a force of 72 dynes to break a surface film of water 1 cm in length [11]. Surface tension varies between liquids and is affected by temperature.

Surface tension is responsible for a broad range of phenomena. Surface tension impacts the shape of droplets, allows for bubble formation, and also enables animals such as the water strider to move on the surface of bodies of water without sinking. A water strider, shown in Fig. 8.4A, relies on the relatively high surface tension of water to enable

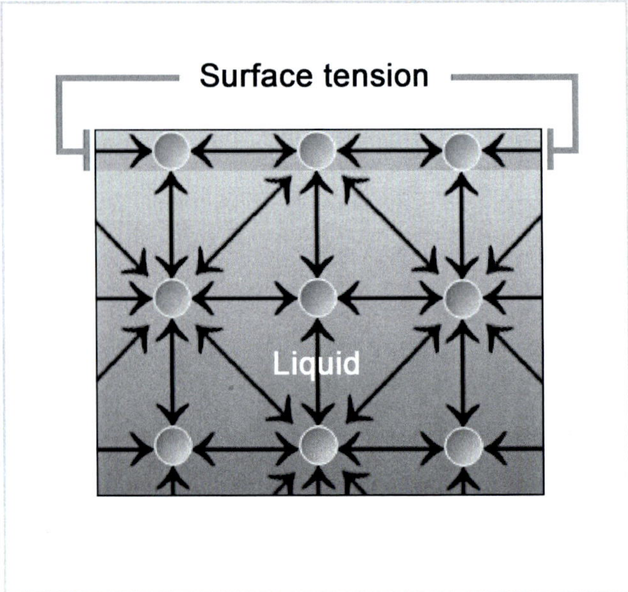

FIG. 8.3 Cohesive forces acting on liquid molecules. At the surface of the liquid, stronger attractive forces exist between neighboring molecules since the molecules are not surrounded on all sides. *Image courtesy of the Ramé-Hart Instrument Co. http://www.ramehart.com/surface_tension.htm (accessed 14.08.18).*

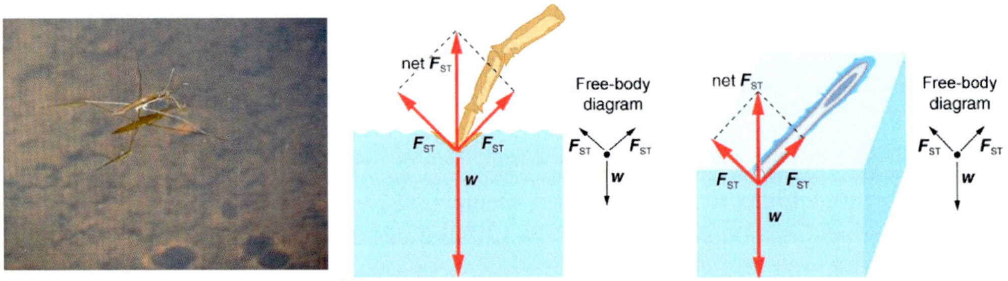

(A) (B)

FIG. 8.4 (A) Water strider on the surface of water. (B) Surface tension forces balance the weight of the insect thereby allowing it to remain on the surface of the water. *(A) Reproduced from https://commons.wikimedia.org/wiki/File: Water_Strider_on_Lily_Lake_(15058406522).jpg (accessed 14.08.18). (B) Reproduced from https://courses. lumenlearning.com/boundless-physics/chapter/bernoullis-equation/.*

it to walk on the surface of the water. If an object is placed on the surface of a liquid, the weight of the object causes the surface to deform. If the downward force due to the weight of the object is balanced by surface tension forces on either side of the object and the object is nonwettable, the object will not sink. A water strider is an extremely lightweight insect with submillimeter diameter legs covered with tiny, superhydrophobic hairs. When placed on the surface of a body of water, the water strider will slightly deform the surface of water as shown in Fig. 8.4B. However, surface tension forces balance the weight of the water strider and, when combined with the effect of the nonwetting hairs, allow the insect to move on the surface of the water without sinking. Recent advances in microrobotics have utilized surface tension effects to create novel devices that can move on the surface of a liquid similar to the water strider [12].

8.3.5 Electrostatic Forces

Electrostatic forces can develop between nonconductive objects resulting from charge buildup, electrical dipoles, or an applied voltage [5]. Forces can also be induced between conductive surfaces when a voltage potential is applied to them. Electrostatic forces typically scale as l^2 and have been utilized to develop novel actuation mechanisms at the microscale. However, the magnitudes of electrostatic forces at the microscale are also great enough to create potential undesired effects on microscale devices. At the microscale, the presence of electrostatic forces can dominate, but are difficult to measure since the forces are complex and often time-variant [13].

At the most fundamental level, Coulomb's law can be used to calculate the electrostatic force between two objects. Coulomb's law states that

$$F = k_e \frac{q_1 q_2}{r^2} \tag{8.13}$$

where k_e is the electrostatic constant dependent on the medium, q_1 and q_2 represent the charges of the first and second objects, respectively, and r is the distance between the objects. At the microscale, the magnitude of this force potentially becomes significant for distances as small as several microns.

For a simple parallel plate capacitor, the electrostatic force normal to the plate surface, in the z direction, can be calculated using

$$F = \frac{\varepsilon_0 \varepsilon_r w l}{2 g_0^2} V_0^2 \tag{8.14}$$

In Eq. (8.14), V_0 is the applied voltage, ε_0 the permittivity of vacuum, ε_r is the permittivity of the dielectric, l and w are the plate dimensions in the x and y directions, and g_0 is the gap between the two parallel plates [13]. If one plate of the capacitor is moved parallel to the other plate, a force in the x or y direction is generated and acts to realign the plates. The forces in the x (F_x) and y (F_y) directions can be calculated using

$$F_x = \frac{\varepsilon_0 \varepsilon_r l}{2 g_0^2} V_0^2 \tag{8.15}$$

or

$$F_y = \frac{\varepsilon_0 \varepsilon_r w}{2 g_0^2} V_0^2 \tag{8.16}$$

The magnitude of electrostatic forces can change if different materials are used, surfaces are roughened, or if different media are used. For example, electrostatic forces can be much smaller in water than in air due to the relative permeabilities of the media [5]. The principle underlying Eqs. (8.10), (8.11) has been used to develop microscale tangential force actuators, such as comb drives, in which several elements are arranged in parallel to increase force [13]. Electrostatic forces have also been used to generate locomotion in microrobots as well as adhesive forces that enable microscale robotic bees to perch to conserve energy. RoboBee, a micro air vehicle, is an example of a robotic device capable of utilizing electrostatic forces to adhere to the underside of a range of materials [14].

8.3.6 Friction at the Microscale

Friction is a force that resists the relative motion of two surfaces. Contact occurs between the tips of surface asperities [15]. The contacting surface asperities deform as the load increases, resulting in an increase in the area of contact with load. At the macroscale, the frictional force due to the increase in area of contact with load is often modeled using Amontons' Law which states that

$$F = \mu L \tag{8.17}$$

where F is the friction force, μ is the coefficient of friction, and L is the normal load. Amontons' law is considered load-controlled since the friction is directly proportional to the applied load. However, in cases where surfaces are adhering, Amontons' law does not always adequately capture the frictional behavior with applied load.

At the microscale, both load-controlled and adhesion-controlled friction must be considered. Adhesion-controlled friction accounts for adhesive interactions between surfaces [15]. As a consequence of the adhesive interactions, asperities on the surface deform mechanically and the area of contact is no longer linearly proportional to the normal load. The area of contact is determined by the elastic properties of the surfaces and the strength of the adhesive interactions [16]. Adhesion-controlled friction results in a nonzero frictional force at zero load and can be represented mathematically as

$$F = F_0 + \mu L \tag{8.18}$$

where $F_0 = \mu L_0$ represents the friction force at zero load.

Friction forces can be particularly problematic at the microscale since friction forces scale with area as l^2. Surface roughness can impact the real contact area between surfaces and can be used to reduce adhesion-controlled friction. If two rough, rigid objects come into contact, the real contact area is less than the apparent contact area due to the presence of the asperities. In cases where the normal contact force is strongly impacted by adhesion, friction can be reduced by adding small microbumps or ridges that act as contact points with a reduced contact area [1]. Friction is of particular concern for micromotors. Wear is also a critical problem for micromechanisms in which sliding interfaces are used. Material selection can be utilized to minimize wear effects.

8.4 Physical Considerations at the Nanoscale

8.4.1 Brownian Motion

Robotic devices at the microscale are strongly impacted by inertial forces. Devices at the nanoscale are also impacted by Brownian motion. Brownian motion is the random movement of small particles suspended in a fluid due to collisions with the fluid molecules. If particles in any medium undergo Brownian movement and there is no preferred direction for the random motions, over a period of time, the particles will become evenly distributed. The rate of Brownian motion increases with temperature, the number of particles, small particle size, and low viscosity of the surrounding medium. Diffusion is the movement of particles from a region of high concentration to low concentration and is a prime example of Brownian motion. When designing devices for operation at the nanoscale, both inertial forces and Brownian motion must be considered. Brownian motion presents a major challenge for locomotion of nanorobotic devices and strategies must be developed to overcome the inherent Brownian motions associated with the surrounding environment [17].

8.5 Approaches to Micro- and Nanoscale Propulsion

The unique physics associated with micro- and nanoscale phenomena create both challenges and opportunities for generating robotic propulsion mechanisms. Many of the micro- and nanorobotic devices created to date utilize synthetic materials to mimic propulsion methods found in nature. Synthetic robots are made of materials such as polymers, magnetic materials, silicon, composites, elastomers, and metals [18]. Other biohybrid robots utilize living organisms, such as cardiomyocytes, bacteria, algae, spermatozoids, and protozoa for propulsion [17, 18].

Propulsion methods can be on-board or off-board. Macroscale robots are typically propelled using an on-board approach. With an on-board approach, the robot is untethered, self-contained, and self-propelled. All of the components are on-board to operate autonomously or with a remote control. Components for on-board propulsion and control are readily available for devices at the milliscale or larger, but are limited at microscale sizes or smaller. An off-board approach relies on external actuation, sensing, control, and power. The majority of micro- or nanorobotic devices created to date rely on off-board approaches. Synthetic propulsion can be accomplished using mechanical, chemical, magnetic, acoustic, or optical methods or a combination of the methods.

8.5.1 Mechanically Powered Propulsion

Microscale swimming biological organisms utilize elaborate mechanisms for swimming. At low Reynolds numbers, viscous forces dominate inertial forces and sophisticated approaches for swimming need to be implemented. Low Reynolds number flows are fully reversible and do not change with time. As a result, reciprocal motions, in which body motions simply alternate between two configurations, such as might be observed with a large-scale organism like a fish, do not result in a net displacement. To compensate, microorganisms rely on a range of propulsion mechanisms, such as rotation of helical bacterial flagella or nonreciprocal beating of a sperm tail [19]. The inspiration for mechanical propulsion for the majority of microscale swimming robots is drawn from nature.

Cilia are active organelles that are used for swimming in microorganisms such as paramecia. Cilia are small hairs or flexible rods approximately 2–15 μm in length that cover the outside surface of the microorganism. During the power stroke of the swimming motion, the cilia are stretched out perpendicular to the flow and move quickly in one direction [19, 20]. During the recovery stroke, the cilia twist and retract to become parallel to the flow [19]. The beating of the cilia forms what is called metachronal wave, which is neither random nor fully synchronous [20]. Until recently, ciliary motions were thought to be too complex to implement in a practical microrobotic system. Kim et al. [21] created a paramecium-like microrobot, shown in Fig. 8.5, that utilized a magnetic field to actuate

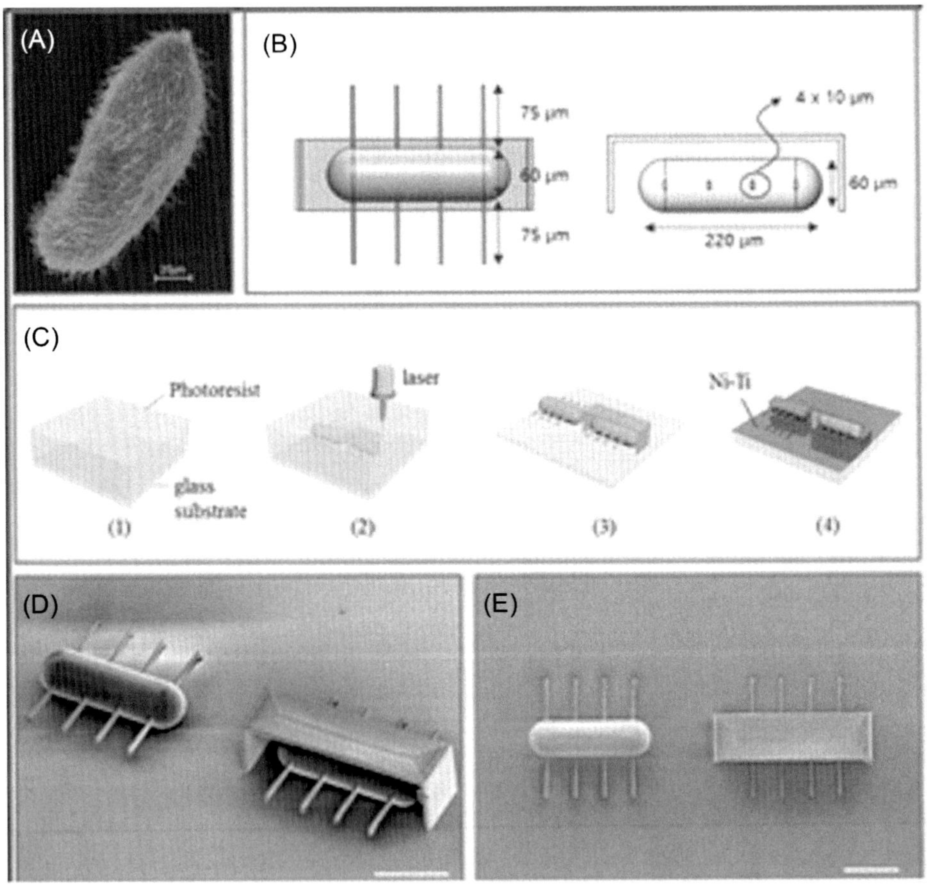

FIG. 8.5 (A) SEM image of a Paramecium. (B) Design of an artificial ciliary microrobot. (C) Microfabrication process for creating the microrobot. (D) and (E) SEM images of the ciliary microrobots with and without a mask structure (the scale bar is 100 μm). *Reproduced from S. Kim, S. Lee, J. Lee, B. Nelson, L. Zhang, H. Choi, Fabrication and manipulation of ciliary microrobots with non-reciprocal magnetic actuation, Sci. Rep. 6 (2016).*

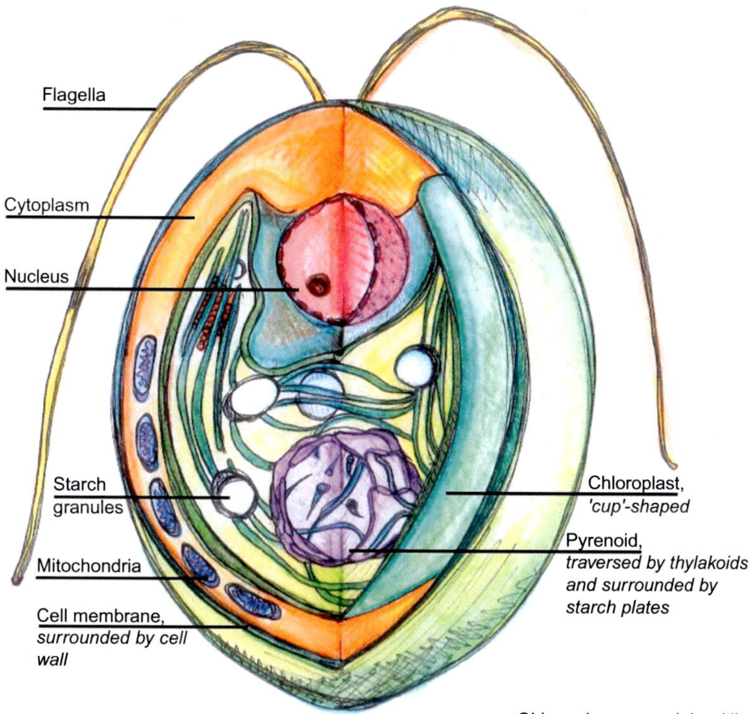

FIG. 8.6 Swimming motions in *Chlamydomonas reinhardtii*. The flagella create breaststroke-like motions for propulsion. *Reproduced from https://commons.wikimedia.org/wiki/File:Cross_section_of_a_Chlamydomonas_reinhardtii_algae_cell,_a_3D_representation.jpg (accessed 15.08.18).*

artificial cilia. The cilia were actuated using a nonreciprocal motion that imparted different actuating forces during the power stroke and recovery stroke.

Flagella are found in both eukaryotic and prokaryotic organisms. Eukaryotic flagella, such as those found in sperm or *Chlamydomonas reinhardtii* deform to create paddling motions [19, 20]. The sperms swim using a sinusoidal traveling wave on the flagellum (Fig. 8.6). The wave creates a snakelike motion as the wave propagates from the head to the tip. MagnetoSperm is a sperm-like microrobot with a magnetic head 42 μm in length and 280 μm flexible tail [22]. The robot utilizes weak oscillating magnetic fields to generate paddling motions using flagella similar to sperm [22].

The green algae *Chlamydomonas reinhardtii* swims using two flagella [20, 23]. The flagella move in a breaststroke-like motion, shown in Fig. 8.6, to generate propulsion and use motions of single flagella to steer [20]. In a study using flagella attached to microbeads, biflagellate robots were shown to be able to propel forward while beads with a single flagella rotated [24]. Development of microrobots that utilize biflagellar propulsion mechanisms represents an area to be further explored in the future.

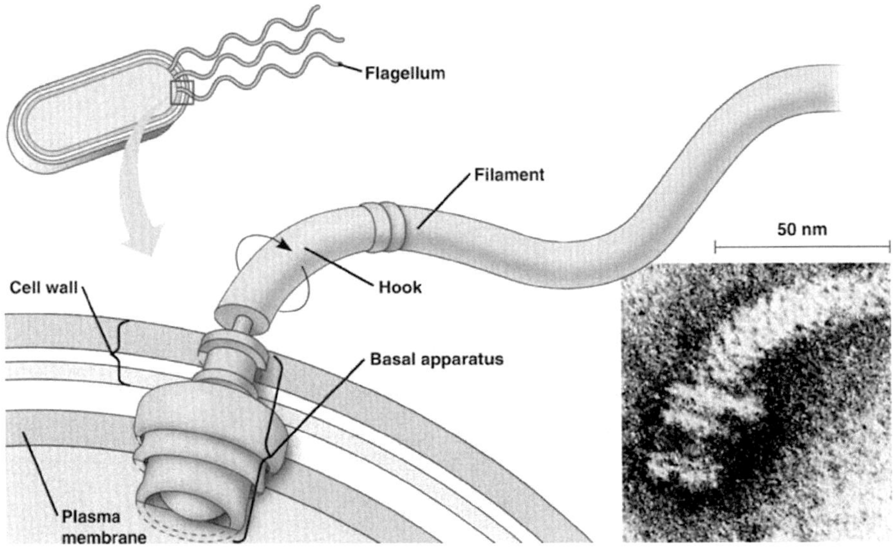

FIG. 8.7 Prokaryotic flagella. The curved hook rotates the attached filament to generate propulsion via a helical wave [25].

Prokaryotic flagella, such as those found in bacteria, utilize a molecular motor embedded in the plasma membrane, to turn the base of the flagella [19]. Fig. 8.7 depicts a prokaryotic flagella. As shown in Fig. 8.7, a curved hook rotates the attached filament thereby allowing directional movements [25]. The bacteria *Escherichia coli* (*E. coli*) swim using molecular motors to generate a helical wave and have served as the inspiration for a broad range of magnetic helical microrobots [26]. Fig. 8.8 shows examples of several magnetic helical micro- and nanorobots. The helical robots utilize rotating magnetic fields to translate rotational motions into translational movements. Helical robots have been described as one of the most promising configurations for biomedical applications, particularly since the devices can be used at both the micro- and nanoscales [26].

8.5.2 Chemical-Powered Propulsion

Chemical reactions can be used to provide a replenishable, remote power supply for micro- and nanoscale robots. Synthetic self-propelled microrobots use synthetic components and available chemical energy from the microenvironment. Chemical micromotors are the most commonly used method for transforming the available energy into motion. Chemical micromotors perform some type of chemical conversion to release energy stored in chemical bonds. The energy sources can be organic, such as glucose, or inorganic, such as hydrogen peroxide, depending on the catalyst that interacts with the fuels [19, 27]. A key challenge with chemical micromotors is ensuring that the chemical conversion is continuous to be able to sustain motion [19].

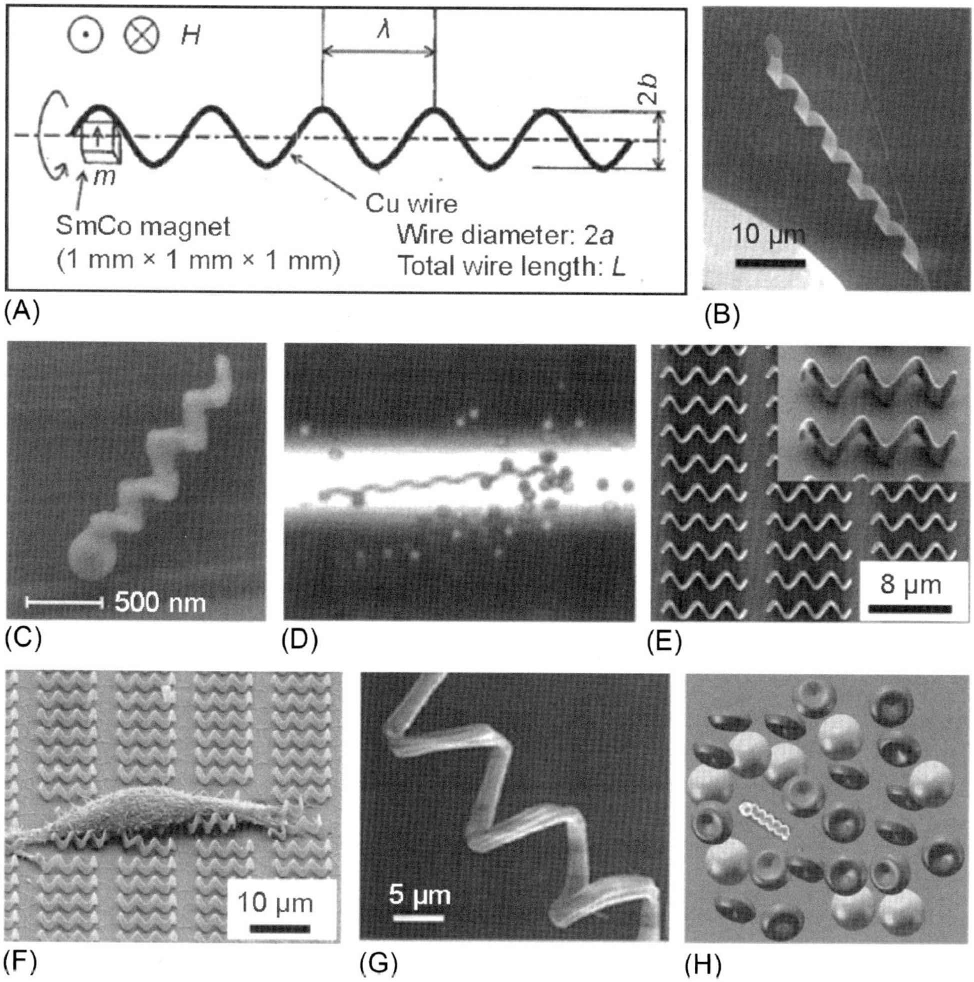

FIG. 8.8 Examples of magnetic helical micro- and nanorobots. (A) Prototype of a centimeter scale magnetic helical robot. (B) The first microscale magnetic helical robot. (C) The first nanoscale helical propeller. (D) Manipulation of fluids. (E) Examples of magnetic helical microrobots fabricated using 3D laser lithography. (F) Other magnetic helical microrobots designed for use with cells. (G) Helical microstructures derived from plants. (H) Nanohelix in blood [26].

Two general strategies have been used to date for the creation of chemical micromotors. The first strategy relies on the continuous formation of bubbles in the cavity of a tubular micromotor and the subsequent expulsion of the bubbles from one end of the tube. These bubble-propelled micromotors rely on the interaction of a catalyst within the micromotor with the surrounding liquid. For example, a platinum-coated micromotor reacting with hydrogen peroxide fuel will yield water and gaseous oxygen that manifests as bubbles. The primary challenge with this approach is finding biocompatible fuels that are efficient and have good endurance [27]. For example, as shown in Fig. 8.9, tubular

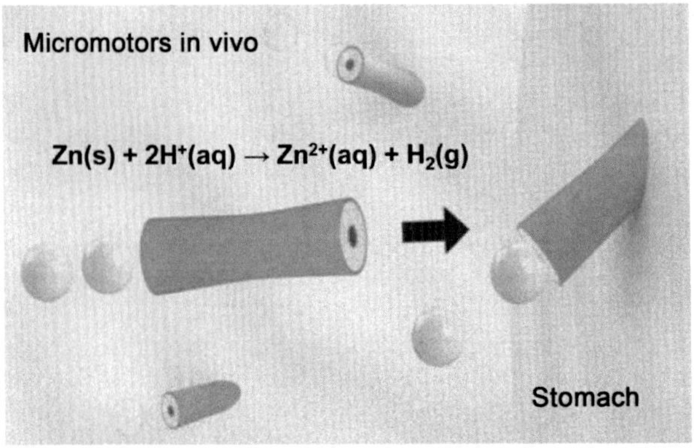

FIG. 8.9 Zinc-based micromotors in gastric acid. Bubbles that propel the motor are generated as a consequence of the reaction between the zinc body and the gastric acid in the stomach. The zinc body of the motor dissolves in the acid so it does not leave behind any potentially harmful chemicals. *Reproduced from H. Wang, M. Pumera, Fabrication of micro/nanscale motors, Chem. Rev. 115(6) (2015) 8704–8735. https://pubs.acs.org/doi/abs/10.1021%2Fnn507097k. All permission requests should be directed to the American Chemical Society.*

micromotors were created using dissolved zinc in gastric acid. The motors show good potential for use as a drug delivery vehicle, however, the efficiency of these motors was limited [27, 28]. Future work is needed to investigate the use of additional fuels that are normally present in the body, such as glucose, urea, or other bodily fluids, and to develop strategies for controlling the chemical-powered motors. Efforts are currently underway to utilize external stimuli, such as light, temperature, and ultrasound, to selectively impact the bubble ejection [19]. The net effect is to tune the magnitude of the momentum generated by the bubble ejection to control the speed of the microswimmer [19]. External stimuli are also being investigated as a mechanism for terminating swimming when the microrobot reaches its target location [19].

An alternate strategy relies on self-phoresis, in which the motor generates its own local concentration gradient or field for propulsion. In self-diffusiophoretic micro- and nanomotors, concentration gradients of solute across the particle interfacial region cause water to flow from regions of low to high solute concentrations [29]. The induced flow generates a thrust force and propels the motor [19, 29]. Self-phoresis is an effective strategy for micromotors smaller than 3 μm since the forces produced are limited [29].

8.5.3 Magnetically Powered Propulsion

Magnetic fields are well suited to remotely power and control microscale robots, particularly in remote, inaccessible spaces [19, 30]. Magnetic fields can penetrate biological tissue and other materials easily and safely. Magnetic fields and torques can be independently applied in a time-varying manner to enable a broad range of actuation options [19]. The magnetic forces and torques are also relatively strong in comparison to other external stimuli, such as light or chemical signals [30]. Magnetic materials are also easily integrated into microfabrication methods [19].

The fundamental mechanism underlying magnetic actuation relies on the use of time-varying magnetic fields and their spatial gradients. In an inhomogeneous magnetic field, the magnetic moment of a microrobot is attracted to the region of greater magnetic flux density [19]. The attraction causes the microrobot to experience a greater magnetic force [19]. If the external field remains invariant, the magnetic force will scale with l^3 as a function of the volume of the material. Magnetic torques are induced when the magnetic moment in an external magnetic field undergoes a restoring torque to align it with the field [19]. Rotating magnetic fields will cause unconstrained microrobots to rotate [19, 30]. Rotating fields can be used to generate the nonreciprocal motions needed for propulsion in low Reynolds number flows. For example, rotating magnetic fields are ideally suited for the rotation of helical propellers. Forces from magnetic torques scale with l^2, thus swimming is more frequently accomplished using torque-induced rotations [19, 30].

The magnetic forces and torques developed on a uniformly magnetized object within an externally imposed magnetic field can be expressed as

$$\vec{f_m} = \left(\vec{M} \cdot \nabla\right) \vec{B} \tag{8.19}$$

and

$$\vec{\tau_m} = \vec{M} \times \vec{B} \tag{8.20}$$

respectively. In Eqs. (8.19), (8.20), f_m is the magnetic force developed, τ_m the magnetic torque developed, $\vec{M}$ the total dipole moment of the magnetic body, and $\vec{B}$ is the flux density of the applied field [31]. The total dipole moment $\vec{M}$ can be calculated as the product of the magnetic body volume and the average magnetization. Eq. (8.14) indicates that, since the magnetic force is proportional to the gradient of the magnetic field, an object in the field will move to a local maximum [31]. Since magnetic torque is proportional to the magnetic field, as indicated in Eq. (8.15), a magnetized object will become aligned with the field [31].

Independent magnetic torques and forces can be generated by controlling both the magnetic field and its gradients [19]. Magnetic fields can be supplied by magnetic coils, such as Helmholtz or Maxwell coils, or using large permanent magnets outside of the microrobots' workspace. Magnetic coils offer the advantage of delivering varying fields with no moving parts [19, 31]. A Helmholtz coil pair consists of two circular coils positioned on the same axis and separated by a distance equal to the radius of the coil [31]. If identical currents are passed through the coils in the same direction, a uniform magnetic field can be generated. If the electromagnets are combined, more complex magnetic fields can be created that enable motions with multiple degrees of freedom [31]. A 3-axis Helmholtz coil system consists of three orthogonal Helmholtz coil pairs. This arrangement, shown in Fig. 8.10A, can generate a uniform rotating magnetic field around any axis in the three-dimensional (3D) space. The 3-axis Helmholtz coils are widely used for the propulsion of helical swimming robots [31].

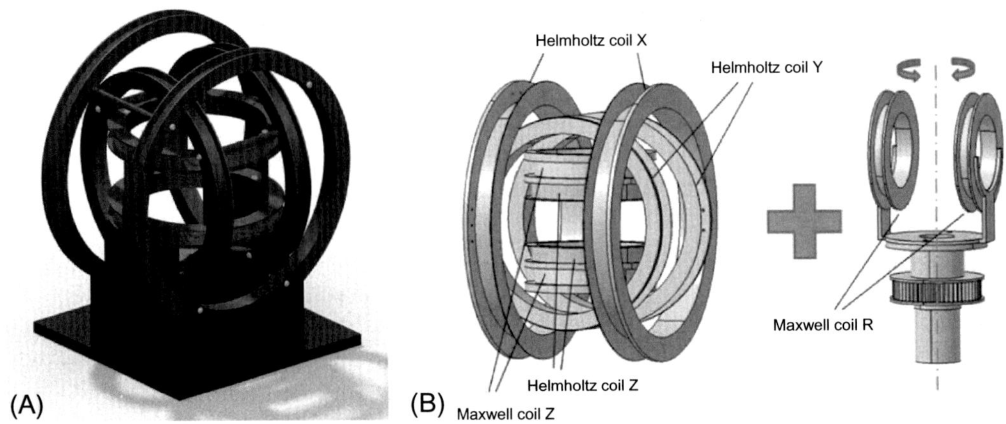

FIG. 8.10 (A) Three-axis Helmholtz coil. (B) EMA system consisting of three pairs of Helmholtz coils and two pairs of Maxwell coils. *(A) Reproduced from http://www.directvacuum.com/shopexd.asp?id=145. (B) Reproduced from C. Yu, J. Kim, H. Choi, J. Choi, S. Jeong, K. Cha, J. Park, S. Park, Novel electromagnetic actuation system for three-dimensional locomotion and drilling of intravascular microrobot, Sensors Actuators A: Phys. 161(1–2) (2010) 297–304.*

A Maxwell coil pair contains coils separated by $\sqrt{3}$ times the radius and currents pass in the opposite direction [31]. Combinations of Maxwell and Helmholtz coils can be used to provide faster control and to reduce mechanical noise in the system [31]. Fig. 8.10B depicts a novel electromagnetic actuation system (EMA) consisting of multiple pairs of Helmholtz and Maxwell coils that can be used to manipulate the position of an intravascular microrobot in three dimensions. The Helmholtz coils align the microrobot in a particular direction while the Maxwell coils propel the robot in the aligned direction [32]. The Helmholtz coils can also be used to rotate the microrobot [32].

Permanent magnets can be used as an alternate approach to generate large fields without the use of high electrical currents [19]. A single, rotating permanent magnet mounted on a linear stage or robotic arm can be used to generate a periodic magnetic field with multiple degrees of freedom. Permanent magnet systems are better suited for in vivo tasks when the organisms are too large to fit into an electromagnetic system [31]. However, the magnetic field falls off at a shorter distance with a permanent magnet system than with an electromagnetic system. In general, the strength of a magnetic field decreases with distance (d) as d^{-3} [30]. Gradients in the magnetic field decrease as d^{-4}. Fig. 8.11 shows examples of rotating permanent magnet systems. Fig. 8.11A depicts a rotating magnetic system mounted on a robotic arm while Fig. 8.11B shows two synchronized rotating magnets mounted on two linear stages.

Magnetically actuated robots capable of crawling, walking, and swimming have been widely investigated [19, 30, 31]. Fig. 8.12 shows a crawling microrobot (Fig. 8.12A) that is actuated by a magnetic navigation system (Fig. 8.12B). The microrobot is capable of crawling forward and backward and can anchor in tubular environments [33]. One particular challenge associated with crawling microrobots is the need to overcome high surface friction and adhesion [30]. Vibrating motions can be used to periodically break the adhesion and enable controlled motions with constant velocity [30]. Fig. 8.13 shows a swimming

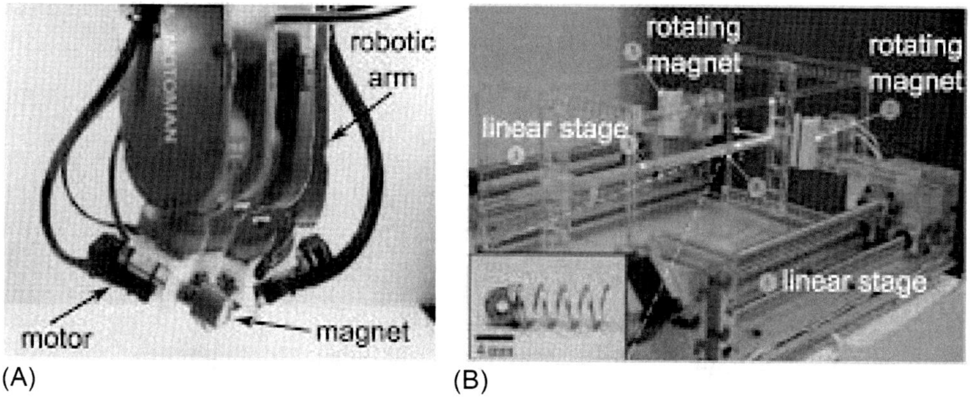

FIG. 8.11 Rotating permanent magnet systems for control of microrobots. (A) Rotating system mounted on a robotic arm. (B) Multiple rotating magnets mounted on two linear stages. *Reproduced from T. Xu, J. Yu, X. Yan, H. Choi, L. Zhang, Magnetic actuation based motion control for microrobots: an overview, Micromachines 6 (2015) 1346–1364.*

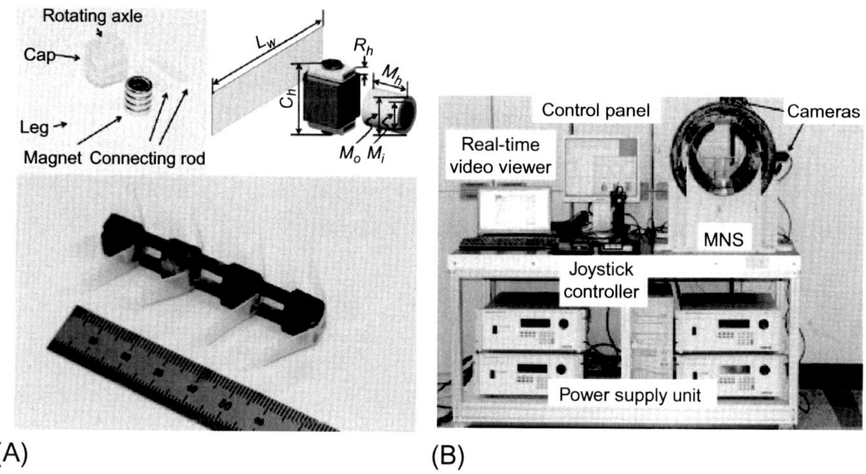

FIG. 8.12 (A) Magnetically actuated crawling microrobot. (B) Helmholtz coil system used for magnetic propulsion. *Reproduced from J. Nam, S. Jeon, S. Kim, G. Jang, Crawling microrobot actuated by a magnetic navigation system in tubular environments, Sensors Actuators A: Phys. 201 (2014) 100–106.*

microrobot actuated by two Helmholtz coil pairs [34]. The microrobot consists of a buoyant rubber body, a permanent magnet, and a fin that moves in response to the alternating magnetic field [34]. Other approaches for microswimming utilize the torque generated in response to a rotating magnetic field to rotate helical structures and propel the robot forward.

8.5.4 Acoustic Propulsion

Magnetic fields are one of the most common methods used to power and propel microrobots. The use of acoustic fields is an emerging method for off-board propulsion and manipulation [19]. An acoustic radiation force occurs when momentum is transferred

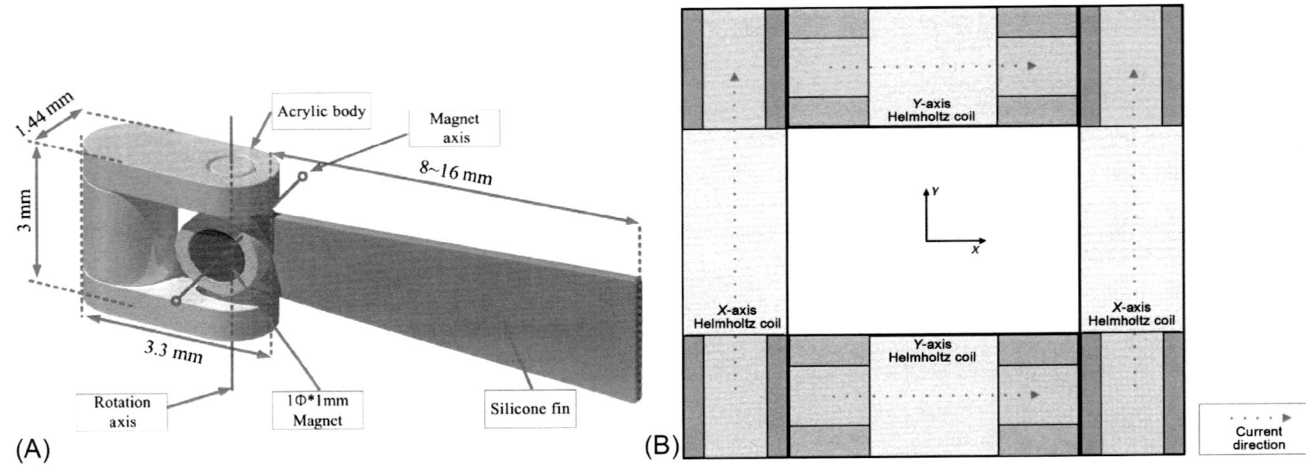

FIG. 8.13 (A) Magnetically powered swimming microrobot. (B) The robot is actuated by two pairs of Helmholtz coils. *Reproduced from D. Byun, J. Choi, K. Cha, J. Park, S. Park, Swimming microrobot actuated by two pairs of Helmholtz coils system (2011).*

from an ultrasound wave to the surrounding medium. In a resonator, force can be generated by a standing wave reflecting back and forth [19]. A hydrodynamic drag force is then created which can be used to drive microobjects, such as a microrobot, to the sound pressure nodes and antinodes. The nodes and antinodes correspond to the minimum and maximum amplitude points, respectively [19]. For microrobotic propulsion, the wave functions could be altered to generate predictable, dynamic movements. The method was used in a swimming robot that utilized an acoustically actuated flagellum for propulsion. An acoustically actuated flagellum was used to propel a swimming microrobot. Small amplitude oscillations of the tail were generated in the presence of traveling acoustic waves [19].

Acoustic streaming can also be used to provide directional propulsion forces in fluids. Acoustic streaming is a nonlinear effect where a steady flow in a fluid is generated in response to the absorption of high amplitude acoustic oscillations. Acoustic streaming was used to propel a microrobotic structure consisting of an array of comb-like cavities. Ultrasound was used to oscillate trapped air bubbles in the cavities at their resonant frequency and to propel the microrobot in the direction opposite to the acoustic streaming [19]. The utilization of acoustic waves for robotic propulsion is still an active area of exploration, but could hold great promise for the future.

8.5.5 Optical/Thermal-Powered Propulsion

Heat or momentum transfer can be used to remotely actuate microrobots. Thermal gradients across air-liquid interfaces will induce fluid flow in the direction of the lower temperature region since surface tension is impacted by temperature [19]. An alternate method of propulsion uses localized heating and thermocapillary effects to propel microbubble robots. Opto-thermocapillary flow-addressed bubble microrobots rely on the use of light patterns to generate thermal gradients in a liquid [35]. A focused light is used to create a heated region on the bubble that can be used to move the bubble toward the light pattern [35]. Opto-thermocapillary manipulation for propulsion of microrobots, however, requires a specialized setup to transduce light into heat and is still in the early phases of investigation [19, 35].

Thermal expansion can also be used to directly propel microrobots. If a focused laser is used to apply an asymmetric excitation, materials can be heated and cooled to undergo repeated periods of expansion and contraction. Thermal expansion was used to generate crawling motions in a microrobot resting on asymmetric, thin metal bimorph legs [19]. As the legs were heated and cooled, expansion and contraction of the joints in the legs occurred and the microrobotic underwent walking-like motions. Varying intensity of the light was used to control step size. A similar approach was used to generate walking motions in a polyamide joint actuator using electrical heating. By using polysilicon heaters, local heating at the joints could be used to thermally expand the joint and generate walking motions [36]. A minimum device size of approximately 5 μm is required since below this size, the entire device would be heated rather than just an individual leg or joint [19].

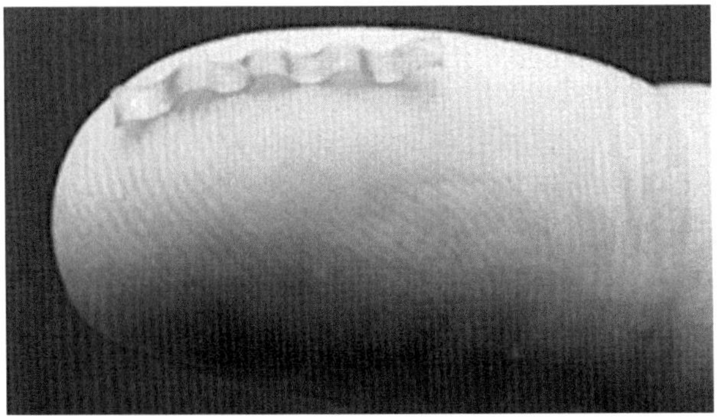

FIG. 8.14 Soft caterpillar robot made from an optomechanical liquid crystalline elastomer. *Reproduced with permission from the Faculty of Physics, University of Warsaw, Poland.*

Modulated traveling light waves can also be used to propel microrobotic devices. A soft caterpillar robot, shown in Fig. 8.14, was created using an optomechanical liquid crystalline elastomer monolith [37]. The body of the robot was made from a light-sensitive elastomer material with patterned molecular alignment. Traveling deformations were induced in the robot as a result of sequential, nonreciprocal illumination of the patterned and unpatterned regions using laser light [37]. The traveling deformation patterns moved along the length of the structure and caused the microrobot to move. Varying deformation pattern was used to implement different gaits on horizontal surfaces [37].

8.5.6 Propulsion Through Integration With Motile Organisms

All of the previously described methods for propulsion of micro- and nanorobots rely on off-board mechanisms for propulsion. Significant efforts have been undertaken recently to use the power of microorganisms for on-board actuation [19]. Microorganisms, such as bacteria, algae, or spermatozoa, and muscle cells are in the size range of 1–20 μm and operate at the same scale as many microrobotic devices. These microorganisms are highly efficient in converting chemical energy from their surroundings to useful mechanical work. Furthermore, microorganisms can respond to forces, strain, and chemicals in their environment via elaborate on-board control pathways. Control of microrobotic devices using off-board propulsion methods is particularly challenging. The biohybrid approach to micro- and nanorobotic propulsion relies on the integration of single cells, tissue, and microorganisms to create fully functional microrobots [19]. Biohybrid systems are already adapted to function in a physiological environment making them ideally suited for use in physiological environments.

Flagellated eukaryotic and prokaryotic species, such as bacteria or spermatozoa, can generate thrust forces of the orders of 10^{-3} and 10^{-1} nN, respectively [19]. Forces of these magnitudes are sufficient to power microrobots, but controlling the actuation can become problematic. Several strategies have been undertaken to use the natural responses of cell or organisms in a physiological environment for directional cues. One approach relies on utilizing natural responses to changes in pH, temperature, chemicals, and light [19]. Flagellated bacteria, such as *E. coli* or *Serratia marcescens*, will move away from strong acidic or alkaline pH environments [38]. The pH-sensing capabilities of these organisms can be exploited to achieve drift control for microrobots. The pH-based control could be particularly useful for targeting cancerous tumors since tumors have been shown to have a lower pH than surrounding tissue [38]. Zhuang et al. [38] demonstrated that pH-sensing, bacteria-propelled microrobots could be synthesized using polystyrene beads and *S. marcescens*. *S. marcescens* is a highly agile, multiflagellated species that swims efficiently in low Reynolds number environments. Spatial pH gradients were used to generate drift motions in the biohybrid microrobots [38].

Biohybrid robots consisting of polystyrene beads and *S. marcescens* were also used to demonstrate on/off control in response to chemical changes [39]. Copper ions were introduced to stop flagellar motions. Ethylenediaminetetraacetic acid was used to resume motion. A similar biohybrid robot configuration was used to demonstrate that localized control could be accomplished by exposing the microrobots to ultraviolet light [40].

Magnetic fields can also be used to control biohybrid robots. Two strategies for utilizing magnetic fields for the control of biohybrid robots rely on the use of magnetotactic bacteria or the addition of microorganisms to magnetic particles. Magnetotactic bacteria utilize flagella for propulsion and contain intracellular membrane-surrounded magnetosomes [41]. Magnetotactic bacteria synthesize chains of nanoscale magnetic particles within the magnetosomes and can passively orient in the direction of the Earth's magnetic field. The bacteria utilize this feature to navigate along geomagnetic field lines and swim in a fixed magnetic direction. MO-1 is a type of magnetotactic bacteria that has seven individual flagella for propulsion and can swim along a fixed magnetic direction making it ideally suited for integration into a biohybrid robot [41]. Biohybrid robots were synthesized by Ma et al. [41] using microbeads and ovoid MO-1 cells. The MO-1 cells were used for both propulsion and navigation. The microrobots were shown to respond to rotating magnetic fields in a specially developed electromagnetic coil device highlighting the utility of using magnetic fields for the control of biohybrid robots.

The addition of microorganisms to magnetic beads can also be used to create biohybrid robots that can be controlled through magnetic fields. Carlsen et al. [42] attached multiple *S. marcescens* bacteria to a superparamagnetic bead to investigate the use of magnetic fields for controlling biohybrid robots. Weak magnetic fields were applied and successfully used to control swimming along two-dimensional (2D) trajectories.

Cells, such as cardiomyocytes or muscle cells, are also being investigated for their use as potential propulsion mechanisms for biohybrid robots. The use of cells represents a bottom-up approach in which cells are assembled together to provide propulsion mechanisms or form robotic actuators [43]. Skeletal and cardiac muscle cells generate a directional tension during contraction of the order of 10^3 to 10^5 nN in magnitude [19]. When seeded onto an artificial substrate, the cells can be used to provide propulsive forces.

Chan et al. [44] demonstrated that poly(ethylene glycol) diacrylate hydrogels and cardiomyocytes could be used to create a walking microrobot. The biohybrid microrobot

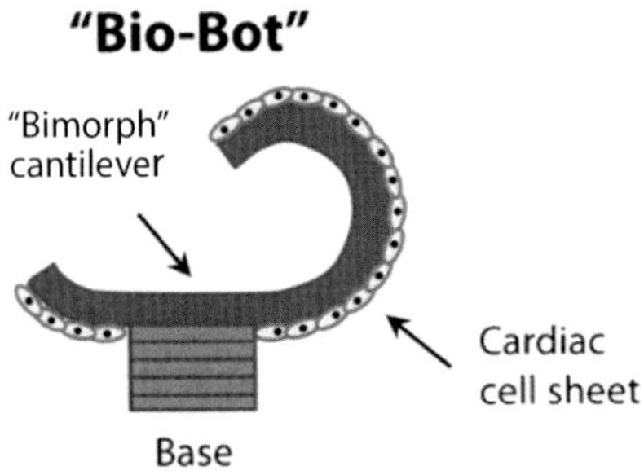

FIG. 8.15 Biohybrid robot powered by cardiomyocytes. Contraction of the cardiomyocyte sheet leads to actuation of the cantilever and production of a power stroke. *Reproduced from V. Chan, K. Park, M.B. Collens, H. Kong, T.A. Saif, R. Bashir, Development of miniaturized walking biological machines, Sci. Rep. 2 (2012) 857.*

"Bio-Bot" consisted of a base structure, a bimorph cantilever structure, and cardiomyocyte sheets [44]. The cardiomyocytes were seeded onto the surface of the cantilevers to generate contractile motions as shown in Fig. 8.15. As the cardiomyocytes contract, the cantilever deforms leading to the production of a power stroke and locomotion [44].

Williams et al. [45] showed that cardiomyocytes could be used in a swimming microrobot. The swimmer consisted of a polydimethylsiloxane filament with a rigid head and tail on which cardiomyocytes were cultured [45]. As the cardiomyocytes contracted, the filament deformed to propel the microswimmer. Fig. 8.16 shows a schematic of the biohybrid swimmer. The contractile cells are used to generate force to propel the microrobot. Similar approaches could be undertaken using skeletal muscle cells to generate swimming or walking motions.

Cardiomyocytes and muscle cells normally respond to chemicals or an electric field, but can be engineered to express light-responsive proteins so that the cells respond optogenetically. Modifying the cells to respond to light makes them inherently easier to

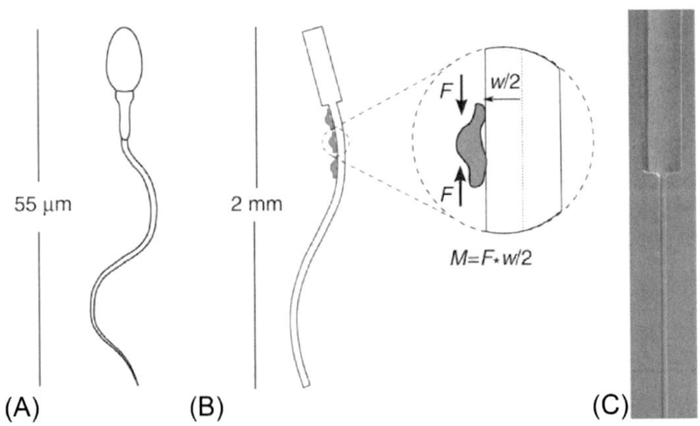

FIG. 8.16 Biohybrid microswimmer that utilizes cardiomyocytes for propulsion. The contractile cells generate a force (*F*) at a distance *w*/2 from the filament's neutral axis. *Reproduced from B. William, S.V. Anand, J. Rajagopalan, M. Taher, A. Saif, A self-propelled biohybrid swimmer at low reynolds number, Nat. Commun. 5 (2014).*

control, particularly for microrobotic applications. Park et al. [43] utilized optogenetic muscle cells to control a biohybrid ray-like robot. The velocity and the direction of the robot were largely controlled by modulating the frequency of light stimuli. Optogenetic modification of cells is an emerging strategy with the potential to improve the control of biohybrid microrobots.

Biohybrid robots have great potential for use in a broad range of applications including drug delivery, tumor targeting, tissue engineering, imaging, and lab-on-a-chip or organ-on-a-chip applications. However, despite their potential, several challenges still remain. Biohybrid robots inherently have a limited lifetime and may have a high sensitivity to environmental conditions. Biohybrid microrobots could also elicit pathogenic responses from the host. Microorganisms also inherently undergo stochastic processes making their control particularly challenging. Despite these many issues, the integration of microorganisms or cells has tremendous potential for providing both propulsion and control of micro- and nanorobots.

8.6 Applications of Micro- and Nanorobots at the Molecular and Cellular Levels

The size of micro- and nanorobots make them ideally suited to perform a range of tasks and functions at the cellular and molecular levels. The following sections describe a broad range of existing and potential applications for these devices.

8.6.1 Imaging

Imaging is one of the most powerful methods for diagnosing diseases. Current methods for imaging utilize flexible endoscopes or catheters, however, they can only be utilized in the body for limited periods of time [12]. Micro- and nanorobots have the potential to provide long-term visual imaging and access small regions of the body that cannot currently be visualized. Pill-sized endoscopes containing an on-board camera, wireless transmission device, and battery currently serve as an alternative to large-scale endoscopes. However, the pill sized endoscopes are passive devices. Capsule-like milli- or microrobots would provide the same functionality as pill-sized endoscopes, but would enable active imaging and could be propelled to hard-to-access regions in the body.

Remote, off-board magnetic actuation is currently being investigated as a means of navigating capsule-like robots in the gastrointestinal (GI) tract. A magnetically actuated soft capsule endoscope with an integrated camera was proposed by Yim and Sitti [12, 46] as a method for actively imaging the lining of the stomach. The capsule was designed and fabricated from an elastomeric material to allow for passive deformation during robot-tissue contact interactions and extra degree-of-freedom actuation [46]. The device was designed to be swallowed to reach the stomach and an external magnet was used to roll it inside the stomach. The feasibility of the device was demonstrated in vitro using a synthetic stomach surface [46].

8.6.2 Diagnosis and Health Monitoring

Biosensors are currently used to monitor a range of physiological factors both inside and outside of the human body. Biosensors can be used intermittently or continuously to measure or detect factors such as glucose, pH, temperature, oxygen, viral or bacterial activity, body motions, blood pressure, respiration, muscle activity, neural activity, and pulse rate [12]. Integrating biosensors with milli-, micro-, or nanorobots could enable active health monitoring that would allow for detection and diagnosis of pathological activity or abnormal conditions [12]. Ergeneman et al. utilized a magnetically controlled microrobot to perform optical oxygen sensing inside the eye [47]. Biosensing microrobots are still in the investigational stage, but represent an opportunity for future development.

8.6.3 Surgical

Microrobots could also be used for minimally invasive surgery. Examples of surgical procedures that could be performed by a microrobot include opening of clogged vessels or other channels, cauterization, treatment of hyperthermia, biopsies, electrical stimulation, injection, cutting, drilling, or biomaterial removal [12]. Several groups to date have investigated the use of integrated biopsy tools for capsule endoscopes to collect tissue samples for diagnostic purposes [48, 49]. Expansion of these capabilities into surgical microrobots would enable more precise targeting of sampling locations using a minimally invasive surgical system. Others have proposed magnetic robots that could be used to puncture or open clogs in blood vessels [50–52]. To date, only a limited number of studies on the use of microrobots for minimally invasive surgery have been undertaken. However, as methods for propulsion and control have improved, microrobots could find a broad range of potential applications throughout the human body [12].

8.6.4 Cell Manipulation

Cell manipulation is important in vitro for single cell analyses, microfluidic studies, drug discovery, or lab-on-a-chip applications. Cell manipulation is traditionally done using micromanipulators. Micromanipulators utilize microscale end effectors connected to macroscale actuators to manipulate single cells [12]. Microrobots are increasingly used for single cell manipulation in vitro since they offer a more compact means of manipulating single cells and do not rely on the dexterity of the operator. Microrobots are likely to become an alternative to traditional micromanipulators. Applications in which microrobots have been used to date include microtransportation for cell culture, delivery of drug compounds to modulate intracellular communication, and arrangement of cells into particular configurations for research [12, 53]. Other studies have sought to use microrobots to stimulate cellular responses to forces, to distinguish abnormal cells, and to perform positioning and cutting procedures on single cells [12, 54]. As control strategies for microrobots improve, the potential applications for single cell manipulations are likely to increase significantly. Swarms of microrobots could be used in cell cultures to study cellular communications. Microrobots could also be released into cell cultures to detect abnormal cells, such as tumor cells. In the long term, microrobots could be introduced in vivo to perform the same types of single cell manipulations in an intact, physiological system.

8.6.5 Cancer Detection and Treatment

The detection and treatment of cancer is one of the most promising applications of microrobots. Microrobots could precisely target cancerous cells or tumors without the need for systemic treatments, such as chemotherapy. The feasibility of several approaches has been demonstrated to date. Hyperthermia induced by microrobots is a promising method to target cancer cells, without causing damage to the surrounding healthy tissue [19]. The feasibility of delivering cancer-targeting drugs using spirulina algae coated with magnetic microparticles is also a promising strategy for cancer treatment [55]. One of the most promising approaches for cancer treatment involves the use of spinning nanorobots to penetrate cell membranes in response to ultraviolet light [56].

8.6.6 Drug Delivery and Targeted Therapy

Micro- and nanorobots hold great potential for delivering therapeutics such as drugs, mRNA, genes, radioactive seeds, imaging contrast agents, stem cells, and proteins to specific targeted regions inside the body [12]. By delivering the therapeutics to specific targeted locations, potential side effects can be minimized and more potent amounts of the substances could be delivered. Small-scale robots have been successfully used for targeted drug delivery in the GI tract and larger blood vessels using either passive or active drug release methods. For example, milliscale robots can be propelled using magnetic field gradients and drug release from capsules or bubbles can be triggered using ultrasound or magnetic fields [12]. Similarly, micro- or nanorobots could be used to deliver drugs and therapeutics to vessels smaller than arterioles (less than 150 μm), however, propulsion and control of these robots is more challenging. Active release of drugs from microrobots is still problematic. Fusco et al. [57] used microswimming robots coated with drugs to deliver them to a target location via passive diffusion. Others, such as Carlsen et al. [42], used magnetically controlled biohybrid robots embedded with drug microparticles for targeted delivery via passive diffusion. Future research into active delivery mechanisms or delivery using microrobot swarms has the potential to transform the way drugs and therapeutics are delivered in vivo.

8.6.7 Tissue Engineering

The capability to deliver differentiated stem cells to precise locations in the body using microrobots could have a profound impact on the field of tissue engineering. Delivery of stem cells could enable tissue regeneration at pathological sites and enable treatment of a broad range of diseases [12]. Microrobots could also be used to manipulate nonfunctionalized microgels or cells in vitro for the synthesis of tissue constructs that can be implanted in vivo. Tasoglu et al. [58] used crawling microrobots to arrange cell laden microgels for the synthesis of tissue constructs. Microrobots could also be used to deliver nutrients or growth factors to target regions in vitro or in vivo in the future. As the field of tissue engineering continues to expand, it is likely that the microrobots will play an increasingly important role in cell delivery, micromanipulation, delivery of therapeutic agents needed for synthesis of tissue constructs, and imaging of engineered tissue in vivo.

8.6.8 Detoxification

Another application in which micro- and nanorobots are finding utility is for destroying potentially toxic bacteria. Direct contact between an *E. coli* bacterium and a silver-coated microrobot was used to disintegrate the membrane of the bacteria [12]. These microrobots are finding applications in the detoxification of water supplies. Microrobots could also be used to heat localized areas to destroy bacteria [12]. Single or swarms of microrobots could thus be used to destroy bacterial colonies. Microrobots could also be used to deliver drugs or therapeutics aimed at reducing bacterial populations.

8.7 Future Perspective

Micro- and nanorobots are likely to have a promising future in bioengineering and healthcare applications. No other devices are capable of actively performing a broad range of tasks at the same scale as cells and molecules. Thus, the potential exists to fundamentally transform the ways in which compounds are delivered to cells and tissues, surgeries are performed, imaging methods are implemented, and tissue is regenerated. Microrobots could be used to develop new transformative strategies for the treatment of chronic disease states such as cancer, cardiovascular disease, liver disease, respiratory diseases, and kidney disease. Microrobots could also be used to treat everyday ailments more cost-effectively and efficiently or to aid in the development of novel devices or therapeutic strategies.

Although many advances have been made to develop devices at the micro- and nanoscales, several challenges still exist. Further advances are needed to effectively provide power at the microscale, to control synthetic robots using on-board control strategies, to microfabricate the devices, and to ensure that the devices can be safely used in humans with no adverse consequences. However, given the rapid rate of advances in microsystems and nanotechnology, the field of micro- and nanorobotics is likely to continue to expand rapidly until these challenges are overcome. The field is becoming increasingly close to what science fiction writers envisioned in the 1966 film “Fantastic Voyage” in which a miniaturized submarine was transported within the human vasculature to deliver a cure to a diseased region of the brain.

References

[1] M. Sitti, Mobile Microrobotics, MIT Press, Cambridge, MA, 2017.

[2] B.S. Yamaan Saadeh, D. Vyas, Nanorobotic applications in medicine: current proposals and designs, Am. J. Robot. Surg. 1 (2014) 4–11.

[3] T. Hsu, MEMS and Microsystems: Design, Manufacture, and Nanoscale Engineering, John Wiley & Sons, 2008.

[4] M.A. Peterson, Galileo's discovery of scaling Laws, Am. J. Phys. 70 (2002) 575.

[5] D. Christensen, E.W. Hawkes, S.A. Suresh, K. Ladenheim, M.R. Cutkosky, μTugs: enabling microrobots to deliver macro forces with controllable adhesives, in: 2015 IEEE International Conference on Robotics and Automation (ICRA), 2015.

[6] Y. Meng, Electrostatic field effects on adhesion, in: Q.J. Wang, Y.W. Chung (Eds.), Encyclopedia of Tribology, Springer, Boston, MA, 2013.

[7] F.L. Leite, C.C. Bueno, A.L. Da Róz, E.C. Ziemath, O.N. Oliveira, Theoretical models for surface forces and adhesion and their measurement using atomic force microscopy, Int. J. Mol. Sci. 13 (10) (2012) 12773–12856, doi:10.3390/ijms131012773.

[8] J. Laven, J.P.C. Vissers, The Hamaker and the Lifshitz approaches for the Van der Waals interaction between particles of composite materials dispersed in a medium, Colloids Surf. A Physicochem. Eng. Asp. 152 (1999) 345–355.

[9] K. Autumn, Y.A. Liang, S.T. Hsieh, W. Zesch, W.P. Chan, T.W. Kenny, R. Fearing, Adhesive force of a single gecko foot hair, Nature 405 (6787) (2000) 681.

[10] W.R. Hansen, K. Autumn, Evidence for self-cleaning in gecko setae, Proc. Natl. Acad. Sci. 102 (2) (2005) 385–389.

[11] J.M. Cimbala, Y.A. Cengel, Essentials of Fluid Mechanics: Fundamentals and Applications, McGraw Hill, New York, 2008.

[12] M. Sitti, H. Ceylan, W. Hu, J. Giltinan, M. Turan, S. Yim, E. Diller, Biomedical applications of untethered mobile milli/microrobots, Proc. IEEE: Inst. Electr. Electron. Eng. 103 (2) (2015) 205–224, doi:10.1109/JPROC.2014.2385105.

[13] F.D. DeBona, E.T. Enikov, Microsystems Mechanical Design, Springer Science & Business Media, 2007.

[14] http://marketbusinessnews.com/robobees-tiny-robots-fly-use-static-electricity-stick-surfaces/135513.

[15] H. Zeng, Polymer Adhesion, Friction, and Lubrication, John Wiley & Sons, Inc., 2013

[16] A. Berman, J. Israelachvili, Control and minimization of friction via surface modification, in: B. Bhushan (Ed.), Micro/Nanotribology and its Applications. NATO ASI Series (Series E: Applied Sciences), vol. 330, Springer, Dordrecht, 1997.

[17] J. Li, B. de Avila, W. Gao, L. Zhang, J. Wang, Micro/nanorobots for biomedicine: delivery, surgery, sensing, and detoxification, Sci. Robot. 2 (4) (2017).

[18] M. Sitti, H. Ceylan, W. Ju, J. Giltinan, M. Turan, S. Yim, E. Diller, Biomedical applications of untethered mobile milli/microrobots, Proc. IEEE 103 (2) (2015) 205–224.

[19] H. Ceylan, G. Giltinan, K. Kozielski, M. Sitti, Mobile microrobots for bioengineering applications, Lab Chip 17 (2017) 1705–1724.

[20] J. Elgeti, R.G. Winkler, G. Gompper, Physics of microswimmers-single particle motion and collective behavior, Rep. Prog. Phys. 78 (5) (2015).

[21] S. Kim, S. Lee, J. Lee, B. Nelson, L. Zhang, H. Choi, Fabrication and manipulation of ciliary microrobots with non-reciprocal magnetic actuation, Sci. Rep. 6 (2016).

[22] I.S. Khalil, H.C. Dijklag, L. Abelmann, S. Misra, MagnetoSperm: a microrobot that navigates using weak magnetic fields, Appl. Phys. Lett. 104 (2014) 223701.

[23] https://commons.wikimedia.org/wiki/File:Cross_section_of_a_Chlamydomonas_reinhardtii_algae_cell,_a_3D_representation.jpg (accessed 15.08.18).

[24] N. Mori, K. Kuribayashi, S. Takeuchi, Artificial flagellates: analysis of advancing motions of biflagellate micro-objects, Appl. Phys. Lett. (8) (2010) 96.

[25] http://bio1151.nicerweb.com/Locked/media/ch27/flagellum.html.

[26] F. Qiu, B. Nelson, Magnetic helical micro- and nanorobots: toward their biomedical applications, Engineering 1 (1) (2015) 021–026.

[27] M. Medina-Sanchez, O.G. Schmidt, Medical microrobots need better imaging and control, Nature 545 (7655) (2017).

[28] W. Gao, R. Dong, S. Thamphiwatana, J. Li, W. Gao, L. Zhang, J. Wang, Artificial micromotors in the mouse's stomach: a step toward in vivo use of synthetic motors, ACS Nano 9 (1) (2015) 117–123, doi:10.1021/nn507097k.

[29] H. Wang, M. Pumera, Fabrication of micro/nanscale motors, Chem. Rev. 115 (16) (2015) 8704–8735.

[30] J.J. Abbott, K.E. Peyer, L. Dong, B.J. Nelson, How should microrobots swim? in: 13th International Symposium of Robotics Research, Hiroshima, Japan, 2007.

[31] T. Xu, J. Yu, X. Yan, H. Choi, L. Zhang, Magnetic actuation based motion control for microrobots: an overview, Micromachines 6 (2015) 1346–1364.

[32] C. Yu, J. Kim, H. Choi, J. Choi, S. Jeong, K. Cha, J. Park, S. Park, Novel electromagnetic actuation system for three-dimensional locomotion and drilling of intravascular microrobot, Sensors Actuators A Phys. 161 (1–2) (2010) 297–304.

[33] J. Nam, S. Jeon, S. Kim, G. Jang, Crawling microrobot actuated by a magnetic navigation system in tubular environments, Sensors Actuators A Phys. 201 (2014) 100–106.

[34] Byun, D., Choi, J., Cha, K., Park, J., & Park, S. (2011). Swimming microrobot actuated by two pairs of Helmholtz coils system.

[35] W. Hu, Q. Fan, A.T. Ohta, Interactive actuation of multiple opto-thermocapillary flow-addressed bubble microrobots, Robot. Biomimet. 1 (1) (2014) 14, doi:10.1186/s40638-014-0014-3.

[36] T. Ebefors, J.U. Mattsson, E. Kalvesten, G. Stemme, A walking silicon micro-robot, in: 10th International Conference on Solid-State Sensors and Actuators, Sendai, Japan, 1999.

[37] M. Rogóż, H. Zeng, C. Xuan, D.S. Wiersma, P. Wasylczyk, Light-driven soft robot mimics caterpillar locomotion in natural scale. Adv. Optical Mater. 4 (2016) 1689–1694, doi:10.1002/adom.201600503.

[38] J. Zhuang, R.W. Carlsen, M. Setti, pH-taxis of biohybrid microsystems, Sci. Rep. 5 (2015).

[39] B. Behkam, M. Setti, Bacterial flagella-based propulsion and on/off motion control of microscale objects, Appl. Phys. Lett. 90 (2007), doi:10.1063/1.2431454.

[40] E. Steager, C. Kim, J. Patel, S. Bith, C. Naik, L. Reber, M.J. Kim, Control of microfabricated structures powered by flagellated bacteria using phototaxis, Appl. Phys. Lett. 90 (2007)doi:10.1063/1.2752721.

[41] Q. Ma, C. Chen, S. Wei, C. Chen, L.-F. Wu, T. Song, Construction and operation of a microrobot based on magnetotactic bacteria in a microfluidic chip, Biomicrofluidics 6 (2) (2012) 024107–024112, doi:10.1063/1.3702444.

[42] R.W. Carlsen, M.R. Edwards, J. Zhuan, C. Pacoret, M. Sitti, Magnetic steering control of multi-cellular bio-hybrid microswimmers, Lab Chip 14 (19) (2014) 3850–3859.

[43] L. Ricotti, B. Trimmer, A.W. Feinberg, R. Raman, K.K. Parker, R. Bashir, M. Sitti, S. Martel, P. Dario, A. Menciassi, Biohybrid actuators for robotics: a review of devices actuated by living cells, Sci. Robot. 2 (12) (2017).

[44] V. Chan, K. Park, M.B. Collens, H. Kong, T.A. Saif, R. Bashir, Development of miniaturized walking biological machines, Sci. Rep. 2 (2012) 857.

[45] B. William, S.V. Anand, J. Rajagopalan, M. Taher, A. Saif, A self-propelled biohybrid swimmer at low Reynolds number, Nat. Commun. 5 (2014).

[46] S. Yim, M. Sitti, Design and rolling locomotion of a magnetically actuated soft capsule endoscope, IEEE Trans. Robot. 28 (2012) 183–194.

[47] O. Ergeneman, G. Dogangil, M.P. Kummer, J.J. Abbott, M.K. Nazeeruddin, B.J. Nelson, A magnetically controlled wireless optical oxygen sensor for intraocular measurements, IEEE Sensors J. (8) (2008) 29–37.

[48] K. Kong, J. Cha, D. Jeon, D. Cho, A rotational micro biopsy device for the capsule endoscope, in: Proc. 2005 IEEE/RSJ Int. Conf. Intell. Robots Syst., 2005 (IROS 2005), Edmonton, AB, Canada, 2005, , pp. 1839–1843.

[49] S. Park, K. Koo, G. Kim, S. Bang, S. Song, J. Byun, C. Chu, A novel micro-biopsy actuator for capsular endoscope using LIGA process, in: Proc. 2007 Solid-State Sensors, Actuators and Microsystems Conf; Lyon, 2007, 2007, p. 209.

[50] F. Ullrich, C. Bergeles, J. Pokki, O. Ergeneman, S. Erni, G. Chatzipirpiridis, S. Pane, C. Framme, B. Nelson, Mobility experiments with micro-robots for minimally invasive intraocular surgery, Invest. Ophthalmol. Vis. Sci. 54 (2013) 2853–2863.

[51] C. Yu, J. Kim, H. Choi, J. Choi, S. Jeong, K. Cha, J. Park, S. Park, Novel electromagnetic actuation system for three-dimensional locomotion and drilling of intravascular micro-robot, Sensor Actuat A – Phys. 161 (2010) 297–304.

[52] P. Miloro, E. Sinibaldi, A. Menciassi, P. Dario, Removing vascular obstructions: a challenge, yet an opportunity for interventional microdevices, Biomed. Microdevices 14 (2012) 511–532.

[53] Z. Lu, C. Moraes, G. Ye, C.A. Simmons, Y. Sun, Single cell deposition and patterning with a robotic system, PLoS ONE 5 (2010), e13542.

[54] E.B. Steager, M.S. Sakar, C. Magee, M. Kennedy, A. Cowley, V. Kumar, Automated biomanipulation of single cells using magnetic micro-robots, Int. J. Robot. Res. 32 (2013) 346–359.

[55] X. Yan, Q. Zhou, M. Vincent, Y. Deng, J. Yu, J. Xu, T. Xu, T. Tang, L. Bian, Y. Wang, K. Kostarelos, L. Zhang, Multifunctional biohybrid magnetite microrobots for imaging-guided therapy, Sci. Robot. 2 (12) (2017).

[56] V. Lopez, F. Chen, L.G. Nilewski, G. Duret, A. Aliyan, A.B. Kolomeisky, J.T. Robinson, G. Wang, R. Pal, J. M. Tour, Molecular machines open cell membranes, Nature 548 (2017) 567–572.

[57] S. Fusco, G. Chatzipirpiridis, K.M. Sivaraman, O. Ergeneman, B.J. Nelson, S. Pane, Chitosan electrodeposition for micro-robotic drug delivery, Adv. Healthc. Mater. 2 (2013) 1037–1044.

[58] S. Tasoglu, E. Diller, S. Guven, M. Sitti, U. Demirci, Untethered micro-robotic coding of three-dimensional material composition, Nat. Commun. 2014 (5) (2014) 3124.

Further Reading

[59] https://www.livescience.com/47307-how-geckos-stick-and-unstick-feet.html.

[60] https://commons.wikimedia.org/wiki/File:Water_Strider_on_Lily_Lake_(15058406522).jpg (accessed 14.08.18).

[61] http://www.ramehart.com/surface_tension.htm (accessed 14.08.18).

[62] https://courses.lumenlearning.com/boundless-physics/chapter/bernoullis-equation/.

[63] http://www.directvacuum.com/shopexd.asp?id=145.

9

Prosthetic Limbs

Philipp Beckerle*,†, Steffen Willwacher‡, Minas Liarokapis§, Matthew P. Bowers¶, Marko B. Popovic¶

*TECHNISCHE UNIVERSITÄT DORTMUND, DORTMUND, GERMANY †TECHNISCHE UNIVERSITÄT DARMSTADT, DARMSTADT, GERMANY ‡GERMAN SPORT UNIVERSITY COLOGNE, COLOGNE, GERMANY §THE UNIVERSITY OF AUCKLAND, AUCKLAND, NEW ZEALAND ¶WORCESTER POLYTECHNIC INSTITUTE, WORCESTER, MA, UNITED STATES

CHAPTER OUTLINE

Biomechatronics. https://doi.org/10.1016/B978-0-12-812939-5.00009-4

9.1 Introduction

A prosthesis (plural: prostheses; from the Ancient Greek word πρόσθεσις that means "addition") is a technical aid for people with limb loss or congenital limb absence that aims at substituting missing body parts. Prosthetic limbs have a long history. For example, in ~200 BCE, the Roman general Marcus Sergius Silus lost his right hand during a battle and he ordered an iron hand that allowed him to hold his shield and continue campaigning [1]. This was the first documented example of use of an upper-limb prosthesis.

When a prosthesis is needed to replace a part of the body, the reason for doing so is either described as traumatic (lost by injury), dysvascular, cancer related, or congenital (missing from birth). For example, most lower-limb amputations occur due to complications of the vascular system (pertaining to the blood vessels), especially from diabetes. These types of amputations are known as dysvascular. They account for approximately 82% of all (lower and upper limb) amputations excluding finger or toe amputations. Out of those dysvascular amputations, 97% are lower limbs and 3% are upper limbs [2–4].

The terms unilateral, bilateral, trilateral, and quadrilateral are used to define the level of amputation depending on the number of amputated limbs.

9.1.1 Demographics and Statistics

Nowadays, the global demand for sophisticated prostheses is growing. Statistics indicate that the number of individuals missing one or more limbs is rising due to demographic change and other factors. In terms of prevalence, there are between 1.7 and 2 million people living with limb loss in the United States (current US population is about 321 million) [2]. About 70% of them are lower-limb amputees and 30% are upper-limb amputees [5]. Causes of upper extremity amputation are 8.9% congenital, 8.2% tumor, 5.8% disease, and 77% trauma [6]. Hence, trauma is a major cause for upper extremity amputation. Patients who receive amputation due to trauma tend to be younger, with 72% being <65 years of age [2]. The trauma-related amputation statistics are also distinctly affected by major military conflicts, since injured soldiers become a significant portion of this population.

In terms of incidence, approximately 185,000 amputations occur each year in the United States [7]. The total number of digit-related amputations performed in the United States each year, resulting in the loss of one or more fingers or toes, is about 65,000. Out of roughly 120,000 other amputations each year, based on hospital discharges, 14% of amputations are upper-limb amputations and 86% are lower-limb amputations [4, 8]. Hence, the number of upper extremity amputations that are not digital amputations is about 17,000 per year. Out of this number, 5% (approximately 850) of cases are hand and wrist area amputations, 59% (approximately 10,000) of cases are amputations below the elbow area, 28% (approximately 4750) of cases are above elbow and elbow disarticulation, and 8% (approximately 1350) of cases are shoulder disarticulation [9].

9.1.2 Passive and Active Prostheses

Upper- and lower-limb prostheses can be made out of biological or synthetic materials and can be either a passive extension of the human body or an active device that performs accordingly to electronic sensory input. Prostheses that actively generate forces or torques and thereby provide motive power are actually robotic systems that closely interact with the user and, in some cases, can even interact with the user's nervous system. These neural prostheses can use various methods of communication, either directly or indirectly, by means of unidirectional or bidirectional invasive or noninvasive interfaces.

Prosthesis can be categorized as active or passive [10]. Active prostheses via their actuators and batteries provide positive work and power to perform manipulation or locomotion task. In difference, passive prostheses do not utilize actuators and batteries and hence they do not provide positive work and power; motive power to perform manipulation or locomotion task is primarily provided by the user. Yet, biomechatronic devices can also be semi-active. A famous example is the Otto Bock C-Leg, which automatically alters its mechanical joint characteristics, specifically friction and damping, and thereby adjust to the gait situation by actively providing typically small amounts of negative power to the user [10].

9.1.3 Engineering Design Challenges

The engineering design of advanced biomechatronic prostheses comprises various challenges since such devices need to recreate biomechanical function of missing parts of the musculoskeletal system, recognize actions intended by the user, and restore the appearance of the lost limb. Hence, not only limb mechanics, but also actuation, sensing, and controls have to be designed. When developing prostheses for humans, also non-technical requirements have to be taken into account, for example, giving users the feeling of body integrity with their artificial limb [11].

The historic development of upper- and lower-limb prostheses begun with replacing the missing part of the body by passive tools, for example, hooks or peg legs. To overcome the functional limitations of these devices, "intelligent" but still passive mechanical solutions such as body-powered grippers or gait-specific knee mechanisms were introduced. Through the introduction of biomechatronic approaches, remarkable progress has been observed in prosthetic technology over the last decades, for example, myoelectric control of upper-limb prostheses [12] and actuated lower-limb prostheses that actively provide forces/torques at the joint [13]. For those systems the detection of user intent and appropriate actuation feedback are of paramount importance. While electromyographic intent detection was introduced to upper-limb prosthetics in the 1960s [12], it took until the 1990s to bring automatic locomotion adaptation via actuators to provide negative power into commercial lower-limb prostheses [10]. Still, these semi-passive devices do not actively support gait, but adjust their dynamic characteristics to match different phases of gait and tackle various gait situations like walking on slopes or stairs. Just in the last 15 years, first commercial knee and ankle devices that are driven by actuators, which provide positive and negative power, appeared [10, 14]. Despite these developments and gains in function and user experience, both, biomechatronic upper- and lower-limb prostheses, are subject to ongoing academic research.

There is still a number of engineering challenges to be solved before prosthesis that can provide full versatility of motion and that can minimize user's energy expenditure is introduced on the market. Currently, only a few lower-limb's running prostheses provide active forces [15], and only a few walking prostheses support lateral motion as during the turning [16]. The clear potential of powered devices is that they can improve walking economy and decrease the user's metabolic energy consumption [17]. However, the design of appropriate actuation and control systems remains challenging [13].

Multidisciplinary process of prosthesis design may comprise of various considerations including: biomechanics and neuroscience, mechatronics and robotics, psychological and aesthetical, health care related, ethical and legal aspects.

The first part of this chapter introduces biomechanical fundamentals and describes how limb loss and prosthesis use influence human biomechanics. Focusing on prosthetics as biomechatronic systems, the main part of this chapter is devoted to engineering aspects in upper- and lower-limb prostheses. It discusses the characteristics and design of components, systems, and their interaction with the user. Despite many amputees already benefit from biomechatronics prostheses, there are still numerous research and development challenges to be addressed and solved. These are related to biomechanical, psychological, and medical aspects and they are discussed throughout the chapter. In the final part, human-oriented methods are presented and future directions are presented.

9.2 Prosthetic Biomechanics

This section introduces fundamental aspects on biomechanics in general and prosthetic biomechanics in particular. The presented knowledge builds the basis for understanding the design requirements and prosthetic systems, which are described subsequently.

9.2.1 Biomechanical Fundamentals

9.2.1.1 Functional Anatomy

The human body is a remarkable "machine," shaped by millions of years of evolution [18]. Its construction is built around a rigid lever system, the skeleton, made up from 206 bones in a grown-up individual. Its main function is to support the body against external forces, to allow for a large variety of movements to be performed in variable settings, and to protect inner organs, but it is also an important calcium reserve and mainly involved in blood cell production. The long bones of the lower and upper extremities show an anisotropic behavior. This implies that the stiffness of these bones is dependent on the direction of loading. Highest stiffness values are obtained for compression along the long axis, followed by bending and finally torsional stiffness. The characteristics (duration, frequency, rate, etc.) of the mechanical stimuli applied on the bone determine its adaptation behavior in response to mechanical loading [19].

Attached to the bony lever system we find both active and passive soft tissues: passive tissues are mostly ligaments and joint capsules, which provide passive stability to joints and guide joint movement. Structurally, ligament tissue is mostly made up from Collagen proteins. Ligaments directly connect two bones with each other and are therefore crucial for controlling joint congruency and motion. The force-deformation behavior of ligaments is following a nonlinear relationship. At low levels of deformation (the "toe region"), relatively low forces are needed to flatten out the crimp contained in unloaded ligament tissue, resulting in a low, nonlinear stiffness of the tissue. Once all crimp has been removed, the force-deformation behavior follows a mostly linear relationship, whose slope is equal to the stiffness of the tissue. The structural and mechanical properties of ligamentous tissues can adapt to their mechanical environment. This adaptational response is mainly governed by the strain experienced by these tissues as a result of the external forces they need to counteract. For an adaptational increase in tissue stiffness high strain amplitudes need to be applied over a considerable duration [19].

Muscles are the most important active tissues in the human locomotor system. They are the main biological actuators within the human body and also serve an important function as dampers when absorbing kinetic energy from the whole-body system and convert it into other forms of energy (mainly heat).

The force generating capacities of muscles are governed by two major relationships [20, 21]. The first one is the force-length relationship, which states that the amount of force a sarcomere can exert is dependent upon its length (see Fig. 9.1 regarding the structure of a muscle, and Fig. 3.26). The force-length relationship of muscles shows a plateau region at the mid-portion of typically occurring lengths and exhibits lower levels of force at higher and lower lengths. Muscle tendon units typically cross the joints with a certain lever arm which changes depending on the joint configuration. Thus, a moment is created by muscle force application. The force-length relationship can therefore be transformed to the joint moment—joint angle relationship when considering the joint perspective. The second law governing the force generating capacities of muscle is the force-velocity relationship, which states that the maximum force a muscle can exert is dependent upon its contraction velocity [21]. Muscles can exert greater forces when being actively stretched (eccentric force) compared to the situation when they are shortening (concentric force). This in turn constitutes the power-velocity relationship when muscles are performing positive (concentric) or negative (eccentric) work. Peak positive power values are typically observed at around 30% of the maximum contraction velocity of muscles. When muscles are performing negative work, they act as dampers, when they perform positive work they are actuators, increasing the kinetic energy levels of their adjoining bone segments.

The force generated by a muscle is furthermore related to its activity level, which is set by the neural system. The activity level is controlled in a complex interplay between central components originating in the brain and peripheral components reflecting the results of inhibitory and excitatory reflex loops [22].

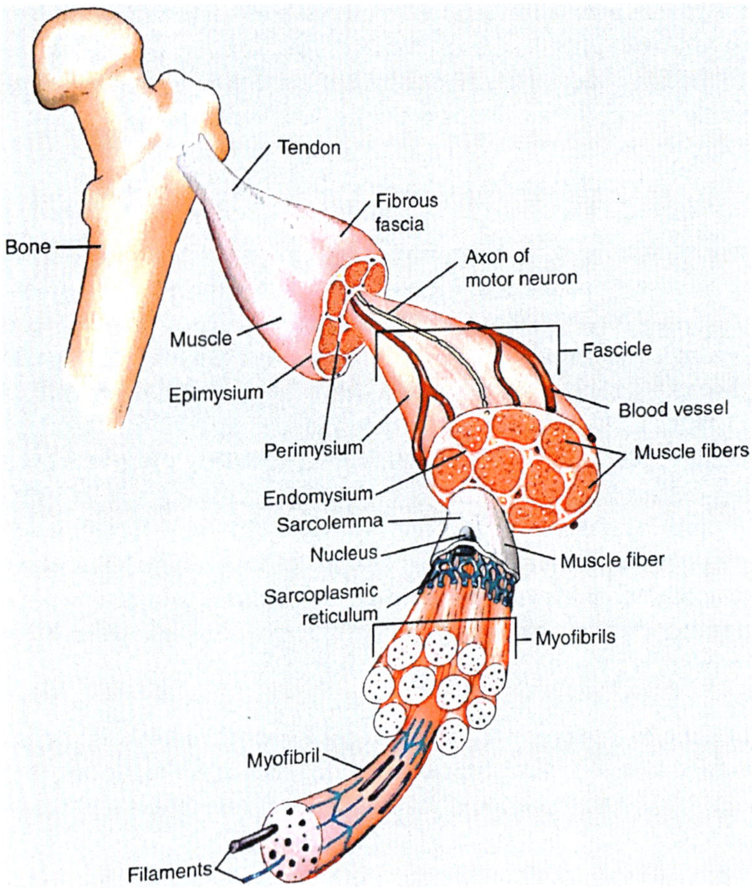

FIG. 9.1 Structure of a human muscle (Fig. 10.7 "Kinesiology—The skeletal system and muscle function" by Joseph E. Muscolino).

9.2.1.2 Locomotion Basics

While the human "machine" is capable of performing a remarkable number of different, highly complex movement tasks, walking and running can be considered the most frequently performed and are the most basic movements that might have played an important role in human evolution [23]. Steady-state walking and running is represented by cyclic movements with similar, but alternating movements of lower and upper extremities. The time from one distinct event (for example the heel strike) to the next occurrence of this event on the ipsilateral side is referred to as a stride cycle or gait cycle. In walking and running this involves a time period when the foot is in contact with the ground (stance period) and a time period when the foot is swung forward not in contact with the ground (swing period). During stance phases, humans can apply forces to the environment in order to accelerate body parts. The average vertical ground reaction forces exerted during one complete gait cycle during steady state walking must equal the body weight of a person. The net horizontal ground reaction forces determine the horizontal acceleration of the center of mass (CoM). Consequently, ground reaction forces are represented as a vector quantity with a direction and length, that is, force amplitude. The ground reaction force vector originates at the point of force application, which can be calculated from measurements taken from state-of-the-art 3D-force platforms. The point of force application coincides with the center of pressure between the foot and the ground. When considering both legs, running is characterized by a flight phase where no contact is existing between foot and the ground. During walking, one leg is always in contact with the ground and there is a short period where both feet are applying forces to the environment, called the double support phase. Vertical force demands are higher during the individual stance phases in running, because the average vertical force requirements need to be satisfied in a shorter period of time, thereby leading to higher stance phase average and peak forces. The point of force application and the orientation of the force vector determine the external lever arm of the ground reaction force for a given leg configuration and can therefore modify the external moments acting at the lower extremity joints. These external moments need to be counteracted by internal moments created by active and passive biological tissues [24]. As the line of action of these structures often passes the joints with a much shorter lever arm, internal forces transmitted, for example, through the Achilles tendon can be much higher than the external ground reaction force [25].

Apart from providing propulsion and support, humans need to control the postural alignment of the parts of their bodies with respect to each other and with respect to their surrounding world in order to stay in a dynamically stable state. Here, a successful sensory input integration and the consideration of biomechanical constraints imposed by the characteristics of the musculoskeletal system are crucial. Without proper dynamic stability control, human movement becomes less efficient and potentially dangerous due to a higher risk of falling. Dynamic stability during locomotion requires properly timed activations of muscles involved in the locomotor task resulting in movement patterns characterized by low whole-body angular momenta [26, 27]. Another very important feature of successful dynamic stability control is locomotor adaptability, which incorporates both reactive and predictive control processes [28].

9.2.2 Modeling and Simulation

The most basic model of the human body is the representation as a point mass. This representation is useful to describe the body's position, velocity, and acceleration in three-dimensional (3D) space. Despite being a dramatic simplification of the true body structure, the CoM concept can be used to determine the energy level of the system with respect to potential and translational kinetic energy [29]. In biomechanics, the work performed on the CoM is referred to as "external work" because it does not include the "internal work" performed by the lower and upper extremities (legs and arms) with respect to the CoM. This work is often not reflected in the external work of the CoM, as the work performed by body parts moving in opposite directions is not addressed. Consider for example the energy of the arms when swinging during walking. The work of the left arm when swinging forward can be equal but with opposite sign compared to the work of the right arm when swinging backwards; therefore no change in CoM energy is observed [30].

Gaining a deeper insight into the specific patterns of power generation and absorption at the joint level requires approaches that are more sophisticated. On the highest level these can be subdivided into two different concepts: The first one aims at understanding underlying principles of human motion by developing basic models that can predict certain features of human motion. Subsequently, these models can be tested against real motions experimentally observed from humans. The second concept starts with a very precise measurement and analysis of human motion, from which underlying mechanisms of motion control are derived. Ideally, both analytic concepts converge toward a detailed understanding of human motion control [31].

Basic features of walking can be reasonably well predicted by modeling the lower extremities as two coupled inverted pendula. This model also allows estimating the energy exchanges between kinetic and potential energies observed at the CoM during the stride cycle. Furthermore, this model predicts considerable amounts of mechanical work performed during the transition from one step to another in order to redirect the CoM motion from one pendular arc to the next [32]. In reality, stance leg behavior during walking is not perfectly stiff, but involves considerable amounts of compliant leg behavior [33]. Modeling compliant legs is essential to obtain the typical double peak ground reaction force behavior in walking, which cannot be predicted using rigid leg inverted pendula models alone [33]. Furthermore, essential parts of the energy involved during the transitions from one step to the other might be stored in the elastic structures of the leg and can therefore be conserved within the moving system.

The CoM behavior during running can also be modeled well with spring-mass models, leading to reasonably well resemblance of the ground reaction force profiles observed in real humans. Elastic leg behavior can also cope well with uneven terrains, in particular at higher running speeds when the time needed to receive and process sensory information might get critical. Therefore, spring-mass models might be useful as templates guiding the control of human locomotion [34].

While the inverted pendulum and spring-mass model are potentially the most prominent representatives of basic models of human locomotion used to test mechanisms of human motion, musculoskeletal models are addressing the analysis of human motion from the opposite direction, by trying to model the detailed loads imposed on the structures composing the human body. These approaches usually take precise measurements of the body using sophisticated motion capture techniques. Furthermore, relevant external forces including, for example, the ground reaction force, are captured using force-sensing equipment. Since these models need to resemble the anthropometrics of the human under investigation, body dimensions are measured as a basis for regression models estimating the body segments masses, moments of inertia, and CoM locations [35]. After obtaining these measurements, computer modeling is applied to estimate the internal loading of biological structures involved in performing the captured motion.

Using inverse dynamics approaches, resulting joint reaction forces and resulting joint moments can be calculated. Sophisticated models even address the issue of calculating forces of individual muscle tendon units. Therefore, muscles are modeled as force actuators. In this task, problems arise from the fact that the number of muscles within the body exceeds its kinematic degrees of freedom (DoF), resulting in muscle redundancy. To solve this issue optimization methods are utilized aiming at movement performance criteria or physiological functions, like, for example, metabolic energy expenditure or the sum of muscle activations [21].

9.2.3 Experimental Studies

9.2.3.1 Studies Addressing Energy Consumption

The main function of an upper limb is to robustly execute dexterous manipulation. The human hand allows us to interact, understand and feel our surroundings using a combination of proprioceptive and tactile feedback. In [36], it was demonstrated that the major goal of upper-limb motor coordination is the minimization of the square of the magnitude of jerk of its motion (rate of change of acceleration), which is equivalent to the execution of the smoothest possible movements by the arm hand system. In particular, when humans perform unconstrained, point-to-point motions on a plane with their upper limbs, these motions are approximately straight and they have bell-shaped tangential velocity profiles. In [37], the authors suggest that the movements of the human arm in 3D space, are performed as "effortlessly" as possible, expending the minimum required amount of energy, while in [38] the authors demonstrate that in certain cases the arm trajectories are chosen so as to minimize metabolic energy costs. A review of the optimality criteria for upper-limb motor control studies is presented in [39]. All these results can be used for optimizing the control of advanced upper-limb prostheses to guarantee minimum effort.

Regarding lower limbs, the main function of the legs during locomotion is to support the weight of the body and to accelerate its mass [29, 30, 32, 40]. In addition, locomotion incurs independent costs for swinging the legs and for maintaining stability [41–43]. Humans normally choose gait patterns that minimize their metabolic energy consumption [44]. Metabolic energy refers to energy that is recruited from food intake and that gets subsequently stored as energy-rich phosphates within the body, which can be converted by muscle fascicles into mechanical energy. The rate of metabolic energy consumption (or metabolic power) can be estimated by measurements of the rate of oxygen consumption at submaximal walking or running speeds [45]. Net metabolic cost is the metabolic cost of locomotion after subtracting the resting metabolic cost, which is typically determined during quiet standing or in a sitting or lying position.

During walking, a study of Grabowski and Kram found that 28% of the net metabolic cost can be attributed to force generation in order to support body weight, while 45% of the net energy is spent for the acceleration of the bodies mass [40]. In particular, it is assumed that a great amount of energy is spent during the transition from one step to another, when the CoM trajectory needs to be redirected from forward downward to forward upward motion [46]. Swinging the legs also incurs a metabolic cost during walking of about 10% of the overall net metabolic cost. The remaining net metabolic costs are related to arm swing and the provision of dynamic stability during walking. The compliant leg behavior during walking and even more during running requires negative and positive work to be performed during the stance phase of gait. During walking, the hip and ankle joints perform most of the positive work during the stance phase. The ankle joint generates clearly more energy compared to the energy it absorbs and is therefore a net energy contributor during walking. An active and properly timed push-off from the ankle joint is in particular important for the reduction of energy losses during the collision-like transition from one step to the subsequent step. Passive elastic ankle-foot prostheses cannot replicate this positive energy generation at the ankle joint during the push-off. They can only perform slightly less positive work than has previously been stored within their passive elastic components. Consequently, increased rates of metabolic energy consumption (+10% to +30%) have been found in humans with amputations utilizing commercially available passive elastic lower-limb replacements [47–49].

On the other hand, when using a bionic (biomechatronic) prosthesis, including a powered ankle joint in addition to passive elastic elements, a clearly more normal gait pattern has been observed which also diminished the differences in metabolic power between participants with below the knee amputation and nonamputees [50]. Furthermore, in walking conditions with greater net positive work demands like stair climbing or upslope walking, gait improvements have been observed in subjects with below the knee amputations [51, 52]. Still, these improvements were less than those observed for less challenging walking tasks like level walking on a smooth surface.

While running, similar tasks as in walking incur metabolic costs. During several experiments, Kram and coworkers identified that the metabolic cost of body weight support and forward propulsion comprise about 80% of the net metabolic cost of running [43]. Furthermore, they estimated that swinging the legs constitutes a metabolic cost of about 7%, while the cost of maintaining balance equals only about 2% of the net metabolic cost [43]. During running, the joints of the lower extremities (in particular the ankle and knee joint) display a more spring-like behavior than during walking. The amount of energy which is passively stored and returned within tendons and ligaments increases with increasing running speed [53]. Therefore athletes with below the knee amputation can achieve sprint performances (in particular in longer sprint events like the 400 m distance) which are very similar to elite nonamputee athletes when using passive, lightweight, running specific prostheses [54]. These prostheses are mainly constructed of carbon fiber spring elements attached to the residual limbs. Despite the remarkable performances of athletes with passive elastic prostheses, there exist strong efforts to develop powered prosthetic feet and knee joints which suffice the higher joint torque demands imposed by the greater force application involved in running [55]. Active prostheses could allow a normalization of the running mechanics in running environments that require net positive joint work, like during uphill running. Furthermore, damping mechanisms might facilitate a more natural downhill running behavior.

9.2.3.2 Studies Addressing Joint Loading

In everyday life, the human upper limbs are used to carry significant loads and to exert significant forces, interacting with dynamic and unstructured environments. During these interactions, the forces are exerted by the dozens of muscles of the upper arm and the forearm and they are transmitted through various flexor and extensor tendons to the human fingertips to grasp or manipulate objects or parts of the environment. The processes of grasping and manipulation are facilitated by the structural compliance of the human arm hand system and the increased dexterity offered by the corresponding musculoskeletal system. However, the human arm hand system effectively performs as a cable-driven structure that has varying force exerting capabilities depending on the attained configurations. An ergonomics study [56] has demonstrated that the upper-limb strength is directly affected by the upper-limb postures and proposes predictive equations that can estimate the maximal forces and torques that can be exerted for given upper-limb postures. Another important characteristic is variable joint stiffness that can be efficiently controlled to facilitate the system response to perturbations [57].

Modern upper-limb prostheses do not have the capabilities and the dexterity of the human arm hand system but they are adequately efficient to execute a series of everyday life activities, including assembly and dexterous manipulation tasks even if they require significant forces [7]. Another drawback of modern prostheses is that they do not exhibit a humanlike, compliant behavior when they face external disturbances. Body-powered prostheses have different force exertion capabilities depending on the system configurations, just like the human arm hand system. To achieve the required tendon/cable displacements they need significant body compensation that requires efforts from the user and can be accomplished in limited body postures.

Regarding lower limbs, every kind of locomotion leads to loading of the biological tissues in the body. During walking and running, this relates most strongly to the tissues that make up the lower extremities and the trunk/spine. Muscles need to create forces to satisfy the weight bearing and acceleration demands put upon the lower extremities. These forces are transmitted by tendons, which experience considerable amounts of stress during each gait cycle. Ligaments and joint capsules provide passive stability to the joints and thereby take-up considerable forces. Cartilage structures provide low friction for a proper movement of the joints while being exposed to high compression and shear forces. Chronic overuse and/or traumatic injury of these structures can lead to severe long-term diseases, like joint osteoarthritis, a degenerative disease of the articular cartilage.

The causes of degenerative changes in the articular cartilage lie in a complex interaction of biological [58–60], mechanical [61–63], and structural [64, 65] mechanisms. Osteoarthritis-related joint alterations initially result from a lack of balance between anabolic and catabolic processes within the cartilage metabolism, which leads to a reduction in the quality of the extracellular matrix of the articular cartilage. The resulting reduced mechanical strength of the cartilage surface leads to progressive wear of the cartilage structures, up to the (local) complete exposure of the underlying bone [66].

Altered and/or increased mechanical stress on articular cartilage plays a key role in the development and progression of osteoarthritis [67, 68], with some mechanical stress being important for the structural integrity and biomechanical functionality of articular cartilage [69, 70].

The most common site of osteoarthritis is in the medial compartment of the tibiofemoral joint (knee joint) [71–73]. Here, the osteoarthritis-related loss of cartilage volume and quality leads to an increase in the local stress at the joint contact surfaces. If a larger cartilage surface is lost and there may already be a pathological remodeling of the underlying bone, it may lead to a structural tilting of the joint and thus to a further increase in stress on the medial compartment of the knee joint [72]. This misalignment of the joint load is one of the major risk factors for joint osteoarthritis [74]. Local inflammation of synovial fluid and cartilage in this situation contributes to the development of pain and progressive destruction of the joint [75].

Humans, which are actively engaged in sports, are typically at higher risk of sustaining traumatic or overuse related injuries or pain of the lower extremities. Distance runners, for example, display injury rates of up to 79%, with the knee being the most frequent injury location, followed by the foot and ankle injuries [76]. Still, these running-related injuries are mostly curable, even though reduction or omission of training sessions is most often required.

Modern prostheses often allow for an active lifestyle, but humans with amputations are often faced with secondary physical conditions that arise from the altered loading conditions of their affected, but also their unaffected legs. People with unilateral amputation put greater load on their nonaffected leg to compensate for the reduced performance of their affected legs [77, 78]. As a consequence, higher resultant joint moments and power output have been observed in people with unilateral amputations on their nonaffected side [79]. Joint moments, determined by inverse dynamics procedures, quantify the load experienced by biological structures that are capable of creating a moment around the joint of interest. Moments can be created by muscle-tendon forces, but also by passive forces, like ligament, joint capsule or bone-to-bone contact forces. Consequently, osteoarthritis is more common in the nonaffected legs of people with lower-limb amputation [80]. As people with amputation avoid loading of their affected leg, they are at a higher risk of sustaining osteoporotic degenerative changes of the skeleton on their affected side [81].

Most of the studies cited above refer to investigations performed on amputees using passive prosthetic systems. Nonetheless, the use of powered ankle joint prostheses can improve the loading asymmetry between affected and nonaffected legs [50, 82, 83], i.e., it can have positive effects on lower extremity load symmetry and associated degenerative changes [80].

9.2.3.3 Effects of the Motion Environment (Shoes, Surface Characteristics, etc.)

Most upper-limb studies focus on oversimplified tasks executed in a low-dimensional setting, for example, two-dimensional (2D) equilibrium point manipulation tasks, analysis of the arm motions on a plane, etc., using simple objects and without considering environmental uncertainties or constraints, for example, external disturbances, and difficult to model phenomena (uncontrolled slipping and rolling). Nevertheless, in reality, upper limbs are responsible for the execution of robust grasping and dexterous manipulation tasks that rely on the selection of appropriate contact points, the exertion of adequate forces and accurate control of the fingers. All these prerequisites can be significantly affected by the aforementioned error sources and uncertainties [84], thus, analysis and modeling of the upper limbs and the upper-limb prostheses should be conducted using more realistic scenarios.

Similarly, the majority of studies on human locomotion are performed on level and regular surfaces. Nonetheless, in the real world, deviations from this typical laboratory setting frequently occur. Locomotion is performed in a great variety of different environments. The main factors that determine the locomotor environment are the stiffness of the surface, as well as its frictional characters. Furthermore, irregularities of the ground challenge locomotion strategies and dynamic stability control [85]. Major adjustments of gait mechanics are further needed when the ground is sloped, either uphill, downhill, or in the mediolateral direction (cross-slope). Footwear builds the interface between human and environment and can modify locomotor mechanics to a considerable extent. All these factors or various interactions between them influence gait mechanics and therefore the load imposed on the joints of the human body. In the next section, a brief introduction is provided into some of the results studies on the effects of surface and footwear variations on gait mechanics have provided.

When walking and running on softer and very viscous grounds like sand, humans need to consume more metabolic energy than when moving on level and hard surfaces. This is partly because additional work needs to be performed to penetrate into and move the sand. On the other hand, the motion becomes less efficient, because the working conditions for muscles and tendons are worse compared to moving on stiffer surfaces. Furthermore, compensating movements are necessary to keep dynamic stability [86]. While walking and running on sand is associated with increased metabolic energy consumption, running on softer, but very elastic surfaces with a high energy return has been shown to reduce energy consumption for a given speed [87, 88].

Locomotion on rigid, but irregular surfaces (surfaces with small height differences, bumps, etc.) is related to an increased energy consumption. Similar to walking on sand this is related to mechanisms aiming at keeping dynamically stable movement patterns. To achieve this, humans tend to increase their overall muscle activity and the co-contraction at their joints. Furthermore, additional work at the knee and hip have been observed, while ankle joint work is decreased [89, 90]. Despite the major changes observed in the locomotion mechanics, the actual control of these responses to irregular ground conditions might necessitate only little central nervous control, but might be governed by the self-stabilizing mechanisms inherent to the spring-mass behavior of the human body and leg during running [91].

Several changes in gait mechanics are also induced when walking or running on up- or downhill slopes. Adaptations include alterations in step length and step frequency. Uphill locomotion is linked with an increased metabolic cost, because additional work needs to be performed to lift the CoM upwards. All joints of the lower extremities need to perform more positive work in order to facilitate the positive work performed on the CoM. Downhill locomotion typically reduces the metabolic costs up to a certain downhill slope. When further increasing the slope, the pronounced impact absorption and overall negative work requirements lead to an increase in metabolic energy consumption compared to less extreme downhill slopes. Walking and running on mediolaterally tilted surfaces requires pronounced adaptations of gait mechanics in particular within the frontal plane of movement [92–94].

Passive elastic prosthetic systems do not allow for a net positive work output at a joint due to missing actuators. Therefore, uphill walking or running is very challenging and requires compensational adaptations of the gait mechanics on the affected and unaffected side [95]. Walking on irregular terrain is also more difficult for any user of a prosthetic system at the moment, not the least because afferent sensory input from biological sensors is missing. Therefore, additional compensational mechanisms are applied, for example, by greater mediolateral movements of the arms [96], in order to keep dynamic stability.

Footwear affects both the mechanics of walking and running. A great variety of footwear types are commonly used for human walking, ranging from thin soles moccasins to high-heeled shoes with sometimes spectacular heel-to-toe drop. People with lower-limb amputation typically choose footwear that works well in combination with their prosthetic device, which narrows down the effects of footwear introduced to their lower extremities. Nonetheless, during higher dynamic movements like running, even small changes in the geometry or density distribution of footwear can affect the kinematics and kinetics of the lower extremities. These are often introduced by technological features aiming at a shift in the point of force application in the anteroposterior or mediolateral direction. This affects the gearing of the external ground reaction force and thereby alters the internal loading experienced by biological structures. Typical footwear design features are heel height and stiffness, midsole bending stiffness or medial postings [97].

9.3 Design Considerations

Upper- and lower-limb prosthetic devices share certain general design considerations. In [98], the main considerations when designing a prosthetic device are discussed.

9.3.1 General Requirements

In summary, according to Popović [98], prosthetic design commonly addresses the following factors:

- *Functionality:* Prostheses are intended to help amputees to regain their lost dexterity and efficiently interact with the environment, so functionality is of paramount importance. Prosthetic devices should inherit many important features of the missing limb. These may include appropriate ranges of prosthesis joint angles, joint angular speeds, and passively or actively generated joint torques. Additionally, active prostheses collect sensory data on the physical state of the body and the interaction with the environment to decipher user intent and command the actuators of the device.

- *Ease of use:* Prosthetic devices are an intimate extension of the human body. They are thus required to perform the desired tasks in a fluid and coordinated manner as if they were biological. Ease of use is extremely important as it increases the prosthesis acceptance by the amputees, as they tend to use "handy" devices for longer periods of time.
- *Comfort:* As prostheses are worn over longer periods of time, the users' stumps might be loaded with large pressures and forces. While pressures are necessary to ensure the attachment between the human body and the prosthesis, users might still experience significant pain, swelling, rash, and even breakdown of the skin even with the best stump-socket interface technologies. Thus, both the prosthesis and the socket need to be carefully designed in order to maximize comfort.
- *Mass:* The mass of prosthesis should be equal or smaller than that of the missing limb. It must be noted though that a very lightweight prosthesis may cause an imbalance between the left and the right side of the human body that may lead to undesired spinal rotation, skeletal strains, or scoliosis, affecting the overall body stability. Prosthetic devices are typically designed to exhibit similar or lesser weight than that of the missing limb in order to minimize fatigue during usage and maximize ease of use. The weight reduction is accomplished through careful selection of the mechanical components and the actuation system.
- *Size:* Dimensions introduce yet another important aspect for prosthesis design. Typically, prosthesis should be of the same size as the missing limb. This puts a strong constraint on the size of robotic prostheses, complicating the selection and packaging of the various components (actuators, battery, etc.) and limiting the prosthesis performance.
- *Visual appearance:* Many amputees prefer a human-like/anthropomorphic appearance for their artificial limbs, achieved through the use of cosmetic gloves, artificial skins, etc.
- *Sound:* The noise produced by the prosthesis should be minimal. This specification affects the actuators selection, since several options are extremely noisy, for example, some types of linear actuators.
- *Energy:* Good energy efficiency of prosthetic devices leads to greater autonomy and improved user experience, for example, maximization of the achievable number of grasps for a prosthetic hand or number of steps of a prosthetic leg.
- *Variability of use:* The prosthesis should be able to perform well in various activities, for example, grasping, writing, carrying, walking/running, slope/stair ascent, cycling, etc., and should work irrespective of the environmental conditions, for example, coefficient of friction, materials of objects, ground surface, etc. Different types of prostheses may have different requirements, for example, being waterproof for bath use.

- *Durability:* Prosthetic devices should be durable enough to minimize the required maintenance, reduce repair costs, and improve the user experience offering increased reliability and robustness. Quality control offers various methods to test failure limits and determine the lifetime of the device, for example, a walking prosthesis could be designed to take 2,000,000 one-second loading cycles (a bit less than time between consecutive stance phases of the same leg in walking) and accordingly tested in a walking simulator.
- *Personalization (see also Size and Visual appearance above):* The prosthetic devices should be able to be personalized to specific users, according to their body mass, body dimensions, and physical condition, for example, development of a prosthetic hand that matches the dimensions of the intact hand of an amputee.
- *Modularity:* Modularity is very important for prosthetic systems as it gives the ability to the user to change, upgrade, or extend the capabilities of a device as well as to easily replace broken or damaged parts.
- *Cost:* While the anticipated cost is a crucial design metric, advanced biomechatronic prostheses tend to require a higher budget than passive prosthesis (several tens of thousand dollars or even close to hundred thousand dollars). This is due to higher costs required for specialized mechanical parts, sophisticated sensors and actuators, complex control software, research and development, and finally, a much smaller market volume.

Such characteristics should be taken into consideration while designing a new prosthetic device to find a good compromise between users' requirements and technical capabilities. Generally, the evaluation of design factors can happen empirically or more systematically by being incorporated as constraints in a multiobjective design optimization formulation (see [4–7] for examples). Moreover, factors such as ease of use, comfort, or visual appearance are distinctly connected to the users' needs and experiences. Hence, current research tries to explore the impact of such factors and first approaches to tackle these factors systematically in the design process have been developed [11] (see Section 9.6.2).

9.3.2 Social Aspects and Device Acceptance

The acceptance rate of a prosthetic device can be assessed considering the amount of time that the amputee spends using the device in everyday life tasks. The lack of comfort and perhaps limited functionality of an advanced upper limb prosthesis may lead user to less frequently use this device; instead, users may prefer more passive tools or simple hooks. When the amputees are involved in the selection/preparation of the prosthesis, for example, replication of an open-source design, studies found the likelihood of prosthesis acceptance is increased eight times [9]. These findings confirm that what amputees need is highly functional, personalized, affordable, and lightweight prostheses that can be easily developed and repaired.

The acceptance rates of prosthetic devices are greater among lower-limb amputees than upper-limb amputees. Lower-limb prostheses are mainly used for locomotion while upper-limb prostheses are typically used for grasping and manipulation tasks that are far more complex and require an increased level of dexterity. It should also be noted that amputees learn over time to execute many tasks one handed or with the portion of the limb(s) remaining. For example, upper-limb amputees often develop the ability to grasp or stabilize large objects between the residual limb and the thorax. Thus, in order to be convinced to use on a daily basis an expensive upper-limb prosthesis the device should be highly dexterous, easy to use and robust. Lower-limb prostheses give individuals with lower-limb amputations the ability to walk independently, and are more commonly used full-time.

9.3.2.1 Aesthetics and Anthropomorphism

The uncanny valley is a hypothesis that human replicas or human body part replicas that appear almost, but not exactly, like real human beings or body parts can create feelings of strangeness and revulsion (or uncanniness) among some observers. The uncanny valley is represented as a dip in the human observer's familiarity toward the replica, a relation that otherwise increases with the replica's human likeness. Examples can be easily found in robotics and prosthetics. For example, if someone shakes an absolutely realistic, human-like prosthetic hand that is hard and cold, they might be surprised by the lack of soft tissue and by the low temperature and the initial familiarity becomes a sense of strangeness. The concept was first postulated by the robotics professor Masahiro Mori in 1970 [99]. A graph presenting the uncanny valley is shown in Fig. 9.2. Mori suggested that designers should take the first peak of the uncanny valley as the goal in building robotic and prosthetic devices rather than the second. Indeed, some amputees prefer

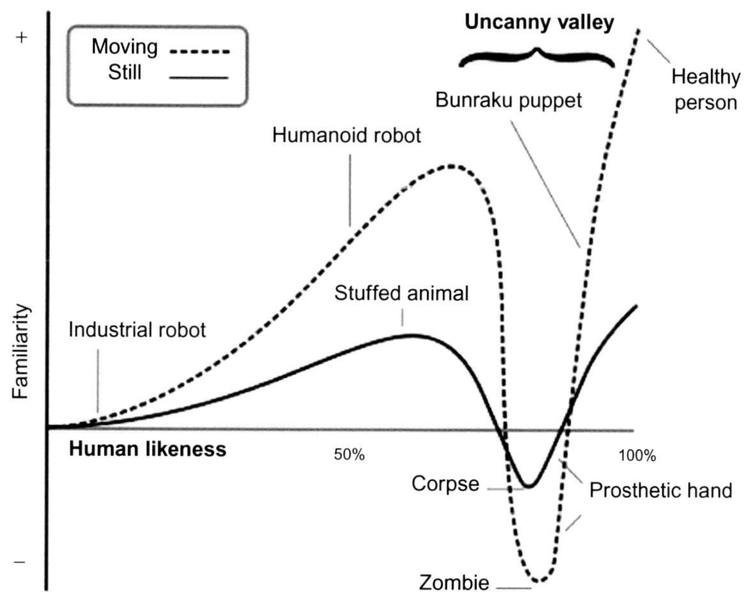

FIG. 9.2 The Masahiro Mori's uncanny valley [99]. *Source: https://commons.wikimedia.org/wiki/File:Mori_Uncanny_Valley.svg.*

robot looking, non-human-like devices to human-like prostheses. An influencing factor to this preference may be the length of time since amputation (people with longstanding amputations might develop a novel and individual understanding of their body shape) and the popular culture, for example, movies and fiction that contain advanced prosthetics with non-human-like appearance.

9.4 Upper-Limb Prostheses

This section focuses on prosthetic devices that help amputees with upper-limb amputations to regain their lost dexterity and improve their ability to efficiently interact with the environment, for example, grasping an object, pushing a button, opening a door, etc. Upper-limb prostheses are mainly developed to substitute a missing shoulder, elbow, wrist, and/or hand. The possible upper-limb amputations are forequarter, shoulder disarticulation (at the shoulder joint), transhumeral (above elbow), elbow disarticulation (at the elbow joint), transradial (below elbow), wrist disarticulation (at the wrist joint), partial hand and finger. An image presenting all the different levels of upper-limb amputations can be found in Fig. 9.3.

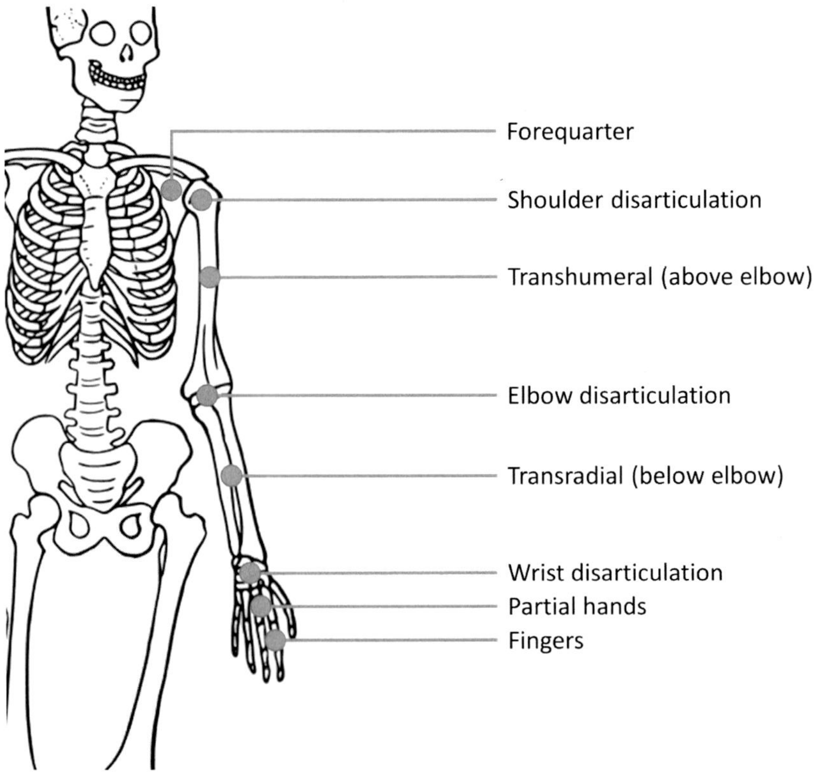

FIG. 9.3 The different levels of upper-limb amputations.

9.4.1 Classes of Upper Limb Prostheses

9.4.1.1 Prosthetic Hands and Tools

Passive prostheses for the replacement of the hand include prosthetic hands and prosthetic tools. Prosthetic tools are limited to the performance of one specific activity or task that needs to be performed bimanually, while prosthetic hands can perform multiple activities and tasks. In addition, prosthetic hands appear human-like in contrast to the mechanical appearance of prosthetic tools, for example, the appearance of a passive prosthetic hook.

9.4.1.2 Adjustable and Static Devices

Both types of passive prostheses can be either static or adjustable. Static prostheses cannot be moved at all. Adjustable prostheses feature an adjustable grasping or manipulation mechanism and parts of the prosthesis can be reconfigured to multiple positions or orientations. Adjustment of the prosthesis is performed by the sound hand or by pushing the prosthesis against the environment, thus by using external environmental constraints. Often little functional value for daily-life tasks is attributed to passive hand prostheses when compared to active prostheses in prosthetic research literature. Yet, user studies show that as much as one out of three amputees uses a passive prosthesis as a terminal device. Examples of the different classes of passive prostheses can be found in Fig. 9.4.

People with a unilateral upper-limb deficiency often only need limited prosthetic function at a time and passive devices are typically very light and hence at least in that sense more comfortable than active prostheses. In addition, the control problem is simplified to the level of functional residual limb. Still further, due to their simplicity these devices are easy to use and can be made more robust and more resistant to wear and tear than an active prosthesis. Finally, most passive devices are low cost, especially in the group of prosthetics tools, and users can typically afford a number of these tools that can be used for various activities and tasks.

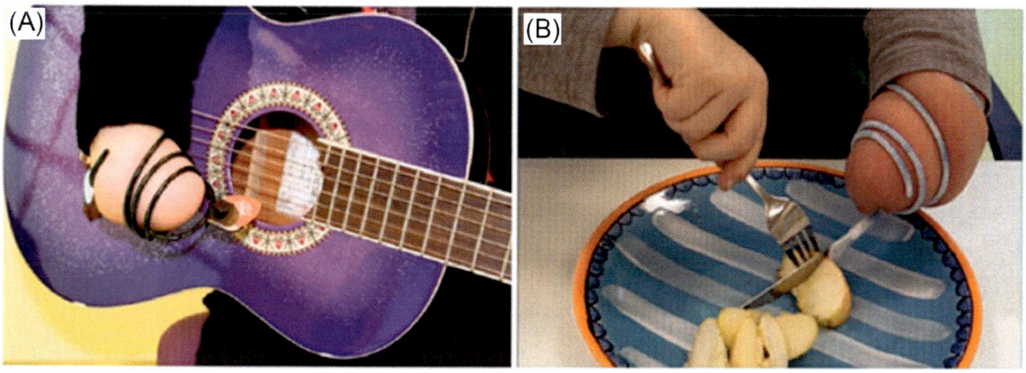

FIG. 9.4 Example of passive prosthetic tools that can be used for specific activities like guitar playing and eating [100].

9.4.1.3 Body-Powered, Electrically Powered, and Hybrid Devices

Active upper-limb prosthetics can be classified into two categories, depending on the means of generating movement at the joints: body powered and electrically powered movement. They have both been in use for over 50 years and each possesses unique advantages and disadvantages. The body-powered prostheses use a body harness and a cable system to provide functional manipulation of the elbow and hand.

Voluntary movement of the shoulder and/or limb stump displaces the cable and transmits the force to the terminal device. Prosthetic hand attachments, which may be claw-like devices that allow good grip strength and visual control of objects or latex-gloved devices that provide a more natural appearance at the expense of control, can be opened and closed by the cable-driven actuation mechanism. Body-powered mechanisms are typically very durable and versatile and they offer intuitive control of the prosthetic device as well as force feedback via the cable tensioning. In the 2016 Cybathlon competition, where different prosthetic technologies competed in daily-life benchmark activities, the winning limb prosthetic arm was not biomechatronic, but body powered [101]. This indicates the complexity of engineering upper-limb prosthetic devices. Possible reasons might be that the system exhibited minimal latency and fine-grained proportional control through the mechanical connection. Yet, body-powered prostheses can have issues such as harness discomfort (particularly the wear temperature), wire failure, and the unattractive appearance. In addition, often the arm only moves accurately when the wearer is standing upright and abnormal body movements are required to make the prosthetic move. Sometimes it may be also hard for weaker amputees to achieve the required cable displacement and make the arm or the prosthetic hand move. Examples of passive tool prostheses and an active body-powered prosthesis are presented in Figs. 9.4 and 9.5, respectively.

Hybrid systems represent combinations of body-powered and myoelectric components that may be used for high-level amputations, that is, at or above the elbow. Those systems allow control of two joints at once, that is, one body-powered and one

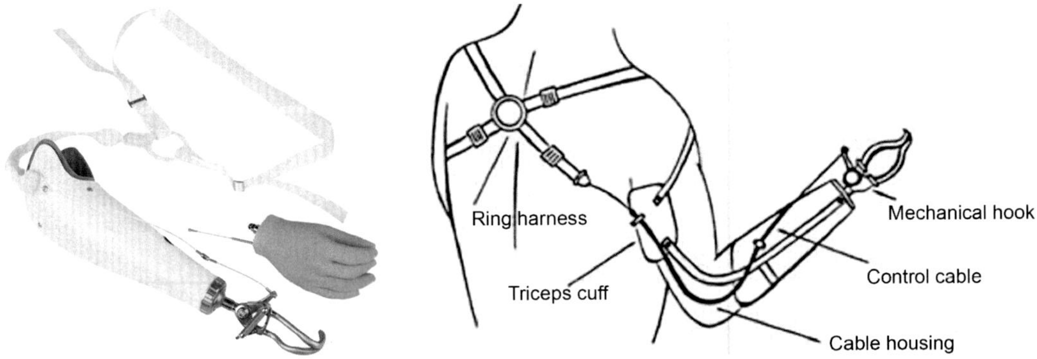

FIG. 9.5 A body-powered arm hand system prosthesis. Pulling or releasing the cable controls closing or opening of the split hook gripper [102]. *Reprinted with permission from Ottobock.*

myoelectric, and are generally lighter and less expensive than a prosthesis composed entirely of EMG controlled components.

9.4.1.4 Advantages and Disadvantages of the Different Solutions

Although myoelectric prostheses offer a more dexterous control of the prosthetic devices, some of their disadvantages are that they are typically heavier (because of the batteries and the motor unit), more expensive than other types of prostheses, and can operate for a limited amount of time, depending on usage and power consumption. Moreover, EMG sensors need to be attached to the skin, and the system sometimes "misreads" the user intent and hence it is not 100% reliable. Still further, at least for the generation of prostheses currently on the market, there is no full feedback loop; the tactile, haptic, and proprioceptive sensory input is not fed back to the user's natural neural pathways (not even noninvasively). In other words, the user does not 'feel' what prosthesis feels and cannot fully interact with the environment. The feedback can be achieved through other modalities as it will be addressed later in this section.

9.4.2 Prosthetic Systems Examples

Historically, the first close-to-biologically accurate prosthetic hand was made more than 55 years ago. The *Belgrade Hand* was developed by Rajko Tomovic in collaboration with Miodrag Rakic at the University of Belgrade, between 1962 and 1964. It was the first robotic five-fingered hand prostheses and it was the first device with a sense of touch. It was anthropomorphic and it had the size of a human hand. The fingertips were equipped with pressure sensors, and when contact with an object was made, a single actuator began to close all five fingers until the pressure was equalized. This made the hand shape adaptable without effort from the wearer.

Fast forwarding across decades, the Luke Arm codenamed after Star Wars' Luke Skywalker's artificial hand was developed by Dean Kamen, the inventor of the Segway and introduced by company DEKA. The FDA announced it has approved the Luke Arm in May 2014 [103]. The result, Luke Arm weighs only 8 lb (3.6 kg), contains many electric motors that give it 18 DoF (the human arm has 22) and pressure control. The original Deka's Luke is controlled by pressure-sensitive pads under the feet and attached to some of the shoulder muscles. The project has been funded since 2006 by the US Defense Advanced Research Project Agency (DARPA). Based on the need for a more functional upper-limb prosthetic system for veterans returning home from the Iraq War, the DARPA launched the Revolutionizing Prosthetics program in 2006 and funded multiple research initiatives to create a better class of advanced upper-limb prosthetic solutions. Several prosthetic arms have been under development including the DEKA arm and the Johns Hopkins' modular prosthetic limb (MPL).

The Johns Hopkins Applied Physics Laboratory (JH APL) MPL (see Fig. 9.6) is capable of effectuating almost all of the movements as a human arm and hand. With more than hundred sensors in the hand and upper arm, it is probably the world's most sophisticated

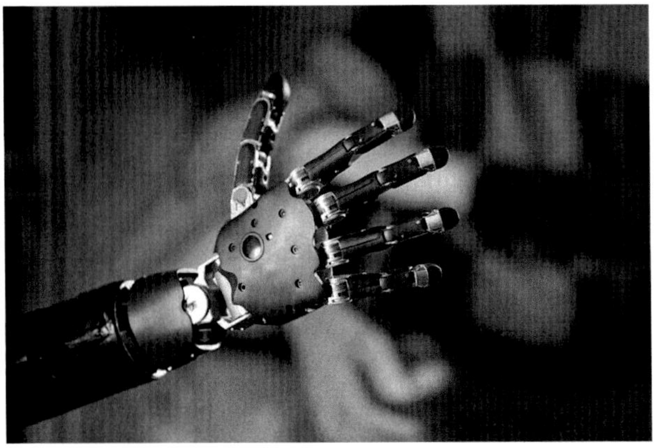

FIG. 9.6 The Johns Hopkins Applied Physics Laboratory (JH APL) modular prosthetic limb (MPL) [104]. *Source: https://commons.wikimedia.org/wiki/File:Brain-Controlled_Prosthetic_Arm.jpg.*

upper-extremity prosthesis [104]. The 10.5 lb (4.8 kg) modular and extensible limb including battery features 3 DoF shoulder, integrated powerful elbow with active extension, 3 DoF wrist assembly, and articulated hand with 10 actuated joints.

Another advanced robotic prosthesis was introduced in 2007 by the Scottish Company Touch Bionics, called the i-LIMB [105]. This robotic hand is shown in Fig. 9.7 and is capable of a variety of unique grip positions that allow the user to balance power and precision as needed. By extending the index finger alone, patients can type on a keyboard or push buttons. The user can also grip a key or dinner plate by rotating the thumb to meet the side of the index finger. The prosthesis is capable of stopping when a sufficient grip is achieved, allowing the patient to grip sensitive objects, for example, a styrofoam cup, without crushing them. The i-Limb offers EMG-based individual finger control to the user that allows him/her to perform dexterous tasks.

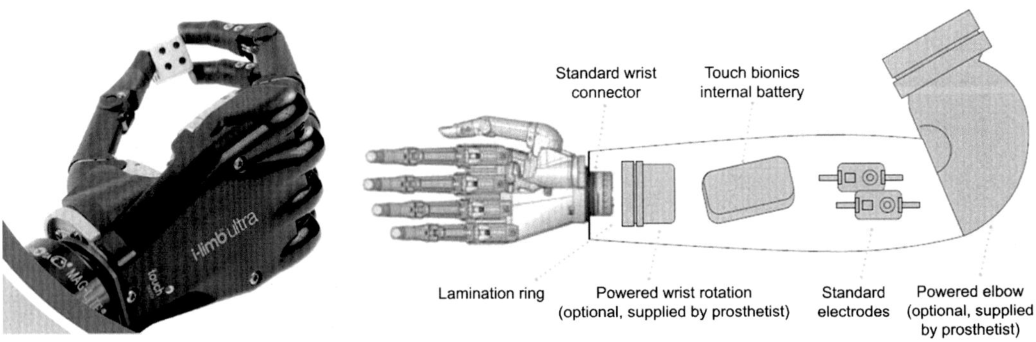

FIG. 9.7 The i-LIMB prosthetic hand of Touch Bionics. *Image adapted based on M.B. Popović, Biomechanics and Robotics, CRC Press, 2013.*

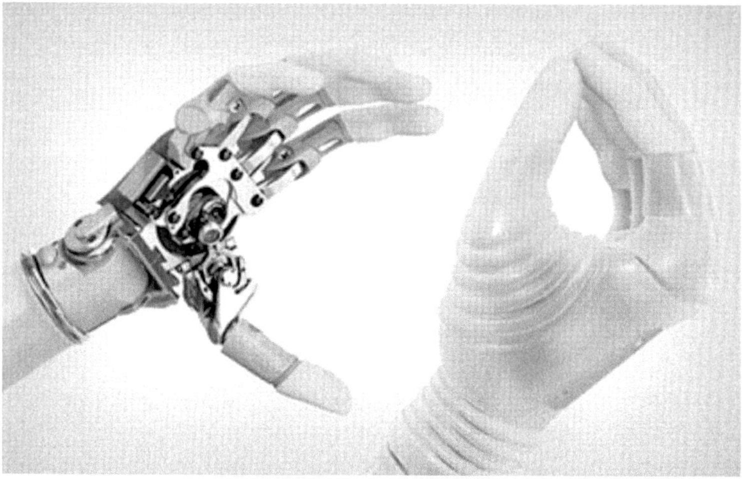

FIG. 9.8 The Ottobock Michelangelo prosthetic hand-wrist system [106]. *Reprinted with permission from Ottobock.*

Ottobock Michelangelo shown in Fig. 9.8 is one of the most popular, commercially available robot hand prostheses. The Michelangelo hand is a multiarticulated hand-wrist system that uses standard myoelectric control (using two surface EMG electrodes) and that can perform seven grasp configurations as well as thumb opposition in order to execute power grip and tripod pinch grasps (with the thumb, and index and middle fingers). The Michelangelo is considered a hand-wrist system as the hand is proximally connected to a passive wrist joint that allows for flexion-extension and pronation-supination. Wrist flexion-extension can be either locked in eight different angles or it can be used in the full range of motion against the resistance of a spring that returns passively the wrist to the neutral position [106].

The SmartHand prosthetic device is a forerunner for the next generation of prosthetics. SmartHand project, which began on November 1, 2006, is a joint effort by several European research institutions, led by Scuola Superiore Sant'Anna, Pisa, Italy, to create a transradial prosthesis that is truly authentic to a biological hand. In addition to myoelectric control, researchers are also investigating a novel peripheral intraneural multielectrode approach for multimovement prosthesis control and for sensory feedback, while assessing cortical reorganization using the reacquired stream of data [107].

The WPI prosthetic hand is a high-fidelity biomimetic device actuated by nine stepper motors packaged within the forearm casing that was manufactured for less than 350 USD, it has 18 mechanical DoF, is 38 cm long and it weighs 2.2 kg [108, 109]. The hand model has 3D-printed replicas of human bones and laser cut tendons and ligaments. The user intent is decoded from EEG (electroencephalography) and EMG (electromyography) signals, obtained by Neurosky and Myoware commercial sensors, respectively. Three distinct EEG patterns trigger pinch, hook, and point actions. EMG signals are used for finer motor control, for example, strength of grip. A pilot test study on three subjects showed that EMG

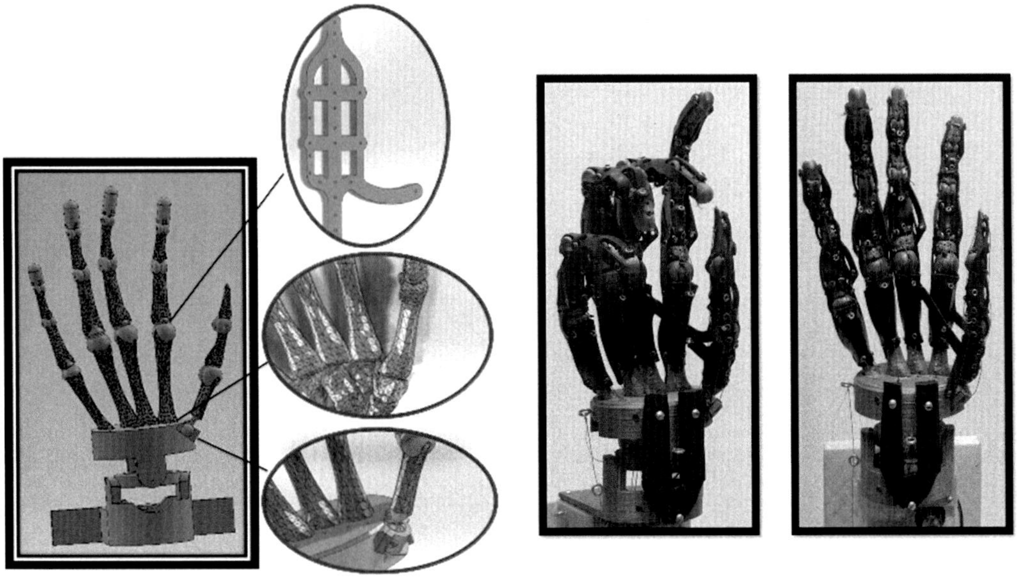

FIG. 9.9 Biologically accurate prosthetic hand, CAD models on the left, physical on the right [108, 109].

can actuate the hand with an 80% success rate, while EEG allows for a 68% success rate. The system proved its robustness at the 2017 Cambridge Science Festival, using EEG signals alone. Out of approximately 30 visitors the majority could generate a "peace" sign after 1–2 min. The WPI biologically accurate prosthetic hand is depicted in Fig. 9.9.

In [110] Dollar et al., evaluated the potential of applying the principles of adaptive robot hands created with the concept of shape deposition manufacturing (SDM) to a prosthetic terminal device. The presented experimental results were quite promising, demonstrating increased robustness, adaptability, durability, and a reliable performance combined with the potential for inexpensive mass production and a realistic appearance. The developed prosthetic hand requires a single actuator that controls eight DoF and can be either body-powered or externally powered. The particular work inspired many other research initiatives that resulted in several open-source designs for both robotics and prosthetics [111].

In [112] the authors presented an open-source design for the development of anthropomorphic, modular, adaptive robot hands of low complexity and cost. The particular hands take advantage of a novel, selectively lockable differential mechanism that can block the motion of each finger independently, allowing the user to select multiple grasping postures and gestures easily and intuitively. The proposed prosthetic hand uses a single actuator in order to control 15 DoF and execute 144 different grasping patterns. The hand costs less than 200 USD, weighs less than 300 g, and is able to carry around a payload of 1.5 kg or to execute assembly tasks with the help of the intact hand.

9.4.2.1 Prosthetic Shoulders, Elbows, and Sockets

The human arm contributes significantly to our stability and efficiency when we walk. This is because the angular momentum of the arms counteracts the angular momentum created by the motion of the lower extremities, in particular the swinging leg. Therefore, proper arm motion is highly relevant for whole-body angular momentum balance [27]. For shoulder disarticulation and forequarter amputees who have lost an entire arm, these dynamics are no longer a part of their biomechanics.

A novel shoulder prosthesis [113–115] relying on a control moment gyroscope (CMG, see Fig. 9.10) was proposed by Worcester Polytechnic Institute (WPI) researchers that focuses on restoring some of the complex dynamics of the arm for whole-arm amputees. The CMG-based prosthesis remains affixed to the shoulder and exerts a moment on the user's trunk similar to that of the arm during walking for dynamic motion assistance. The size, ease of use, and relatively low cost to manufacture (<$1000) of the proposed device makes it an attractive complement or alternative to standard prostheses, especially for amputees who pursue rigorous or prolonged physical activity. The prototype actuator has successfully produced moments in excess of those exerted by the arm during walking, at a similar speed. The disk weighs approximately 2 lb (∼0.9 kg), considerably less than the average human arm, meaning that a final device has room to be customized for each user to match the weight of the missing arm.

Although prosthetic devices are an essential improvement for the lifestyle of many amputees, their use can still lead to many serious long-term health issues such as deep tissue necropathy, skin lesions, and vascular occlusion. Short-term discomfort can also occur because of swelling, perspiration, and bruising on the limb within the socket. Thus, sockets need to keep artificial limbs securely tightened onto residual limb. To address comfortable yet robust body attachment to support the range of movement and load capacity of the MPL for varying amputation levels, JH APL researchers studied multiple volume accommodating and dynamic shape-changing socket methods. These included pneumatic or air-filled bladders, hydraulic or fluid-filled bladders, vacuum-attachment

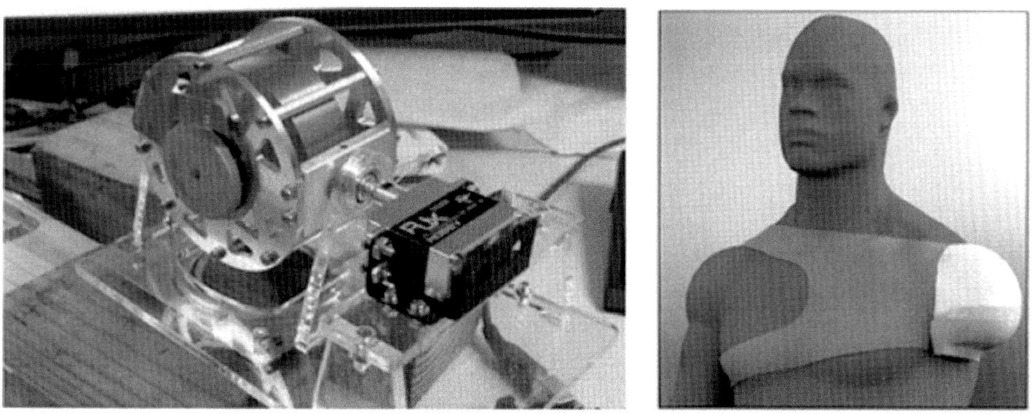

FIG. 9.10 WPI shoulder prosthesis: control moment gyroscope, left, and device concept rendering, right.

methods, electro-active polymers, and shape-changing material structures. Additional socket design challenges included the need to provide space for controllers, for the tactor or other afferent devices, and for peripheral control transduction elements [implantable myoelectric sensor (IMES) and Utah Slanted Electrode Array (USEA) by Blackrock Microsystems].

Most of the described robotic and biologically accurate prosthetic projects were result of extensive, multimillion-dollar funding by various government agencies and potential realizations in the form of commercial devices are typically expected to be on the order of US$10,000 or more. However, recent advancements in rapid prototyping and introduction of novel concepts in particular that related to soft robotics may soon change this situation and drive costs down. One of the authors (MP) led several projects in recent years that nicely illustrate this point including Robotic Socket for Transtibial Amputees [116, 117]. This novel prosthetic socket reduces stress on the skin and soft tissues of the limb by automatically redistributing the pressures between the limb and the socket. The final design consisted of a system of soft bladders and servo-actuated valves controlled using input from pressure sensors. All design specifications were met and pressures were successfully redistributed across the limb in less than 100 ms even under a 270 lb (122 kg) load. This advanced robotic prosthetic socket was built for less than $250 USD.

9.4.3 Interfaces and Control of Prosthetic Devices

Several control modalities have been used in upper-limb prosthetics utilizing various sensory inputs including pressure-sensitive pads under the feet or attached to the shoulder muscles, noninvasive EMG (see Fig. 9.11) measuring muscle activity in residual limb and EEG sensors measuring brain activity with surface electrodes, invasive EMG sensors, and finally direct neural interfaces. Most common are the prostheses that use EMG signals (recorded with double differential surface EMG sensors) to decode based on the muscle activities of the residual limb the desired human movement or the human intention.

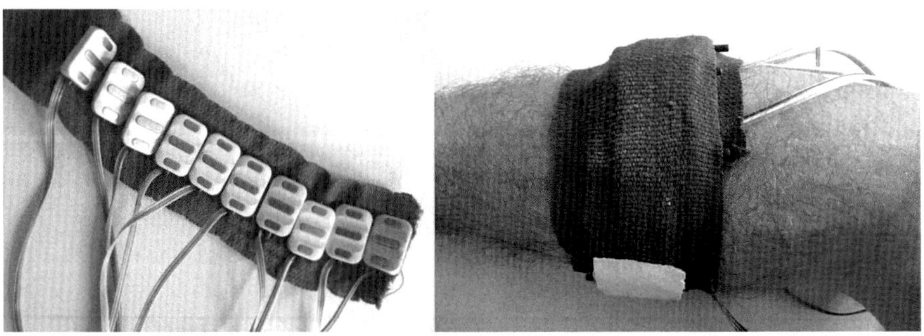

FIG. 9.11 Example of an Ottobock EMG system that is used for the control of the Ottobock prosthetic devices. *Reprinted with permission from A. Gijsberts, R. Bohra, D. Sierra González, A. Werner, M. Nowak, B. Caputo, M.A. Roa, C. Castellini, Stable myoelectric control of a hand prosthesis using non-linear incremental learning, Front. Neurorobot. 8 (2014) 8. https://doi.org/10.3389/fnbot.2014.00008.*

The control process involves recording of the EMG signal from the surface electrodes, amplification and filtering of the captured myoelectric activations, and formulation of a mapping problem, for example, a regression problem, that will map the recorded myoelectric activations of the muscles of the forearm and the upper arm to appropriate joint space motions or motor trajectories for the prosthesis.

A second version of the Luke arm (presented in Section 9.4.2) uses targeted reinnervation and nerve-muscle grafts, developed by Dr. Todd Kuiken from the Rehabilitation Institute of Chicago. Dr. Kuiken put into practice a new surgical technique called *targeted muscle reinnervation* (TMR) that allows for better control of an externally powered artificial arm. The surgical technique involves grafting some of the still existing hand nerves to muscles in the arm or elsewhere so that these muscles contract when the subject thinks about using their amputated hand. Electric signals from these muscles can then be used to control the artificial hand. In this way, the subject is able to control both an artificial hand and elbow at the same time with the muscles in their residual arm. Hence, in addition to EMG sensory data obtained from existing muscle tissue in the residual limb, the neural signal from amputated nerves which have been disconnected from original muscle tissue can still be used by EMG sensing and TMR techniques.

Cortical level brain machine interfaces were used to decipher user intent and control robotic arm in pioneering Brain Gait project already between 2004 and 2006. The sense of touch has also been fed back directly to cortex by utilizing electrocorticography (ECoG) as reported in October 2016, by the University of Washington researchers working at the US National Science Foundation Center for Sensorimotor Neural Engineering (CSNE) (see Chapter 6 for details).

The neural interface encompasses the most significant technical challenges—to directly interface with the body's nervous system for limb control and sensory feedback. Several types of recording devices are used to record various biological signals from muscles, peripheral nerves, and the cortex. Implanted intramuscular electrodes and surface electromyogram electrodes are used to record muscle activity; implantable peripheral nerve electrode intercept signals, propagating along peripheral nerves; and implantable cortical electrode capture spike and local field potentials, near their origins in the primary motor, premotor, and posterior parietal cortices. Collecting these signal modalities provides complementary information as well as a certain level of redundancy that maintains long-term high levels of fidelity and provides modularity.

The sensory feedback has been demonstrated in summer 2016 by the Johns Hopkins Advanced Physics Laboratory researchers utilizing *targeted sensory reinnervation* (following surgical methods developed earlier by Dr. Kuiken) to transfer patient hand's nerves associated with touch to patient upper arm. Now, while wearing a sensory sleeve over new nerve endings, a patient is able to receive signals from a prosthetic device, including the sense of pressure needed to grasp for example a ball. In other words based on device sensory input the sensory sleeve affects the muscle tissue which then affects the nerve ending associated with touch sensation of missing limb. The virtual integration environment is used to visualize and monitor the performance of various design approaches, pilot neural signal analysis algorithms, simulate emerging mechatronic elements, train end users to control real or virtual neuroprosthetic devices, and configure and customize clinical and take-home devices.

9.5 Lower-Limb Prostheses

The paramount aim of lower-limb prostheses is to recreate the locomotion capabilities of amputees and thereby improve their ability to participate in society. Amputations of the lower extremity occur on different levels, which are displayed in Fig. 9.12: hemipelvectomy (transpelvic), hip disarticulation (at the hip joint), transfemoral (above knee), knee disarticulation (at the knee joint), transtibial (below knee), ankle disarticulation (at the ankle joint), partial foot and toe. With increasing level of lower-limb amputation, that is, with increasing loss of parts of the biological limb, walking speed is reduced and energy expenditure is increased [118].

Fundamentally, lower-limb prosthetics are designed to increase mobility by allowing an amputee to ambulate normally. However, prosthetic limb characteristics differ from those of real limbs, for example, regarding inertia and the ability to generate work. Hence, motor functionality and economy can be significantly reduced in prosthetic users.

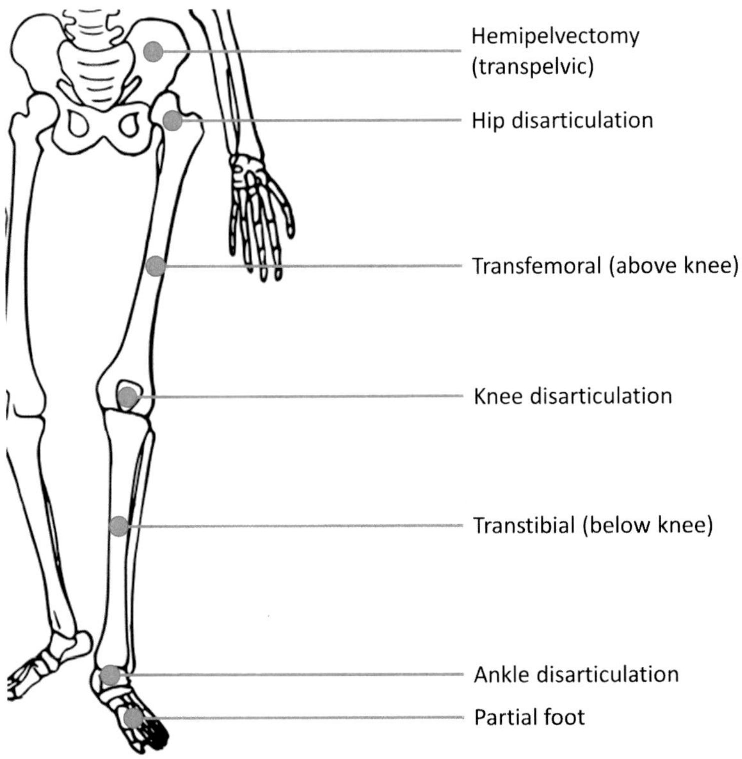

FIG. 9.12 The different levels of lower-limb amputations.

9.5.1 Contemporary Lower-Limb Prosthetic Devices

All prostheses used above ankle level include a foot-and-ankle components. If the amputation is located above the knee, devices replacing the knee are added. As mentioned in Section 9.1, passive, semiactive, and active devices exist, which are categorized based on if they include mechatronic components and if they introduce forces/torques to locomotion. Fig. 9.13 gives an overview of systems that are available on the market.

Passive prosthetic knees such as the Össur Mauch SNS or the ottobock 3R80 contain mechanical linkages and hydraulics to provide damping characteristics. While these passive devices are optimized to certain gait speeds, more modern microprocessor controlled knees, such as the ottobock C-Leg, the Freedom Innovations Plié3, or the Blatchford/endolite Orion (part of the Linx system) are semi-active and can adapt their mechanical properties through actuators. For instance, the C-Leg varies joint stiffness and damping by altering the valve opening of its hydraulic component. Among commercial knee products, only the Össur Power Knee is capable of providing additional power to locomotion and it is thus categorized as an active knee prosthesis.

Passive prosthetic feet are usually based on carbon springs that store energy during ground contact and release it during push-off. This basic design is also used in most

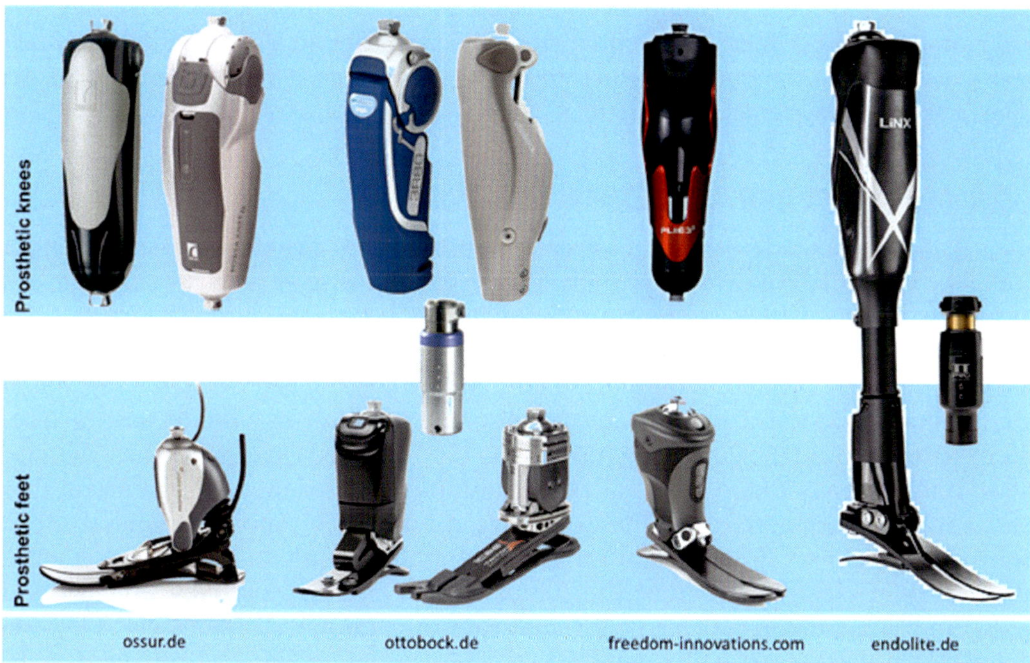

FIG. 9.13 Contemporary, commercial lower-limb prosthetic components from different manufacturers. Knees (top): Össur Mauch SNS & Power Knee, ottobock 3R80 & C-Leg, Freedom Innovations Plié3. Torsion adaptors (middle): ottobock DeltaTwist, Blatchford/endolite TTPro. Feet/ankles (bottom): Össur ProprioFoot, Ottobock empower and triton, Freedom Innovations Kinnex. Integrated limb system (left): Blatchford/endolite Linx.

semi-active and active components. Semi-active devices such as the Össur ProprioFoot, the ottobock triton, the Freedom Innovations Kinnex, or the Blatchford/endolite elan (part of the Linx system) use actuators to provide ground clearance during swing phase or align to slopes. Only one active ankle device exists on the market, namely the Empower by Ottobock. It is based on research outcomes of Prof. Hugh Herr and Biomechatronics group from the Massachusetts Institute of Technology (see [17, 50] for instance).

Most existing components are modular and can be combined with those of other manufacturers using to standardized connectors. Beyond purely structural adapters that assure the correct length of the leg, for example, by covering the distance between foot/ankle and knee devices, there are torsional adaptors that enable rotation around the shank axis as in the real leg. This functionality makes them attractive to users who turn a lot, for example, for maneuvering in confined spaces or playing Golf. Despite the modular approach, integrated limb systems like the semi-active Blatchford/endolite Linx are emerging since they promise improved locomotion capabilities through data exchange between ankle/ft and knee devices.

The remainder of this section reviews design aspects of lower-limb prosthetics and takes a strong focus on active, that is, powered, devices, and systems. It considers how powered prosthetic limbs could restore biomechanical functionality and user experience, which enables design implications for future, biomechatronic systems. Thereby, mechanics, actuators, sensors, and controls as well as the socket, which implements the mechanical human-machine interface, are discussed. The text relies on a systematic review [13], which discussed 21 different active prostheses (8 above-knee, 9 below-knee, and 4 combined knee-ankle systems).

9.5.2 Mechanics and Kinematics

Over the last years, a few users of lower-limb prostheses have drawn major attention within popular media due to the remarkable performances achieved by some of the best athletes with transtibial lower-limb amputations. Oscar Pistorius, a South African 400 m sprinter, was the first athlete with lower amputations to perform with nonamputee athletes at the 2012 Olympic Games in London. More recently, Markus Rehm, a German long jumper achieved a personal record distance (8.40 m) that would have allowed him to win the Gold medal in long jumping in the last three Olympic Games. Actually, he won the German Championships of nonamputee athletes in 2014. The use of their passive elastic, carbon fiber prostheses has raised a discussion about a potential advantage of passive lower-limb prostheses in certain athletic competitions. Biomechanical studies could clearly show that the motor control strategy used by athletes with lower-limb amputations is clearly different from nonamputee athletes [54, 119]. The use of highly elastic carbon fiber prostheses seems to offer potential advantages for running longer distances with constant and high running speeds. This is due to the large quantities of stored and returned energy within and from the prostheses and due to the reduced mass and moment of inertia of these artificial lower-limb replacements compared to biological legs [54, 120]. During the long jump take-off, the generation of large amounts of positive work from energy storage and return from a carbon fiber prosthesis offers a performance advantage for transtibial unilateral amputees taking-off from their prosthetic limb [119]. Despite these advantages for high, constant speed running and jumping for distance after a high-speed approach run, clear disadvantages evoked by the use of passive elastic, running specific prostheses have been reported, in particular for the sprint start and acceleration phases [121, 122]. Sprinting through a curve is also more challenging for athletes with lower-limb amputation, in particular if amputated above the knee and if the affected leg is inside the curve [123]. The lower acceleration performance can be nicely explained by the missing muscle fascicle actuators in the limb parts replaced by the passive elastic prostheses. Passive elastic components cannot create mechanical energy de novo, which inhibits the increase in kinetic energy, that is, sprinting speed, in athletes with lower-limb amputations. The competition rules of the International Paralympic Committee (IPC) prohibit the use of actively powered prosthetic systems. Nonetheless, their development and use will clearly enhance the biomechanics of walking and running, especially in situations when net positive energy output is required from the replaced joint, like for example, during uphill locomotion or during acceleration tasks.

Typically, the mechanical designs of powered knee and ankle prostheses use rotational single-axis joints while very few knees rely on multibar linkages (see [13] and references thereby). The latter is due to the kinematics of the human knee, which exhibits a rolling-sliding motion and motivates polycentric knee joints using four-bar linkages to increase stability during stance phase.

Commonly, the applied actuators, for example, electric motors, are connected to the joint via a ball screws, belt transmissions, or slider-crank mechanisms. Moreover, the addition of clutches to elastic actuators is currently investigated as it can allow motor to be completely disengaged, when needed, from the load and hence transition to actively controlled nonpowered system that can store and release elastic energy along the lines of the One-To-Many (OTM) principle [98]. Since energy efficiency results in long operating times and is thus a key design goal for powered prosthetics, mechanisms to exchange energy between ankle and knee, similar to the human Musculus gastrocnemius, receive increased interest. Due to the dissipative behavior of the knee during level walking, such kinematic approaches can reduce power and energy requirements regarding the ankle joint. There is still no commercially available device nor there are many examples of academic research of two-joint prosthetic systems with powered ankle and knee.

9.5.3 Actuation

Designing actuators that meet the requirements of powered lower-limb prosthetics is challenging, especially since the demanded performance is hard to implement while keeping within weight and size constraints. These issues are currently addressed by research groups affiliated with both industry and academia. Fig. 9.14 presents

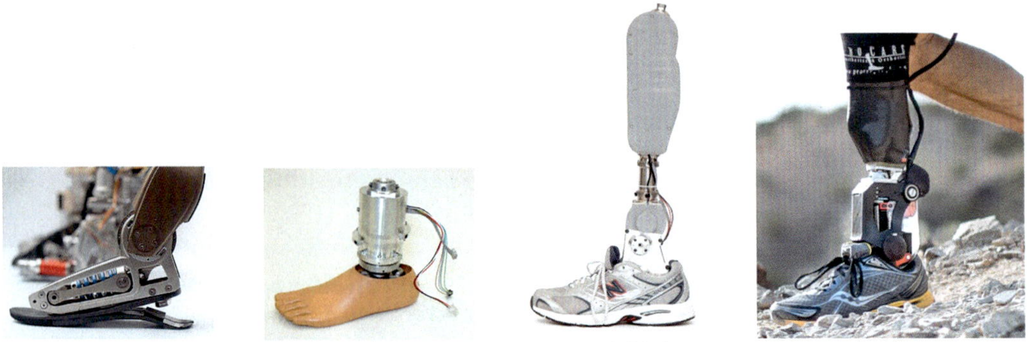

FIG. 9.14 Precommercial active lower-limb prosthetic devices stemming from long-term academic research projects. AMP foot (left, Vrije Universiteit Brussel and Axiles Bionics, Belgium), powered torsion adaptor (middle left, Technische Universität Darmstadt, Germany), multijoint robotic prosthesis (middle right, Vanderbilt University, USA), and prosthetic ankle (right, Arizona State University and SpringActive, USA).

four-powered prosthetic devices of which some are currently in the transition from academic research to commercial products. As the ottobock empower and the Össur Power Knee that are shown in Fig. 9.13, all of them are based on electromechanical actuators.

Besides electromechanical actuators, pneumatic, and hydraulic actuation units are used to tackle this challenge. Most electromechanical motors are incorporated in elastic actuators, some of them with variable impedance, to yield characteristics that are closer to the biomechanical role model. Another approach to tackle this aim is with fluid powered artificial muscle, which contract like human muscles and can vary stiffness as well, see Chapter 3; while McKibben muscle has limited strain, that is much smaller than its biological counterpart, the Hydro Muscle can match the properties of the real muscle. However, pneumatically and hydraulically actuated prostheses need to integrate a lot of auxiliary equipment like pump, tubes, and valves.

Specifically, elastic and variable impedance actuators are able to buffer energy in the elastic element and return in subsequent gait phases, for example, in the ottobock empower or the AMP foot (see Figs. 9.13 and 9.14). Thereby, they can reduce the peak forces, required peak powers, and energy consumption of the actuation system. In addition, this might lead to a downsizing of the actuator itself, which again increases efficiency and is beneficial for the size and weight of the overall system. Besides this positive effect on energy balance, adapting the variable impedance to the properties of the physiological joint can make a prosthesis feel more natural and yield a more predictable behavior for the users.

To optimize actuator designs, gait analysis via motion capturing and biomechanical simulations are very helpful tools. Based on the acquired kinematic and kinetic data of human walking, required actuator powers can be determined [15] and compared considering different spring configurations, for example, serial versus parallel [124], or actuator implementations, for example, elastic versus variable elastic.

9.5.4 Sensors

The control unit for biomechatronics prostheses typically heavily relies on motion sensors, for example: incremental encoders, accelerometers, inertial sensors, and potentiometers. Moreover, mechanical loads of joints or actuators can be acquired to assess ground contact behavior or aspects such as the body-socket interface, for example, using force-sensitive resistors, Hall effect sensors, or pneumatic bladders. Finally, noninvasive EMG and EEG methods are also utilized to improve assessment of user intent.

9.5.5 Control

The advanced control algorithms are necessary for semiactive and active prosthesis. As powered prostheses are capable to generate forces/torques and aim at generating a desired movement, those algorithms are of paramount importance to achieve appropriate and intuitive interaction with the user [13].

Due to the complexity and variety of motion patterns in locomotion, transition, and other lower-limb movements, prosthetic control faces rather challenging requirements and includes the detection of user intent and environmental conditions. To cope with this, control approaches with multiple levels are common [13]. These methods combine (feedback) control with techniques to switch between different modes, for example, state machines, and trajectory generators, for example, central pattern generators. The highest level controller detects user intentions and environmental conditions to select and generate the desired motion patterns. These patterns are transformed into position or torque requests to be set by a lower level controller that directly commands the actuators.

9.5.5.1 High-Level Control

Two broadly applied high-level control schemes are echo control and intent/gait recognition [13]. Echo control acquires sensory data about the motions of the healthy leg and transfers it to the prosthetic controller delayed by a half gait cycle. Despite the simple implementation, echo control has some crucial drawbacks. Tracking the sound-side trajectories requires a rather high mechanical impedance, asymmetric locomotion requires the implementation of auxiliary algorithms, and the whole approach is limited to unilateral amputees [13]. Due to the high mechanical impedance, the prosthesis might feel less natural and be constrained in interacting with the environment.

Control methods based on intent and/or gait mode recognition rely on sensory input from the user and the environment, for example, ground reaction forces, sEMG signals, angular positions, or forces applied by the amputee, and detects the current user intention and gait scenario. Depending on these classifications, the prosthesis selects a locomotive mode and commands the actuators. While finite-state machines are often used for high-level decision-making, impedance control with adaptable spring and damper characteristics commonly serves as the low-level controller. The apparent impedance of the prostheses is practically realized physically by varying the mechanics of a variable impedance actuator, rendered virtually through the actuator and controller, or through a combination of both. To adapt the apparent impedance behavior of the prosthesis appropriately, the gait cycle is separated into phases, for example, stance and swing, and gait scenarios such as straight walking, stair climbing, and turning are differentiated. The corresponding spring and damper parameters are mostly set similarly to the ones physiological joint, which seems to yield more intuitive control than conventional position controllers [13].

While being an established method in upper-limb prosthetics, myoelectric control of lower-limb prosthetics started gaining more interest in the research community [13]. Data about the muscle activity in the residual limb are acquired through electrodes embedded in the prosthetic socket and facilitates users to volitionally control their prosthetic device, for example, by commanding torques by varying muscle contraction. Despite these advantages, myoelectric control exhibits several challenges such as noise, latency, and additional training required from the users.

Combinations of echo control and myoelectric control can resolve limitations to symmetric locomotion patterns. However, intent recognition and environmental monitoring using finite-state machines is applied in most recent devices [13]. Moreover, methods using artificial intelligence or bioinspired central pattern generators are introduced into high-level control algorithms.

9.5.5.2 Low-Level Control

Feedback control is a common technique utilized for the low-level control of the actuators. The controlled quantities comprise joint positions as well as actuator currents and forces [13]. Recently, the prevalent and most promising approach seems to be impedance control. It adjusts the mechanical compliance, damping, and inertial behavior of the limb and can be used to inject power to stimulate and support locomotion. Instead of controlling positions, torques are generated to let the prosthetic device act like biologically inspired system composed of passive springs and dampers, mimicking properties of biological muscles, tendons, and ligaments. This also leads to a more predictable behavior, which improves human-machine interaction.

9.5.6 Sockets

Generally, prosthetic limbs connect to the residual limb of the amputee via a socket (see Fig. 9.14). These are attached through belts and cuffs or through suction either directly or using an intermediate part, that is, the liner. Liners are fixed within the sockets by vacuum or a pin lock. While most liners are manufactured in series, sockets are usually handcrafted and patient tailored. This section gives a broad overview on this mechanical human-machine interface based on Popović [98].

To achieve a good fit between residual limb and socket, various biomechanical factors such as the geometry or soft tissue behavior have to be considered. While this is mostly done by manual plaster modeling, methods like laser scanning and MRT are introduced to support the acquisition of the residual limb geometry and tissue properties. The positive mold of the residual limb serves as the basis to from a thermoplastic socket model. Therefore, a sheet of clear thermoplastic is heated in an oven and vacuum formed around the positive mold. The resulting thermoplastic test socket is transparent and allows the orthopedic technician to assess the fit, which is subsequently elaborated in a discussion with the patient. Besides laser scanning and MRT, motion capturing and force measurement are used for gait analysis to examine the fit of the prosthesis as a whole and the socket in particular. A very recent development is robotic devices that measure the surface stiffness. The implementation of a permanent socket usually relies on polypropylene, which can be vacuum formed over a mold or used in injection molding.

Liners are commonly made from materials such as gel or silicone. Gel liners as shown in Fig. 9.15 can increase comfort and protect the skin while possible reducing proprioception and thereby the control over the prosthetic limb. To enable secure suspension, gel liners can be attached to the socket with a pin lock or a lanyard cord. The main causes of suspension when using gel liners are suction and tackiness between the skin and the liner. Additionally, suction between the liner and the socket might improve the overall suspension. For improved fitting, elevated vacuum socket use a pump to create a vacuum between the liner and the socket wall, which can also reduce perspiration inside the socket.

Beyond sockets and liners, novel approaches to realize the mechanical human-machine interface of the prosthesis are researched and enter clinical application. Osseointegration is a very promising approach, where a titanium connector is inserted into the

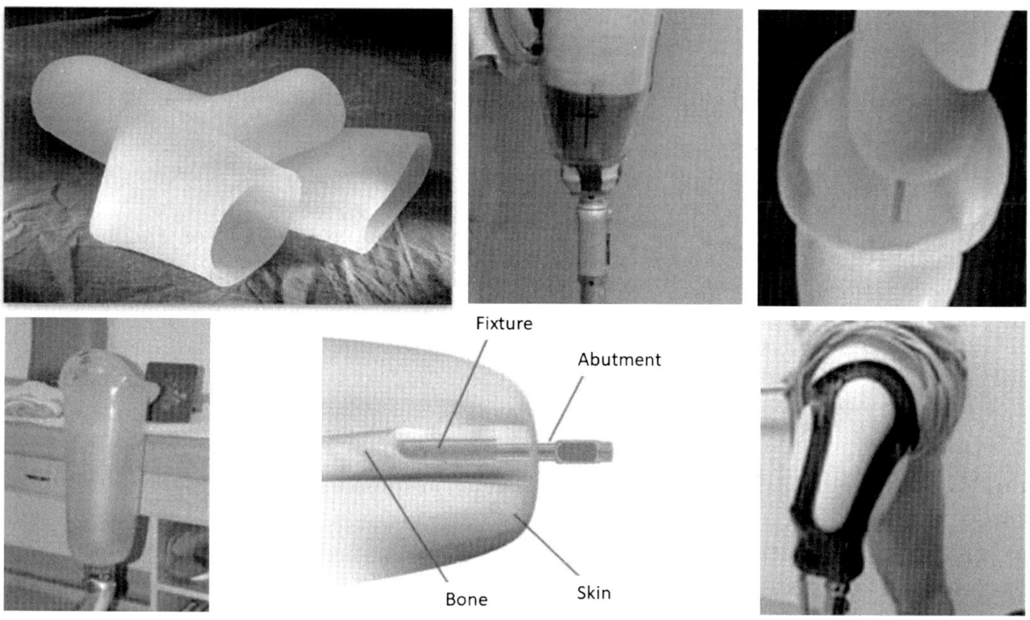

FIG. 9.15 Types of lower-limb prosthetic interfaces. Top (left to right): gel liners, elevated vacuum socket, pin lock system. Bottom (left to right): temporary transparent socket, osseointegration, Hi-Fi (high-fidelity) stump-brace interface.

bone of the residual limb (see Fig. 9.14). This connector interfaces to the prosthetic limb and the bone attaches itself to it after several months. However, osseointegration is not fully established in clinical practice due to potential medical issues such as infections at the passage through the skin or bone fractures. Another recent trend is High-Fidelity (HiFi) interfacing, which distributes the forces and pressures more functionally localized at particular leg areas. By targeted and patient-tailored compression and relieving of soft tissue, compression is increased where it has the highest effect without affecting more sensitive areas as presented in Fig. 9.15.

In the future, biomechatronic approaches might bring completely novel approaches to the field of body-prosthesis interfaces, for example, using advanced materials, sensors, and microscale actuators. One of the authors (MP) recently led the Robotic Socket for Transtibial Amputees project [116,117]. This novel prosthetic socket reduces stress on the skin and soft tissues of the limb by automatically redistributing the pressures between the limb and the socket. The final design consisted of a system of soft bladders and servo-actuated valves controlled using input from pressure sensors. As a result pressures are successfully redistributed across the limb in less than 100ms even under a 270 lb (122 kg) load.

9.6 Future Directions

This section gives an outline of future directions and developments in prosthetic engineering.

9.6.1 Considering Psychological Aspects in Prosthetic Engineering

Besides physical effects, losing a limb has an enormous psychological impact [125] and relates to social and personal aspects [126]. Hence, various studies suggest to focus on the needs and experience of the human user in the medical [127] as well as the engineering domain [11]. While user requirements have been taken into account for decades, for example, ease of use, comfort, or variability of use (see Section 9.3), systematic procedures to consider and assess user needs and experience in engineering design are scarce [128–130].

Psychological effects in prosthetic use are commonly evaluated with questionnaires and interviews, while biomechanical aspects are tackled using rehabilitation measures. For instance, the prosthesis evaluation questionnaire (PEQ) [131] and the trinity amputation and prosthesis experience scales (TAPES) [132] are widely applied to understand lower-limb prosthetic function, mobility, and psychosocial aspects. For the upper limb, questionnaires are similarly commonly applied to assess user satisfaction [133, 134]. Moreover, specific issues are assessed by specialized questionnaires, for example, by the amputee body image scale (ABIS) [135].

Gauthier-Gagnon et al. [136] distinguish enabling, predisposing, and reinforcing factors that influence prosthetic design and its user experience. According to their categorization, enabling factors are those that can be influenced by technical development, while predisposing ones are related to personal traits or attitudes and reinforcing ones comprise aspects such as the psychosocial situation of the amputee, for example, availability of family support. To systematically investigate and consider the influence of user needs and experience on prosthetic design, connections between enabling and predisposing factors might be investigated [11].

Fig. 9.16 suggests a potential design procedure to consider user needs and experience in the development of future biomechatronic prostheses, which is based on Beckerle et al. [11]. While the concrete implementation will differ with respect to the particular design task, it implies which steps could help to integrate user needs and experience in engineering design. In parallel to common technical requirement analysis relying on biomechanical data in prosthetics, user needs and experience are examined. Therefore, user and expert data is surveyed by questionnaires, interviews, and similar techniques. From this data, user needs and experience are identified by statistical analysis, verbalized for a joint understanding in the design team, and ranked if appropriate. Using a transfer method, the results from both domains are fused, for example, by prioritizing technical requirements according to their relation to user needs and experience, to determine a human-oriented development focus. Eventually, this focus and the corresponding and weighted technical requirements are forwarded to common mechatronic systems engineering of the biomechatronic limb as indicated in Fig. 9.16. Systems engineering can be implemented by established methods such as V model design, which is commonly used in Germany [137].

The consideration of user needs and experience may facilitate the development of a new generation of prosthetic devices that will allow users to "feel" the device and fluidly

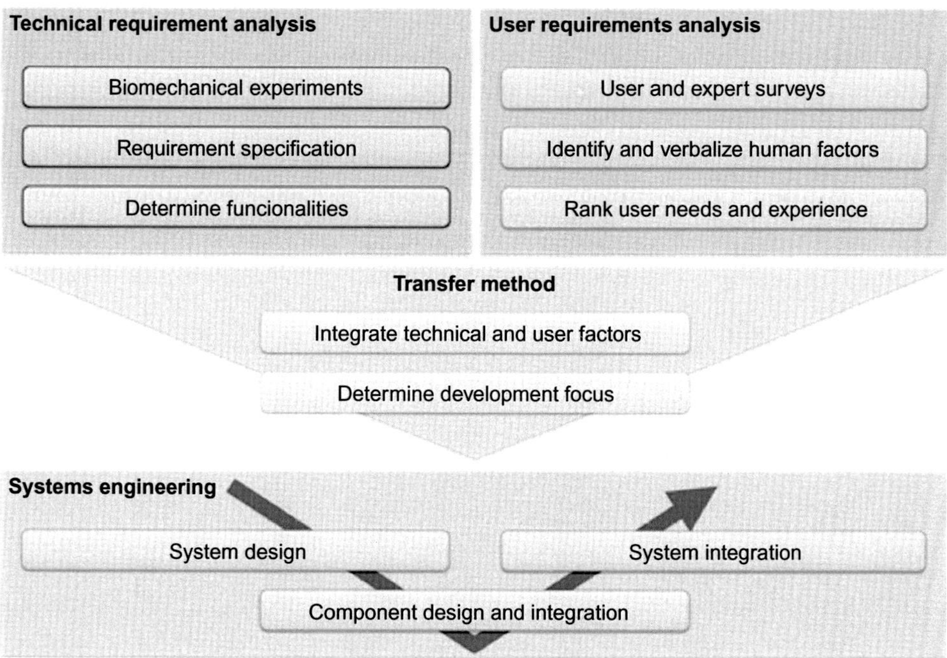

FIG. 9.16 Suggestion for a design procedure that considers user needs and experience in biomechatronic prostheses.

communicate with users through bidirectional interfaces [12, 138–140]. If engineered well, those devices have the potential to not only assist or augment their users' capabilities, but might also improve the experience of being part of their users' own bodies [11, 138, 140].

9.6.2 Long-Term Visions

The future direction of the prosthetics field is almost certainly focused on the development of even more biomimetic synthetic limbs. No amputee would ever object having a comfortable and fully functional biomimetic limb that feels like their own. How far are we from that goal?

From an engineering standpoint technological readiness is very close to that objective and the first highly biomimetic limbs are expected on market during the following decade. Of course, the notion of biomimicry in this context refers only to macro shape, dimensions, mass, and function rather than to details of substructure all the way to the level of sarcomere. Biomimetic limbs that will closely match biological ones even at the microscale are expected by the end of 21st century and there is a good chance that they will be engineered either by using biological cells or maybe even by their augmented, synthetic counterparts.

This optimistic point of view is supported by the advent of more bioinspired actuators, discussed in Chapter 3, that could exactly match the biological muscle function both in terms of passive and active properties, novel materials that could easily mimic ligaments and tendons, rapid prototyping capabilities that could for example allow 3D printing of bones obtained from scans of nonamputated side, inexpensive sensors, discussed in Chapter 4, that could be embedded in muscles as well as in synthetic skin, better battery energy and mass densities, tiny control units with powerful computational capabilities, smart socket technologies, and finally, bidirectional communication between the user and prosthesis. Hence, all these elements are all within reach and it seems to be a matter of dedicated integration and the creation of market ready commercial product that will most likely take place in the next 10 years.

References

[1] H. Rackham, W.H.S. Jones, D.E. Eichholz, Pliny: Natural History, vol. 7, Harvard University Press, 1947.

[2] K. Ziegler-Graham, E. MacKenzie, P. Ephraim, T. Travison, R. Brookmeyer, Estimating the prevalence of limb loss in the United States: 2005 to 2050, Arch. Phys. Med. Rehabil. 89 (2008) 422–429.

[3] P.F. Adams, G.E. Hendershot, M.A. Marano, Current estimates from the National Health Interview Survey, 1996, Vital Health Stat. 10 (1999) 200.

[4] T.R. Dillingham, L.E. Pezzin, E.J. MacKenzie, Limb amputation and limb deficiency: epidemiology and recent trends in the United States, South. Med. J. 95 (8) (2002) 875–884.

[5] A.S. Sarvestani, A.T. Azam, Amputation: a ten-year survey, Trauma Mon. 18 (3) (2013) 126.

[6] E.J. Davies, B.R. Friz, F.W. Clippinger, Amputees and their prostheses, Artif. Limbs 14 (2) (1970) 19–48.

[7] M.F. Owings, L.J. Kozak, Ambulatory and inpatient procedures in the United States, 1996, Vital Health Stat. 13 (139) (1998) 1–119.

[8] A. Esquenazi, Amputation rehabilitation and prosthetic restoration. From surgery to community reintegration, Disabil. Rehabil. 26 (14/15) (2004) 831–836.

[9] L. Cunningham, H. Lax, M. Nielsen, Amputee Rehabilitation (Online), Available from:http://www.mageerehab.org/wp-content/uploads/2016/04/Amputee-Rehab.pdf, 2016.

[10] D. Berry, Microprocessor prosthetic knees, Phys. Medica 17 (2006) 91–113.

[11] P. Beckerle, O. Christ, T. Schürmann, J. Vogt, O. von Stryk, S. Rinderknecht, A human–machine-centered design method for (powered) lower limb prosthetics, Robot. Auton. Syst. 95 (2017) 1–12.

[12] C. Castellini, et al., Proceedings of the first workshop on peripheral machine interfaces: going beyond traditional surface electromyography, Front. Neurorobot. 5 (22) (2014) 1–17.

[13] M. Windrich, M. Grimmer, O. Christ, S. Rinderknecht, P. Beckerle, Active lower limb prosthetics: a systematic review of design issues and solutions, Biomed. Eng. Online 15 (3) (2016) 5–19.

[14] R. Versluys, et al., From conventional prosthetic feet to bionic feet: a review study, in: IEEE International Conference on Biomedical Robotics and Biomechatronics, 2008.

[15] M.A. Holgate, J.K. Hitt, R.D. Bellman, T.G. Sugar, K.W. Hollander, SPARKy (spring ankle with regenerative kinetics) project: choosing a DC motor based actuation method, in: IEEE International Conference on Biomedical Robotics and Biomechatronics, 2008.

[16] F. Stuhlenmiller, J. Schuy, P. Beckerle, S. Rinderknecht, A user-specific human-machine interaction strategy for a prosthetic shank adapter, Curr. Dir. Biomed. Eng. 3 (2) (2017) 493–496.

[17] S.K. Au, J. Weber, H. Herr, Powered ankle-foot prosthesis improves walking metabolic economy, IEEE Trans. Robot. 25 (1) (2009) 51–66.

[18] R.M. Alexander, The Human Machine, Columbia University Press, 2010.

[19] B.M. Nigg, W. Herzog, Biomechanics of the Musculo-Skeletal System, Wiley, 2007.

[20] W. Herzog, Skeletal muscle mechanics: questions, problems and possible solutions, J. NeuroEng. Rehabil. 14 (Sep. 2017) 98.

[21] Biomechanics of Skeletal Muscles—Vladimir Zatsiorsky, Boris Prilutsky, human-kinetics, 2011. (Online). Available from: http://www.humankinetics.com/products/all-products/biomechanics-of-skeletal-muscles (accessed 12.01.18).

[22] C. Schumacher, A. Seyfarth, Sensor-motor maps for describing linear reflex composition in hopping, Front. Comput. Neurosci. 11 (2017).

[23] D.M. Bramble, D.E. Lieberman, Endurance running and the evolution of Homo, Nature 432 (7015) (2004) 345–352.

[24] D.A. Winter, Biomechanics and Motor Control of Human Movement, John Wiley & Sons, 2009.

[25] P.V. Komi, Relevance of in vivo force measurements to human biomechanics, J. Biomech. 23 (Suppl. 1) (1990) 23–34.

[26] M. Popovic, A. Hofmann, H. Herr, Angular momentum regulation during human walking: biomechanics and control, in: 2004 IEEE International Conference on Robotics and Automation, 2004. Proceedings. ICRA'04, vol. 3, 2004, pp. 2405–2411.

[27] H. Herr, M. Popovic, Angular momentum in human walking, J. Exp. Biol. 211 (4) (Feb. 2008) 467–481.

[28] S. Bohm, L. Mademli, F. Mersmann, A. Arampatzis, Predictive and reactive locomotor adaptability in healthy elderly: a systematic review and meta-analysis, Sports Med. 45 (12) (Dec. 2015) 1759–1777.

[29] G.A. Cavagna, H. Thys, A. Zamboni, The sources of external work in level walking and running, J. Physiol. 262 (3) (1976) 639–657.

[30] P.A. Willems, G.A. Cavagna, N.C. Heglund, External, internal and total work in human locomotion, J. Exp. Biol. 198 (Pt 2) (1995) 379–393.

[31] J.E.A. Bertram, Considering gaits, in: J.E.A. Bertram (Ed.), Understanding Mammalian Locomotion, John Wiley & Sons, Inc., 2016, pp. 27–50

[32] A.D. Kuo, J.M. Donelan, A. Ruina, Energetic consequences of walking like an inverted pendulum: step-to-step transitions, Exerc. Sport Sci. Rev. 33 (2) (2005) 88–97.

[33] H. Geyer, A. Seyfarth, R. Blickhan, Compliant leg behaviour explains basic dynamics of walking and running, Proc. R. Soc. Lond. B Biol. Sci. 273 (1603) (2006) 2861–2867.

[34] R. Blickhan, A. Seyfarth, H. Geyer, S. Grimmer, H. Wagner, M. Günther, Intelligence by mechanics, Philos. Trans. R. Soc. Lond. Math. Phys. Eng. Sci. 365 (1850) (2007) 199–220.

[35] V.M. Zatsiorsky, Kinetics of Human Motion. Human Kinetics, (2002).

[36] T. Flash, N. Hogan, The coordination of arm movements: an experimentally confirmed mathematical model, J. Neurosci. 5 (7) (1985) 1688–1703.

[37] J.F. Soechting, C.A. Buneo, U. Herrmann, M. Flanders, Moving effortlessly in three dimensions: does Donders' law apply to arm movement? J. Neurosci. 15 (9) (1995) 6271–6280.

[38] R.M. Alexander, A minimum energy cost hypothesis for human arm trajectories, Biol. Cybern. 76 (2) (1997) 97–105.

[39] T. Flash, N. Hogan, Optimization principles in motor control, in: The Handbook of Brain Theory and Neural Networks, MIT Press, Cambridge, MA, 1998, pp. 682–685.

[40] A. Grabowski, C.T. Farley, R. Kram, Independent metabolic costs of supporting body weight and accelerating body mass during walking, J. Appl. Physiol. 98 (2) (Feb. 2005) 579–583.

[41] R.L. Marsh, Partitioning the energetics of walking and running: swinging the limbs is expensive, Science 303 (5654) (2004) 80–83.

[42] C.J. Arellano, R. Kram, The energetic cost of maintaining lateral balance during human running, J. Appl. Physiol. 112 (3) (2011) 427–434.

[43] C.J. Arellano, R. Kram, Partitioning the metabolic cost of human running: a task-by-task approach, Integr. Comp. Biol. 54 (6) (2014) 1084–1098.

[44] R.M. Alexander, Optimization and gaits in the locomotion of vertebrates, Physiol. Rev. 69 (4) (1989) 1199–1227.

[45] W.D. McArdle, F.I. Katch, V.L. Katch, Exercise Physiology: Nutrition, Energy, and Human Performance, Lippincott Williams & Wilkins, 2010.

[46] J.M. Donelan, R. Kram, A.D. Kuo, Mechanical work for step-to-step transitions is a major determinant of the metabolic cost of human walking, J. Exp. Biol. 205 (23) (2002) 3717–3727.

[47] M.-J. Hsu, D.H. Nielsen, S.-J. Lin-Chan, D. Shurr, The effects of prosthetic foot design on physiologic measurements, self-selected walking velocity, and physical activity in people with transtibial amputation, Arch. Phys. Med. Rehabil. 87 (1) (2006) 123–129.

[48] L. Torburn, C.M. Powers, R. Guiterrez, J. Perry, Energy expenditure during ambulation in dysvascular and traumatic below-knee amputees: a comparison of five prosthetic feet, J. Rehabil. Res. Dev. 32 (2) (1995) 111.

[49] R.L. Waters, S. Mulroy, The energy expenditure of normal and pathologic gait, Gait Posture 9 (3) (1999) 207–231.

[50] H.M. Herr, A.M. Grabowski, Bionic ankle–foot prosthesis normalizes walking gait for persons with leg amputation, Proc. R. Soc. B 279 (1728) (2012) 457–464.

[51] J.M. Aldridge, J.T. Sturdy, J.M. Wilken, Stair ascent kinematics and kinetics with a powered lower leg system following transtibial amputation, Gait Posture 36 (2) (2012) 291–295.

[52] F. Sup, H.A. Varol, M. Goldfarb, Upslope walking with a powered knee and ankle prosthesis: initial results with an amputee subject, IEEE Trans. Neural Syst. Rehabil. Eng. Publ. IEEE Eng. Med. Biol. Soc. 19 (1) (2011) 71–78.

[53] A. Lai, A.G. Schache, Y.-C. Lin, M.G. Pandy, Tendon elastic strain energy in the human ankle plantar-flexors and its role with increased running speed, J. Exp. Biol. 217 (17) (Sep. 2014) 3159–3168.

[54] G.-P. Brüggemann, A. Arampatzis, F. Emrich, W. Potthast, Biomechanics of double transtibial amputee sprinting using dedicated sprinting prostheses, Sports Technol. 1 (4–5) (2008) 220–227.

[55] M. Grimmer, et al., A powered prosthetic ankle joint for walking and running, Biomed. Eng. Online 15 (Suppl. 3) (2016).

[56] D. Roman-Liu, T. Tokarski, Upper limb strength in relation to upper limb posture, Int. J. Ind. Ergon. 35 (1) (2005) 19–31.

[57] F.A. Mussa-Ivaldi, N. Hogan, E. Bizzi, Neural, mechanical, and geometric factors subserving arm posture in humans, J. Neurosci. 5 (10) (1985) 2732–2743.

[58] S. Maniwa, T. Nishikori, S. Furukawa, K. Kajitani, M. Ochi, Alteration of collagen network and negative charge of articular cartilage surface in the early stage of experimental osteoarthritis, Arch. Orthop. Trauma Surg. 121 (4) (2001) 181–185.

[59] L.S. Lohmander, M. Ionescu, H. Jugessur, A.R. Poole, Changes in joint cartilage aggrecan after knee injury and in osteoarthritis, Arthritis Rheum. 42 (3) (1999) 534–544.

[60] I.G. Otterness, A.C. Swindell, R.O. Zimmerer, A.R. Poole, M. Ionescu, E. Weiner, An analysis of 14 molecular markers for monitoring osteoarthritis: segregation of the markers into clusters and distinguishing osteoarthritis at baseline, Osteoarthr. Cartil. 8 (3) (May 2000) 180–185.

[61] G.S. Beaupre, S.S. Stevens, D.R. Carter, Mechanobiology in the development, maintenance, and degeneration of articular cartilage, J. Rehabil. Res. Dev. 37 (2) (2000) 145.

[62] A.J. Grodzinsky, M.E. Levenston, M. Jin, E.H. Frank, Cartilage tissue remodeling in response to mechanical forces, Annu. Rev. Biomed. Eng. 2 (1) (2000) 691–713.

[63] V.C. Mow, A. Ratcliffe, A. Robin Poole, Cartilage and diarthrodial joints as paradigms for hierarchical materials and structures, Biomaterials 13 (2) (1992) 67–97.

[64] F. Eckstein, et al., Functional adaptation of human joints to mechanical stimuli, Osteoarthr. Cartil. 10 (1) (2002) 44–50.

[65] M.F. Koff, O.F. Ugwonali, R.J. Strauch, M.P. Rosenwasser, G.A. Ateshian, V.C. Mow, Sequential wear patterns of the articular cartilage of the thumb carpometacarpal joint in osteoarthritis1 1No benefits in any form have been received or will be received from a commercial party related directly or indirectly to the subject of this article, J. Hand Surg. 28 (4) (2003) 597–604.

[66] T.P. Andriacchi, A. Mündermann, R.L. Smith, E.J. Alexander, C.O. Dyrby, S. Koo, A framework for the in vivo pathomechanics of osteoarthritis at the knee, Ann. Biomed. Eng. 32 (3) (2004) 447–457.

[67] C.C. Prodromos, T.P. Andriacchi, J.O. Galante, A relationship between gait and clinical changes following high tibial osteotomy, JBJS 67 (8) (1985) 1188.

[68] T. Miyazaki, M. Wada, H. Kawahara, M. Sato, H. Baba, S. Shimada, Dynamic load at baseline can predict radiographic disease progression in medial compartment knee osteoarthritis, Ann. Rheum. Dis. 61 (7) (2002) 617–622.

[69] D.R. Carter, G.S. Beaupré, N.J. Giori, J.A. Helms, Mechanobiology of skeletal regeneration, Clin. Orthopaed. 355 (1998) S41–S55.

[70] D.R. Carter, M. Wong, Modelling cartilage mechanobiology, Philos. Trans. R. Soc. Lond. Ser. B Biol. Sci. 358 (1437) (Sep. 2003) 1461–1471.

[71] T.E. McAlindon, S. Snow, C. Cooper, P.A. Dieppe, Radiographic patterns of osteoarthritis of the knee joint in the community: the importance of the patellofemoral joint, Ann. Rheum. Dis. 51 (7) (1992) 844–849.

[72] D.T. Felson, Epidemiology of hip and knee osteoarthritis, Epidemiol. Rev. 10 (1988) 1–28.

[73] D.T. Felson, Osteoarthritis of the knee, N. Engl. J. Med. 354 (8) (2006) 841–848.

[74] L. Sharma, J. Song, D.T. Felson, S. Cahue, E. Shamiyeh, D.D. Dunlop, The role of knee alignment in disease progression and functional decline in knee osteoarthritis, JAMA 286 (2) (Jul. 2001) 188–195.

[75] J.-P. Pelletier, J. Martel-Pelletier, S.B. Abramson, Osteoarthritis, an inflammatory disease: potential implication for the selection of new therapeutic targets, Arthritis Rheum. 44 (6) (Jun. 2001) 1237–1247.

[76] R.N. van Gent, D. Siem, M. van Middelkoop, A.G. van Os, S.M. Bierma-Zeinstra, B.W. Koes, Incidence and determinants of lower extremity running injuries in long distance runners: a systematic review, Br. J. Sports Med. 41 (8) (2007) 469–480. discussion 480.

[77] J.R. Engsberg, A.G. Lee, K.G. Tedford, J.A. Harder, Normative ground reaction force data for able-bodied and below-knee-amputee children during walking, J. Pediatr. Orthop. 13 (2) (1993) 169–173.

[78] J.R. Engsberg, A.G. Lee, J.L. Patterson, J.A. Harder, External loading comparisons between able-bodied and below-knee-amputee children during walking, Arch. Phys. Med. Rehabil. 72 (9) (1991) 657–661.

[79] L. Nolan, A. Wit, K. Dudziñski, A. Lees, M. Lake, M. Wychowañski, Adjustments in gait symmetry with walking speed in trans-femoral and trans-tibial amputees, Gait Posture 17 (2) (Apr. 2003) 142–151.

[80] S. Farrokhi, B. Mazzone, A. Yoder, K. Grant, M. Wyatt, A narrative review of the prevalence and risk factors associated with development of knee osteoarthritis after traumatic unilateral lower limb amputation, Mil. Med. 181 (S4) (2016) 38–44.

[81] R. Gailey, K. Allen, J. Castles, J. Kucharik, M. Roeder, Review of secondary physical conditions associated with lower-limb amputation and long-term prosthesis use, J. Rehabil. Res. Dev. 45 (1) (2008) 15–29.

[82] K.R. Kaufman, S. Frittoli, C.A. Frigo, Gait asymmetry of transfemoral amputees using mechanical and microprocessor-controlled prosthetic knees, Clin. Biomech. Bristol Avon 27 (5) (2012) 460–465.

[83] G.R. Hurley, R. McKenney, M. Robinson, M. Zadravec, M.R. Pierrynowski, The role of the contralateral limb in below-knee amputee gait, Prosthetics Orthot. Int. 14 (1) (1990) 33–42.

[84] W.S. Howard, V. Kumar, On the stability of grasped objects, IEEE Trans. Robot. Autom. 12 (6) (1996) 904–917.

[85] A. Santuz, A. Ekizos, N. Eckardt, A. Kibele, A. Arampatzis, Challenging human locomotion: stability and modular organisation in unsteady conditions, Sci. Rep. 8 (1) (2018) 2740.

[86] T.M. Lejeune, P.A. Willems, N.C. Heglund, Mechanics and energetics of human locomotion on sand, J. Exp. Biol. 201 (13) (1998) 2071–2080.

[87] T.A. McMahon, P.R. Greene, Fast running tracks, Sci. Am. 239 (6) (1978) 148–163.

[88] A.E. Kerdok, A.A. Biewener, T.A. McMahon, P.G. Weyand, H.M. Herr, Energetics and mechanics of human running on surfaces of different stiffnesses, J. Appl. Physiol. 1985 92 (2) (2002) 469–478.

[89] A.S. Voloshina, A.D. Kuo, M.A. Daley, D.P. Ferris, Biomechanics and energetics of walking on uneven terrain, J. Exp. Biol. 216 (21) (2013) 3963–3970.

[90] A.S. Voloshina, D.P. Ferris, Biomechanics and energetics of running on uneven terrain, J. Exp. Biol. 218 (5) (2015) 711–719.

[91] M.A. Daley, Biomechanics: running over uneven terrain is a no-brainer, Curr. Biol. 18 (22) (2008) R1064–R1066.

[92] S. Willwacher, K.M. Fischer, R. Benker, S. Dill, G. Brüggemann, Kinetics of cross-slope running, J. Biomech. 46 (16) (2013) 2769–2777.

[93] M. Damavandi, P.C. Dixon, D.J. Pearsall, Ground reaction force adaptations during cross-slope walking and running, Hum. Mov. Sci. 31 (1) (2012) 182–189.

[94] M. Damavandi, P.C. Dixon, D.J. Pearsall, Kinematic adaptations of the hindfoot, forefoot, and hallux during cross-slope walking, Gait Posture 32 (3) (2010) 411–415.

[95] A.H. Vrieling, et al., Uphill and downhill walking in unilateral lower limb amputees, Gait Posture 28 (2) (2008) 235–242.

[96] C. Curtze, A.L. Hof, K. Postema, B. Otten, Over rough and smooth: amputee gait on an irregular surface, Gait Posture 33 (2) (2011) 292–296.

[97] S. Willwacher, B. Müller, S.I. Wolf, G.-P. Brueggemann, Z. Deng, A. McIntosh, F. Miller, W.S. Selbie (Eds.), Running shoes: injury protection and performance enhancement, in: Handbook of Human Motion, Springer International Publishing, 2017, pp. 1–16.

[98] M.B. Popović, Biomechanics and Robotics, CRC Press, 2013.

[99] M. Mori, The uncanny valley, Energy 7 (4) (1970) 33–35.

[100] B. Maat, G. Smit, D. Plettenburg, P. Breedveld, Passive prosthetic hands and tools: a literature review, Prosthet. Orthot. Int. (2017). 309364617691622.

[101] T.R. Makin, F. de Vignemont, A.A. Faisal, Neurocognitive barriers to the embodiment of technology, Nat. Biomed. Eng. 1 (2017) 0014.

[102] A.L. Muilenburg, M.A. LeBlanc, Body-powered upper-limb components, in: Comprehensive Management of the Upper-Limb Amputee, Springer, 1989, pp. 28–38.

[103] S. Adee, Dean Kamen's' Luke arm' prosthesis readies for clinical trials, IEEE Spectr. 2 (2008) 34.

[104] M.S. Johannes, J.D. Bigelow, J.M. Burck, S.D. Harshbarger, M.V. Kozlowski, T. Van Doren, An overview of the developmental process for the modular prosthetic limb, Johns Hopkins APL Tech. Dig. 30 (3) (2011) 207–216.

[105] C. Connolly, Prosthetic hands from touch bionics, Ind. Robot Int. J. 35 (4) (2008) 290–293.

[106] M. Luchetti, G. Verni, R. Sacchetti, N. Rossi, Impact of michelangelo prosthetic hand: findings from a crossover longitudinal study, J. Rehabil. Res. Dev. 52 (5) (2015) 605.

[107] C. Cipriani, M. Controzzi, M.C. Carrozza, The SmartHand transradial prosthesis, J. Neuroeng. Rehabil. 8 (1) (2011) 29.

[108] E. Saint-Elme, M.A. Larrier Jr., C. Kracinovich, D. Renshaw, K. Troy, M. Popovic, Design of a biologically accurate prosthetic hand, in: Extended Abstracts Proceedings of the International Symposium on Wearable Robotics (WeRob 2017), Houston, Texas (USA), November 5–8, 2017.

[109] Brain Controlled Accurate Prosthetic Hand, WPI Popovic Labs http://users.wpi.edu/~mpopovic/pages/AccurateProstheticHand.html (accessed 17.03.18).

[110] A.M. Dollar, R.D. Howe, The SDM hand as a prosthetic terminal device: a feasibility study, in: IEEE International Conference on Rehabilitation Robotics, 2007, pp. 978–983.

[111] M.V. Liarokapis, A.G. Zisimatos, C.I. Mavrogiannis, K.J. Kyriakopoulos, Openbionics: an open-source initiative for the creation of affordable, modular, light-weight, underactuated robot hands and prosthetic devices, in: 2nd ASU Rehabilitation Robotics Workshop, 2014.

[112] G.P. Kontoudis, M.V. Liarokapis, A.G. Zisimatos, C.I. Mavrogiannis, K.J. Kyriakopoulos, Open-source, anthropomorphic, underactuated robot hands with a selectively lockable differential mechanism: towards affordable prostheses, in: 2015 IEEE/RSJ International Conference on Intelligent Robots and Systems (IROS), 2015, Hamburg, pp. 5857–5862.

[113] M. Muller, M. Popovic, Shoulder mounted gyroscopic prosthesis for assisting arm amputees during walking, in: Northeast Bioengineering Conference 2017, Newark, New Jersey, USA, April 1–2, 2017.

[114] M. Muller, M. Popovic, Shoulder mounted gyroscopic prosthesis for assisting arm amputees during walking, in: Dynamic Walking Conference 2017. Mariehamn, Finland, June 4–9 2017, 2017.

[115] Gait Assistive Shoulder Prosthesis: Shoulder Control Moment Gyroscope (SCMG) WPI Popovic Labs. http://users.wpi.edu/~mpopovic/pages/scmg.html (accessed 17.03.18).

[116] D. Felix, S. Tanreverdi, C. Trivedi, E. Wenzlaff Designing a Dynamic Prosthetic Socket for Transtibial Amputees. MQP, Primary Advisor: Marko Popovic (PH/RBE/BME), Co-advisors: Selcuk Guceri and Yitzhak Mendelson, WPI, 2016. https://web.wpi.edu/Pubs/E-project/Available/E-project-042816-151043/unrestricted/Designing_and_Testing_a_Dynamic_Prosthetic_Fitting_for_Transtibial_Amputees.pdf (accessed 17.03.18).

[117] Design and Testing of Robotic Socket for Transtibial Amputees WPI Popovic Labs. http://users.wpi.edu/~mpopovic/pages/roboticlegsocket.html (accessed 17.03.18).

[118] R.L. Waters, J. Perry, D. Antonelli, H. Hislop, Energy cost of walking of amputees: the influence of level of amputation, J. Bone Joint Surg. Am. 58 (1) (1976) 42–46.

[119] S. Willwacher, et al., Elite long jumpers with below the knee prostheses approach the board slower, but take-off more effectively than non-amputee athletes, Sci. Rep. 7 (1) (2017) 16058.

[120] P.G. Weyand, M.W. Bundle, Point: artificial limbs do make artificially fast running speeds possible, J. Appl. Physiol. 1985 108 (4) (2010) 1011–1012.

[121] S. Willwacher, et al., Sprint start kinetics of amputee and non-amputee sprinters, PLoS ONE 11 (11) (2016) e0166219.

[122] G. Strutzenberger, et al., First and second step characteristics of amputee and able-bodied sprinters, Int. J. Sports Phys. Performance (2017) 1–22.

[123] J. Funken, et al., Leg amputation side determines performance in curve sprinting: a case study on a paralympic medalist, Sports Biomech. (2017) 1–13.

[124] M. Grimmer, A. Seyfarth, Stiffness adjustment of a series elastic actuator in a knee prosthesis for walking and running: the trade-off between energy and peak power optimization, in: IEEE/RSJ International Conference on Intelligent Robots and Systems, 2011.

[125] C.D. Murray, J. Fox, Body image and prosthesis satisfaction in the lower limb amputee, Disabil. Rehabil. 24 (17) (2002) 925–931.

[126] E.M. Schaffalitzky, Optimising the Prescription and Use of Lower Limb Prosthetic Technology: A Mixed Methods Approach, Dublin City University, 2010.

[127] R. Baumgartner, P. Botta, Amputation und Prothesenversorgung, Thieme, 2008.

[128] O. Christ, P. Beckerle, Towards active lower limb prosthetic systems: design issues and solutions, Biomed. Eng. Online 15 (3) (2016) 139.

[129] C. Castellini, R.M. Bongers, M. Nowak, C.K. van der Sluis, Upper-limb prosthetic myocontrol: two recommendations, Front. Neurosci. 9 (496) (2015).

[130] P. Beckerle, Commentary: proceedings of the first workshop on peripheral machine interfaces: going beyond traditional surface electromyography, Front. Neurorobot. 11 (2017) 32.

[131] M.W. Legro, G.D. Reiber, D.G. Smith, M. del Aguila, J. Larsen, D. Boone, Prosthesis evaluation questionnaire for persons with lower limb amputations: assessing prosthesis-related quality of life, Arch. Phys. Med. Rehabil. 79 (1998) 931–938.

[132] P. Gallagher, M. MacLachlan, Development and psychometric evaluation of the trinity amputation and prosthesis experience scales (TAPES), Rehabil. Psychol. 45 (2) (2000) 130–154.

[133] J. Siller, S. Silverman, Studies of the upper-extremity amputee; VII. Psychological factors, Artif. Limbs 5 (2) (1958) 88–116.

[134] F. Cordella, et al., Literature review on needs of upper limb prosthesis users, Front. Neurosci. 10 (2016) 209.

[135] J.W. Breakey, Body image: the lower-limb amputee, J. Prosthet. Orthot. 9 (2) (1997) 58–66.

[136] C. Gauthier-Gagnon, M.C. Grisé, D. Potvin, Enabling factors related to prosthetic use by people with transtibial and transfemoral amputation, Arch. Phys. Med. Rehabil. 80 (6) (1999) 9+–713.

[137] Verein Deutscher Ingenieure, VDI 2206, Design Methodology for Mechatronic Systems, 2004.

[138] O. Christ, M. Reiner, Perspectives and possible applications of the rubber hand and virtual hand illusion in non-invasive rehabilitation: technological improvements and their consequences, Neurosci. Biobehav. Rev. 44 (2014) 33–44.

[139] N. Jiang, S. Dosen, K. Müller, D. Farina, Myoelectric control of artificial limbs—is there a need to change focus? IEEE Signal Process. Mag. 29 (5) (2012) 148–152.

[140] T.R. Makin, F. de Vignemont, A.A. Faisal, Neurocognitive barriers to the embodiment of technology, Nature 1 (2017).

10

Powered Orthotics: Enabling Brace Technologies for Upper and Lower Limbs

Marko B. Popovic

WORCESTER POLYTECHNIC INSTITUTE, WORCESTER, MA, UNITED STATES

CHAPTER OUTLINE

10.1 Introduction

In contrast to prosthetic technologies that completely replace limbs, the goal with orthotics, exoskeletons, and exomusculatures is to assist (in sense of assistance with activities of daily life), help regain (as part of rehabilitation treatment), and protect or augment the functionalities of existing limb(s). Again, any device that actively affects dynamics and intimately interacts (via sensory-control system) with the human body is considered to be a robotic [1] or more specifically biomechatronic device.

Here, exoskeletons are differentiated from rigid braces and general orthoses based on the number of human joints that they interface with. In contrast to exoskeletons, addressed in Chapter 11, defined as rigid structures that interface with a plurality of human joints, here braces or orthoses are assumed to interface with just a single human joint. This line is somewhat blurred in case of a hand with multiple joints.

In this chapter a number of orthotic systems for hip, ankle, shoulder, elbow, wrist, and hand are briefly reviewed with a goal to provide the reader a succinct "big picture" of the state-of-the-art technologies and research directions. By no means is this review complete; this field is rapidly advancing with a number of larger and smaller companies

Biomechatronics. https://doi.org/10.1016/B978-0-12-812939-5.00010-0

actively performing R&D as well as with numerous university labs focused on fundamental research resulting in proof of concepts and variety of noncommercial prototypes.

10.2 Powered Hip Braces, Waist Assist, and Lumbar Support

Hip torque values normalized with body mass are typically slightly less than 1 Nm/kg for both leg flexion and extension during regular gait at self-selected speed [1]. Similar maximal values of extension torque are required for unobstructed sit-to-stand and (slow) stand-to-sit action, clearly depending on the chair height and legs' dimensions. However, the factor 2–3 times larger torques are required for stair climbing, and even substantially larger hip torques might be required for weight lifting, details clearly depending on the mass being lifted and body's geometry during these specific weight lifting movements.

On similar lines, typical hip and lumbar area related injuries are caused by either frequent high loading or singular but sudden, jerky application of very large force.

Clearly, there is a well-motivated need for hip and lumbar support and active hip actuation for those with weak hip joint muscles, for example, a large elderly population as well as for the able bodied population who might be lifting heavy weights on a frequent basis and are hence more prone to injuries.

In September 2017, Dr. Yoshiyuki Sankai, a professor at the University of Tsukuba, also the President and CEO of CYBERDYNE Inc., announced the commercialization, starting in October 2017, of "HAL Lumbar Type to Promote Independent Living (tentative name)" (FB02) (Fig. 10.1) marketed to "maintain and improve the weakened bodily functions of a user in the core and lower body, supporting both care givers and care receivers" [2]. Similar to other hybrid-assisted limb (HAL)-like products, this 3.1 kg heavy (including

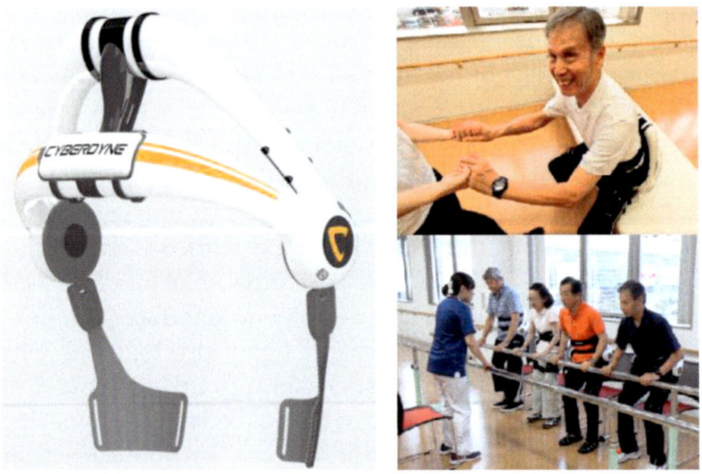

FIG. 10.1 CYBERDINE's "HAL Lumbar Type to Promote Independent Living (tentative name)" (FB02), on the left, and patients performing sit-to-stand exercise with this device, on the right. *Reprinted with permission from CYBERDINE.*

lithium polymer battery), actuated brace, utilizes EMG signals (see Chapter 4) to understand user's intents and provide assistive hip torques and reduce the load and stress on the lower back. EMG sensors are placed either in the form of individual patches on lumbar area skin or incorporated within belt solution. The brace is designed to operate for 3 h on a single charge. The level of assistance could be adjusted with only two buttons.

The same company, CYBERDINE Inc., introduced the "HAL Lumbar Type for Labor Support" (LB03) in December 2017, which is intended to support outdoor use for labor workers who frequently lift weights (Fig. 10.2) [3]. It is both waterproof and dustproof, utilizes advanced 4G (5G) reinforced communication functions, and promotes network with IoH/IoT (Internet of humans/Internet of things). There are only three operating buttons and five levels of assistance available for various tasks. This wearable robot has been approved using international safety standard ISO13482. It is suggested that test results have shown a measurable reduction in lumbar and intervertebral disk loads [4]. The entire device weighs only 3 kg (6.6 lb) and is designed to operate for 3 h on a single charge.

In November 2016, the same company launched a similar waterproof, actuated hip and lumbar support brace CB02 intended for health-care workers, for example, a nurse could lift a patient with more ease. According to review [3], as of September 2017, about 800 of these braces are already in use.

Although assistive torque specifications have not been publicly provided, or at least they are not known by the authors, based on mass and size of this device, authors estimate that each hip motor provides maximally ~15 Nm torques, and probably less than that.

FIG. 10.2 CYBERDINE's "HAL Lumbar Type for Labor Support" (LB03), on the left, and labor worker lifting weight, on the right. *Reprinted with permission from CYBERDINE.*

Normalized by body mass of, for example, 60 kg user that translates to maximally ~0.25 Nm/kg. Clearly, this torques cannot fully exchange needed hip torque, but could certainly provide substantial and possibly, for a given person and task, critical support.

Interestingly, in terms of physical interface with the body, CYBERDINE's devices apply extension forces on front thighs through rigid interface and also stomach area through soft straps. Unfortunately, forces directed onto stomach might be uncomfortable if larger extension forces are applied.

Primarily, shoulder-based interface is utilized by another brace [5, 6] intended to help labor workers.

Muscle suit by Innophys Inc., Japan (Fig. 10.3) which is a spin-off from the Tokyo University of Science, is another approach to the generation of hip torques and lumbar support. Associate Professor Dr. Kobayashi Hiroshi has been developing this approach since 2001. Waist assistance Muscle suit standard model has been launched in the market in 2014. According to Dr. Hiroshi, more than 3400 Muscle suits have been sold in Japan as of summer 2018.

The suit exploits McKibben type of artificial muscles (see Chapter 3) which are 1.5 in. in diameter, contracts maximally 30%, and operates at maximal 5 atm air pressure (by utilizing either compressed air bottle or compressor on a side). The brace is directly controlled and operated by the user by an interface consisting of mouth exhalation switch and touch switch for chin (hence no hands are required), and possibly wireless finger switch.

If instead of McKibben muscle, air operated Hydro Muscle (see Chapter 3) had been used, the maximal force, ignoring friction, could be estimated roughly as product of pressure and area or about 48 kgf. In case of McKibben muscle, as expected, these forces are

FIG. 10.3 Muscle suit by Innophys Inc. equipped with McKibben muscles, shown on the left, and labor worker utilizing this brace that is, waist assist device, on the right. *Reprinted with permission from Innophys. Courtesy of Dr. Kobayashi Hiroshi.*

somewhat smaller for the same pressures and as reported on company website, the generated forces are between 25 and 35 kgf per muscle (range of force ~250–350 N).

The company offers two different versions of suit, one operated with two muscles (range of force ~ 500–700 N) and one operated with four muscles (range of force ~ 1000–1400 N). Hence, for a moment arm of about 0.1 m about the hip joint the heavy lifting option of the muscle suit, which has a mass of 6.6 kg, has torque range up to ~ 100–140 Nm combined for both the hips. Normalized by body mass of, for example, 60 kg user that translates to maximal range of ~0.83–1.16 Nm/kg per hip (or ~0.75–1.05 Nm/kg if mass of device is also taken into account). This is about factor 2–3 times larger torque than previously estimated torque for CYBERDINE's devices for lumbar support.

Power assist suits [7, 8] for industrial use have also been developed by Atoun Inc., located at Nara City, Japan, formally Activelink Co., Ltd., one of Panasonic's in-house venture companies. To help with lifting and carrying heavy loads, Panasonic has introduced the 6.7 kg heavy AWN-03 (Fig. 10.4) an assist suit designed specifically to provide lower back support. Suit can operate for 8 h using a 48.1 V lithium-ion battery.

Note that similar to Innophys suit, this suit also exploits shoulder-based interface to help labor workers.

Here manual switching is unnecessary between four (assist or lifting up, hold, walk, and break or putting down) action modes. Model A switches on its own by following the user's movements. By sensing the wearer's motion when lifting or holding heavy objects, the suit sends a signal to its motors. Hence, the gears spin according to the user's movements.

By raising the user's upper body while simultaneously pushing on their thighs, the suit operated with two high torque alternating current (AC) motors, alleviates burden on the waist. It is marketed by a promise "to reduce stress on the lower back by 15 kg".

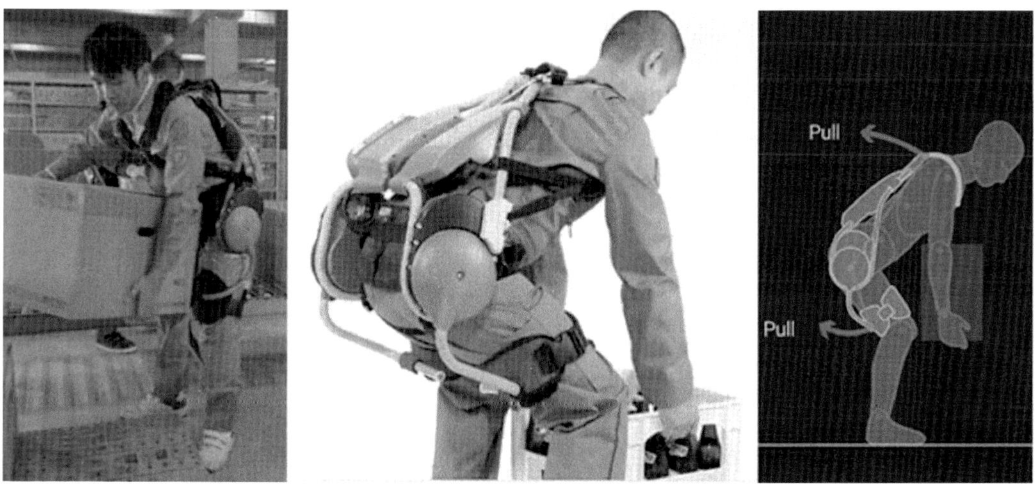

FIG. 10.4 AWN-03, an assist suit designed specifically to provide lower back support, developed by Atoun Inc., Nara City, Japan, formally Activelink Co., Ltd., one of Panasonic's in-house venture companies.

Unfortunately, torque information is not publicly available (or at least it is not known by the authors); however, if a moment arm of 0.3–0.4 m in respect to hip joint is assumed, then 15 kgf suggests about 45–60 Nm, that is, 23–30 Nm per hip, or normalized by body mass of, for example, 60 kg user that translates to range of ~0.38–0.50 Nm/kg per hip (or ~0.35–0.45 Nm/kg if mass of device is also taken into account).

The GEMS [9–13], developed by the Samsung Advanced Institute of Technology, is another example of a robotic brace that assists motion directly at the hip joints (Fig. 10.5). The 2.9 kg device is worn around the waist and utilizes a set of belts with Velcro to fasten its power source in the form of a rechargeable lithium battery; electronics, actuation, and computation are located in the backpack of the device. The device can operate for 1.3 h when fully charged.

Unfortunately, similar to CYBERDINE's suits GEMS also applies extension forces onto the stomach which might be uncomfortable if larger forces are applied.

The device generates assist torque to the hip joints through two 70-W brushless DC motors mounted near the joints. The torque generated by these actuators is channeled to each hip joint through a 75:1 multistage gear system capable of generating up to 14 Nm. If normalized by, for example, a body mass of a 60 kg user that is equal to about 0.23 Nm/kg per hip.

Each joint has one passive degree of freedom (DOF) for abduction and adduction in the frontal plane and one active DOF for extension and flexion in the sagittal plane.

To fit various sizes and to provide comfort, a pair of special thigh frames conforms to a user's thigh contours frames when transmitting assist power from the actuators to the thighs. These thigh frames are available in three sizes (large, medium, and small) for different leg lengths. The GEMS is also available in two sizes to fit various waist/hip sizes, small (for hip circumferences 70–90 cm) and medium (for hip circumferences

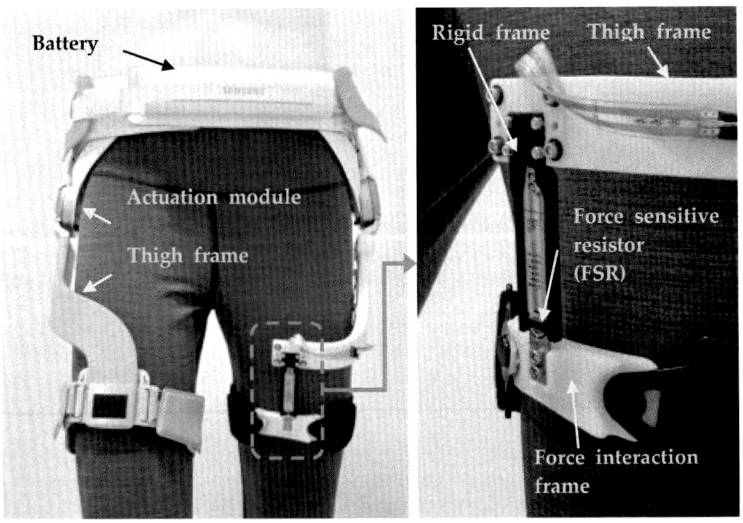

FIG. 10.5 Gait enhancing mechatronic system (GEMS) developed by the Samsung Mechatronics Lab.

90–100 cm). The difference between the two sizes is the size of the two-side hip braces. Additionally, to fit individual body size within the circumference range, the width of each version can be further adjusted.

To measure the hip joint angles, the potentiometer sensors at the hip joints and encoders in the motors are used. Also, there is an inertial measurement unit (IMU) in the backpack to measure the acceleration and rotation of the device.

According to the new research [13] presented at the 15th IEEE International Conference on Rehabilitation Robotics, ICORR 2017 held in London, the GEMS can bestow as much as 21% reduction in metabolic cost (with vs without exo condition) when walking on a treadmill, but only when the control timing is perfectly optimized. For more realistic situations such as changing terrain, slope, and surface conditions, future control systems will have to be able to change motor output timing on the fly. The Samsung team is currently working on developing the GEMSv2, optimized to be used on slopes, which will help solve the problem of a changing slope.

In July 2015, Honda Motor Co., Ltd. announced that it will begin lease sales of the Honda Walking Assist Device [14–16] (Fig. 10.6) to businesses in Japan in November 2015. This is an assistive device used in the training of walking.

This 2.7 kg (including battery) heavy, nonwaterproof device provides very small torques, ~4 Nm, which translate to only 0.07 Nm/kg normalized hip torque (for 60 kg person). However, the device is very fast and can be utilized to assist walking even for very fast walking speeds.

The study on 14 elderly people, average age 78 (Fig. 10.6, top right) suggests that after 3 months of training with this devices walking without assistive device stride length is increased by more than 7% and speed is increased by more than 15%.

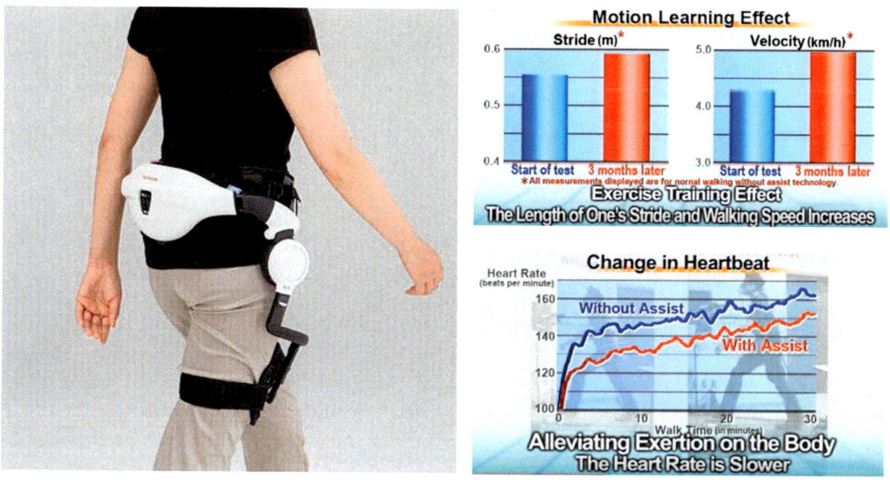

FIG. 10.6 Honda Walking Assist Device (left); study on stride and speed (top right); study on heartbeat (bottom right). *Adapted from Honda Walking Assist Devices, YouTube, Honda, May 24, 2016. https://www.youtube.com/watch?v=lo3MBpzmVJc (retrieved 17.03.18).*

In another study it is shown that walking with assistive device can lower the heartbeat by 10% compared to walking without assistive device (Fig. 10.6, bottom right).

Honda has been optimizing the functions of this device since 2013, with the cooperation of approximately 50 hospitals and other facilities in Japan. Honda's walking assist device has the functionality to guide the user to achieve efficient walking based on the inverted pendulum model.

Three training modes are included: (1) following mode: influence the user's walking motions based on the walking pattern of the user, (2) symmetric mode: device influences the user to achieve bilaterally symmetric motions such as bending and extending both legs, and (3) step mode: device influences the user's steps to enable smooth shifting of weight from heel to toe.

10.3 Powered Knee Braces

Ottobock's C-Brace [17], shown in Fig. 10.7, was launched in the United States in 2014 and in 2015 it was placed on sale (for £60,000) in Britain [18]. This robotic support, designed to wrap around the lower leg, has been developed to help people with disabilities and

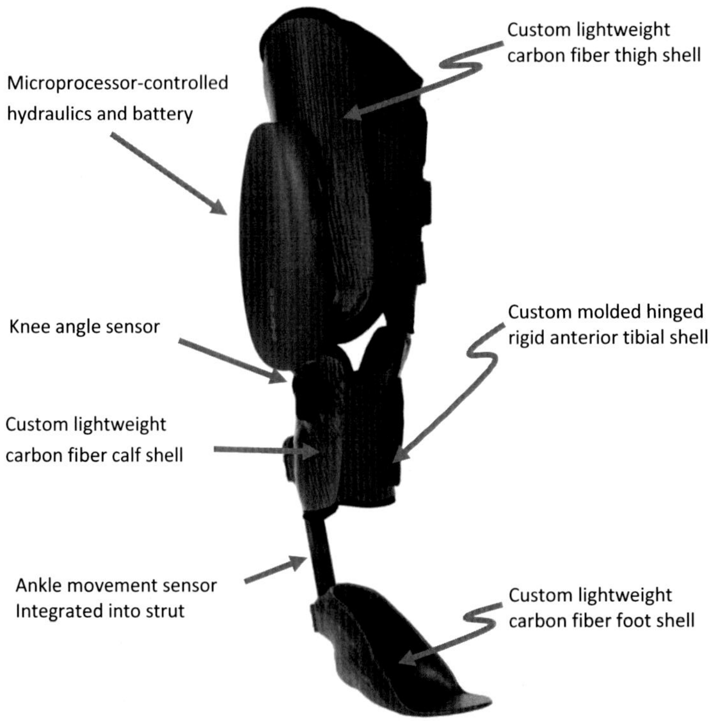

FIG. 10.7 Ottobock's C-Brace. *Reprinted with permission from Ottobock.*

improve the way they walk or play sports, it is the first exoskeletal brace marketed for leisure activities such as walking, cycling, and even roller blading.

The brace reduces strain in the joint, enabling people with minor movement disabilities—such as those experienced after a stroke—to employ a full range of actions by generating the power needed for movement.

For the brace to carry out actions needed for a particular activity, the user selects the desired activity on a remote control which transmits to the leg via wireless Bluetooth technology. The brace then uses computers and sensors on the upper leg to sense the movement of the thigh and transfer them to the robotic knee support.

According to Ottobock: "C-Brace is a computer controlled knee ankle foot orthosis (KAFO) that provides an entirely new approach to walking for individuals who have lower-limb mobility issues due to partial paralysis, incomplete spinal cord injury, post-polio syndrome, quadriceps weakness, and post-stroke." [17] And it is also claimed that: "C-Brace is the only KAFO on the market that provides dynamic control of the entire gait cycle*—stance and swing phases—in real time." [17]

C-Brace has microprocessor-controlled hydraulics and an integrated battery. Its main function is not to walk on its own while moving patient, but to help normalize the patient's gait. In the same way that C-Leg operates, C-Brace is also an intelligent braking system; it provides damping and resists undesired loading. C-Brace does not provide active torque as it essentially represents an actively controlled damping and resistive braking element (Fig. 10.8).

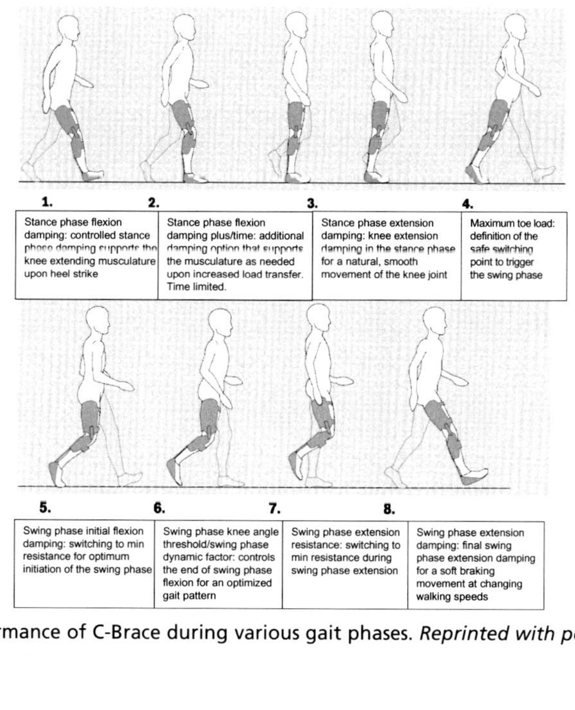

FIG. 10.8 Different performance of C-Brace during various gait phases. *Reprinted with permission from Ottobock.*

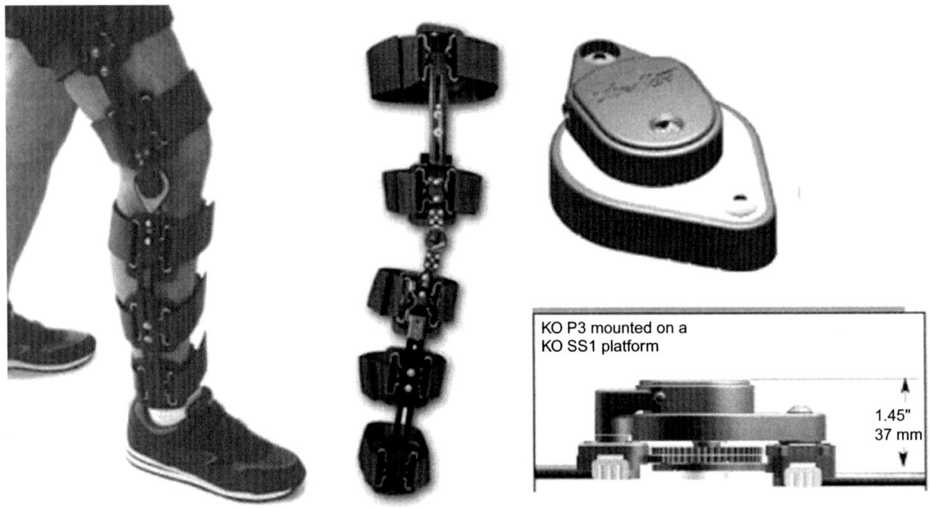

FIG. 10.9 Ultraflex Solution: Knee Orthosis Custom Fit (KO CF) with OPTIONAL extension, flexion, or bidirectional power unit by UltraflexSystems, Inc. *Courtesy of Ultraflex Systems.*

Another company Ultraflex Systems Inc. located in Pennsylvania, US, offers an optional active torque element for their knee braces (Fig. 10.9) [19, 20]. Less than 4-cm-thick motor solution can provide up to 10 Nm assistive torque to assist in achieving full range of knee joint motions. Normalized by body mass this translates to about 0.16 Nm/kg for a 60 kg person. Substantially less than the observed normalized knee torques in walking which are about 0.60 Nm/kg, however, possibly still very important assistive torque.

Another potential practical approach to active generation of less than 10% required torque for walking is the concept of a soft exosuit.

In May 2016, the Wyss Institute for Biologically Inspired Engineering at Harvard University entered into collaboration with ReWalk Robotics Ltd., to accelerate the development of the Institute's lightweight, wearable soft exosuit technologies for assisting people with lower-limb disabilities. The agreement with ReWalk is anticipated to help speedup the design of assistive exosuits that could help patients suffering from stroke and multiple sclerosis (MS) regain mobility. [21]

According to the Wyss Institute: "Using form-fitting, fabric-based designs that are lightweight and non-restrictive, the Wyss Institute's soft exosuit uses compact, powerful actuators packaged in a belt to provide assistance to the wearer's legs in a physiologically relevant manner. These enhanced movements have the potential to assist wearers in walking with greater stability and metabolic efficiency, which could prevent injury and reduce fatigue." [21]

In addition to a license for intellectual property and additional funding from the ReWalk for the continued development of the exosuit at the Institute, this project has also been heavily funded by the National Science Foundation, the Harvard University Star

Family Challenge, and the Defense Advanced Research Projects Agency's Warrior Web program (whose purpose is to develop technologies to mitigate musculoskeletal injuries among military service members while improving performance).

In spring 2010, Peter Dilworth, who started his roboticist and inventor career at the MIT Leg lab in the early 1990s, and MP, one of the authors of this book, who joined the same lab in the early 2000s, were brainstorming on new robotics technologies while enjoying coffee at one of the famous Cambridge coffee shops. Suddenly, Peter jumped in his seat and said: "Let's use bicycle solution!" Bicycle solution referred to Bowden cable used for standard bicycle brakes (see Chapter 3). Peter envisioned Bowden cable added from inside with small wheels that could serve to bend cable at various points and in various directions. It was quite ingenious. MP envisioned somewhat simpler but more biologically inspired approach with Bowden cable in series with elastic element simply used to transfer securely forces from belt or backpack to various joints in muscle like fashion (Fig. 10.10, right). Both Peter Dilworth and MP were MIT alumni and looking for a new research lab home. The next day, they met with Harvard Professor Rob Howe, acting area Dean for Bioengineering at Harvard, and presented him with these exciting concepts. Rob liked MP's idea better and opened the doors of his lab for collaboration. MP got a job at WPI in the summer of 2010. During 2010–12, the author's group at WPI and Prof. Robert Howe's group at Harvard University collaborated on the soft robotic brace for stroke

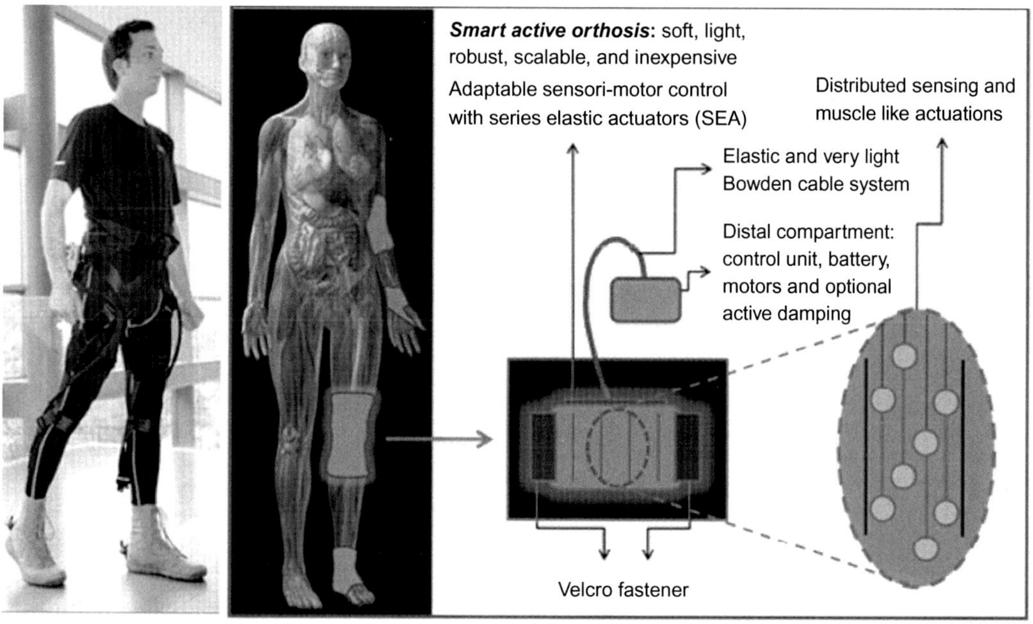

FIG. 10.10 Ignacio Galiana, one of the researchers from the Wyss Institute for Biologically Inspired Engineering at Harvard University, wearing a soft exosuit [22, 23] advanced in Connor Walsh's Biodesign Lab (left). Concept design of exosuit by one of the authors (MP) dating back to October 2010 (right).

rehabilitation as an example of practical exomusculature and published two papers [24, 25] that explored fundamental concepts of exosuit applied to shoulder. In Fall 2010, they also applied to EUREKA program together with Paolo Bonato, then at Spaulding Rehabilitation Hospital. Ignacio Galiana (Fig. 10.10, left), at that time still with Centre for Automation and Robotics UPM-CSIC, Madrid, Spain joined these efforts in summer 2011. Shortly after, the Wyss Institute has been formed and new professor Connor Walsh picked up these interesting ideas and applied them to lower limbs. It is rather interesting how good ideas have their own life and eventually find way to realization for benefit of those in need.

While very promising, the concept of an exosuit still has to be substantially improved (which might require even going back to some of the original ideas of Peter Dilworth). One of the main problems is related to the soft, cloth-like interface. Unless the cloth is extremely tightly positioned on the body, which might be very uncomfortable for the wearer, the actuating forces necessary are very limited and could only account for less than 10% of required torques for walking. Otherwise the attachment points of cable system would move together with skin which could be especially troublesome for the more elderly population.

The AlterG Bionic Leg is another powered knee brace that can be used both for rehabilitation and for everyday assistance (Fig. 10.11) [26, 27]. From the company's description it is "the first wearable, mobile robotic exoskeleton for lower extremity physical therapy."

As specified by The AlterG Bionic Leg, it provides patient initiated motor assistance during sit-to- stand exercises, over ground walking, and stair climbing; this wearable,

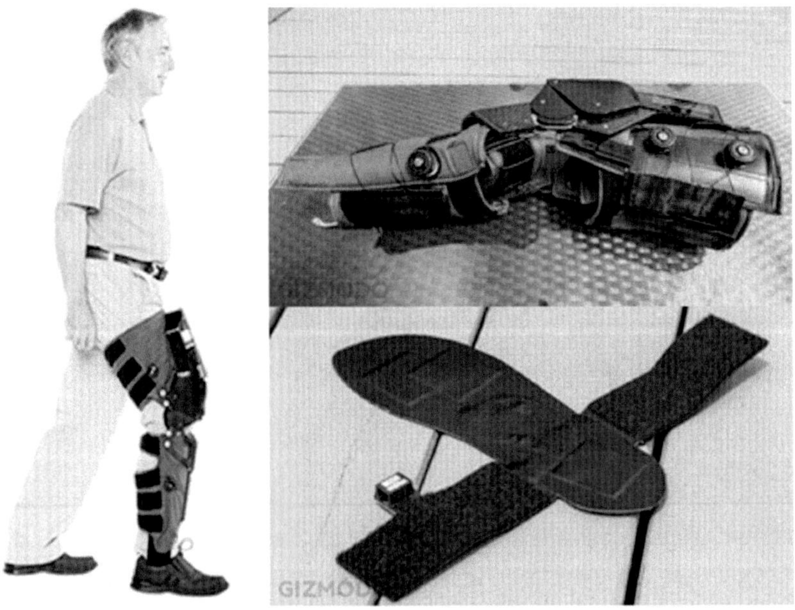

FIG. 10.11 Alter G Bionic Leg. *Reprinted with permission from AlterG.*

robotic rehabilitation device enables neuro and ortho physical therapy patients to: (1) reach higher functional levels, (2) increase balance, (3) improve gait mechanics and endurance, and (4) strengthen weak muscles.

Sensory information from four separate pressure sensors embedded within a thin shoe insole in tandem to the information on the effort exerted by user's knee is utilized to decipher user intention. The 8 lb. leg brace acting as an additional quadriceps muscle has a maximum walking speed of just over 2 miles/h. It can provide approximately 30% assistance to the required knee effort.

10.4 Powered Ankle Brace

One of the first projects within the, at the time, newly formed MIT Biomechatronics group (formed by the MIT Leg Lab members at the time, including one of the authors MP) was the MIT Active Ankle Foot Orthosis (AFO) project (c. 2002) [28, 29] led by then master student Joaquin Blaya co-advised by Hugh Herr and Dava Newman (Fig. 10.12).

The AFO was geared to assist drop-foot patients to regain better balance and stability as well as to increase their gait speed. This simple system had one "artificial muscle-tendon" and a few basic sensors measuring ankle angle and time when ground contact has been made. One prototype also had strain gauges to estimate force.

The "artificial muscle-tendon" system was mimicked by a series elastic actuator (SEA) [30, 31]; technology that was introduced by Gill Pratt who directed Leg lab after Marc Raibert. In a SEA, the actuator is connected in series with a spring. Even if the motor is only position controlled, the spring introduces slack that is necessary for safe human-

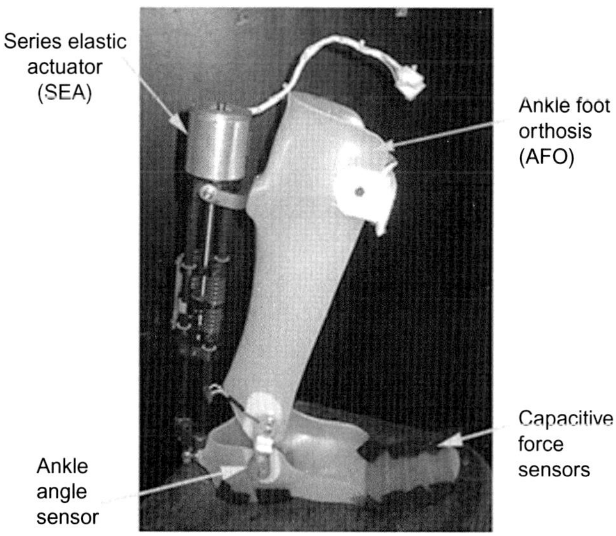

FIG. 10.12 The MIT active ankle foot orthosis (AFO).

robot interaction; if there is no spring, then the motor may apply very large forces directly onto the load if a large position error was to occur. Moreover, the spring force can be easily measured with SEA and hence this technology naturally addresses both position and force control—variable impedance control!

Preliminary tests with drop-foot patients were performed and the results were formidable; patients felt that their balance and stability substantially improved and their self-selected walking speeds significantly increased. The patent for this technology was issued in 2013. The author is not aware if the AFO or a similar device has been commercialized since.

An active autonomous ankle exoskeleton utilizing winch on the shin [32] was yet another successful project by the Biomechatronics group led by Hugh Herr (c. 2014).

The active autonomous ankle exoskeleton uses a winch actuator on the shin to actuate the proximal ends of fiberglass struts attached to the boot. The winch actuator then implements a brushless DC motor and pulley transmission to wind a high-strength cord. Reflective markers on the strut were used to measure the deflection of the struts and the applied exoskeletal torque. The batteries and motor controllers are worn around the chest and waist.

This is a quite inventive and practical solution for the generation of large ankle assistive torques for walking. The body mass normalized peak ankle torque is about 1.7, 1.8 Nm/kg; it is substantially larger than the same quantity for the knee or for the hip during regular gait. Off-the-shelf winch technology has already been advanced by many researchers such that a very light mechanism can provide high power, that is, very forceful and fast actuation. The concept with fiberglass strut is very nonstandard but in the given context extremely useful; it provides a very large moment arm, potentially much longer than boot or shoe length that when combined with large tension from the winch cord finally gives a very large moment about biological ankle joint.

From Fig. 10.13, it appears that the moment arm perpendicular to cord is roughly about 0.2 m. According to the authors of this study, the winch consisted of a 200 W brushless motor (model: 305015, Maxon Motor, Sachseln, CH) with 92.9 mNm torque actuating an 8 mm diameter spool through a single-stage pulley transmission with a speed reduction of 44:14. Hence, by ignoring friction, one may estimate maximal tension in the cord as 92.9 mNm 44,141,000 8 mm = 36.5 N. This translates to maximal assistive ankle torque of about 36.5 N 0.2 m = 7.3 Nm and this still further translates to body normalized assistive ankle torque of about 0.121 Nm/kg per 60 kg person. If entire 3.6 kg mass of device was included, one would obtain 0.115 Nm/kg. This corresponds to approximately 7% torque assistance in walking.

In a study with six participants walking on a level treadmill at 1.4 m/s, the metabolic, kinetic, and kinematic effects of wearing a battery powered bilateral ankle exoskeleton were quantified. The net metabolic cost of walking with the active exoskeleton (3.28 ± 0.10 W/kg) was an $11 \pm 4\%$ ($P = .019$), that is, 7%–15% reduction compared to the cost of walking without the exoskeleton (3.71 ± 0.14 W/kg).

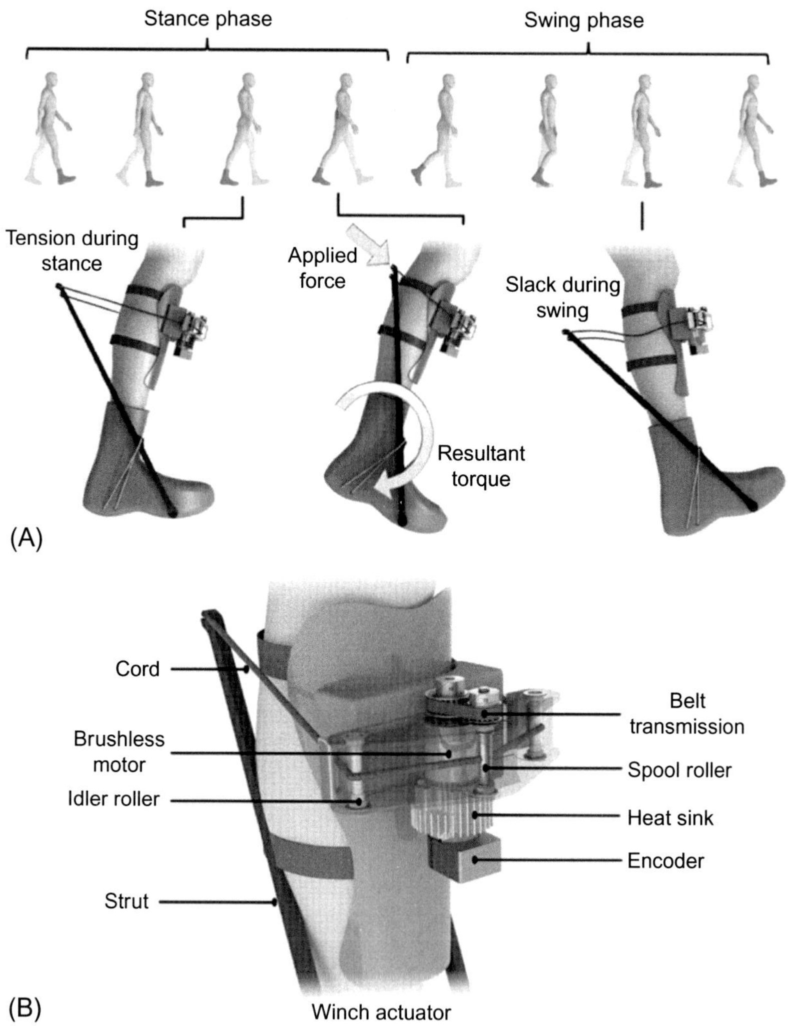

FIG. 10.13 The active autonomous ankle exoskeleton by Center for Extreme Bionics, MIT Media Lab, Massachusetts Institute of Technology.

The study, "Design and control of a bio-inspired soft wearable robotic device for ankle–foot rehabilitation" [33], led by Yong-Lae Park from Robotics Institute, Carnegie Mellon University, published in Bioinspiration and Biomimetics in 2014, provides yet another very innovative way to think about actuated ankle brace (Fig. 10.14).

While soft wearable device was primarily intended for use in rehabilitation as it utilized pneumatic artificial muscles (PAM) of McKibben type, Chapter 3, requiring huge compressed air source, the proposed concept can certainly well advise slightly different

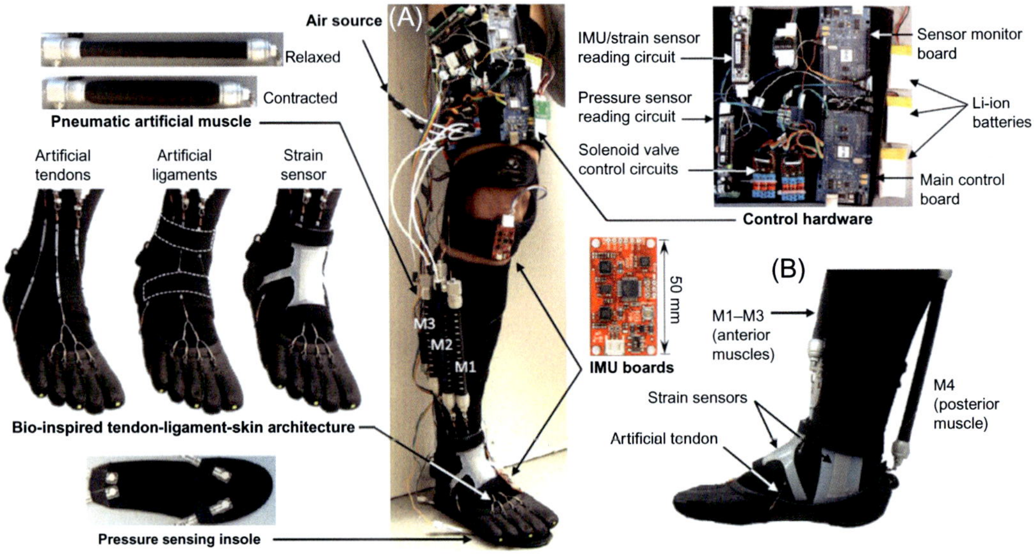

FIG. 10.14 Main design of the active soft orthotic device (highlighting key components). (A) Front view of the prototype integrated on a human leg showing the three anterior muscles (M1: tibialis anterior, M2: extensor digitorum longus, and M3: peroneus tertius). (B) Side view of the prototype showing the posterior muscle (M4: gastrocnemius) and the strain sensors. *Adapted from Y.E. Park, B.-R. Chen, N.O. Pérez-Arancibia, D. Young, L. Stirling, R.J. Wood, E.C. Goldfield, R. Nagpal, Design and control of a bio-inspired soft wearable robotic device for ankle–foot rehabilitation, Bioinspir. Biomimet. 9(1) (2014) 016007. Courtesy of Yong-Lae Park.*

approaches; for example, using wearable closed system based on Hydro Muscles, Chapter 3, which then can be used for everyday assistance.

The design mimics the morphology and functionality of the muscle-tendon-ligament structure found in the biological musculoskeletal system of the human foot and lower leg. The soft structure of the device provides active assistance without restricting natural DOF at the ankle joint (Fig. 10.15). For the subject tested, the prototype is capable of generating an ankle range of motion of 27° (14° dorsiflexion and 13° plantarflexion). Four pneumatic artificial muscles assist dorsiflexion and plantarflexion as well as inversion and eversion. The prototype is also equipped with various embedded sensors for gait pattern analysis.

According to this study: "Based on the configurations of the artificial tendons and the dimensions of the prototype, the maximum torques the system can apply are 110 Nm, 53 Nm, 20 Nm, and 21 Nm for dorsiflexion, plantarflexion, inversion, and eversion, respectively." [33]

This is a quite impressive result as 110 Nm devices can completely exchange biological ankle torques required for walking for a 60 kg person. And all of this is achieved with only 570 kPa (~85 PSI) air pressure (!), that is, low pressure equal to the water pressure in many households.

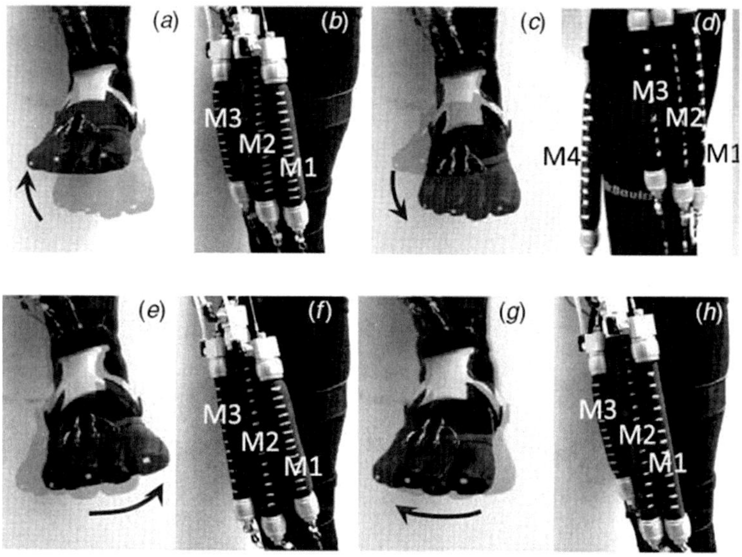

FIG. 10.15 Actively assisted ankle motions: (A) dorsiflexion (B) with all three anterior muscles contracted; (C) plantarflexion (D) with only muscle 4 contracted; (E) inversion (F) with only muscle 1 contracted; (G) eversion (H) with only muscle 3 contracted. *Adapted from Y.E. Park, B.-R. Chen, N.O. Pérez-Arancibia, D. Young, L. Stirling, R.J. Wood, E.C. Goldfield, R. Nagpal, Design and control of a bio-inspired soft wearable robotic device for ankle–foot rehabilitation, Bioinspir. Biomimet. 9(1) (2014) 016007. Courtesy of Yong-Lae Park.*

10.5 Powered Shoulder Brace

The human shoulder has a very complex anatomical structure with approximately 5 DOF actuated by a large number of muscles.

The soft-robotics exomusculature [24, 25, 34], proposed by one of the authors (MP), is a network of a large number of cable-driven artificial muscle fibers embedded in a suit (or brace). All heavy parts (e.g., motors, controller, batteries, etc.) are located in a separate compartment and force transmission is provided via Bowden-type cable (Fig. 10.16).

Clearly, cable-driven artificial muscle can be extremely light. Moreover, they may be driven only by a single (or several) motor/s within the one-to-any (OTM) system architecture [34–36]. The OTM system utilizes active clutches to store and release energy through passive springs. The same muscle consists of many fibers, where each fiber is an independently actuated clutch-spring system. Hence, the muscle's stiffness is varied by recruiting different fibers, exactly as in the human body. The advantages of the OTM-assisted exomusculature system, with a very small number of conventional electric motors and a large number of actuated DOF, are that it can be highly dexterous, wearable, energy efficient, compliant, lightweight, adaptive, and cost-effective. Due to the poor energy density of pneumatic systems, the battery-powered electric motors may be more suitable for applications requiring full mobility.

During 2010–12, the MP's group at WPI and Prof. Robert Howe's group at Harvard University collaborated on the soft robotic brace for stroke rehabilitation and possibly

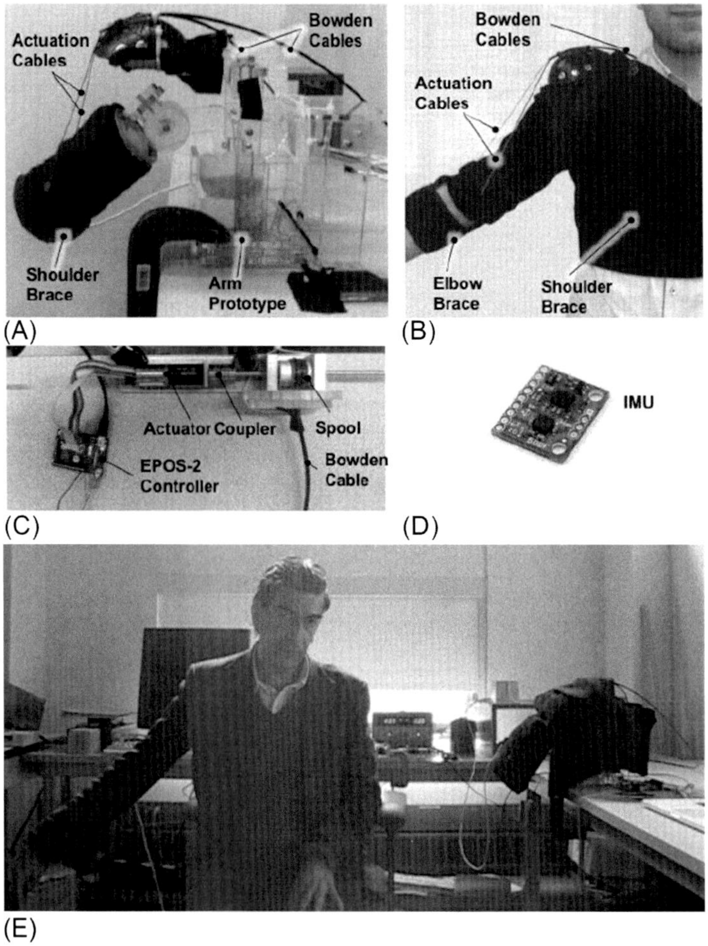

FIG. 10.16 The example of the soft robotics exomusculature (A–D) [24, 25], the exomusculature shoulder brace, on the right, attached to a mannequin arm with added weights is teleoperated by Prof. Popovic (E).

for daily assistance as an example of practical exomusculature [24, 25]. Ignacio Galiana from Centre for Automation and Robotics UPM-CSIC, Madrid, Spain joined these efforts in summer 2011.

While very little work has been done to extend this work since then and no commercialization took place, it would be worth trying to incorporate closed hydraulic system possibly based on Hydro Muscles [37–39] and also explore integration with EMG or EEG type of control.

10.6 Powered Elbow and Wrist Brace

The MIT spin-off Myomo is used by individuals with neuromuscular conditions who have lost movement in their hand and arm to perform everyday activities. MyoPro [40]

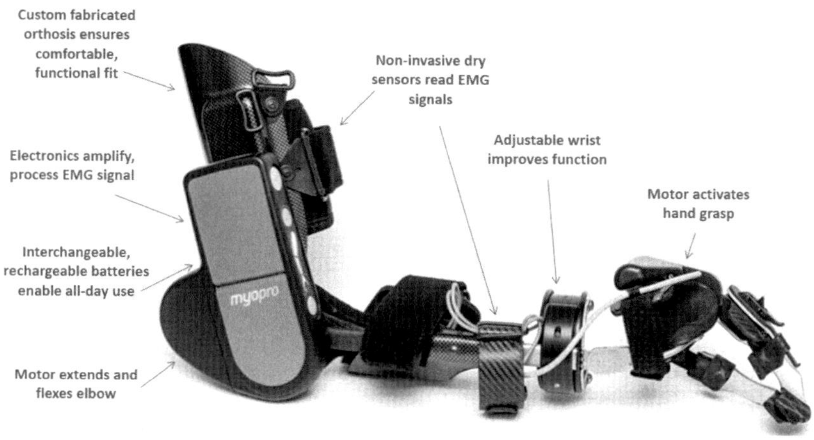

FIG. 10.17 MyoPro a myoelectric elbow/wrist/hand orthosis (powered brace) by Myomo. *Reprinted with permission from Myomo.*

(Fig. 10.17) by Myomo, is a myoelectric elbow/wrist/hand orthosis (powered brace) used to support the weak arm and enable patients to move an impaired hand and arm. According to company's website, MyoPro is the only product of its kind for people who suffer from debilitating neurological disorders such as brachial plexus injury, brain or spinal cord injury, CVA stroke, and MS or amyotrophic lateral sclerosis (ALS).

The MyoPro is a custom fabricated, wearable powered brace incorporating noninvasive myoelectric technology that can support and move an impaired hand and arm. By the use of sensors in the brace, weak muscle signals can be detected, powering motors to move the affected hand and arm. The user is in complete control of the MyoPro at all times.

Essentially, the brace amplifies force output based on user's intent communicated via EMG signals, Chapter 4.

10.7 Powered Hand and Finger Braces, Robotic Gloves

In case of hand which has a number of joints, the line between powered braces, gloves, and exoskeletons, based on definition introduced earlier, is somewhat blurred.

NASA's 'Robo-Glove' [41, 42] is result of almost decade long NASA and General Motors (GM) collaboration on humanoid robot Robonaut 2 (R2). It was originally utilized as add-on equipment for haptic teleoperation and robot learning by demonstration (Fig. 10.18).

> *"When engineers, researchers and scientists from GM and NASA began collaborating on R2 in 2007, one of the design requirements was for the robot to operate tools designed for humans, alongside astronauts in outer space, and factory workers on Earth. The team achieved an unprecedented level of hand dexterity on R2 by using leading-edge sensors, actuators and tendons comparable to the nerves, muscles and tendons in a human hand. Research shows that continuously gripping a tool can cause fatigue in hand muscles within a few minutes, but initial testing of the Robo-Glove indicates the wearer can hold a grip longer and more comfortably."* [41]

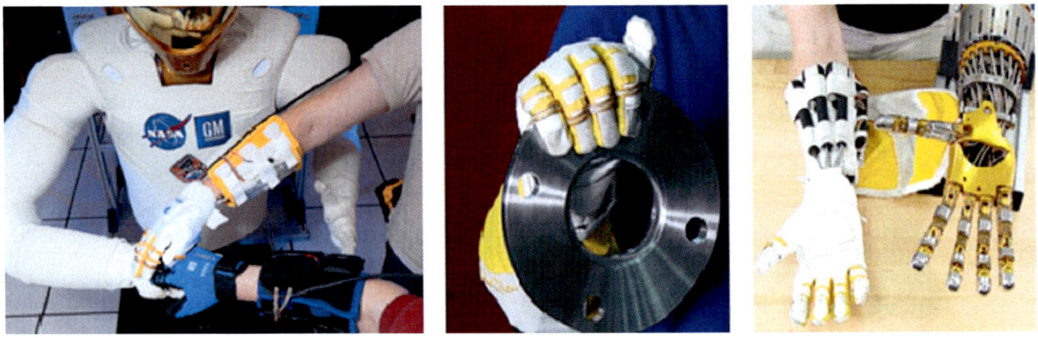

FIG. 10.18 NASA Robo-Glove.

Touch sensors are embedded in glove's fingertips. When the user grasps the object, the input from sensors causes actuators, attached onto lower arm/wrist area, to pull the cables and apply tension in the "tendons" placed laterally on both side of each finger. There are three motor units (one for the thumb, one for the index and the middle fingers, and one for the ring and the little fingers). This action introduces additional grasping force. The mass of device on arm is about 900 g. It also includes control electronics and a small display for programming and diagnostics. The battery is separately worn on the belt.

It is reported that glove can add 10 pounds (~44 N) of extra force.

In 2016, GM and Swedish medical company Bioservo partnered to create commercial version of this device using Bioservo's SEM (soft extra muscle) patented technology. GM will integrate Bioservo's strength-boosting patented (soft extra muscles) SEM technology into the NASA Super-Glove [43].

SEM technology actually refers to artificial tendons pulled by motorized units.

SEM Glove [44] is a commercial product by Swedish company Bioservo (Fig. 10.19). It is designed for people with impaired hand function or for people with grip-intensive occupations. It operates somewhat similar to NASA Robo-Glove except that actuators are placed proximally in unit worn on the belt. The total mass of device including batteries is 700 g. Forces are transmitted by Bowden cable system that is attachable/detachable from connector close to wrist. Similar to Robo-Glove, the SEM Glove has three motor units that actuate the thumb, the middle finger, and the ring finger which are only partially covered by glove's textile material. The index finger and the little finger are not actuated and they are not covered by glove's textile material.

SEM Glove is built so it can be adapted to each user's individual needs. For example, the length of SEM Glove fingers can be adjusted to suit every individual. Even the sensors can be slightly moved to suit the wearer. For ease of use, three grip profiles can be also defined. Different strength and sensibility levels can be preset here tailored to the wearer's needs.

Carbonhand is another commercial product by Swedish company Bioservo. It is essentially slightly improved version of SEM Glove.

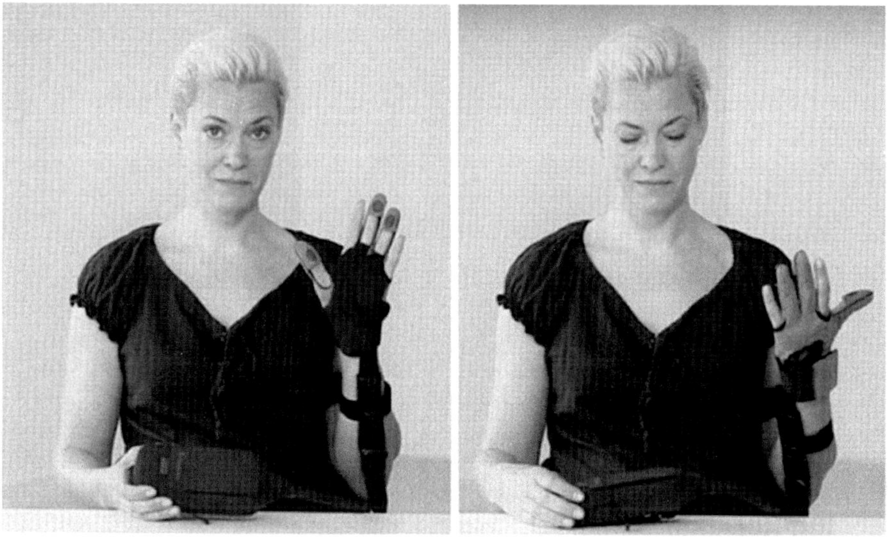

FIG. 10.19 Bioservo's SEM Glove [44]. *Reprinted with permission from Bioservo.*

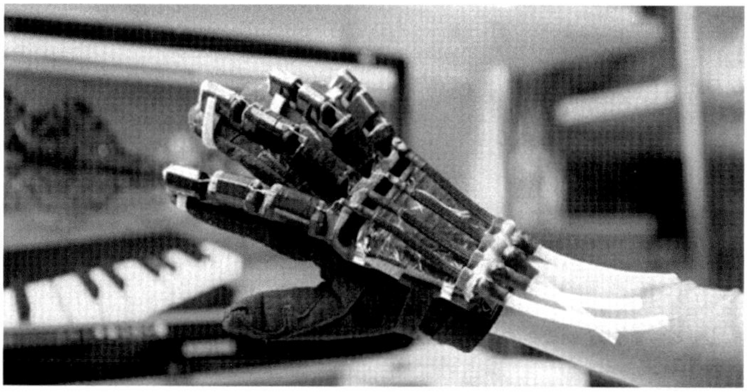

FIG. 10.20 Prototype of haptic virtual reality (VR) glove, Jacobs School of Engineering/UC San Diego.

Prototype of haptic virtual reality (VR) glove [45] shown in Fig. 10.20 was recently introduced by researchers affiliated with Jacobs School of Engineering/UC San Diego. To mimic interaction with physical objects the haptic VR glove is applying resistive forces that cause finger extension.

The key soft-robotic component of the new glove is a version of the McKibben muscle using soft latex chambers covered with braided fibers, already discussed in Chapter 3. To apply tactile feedback when the user moves their fingers, the muscles respond like springs. The board controls the muscles by inflating and deflating them. The muscles then pull cords that act as tendons.

The researchers have used a 3D-printed mold to make the glove's soft exoskeleton. They used silicone rubber for the exoskeleton with velcro straps embedded at the joints.

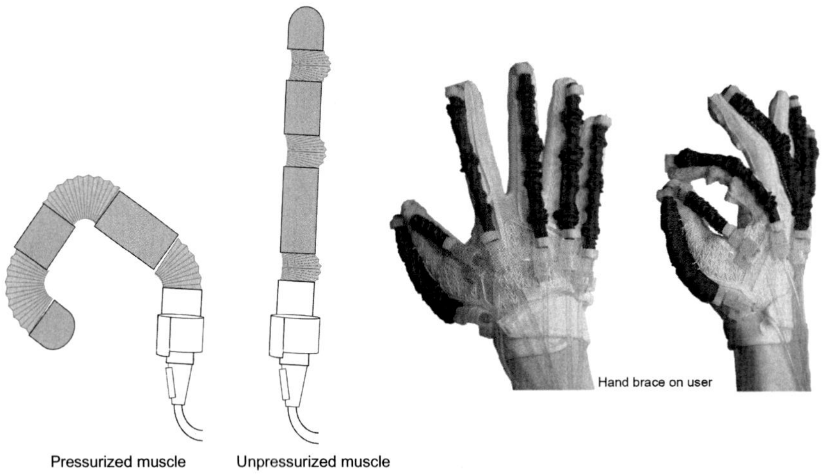

FIG. 10.21 Hydro Muscle Hand Brace, WPI Popovic Labs. Pressurized miniature version of slightly redesigned Hydro Muscle, Chapter 3, expands where slack exists in the outer sleeve, i.e., within bunched fabric located specifically at the knuckles hence causing joint flexion (left). The lightweight and flexible brace successfully performed the "Key Grasp" and the Palmar Prehension gestures with the fingers exerting a force of 5 N, adequate for pinching and lifting standard objects (right).

New power glove called Hydro Muscle Hand Brace shown in Fig. 10.21 was recently introduced by WPI Popovic Labs and demonstrated at the 2018 Cambridge Science Festival [46, 47]. This glove is actuating fingers by both pushing and pulling actions. It is opened on palmar side with hydraulic miniature tubes anchored on the dorsal side of the hand in order not to obstruct object grasping or haptic tactile feeling. Five tiny curving Hydro Muscles provides extension/flexion of the thumb and fingers and one more linear Hydro Muscle is used for thumb's adduction/abduction.

References

[1] Popovic, M. (2014) "Biomechanics and Robotics", 364 pages, Book, Copyright © 2014 Pan Stanford Publishing Pte. Ltd., Singapore, ISBN 978-981-4411-37-0 (Hardcover), 978-981-4411-38-7 (eBook).

[2] Cyberdyne, Inc., press release, https://www.cyberdyne.jp/wp_uploads/2017/09/20170925_FB02_eNG.pdf, 2017 (retrieved 17.03.18).

[3] Cyberdyne, Inc., press release, https://www.cyberdyne.jp/wp_uploads/2017/11/20171114_kessanhosokusetsumei_ENG.pdf, 2017 (retrieved 17.03.18).

[4] HAL for Lumbar Support. Exoskeleton Report, https://exoskeletonreport.com/product/hal-for-lumbar-support/ (retrieved 17.03.18).

[5] What is 'Muscle Suit'? Innophys, https://innophys.jp/about/ (retrieved 17.03.18).

[6] Muscle Suit. Exoskeleton Report, https://exoskeletonreport.com/product/muscle-suit/ (retrieved 17.03.18).

[7] No More Power Barriers With Panasonic Assist Robots, Panasonic Press release, http://news.panasonic.com/global/stories/2016/44969.html, 2016 (retrieved 17.03.18).

[8] AWN-03. Exoskeleton Report, https://exoskeletonreport.com/product/awn-03/ (retrieved 17.03.18).

[9] The Increasing Importance of Exoskeleton Software. Exoskeleton Report, https://exoskeletonreport.com/2017/08/increasing-importance-exoskeleton-software/ (retrieved 17.03.18).

[10] L. Younbaek, C. Byungjung, L. Jongwon, L. Minhyung, R. Se-Gon, K. Jeonghun, C. Hyundo, K. Yong-Jae, Flexible sliding frame for gait enhancing mechatronic system (GEMS), Conf. Proc. IEEE Eng Med Biol Soc. 2016 (2016) 598–602.

[11] Y. Lee, B. Choi, J. Lee, M. Lee, S.-G. Roh, J. Kim, H. Choi, Y.-J. Kim, Flexible sliding frame for gait enhancing mechatronic system (GEMS), in: 2016 IEEE 38th Annual International Conference of the Engineering in Medicine and Biology Society (EMBC), IEEE, 2016, pp. 598–602.

[12] S.H. Lee, H.J. Lee, W.H. Chang, B.O. Choi, J. Lee, J. Kim, G.H. Ryu, Y.H. Kim, Gait performance and foot pressure distribution during wearable robot-assisted gait in elderly adults, J. Neuroeng. Rehabil. 14 (1) (2017) 123, doi:10.1186/s12984-017-0333-z.

[13] J. Lee, K. Seo, B. Lim, J. Jang, K. Kim, H. Choi, Effects of assistance timing on metabolic cost, assistance power, and gait parameters for a hip-type exoskeleton, in: July (Ed.), 2017 International Conference on Rehabilitation Robotics (ICORR), IEEE, 2017, pp. 498–504.

[14] Walking Assist, Honda, http://world.honda.com/Walking-Assist/ (retrieved 17.03.18).

[15] Freedom of Motion: The Story of Honda Walking Assist. Honda, https://www.honda.com/mobility/walking-assist-technology (retrieved 17.03.18).

[16] Honda Walking Assist Devices. YouTube, Honda, 2016. https://www.youtube.com/watch?v=Io3MBpzmVJc (retrieved 17.03.18).

[17] C-Brace High-Tech Leg Brace. Ottobock, https://www.ottobockus.com/orthotics/solution-overview/orthotronic-mobility-system-c-brace/ (retrieved 17.03.18).

[18] Victoria Ward, Bionic Leg: Golfer Becomes First to Wear "Bluetooth" Brace Allowing him to Walk, Cycle and Tackle the Stairs. The Telegraph, https://www.telegraph.co.uk/news/science/11388954/Bionic-leg-Golfer-becomes-first-to-wear-Bluetooth-brace-allowing-him-to-walk-cycle-and-tackle-the-stairs.html, 2015 (retrieved 17.03.18).

[19] Ortho Rehab – Knee Orthosis Custom Fit (KO CF), Ultraflex, http://ultraflexsystemsinc.com/bracedesigns/PostOpKneePrefab.htm (retrieved 17.03.18).

[20] Power Units, Ultraflex, http://ultraflexsystemsinc.com/components/PowerUnits.htm (retrieved 17.03.18).

[21] Wyss Institute collaborates with ReWalk Robotics to develop wearable exosuits for patients with limited mobility, Harvard University, 2016. Wyss Institute Press release https://wyss.harvard.edu/wyss-institute-collaborates-with-rewalk-robotics-to-develop-wearable-exosuits-for-patients-with-limited-mobility/ (retrieved 17.03.18).

[22] Walsh, C., A.T. Asbeck, I.G. Bujanda, Y. Ding, R.J. Dyer, A.F. Larusson, B.T. Quinlivan, K. Schmidt, D. Wagner, and M. Wehner. Soft exosuit for assistance with human motion. U.S. Patent 9,351,900, issued May 31, 2016.

[23] A.T. Asbeck, S.M.M. De Rossi, I. Galiana, Y. Ding, C.J. Walsh, Stronger, smarter, softer: next-generation wearable robots, IEEE Robot. Autom. Mag. 21 (4) (2014) 22–33.

[24] S.B. Kesner, L. Jentoft, F.L. Hammond, R.D. Howe, M.B. Popovic, Design considerations for an active soft orthotic system for shoulder rehabilitation, in: 33rd Annual International IEEE EMBS Conference, August 30–September 02, 2011, Boston, USA, 2011.

[25] I. Galiana, F.L. Hammond, R.D. Howe, M.B. Popovic, Wearable soft robotic device for post-stroke shoulder rehabilitation: identifying misalignments, in: 2012 IEEE/RSJ International Conference on Intelligent Robots and Systems, October 7–12, 2012, Portugal, 2012.

[26] AlterG Bionic Leg, AlterG, https://www.alterg.com/products/bionic-leg/consumer (retrieved 17.03.18).

[27] Brent Rose "I Wore a Bionic Leg, And I Never Wanted To Take It Off Again," GIZMODO 7/10/13, https://gizmodo.com/i-wore-a-bionic-leg-and-i-never-wanted-to-take-it-off-726536822 (retrieved 17.03.18).

[28] J.A. Blaya, H. Herr, Adaptive control of a variable-impedance ankle-foot orthosis to assist drop-foot gait, IEEE Trans. Neural Syst. Rehabil. Eng. 12 (1) (2004) 24–31.

[29] H. Herr, J. Blaya, G.A. Pratt, Active ankle foot orthosis. U.S. Patent 8,075,633, issued December 13, 2011.

[30] G.A. Pratt, M.M. Williamson, Series elastic actuators, in: IEEE International Conference on Intelligent Robots and Systems, vol. 1, 1995, pp. 399–406.

[31] M.M. Williamson, Series Elastic Actuators, Massachusetts Institute of Technology, 1995.

[32] L.M. Mooney, E.J. Rouse, H.M. Herr, Autonomous exoskeleton reduces metabolic cost of human walking during load carriage, J. Neuroeng. Rehabil. 11 (1) (2014) 80.

[33] Y.-L. Park, B.-R. Chen, N.O. Pérez-Arancibia, D. Young, L. Stirling, R.J. Wood, E.C. Goldfield, R. Nagpal, Design and control of a bio-inspired soft wearable robotic device for ankle–foot rehabilitation, Bioinspir. Biomimet. 9 (1) (2014), 016007.

[34] T. Hunt, C. Berthelette, G.S. Iannacchione, S. Koehler, M.B. Popovic, Soft robotics variable stiffness exo-musculature, one-to-many concept, and advanced clutches, in: IEEE ICRA 2012 WORKSHOP: Variable Stiffness Actuators Moving the Robots of Tomorrow, St. Paul, Minnesota, May, vol. 14, 2012.

[35] T. Hunt, C. Berthelette, M.B. Popovic, Linear one-to-many (OTM) system, in: 2013 IEEE International Conference on Technologies for Practical Robot Applications (TePRA), IEEE, 2013, pp. 1–6.

[36] M. B. Popovic, C. J. Berthelette, M. DiPinto, J. D. Sareault, and T. R. Hunt Actuation Systems and Methods. U.S. Patent Application No. 14/903,126 Filed January 6, 2016. U.S. Patent No. 9,879,767 Issued January 30, 2018.

[37] M.B. Popovic, C. Onal, G. McCarthy, N. Corso, D. Effraimidis, and B. Jennings (2015) Actuators and Methods of Use (Hydro Artificial Muscles). United States Patent and Trademark Office, Assignee Worcester Polytechnic Institute. Serial No.: 62/011,830. Filed: June 13, 2014. US Patent 20,150,359,698, 2015.

[38] G. McCarthy, D. Effraimidis, B. Jennings, N. Corso, C. Onal, M.B. Popovic, Hydraulically actuated muscle (HAM) exo-musculature, in: Robot Makers: The Future of Digital Rapid Design and Fabrication of Robots (RoMa) Workshop, the 2014 Robotics: Science and Systems Conference, Berkeley, CA, July 12, 2014, 2014.

[39] S. Sridar, C.J. Majeika, P. Schaffer, M. Bowers, S. Ueda, A.J. Barth, J.L. Sorrells, J.T. Wu, T.R. Hunt, M. Popovic, Hydro muscle—a novel soft fluidic actuator, in: 2016 IEEE International Conference on Robotics and Automation (ICRA), 2016, pp. 4014–4021.

[40] What is a MyoPro orthosis? Myomo, http://myomo.com/what-is-a-myopro-orthosis/ (retrieved 17.03.18).

[41] NASA, GM Jointly Developing Robotic Gloves For Human Use, NASA Press release J12-006, https://www.nasa.gov/centers/johnson/news/releases/2012/J12-006.html#.Wq0Ky-jwbIV, 2012 (retrieved 17.03.18).

[42] Partnerships to Move Humanity Forward. General Motors, http://www.gm.com/mol/GM-and-NASA-partner-together.html (retrieved 17.03.18).

[43] The Glove that Gives you Super-Human Strength: Nasa Device Could DOUBLE your Gripping Power, The Daily Mail, 2016. http://www.dailymail.co.uk/sciencetech/article-3676897/The-glove-gives-super-human-strength-Nasa-device-DOUBLE-gripping-power.html (retrieved 17.03.18).

[44] SEM Glove™. Bioservo, http://www.bioservo.com/healthcare/product-portfolio/sem-glove/ (retrieved 17.03.18).

[45] P. Polygerinos, N. Correll, S.A. Morin, B. Mosadegh, C.D. Onal, K. Petersen, M. Cianchetti, M.T. Tolley, R.F. Shepherd, Soft robotics: review of fluid-driven intrinsically soft devices; manufacturing, sensing, control, and applications in human-robot interaction, Adv. Eng. Mater. (2017) e201700016.

[46] M. McCormick, "Hydro Muscle Hand Brace," A Major Qualifying Project (MQP), Primary Advisor Marko Popovic, Worcester Polytechnic Institute, Worcester, Massachusetts, USA, https://web.wpi.edu/Pubs/E-project/Available/E-project-060118-153854/unrestricted/MQP_report_Final.pdf.

[47] WPI Popovic Labs, Cambridge Science Festival, 2018, YouTube video:https://www.youtube.com/watch?v=LbR6QASKtHc.

11

Exoskeletons, Exomusculatures, Exosuits: Dynamic Modeling and Simulation

Yunus Ziya Arslan*, Derya Karabulut*, Faruk Ortes*, Marko B. Popovic†

*ISTANBUL UNIVERSITY, ISTANBUL, TURKEY †WORCESTER POLYTECHNIC INSTITUTE, WORCESTER, MA, UNITED STATES

CHAPTER OUTLINE

11.1 Introduction to Wearable Exoskeletons, Exomusculatures, and Exosuits

Exoskeletons are rigid wearable structures that are differentiated from rigid braces or orthoses based on the number of human joints that they interface with. In contrast to rigid braces or orthoses, addressed in Chapter 10, that interface with just a single human joint, exoskeletons are defined as rigid structures that interface with a plurality of human joints. Clearly, the exoskeleton may consist of several rigid braces or orthoses that are integrated

Biomechatronics. https://doi.org/10.1016/B978-0-12-812939-5.00011-2

into single, rigidly connected, functional entity. Terminology in common use is the lower body or single leg exoskeleton, upper body as well as single arm and back exoskeleton, and finally, whole body exoskeleton.

The first exoskeleton was proposed by Schulte [1] and appeared in the late 1960s [2]. The main goal of exoskeletons has changed since the first one was built. Initially, they were thought to empower soldiers, allowing them to carry heavy loads. Nowadays, assistive exoskeletons are developed to improve rehabilitation interventions for individuals with muscle disease [3], compensate disabilities caused by strokes [4], and contribute to recovery and movement after spinal cord injuries [5]. Also, new exoskeleton applications are anticipated within the physical labor intensive occupations like construction, moving and transportation, manufacturing, etc.

The exoskeletons can be completely passive mechanical devices, which are a bit less interesting from the biomechatronic point of view but still potentially very useful, then mostly passive devices with sensors and, for example, electrically operated clutch or brake mechanisms, and finally partially or fully sensitized and actuated devices. Sometimes the terminology powered exoskeleton and active exoskeleton are also used (Fig. 11.1).

Advantages of passive devices are that they do not require batteries, actuators, sensors, or control units that can "misbehave," they are typically very sturdy, lightweight and also

FIG. 11.1 Examples of exoskeleton technologies in period 2000–12 discussed in book "Biomechanics and Robotics" [6]. From left to right, top to down: MIT Biomechatronics Group exoskeleton (3 figures), BLEEX exoskeleton (1 figure), HULC exoskeleton (2 figures), Sarcos exoskeleton (1 figure), XOS exoskeleton (1 figure), HAL exoskeleton (1 figure), ReWalk exoskeleton (2 figures), and Tek RMD wheelchair-exoskeleton (2 figures).

inexpensive and they could still assist, for example, workers (e.g., Ford manufacturing facility personnel [7]) or military personnel (a nice example is Lockheed Martin's FORTIS exoskeleton [8]) for numerous specific tasks. These devices can be especially useful if they are designed for a very specific activity. Some of these fully manually operated devices can even store and release elastic energy.

Advantages of mostly passive sensitized exoskeleton with clutch or brake mechanism are that they can more efficiently store and release elastic energy and also react faster and in a more sophisticated manner than manually operated exoskeletons, a feature that can prove critically important for injury or fall prevention. Finally, they can still be lightweight and relatively inexpensive. By simply invoking different control modes they can be very useful for more than a single specific activity, hence in practice, they can be more universal than completely passive devices.

Advantages of partially or fully sensitized and actuated exoskeletons are that they do not depend on the previous history of the system (e.g., whether enough elastic energy was stored or not), they can provide full actuation at any moment in time independent of, for example, posture or joint speeds, and in principle, they may not even require any choice of control mode targeting some specific activity as they can perform with the user in the feedback loop.

In the context of man-machine mergence and biomechatronics feedback loop, the actuated exoskeleton can be fully feedforward, then partially looped, that is, able to decipher user intent and act accordingly, or fully looped, that is, it can also transfer some of its own sensory inputs back to the user.

For example, consider exoskeletons for those paralyzed below the waist. If using the fully feedforward lower body exoskeletons, the user can in advance choose speed and stride length, press the start button and then navigate path using crutches (e.g., NASA-IHMC X1 Mina Exoskeleton [9]). If using the partially looped exoskeleton, the user can, for example, navigate each foot three-dimensional (3D) path by hand-held joysticks or sticks physically attached to feet (e.g., New Jersey TREKKER Exoskeleton [10, 11]). This approach can be particularly useful for walking over obstacles or ascending/descending stairs. Finally, the fully looped exoskeleton communicating its own sensory data back to the user can also be very useful; for example, if the user does not feel the location of the center of pressure (CoP) (or Zero-Moment Point, ZMP) in soles, the fully looped exoskeleton may communicate CoP (ZMP) location to the user, for example, by vibro-tactile device attached to the upper body to minimize the risk of fall.

The user intent can be communicated through mechanical actions, such as joint speed and force applied to exoskeleton, or noninvasive surface electromyography (EMG), electroencephalography, EEG signals, discussed in Chapters 4 and 6, invasive EMG, electrocorticography (ECoG), or direct intracortical brain-machine interfaces (BMI) all discussed in Chapter 6.

The research on powered human exoskeleton devices began in the late 1960s, almost in parallel between a number of research groups in the United States and in the former Yugoslavia, particularly within the Institute Mihailo Pupin in Belgrade, Serbia. Very detailed and comprehensive review of these and other technologies until approximately 2006 is written by Dollar and Herr [12]. Exoskeleton technologies in the period 2000–12 are reviewed in more detail in book "Biomechanics and Robotics" [6].

Here we provide a few examples of novel exoskeletons and exoskeleton-exomusculature technologies that surfaced in the last 5 years.

The exoskeletons introduced by two California based companies, EksoBionics and SuitX, both initially cofounded by UC Berkeley Professor Homayoon Kazerooni, illustrate well all three classes of exoskeletons previously addresses, see Fig. 11.2 [7].

Variable rigidity HydroBone exoskeleton [13–15] invented and developed in WPI Popovic Labs was first unveiled to the general public at the Cambridge Science Festival in 2015 (Fig. 11.3). It represents a wearable structure that can transition from a fully rigid exoskeleton to pure soft exomusculature and vice versa in less than 100 ms. Moreover, it is also possible to fine-tune rigidity level within the entire range in a controlled manner. This wearable device consists of the so-called HydroBones [13–15] utilized as support "skeletal" elements described below. Biologically inspired Hydro Muscle [14, 16, 17] were utilized for joint actuation for the first prototype, but this is likely of less importance here.

The HydroBone is a variable rigidity load-bearing element controlled by fluid pressure. The HydroBone consists of an inner layer, latex tube filled with pressurized fluid, and an outer layer, granular media enclosed by nonstretchable sheathing. The axial compression experiments verified that the HydroBone's stiffness and critical buckling force vary with fluid pressure. In order to avoid undesirable buckling, fluid pressure has to be greater than a certain minimal value that depends on the external load. In wearable devices, such as an

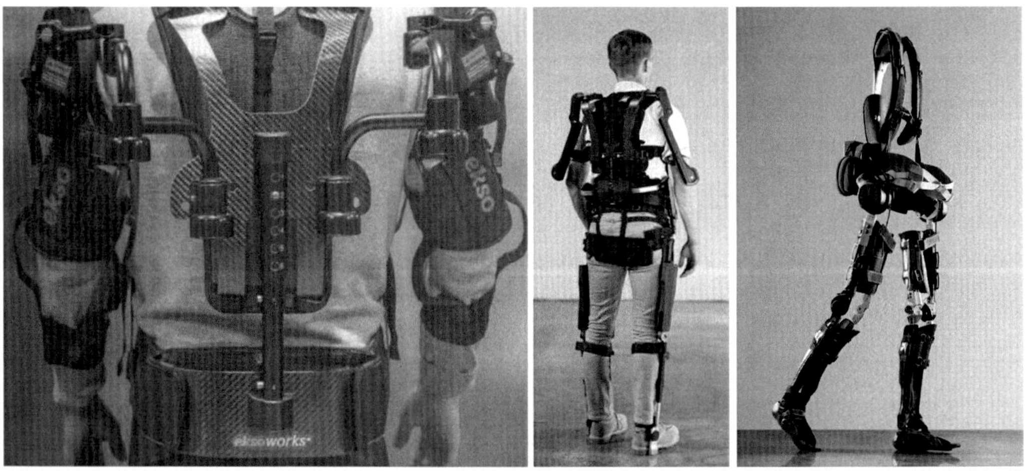

FIG. 11.2 Passive 9 pounds back shoulder exoskeleton Ekso vest introduced by California based EksoBionics currently tested by Ford to assist workers and prevent injuries (left). Mostly passive sensitized exoskeleton with clutch or brake mechanism lower body exoskeleton that can intelligently store and release energy (middle) and fully actuated lower body Phoenix exoskeleton (right) both by California based SuitX company [7].

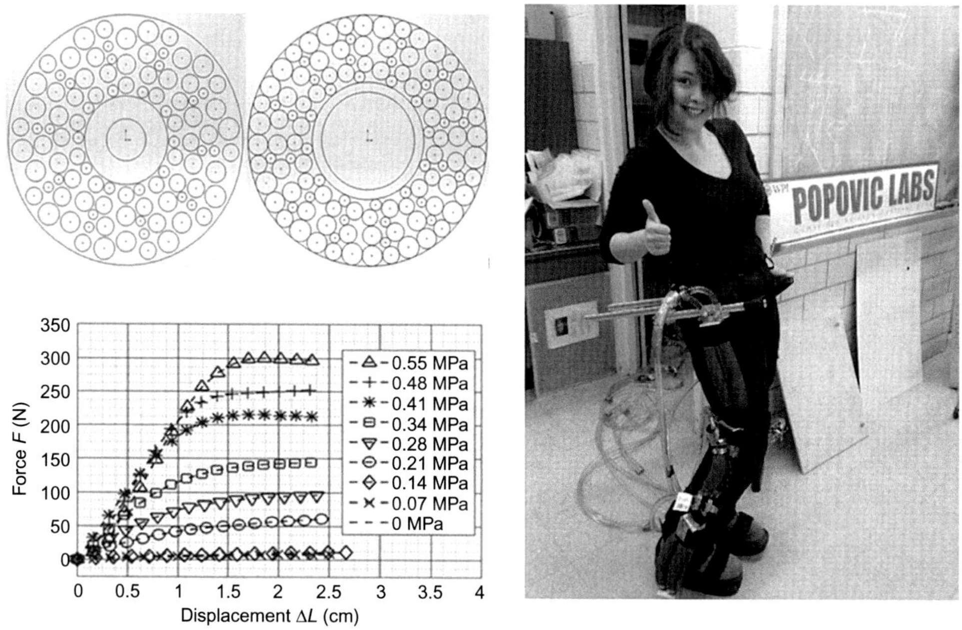

FIG. 11.3 Variable rigidity HydroBone exoskeleton [13–15]. Cross sectional view of granular media in unpressurized and pressurized HydroBone (top left). Result of compressive testing under an Instron 5567A system running the Blue Hill 2 software—force displacement properties of HydroBone with EPDM rubber granules pressurized at specified hydraulic pressure levels (bottom left). Researcher in Popovic Labs testing Variable rigidity HydroBone exoskeleton (right).

exoskeleton, rigid elements are often uncomfortable and may obstruct movements. To address both the user comfort and the necessity for load-bearing rigidity, several variable rigidity devices controlled in an automatic manner were proposed, designed, and the prototypes were tested. Based on the pressure and the external load the axial force fluid pressure is adjusted such that the critical buckling force is outside of the predetermined safety margin for the force. The HydroBone automatically stiffens when the load increases, and softens when the load decreases. The potential applications of this technology range across numerous disciplines, including wearable assistive technologies and civil engineering. The wearable applications are illustrated with HydroBone lower limb variable stiffness exoskeleton.

A common problem with all exoskeletons is that interconnected rigid joints unavoidably limit the possible human body motions to some extent. Even the best possible exoskeleton architecture cannot exactly match the human body degrees of freedom (DoF). Exoskeleton is not attached to human skeleton directly. Due to skin and tissue movement, the bony structure of the human skeleton is continuously changing its position and orientation with respect to the exoskeleton. As a result of this, biological and artificial joints are inevitably dislocated to some extent. For slow movements, small force transmission, and overall less ambitious activities, this may not be of

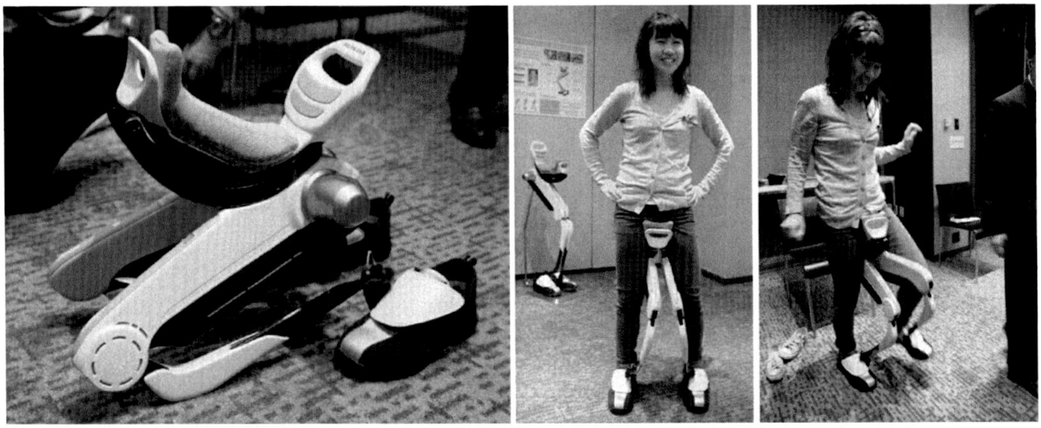

FIG. 11.4 Honda's Bodyweight Support Assist [18].

serious concern. The relative motion of two systems can even be advantageous. However, for fast movements, large force transmission, and unrestricted ranges of desired motion, especially for leg exoskeleton, this might be a problem. For example, users of leg exoskeletons for augmentation often complain about discomfort. This can be addressed with a biologically inspired soft robotic exomusculature [6] and exosuits already introduced in Chapter 10.

Exosuit concept is currently pursued by a group at Harvard in partnership with ReWalk, see Chapter 10, as well as in an independent effort by Silicon Valley-based company Seismic, formerly known as SuperFlex, which is also looking to make an entrance into the market for powered clothing [7]. Their early prototypes look more like gym clothes, with built-in motors and sensors. As discussed in Chapters 3 and 10, the issue with soft exosuits due to their inherent softness is that they can only provide very limited joint torques typically less than 10% of those naturally generated by the human body.

In difference to exoskeleton, soft exomusculatures do not have "bones," that is, rigidly interconnected joints. At the local joint level, the exomusculature can also consist of rigid or semi-rigid brace but only local to joint. This substantially relaxes the problem of multi-joint dislocation and user's discomfort. The disadvantage, however, is that pure exomusculature cannot transmit large load bearing forces through the external structure, that is, the large forces inevitably are transferred through human bones. One of a few device concepts till now that can effectively deal with both inherent discomfort and load bearing capability is variable rigidity exoskeleton. The other successful concept may be the rigid robotics legs that exactly follow the user, but the user is not directly strapped onto robotic legs throughout the legs; one example of that technology is Honda's Bodyweight Support Assist with foot attachments and saddle-like seat introduced in 2009 (Fig. 11.4) [18].

11.2 Dynamic Modeling and Simulation of the Human Musculoskeletal System for Exoskeleton Designs

The safe and comfortable interaction between the exoskeleton and the human body remains a major challenge. Great efforts in developing novel actuators, control methods, and ergonomic mechanical designs are made to facilitate the adaptation of the users to the exoskeletons. However, obtained results are still not satisfactory. It is clear that a prior analysis of the interaction between the human body and the exoskeleton before the construction of the exoskeleton would provide a significant feedback to the developers to designate the optimum parameters used in the electromechanical design of the exoskeleton. At this point, musculoskeletal modeling and simulation integrated with the exoskeleton dynamics gain an important role in the decision-making process in terms of the design and production of the exoskeleton systems.

Computational modeling and simulation of musculoskeletal systems provide the investigation of the mechanism underlying the movement of individuals. Understanding how muscular, skeletal, and neural systems behave and influence the locomotor system is an important issue not only for improving the quality of human life, but also for gaining new insights into the individuals' locomotor behaviors. Simulations of several musculoskeletal models were carried out to predict muscle forces [19], muscle contraction velocity [20], bone-to-bone contact force [21], and muscle-tendon unit (MTU) length change [22], to simulate the effects of surgeries [23], to determine the causes of abnormal gait [24], and to develop devices such as exoskeletons [25].

11.2.1 Construction of the Musculoskeletal Model

A musculoskeletal model represents the physical model of the human body which is a mechanical multi-body system actuated by the individual muscles. Design criteria of musculoskeletal model depend on the intended use of the model. To obtain more accurate and reasonable results from simulations, musculoskeletal models should reflect the anatomical, physiological, and functional characteristics of the human body [26]. Fig. 11.5 shows the general aspects of generating a musculoskeletal model.

11.2.1.1 Modeling of the Skeletal Elements

To model the skeletal system, it is necessary to specify a set of rigid bodies which represents the components of the skeletal system of the human body. Models of the skeletal structures are often created by converting two-dimensional (2D) computed tomography (CT) images into three-dimensional (3D) rigid body models using a set of image processing software [27, 28]. By means of these software packages, skeletal structures can be isolated from the other tissues such as muscles, fat, and tendon. Different biologic tissues absorb radiation at different rates. The different radiation absorption reveals itself as a visual difference which was visualized through tomography software, thereby enabling the separation of different tissues from each other based on the Hounsfield Scale, which

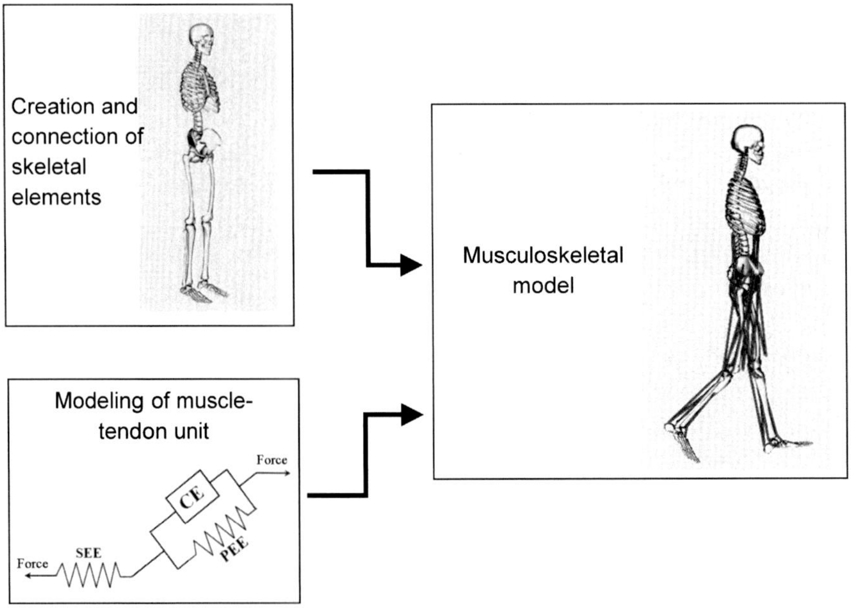

FIG. 11.5 A general perspective in the musculoskeletal modeling. Human body is modeled as a multi-body system consisting of bony structures actuated by the individual muscle-tendon units which are modeled by the three-element Hill-type muscle model. More information regarding Hill-type muscle models is given in Section 11.2.2. Modeling of the muscle-tendon unit.

quantitatively defines the different structures on the CT data by describing radiation density. By setting an appropriate threshold level, it is possible to isolate bony structure from other neighboring soft tissues.

In order to obtain 3D reconstruction of bone model from CT data, various open-access software such as Osirix (Pixmeo, Geneva, Switzerland), Seg3D (University of Utah Center for Biomedical Computing, USA), or Slicer3D (Fortner Research, Sterling, VA) can be used besides commercial image-processing software including Mimics (Materialise, Leuven, Belgium), 3D Doctor (Able Software Corp, Lexington, MA, USA), Amira (Mercury Computer Systems GmbH, Berlin, Germany), and Avizo (Mercury Computer Systems, Chelmsford, MA). The surface of each bone can be generated using computer-aided design (CAD) software such as Solidworks (Solidworks Corp., Dassault Systemes, Concord, MA, USA), Catia (Dassault System, Paris, France), 3D Builder (3D Construction Company, USA), Autodesk 123D Design (Autodesk Inc., San Rafael, CA, USA), and DraftSight (Dassault Systemes, USA). Image processing of bone reconstruction is briefly present in Fig. 11.6.

Motion capability of each body and their interactions with other rigid bodies are identified and assigned to the model. Joints are employed to define the range of motion and DoF of the rigid bodies. The minimum number of independent parameters necessary to specify the position and orientation of a body in space defines number of DoF. In simulation studies, joint models are kept as simple as possible to minimize complexity.

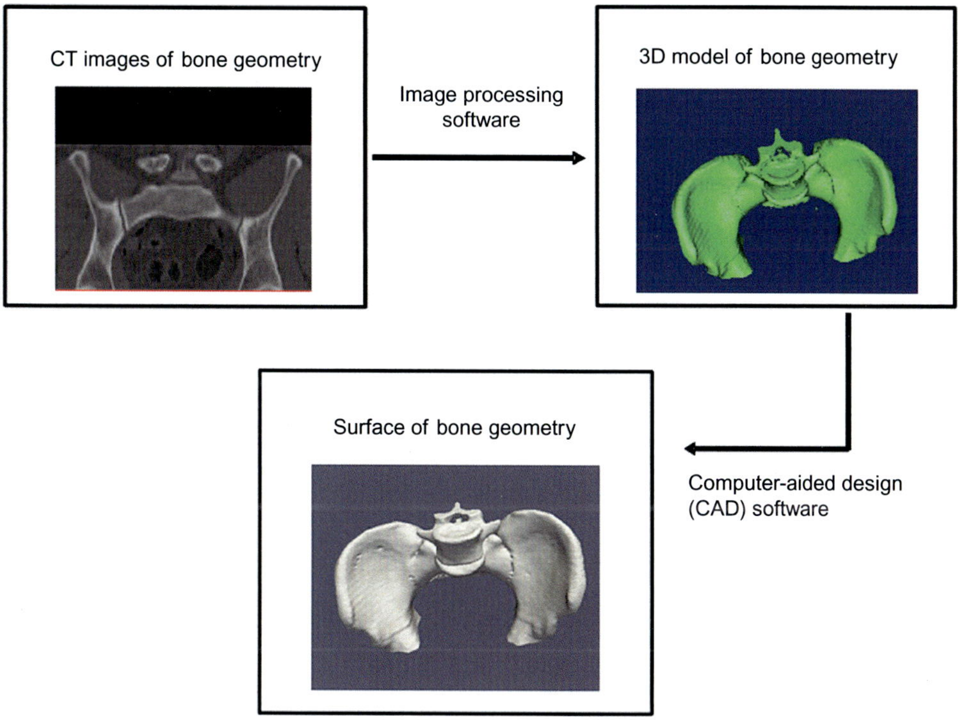

FIG. 11.6 Creation of skeletal model from computed tomography (CT) data using image processing and computer-aided design software.

To accurately mimic the motion of lower extremity of the human body and provide a realistic simulation of the trajectory of the whole body center of mass, six gait determinants, namely pelvic rotation, pelvic tilt, stance knee flexion, foot and ankle mechanisms, and tibiofemoral angle should be taken into account [29].

A typical 3D skeletal model is presented in Fig. 11.7 [30]. This model is represented by 10 segments: upper extremity (head, arms, and torso were modeled as a combined and unique rigid body), pelvis, thigh (x2), shank (x2), hindfoot (x2), and the toe segments (x2). There are six global DoF (three for position and three for orientation). Upper extremity articulates with the pelvis via a three-DoF ball-and-socket joint. The hip, knee, ankle, and metatarsal joints are modeled as three, one, two, and one DoF joints, respectively. Since the total DoF of all joints is 17 and this model is freely movable, the total DoF of the model is $17+6=23$.

11.2.2 Modeling of the Muscle-Tendon Unit

In a human musculoskeletal system, muscles act as force generators. Muscle modeling approaches can be classified into two groups, namely structural (or molecular) and phenomenological models. In structural models, which are also known as Huxley-type models [31], the internal mechanical and chemical events in the sarcomeres are taken into

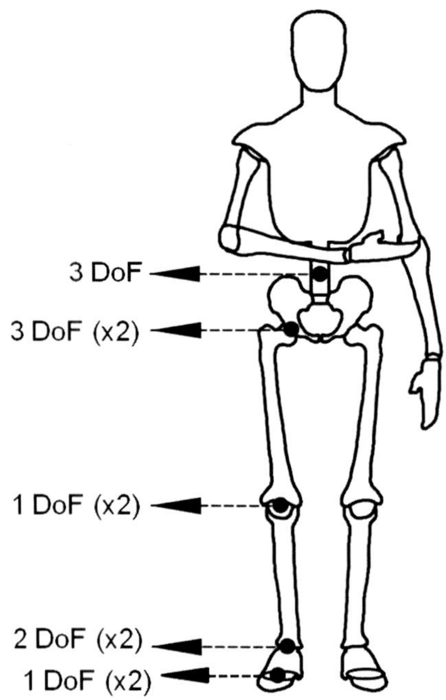

FIG. 11.7 A skeletal model with 10 segments and 23 degrees of freedom.

account. The number of cross-bridges between myofilaments (actin and myosin) and the involvement of titin determine the force production capability of the models. However, the classical structural models represent the high complexity and they are still far from the practical use in the human motion analysis. On the other hand, phenomenological (Hill-type) models are widely used to mimic the biomechanical behaviors of the MTUs by means of rheological elements [32, 33] (Fig. 11.8). Hill-type muscle models consist of rheological elements (spring and/or dashpot) and a controlled black box representing the muscle contractile properties such as force-length and force-velocity relationships. Even though Hill-type models are not based on realistic muscle architecture and internal

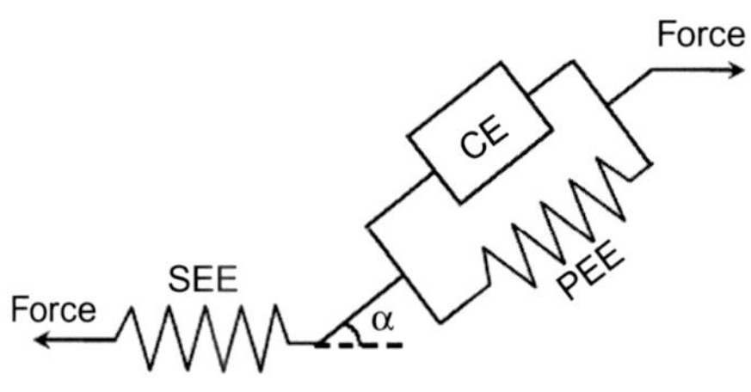

FIG. 11.8 The most commonly used three-element Hill-type muscle model [33]. CE: The contractile element of muscle representing the interaction between actins and myosins and is responsible for active force generation. SEE: The series elastic element of muscle representing the elasticity of the myofilaments, tendon, and aponeuroses. PEE: The parallel elastic element of muscle representing the connective tissue surrounding the muscle fibers. α is the pennation angle which is defined as the angle between the direction of muscle fibers and the line of action of the muscle force.

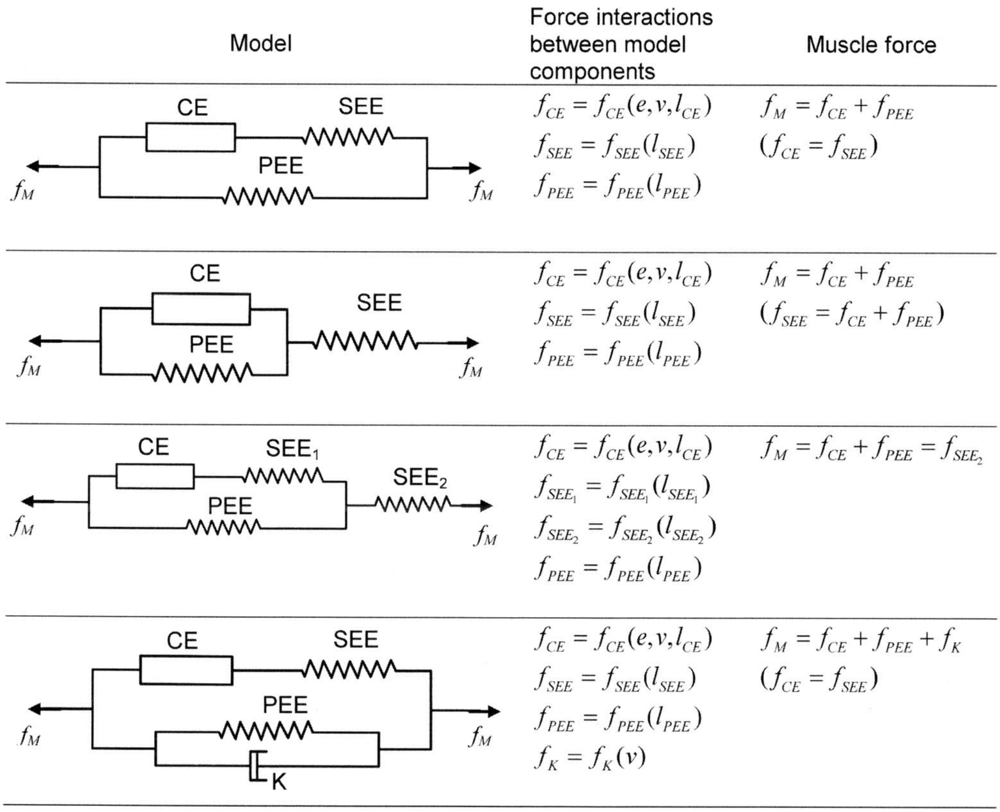

FIG. 11.9 Various Hill-type muscle models. f_M = muscle force, f_{CE} = contractile element force, f_{SEE} = series elastic element force, f_{PEE} = parallel elastic element force, f_K = force acting along damping element, l_{CE} = contractile element length, l_{SEE} = series elastic element length, l_{PEE} = parallel elastic element length, e = activation level of muscle, v = contractile velocity of muscle.

sequential events, the gross biomechanical behavior of MTU could be practically simulated by using this model with a low computational burden.

A few further typical Hill-type models are shown in Fig. 11.9. Interactions of the active (contractile) and passive (noncontractile) muscle components are provided mathematically. Passive elements are considered either as pure elastic (spring) or viscoelastic elements (spring + dashpot), since muscles exhibit hysteresis, force relaxation, and creep properties [34, 35].

Three major parameters including instantaneous muscle length (l), instantaneous muscle velocity (v), and activation level determine the force production capability of the muscles. The maximal muscle force at instantaneous muscle length ($f(l)$) can be mathematically characterized by a Gaussian curve (Fig. 11.10A) around the optimal length (l_o) for which the maximum isometric muscle force (f_o) is observed [36]. In addition, the maximal muscle force at instantaneous muscle contraction velocity ($f(v)$) has a hyperbolic shape (Fig. 11.10B), which is mathematically expressed as [37]

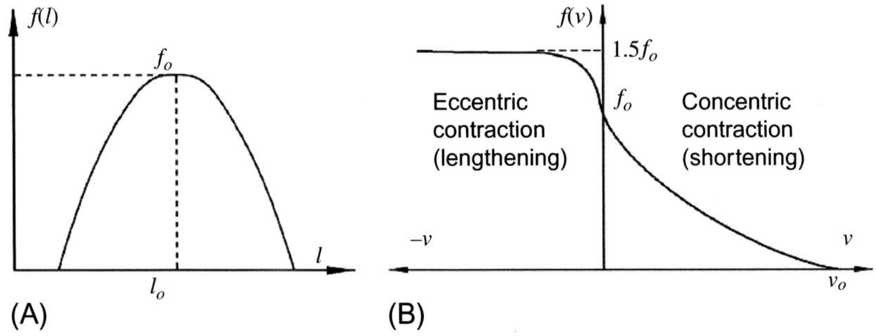

FIG. 11.10 (A) Force-length and (B) force-velocity relationships of muscles.

$$f(v) = \begin{cases} 0, & v_o < v \\ \dfrac{f_o b - av}{v + b}, & 0 < v \leq v_o \\ 1.5f_o - 0.5\dfrac{f_o\frac{b}{2} + av}{-v + \frac{b}{2}}, & -\dfrac{v_o}{2} < v \leq 0 \\ 1.5f_o, & v \leq -\dfrac{v_o}{2} \end{cases} \tag{11.1}$$

while a and b are Hill's thermodynamic constants [32].

Even though the Hill-type muscle models provide a practical way of simulation of the muscle contraction mechanism, the identification of the model parameters, which should be in agreement with the experimental findings, can be challenging task [38]. One of the main drawbacks of the Hill-type models is that they do not account for the history dependency of the skeletal muscle force production, which is the effect of the "recent" contractile conditions of muscle on the produced force [39]. Despite these shortcomings, the Hill-type muscle model is widely used in musculoskeletal modeling and simulation platforms due to its simplicity and practicability.

11.2.3 Model Validation

Model validation is the process of defining how well the model reflects the experimentally observed reality. To validate the musculoskeletal model, the most preferred method is to compare the theoretical muscle moment arms with those reported in the literature and adjust the muscle paths according to moment arms [26].

11.2.3.1 Muscle Moment Arm

Muscles generate forces and cause moments around joints. Joint moment production capability of a muscle is directly related to the moment arm which can be defined as

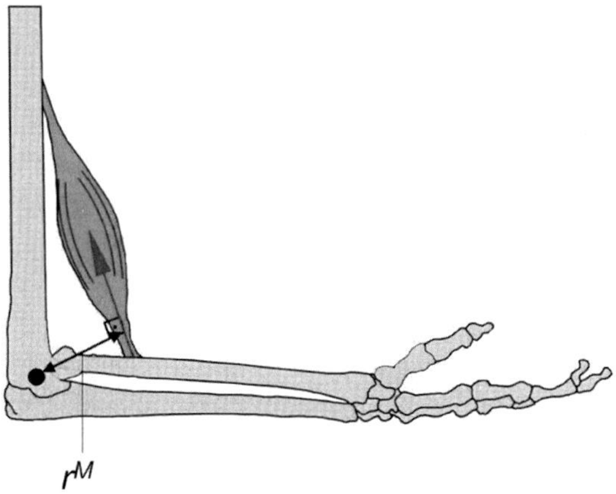

FIG. 11.11 The moment arm (r^M) of a muscle, which is the perpendicular distance between the line of action of the muscle and the center of rotation of the joint.

the distance between the joint rotation center and the line of action of muscle (Fig. 11.11). Determination of the muscle moment arm requires a kinematic characterization of the anatomical structures of the muscle and joint of interest. Muscle moment arm can be directly measured from cadavers or can be determined from MRI or CT data [40].

11.2.3.2 Modeling of the Muscle Path

To model the muscle paths, two different methods are commonly used, namely (i) the straight-line method and (ii) the centroid-line method. In the straight-line method, muscle path is defined as a straight line connecting the centers of the muscle attachment sites. This method is easy to apply, but it may not provide meaningful results [41, 42]. In the centroid-line method, the path of muscle is defined by a line passing through the cross-sectional center of the muscle. Using this method, the muscle line of action can be represented more accurately. However, it is difficult to implement this approach since (a) obtaining the locations of muscle cross-sectional centroids may not be available for even a single position of the body and (b) even if a path of muscle center is determined, it is difficult to define how the muscle path changes as body moves [41]. To deal with these constraints, effective attachment sites or via points at specific locations can be used. Muscle models include muscle paths determined according to the corresponding muscle's origin and insertion. These muscle attachment sites are combined with wrapping surfaces and via points to prevent muscles from passing through bones. To provide consistency with the experimentally measured moment arms, it is recommended to use via points and wrapping surfaces [43]. To represent muscle-tendon dynamics precisely, muscle path geometry should be modeled in detail. Origins and insertions, via points, and wrapping surfaces can be determined from cadaver dissections, magnetic resonance imaging (MRI), or CT data [44].

11.2.4 Motion Analysis

Motion capture systems are used to obtain 3D human body kinematics (position, velocity, and acceleration) and force plate systems to obtain kinetic data (directly measured ground reaction force—GRF, and indirectly deduced joints' moment and power), enabling the researchers to quantitatively evaluate, identify, and interpret the movement. The fundamental aim of the motion analysis is to evaluate the functional motor capabilities of individuals and determine the presence of impairments in gait pattern, if any. Motion analysis is also used for different purposes such as determination of the treatment and rehabilitation protocols for abnormal gait, investigation of the effectiveness of orthotics, prosthesis, or exoskeletons, and identification of sports injury risks in a reliable and quantitative manner. The traditional motion analysis systems consist of specific cameras which perceive reflective markers' positions and force plates which record the GRF (Fig. 11.12). Moreover, with the incorporation of the EMG recordings, it is also possible to gain insight into the activation levels of muscles during movement.

11.2.4.1 Optical Systems

Kinematic analysis is utilized to define the time course of changes in position and orientation of the body segments and the geometry of motion in terms of displacements, velocities, and accelerations without taking into account the kinetics behind the generation of the motion. Kinematics data can be obtained using infrared camera systems which record the 3D position of markers placed onto the body surface. Data are recorded at a series of

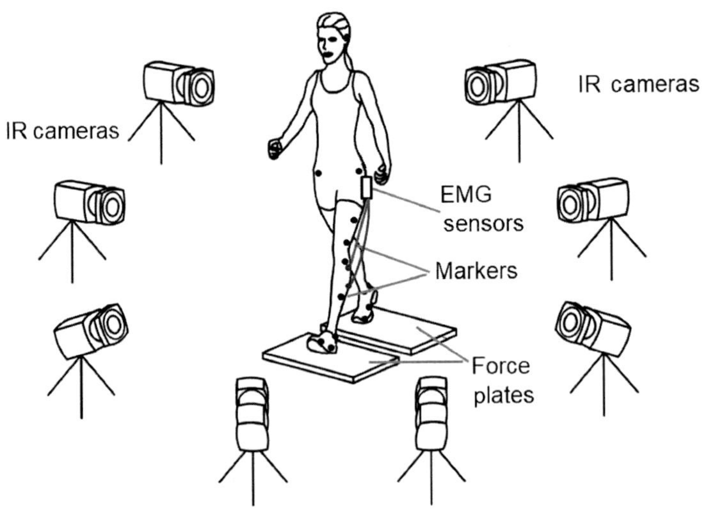

FIG. 11.12 Traditional motion analysis system including infrared (IR) cameras, reflective markers, EMG recording system, and force plates.

time intervals. Kinematic systems require calibration process in order to provide accuracy in the recording sessions.

11.2.4.2 Markers

The 3D motions of markers placed onto the specific anatomical regions of subjects can be recognized by camera system. The objective of recording the markers' motions is to track the motion (or movement patterns) of the body segments. In order to define the spatial location of each body segment, at least three markers are required. When a subject walks in the volume that the cameras can detect, 3D positions of markers are computable as long as they are visible to at least two cameras. If a marker can be seen by only one camera, its 3D position cannot be calculated. However, in such cases, the position could be estimated using data from earlier and later frames by "interpolation."

Another significant error affecting the estimation of angles and moments around joints is the miscalculation of locations of the joint centers from surface markers [45]. Marker placement protocol must be simple, but meticulous enough to define the relative motion of the body segments [46]. The most commonly used marker sets are Helen Hays Hospital Markers Set, Helen Hayes Modified Markers Set, Vaughan Marker Set, and Vaughan Modified Marker Set [46–48].

11.2.4.3 Force Platform

Force platform, which is also known as force plate, is used to measure the GRF exerted by the ground on a body in contact with it. Force plates provide 3D GRF components, the center of pressure components defining the location of the force vector, and orthogonal moment components.

A force platform should be located below floor level so that it is not discernible from the ground. The subject should not be informed of the plate's presence. To obtain "natural" gait data representing the real walking patterns, entire sole of the subject's foot must be placed on the force platform. This may require a number of walking trials to be performed [49]. By using the GRF and kinematic data with the inverse dynamic analysis, the resultant joint forces and moments can be calculated.

11.2.5 Exoskeleton

Exoskeletons are wearable biomechatronics devices that assist and/or augment [50] physical performance of the user. These devices fit to user's body are developed for (i) increasing the load-carrying capacity, human endurance, and joint strength, (ii) augmenting walking and running, and (iii) reducing energy consumption, joint pain, and impact effects coming from the ground during locomotion. In addition, the rehabilitation applications (i.e., recovery after stroke) exist benefitting from the exoskeletons [51]. An exoskeleton might not meet all of these functional requirements at the same time, however, all these kinds of wearable machines are designed to respond one or combinations of above-listed demands. Exoskeletons are devices consisting of not only mechanical

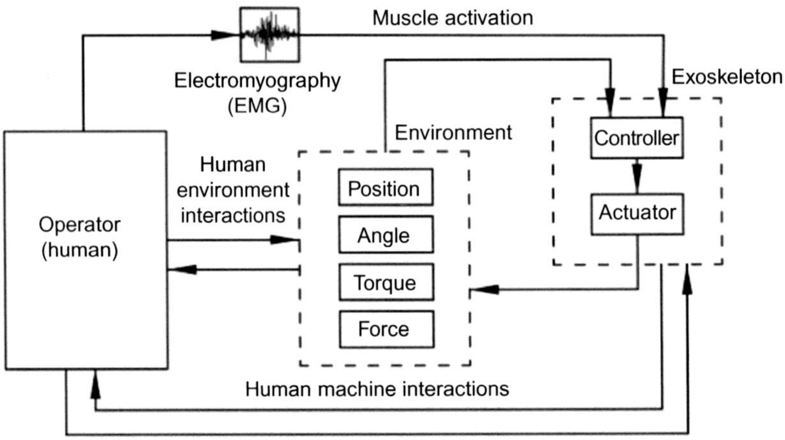

FIG. 11.13 A sample exoskeleton control mechanism. The exoskeleton is in interaction with the human body of the user and environment to implement specific tasks.

parts integrated into each other, but also a system for sensing and processing data and controlling motion in the direction of the user's intention. Therefore, control of exoskeletons is an important issue besides the mechanical compatibility. An EMG-driven control mechanism for an exoskeleton is presented in Fig. 11.13 [52].

As a wearable device, electromechanical structure of the exoskeleton should be compatible with the user's body. Interaction points between the human body and exoskeleton are important in terms of the effective force transmission and comfortable usage [53]. Lightness and strength of the exoskeleton's hardware affect the mobility capacity, endurance, and energy consumption. Moreover, the determination of the structure of joints (e.g., degrees of freedom, the range of motion), actuators (e.g., electrical motors, artificial muscles), and control methods is also a challenging task for stability and performance of these robotic systems [4, 54]. To figure out such challenges, musculoskeletal modeling and simulation platforms integrated with the exoskeleton dynamics can be considered as complementary tools, enabling to predict the joint contact force, muscle power, and muscle force, as well as the muscle contraction velocity, length change, and activation level.

11.3 Computational Musculoskeletal Modeling and Simulation

In addition to experimentally obtained kinematic and kinetic data, computational musculoskeletal dynamic modeling, and simulation environments can also provide further information regarding the muscle contractile properties and joint dynamics such as joint contact and muscle forces (Fig. 11.14). Musculoskeletal modeling and simulations of movement have a potential to give insights into the dynamics of the musculoskeletal

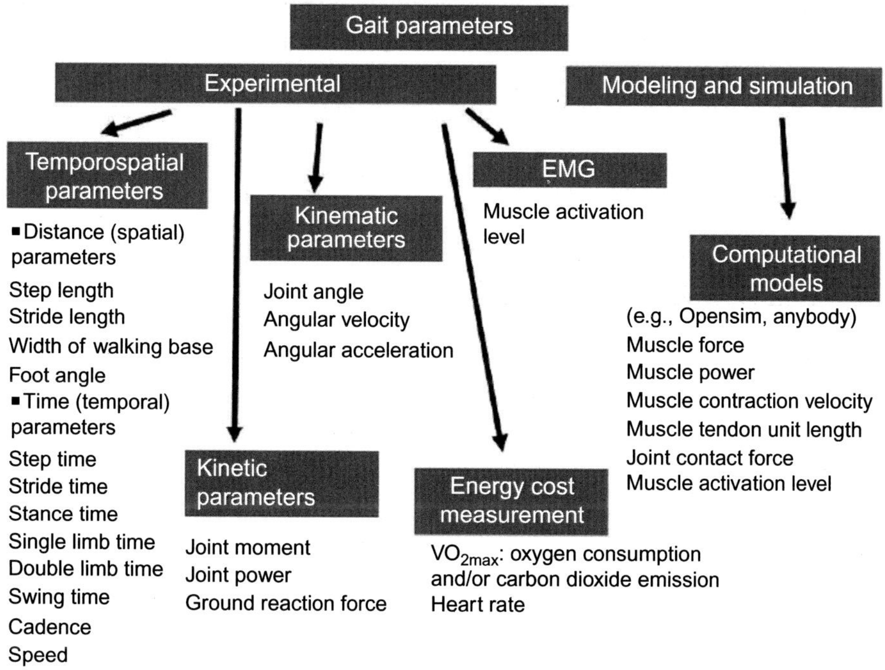

FIG. 11.14 Gait parameters obtained from both experimental gait analysis and computational modeling and simulation platforms.

systems [55], and there has been a growing development of musculoskeletal models ranging from the study of muscle contributions of movement [56] to the development of human-machine interfaces [57].

The knowledge of all joints' states and torques is still limited in sense that: (1) muscles' forces need yet to be determined, and (2) it is not trivial task to create a cause-effect relationship between action of muscles and observed body movements. Hence it is a big challenge to determine how individual muscles accelerate body segments and contribute to observed motion [33, 58].

Early simulations of the movement were limited primarily by the performance of the computers [59]. Interest in using musculoskeletal models to determine movement has been encouraged by the ever-increasing performance of the computers. Computer modeling and simulation, which allow noninvasive prediction of the joint loads and muscle forces, can provide quantitative explanations of how the musculoskeletal systems interact to produce movement [26, 60]. In computer simulations, a theoretical framework is needed to explore the biomechanics of movement and the mechanics of multi-joint movement. A dynamic simulation of motion combining models describing the anatomical and physiological components of the musculoskeletal system would supply such a framework.

Although dynamic simulations of movement are broadly adopted [30, 61, 62], simulation development of movements is quite challenging because of the complexity of the biological systems. Over the last two decades, a significant effort has been made to develop open-access and commercial software that allow dynamic simulations of the musculoskeletal models combined with optimization theory for a wide variety of movements. OpenSim [58], Anybody [63], Human Body Model [64], and Biomechanics of Bodies [65] are widely used simulation programs for musculoskeletal systems. Their computational capabilities present an important tool to solve the redundancy (muscle-force sharing or load distribution) problem, which arises since a number of muscles crossing a joint exceeds the number of DoF of the joint of interest. To be able to run these simulation environments, it is necessary to feed the kinetic and kinematic data obtained from motion analysis systems.

11.3.1 Scale Tool

Scale tools are used to adapt the anthropometry of generic musculoskeletal models existing in the musculoskeletal simulation platforms to the anthropometry of the subject of interest. Scale tools are implemented based on the comparison of the experimental marker data and virtual markers located on the model, with a combination of manually specified scale factors and measured distances between experimental marker locations. Scale tools also scale both mass and inertia properties of the models as well as the dimensions of the body segments. Fig. 11.15 shows the generic musculoskeletal model of OpenSim scaled according to the anthropometric data of the human subject of interest.

11.3.2 Inverse Kinematics

Inverse kinematics is used to determine the generalized coordinates (joint angles and positions) of the model by minimizing the error value between the experimental and theoretical markers for each time frame. The inverse kinematics is performed by minimizing the weighted squared error which is formulated as follows:

$$\sum_{i=1}^{n} w_i \left(x_i^{\exp} - x_i^{subject} \right)^2 + \sum_{j=1}^{m} w_j \left(\theta_j^{\exp} - \theta_j^{subject} \right)^2 \tag{11.2}$$

where n and m are the number of markers and joint angles, respectively. $x_i^{\exp}$ is the experimental marker position. $x_i^{subject}$ is the position of the simulated model. $\theta_j^{\exp}$ and $\theta_j^{subject}$ are the values of the jth joint angle for the experimental and simulated model, respectively. w_i and w_j are the weight values that provide markers and joint angles to be weighted differently. The weight value affects to what degree a match should be satisfied.

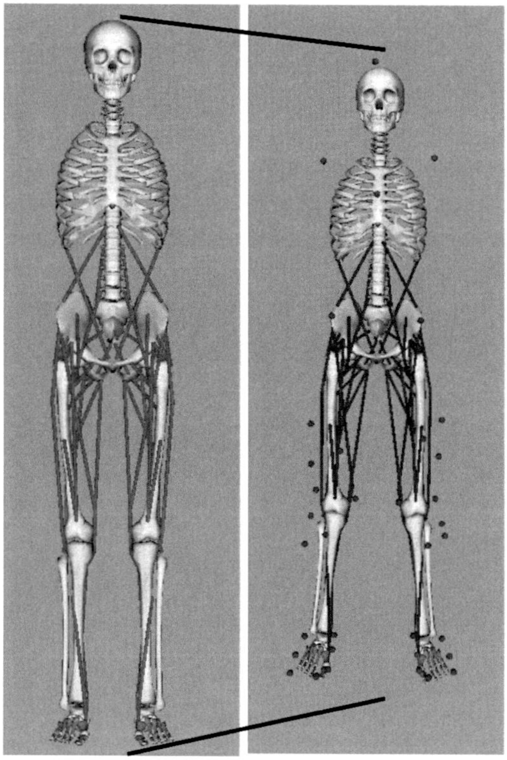

FIG. 11.15 Scaling of the OpenSim generic musculoskeletal model according to the experimental marker data.

11.3.3 Inverse Dynamics

Inverse dynamics derives the generalized forces (net forces and torques) at each joint for targeted movement. It is performed based on a mass-dependent relationship between force and acceleration (Newton's second law of motion). The general form of the equation of motion, which describes the kinematic and kinetic behaviors of the musculoskeletal model in terms of its motion as a function of time, obtained by inverse dynamics can be expressed in the following form:

$$\mathbf{A}(\boldsymbol{\theta})\ddot{\boldsymbol{\theta}} + \mathbf{b}\left(\boldsymbol{\theta}, \dot{\boldsymbol{\theta}}\right) + \mathbf{g}(\boldsymbol{\theta}) = \boldsymbol{\tau} \tag{11.3}$$

where $\boldsymbol{\theta}, \dot{\boldsymbol{\theta}}$, and $\ddot{\boldsymbol{\theta}}$ are the vectors of generalized coordinates, velocities, and accelerations, respectively. $\mathbf{A}(\boldsymbol{\theta})$ is the mass (or inertia) matrix of the model, $\mathbf{b}\left(\boldsymbol{\theta}, \dot{\boldsymbol{\theta}}\right)$ is the vector including the Coriolis and centrifugal terms, $\mathbf{g}(\boldsymbol{\theta})$ the vector of the gravity terms, and $\boldsymbol{\tau}$ is the vector of the generalized forces. An equation of motion is a second-order differential equation and one can calculate the generalized coordinates as a function time by solving this differential equation using either analytical or numerical technique which is the more common and practical way.

11.3.4 Static Optimization

Theoretical calculation of muscle forces depends on solving the redundancy problem [66]. A general approach to overcome the redundancy problem is to formulate an optimization strategy in which a physiologically based cost function is solved [67]. Since it is easy to design and computationally efficient, static optimization in which the unknown muscle forces are solved at each instant of time is the most widely used optimization technique to calculate the muscle forces and activations which satisfy the kinematic and kinetic data recorded experimentally during locomotion [68]. Typical objective functions have the form of a weighted sum of muscle forces (or activations) raised to some nonlinear power [67].

11.3.5 Integration of the Exoskeleton Dynamics to the Simulation Software

The dynamic model of exoskeletons could be integrated into a musculoskeletal model to identify the influences of these external devices on the kinematics and kinetics of the human body, as well as on the motor coordination, energy expenditure, and ergonomics [25]. Once the concept design of an exoskeleton is completed in accordance with the functional requirements, the physical model of the device is imported to a musculoskeletal modeling and simulation program [69]. Contact points between the exoskeleton and human body parts are determined and DoFs and range of motions of the exoskeleton joints were designated in order to release or restrict the motions. Kinematics and GRF data obtained from a subject wearing an exoskeleton are fed to the simulation software to calculate the physiological properties of the exoskeleton-integrated human body such as muscle force, metabolic energy consumption, and joint reactions forces. A basic flowchart illustrating the successive steps of obtaining muscle contractile properties and joint contact forces of a representative exoskeleton-integrated human body is presented in Fig. 11.16.

Integrating 3D dynamic model of an exoskeleton into the simulation imposes a series of changes in operating conditions of the model. Due to the weight of the device, the model will be exposed to higher GRFs and the changes also likely occur in muscle coordination in order to keep the stability and balance and preserve the so-called normal gait characteristics of the human body [70]. Muscle contraction velocity and muscle length are also expected to be affected by the exoskeleton assembly since Hill-type muscle models operate force-velocity and force-length relations. Moreover, the joint moments and joint powers in the model could show differences because of the additional torques coming from the actuators of the device [71]. Exoskeleton may have an influence on the range of motions of joints and limbs such as avoiding excessive knee bending [72]. Therefore, exoskeletons bring about a series of change in the contractile and mechanical properties of the musculoskeletal models consistent with their designs and functions.

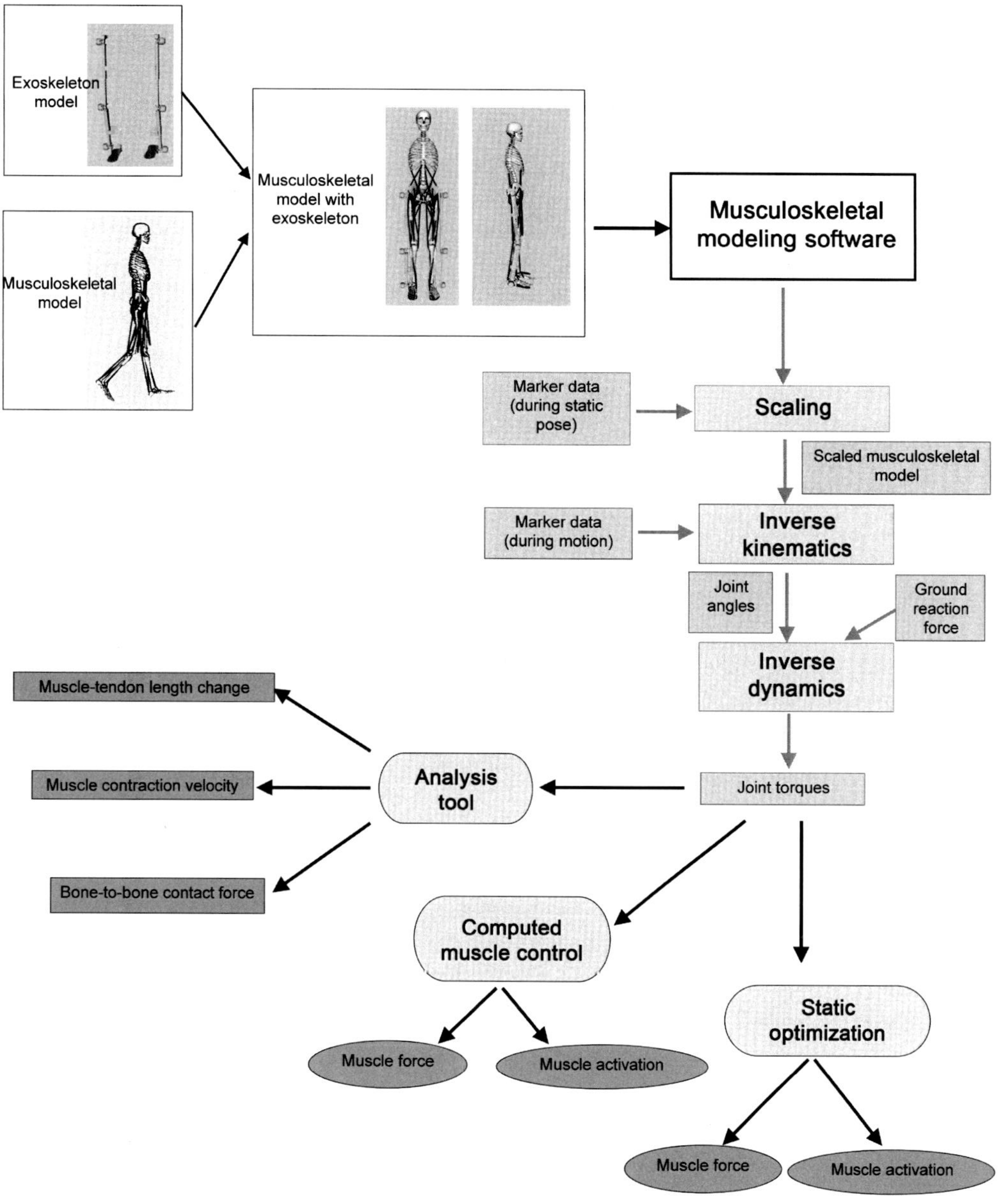

FIG. 11.16 A flowchart representing the successive steps of obtaining muscle contractile properties and joint contact forces with the use of a musculoskeletal modeling and simulation platform including the dynamic model of an exoskeleton.

To develop an exoskeleton as an assistive device in the daily life, many researchers continue to study the optimal modeling and design with the help of simulations of the musculoskeletal system. It is desired that

- the joint angles should be in the anatomical range of motion during movement,
- the exoskeletons should compensate the joint moments and velocities,
- the exoskeletons should predict the intended motion before the actual occurrence.

Exoskeleton takes the form of connected anatomy and follows the natural movement of the body. It is also connected to the individual's body at certain attachment points. One of the corrective tasks of the exoskeleton is to reduce the loads acting on the human body. It is possible to facilitate the joint action by providing assistance power and physical load transfer to the body [73]. Also, it is known that complexity of muscle-tendon mechanics constitutes an impediment to the design of exoskeletons. Several experiments and simulations of musculoskeletal models have shown that wearing elastic exoskeleton during bilateral hopping considerably reduce the plantar flexor muscle forces [74, 75]. Even though a high amount of reductions in the whole-body metabolic rate is observed, predicted metabolic energy consumed from plantar flexor muscles is not considerably decreased, probably because of negative changes occurred in the lengths and velocities of the muscle fibers [75]. Thus it is important to obtain time-dependent muscle contractile properties during locomotion. Determination of muscle fiber length changes can be possible using ultrasound imaging, but the number of muscles that can be imaged is limited. Experimental determination of muscle forces, activations, and MTU length changes also require invasive methods. At this point, computational musculoskeletal models can provide valuable information.

To be able to assess the effectiveness of the exoskeleton during movement in different environments and decide the model parameters to obtain optimal design, computer simulations provide an alternative way. Due to variations in the velocities and torques in human joints during movement, exoskeletons are designed relying on maximal values of joint velocities and torques [76]. But the design of an exoskeleton based on maximal torques and velocities may cause the motors to be oversized and hence, exoskeleton's size, weight, and cost may increase. Therefore it should be aimed to have an optimal design which is developed by analyzing the trade-off between the volume/size and required torques [77]. For the selection of the actuators to be used in exoskeletons, it is required to determine the minimum torques and velocities per articulation, as shown in Fig. 11.17.

Some exoskeleton models are aimed to predict the intended motion of the user using EMG signals [52]. It is also aimed to use the activations of muscles crossing the joint to predict the required moment around the joint of interest. With the obtainment of the joint moment, the controller is fed to allow the exoskeleton to realize the motion. The essential advantage of EMG-based exoskeletons is the potential to predict muscle forces before the actual event of mechanical contractions [25]. Computational musculoskeletal models also have the convenience to determine optimal parameters of the controller of an exoskeleton. It can be easily observed that how the kinetic and kinematic results change depending on the alterations of parameters.

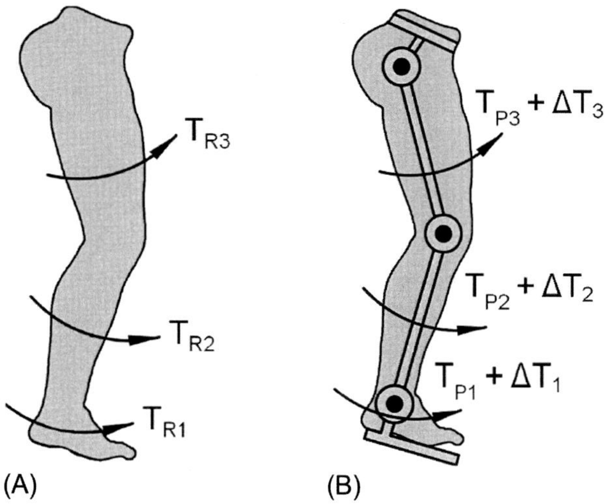

FIG. 11.17 (A) Joint moments of a healthy subject, where T_{Ri} is the required moment to rotate the joint. (B) Joint moments of a patient with motor dysfunction in lower limbs, where T_{Pi} is the joint moment barely generated by the patient and $\Delta T_i = T_R - T_P$ is the difference between the joint torques of the healthy and patient subjects, which is supposed to be produced by the exoskeleton. It is assumed that the healthy and patient subjects are in need of the same amount of joint moments for having a "normal" gait and exoskeleton dynamics is not included in the calculation of ΔT_i.

References

[1] R. Schulte, The characteristics of the mckibben artificial muscle, Appl. External Power Prosthet. Orthet. 874 (1962) 94–115.

[2] M. Vukobratovic, D. Hristic, Z. Stojiljkovic, Development of active anthropomorphic exoskeletons, J. Med. Biol. Eng. 12 (1974) 66–80.

[3] S. Castro, E. Lugo, P. Ponce, A. Molina, Assistive robotic exoskeleton for helping limb girdle muscular dystrophy, in: International Conference on Mechatronics, Electronics and Automotive Engineering, ICMEAE, 2013, pp. 27–32.

[4] M. Dollar, H. Herr, Lower extremity exoskeletons and active orthoses: challenges and state-of-the-art, IEEE Trans. Robot. 24 (2008) 144–158.

[5] P. Beyl, M. Van Damme, R. Van Ham, D. Lefeber, Design and control concepts of an exoskeleton for gait rehabilitation, in: Second IEEE Ras & Embs International Conference on Biomedical Robotics and Biomechatronics, BIOROB, 2008, pp. 103–108.

[6] Popovic M. B. Biomechanics and Robotics. 2014. 351 pages, Copyright © Pan Stanford Publishing Pte. Ltd., Singapore, ISBN 978-981-4411-37-0 (Hardcover), 978-981-4411-38-7 (eBook).

[7] L. Goode, Are Exoskeletons the Future of Physical Labor? Ford and Others are Experimenting With Wearable Robotics, The Verge, 2017. https://www.theverge.com/2017/12/5/16726004/verge-next-level-season-two-industrial-exoskeletons-ford-ekso-suitx (accessed 10.03.18).

[8] Lockheed Martin Exoskeleton Technologies: Media, https://www.lockheedmartin.com/us/products/exoskeleton/media.html (accessed 10.03.18).

[9] NASA-IHMC X1 Mina Exoskeleton, http://robots.ihmc.us/x1-mina-exoskeleton/ (accessed 10.03.18).

[10] G.J. Androwis, K.J. Nolan, Evaluation of a robotic exoskeleton for gait training in acute stroke: a case study, in: Wearable Robotics: Challenges and Trends, 2017, pp. 9–13.

[11] K. Karunakaran, G. Androwis, R. Foulds, Natural user-controlled ambulation of lower extremity exoskeletons for individuals with spinal cord injury, in: Wearable Robotics: Challenges and Trends, Springer International Publishing, 2017, pp. 121–125.

[12] A. Dollar, H. Herr, Lower extremity exoskeletons and active orthoses: challenges and state-of-the-art, IEEE Trans. Robot. 24 (1) (2008).

[13] Popovic M, Sridar S, Majeika C, Deisadze N, Giancarlo E. Variable Stiffness Devices and Methods of Use. United States Patent and Trademark Office, Assignee Worcester Polytechnic Institute, US Patent App. 15/154,443, 2013.

[14] S. Sridar, C.J. Majeika, P. Schaffer, M. Bowers, S. Ueda, A.J. Barth, J.L. Sorrells, J.T. Wu, T.R. Hunt, M. Popovic, Hydro muscle—a novel soft fluidic actuator, in: IEEE International Conference on Robotics and Automation (ICRA), 2016, pp. 4014–4021.

[15] M. Bowers, C. Harmalkar, S. Sridar, C. Majeika, C. Kaan, G. Iannacchione, M. Popovic, An approach to HydroBone and other variable stiffness structures, in: Proceedings of the 20th International Conference on Composite Structures (ICCS20) in Paris, France, 2017.

[16] Popovic MB, Onal C, McCarthy G, Corso N, Effraimidis D, Jennings B. Actuators and Methods of Use (Hydro Artificial Muscles). United States Patent and Trademark Office, Assignee Worcester Polytechnic Institute. 2015. Serial No.: 62/011,830. Filed: June 13, 2014. US Patent 20,150,359,698.

[17] G. McCarthy, D. Effraimidis, B. Jennings, N. Corso, C. Onal, M.B. Popovic, Hydraulically actuated muscle (HAM) exo-musculature, in: Robot Makers: The Future of Digital Rapid Design and Fabrication of Robots" (RoMa) Workshop, the 2014 Robotics: Science and Systems Conference, Berkeley, CA, July 12, 2014.

[18] O. Lee, How It Feels To Walk With Honda's Cyborg Legs, https://gizmodo.com/5212161/how-it-feels-to-walk-with-hondas-cyborg-legs (accessed 10.03.18).

[19] Y.C. Lin, T.W. Dorn, A.G. Schache, M.G. Pandy, Comparison of different methods for estimating muscle forces in human movement, Proc. Inst. Mech. Eng. Part H J. Eng. Med. 226 (2012) 103–112.

[20] R.W. Jackson, C.L. Dembia, S.C. Delp, S.H. Collins, Estimated Changes in Muscle-Level Dynamics and Energetics Under Different Levels of Exoskeleton-Applied Work and Torque, (2015).

[21] K.M. Steele, M.S. DeMers, M.H. Schwartz, S.L. Delp, Compressive tibiofemoral force during crouch gait, Gait Post. 35 (4) (2012) 556–560.

[22] E.M. Arnold, S.R. Hamner, A. Seth, M. Millard, S.L. Delp, How muscle fiber lengths and velocities affect muscle force generation as humans walk and run at different speeds, J. Exp. Biol. 216 (2013) 2150–2160.

[23] R. Balasubramanian, J. Montgomery, K. Mardula, C. Allan, Implanted miniature engineering mechanisms in tendon-transfer surgery improve robustness of post-surgery hand function, in: The 6th Hamlyn Symposium on Medical Robotics, 2013.

[24] B.A. Knarr, D.S. Reisman, S.A. Binder-Macleod, J.S. Higginson, Understanding compensatory strategies for muscle weakness during gait by simulating activation deficits seen post-stroke, Gait Post. 38 (2013) 270–275.

[25] I.S. Dhindsa, R. Agarwal, H.S. Ryait, Principal component analysis-based muscle identification for myoelectric-controlled exoskeleton knee, J. Appl. Stat. 44 (10) (2016) 1707–1720.

[26] M.G. Pandy, R. Barr, M. Kutz, Biomechanics of the musculoskeletal system, in: Standard Handbook of Biomedical Engineering and Design, The McGraw-Hill Companies, Inc., USA, 2003 (Chapter 6).

[27] M. Wesseling, F. De Groote, L. Bosmans, W. Bartels, C. Meyer, K. Desloovere, I. Jonkers, Subject-specific geometrical detail rather than cost function formulation affects hip loading calculation, Comput. Methods Biomech. Biomed. Eng. 19 (2016) 1475–1488.

[28] Z. Lerner, B.G. Gadomski, K. Haussler, C.M. Puttlitz, R.C. Browning, Modulating tibiofemoral contact forces in the sheep hindlimb via treadmill walking: predictions from an OpenSim musculoskeletal model, J. Orthop. Res. 33 (2014) 1128–1133.

[29] J.B. Saunders, V.T. Inman, H.D. Eberhart, The major determinants in normal and pathological gait, J. Bone Joint Surg. Am. 35-A (1953) 543–558.

[30] M.G. Pandy, Computer modeling and simulation of human movement, Ann. Rev. Biomed. Eng. 3 (2001) 245–273.

[31] A.F. Huxley, Muscle structure and theories of contraction, Prog. Biophys. Biophys. Chem. 7 (1957) 255–318.

[32] A.V. Hill, The heat of shortening and the dynamic constants of muscle, Proc. R. Soc. B 126 (1938) 136–195.

[33] F.E. Zajac, Muscle and tendon: properties, models, scaling, and application to biomechanics and motor control, Crit. Rev. Biomed. Eng. 11 (1989) 359–411.

[34] S.P. Magnusson, E.B. Simonsen, P. Aagaard, M. Kjaer, Biomechanical responses to repeated stretches in human hamstring muscle in vivo, Am. J. Sports Med. 24 (1996) 622–628.

[35] M.L. Audu, D.T. Davy, The influence of muscle model complexity in musculoskeletal motion modeling, J. Biomech. Eng. 107 (1985) 147–157.

[36] A.M. Gordon, A.F. Huxley, F.J. Julian, The variation in isometric tension with sarcomere length in vertebrate muscle fibres, J. Physiol. 184 (1966) 170–192.

[37] M. Epstein, W. Herzog, Theoretical Models of Skeletal Muscle: Biological and Mathematical Considerations, John Wiley and Sons, New York, 1998.

[38] M. Blümel, S.L. Hooper, C. Guschlbauer, W.E. White, A. Büschges, Determining all parameters necessary to build Hill-type muscle models from experiments on single muscles, Biol. Cybern. 106 (2012) 543–558.

[39] W. Herzog, History dependence of skeletal muscle force production: implications for movement control, Hum. Mov. Sci. 23 (5) (2004) 591–604.

[40] K.N. An, K. Takahashi, T.P. Harrigan, E.Y. Chao, Determination of muscle orientations and moment arms, ASME J. Biomech. Eng. 106 (3) (1984) 280–282.

[41] R.H. Jensen, D.T. Davy, An investigation of muscle lines of action about the hip: a centroid line approach vs the straight line approach, J. Biomech. 8 (1975) 103–110.

[42] M.R. Pierrynowski, Analytic representation of muscle line of action and geometry, in: P. Allard, I.A.F. Stokes, J.P. Blanchi (Eds.), Three-Dimensional Analysis of Human Movement, Human Kinetics, Champaign, 1995, pp. 214–256.

[43] E.M. Arnold, S.R. Ward, R.L. Lieber, S.L. Delp, A model of the lower limb for analysis of human movement, Ann. Biomed. Eng. 38 (2) (2010) 269–279.

[44] M.D.K. Horsman, H.F.J.M. Koopman, F.C.T. van der Helm, L.P. Prose, H.E.J. Veeger, Morphological muscle and joint parameters for musculoskeletal modelling of the lower extremity, Clin. Biomech. 22 (2) (2007) 239–247.

[45] S.R. Simon, Quantification of human motion: gait analysis benefits and limitations to its application to clinical problems, J. Biomech. 37 (2004) 1869–1880.

[46] M.P. Kadaba, H.K. Ramakrishnan, M.E. Wootten, Measurement of lower extremity kinematics during level walking, J. Orthop. Res. 8 (3) (1990) 383–392.

[47] R.B. Davis, S. Ounpuu, D. Tyburski, J.R. Gage, A gait analysis data collection and reduction technique, Hum. Mov. Sci. 10 (1991) 575–587.

[48] J. O'Connor, C.L. Vaughan, B.L. Davis (Eds.), Dynamics of Human Gait, second ed., Kiboho Publishers, Cape Town, SA, 1999.

[49] A. Cappozzo, Gait analysis methodology, Hum. Mov. Sci. 3 (1984) 27–50.

[50] H. Herr, Exoskeletons and orthoses: classification design challenges and future directions, J. Neuroeng. Rehabil. 6 (2009).

[51] H.S. Lo, S.Q. Xie, Exoskeleton robots for upper-limb rehabilitation: state of the art and future prospects, Med. Eng. Phys. 34 (3) (2012) 261–268.

[52] J. Rosen, M. Brand, M.B. Fuchs, M. Arcan, A myosignal-based powered exoskeleton system, IEEE Trans. Syst. Man Cybern. Syst. Hum. 31 (2001) 210–222.

[53] J.C. Perry, J. Rosen, S. Burns, Upper-limb powered exoskeleton design, IEEE/ASME Trans. Mechatron. 12 (4) (2007) 408–417.

[54] S. Viteckova, P. Kutilek, M. Jirina, Wearable lower limb robotics: a review, Biocybern. Biomed. Eng. 33 (2) (2013) 96–105.

[55] B.J. Fregly, M.L. Boninger, D.J. Reinkensmeyer, Personalized neuromusculoskeletal modeling to improve treatment of mobility impairments: a perspective from European research sites, J. Neuroeng. Rehabil. 9 (2012) 1–11.

[56] C.R. Winby, D.G. Lloyd, T.F. Besier, T.B. Kirk, Muscle and external load contribution to knee joint contact loads during normal gait, J. Biomech. 42 (2009) 2294–2300.

[57] C. Fleischer, G. Hommel, A human–exoskeleton interface utilizing electromyography, IEEE Trans. Robot. 24 (2008) 827–882.

[58] S.L. Delp, F. Anderson, A.S. Arnold, J.P. Loan, A. Habib, C. John, E. Guendelman, D.G. Thelen, OpenSim: open-source software to create and analyze dynamic simulations of movement, IEEE Trans. Biomed. Eng. 54 (11) (2007) 1940–1952.

[59] H. Hatze, The complete optimization of a human motion, Math. Biosci. 28 (1976) 99–135.

[60] M.A. Marra, V. Vanheule, R. Fluit, B.H. Koopman, J. Rasmussen, N. Verdonschot, A subject-specific musculoskeletal modeling framework to predict invivo mechanics of total knee arthroplasty, J. Biomech. Eng. 137 (2015).

[61] T.S. Buchanan, D.G. Lloyd, K. Manal, T.F. Besier, Neuromusculoskeletal modeling: estimation of muscle forces and joint moments and movements from measurements of neural command, J. Appl. Biomech. 20 (2004) 367–395.

[62] S.J. Piazza, Muscle-driven forward dynamic simulations for the study of normal and pathological gait, J. Neuroeng. Rehabil. 3 (5) (2006) 1–7.

[63] M. Damsgaard, J. Rasmussen, S.T. Christensen, E. Surma, M. de Zee, Analysis of musculoskeletal systems in the AnyBody modeling system, Simul. Model. Pract. Theory 14 (8) (2006) 1100–1111.

[64] A.J. van den Bogert, T. Geijtenbeek, O. Even-Zohar, F. Steenbrink, E.C. Hardin, A real-time system for biomechanical analysis of human movement and muscle function, Med. Biol. Eng. Comput. 51 (10) (2013) 1069–1077.

[65] J. Shippen, B. May, Calculation of muscle loading and joint contact forces in Irish dance, J. Dance Med. Sci. (2010).

[66] Y.Z. Arslan, A. Jinha, M. Kaya, W. Herzog, Prediction of muscle forces using static optimization for different contractile conditions, J. Mech. Med. Biol. 13 (2013) 1–13.

[67] R.D. Crowninshield, R.A. Brand, A physiologically based criterion of muscle force prediction in locomotion, J. Biomech. 14 (11) (1981) 793–801.

[68] A. Erdemir, S. McLean, W. Herzog, A.J. van den Bogert, Model-based estimation of muscle forces exerted during movements, Clin. Biomech. 22 (2) (2007) 131–154.

[69] O. Ashkani, A. Maleki, N. Jamshidi, Design, simulation and modelling of auxiliary exoskeleton to improve human gait cycle, Aust. Phys. Eng. Sci. Med. 40 (1) (2016) 137–144.

[70] R.W. Jackson, C.L. Dembia, S.L. Delp, S.H. Collins, Muscle-tendon mechanics explain unexpected effects of exoskeleton assistance on metabolic rate during walking, J. Exp. Biol. (2017).

[71] H. Kawamoto, Y. Sankai, Comfortable power assist control method for walking aid by HAL-3, in: IEEE International Conference on Systems, Man and Cybernetics, 4, 2002, p. 6.

[72] B. Weinberg, J. Nikitczuk, S. Patel, B. Patritti, C. Mavroidis, P. Bonato, P. Canavan, Control and human testing of an active knee rehabilitation orthotic device, in: IEEE International Conference on Robotics and Automation, 2007, pp. 4126–4133.

[73] V. Samadi, S. Talebian, A. Aliai, Examining the Effect of Lumber Sacral Support on Pattern Moving Control While Combined Move Trunk, Tehran Medical Science University, 2008, pp. 54–62.

[74] D.J. Farris, B.D. Robertson, G.S. Sawicki, Elastic ankle exoskeletons reduce soleus muscle force but not work in human hopping, J. Appl. Physiol. 115 (2013) 579–585.

[75] D.J. Farris, J.L. Hicks, S.L. Delp, G.S. Sawicki, Musculoskeletal modelling deconstructs the paradoxical effects of elastic ankle exoskeletons on plantar-flexor mechanics and energetics during hopping, J. Exp. Biol. 217 (2014) 4018–4028.

[76] M. Cenciarini, A. Dollar, Biomechanical considerations in the design of lower limb exoskeletons, in: IEEE International Conference on Rehabilitation Robotics, ETH Zurich, 2011, pp. 1–6.

[77] A. Zoss, H. Kazerooni, Design of an electrically actuated lower extremity exoskeleton, Adv. Robot. 20 (2006) 967–988.

12

Physical Therapy and Rehabilitation

Adam D. Goodworth*, Michelle J. Johnson†, Marko B. Popovic‡

*UNIVERSITY OF HARTFORD, HARTFORD, CT, UNITED STATES †UNIVERSITY OF PENNSYLVANIA, PHILADELPHIA, PA, UNITED STATES ‡WORCESTER POLYTECHNIC INSTITUTE, WORCESTER, MA, UNITED STATES

CHAPTER OUTLINE

12.1 Introduction

Imagine you just invented a lightweight high-powered exoskeleton device. The device has an on-off switch and variable assistance control located behind one heel. You test this device in the lab with patients recovering from a spinal cord injury (SCI) and find major increases in their ability to generate a stepping pattern. Excited, you begin marketing this device to all patients with known balance or gait disorders. Moreover, in your zeal, you find sponsors

Biomechatronics. https://doi.org/10.1016/B978-0-12-812939-5.00012-4

willing to help you export your device to clinics in remote parts of developing countries so that even more people will benefit. Years later, a follow-up study is conducted to see the impact of your product. The results are rather disappointing. Patients with musculoskeletal injuries were not able to reach the variable control button because their extreme muscle tightness limited their ability to reach the control. Youth with cerebral palsy (CP) found the device too bulky and uncomfortable with their lower-limb spasticity and it exasperated some of their balance problems. In the developing countries, many exoskeletons were broken and due to the requirement of specialized parts, the local clinic did not have monetary resources to repair it. Finally, some patients also felt like the device brought too much attention to their disability or they were more content with using a wheelchair.

The above story illustrates the need to consider the target population's abilities, needs, and resources when integrating technology into the lives of people. Ideally, biomechatronic solutions should be appropriate to the patient's abilities and align with goals of the clinician stakeholders, caregivers, and the patient. These goals may include multiple determinants of function, such as endurance, balance, gait speed, comfort, cosmetics, low cognitive load, etc. The sections below include a description of common patients needing physical rehabilitation, design strategies used to develop biomechatronic solutions, and then highlight the key biomechatronic devices for therapy and assessment for upper extremity, lower extremity, and balance. Various products are described in this chapter to illustrate design considerations, but the reader should remember these products are just a sample of those available. Biomechatronics in rehabilitation is an active area of research and product design with new developments every year.

12.2 Learning Objectives

At completion of this chapter, students will be able to

1. Understand how different target populations require different technology and approaches.
2. Understand how a human-centered design approach can improve the fit of products for people with disabilities.
3. Describe the current treatment strategies for addressing populations (injured, disabled, and elderly) with some type of need for therapy—in a wide variety of spaces.
4. Describe technical examples that span the therapy field for functional recovery.
   a. upper-limb therapy,
   b. lower-limb therapy, and
   c. balance therapy.

12.3 Target Population, Design, and Treatment Strategies

The US Center for Disease Control (CDC) indicates that 22% of adults in the United States have some form of disability [1, 2]. According to the US Disability Statistics and

Table 12.1 Definition of Disabilities Directly From Government American Community Survey [3]

Disability Type (American Community Survey)	Questions Used in the Survey
Hearing disability (asked of all ages)	Is this person deaf or does he/she have serious difficulty hearing?
Visual disability (asked of all ages)	Is this person blind or does he/she have serious difficulty seeing even when wearing glasses?
Cognitive disability (asked of persons ages 5 or older)	Because of a physical, mental, or emotional condition, does this person have serious difficulty concentrating, remembering, or making decisions?
Ambulatory disability (asked of persons ages 5 or older)	Does this person have serious difficulty walking or climbing stairs?
Self-care disability (asked of persons ages 5 or older)	Does this person have difficulty dressing or bathing?
Independent living disability (asked of persons ages 15 or older)	Because of a physical, mental, or emotional condition, does this person have difficulty doing errands alone such as visiting a doctor's office or shopping?

Demographic Report [3], in 2015, 12.6% (approximately 40 million people) of the US population (approximately 316.45 million) reported having a serious disability, which were nonexclusively categorized as hearing (3.6%), vision (2.3%), cognitive (4.8%), ambulatory (6.6%), self-care (2.5%), and independent living (4.5%) disabilities (see Table 12.1 for definitions taken from Ref. [3]). The rates of disability increase with age with the highest percentages of people with disabilities in the US population being ages 65 and over; more than one-third of this age group (35.4%) had a disability with the most common disabilities being in self-care (8.2%), ambulation (22.6%), and independent living (14.9%). As we age, we become more at risk for a disability. In 2015, 50% of people with disabilities are of working age, 41.2% are 65 and older, and 7.2% are children and youth between the ages 5 17 years. Only 0.4% are less than 5 years old. People with disabilities are more likely to earn less than those without a disability and are more likely to be living in poverty (20% chance). Disabilities such as those described in Table 12.1 are often a result of communicable and noncommunicable diseases and are a direct result of one or more impairment in some body part or organ. By the World Health Organization (WHO) International Classification of Function (ICF), an impairment of some body part or organ will lead to disabilities that would ultimately affect a person's ability to easily participate in their community [4]. For example, brain injuries such as those due to a stroke and blunt force trauma and peripheral motor injuries such as those due to a spinal nerve injury are common reasons for motor impairment in the upper and lower limbs. Impairments can then affect the ability of the person to exert the voluntary control over the muscles needed to do common activities such as walk or drink from a glass. The following describes some of these target populations and their common motor impairments requiring physical rehabilitation.

12.3.1 Target Populations

Elders and frail elders. Age is a main factor in the number of persons with disability and it is worth considering the world's growing population and increasing life expectancy [5, 6]. Some of the world's oldest nations in terms of median age of their population are Japan and Germany sharing first place with a median age of 46.1 years, then Italy with 44.5, Switzerland with 44.3, Greece and Austria both with 43.5, Finland with 43.2, Belgium with 43.1, etc. In many of these countries, the number of elderly people requiring continuous close assistance for activities of daily living (ADL) may soon outmatch the number of potential caregivers. The United States is a relatively young nation with a median age of 37.8 years in 2015. It is projected that by 2030 the US population over 65 will double to about 71 million older adults, or one in every five Americans resulting in higher demands for health care [7].

Stroke. A stroke (cerebrovascular accident) occurs when blood flow to the brain is significantly reduced due to ischemia or hemorrhage [8, 9]. With less blood flow, oxygen is limited and brain cells die, which typically damages one's ability to control body movements. In the United States, a stroke occurs about every 40 s with 1 in 20 persons in the United States dying because of it [10]. About 795,000 Americans suffer a stroke each year. Approximately 610,000 of these incidents are first-time strokes, and 185,000 are recurrent attacks. Stroke is the leading cause of serious long-term disability in the United States. In 2013, stroke prevalence was 25.7 million worldwide with about 5.8–6.5 million the US adults living with the effects of a stroke [11]. Strokes range from mild temporary weakness in a limited set of muscles to permanent paralysis on one side of the body. Typically, a stroke is categorized as mild, moderate, or severe using the Fugl-Meyer upper and lower extremity motor scale which rates per limb, a person's control over the one or more joints [12]. Rehabilitation typically focuses on relearning independent living, such as reaching, grasping, bathing, eating, and walking. Because a stroke can impact a very specific set of limb movements, to maximize improvement, clinicians employ mass repetition of focused movements and the application of adaptive forces to support the limb movement as needed. Clinicians often motivate patients to use their injured body segment, which may otherwise be neglected. Therefore, when designing devices, engineers should consider the importance of targeted massed repetition, motivation, and the tendency to neglect use of injured limbs.

Traumatic brain injury (TBI). TBI occurs when a sudden jolt or external force to the head causes brain damage [13]. TBI is typically suspected if the jolt or external forces causes alteration or loss of consciousness at the time of injury. Common causes of this sudden external force to the head could be due to a fall (40.3%), an assault (10.5%), a motor vehicle accident (14.3%), and being struck by or against something such as in a sports-related concussion or a blast explosion due to a bomb (15.5%) [14]. Damage to the brain commonly affects motor, cognition (e.g., memory and problem solving), and sensory processing. Typically, TBI is categorized as mild, moderate, or severe using the Glasgow Coma Scale, which rates a person's visual, verbal, and motor responses immediately after injury [15]. For 2010, 2.5 million ER visits, hospitalizations, and deaths were associated with TBI in the United States. Approximately 1.7 million Americans experience a TBI each year, of which 1.365 million survive with mild injury and 250,000 with severe one [2]. It is estimated that 3.5–5.3 million people live with the consequences of TBI. In the United States, a person experiences a TBI every 19 s. About half of TBI patients require surgery to remove or repair damage in the brain. Similar to stroke, therapy with TBI patients is aimed at restoring function. The public has been made more aware of TBI from hearing about soldiers in the wars of Iraq and Afghanistan having concussions due to blast injuries and football players having concussions from repeated hits to the head. Engineering efforts have focused on cognitive and motor therapies aimed at improving memory, motor skills, and balance.

Cerebral palsy. CP is the leading motor disability in children with an estimated 1 out of 323 children in the United States being diagnosed with CP [12, 16–18]. CP is associated with impairments in motor, cognitive, and sensory systems. CP is a motor disorder caused by brain lesions that occur prenatally or, in some cases, before the age of 2 years. According to the United Cerebral Palsy Association, it is estimated that more than 700,000 Americans have CP [18]. Characteristic symptoms of CP include spasticity, muscle weakness, rigidity, and loss of selective motor control. The most commonly used classification system is based on gross motor ability. Patients with high gross motor ability can walk independently and may have excessive tightness in a one or more muscle (spasticity) resulting in higher energy expenditure, slower gait, and poor balance. In contrast, patients with low gross motor ability spend most of their day in a wheelchair and remain dependent on caregivers for life. Engineering technologies for patients with CP depend widely on the severity of CP and include facilitation and training of upper extremity, gait, and balance. The goal of management of CP is not to cure or to achieve normalcy but to increase functionality, improve capabilities, and sustain health in terms of locomotion, cognitive development, social interaction, and independence.

Spinal Cord Injury. There are 250,000–400,000 people in the United States with SCI or dysfunction. About 12,000 people in the United States suffer a traumatic SCI each year [2]. The main cause of SCI is due to motor vehicle accidents (39.2%), falls (28.3%), or violence (14.6%) [19]. In the elderly, the main cause of SCI is due to falls. In contrast to stroke and TBI, a SCI primarily affects motor function associated with nerves below the level of injury. An SCI injury is typically classified according to the motor, sensory, and neurologic levels of injury along the spinal cord, the ASIA impairment scale designations including an assessment of complete or incomplete spinal injury [20]. If the spinal cord is injured above the thoracic vertebrae (T1), then the person is considered tetraplegic or quadriplegic with impairments across legs, trunk, and arms. Those with injury to the cervical level (C1–C4: neck area), will be most severe and may be fully dependent on caregivers. Persons with a SCI to lower levels of the spinal cord (below T1 level including the remaining thoracic and lumbar levels of the spinal cord) will be considered paraplegic. Thus, patients with injury to lower levels of the spinal cord may have loss of sensation in lower legs but retain control of many actions, including some level of walking. Given the complexity of impairments, biomechatronic approaches will depend on the level of injury and range from improving or enabling gait, enabling standing balance, increasing mobility by strengthening upper and lower-limb muscles, and increasing independence through the development of specialized robotic technologies such as feeders.

Parkinson's Disease (PD). PD is a neurodegenerative disorder. Neurodegenerative disorders are diseases that get progressively worse over time. In this case, Parkinson's is a disease in which brain cells progressively die due to the brain's inability to generate dopamine in the basal ganglia circuit, a major part of the brain that controls movement. Estimates suggest that about 50,000 people are diagnosed with PD each year and about 500,000 are living with the disease [21]. PD occurs in about 1% of population by age 60 and 4% of the population by age 80 [2]. It is the second most common neurodegenerative disease (Alzheimer's is the most common). PD is associated with some unique characteristics. Symptoms often include tremor, rigidity, stiffness, impaired balance, and slow movements. Another unique clinical phenomenon is called freezing of gait, which is characterized by brief episodes of inability to step or extremely short steps that typically occur on initiating gait or turning while walking [22]. Medications, deep brain stimulation, and rehabilitation techniques can alleviate some of the symptoms of PD, but other symptoms such as freezing of gait may persist in the most severely impaired persons with PD [22, 23]. Therefore, clinicians and researchers have focused on motivating patients to make large, high-force movements, and auditory cues to effectively reduced freezing patients with PD. Engineering technologies are focused on supporting improved gait and improved motor coordination and control.

In conclusion, some engineering devices and therapies are applicable across more than one population, while other engineering devices and therapies are highly specific to a particular condition.

12.3.2 Service Locations and Stakeholders

Physical therapy and rehabilitation takes place in several locations. Inpatient rehabilitation refers to the treatment and services provided in a hospital or in a free-standing acute care rehabilitation setting. Outpatient rehabilitation typically refers to treatment received after discharge from an inpatient facility and may take place inside or outside the hospital environment, such as within clinics or adult rehabilitation day centers. If patients are unable to be discharged home or to a community-based home or assisted living facility, they may be discharged to a nursing home or a skilled nursing facility where specialized medical and rehabilitation services that are typically delivered in a hospital can be provided. Typically, more than 40% of stroke survivors will spend considerable time in a skilled nursing facilities and nursing homes. In most of these environments, the number of therapists and nurses is insufficient for the number of residents, affecting the quality of rehabilitation and daily care. This insufficiency is an opportunity for engineers to contribute to physical therapy and rehabilitation.

In general, rehabilitation is delivered by a team of rehabilitation clinicians, which includes physiatrists (rehabilitation medical doctors or PM&R physicians), therapists (speech, occupational, and physical), nurses, and caregivers (Fig. 12.1). Rehabilitation

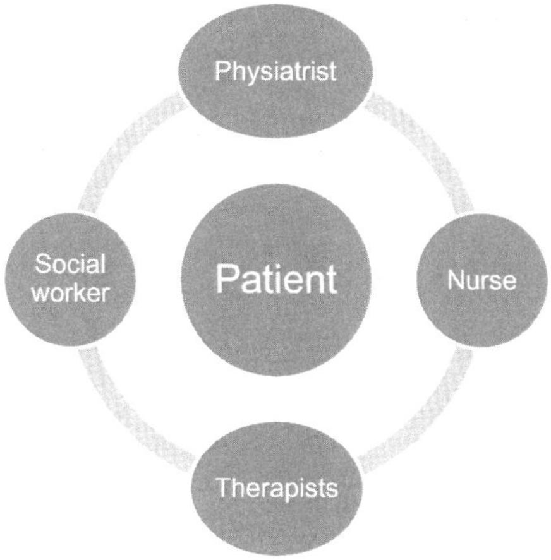

FIG. 12.1 The rehabilitation team.

medical doctors often prescribe therapy and treatments, but therapists and their assistants deliver the treatment. Therapy goals often include reducing pain, reducing impairment, improving functional activities, or normalizing structural properties of the musculoskeletal system. These goals can overlap. Reducing pain may improve sensory processing and can free patients to move limbs farther and faster, potentially improving functional activities. Normalizing structural properties (such as minimizing contractures) can have a similar positive effect on functional activities. Then, as patients become more active, musculoskeletal structures can begin normalizing because patients are using more typical positioning and loading patterns.

In the developed countries like the United States, physical therapists usually specialize in training the lower-limb mobility, joint-based movement, posture, and balance, while occupational therapist specializes in training the upper limb for function, ADLs related to reaching, grasping, and manipulation. In the less-developed countries, these distinctions are not made, and therapists treat both upper and lower limbs. In 2008, there were nearly 200,000 physical therapy jobs and in 2016, there were over 130,000 occupational therapy jobs in the United States [24, 25]. Occupational and physical therapy are two of the fastest growing graduate fields (predicted 21% and 25%, respectively, growth through 2026) and ranked high among health care jobs with an unemployment rate less than 0.3% and 1%, respectively, in 2018 [24, 25]. An entry-level physical therapist must now have a three-year graduate degree called a Doctorate of Physical Therapy while occupational therapists must now have a Master's degree to practice. Unfortunately, it is predicted that there will be a shortage of these rehabilitation clinicians, leading to a common motivation for the use of technology to assist therapists to provide care and support for disabled populations.

It is common for therapists to have a large room for rehabilitation with patients that include assistive technologies to help to reduce impairment, and train balance, mobility, and functional tasks. An assistive technology (AT) is an umbrella term that includes

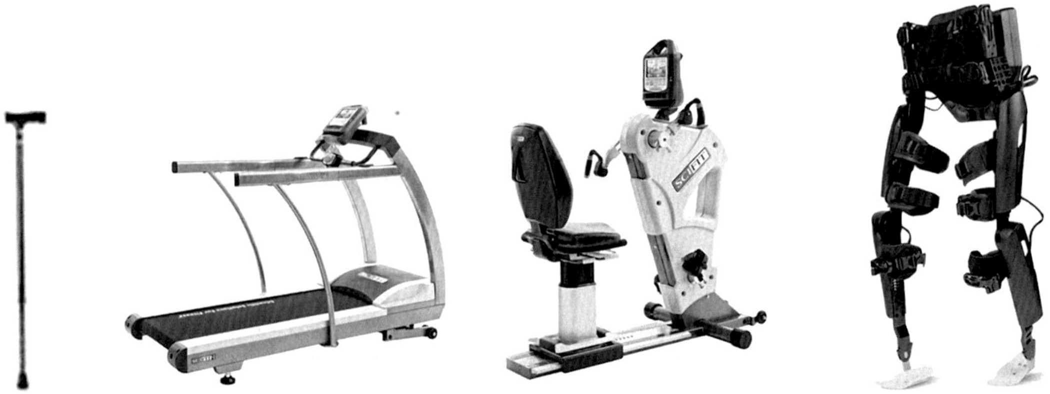

FIG. 12.2 Sample of assistive devices and rehabilitative devices (devices seen on AbleData website [27]). *Used with permission from Scifit and Rewalk Robotics, Ltd.*

assistive, adaptive, and rehabilitative devices for people with disabilities. There is an important process used in selecting, locating, and using AT [26]. AT should promote greater independence, enabling persons with disability to perform tasks they were formerly unable to accomplish (or had great difficulty accomplishing). Often individual modifications to AT are needed to achieve this goal.

Typically, the term assistive device refers to a device used in daily life to augment function, while a rehabilitation device refers to tools used to treat persons, to reduce impairment, and to increase function. A common classification of devices would be low-tech (e.g., a cane), medium-tech (e.g., treadmill), high-tech (e.g., powered wheelchair—see Chapter 13), powered cycle or robotic [27, 28] (Fig. 12.2). Depending on the design, a biomechatronic device can be considered an assistive device, rehabilitation device, or both. Biomechatronic devices are typically considered medium-to-high tech or robotic. A device is robotic if it is reprogrammable and includes mechanical-electrical components such as motors, electrical components such as sensors to measure information about the device's state (e.g., position, orientation, etc.) and its environment (e.g., obstacle or light), and a microprocessor or computer to control the actions and reactions of the device.

12.3.3 Human-Centered Design Considerations for Biomechatronic Devices for Functional Restoration

How should engineers design assistive technologies (whether an assistive device or a rehabilitative device) that can optimally support the patient? There are several accepted design methodologies such as universal design and human-centered design that place the patient/human user at the center of the product design process. The analysis of user needs is a critical first phase and includes a consideration of: (1) the real clinical problem; (2) all the human users; and (3) the user activity context and goals. In the assessment of a disabled user needs, a systematic approach is recommended such the Human Activity-Assistive Technology (HAAT) model, [28]. This model asks the designer/engineer to critically assess the following: (1) the human stakeholders, (2) the activity or activities the human desires the device to do, (3) the context/environment in which the human desires to do these activities, and (4) the AT device requirements implied after assessing humans, activity, and context. If working with a patient, if possible, the engineer interviews the user, therapists, and family who will support patient user. The interview process should uncover status of the users' disability including cognition, motor, aural, visual, etc., request any information about levels of function (low, moderate, or high function) in each category, current complaints or problems with existing assistive device or rehabilitation device, patient's goals and desired activities, therapist treatment goals for the patients, as well as the family's ability to support the treatment process. In addition, it is important to understand where the device will be used and under what conditions (inside or outside, in a clinic or at home, in a high-resource setting or in a low-resource one). The family's input would be especially important if the device will be used for treatment in the home and away from the therapist. Since people with disabilities and the elderly often require assistance with self-care and ambulatory and independent living tasks such as feeding, personal hygiene tasks such as bathing, brushing teeth, using restroom, combing hair, shaving, etc., these tasks are often supported at home by caregivers. Many of these assistive tasks can be uncomfortable both to the people with reduced motor abilities and to their caregivers. It may take a lot of effort and skill, patience, and compassion on the side of caregiver to fully understand the person's preferences and needs and to devise adequate strategies to successfully perform these tasks. In addition to accomplishing these tasks, it is also critical not to injure one's sense of dignity and privacy or to cause painful, unpleasant, or uncomforting sensations. Depending on the level of need and type of assistance, having a personal caregiver may be quite expensive and hence it may cause a huge financial burden on care-receivers and their families. Given this, the introduction of an AT into the dynamic is not trivial and its role and support to the person and caregiver should be carefully considered.

Consider the following sample case:

Zinabu is a 13-year old female who was diagnosed with Hemiplegic Cerebral Palsy at age one. She lives in New York city in a two-bedroom apartment with her parents and an older brother. She goes to a school for children with disabilities and is in Grade 6. She has moderate functioning in terms of motor and cognition and is described as easily distractible. Zinabu is right hand dominant. She uses a manual wheelchair that must be pushed by a caregiver. She is weak in both limbs. Her caregiver desires her to strengthen her limbs to the point that she would be able to help. She would like to be able to push herself. Based on the box and block tests, she can use her right limb. Specifically, she is able to pick up 30 blocks with the left arm in 1 min and only 3 within 1 min with her right arm. A healthy 13-year old would pick up greater than 65 blocks within 1 min with either hand.

To start, describe what is known about the patient. In this case, hemiplegic CP implies that Zinabu has one impaired limb and her language and cognitive skills are moderate.

Second, describe her functional need. In this case, she desires to strengthen her weak limbs to push a wheelchair. Wheelchair ergonomics must be considered. Should the elbow flexors and extensors be strengthened and/or the shoulder complex and what rehabilitation device could accomplish this in an engaging way? But if further examination reveals the main goal is increased mobility with decreased burden on the caregiver, then is a powered wheelchair viable option or a wearable powered upper extremity exoskeleton? Third, describe her context. Using a wheelchair in a city could be very challenging with curbs, crowds, and transportation. Finally, describe her needs in light of her condition, goals, and context. Once needs are defined, additional questions include: (1) Does the technology already exist or does it have to be created? (2) Is the device load/weight bearing (implications for pressure ulcers and comfort)? (3) What cognitive ability is needed to operate the device? (4) How can it be made to be safe? and (5) How can it measure her performance and provide information to her therapists about outcomes? Once user needs are adequately considered, then information about the device requirements should become clearer, with a higher probability of meeting the user's goals.

12.3.4 Current Trends in Practices and Treatment Strategies for Functional Restoration

As noted above, mechanical and robotic devices developed to improve function can (1) help to enable the patient to complete tasks outside the clinic (using an AT) or (2) help the patient's neural system to change by practicing skills that improve motor control and strength (rehabilitation device). Important considerations for designing ATs were described in the previous section. This section describes the neural control system and how concepts of motor learning can be incorporated into rehabilitation devices.

Most rehabilitation devices seek to improve muscle strength and motor learning. Motor learning includes the neural plasticity associated with changes in neural connections, perception, sensory integration, and muscle activations following a treatment [29– 31]. Some treatments can evoke motor learning within minutes. For example, procedural learning can be quickly understood and maintained for a very long time. Learning the correct sequencing of how to use a cane can be understood and practiced within a few minutes with the skill acquisition maintained for a long time. But other types of motor learning require weeks, months, or years of practice. These types of motor learning typically involve less cognitive control and more central neural changes such as adjustments to the scaling of muscle outputs, development of muscle synergies, or interpretation of multiple sensory cues. Research suggests engaging activities that require large amplitudes of motor outputs and inclusion of sensory feedback is best to evoke motor learning. However, learning is heavily influenced by a patient's type of neural or biomechanical injury, age, and previous experience. Engineers must design rehabilitation devices with these aspects of motor learning and their patient population in mind.

Motor learning cannot be directly measured. Therapists infer motor learning has occurred by looking at their patient's (1) improvement in a particular skill, (2) consistency in the improvement (i.e., across repeated repetitions in the same therapy session), (3) persistence in the improvement (i.e., across a time scale of weeks, months, or years), and (4) adaptability [30]. Adaptability refers to the ability of a patient to improve performance of the skill in a variety of contexts and to improve variations of the skill. For example, improvements in reaching and grasping should be maintained in both a kitchen setting where one stands up reaching for a cup full of water and in a restaurant setting where one is sitting down reaching for a fork. Again, engineers must consider aspects of measuring motor learning in their design to track improvements, steer future interventions, and justify treatments. It is also worth noting that many of the machines and devices created by engineers to improve function can also be used to quantify learning and provide a baseline assessment measure. For this reason, some of the technologies described in this chapter are similar between treatment and assessment.

As an overview, biomechatronic devices typically elicit motor learning by (1) increasing engagement and brain activity during the practiced skill, (2) providing biofeedback, (3) providing assisted movements, or (4) requiring very specific targeted movements for the patients. In reality, most devices do a combination of these four categories. In category one, some of the common approaches include brain stimulation (e.g., transcranial direct current stimulation) and the use of video games or virtual reality. In category two, body segment kinematics, kinetics, or muscle activation levels can be fed back to subjects on a visual screen or in the form of vibration. This feedback can be integrated into a game to enhance engagement and brain activity. Chapter 16 provides a more detailed description of the benefits of video games and virtual reality in motor learning. In category three, robotic devices can deliver torque around joints to increase amplitude or speed of movements. These facilitated movements increase sensory feedback and can be integrated within a biofeedback system. Finally, in category four, mechanical devices are designed such that "success" in a given activity is only accomplished through a specific kinetic, kinematic, or electromyography pattern. For example, a patient may use an upper extremity motorized exoskeleton device to practice reaching. The device may present resistance only when a patient's reaching kinematics is outside of a predetermined path, thus requiring a very specific movement pattern. In addition, a device like this could integrate additional design categories by simultaneously providing biofeedback or brain stimulation, integrating reaching kinematics into a game, or facilitating movement through a powered joint wrist, elbow, or shoulder joint.

12.4 Upper-Limb Therapy

Many ADLs require the upper limbs. The ability to control the position of the upper extremity while reaching toward an object, the ability to control the orientation of the hand to grasp the object, and finally the ability to control the fingers to grasp the object securely are key skills needed to eat or drink. The ability to swing the upper limb to support balance during walking or in response to a balance perturbation is also important. Interlimb and intralimb muscle coordination, strength, and sensory feedback are key to performing these tasks.

12.4.1 Treatment Strategies

Everyone who ever trained for a sport or worked as coach knows that the process of improving physical performance is rather challenging and that there is no unique or optimal way that works equally well for everyone. It really depends on many factors. Highly skilled, very experienced coaches typically first get to know their trainees well and then carefully gauge when is the right moment to put more attention on certain type of exercise and what is the right level of that activity. Even with best effort on both sides, this endeavor may or may not be fruitful.

Similarly, in the field of physical therapy and rehabilitation, there is no unique or optimal way that works equally well for all patients at any moment in time. Rather, there is variety of treatments that can be more or less beneficial depending on patient physical condition, cognitive ability, and motivation. The treatment for a low-motor functioning, elder stroke survivor whose goal is retirement may look different from that of a moderate-motor functioning young stroke survivor whose goal is going back to work. For example, the level of motor functioning may dictate the type of force control algorithm (e.g., assistive or resistive) to employ during therapy. On the other hand, the age and living goals, may dictate the intensity of therapy and the stopping points to the therapy.

Consider robotic manipulator that can apply end effector forces to a patient's hand holding a handle as shown in Fig. 12.3. Here, it is assumed that desired location of end effector is known (e.g., it may be represented by cursor on screen). Depending on the type of therapy, the end effector may apply attractive force assisting patient to reach desired location or equilibrium point, see Fig. 12.3 on the left, or it may apply repulsive force resisting patient trying to reach desired location or equilibrium point, see Fig. 12.3 on the right.

One can design proportional laws defining end effector forcing by introducing concept of effective potential centered at desired location, that is, the equilibrium point, and then express force as Eq. (12.1), where the force is the gradient of a potential energy function.

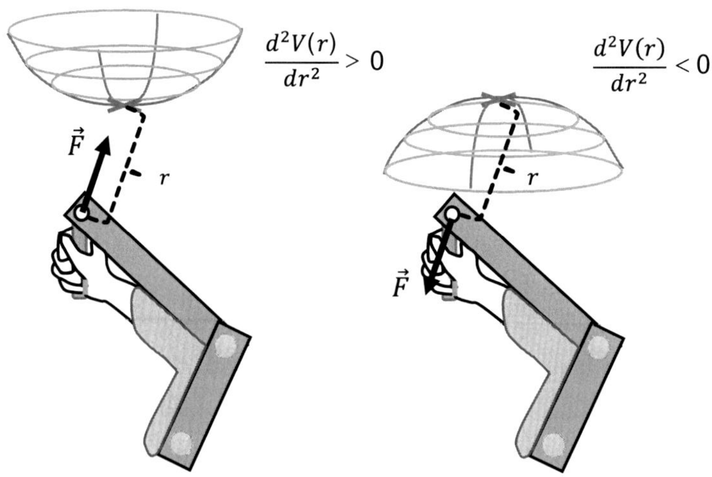

FIG. 12.3 Two robot-aided therapy strategies: assistive (left) and resistive (right). *Courtesy of Dr. Marko Popovic.*

$$\vec{F} = -\vec{\nabla} V = -\left(\hat{i}\frac{\partial}{\partial x} + \hat{j}\frac{\partial}{\partial y} + \hat{k}\frac{\partial}{\partial z}\right) V = -\hat{r}\frac{\partial V}{\partial r} \tag{12.1}$$

If the effective potential is quadratic function of distance between actual and desired location

$$V = \pm\frac{kr^2}{2}, \tag{12.2}$$

then the force law is

$$\vec{F} = \mp k\vec{r} \tag{12.3}$$

with k being interpreted as stiffness.

In the context of stroke, which is by far the main cause of movement impairment, it was shown that depending on the type of activity and functional level both assistive force field [32] and resistive or disturbing force field [33] strategies can be beneficial. The assistive strategies have been shown to help stroke survivors regain/relearn their basic motor skills just following stroke and utilize their brain plasticity to match the proprioceptive and visual pathways of the arm; or help the patient use standard tools to accomplish everyday tasks, for example, use kitchen utensils or use computer mouse/keyboard, etc. The resistive strategies may enhance patient performance by requiring application of an extra level of control and physical force to accomplish a specific task in the environment characterized with resistive and/or disturbance fields. By analogy to athletic training, this would be like doing an exercise with weights attached to the ankle while running.

12.4.2 Treatment Technologies

One recent review identified over 120 upper-limb rehabilitation robotic devices that have been discussed in literature or are commercial products [34]. According to this review, these devices can be classified according to several categorizations, including but not limited to the following: (1) the type of treatment strategy whether active, passive, haptic, or coaching; (2) the involvement of the user whether active exercise or passive exercise; (3) the part of the arm supported whether at for the hand, elbow, wrist, and whole arm; (4) the type of activities trained whether training a specific joint motion, force generation, whole arm reaching, gross or fine grasping, ADLs such as eating and drinking or manipulation; (5) mechanical design such as whether they were designed as an end-effector-based design, exoskeletal-based design, planar, backdrivable, modular, or re-configurable; and (6) the type of input control signals used (dynamic, kinematic, or triggered) or feedback to the user (among them: visual, tactile, audio, and in the form of electrical stimulation). For illustration, several technologies are described below: the InMotionARM (Interactive Motion Technologies) [33, 35, 36], the Activities of Daily Living Exercise Robot (ADLER) [37], mirror-image motion enabler (MIME) [38], ArmIn [39], Armeo (Hocoma) [40, 41], Myomo [42, 43], and Haptic Theradrive [44, 45]. In general, research supports robot-assisted therapy and its ability to be at least as good as conventional therapy [46].

The "InMotionARM" are commercial interactive therapy systems that are designed for use in a clinical setting to help patients practice and improve motor control with the goal of improving function outside the clinic. The patient grasps a joystick and moves his/her arm, wrist, and hand to play a game on a screen in front of them. The games are designed to improve strength, coordination, and speed. While moving, resistance or facilitation can be provided. In a given practice session, patients may use hundreds of repetitive movements focused on their individual impairment and scores are tracked. These products have been suggested for use in stroke, CP, SCI, TBI, PD, and other movement disorders.

The InMotionARM is one of the early examples of a planar two-degree of freedom (DOF) robot manipulator (Fig. 12.4), widely used in clinical trials to assist shoulders, elbow, and wrist movements. Classified as an end-effector system, it permits training through planar point-to-point movements and can be re-configured to apply therapy to the whole arm or just the wrist. The system was originally developed in MIT's Newman Laboratory for Biomechanics and Human Rehabilitation by Dr. Hermano Igo Krebs and Professor Neville Hogan. The goal of the project was to develop, implement, and test a robotic system for physical therapy and neurological rehabilitation.

The MIT-Manus device is a computer-driven robot "arm" that emulates the movements a physical therapist might make while working with a patient to redevelop arm and hand movement and coordination. The robot arm holds the patient's own hand and arm, and either manipulates them according to a preset adaptive assist-as-needed program or records the patient's movements for later analysis.

In the *New England Journal of Medicine* study on post-stroke rehabilitation, researchers at VA hospitals in Baltimore, Seattle, West Haven, CT, and Gainesville, FL, compared the MIT-Manus system to a high-intensity rehab program delivered by a human therapist, which was designed specifically for this study (it essentially repeated the same number of movements but guided by nonrobotic, that is, human therapist) [35]. Each group included about 50 patients, who were also compared with a group of 28 stroke patients who received so-called "usual care"—general health care and 3 h per week of traditional physical therapy for their stroke-damaged limb. At the end of 36 week, both the

FIG. 12.4 InMotion ARM formally known as MIT-MANUS (used with permission from Dr. Herman Igo Krebs) and ADLER (used with permission from Dr. Michelle Johnson).

robot therapy group and intensive human-assisted therapy group showed improvement in arm movement and strength, everyday function and quality of life compared with the usual-care group. Overall health care per-patient costs were not very different over the total 36-week study period—$15,600 per patient for robot-assisted therapy and for intensive nonrobotic therapy, and $14,300 for usual care.

In contrast, ADLER [37] allows training in three-dimensional (3D) space and is focused on training ADLs using real objects (Fig. 12.4). Both systems would be considered an end-effector-based device with the ability to apply both active and haptic assist, but within diverse training strategies. It is typical to have a biomechatronic therapy device capable of providing multiple ways of assisting a patient so that adjustments can be made throughout the patient's recovery process. For example, a stroke survivor with low function may need an active device with motors that enable the device to move the patient's limb autonomously and participate in passive exercise. In contrast, a stroke survivor with higher function, who is typically able to engage in active exercise and move their limbs under their own muscle power, may need a device that can adaptively assist with varying levels of forces.

Another comparison point between the InMotion and ADLER is the type of control. The ADLER system is built using the HapticMaster robot platform (MOOG, Inc.), which uses admittance control (i.e., force in and position out) while the InMotion2 is under impedance control (i.e., position in and force out). Finally, the ADLER uses physical objects to provide visual cues while the InMotion2 relies on tracking games to provide visual and aural cues to guide the user. Similar systems that train both arms are the MIME robot [38], and the Bi-Manu-Track (Reha-Stim) [47]. These devices allow for an additional control mode where the intact/less-impaired arm is able to assist in the rehabilitation. Finally, some devices provide users feedback via a haptic environment requiring more sensory integration and variety in feedback.

As seen in some of the products described above, motivation and ability to self-guide and monitor self-progress on continuous basis are important factors for successful therapy. Hence, a number of researchers trust that either game environment or some type of success score might be very beneficial for patients undergoing robot-aided therapy (see Chapter 16 for more detail on video games in therapy).

Treatments can also be applied through wearable exoskeleton devices such as the ArmIn and the commercial powered version of it called the Armeo Power (Fig. 12.5). These systems apply forces directly to a given joint. The "ARMIn" and the Armeo Power are semi-exoskeleton robots that are fixed to the wall while the patient is seated beneath the frame. Both are four DOF exoskeletons that move the lower arm and upper arm [39–41].

The "Myomo" is an upper extremity exoskeleton that can be classified as both an assistive device and a rehabilitation device. It uses surface electromyography (sEMG) to control small motors to move the finger, wrist, and elbow joints (Fig. 12.6) [42, 43]. The sEMG is measured with surface electrodes on the muscles. sEMG signals generated naturally by the users when his/her brain sends a signal to move the upper extremity or hand. Thus, the device is especially designed for patients with intact neural signals to the upper extremity but lack adequate control or strength, commonly stroke, SCI, and multiple sclerosis.

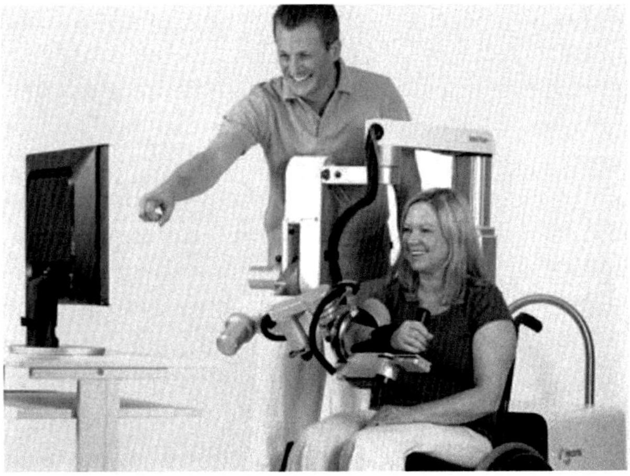

FIG. 12.5 Image of Armeo Power. *Picture: Hocoma, Switzerland.*

FIG. 12.6 Image of Myomo wearable upper extremity. *Used with permission from Myomo.*

One potential advantage of robot-aided therapy is more intense training if devices can accompany patients at their homes. If patients can exercise 3 h or more per day compared to standard level of care, this might have a tremendous effect on overall progress. The physical therapist can still be involved observing exercises and scores online and proposing new set of exercises and levels of difficulty. One barrier to widespread use of robot-aided rehabilitation and assistive devices at home and in low-resource settings is cost. Unfortunately, most robot-assisted therapy devices are expensive with cost greater than 50,000 USD for a single unit and are unable to be used in low-resource settings such as in the home or in the developing countries. There have been some attempts to address this need. More recent affordable systems are being proposed for the home and

FIG. 12.7 Interactive robots. Socially assisted robots for elderly built from Vgo and Nao [51, 52].

rehabilitation facilities use such as Reha-Stim line of devices [47], TyroMotion's Pablo [48], Hand Mentor Pro [49], and Haptic Knob [50] among others. Affordable technology solutions for rehabilitation may provide the means to bridge the gap between the scarcity of rehabilitation providers and their growing disabled population in low-resource settings. The Haptic Theradrive is another good example of a one DOF haptic robot being developed for upper-limb therapy where users engage in passive or active exercised while they play games [44, 45]. The system is geared toward a price of less than $5000 USD, roughly ten times less than many current systems. This robot can be used alone or networked for group play.

Finally, another recent direction in upper extremity robots is toward creating coaching or socially assistive robots. These robots are designed to provide exercise coaching in person or remotely and are usually humanoid (Fig. 12.7). Humanoid and mobile robot research efforts with stroke patients and elderly patients have typically focused on using the humanoid mobile robot as an exercise coach or helper for performing daily activities [51–54]. These systems have been shown effective in motivating stroke survivors to pursue exercise and activities in under-supervised environments with limited caregiver oversight. Also, there have been many efforts to explore robots as tools in therapy for children living with Autism spectrum disorder [55] as these children may find humanoid robots appealing, more than typically developing children.

12.4.3 Assessment Technologies

Clinical and standard metrics are used to judge usefulness of certain robotic-aided rehabilitation method [31–46]. These metrics are typically applied to patient performance before and after robotic-aided therapy. Some metrics include patient-reported outcomes, visual inspect of movement quality (e.g., smoothness), speed of reaching, or accuracy of reaching. While these assessments offer a beneficial global view of reaching and

Table 12.2 Summary of Metrics Used in Upper Extremity Function

Metrics		Definition	Movement
Velocity	Mean velocity	Average velocity across whole movement	Reach, grasp
	Peak velocity	Max velocity across whole movement	
Accuracy	Direction error	Difference between the initial movement direction and the targeted direction	Reach
	Root mean square error	Square root of mean squared distance from acquired position to target position	Reach
	Percent time in target Dwelling percent time in target	Percentage of the time subject stayed within the target window	Tracking tasks
Efficiency	Path length ratio	The line integral of the trajectory/the time taken to reach the target, calculated by the distance between two consecutive points of the patient's path and normalized to the straight line distance between the starting point of the task and the target	Reach
Smoothness	Segmentation	Number of speed peaks that appears in the entire movement	Reach, grasp
	Speed smoothness	Mean velocity/the peak velocity	Reach, grasp
Maximal movement	Max joint angle	Maximal joint angle moved	Reach
	Max grasp aperture	Max distance between the tip of index finger and tip of thumb	Grasp
Time	Reaction time	Time between beginning of task to the first significant movement of the subject	Reach, grasp
	Movement time	Time between the first significant movement and the last significant movement	Reach, grasp
	Time to max movement	Time between beginning of the movement to the max joint angle/grasp aperture	Reach, grasp
	Time to peak velocity	Time between beginning of the movement to the max velocity	Reach, grasp
	Relationship between time to reach peak velocity and grasp peak velocity	Percent of grasp movement when reach movement reaches its max angle/velocity Percent of reach movement when grasp movement reaches its max angle/velocity	Reach to grasp

coordination, additional information can be obtained with engineering-based analyses. Many of the devices have the capability to deliver assessments and treatment. Table 12.2 offers a list of various kinematic metrics used in upper-limb assessment that have been derived from kinematic and kinetic measurements from Biomechatronic devices. Most of these metrics have been successfully correlated with clinical measure of upper-limb motor function and control such as the Fugl-Meyer [12], the Box and Block Test [56], and the Jebsen Taylor Test of Hand Function [57].

To calculate some of the above metrics, it is oftentimes helpful to quantify joint kinematics using motion capture systems (Chapters 2, 4, and 16). Fig. 12.8 shows kinematic plots and two ways of assessing a drinking task [58, 59]. From these plots, information

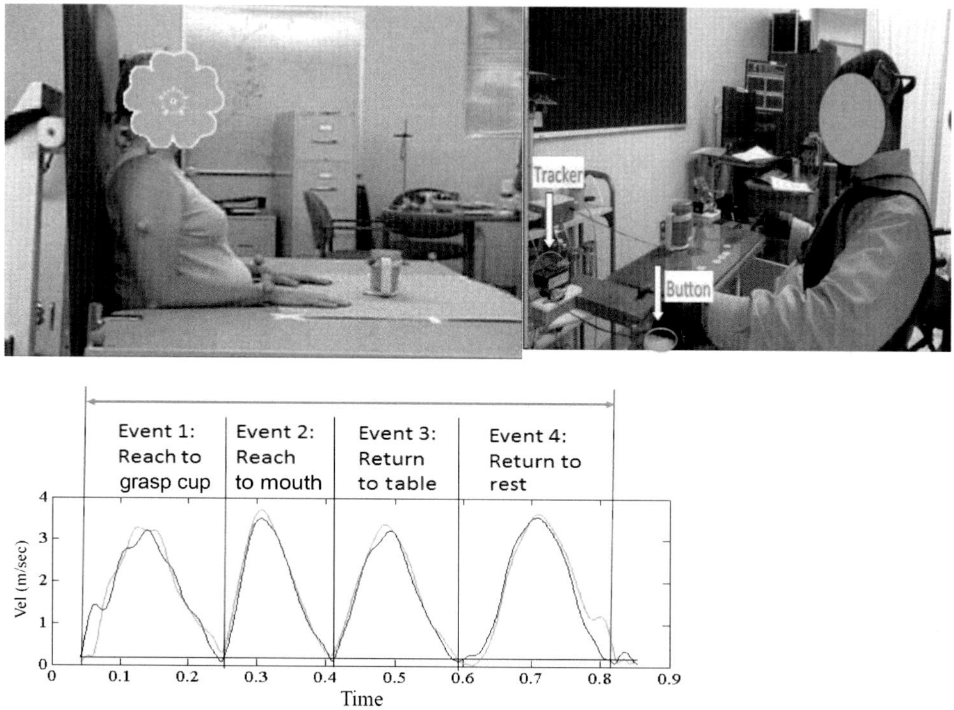

FIG. 12.8 Images of assessing reaching with kinematic collection. *Courtesy of K.J. Wisneski, M.J. Johnson, Quantifying kinematics of purposeful movements to real, imagined, or absent functional objects: implications for modelling trajectories for robot-assisted ADL tasks, J. NeuroEng. Rehabil. 4(1) (2007) 7; M.J. Johnson, S. Wang, P. Bai, E. Strachota, G. Tchekanov, J. Melbye, J. McGuire, Bilateral assessment of functional tasks for robot-assisted therapy applications, Med. Biol. Eng. Comput. 49(10) (2011) 1157.*

such as mean velocity and smoothness can be obtained. It is important to design standard assessment tasks used to measure the users progress as treatment goes on. In many cases, the robots themselves are equipped with the sensors and therefore are able to provide information on kinematics and kinetics during treatment or assessment sessions.

Below is a more detailed description of how assessment can occur for any general goal-directed upper extremity movement.

Imagine a simple experiment or "game" involving two students standing on opposite sides of a glass door, facing each other and moving their pointer fingers along the transparent surface (Fig. 12.9). One student, preferably blindfolded, may move the "target" finger randomly at will and the other student, preferably incentivized to perform well, may try to align his/her finger with the one of his/her colleague based on visual input.

The exact desired value at each instance is defined by the target position. The error at each instance is a two-dimensional (2D) vector

$$\vec{\epsilon} = \vec{r}_{desired} - \vec{r}_{actual} \tag{12.4}$$

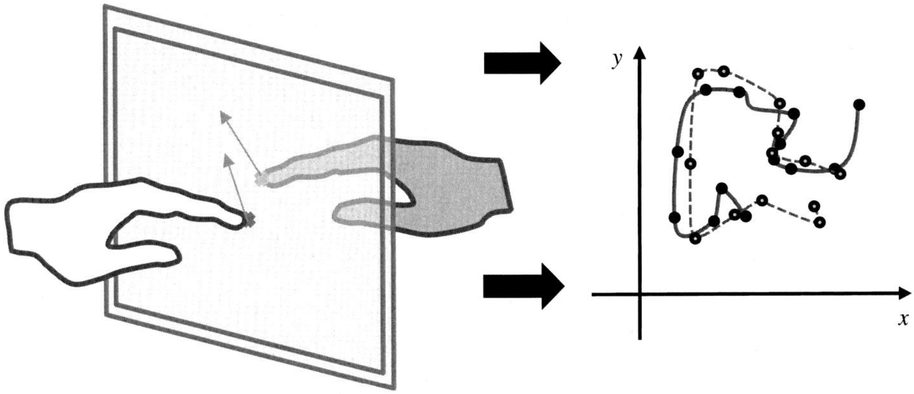

FIG. 12.9 Incentivized student trying to align pointer finger atop pointer finger of blindfolded student on the other side of glass door (left) and their time labeled 2D trajectories (right). *Courtesy of Dr. Marko Popovic.*

If the target is moving slow and along a smooth trajectory, the task to align fingers can be accomplished quickly. However, if the target is moving very fast in an unpredictable manner (e.g., frequent jerky movements), it could be very difficult to accomplish the task. In motor control research, there is a well-known trade-off between speed of reaching and accuracy called Fitt's law [60].

So what can we learn from this experiment?

A lot. Imagine we have a model we trust could explain most of the features of the actual trajectory. For example, imagine that actual acceleration is modeled as simple time delayed PD law

$$\vec{a}_{actual}(t+\Delta t) = \tilde{K}\begin{pmatrix} \vec{\epsilon}\,(t) \\ \dot{\vec{\epsilon}} \end{pmatrix}(t) \cong \tilde{K}_P\,\vec{\epsilon}\,(t) + \tilde{K}_D\dot{\vec{\epsilon}}(t) \cong \begin{pmatrix} K_{Px} & 0 \\ 0 & K_{Py} \end{pmatrix}\vec{\epsilon}\,(t) + \begin{pmatrix} K_{Dx} & 0 \\ 0 & K_{Dy} \end{pmatrix}\dot{\vec{\epsilon}}(t) \qquad (12.5)$$

Now, we can use data collected in our experiment to find best-fit values for five parameters Δt, K_{Px}, K_{Py}, K_{Dx}, K_{Dy}.

Clearly we could also address a slightly more complex, potentially more accurate model, by considering possibility that above parameters might change from initial value Δt_0, K_{Px0}, K_{Py0}, K_{Dx0}, K_{Dy0} and saturate to steady-state values Δt_∞, $K_{Px\infty}$, $K_{Py\infty}$, $K_{Dx\infty}$, $K_{Dy\infty}$ after some initial times characterized by time periods T_t, T_{Px}, T_{Py}, T_{Dx}, T_{Dy} due to fatigue of sensory-motor-control system, including muscle fatigue and maybe increasing attention deficit, or other factors. In other words, each initial parameter f_i is now tripled $f_i \to f_{i0}, f_{i\infty}, T_i$ and expressed in the time-dependent form, for example, as

$$f_i(t) = f_{i0} + \left(1 - e^{-\frac{t}{T_i}}\right)(f_{i\infty} - f_{i0}) \qquad (12.6)$$

Still further, we can address this exercise in the context of even more complex but also more "natural" model based on human anatomy and individual muscle model, for example, some variant of Hill's model (see Chapter 3).

We can remove effects of gravity by considering horizontal contact surface in the form of screen and we can also add passive support for the participant's arm. Moreover, we can remove the "target" pointer and blindfolded student altogether and introduce cursor on the screen instead. Finally based on arm state and joints' accelerations we could deduce joint torques (by following methods addressed in Chapter 2) and subsequently deduce muscle active and passive forces as well as activation signal (by following methods addressed in Chapters 3 and 6).

To verify obtained results, we can independently estimate muscle activation via either surface, fine needle based, or implanted EMG sensor electrodes (see Chapters 4 and 6). And, if we require stronger, that is, more pronounced EMG signal for an independent verification procedure, we could also add an end effector forcing to stipulate continuously modulated joint torques allowing easier extraction of useful EMG signal beyond noise. Moreover, more detailed intersegmental arm kinematics may be obtained through video capture in the lab and/or wearable sensors may be taken home with patients to track reaching activity and practice in the home.

To reliably interpret these results, it would be beneficial to study performance of number of able-bodied participants (to correctly decipher effects of different anatomical features, sex, age, etc.) as well as number of participants with different types of movement impairment (e.g., due to damaged muscle, nerves, demyelinated axons, multiple sclerosis, stroke, Parkinson, semi-paralysis, etc.).

For example, for a specific patient, our data analysis may reveal that neural signals are within healthy ranges both in terms of signal quality and in terms of time delays, whereas a specific muscle group performs maximally up to 30% of baseline active forces. Or, for another patient, health professional can deduce that while active force is within able-bodied range, the activation signal is substantially delayed or characteristically deformed.

Hence, this simple exercise can provide helpful tools for researchers and health professionals to deduce the physical condition of various muscle groups and individual muscles as well as the quality and time delays of activation signals. These research steps can be used to reliably diagnose and monitor health condition of patients subject to a variety of movement impairments. While there is still a lot of ongoing research and development, many hospitals have already implemented one or another variant on these lines.

12.5 Lower-Limb Therapy

12.5.1 Treatment Strategy

Self-directed mobility is an important experiential component of being human. For most humans, a sense of well-being is related to one's ability to engage in the world and self-direct. Wheelchairs and other mobility systems are described in Chapters 13. The current section describes technologies associated with walking (often referred to as robot-assisted walking) that can be either an AT or rehabilitation device. Common conditions that take advantage of robot-assisted gait include SCI, PD, TBI, stroke, multiple sclerosis, and CP.

For most gait training rehabilitation devices, improvements rely on motor learning concepts massed practice and a central pattern generator (CPG). Massed practice refers to the neural plasticity and improvement in coordination that occurs with extensive repetition of hip and knee movements. A CPG is a neural circuit that can generate rhythmic neural output when activated. CPGs have been documented through neural recording in lampreys and other animals, but the extent of their influence in humans is not fully agreed upon.

For all robot-assisted walking technologies, a few considerations are needed. *First, what level of impairment is the technology targeting?* A robot-assisted device for a patient lacking posture control and the ability to generate a gait pattern will be different than a device designed for a patient who can control leg muscles but has weakness. *Second, is the device intended for clinical use to improve motor control (rehabilitation device) or is the device intended for more permanent use at home as an aide to everyday activities (assistive device), or both?* A device designed for clinical use should be able to elicit motor learning in patients, can be more expensive, and is often larger and more stationary. Conversely, a device designed for permanent at-home use would typically be less expensive, more mobile and light (wearable), and needs to directly help with gait but need not necessarily improve gait when the device is removed (although improvements with the device off are always desired). *Third, all robot-assisted gait training devices must consider the extent to which they interfere with the patients' natural movements and comfort.* Size, weight, DOFs, comfort, and cosmetics are thus all considerations. Below is a description of four different types of technologies, each focused on different walking abilities, and each designed specifically for either clinical or at-home use. Chapter 11 also includes descriptions of exoskeletons.

12.5.2 Treatment Technologies

Lokomat. The Lokomat is a rehabilitation device and was the first major technological breakthrough to capture the field of gait training (Fig. 12.10). The Lokomat was developed by the Swiss company Hocoma AG, has been available in the market since 2001, and has included over 250 publications [46, 61]. The Lokomat is a robotic-driven gait orthosis that is used in the clinic for patients with severely impaired gait. By practicing the basic gait patterns, patients can improve their ability for self-generated gait outside the clinic.

Patients are strapped into the orthosis through padded straps at the thighs and shanks. The orthosis provides a gait-like pattern for the lower extremity through four motorized joints, one motor at each hip and each knee, run with direct current motors. Although the actuators generate motion in the sagittal plane only, joints allows additional movements of the legs in other segments in the frontal and transverse plane. The Lokomat is a stationary device integrated into a split belt treadmill. A patient can have vertical translation of the torso and lateral stabilization. Moreover, the patient's body weight can be unloaded as needed through an overhead harness. With both stability and body weight support, the Lokomat can be used in populations with severe motor control impairments. Recent modifications of the Lokomat have been made to

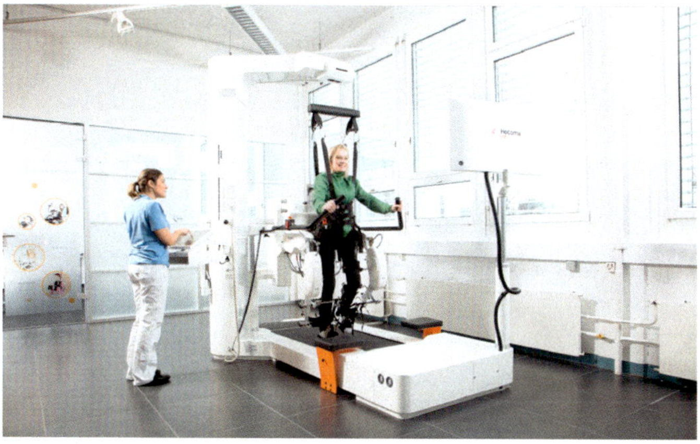

FIG. 12.10 Image of Lokomat. *Picture: Hocoma, Switzerland.*

accommodate the unique needs of children. One limitation is that the Lokomat is not designed for turning or any balance training.

ICARE. The success of the Lokomat has motivated others to develop a less expensive alternative that can accomplish similar improvements in gait (Fig. 12.11). A research group at the Madonna Rehabilitation Hospital in Lincoln, Nebraska, USA, discovered that some elliptical machines also generate gait-like patterns in the lower extremity. These researchers modified the basic components of an elliptical machine to better match human gait, integrated a motor to assist lower extremity movements (as needed), and provided body weight support to ultimately create the "intelligently controlled assistive rehabilitation elliptical (ICARE)" [62]. The patented ICARE integrates a motor to smoothly cycle the legs in a gait-like movement pattern if the patient's voluntary stepping is below a velocity threshold. If the patient's voluntary stepping exceeds the threshold, then the motor no longer assists. The therapist who is working with the patient sets the velocity threshold. Similar to the Lokomat, the ICARE is typically used in a clinical setting but is a simpler and less expensive alternative to the Lokomat. The ICARE has contributed to documented improvements in gait [63]. ICARE has been in use in clinics since 2012.

KineAssist. The iCARE and Lokomat described above enable patients with very little voluntary control to have partially or fully assisted leg movements via a motor. However, some patients with higher levels of voluntary control can move their legs in a gait pattern but still have major impairment in coordination and balance. Motor learning concepts imply that patients learn best when they are challenged to use their own sensory feedback and muscle activations, if possible. The KineAssist (Fig. 12.12) is a device that primarily helps patients to maintain balance while they practice walking [64]. The KineAssist provides a six DOF waist or trunk support while a patient walks on a treadmill. Thus, the patient actively generates all the leg movements and can do so in all planes of motion. A therapist typically assists the patient through the exercise in the clinic.

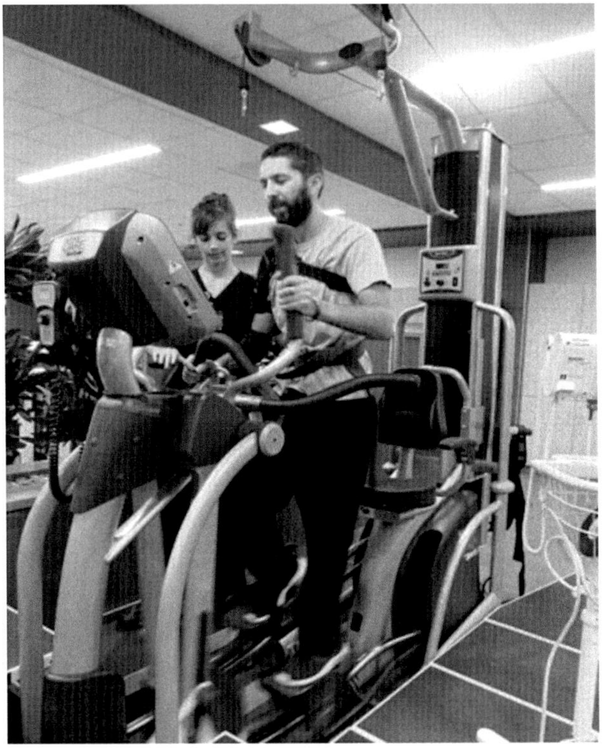

FIG. 12.11 Stroke survivor uses ICARE during therapy to regain walking and endurance. *Photo courtesy Madonna Rehabilitation Hospital.*

Wearable exoskeletons. For higher-functioning patients, the ReWalk [65, 66] and Ekso Bionics [67] take the concept of assisted walking to the next level by enabling patients to walk over ground. These devices are not limited to a treadmill and are appropriate for use in the clinic or at home. Patients wear an exoskeleton around the lower extremities with powered joints that can assist gait. Patients should have moderate balance control during gait or ability to use crutches. The ReWalk has various modes of operation with specific motor output unique to standing (providing stability), walking (providing torque to joints in the hip and knee in the sagittal plane), or sit to stand (proving torque to help to accelerate the body against gravity) (Fig. 12.13). The ReWalk is FDA approved for use at home whereas the Ekso Bionics is a class II FDA-approved device for rehabilitation in the clinic. Finally, some people with adequate balance and mobility do not need extensive support but their muscle weakness or fatigue warrants minimal robot aid. For these individuals, lower-profile devices exist. The Honda Stride Management Assist and Body Support Assists are good examples of simple single and multi-joint devices that can be quickly fit onto a person to augment a person's existing movements and posture [68].

It is remarkable to see how biomechatronic devices have transformed gait rehabilitation. Each device described above has been designed to meet the needs of specific levels of

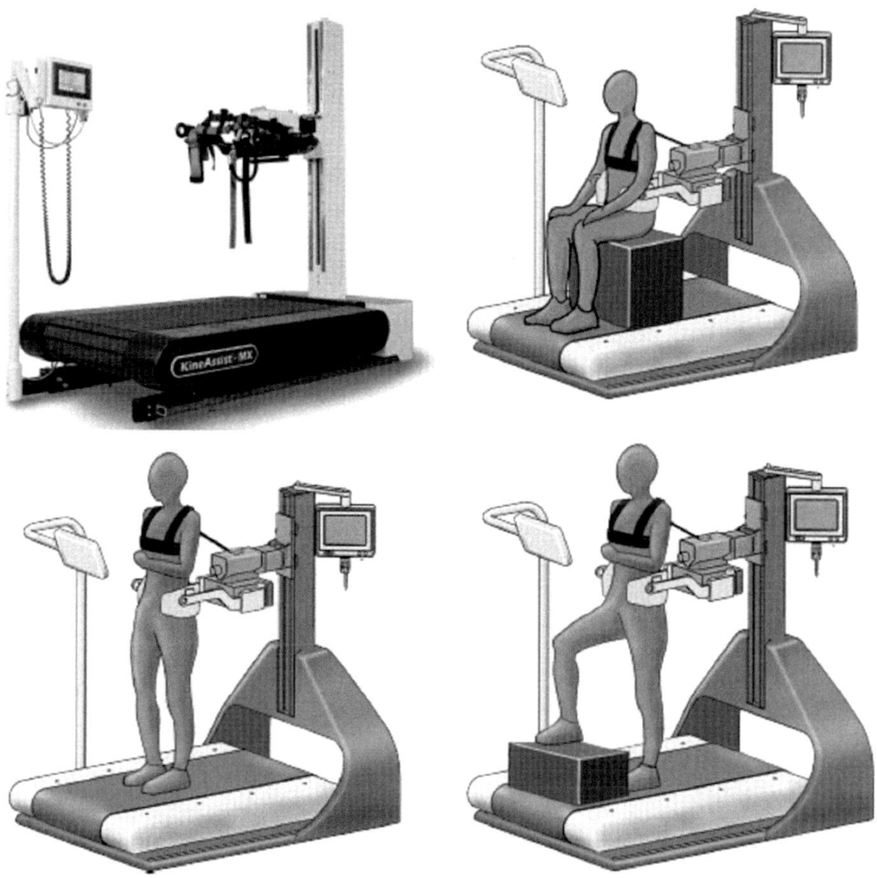

FIG. 12.12 Schematic of KineAssist (HDT Global website). *Used with permission from HDT Global.*

impairments with some devices used exclusively in clinics and others designed for home use. Future work will continue to investigate effectiveness, improve human interfaces, control, and lower costs.

12.5.3 Assessment Technologies

Similar to upper extremity training, assessing the success of gait training often takes place by comparing pretraining versus posttraining performance based on a variety of metrics. Moreover, many of locomotion robotic training devices described above automatically obtain and track metrics so that information can be collected during the training session. Common metrics of the body include speed, distance, stride length, step width, cadence, symmetry, number of steps during turning, and lower extremity kinematics and torque generation. Common metrics captured in robotic devices include speed, degree of body weight support, time of use, and assistance from the motor.

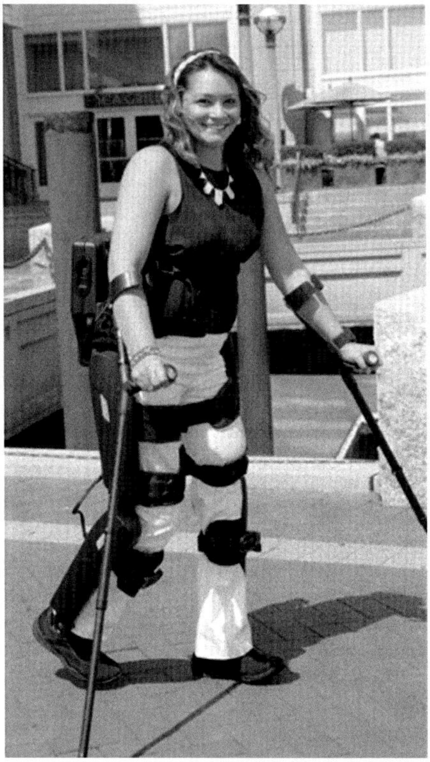

FIG. 12.13 Image of ReWalk wearable exoskeleton. *Used with permission from ReWalk.*

If a patient is progressing, typically one expects greater speeds, distance, larger stride lengths, more symmetry, few steps during turning, higher torque generation, less body weight support, more time of use, and less motorized support required. During over-ground walking, these changes can be easily measured in a standard gait lab with motion tracking camera (Chapter 4). However, these labs are relatively fixed in place, relatively expensive, and time consuming to set up. Less-expensive technologies include wearable sensors [69] or instrumented mats. Instrumented mats automatically calculate metrics like cadence, stance width, step length, and symmetry. These mats can be transported easily and provide basic but important outputs. Numerous wearable sensors exist. One system of the easiest systems to wear is the G-WALK [70]. Subjects wear the G-WALK around their waist like a belt and can walk anywhere. The device is embedded with sensors to capture a few of the prominent metrics in gait and other activities. However, most wearable sensor systems measure more than one body segment to provide more resolution and detail [71–73]. Obtaining meaningful and reliable measures from wireless sensors is an active and ongoing area of research.

Finally, clinics and researcher may have access to advanced medical treadmills that can measure oxygen levels used during locomotion, CO_2 gas exchangers, EEG to examine brain activity, EMG to capture muscle activation, and a split-belt treadmill. The split belt refers to two independently controlled treads, one under each foot. The treads can move at different speeds to evoke adaptions or suddenly change speeds to challenge balance. With improvements in gait, one anticipates lower levels of cardiovascular efforts, more symmetrical and normalized patterns of EMG, and better responses to challenges in balance. However, balance is a complex topic in itself and is therefore the topic of the following section.

12.6 Balance Therapy

12.6.1 Treatment Strategies

Falls are an important societal topic. Approximately one out of four adults over age 65 falls per year [2]. In 2014, older Americans had 29 million falls resulting in seven million injuries which together cost over $30 billion in Medicare costs [2]. Falls can lead to bone fracture, concussion, reduced mobility, fear, and even death. However, age alone is not a determinant of falls, as there is a wide range of balance abilities across older adults. Various factors lead to decreased balance control, muscle strength, sensory feedback, cognitive function, and biomechanical constraints [74]. In populations with certain pathologies, such as diabetes, vestibular impairment, PD, and CP, the risk of falling is increased. Importantly, vestibular dysfunction is estimated to be in approximately one out of three adults over the age of 40 years [75]. Because vestibular cues provide the brain with information about where the body is with respect to the gravitational field, vestibular dysfunction can lead to poor balance, increased likelihood of falls, and decreased mobility.

The reason falls are so critical is because balance underlies most voluntary ADLs. For example, reaching for a cup while standing or walking to one's car requires a person to maintain stability of their body again gravity and to maintain stability in response to internal and external perturbations. Internal perturbations include those generated by oneself when muscles are activated and when one body segment accelerates relative to another. Typically, people learn to anticipate the effects of internal perturbations and are rarely aware of their presence. However, in certain pathologies, the anticipation of internal perturbations is not fully learned. External perturbations include gravity and disturbance from the surroundings, such as a sudden change of surface or another person bumping into you. External perturbations are often not anticipated and require appropriate reactive strategies for activating the appropriate muscles with precise force and timing. Patients with balance impairments often react to external perturbations with the wrong muscles and poor-scaled responses that can put them at greater risk of falling.

From the standpoint of control systems engineering, to stabilize the body, balance is considered inherently unstable because a small deviation from upright results in gravitation forces that further accelerate the body away from upright. Corrective torques must be generated in proportion to body sway position and body sway velocity. Representing the body as a single link inverted pendulum rotating about the ankle joint, the torque due to gravity (T_g) varies in direct proportion to mass (m), gravitational constant (g), center of mass (CM) height above the ankle joint (h), and the sine of body sway angle with respect to upright (Fig. 12.14). Corrective torque is based on the information from sensory feedback and is subject to neural time delays.

A simple linear negative feedback model can be developed to describe the body sway (Fig. 12.15). In this model, s represents the Laplace variable, which enables a characterization of body sway across any frequency. J is the moment of inertia and τ is the neural time delay. Corrective torque (T_c) is assumed to drive the body toward a kinematic stimulus. If the stimulus is zero, T_c is in the opposite direction of body tilt to offset the torque due to gravity. Corrective torque is based on position gain (K_p) plus velocity gain (K_d) plus integral gain (K_I). These gains represent the responsiveness of the system to a stimulus and must be tightly controlled. If K_p and K_d are too low, the body will not be stabilized against gravity or will become resonant. If K_p, K_d, and K_I are too high, the system will be "overcorrecting" and this is incompatible with stability when time delay is included in a feedback system.

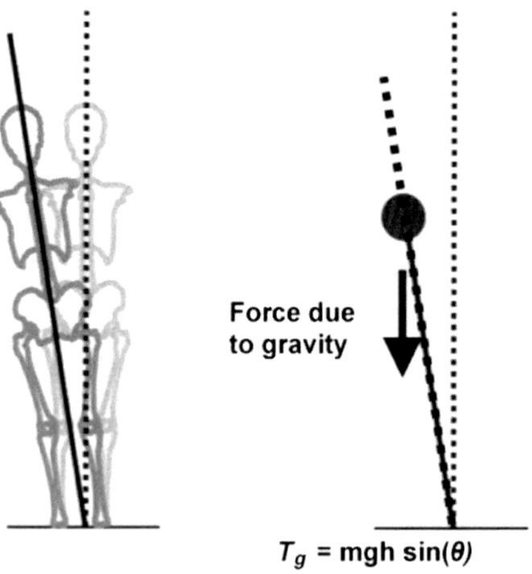

FIG. 12.14 Inherently unstable body where torque due to gravity (T_g) accelerates the body away from upright in proportion to mass (m), gravity constant (g), center of mass height (h), and sine of body sway (θ).

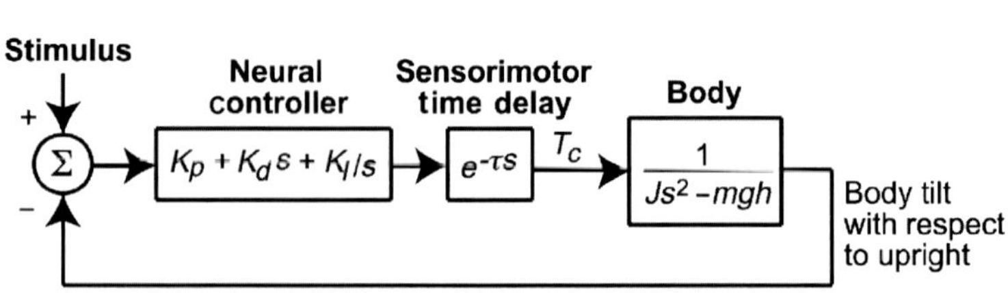

FIG. 12.15 Simple feedback control model of human posture approximated as a single link. *Based on published model by R.J. Peterka, Sensorimotor integration in human postural control, J. Neurophysiol. 88(3) (2002) 1097–1118.*

The dynamic relation between the single link body tilt and the stimulus can written as an equation:

$$\frac{Bodysway}{Stimulus} = \frac{\left(K_p + K_d s + K_I/s\right)e^{-\tau s}}{Js^2 - mgh + \left(K_p + K_d s + K_I/s\right)e^{-\tau s}} \tag{12.7}$$

Using various types of stimuli, biomechanical and neural (such as sensory feedback from visual, vestibular, and somatosensory) systems contribute to corrective torque have been investigated and results provide extensive insights into the analysis of complex posture data in sitting, standing, and walking. For example, an important variant of Eq. (12.7) was developed to include separate channels for each sensory system that could orient the body vertical (vestibular) or toward a stimulus (visual or proprioceptive), depending on the stimulus modality. This model accurately described sensory reliance in adults with normal and bilateral vestibular dysfunction when standing on a tilting surface across a wide bandwidth of stimulus frequencies [76]. While the above examples pertain to human balance, it is worth noting that most robotic solutions to balance instability use a feedback control framework.

With feedback control in mind, some approaches to target specific contributions to balance (analogous to individual parameters in the feedback model) while other approaches target the posture system more globally (all parameters in the feedback model). Therapies for older adults who are not diagnosed with a specific condition are typically more global, centered on mobility, coordination, and strength, consistent with the principle of "use it or loose it." In contrast, vestibular balance treatments are more focused; typically aimed at improving vestibular integration (and reducing symptoms of dizziness by repositioning crystals in the vestibular apparatus) or aimed at improving a patient's ability to make corrective torque. For patients with more severe posture impairments, balance treatments may focus on the basic task of upright sitting. There is increased emphasis on the value of using a stimulus (termed perturbation training) in treatment for all populations. Unexpected perturbations require reactive balance response and predictive perturbations help patients practice anticipatory responses. Delivering repeatable and safe perturbations almost always involves biomechatronics technology.

12.6.2 Treatment Technologies

Many current balance techniques include activities like Tai Chi, dancing, aquatic exercises, and rapid movements that build muscle strength and require anticipatory posture control. These activities do not often include a high level of technology. When technology is used, the most common include Wii sports in clinics (which include biofeedback of arm movements or forces generated under the feet) and wearable sensors such as FitBits that can detect general activity levels. Many of these approaches are used in patients with SCI, TBI, PD, and stroke. Similarly, for children with disabilities such as mild-to-moderate CP, balance training may include perturbation training and the inclusion of video games. Video games can increase motivation and offer the potential for greater specificity of training and feedback.

For those with milder impairments, off the shelf systems such as Wii and Kinect have been used to increase muscle strength and coordination and endurance [77]. For those with more severe CP, typically customized engineering is required.

Other technologies used for practicing movements include vibration biofeedback [78] and virtual reality. Vibrotactile feedback includes sensors that monitor body sway characteristics and active vibrators (typically located on the torso or waist) to provide patients with augmented sensory feedback of their orientation. Virtual reality has been used in habituation by presenting patients with a visual stimulus to practice integrating visual feedback with vestibular stimuli. For habituation, patients are often directed to move their head (right, left, up, and down) and body (walking and turning) at slightly uncomfortable levels so that they habituate to the subnoxius stimulus from vestibular feedback. One system, at the University of Pittsburgh, has a virtual reality dome where patients walk on the treadmill while looking into a moving scene representing a supermarket. Patients practice in a safe environment turning their head to the right and left, and naming off items in the virtual store to simulate the real experience. The most elaborate system for practicing balance and mobility is the CAREN system, a fully enclosed virtual reality visual surround with movable split belt treadmill (explained in more detail in Chapter 16). The CAREN system is used in academic and military centers for research and rehabilitation (especially for veterans using a prosthesis, see Chapter 9 for information on prosthetics).

When perturbations are included with in devices, they typically include an external force (push or pull), surface motion (moving platform or sudden change in acceleration on treadmill), or a moving visual stimulus. These perturbation modalities are sometimes combined with video game or virtual reality systems (e.g., CAREN system). Some common perturbations (such as moving platforms) are covered in the Balance Assessment section because historically perturbations were often used as an assessment tool. However, any system that requires balance responses can be considered perturbation training. One example is hippotherapy (training motor control on a horse) (Fig. 12.16) and simulated hippotherapy (training on a mechanical horse). Both have been shown to improve the inherent balance system, gait, coordination, and strength. Mechanical horses are used in a variety of populations—stroke, TBI, SCI, etc.—ranging from mild to severe disabilities. However, it is noteworthy that for children with CP, hippotherapy and simulated hippotherapy have provided some of the strongest evidence for improvement [77, 79]. In mechanical horses, these improvements may be attributed to (1) the similarity in pelvis motion between human gait and riding a horse [80], and (2) the high number of balance corrections practiced during a typical riding session. The motivations for using a mechanical horse include convenient access to riding motions, indoor use regardless of weather, reduced risk of allergies or fear, improved safety, and convenience for increased frequency of riding sessions. To date, the most accurate mechanical horse available for replicating the 3D horse motion were developed by a researcher at the Baylor University in 2006, now sold through Chariot Innovations. Interestingly, this mechanical horse is able to generate complex 3D horse patterns through a single motor with a series of cams connected to mechanical linkages.

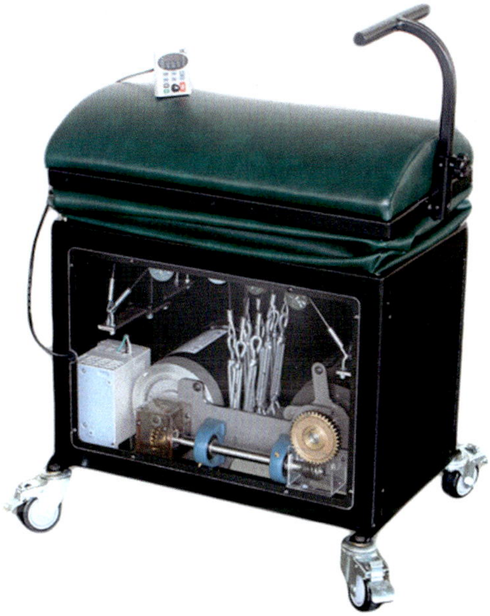

FIG. 12.16 Image of mechanical horse created by Chariot Innovations. *Used with permission from Chariot Innovations.*

Finally, while balance therapy is typically associated with standing or walking, the basic skill of independent sitting with respect to gravity is not something everyone can accomplish. People with neurological disorders or severe SCI may have an inability to sit independently or control one's head against gravity. For these individuals, fewer treatments are available, and these treatments require different technologies. For children with impaired trunk and head control, research has demonstrated improvement in function when seating devices are implemented [79]. Mechanical engineering is involved in creating seating systems that are comfortable and biomechanically effective. For populations with severe balance impairments, it may be necessary to combine mechanical trunk support with robotic technologies.

12.6.3 Assessment Technologies

Assessment of balance can be categorized across three activities: spontaneous sway in standing or sitting, perturbed stance or sitting, and perturbed gait. In the modern era of posture research, technologies have been instrumental in each of these assessments [81]. While assessment of balance is treated as a separate section in this chapter, we note than many of assessment technologies overlap with balance treatments (Fig. 12.17, [82]). For example, robotic-derived perturbations are used in training and assessment. In addition, balance assessments are important to track progress both balance training and gait training because functional gait requires balance as an underlying system. Shirota et al. reviews many of these considerations for robotic balance and gait devices [81].

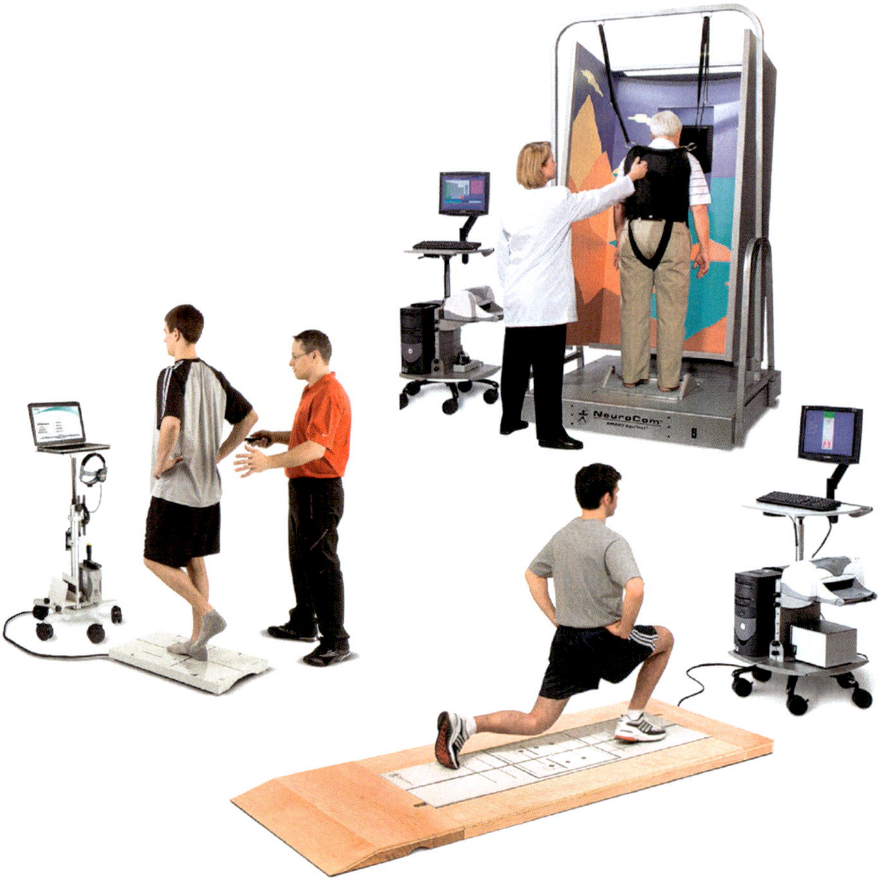

FIG. 12.17 Image from Natus. *Used with permission from Natus.*

Spontaneous sway. Even when sitting or standing still, there is movement across body segments due to variable muscle activations associated with (1) muscle tone (to offset gravity and maintain posture), (2) breathing, and (3) sensorimotor noise (activation of sensory receptors, variability in activation of motor units, and uncertainty in sensory integration). Abnormalities in spontaneous sway can be linked to various pathologies where balance is impaired. Force plates combined with signal processing have been the most commonly used technology to capture spontaneous sway. Using a force plate is very simple. A patient stands or sits on a plate for about 30–60 s. Common test conditions include standing on one or two legs, with or without eyes open. During the test, the patient maintains balance through muscle activations that ultimately change the location and magnitude of forces under his/her feet. Forces are measured in three planes and often characterized in terms of torque and center of pressure (CoP). Once CoP data are captured, typical processing includes measures in the time domain, such as root mean square of CoP position and velocity, peak-to-peak, mean velocity, and path length. Frequency domain measures typically include power spectra and centroid frequency.

One clever way to make spontaneous sway more challenging and to examine sensory integrity with more resolution is to alter the surface a patient is standing or sitting upon. To better understand the contribution of each sensory system, the Neurocom system was created by Lewis Nashner in the 1970s. In the Neurocom, a patient stands on a force plate and looks at a visual dome (Fig. 12.17, upper right). Both the force plate and visual dome can rotate via a servomotor about the ankle joint. Body tilt is continuously captured by low pass filtering the CoP data. In some tests (termed "surface sway referencing"), the surface rotates in proportion to the body tilt. Thus, even though the body tilt is away from upright, the ankle joint remains approximately neutral (i.e., shank relative to foot does not change). Proprioception originating from ankle joint motion is therefore very small, forcing patients to rely upon vision and vestibular in eyes open conditions or only vestibular information in eyes closed conditions. Impairments in visual and/or vestibular function result in higher posture sway. Another similar condition is "visual sway referencing" where the visual dome rotates in proportion to body sway and therefore minimizes visual feedback, forcing subjects to rely upon vestibular and proprioception. Finally, by combing both surface and visual surround rotation, subjects must use a very high degree of vestibular feedback. Impairments in vestibular function result in higher posture sway and sometimes an inability to stand up.

Because of the prominent role of vestibular feedback in human stance control, standing posture tests are often combined with a more direct measure of vestibular function. The gold standard test for vestibular function is the rotation chair. A patient sits on chair while wearing goggles with camera embedded. The chair then rotates at various speeds and directions while the subject looks forward. Eye movement is recorded and related to the rotation chair motion (Fig. 12.18, [83]). The vestibular ocular reflex and various other measures related to the gain between eye motion and chair rotation motion are compared with established norms to quantify the integrity of vestibular function.

Perturbed stance or sitting. As noted above, perturbations are a powerful probe into the posture system. In response to a perturbation, EMG of leg and trunk muscles can detect onset and neural time delays following sudden transient stimuli (such a surface translation or push). Kinematics and kinetics can determine scaling between body motion and torque generation giving key insight into important neural processes. Kinematics has also been analyzed in the frequency domain [often combined with feedback modeling, e.g., variants of Eq. (12.5)] to quantify sensorimotor noise and sensory reliance in standing and sitting [76, 84].

Walking balance. Most falls do not occur during quiet standing or sitting, but rather occur during walking or turning. Therefore, many of the concepts underlying perturbed stance control have been applied to study walking balance. A few examples include delivering (1) a sudden surface translation when a subject walks over a movable plate, (2) a transient or oscillating treadmill that a participants walks on, (3) a sudden change in tread speed on a treadmill (single or split-belt), (4) and an external force applied to the body while walking. Common metrics for walking balance include trips, falls, and magnitude of posture sway. However, we describe in detail another recently developed metric based on physical principles.

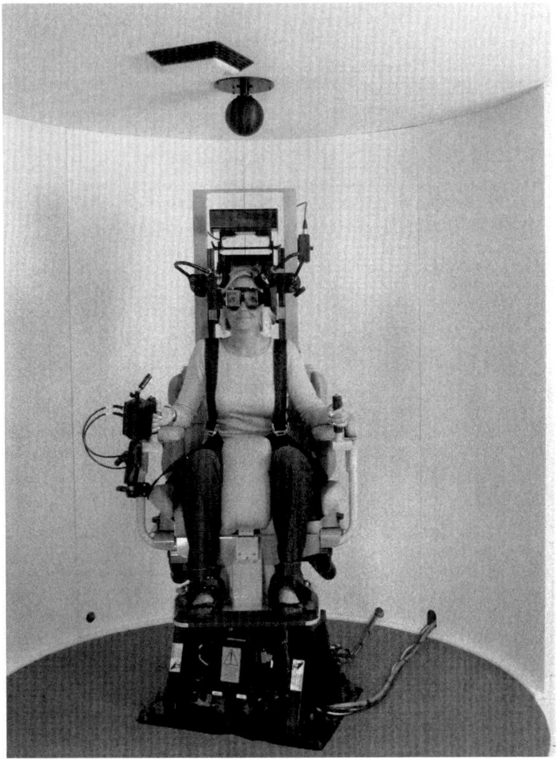

FIG. 12.18 Rotation chair test by Neurokinetics. *Used with permission from Neurokinetics.*

In physics, the notion of equilibrium states that both net external forces and torques must be zero. Thus, ground reaction force must exactly balance body weight. In equilibrium, the zero-moment point (ZMP) [85, 86] from which ground reaction force originates, must coincide with vertical projection of *CM* onto ground, see Fig. 12.19A. The most important part of the ZMP concept, applicable for both single and multi-leg ground support phases, is that it resolves the ground reaction force distribution to a single point such that horizontal component of moment about *CM* can be expressed as

$$\vec{\tau}\big|_{horizontal} = \left[\left(\vec{r}_{ZMP} - \vec{r}_{CM}\right) \times \vec{F}_{GR}\right]_{horizontal} \tag{12.8}$$

The moment is also equal to rate of change of spin angular momentum, that is, rate of change of angular momentum about *CM*,

$$\vec{\tau} = \frac{d\vec{L}_{CM}}{dt} \tag{12.9}$$

Although for flat horizontal ground surfaces, the ZMP is equal to the CoP, the points are distinct for irregular ground surfaces [86]. In general, ZMP or/and CoP must be located within foot ground support. This, in turn constrains the moment about *CM*, as emphasized in Fig. 12.19B and C.

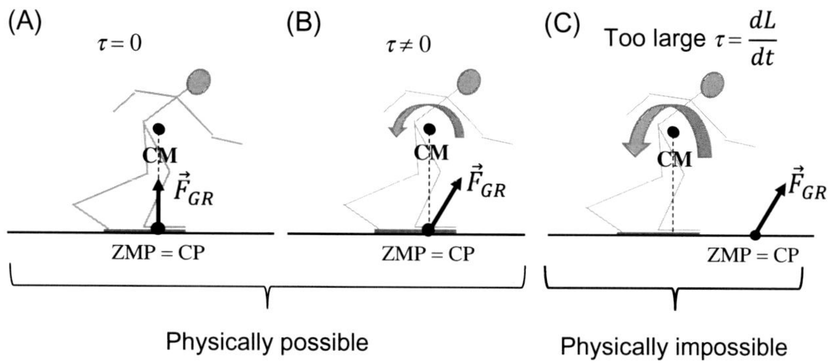

FIG. 12.19 Relationship between ZMP and moment about *CM*. *Courtesy of Dr. Marko Popovic.*

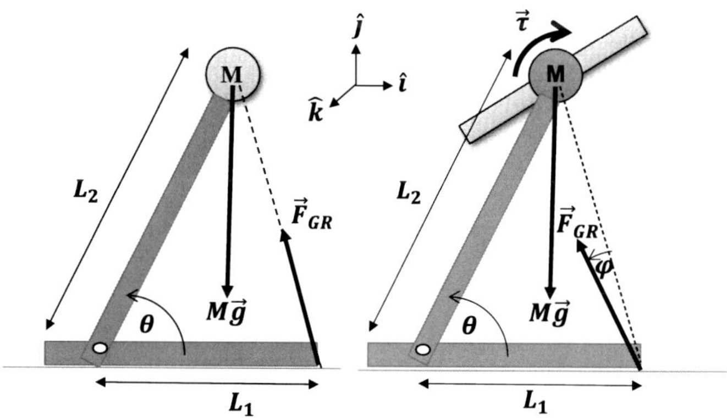

FIG. 12.20 Simplified system that can exhibit only zero-moment balance strategy (left) and system that can utilize both zero-moment and moment balance strategies (right). *Courtesy of Dr. Marko Popovic.*

For static balance, the vertical projection of *CM* onto ground must be within foot ground support.

In gait, the vertical projection of *CM* is inside the support base only during the double-support phase (approximately 20% of the gait cycle for younger population and about 30% for older population). Clearly, the condition for dynamic balance is very different from condition for static balance.

Within this framework, two walking balance strategies emerge: zero-moment and moment balance [86, 87]. To illustrate, consider a simple system (Fig. 12.20) with massless foot (L_1), ankle joint (θ), and leg (L_2). A massive link (mass, M) is rotating about its *CM* with rate of change of spin angular momentum, denoted by moment $\vec{\tau}$. Generally, for a given system supported by a horizontal contact surface, one can express horizontal force as

$$F_{GRx} = M\left(a_{CMy} + g\right)\frac{x_{CM} - x_{ZMP}}{y_{CM}} - \frac{\tau_z}{y_{CM}} \tag{12.10}$$

The zero-moment balance term is designated as

$$F_{GRx\ zero-momene} = M\left(a_{CMy} + g\right)\frac{x_{CM} - x_{ZMP}}{y_{CM}} \tag{12.11}$$

and moment balance term is designated as

$$F_{GRx\ moment} = -\frac{\tau_z}{y_{CM}} \tag{12.12}$$

For any given state and *CM* vertical acceleration, one can manipulate horizontal force by repositioning ZMP or CoP (zero-moment strategy), or changing the moment (moment balance strategy). Stepping can potentially affect either of these strategies (Fig. 12.20).

Numerous simplified model of walking utilize zero-moment strategy either by addressing inverted pendulum dynamics or some variant of it. Experimental gait studies also indicate that both spin angular momentum and moment about *CM* are very small. Clearly the zero-moment strategy is more energy cost effective and it can be continuously used for a long time. However, the moment balance strategy is often necessary for dynamic movements and to correct sudden large dynamic perturbations. Thus, it can easily dominate the system for at least short period of time. For example, if a stance foot slips, the rest of body produces short but very fast movement, which resembles motion of the heavy link in the figure above. Note that vertical projection of *CM* with zero speed of *CM* can be well outside the support base and while keeping the support base unchanged the moment balance strategy can restore *CM* back to statically balanced state above the support base. Consider the act of tight rope walking and the lateral stability that is enforced by moment balance strategy. For additional discussion on balance, see Chapter 2.

Clinicians and engineers of robotic devices may take advantage of these physical and control-based metrics. In summary, physics of static and dynamic balance provides detailed insight into body motion and that may play a key role delineating underlying causes of impairments and thus help target rehabilitation solutions.

12.7 Conclusion

For engineers, it is exciting to see biomechatronics in so many areas of physical therapy and rehabilitation. This chapter covered design strategies, treatments and assessments in upper extremity, lower extremity, and balance. Many times, a device used to treat a patient can also be used as an assessment. Devices typically either aim to enable an individual to perform a daily activity or aim to elicit permanent motor learning so that the individual could perform daily activities with less assistance. Also, devices can be distinguished based on their intended location of use, at home or in the clinic—each intended location having unique considerations. Finally, this chapter emphasized that devices must consider the needs of the individual patient, resources available, and clinical goals.

References

[1] https://www.cdc.gov/media/releases/2015/p0730-US-disability.html (accessed February 2018).

[2] https://www.cdc.gov/nchs/data/series/sr_10/sr10_255.pdf (accessed February 2018).

[3] L. Kraus, 2016 Disability Statistics Annual Report, University of New Hampshire, Durham, NH, 2017.

[4] World Health Organization, International Classification of Functioning, Disability and Health: ICF, World Health Organization, 2001.

[5] J.M. Guralnik, L.P. Fried, M.E. Salive, Disability as a public health outcome in the aging population, Annu. Rev. Public Health 17 (1) (1996) 25–46.

[6] https://www.disabled-world.com/disability/statistics/ (accessed February 2018).

[7] https://www.disabled-world.com/health/aging/ (accessed February 2018).

[8] NIH Fact Sheet—Stroke, Available from: http://www.ninds.nih.gov/.

[9] https://stroke.nih.gov/materials/rehabilitation.htm (accessed February 2018).

[10] E.J. Benjamin, M.J. Blaha, S.E. Chiuve, M. Cushman, S.R. Das, R. Deo, S.D. de Ferranti, J. Floyd, M. Fornage, C. Gillespie, C.R. Isasi, M.C. Jimenez, L.C. Jordan, S. E. Judd, D. Lackland, J.H. Lichtman, L. Lisabeth, S. Liu, C.T. Longenecker, R. H. Mackey, K. Matsushita, D. Mozaffarian, M.E. Mussolino, K. Nasir, R.W. Neumar, L. Palaniappan, D. K. Pandey, R.R. Thiagarajan, M.J. Reeves, M. Ritchey, C.J. Rodriguez, G.A. Roth, W. D. Rosamond, C. Sasson, A. Towfighi, C.W. Tsao, M.B. Turner, S.S. Virani, J.H. Voeks, J.Z. Willey, J. T. Wilkins, W. JHY, H.M. Alger, S.S. Wong, P. Muntner, onbehalf of the American Heart Association Statistics Committee and Stroke Statistics Subcommittee, Heart disease and stroke statistics—2017 update: a report from the American Heart Association, Circulation (2017), doi:10.1161/CIR.0000000000000485.

[11] www.cdc.gov.

[12] A.R. Fugl-Meyer, L. Jääskö, I. Leyman, S. Olsson, S. Steglind, The post-stroke hemiplegic patient. 1. A method for evaluation of physical performance, Scand. J. Rehabil. Med. 7 (1) (1975) 13–31.

[13] https://www.ninds.nih.gov/Disorders/All-Disorders/Traumatic-Brain-Injury-Information-Page (accessed February 2018).

[14] A. Wagner, et al., Chapter 49: Traumatic brain injury, in: C. David (Ed.), Braddom's Physical Medicine and Rehabilitation, Elsevier, USA, 2015.

[15] G. Teasdale, B. Jennett, Assessment of coma and impaired consciousness. A practical scale, Lancet 11 (1974) 81–84.

[16] https://www.cerebralpalsyguidance.com/cerebral-palsy/.

[17] M. Oskoui, F. Coutinho, J. Dykeman, N. Jetté, T. Pringsheim, An update on the prevalence of cerebral palsy: a systematic review and meta-analysis, Dev. Med. Child Neurol. 55 (6) (2013) 509–519.

[18] http://ucp.org/ (accessed February 2018).

[19] Annual report of National Spinal Cord Injury Statistical Center, Birmingham, Alabama.

[20] Spinal Cord Medicine, in: D.I. Campagnolo, S. Kirshblum, M.S. Nash, R.F. Heary, P.H. Gorman (Eds.), Chapter 17: Rehabilitation of Spinal Cord Injury, Lippincott Williams & Wilkins, 2011.

[21] NIH Fact Sheet—Parkinson's, Available from:http://www.ninds.nih.gov/.

[22] J.G. Nutt, B.R. Bloem, N. Giladi, M. Hallett, F.B. Horak, A. Nieuwboer, Freezing of gait: moving forward on a mysterious clinical phenomenon, Lancet Neurol. 10 (8) (2011) 734–744.

[23] G. Deuschl, C. Schade-Brittinger, P. Krack, J. Volkmann, H. Schäfer, K. Bötzel, et al., A randomized trial of deep-brain stimulation for Parkinson's disease, N. Engl. J. Med. 355 (9) (2006) 896–908.

[24] https://money.usnews.com/careers/best-jobs/physical-therapist.

[25] https://www.bls.gov/ooh/healthcare/occupational-therapists.htm.

[26] R.A. Cooper, H. Ohnabe, D.A. Hobson (Eds.), An Introduction to Rehabilitation Engineering, CRC Press, 2006.

[27] https://abledata.acl.gov (accessed February 2018).

[28] A.M. Cook, J.M. Polgar, Cook and Hussey's Assistive Technologies: Principles and Practice, third ed., Mosby Elsevier, St. Louis, MO, 2007.

[29] A. Shumway-Cook, M.H. Woollacott, Motor Control: Translating Research into Clinical Practice, Lippincott Williams & Wilkins, 2007.

[30] R.A. Magill, Motor Learning and Control: Concepts and Applications, 10th ed., McGraw-Hill, Boston, 2014.

[31] D.J. Reinkensmeyer, J.L. Emken, S.C. Cramer, Robotics, motor learning, and neurologic recovery, Annu. Rev. Biomed. Eng. 6 (2004) 497–525.

[32] H.I. Krebs, N. Hogan, B.T. Volpe, M.L. Aisen, L. Edelstein, C. Diels, Overview of clinical trials with MIT-MANUS: a robot-aided neuro-rehabilitation facility, Technol. Health Care 7 (6) (1999) 419–423.

[33] J.L. Patton, M.E. Stoykov, M. Kovic, F.A. Mussa-Ivaldi, Evaluation of robotic training forces that either enhance or reduce error in chronic hemiparetic stroke survivors, Exp. Brain Res. 168 (3) (2006) 368–383.

[34] P. Maciejasz, J. Eschweiler, K. Gerlach-Hahn, A. Jansen-Troy, S. Leonhardt, A survey on robotic devices for upper limb rehabilitation, J. Neuroeng. Rehabil. 11 (1) (2014) 3.

[35] A.C. Lo, P.D. Guarino, L.G. Richards, J.K. Haselkorn, G.F. Wittenberg, D.G. Federman, et al., Robot-assisted therapy for long-term upper-limb impairment after stroke, N. Engl. J. Med. 362 (19) (2010) 1772–1783.

[36] http://bionikusa.com/healthcarereform/upper-extremity-rehabilitiation/inmotion2-arm/.

[37] M.J. Johnson, K.J. Wisneski, J. Anderson, D. Nathan, R.O. Smith, Development of ADLER: the activities of daily living exercise robot, in: The First IEEE/RAS-EMBS International Conference on Biomedical Robotics and Biomechatronics, 2006. BioRob 2006, IEEE, 2006, pp. 881–886.

[38] P.S. Lum, C.G. Burgar, M. Van der Loos, P.C. Shor, MIME robotic device for upper-limb neurorehabilitation in subacute stroke subjects: a follow-up study, J. Rehabil. Res. Dev. 43 (5) (2006) 631.

[39] T. Nef, M. Mihelj, R. Riener, ARMin: a robot for patient-cooperative arm therapy, Med. Biol. Eng. Comput. 45 (9) (2007) 887–900.

[40] V. Klamroth-Marganska, J. Blanco, K. Campen, A. Curt, V. Dietz, T. Ettlin, et al., Three-dimensional, task-specific robot therapy of the arm after stroke: a multicentre, parallel-group randomised trial, Lancet Neurol. 13 (2) (2014) 159–166.

[41] https://www.hocoma.com/us/solutions/armeo-power/#product.

[42] http://myomo.com/.

[43] Powered Arm Brace For Restored Movement—Stroke Rehabilitaion, Portable Robotic Arm Brace, NeuroRobotic Technology. myomo@home program, http://www.myomo.com/ (retrieved July 2012).

[44] A. Theriault, M. Nagurka, M.J. Johnson, Design and development of an affordable haptic robot with force-feedback and compliant actuation to improve therapy for patients with severe hemiparesis, IEEE Trans. Haptics 7 (2) (2014) 161–174.

[45] M.J. Johnson, R. Rai, S. Barathi, R. Mendonca, K. Bustamante-Valles, Affordable stroke therapy in high-, low-and middle-income countries: from theradrive to rehab CARES, a compact robot gym, J. Rehabil. Assist. Technol. Eng. 4 (2017). 2055668317708732.

[46] U. Dundar, H. Toktas, O. Solak, A.M. Ulasli, S. Eroglu, A comparative study of conventional physiotherapy versus robotic training combined with physiotherapy in patients with stroke, Top. Stroke Rehabil. 21 (6) (2014) 453–461.

[47] http://www.reha-stim.de/cms/index.php?id=60.

[48] http://tyromotion.com/en/products/pablo.

[49] http://motusnova.com/products/hand-mentor-pro/.

[50] O. Lambercy, L. Dovat, R. Gassert, E. Burdet, C.L. Teo, T. Milner, A haptic knob for rehabilitation of hand function, IEEE Trans. Neural Syst. Rehabil. Eng. 15 (3) (2007) 356–366.

[51] R. Wilk, M.J. Johnson, Usability feedback of patients and therapists on a conceptual mobile service robot for inpatient and home-based stroke rehabilitation, in: 5th IEEE RAS & EMBS International Conference on Biomedical Robotics and Biomechatronics 2014, IEEE, 2014, pp. 438–443.

[52] https://www.ald.softbankrobotics.com/en.

[53] J. Fasola, M. Mataric, A socially assistive robot exercise coach for the elderly, J. Human-Robot Interact. 2 (2) (2013) 3–32.

[54] M. Matarić, A. Tapus, C. Winstein, J. Eriksson, Socially assistive robotics for stroke and mild TBI rehabilitation, Adv. Technol. Rehabil. 145 (2009) 249–262.

[55] J.J. Diehl, L.M. Schmitt, M. Villano, C.R. Crowell, The clinical use of robots for individuals with autism spectrum disorders: a critical review, Res. Autism Spectr. Disord. 6 (1) (2012) 249–262.

[56] V. Mathiowetz, G. Volland, N. Kashman, K. Weber, Adult norms for the box and block test of manual dexterity, Am. J. Occup. Ther. 39 (6) (1985) 386–391.

[57] R.H. Jebsen, An objective and standardized test of hand function, Arch. Phys. Med. Rehabil. 50 (6) (1969) 311–319.

[58] K.J. Wisneski, M.J. Johnson, Quantifying kinematics of purposeful movements to real, imagined, or absent functional objects: implications for modelling trajectories for robot-assisted ADL tasks, J. NeuroEng. Rehabil. 4 (1) (2007) 7.

[59] M.J. Johnson, S. Wang, P. Bai, E. Strachota, G. Tchekanov, J. Melbye, J. McGuire, Bilateral assessment of functional tasks for robot-assisted therapy applications, Med. Biol. Eng. Comput. 49 (10) (2011) 1157.

[60] P.M. Fitts, The information capacity of the human motor system in controlling the amplitude of movement, J. Exp. Psychol. 47 (6) (1954) 381.

[61] https://www.hocoma.com/.

[62] https://www.madonna.org/research/icare-movement-and-neuroscience-center-research.

[63] S.L. Irons, G.A. Brusola, T.W. Buster, J.M. Burnfield, Novel motor-assisted elliptical training intervention improves six-minute walk test and oxygen cost for an individual with progressive supranuclear palsy. Cardiopulmonary Phys. Therapy J. 26 (2) (2015) 36–41, doi:10.1097/CPT.0000000000000007.

[64] http://www.hdtglobal.com/product/kineassist/.

[65] http://rewalk.com/.

[66] A. Esquenazi, M. Talaty, A. Packel, M. Saulino, The ReWalk powered exoskeleton to restore ambulatory function to individuals with thoracic-level motor-complete spinal cord injury, Am. J. Phys Med. Rehabil. 91 (11) (2012) 911–921.

[67] https://eksobionics.com/.

[68] http://asimo.honda.com/innovations/feature/body-weight-support-assist/.

[69] S. Chen, J. Lach, B. Lo, G.Z. Yang, Toward pervasive gait analysis with wearable sensors: a systematic review, IEEE J. Biomed. Health Inform. 20 (6) (2016) 1521–1537.

[70] http://www.btsbioengineering.com/products/g-walk/.

[71] https://www.apdm.com/wearable-sensors/.

[72] https://www.xsens.com/.

[73] L. De Vito, O. Postolache, S. Rapuano, Measurements and sensors for motion tracking in motor rehabilitation, IEEE Instrum. Measur. Mag. 17 (3) (2014) 30–38.

[74] F.B. Horak, Postural orientation and equilibrium: what do we need to know about neural control of balance to prevent falls? Age Ageing 35 (Suppl. 2) (2006) ii7–ii11.

[75] Y. Agrawal, J.P. Carey, C.C. Della Santina, M.C. Schubert, L.B. Minor, Disorders of balance and vestibular function in US adults: data from the National Health and nutrition examination survey, 2001–2004, Arch. Intern. Med. 169 (10) (2009) 938–944.

[76] R.J. Peterka, Sensorimotor integration in human postural control, J. Neurophysiol. 88 (3) (2002) 1097–1118.

[77] R. Dewar, S. Love, L.M. Johnston, Exercise interventions improve postural control in children with cerebral palsy: a systematic review, Dev. Med. Child Neurol. 57 (6) (2015) 504–520.

[78] C. Wall III, D.M. Merfeld, S.D. Rauch, F.O. Black, Vestibular prostheses: the engineering and biomedical issues, J. Vestib. Res. 12 (2, 3) (2002) 95–113.

[79] S.R. Harris, L. Roxborough, Efficacy and effectiveness of physical therapy in enhancing postural control in children with cerebral palsy, Neural Plast. 12 (2–3) (2005) 229–243.

[80] B.A. Garner, B.R. Rigby, Human pelvis motions when walking and when riding a therapeutic horse, Hum. Mov. Sci. 39 (2015) 121–137.

[81] C. Shirota, E. Van Asseldonk, Z. Matjačić, H. Vallery, P. Barralon, S. Maggioni, et al., Robot-supported assessment of balance in standing and walking, J. Neuroeng. Rehabil. 14 (1) (2017) 80.

[82] http://www.natus.com/index.cfm?page=products_1&crid=271.

[83] http://www.neuro-kinetics.com/index.aspx.

[84] A.D. Goodworth, Y. Wu, D. Felmlee, E. Dunklebarger, S. Saavedra, A trunk support system and approach to study posture control in populations lacking full sitting ability, IEEE Trans. Neural Syst. Rehabil. Eng. 25 (1) (2017) 22–30.

[85] M. Vukobratovic, A.A. Frank, D. Juricic, On the stability of biped locomotion, IEEE Trans. Biomed. Eng. (1) (1970) 25–36.

[86] M.B. Popovic, A. Goswami, H. Herr, Ground reference points in legged locomotion: definitions, biological trajectories and control implications, Int. J. Robot. Res. 24 (12) (2005) 1013–1032.

[87] H. Herr, M. Popovic, Angular momentum in human walking, J. Exp. Biol. 211 (4) (2008) 467–481.

13

Wheelchairs and Other Mobility Assistance

Dmitry A. Sinyukov*, Karen L. Troy†, Matthew P. Bowers†, Taskin Padir*

*NORTHEASTERN UNIVERSITY, BOSTON, MA, UNITED STATES †WORCESTER POLYTECHNIC INSTITUTE, WORCESTER, MA, UNITED STATES

CHAPTER OUTLINE

Biomechatronics. https://doi.org/10.1016/B978-0-12-812939-5.00013-6

13.1 Introduction

Almost every eighth person in the United States has a certain degree of disability [1]. For 130,000 people suffering from paraplegia (severe or complete loss of motor function in the lower extremities and lower portions of the trunk) and almost 120,000 people suffering from quadriplegia (when all four limbs are affected) in the United States [2], the use of wheelchairs is typically necessary in order to perform activities of daily living (ADLs). When quadriplegia is combined with anarthria (loss of speech), but consciousness is preserved, the condition is known as locked-in syndrome (LIS) [3].

Table 13.1 maps health conditions to the corresponding levels of physical capabilities and provides the prevalence data for the United States. Each health condition imposes specific constraints on the type of mobility aid that can be used by the patient. People whose lower body is affected can resort to walking aids, such as canes, crutches, and walkers. Paraplegia requires a manual or an electric wheelchair. If the upper body is partially or completely paralyzed, standard joystick control interfaces cannot be used, and then alternative human-machine interfaces (HMIs), such as brain-computer interfaces (BCI), single-switch, silent speech interfaces, and others must be employed.

13.1.1 Learning Objectives

At completion of this chapter, students will learn about the following:

- Manual wheelchairs and wheelchair propulsion biomechanics
- Electric wheelchairs: modeling and control
- Wheelchairs with low-throughput HMIs: modeling, shared steering control, and shared position control
- Stair-climbing wheelchairs
- Assisted walking

13.2 Manual Wheelchairs

13.2.1 History of Wheelchair Design

The first wheelchairs were, quite literally, chairs with wheels. While these allowed a person, who was unable to walk, to be transported more efficiently than being carried, there were many practical problems with the devices. These included a high center of gravity (COG), lack of stability, and lack of independence of the user. Self-propelled wheelchairs were first recorded in Europe in the 17th century. Initially, a variety of forms (three- and four-wheeled) and propulsion methods were employed [7].

The "Bath chair" (Fig. 13.1), invented by James Heath in the 1750s, became a popular means of transport for people who were injured, or sick. This chair was similar to a modern-day jogging stroller, in that it had three wheels and a large seat that could recline. Similar to earlier chairs, users lacked the ability to independently transport themselves, an assistant had to push the wheelchair [8].

Table 13.1 Classification of Disabilities and Their Prevalence in the US

Body function	Normal		Paraplegia	Quadriplegia	Locked-in syndrom (LIS)			Vegetative state	Coma
					Incomplete	Classic	Total		
Lower body mobility	+	Limited	Severe or complete loss		Remnants of voluntary movement	–	–	–	–
Upper body mobility	+	+	+			–	–	–	–
Neck movement	+	+	+	+		–	–	–	–
Speech	+	+	+	+		–	–	–	–
Facial expressions	+	+	+	+		–	–	–	–
Eye movement	+	+	+	+		Vertical	–	–	–
Awareness	+	+	+	+		+	+	–	–
Arousal	+	+	+	+		+	+	+	–
Prevalence(a) in the US	99,921 (b)		40.6(b)	36.7(b)	< 0.1(c)–1.0(d)			0.2–3.4(e)	?(f)
Type of mobility aid	-	Canes, crutches walkers	Manual and electric wheelchairs	Electric wheelchairs with low throughput human-machine interfaces				-	-

(a) Estimated number of people managing the condition at any given time per 100,000 of population.
(b) Based on Ref. [2] and US population estimated at 320 million.
(c) According to Ref. [4].
(d) Assuming "several thousand" people in the US as reported by gocognitive [5].
(e) According to Pisa et al. [6].
(f) No data available.

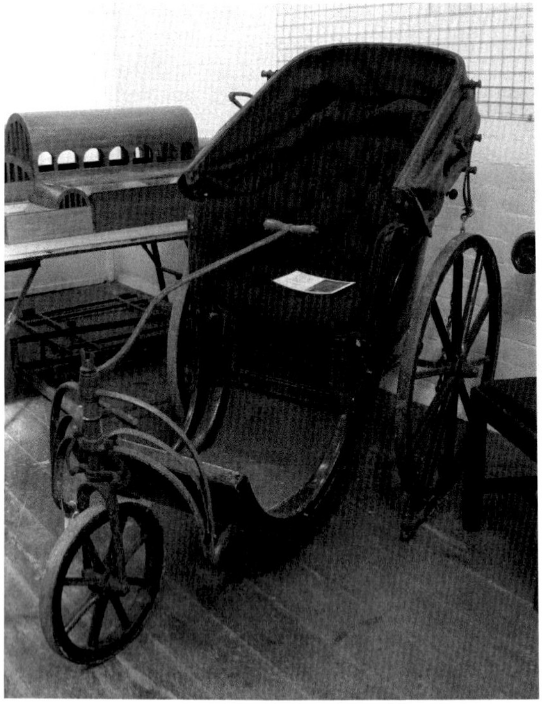

FIG. 13.1 An early 18th-century "Bath chair" from the British Museum collection. *From https://en.wikipedia.org/wiki/Bath_chair#/media.*

After the Civil War in the United States, wheelchairs became used more frequently by people with injuries that prevented them from walking. Improvements in wheel and chair design increased their utility. However, the modern wheelchair was not invented until 1932, when the first folding chair, made from tubular steel, was invented by two engineers, Herbert Everest and Harry Jennings [7, 9].

After World War II, the demand for better wheelchairs increased, leading to the development of more lightweight manual and motorized chairs. In all wheelchairs, improvements in ergonomics and overall design have led to more comfortable, maneuverable, and practical systems. For manual wheelchairs, the primary innovations have been in making chairs substantially more lightweight and customized to fit the user's body. These innovations are important in reducing the very high incidence of shoulder injury among manual wheelchair users [9]. For electric motorized models, the major improvements have been in stability, maneuverability, and the variety of control systems that can be used to direct the chairs.

13.2.2 Modern Manual Wheelchairs

There are approximately 65 million people worldwide who need a wheelchair. In the United States, roughly 3.3 million of these people are manual wheelchair users. Manual wheelchairs: (1) do not require a power or battery source, (2) are lightweight and portable, and (3) can be used in both low- and high-resourced regions. Hence, they are the primary choice for individuals who are able to use their upper body for propulsion [9, 10].

Modern wheelchairs consist of a seat cushion and back with adjustable tilt, depth, and height, customizable rear wheel and caster wheel location and configuration, and front wheels. Currently, many ultralight wheelchairs are available with a mass of <10 kg. Much of this weight savings comes from customizing the chair itself, rather than building adjustability and ranges into the chair. The Rehabilitation Engineering Society of North America (RESNA) recommends that manual wheelchairs are as light as possible, and fully customized for the user [9]. Specifically, they recommend that the following parameters are adjusted, based on individual measurements:

- Seat surface height at the front and rear edge
- Seat plane angle, width, and depth
- Back support height and angle
- Foot support to seat length
- Leg to seat surface angle
- Horizontal and vertical position of the rear wheel axle
- Rear wheel camber, type, and size
- Caster type and size

These adjustments affect wheelchair performance and function, including rolling resistance and propulsion efficiency, the tendency of the chair to turn downhill, yaw and pitch axis control, which affect maneuverability and turning, as well as stability, transportability, and overall footprint [10].

Making these adjustments is important because manual wheelchair users experience a high incidence of upper extremity pain and dysfunction. This includes carpal tunnel syndrome, rotator cuff injuries at the shoulder, and other disorders. The upper extremities did not evolve to perform weight-bearing and high-force tasks such as wheelchair propulsion and transfers. To reduce the number of secondary conditions related to upper extremity injuries in manual wheelchair users, biomechanics research has been directed at better understanding the factors that affect joint and muscle loading during wheelchair propulsion and transfers.

13.2.3 Overview of Wheelchair Biomechanics

Wheelchair biomechanics is treated in a similar manner as gait biomechanics, in that both activities are cyclic in nature. Gait cycle is typically divided into stance phase and swing phase. Similarly, manual wheelchair propulsion can be broadly divided into two phases: a push phase and a recovery phase. During the push phase, the hands of the wheelchair user are in contact with the push rim. The muscles in the shoulder, elbow, and torso are used to drive the rim forward until the hand releases the rim. During the recovery phase, the arms freely swing back into position to begin another push phase. As with gait biomechanics, wheelchair propulsion is typically expressed as a normalized propulsion cycle (0%–100%)

beginning and ending with a hand-rim contact during the push phase. Typical propulsion kinematics for the shoulder are shown in Fig. 13.2.

13.2.4 Kinematic Measurement

Wheelchair propulsion kinematics are typically measured using a motion capture system. The marker set that is used to define upper extremity motion must provide enough detail to resolve the three-dimensional motion of each segment: hand, forearm, upper arm, trunk, and possibly the scapula and clavicle. The International Society of Biomechanics has established standards for measuring and reporting upper extremity motion [11].

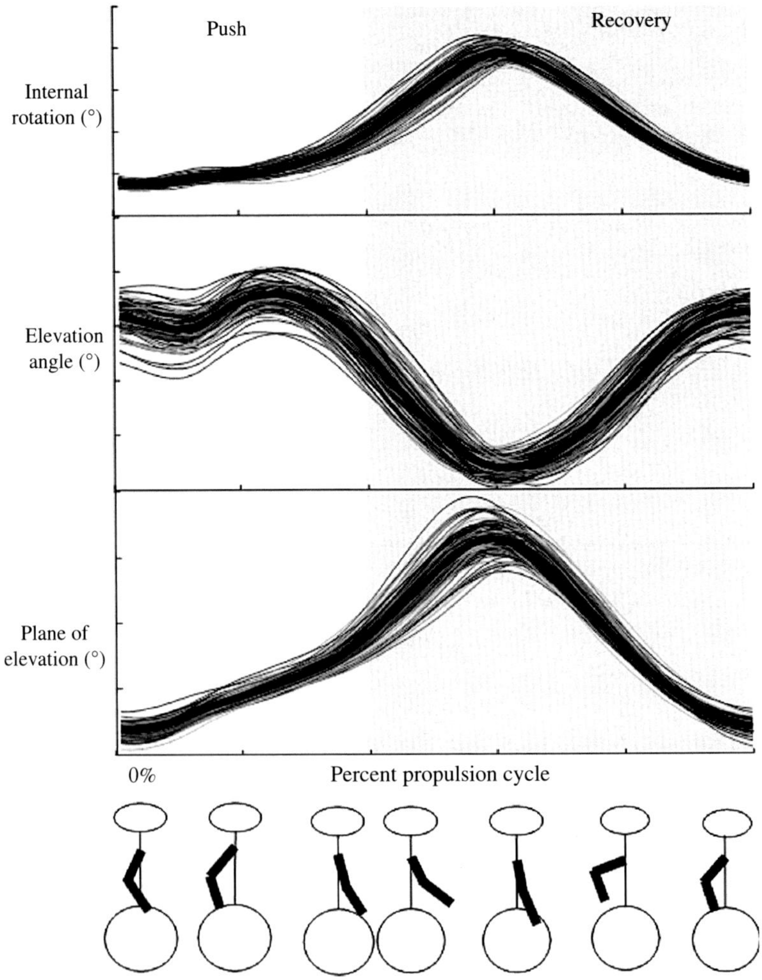

FIG. 13.2 Time-normalized propulsion cycles for 2-min trial at a medium speed for a representative subject. Unshaded area: push phase (max elbow flexion to max elbow extension). Shaded area: recovery phase (max elbow flexion to max elbow flexion). Thin lines: individual propulsion cycles. Thick line: mean of individual propulsion cycles.

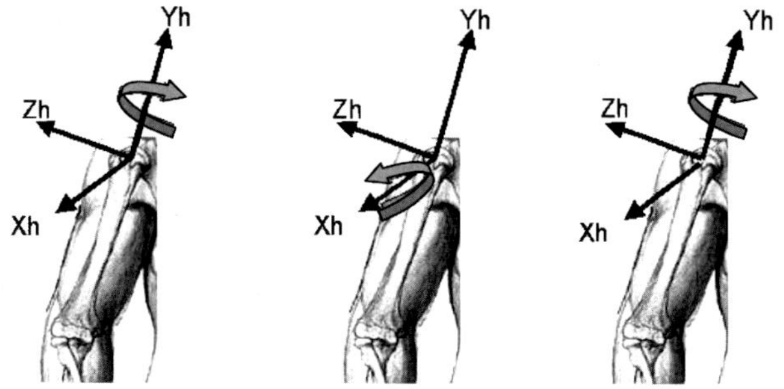

FIG. 13.3 Definition of rotation of the humerus about the thorax, using a *Y-X-Y* Euler rotation sequence [11].

This includes detailed definitions of local segment coordinate systems, relative to anatomic landmarks, and Euler sequences of rotation for defining relative motion between segments as in Fig. 13.3.

13.2.5 Kinetic Measurement of Wheelchair Propulsion

Several approaches have been taken to measure the forces and moments produced during manual wheelchair propulsion. Ultimately, this requires the use of an instrumented wheel, instrumented glove, or another method to measure forces and moments generated through hand/rim contact. It also requires clear definitions of upper extremity kinetics, which have the potential to be ambiguous depending on the frame of reference [12]. Similar to the ISB recommendations on defining and reporting joint motion, joint kinetics should be reported with reference to local joint coordinate systems and using Euler sequences of rotation to define moments [12].

A commercially available SMARTWheel (Fig. 13.4) is now widely used for these measurements. The SMARTWheel was first developed in 1989 by Cooper and Cheda to provide users with feedback to improve wheelchair racing performance, but has since evolved into an important clinical tool [14]. Initial designs of the SMARTWheel included strain-gauged spokes, combined with an encoder to measure wheel position. This design was refined and improved over the following decade and commercialized in 2000. This device can resolve anterior-posterior, medial-lateral, and superior-inferior forces on the wheelchair rim. The wide availability of this measurement tool since its commercialization has contributed significantly to the understanding of wheelchair propulsion biomechanics. Although it remains an important research tool, it has also been adapted to be useful for giving wheelchair users clinically relevant biomechanical feedback.

Around the same time the SMARTWheel was being developed, other research groups developed similar approaches to calculating hand/rim forces. One approach was to strain gauge the handrim mounts (to the wheels). This method allowed force to be measured over time, but without a precise wheel position [15]. Later, Sabick et al. used a combination of load

FIG. 13.4 Commercially available SMARTWheel, which can measure 6 DoF forces and moments during wheelchair propulsion [13].

cells and optical tracking to resolve push-rim forces and to obtain the direction of the forces [16]. Regardless of the specific technology used, measurement of the magnitude and direction of the forces applied to the push-rim during wheelchair propulsion, in combination with the position of the body, is needed to fully quantify the biomechanics of the task.

13.3 Electric Wheelchairs

13.3.1 Modeling Electric Wheelchairs

Model-based development lies at the core of modern engineering. In wheelchair control systems, a mathematical model is necessary for simulation and control system design. A typical electric wheelchair is a differentially driven platform: it has two independently driven wheels placed on either side of the robot body. Additionally, it is normally equipped with two or four caster wheels to maintain the balance. Such kinematic scheme simplifies the mechanical design, but locks 1 degree of freedom (DoF) (lateral motion) making the system nonholonomic.

Several wheelchair mathematical models have been proposed by the research community. They differ in the amount of physical detail. Here, we present a model that is based on Ref. [17]. The model is built with the following assumptions: ground surface is flat, COG location is arbitrary, caster wheels, air drag, frictional moments in the wheel/surface contact region are neglected, linear models are assumed for wheel rolling resistance, axle friction, and direct current (DC) motor torques. Based on these assumptions, we now develop kinematic and dynamic models of an electric wheelchair.

13.3.1.1 Kinematic Model

In mechanical systems, kinematic models relate actuator velocities to the object velocities. In our case, forward kinematic equations express the dependency of the wheelchair linear and angular velocities V and ω on the angular velocities $\dot{\theta}_1$ and $\dot{\theta}_2$ of the right and left wheels, respectively (refer to Fig. 13.5):

$$\begin{bmatrix} V \\ \omega \end{bmatrix} = \frac{R}{2r}\begin{bmatrix} r & r \\ 1 & -1 \end{bmatrix}\begin{bmatrix} \dot{\theta}_1 \\ \dot{\theta}_2 \end{bmatrix} \quad (13.1)$$

where R is the radius of the wheel and r is half of the wheelbase width.

Since our model permits an XY-displacement of the COG C, for the dynamic model, we will need to calculate the COG velocity:

$$V_{\vec{C}} = \vec{V} + \vec{\omega} \times \vec{d} = \begin{bmatrix} 0 \\ V \end{bmatrix} + \begin{bmatrix} \omega\overline{y} \\ \omega\overline{x} \end{bmatrix} = \begin{bmatrix} 0 & -\overline{y} \\ 1 & \overline{x} \end{bmatrix}\begin{bmatrix} V \\ \omega \end{bmatrix} \quad (13.2)$$

where $(\overline{x}, \overline{y})$ is the displacement of the COG in the $x'y'$-coordinate system.

The COG acceleration $a_{\vec{C}}$ can, thus, be expressed as

$$a_{\vec{C}} = \begin{bmatrix} a_{Cx} \\ a_{Cy} \end{bmatrix} = \dot{\vec{V}}_C = \frac{R}{2r}\begin{bmatrix} 0 & -\overline{y} \\ 1 & \overline{x} \end{bmatrix}\begin{bmatrix} r & r \\ 1 & -1 \end{bmatrix}\begin{bmatrix} \ddot{\theta}_1 \\ \ddot{\theta}_2 \end{bmatrix} = \frac{R}{2r}\begin{bmatrix} -\overline{y} & \overline{y} \\ r+\overline{x} & r-\overline{x} \end{bmatrix}\begin{bmatrix} \ddot{\theta}_1 \\ \ddot{\theta}_2 \end{bmatrix} \quad (13.3)$$

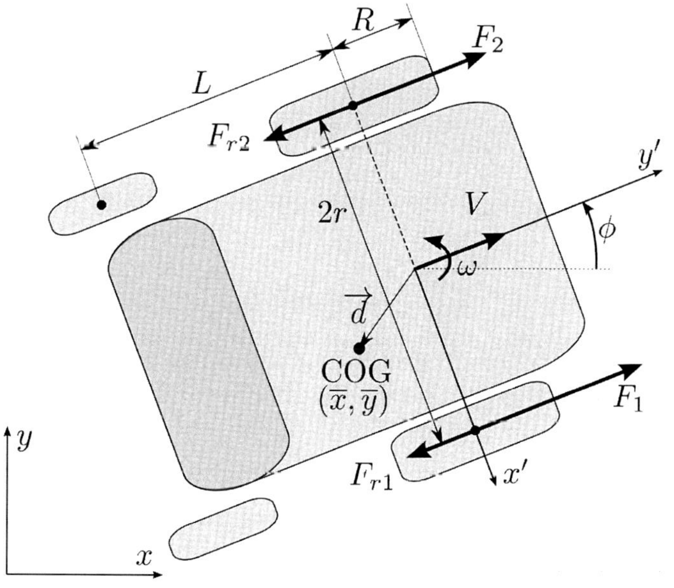

FIG. 13.5 Differentially driven robot (wheelchair) with an XY-displacement of center of gravity.

13.3.1.2 Dynamic Model

By summing forces along the y'-axis, we can write:

$$ma_{Cy} = F_1 + F_2 - F_{r1} - F_{r2} \tag{13.4}$$

where m is the sum of the masses of the wheelchair frame and the operator, F_1 and F_2 are wheelchair driving forces of the right and left wheels, respectively, F_{r1} and F_{r2} are resistive forces of the right and left wheels which are a result of wheel rolling resistance and axle friction.

To include the rotational component, we will apply D'Alembert's principle and take moments about the origin of XY. This yields

$$I_0\dot{\omega} + \overline{x}\cdot ma_{Cy} - \overline{y}\cdot ma_{Cx} = r(F_1 - F_{r1} - (F_2 - F_{r2})) \tag{13.5}$$

where I_0 is the moment of inertia of the frame.

By substituting (13.1), (13.3) into (13.4), (13.5), we will have

$$M^{2\times2}\begin{bmatrix}\ddot{\theta}_1\\ \ddot{\theta}_2\end{bmatrix} + \begin{bmatrix}1 & 1\\ r & -r\end{bmatrix}\begin{bmatrix}F_{r1}\\ F_{r2}\end{bmatrix} = \begin{bmatrix}1 & 1\\ r & -r\end{bmatrix}\begin{bmatrix}F_1\\ F_2\end{bmatrix} \tag{13.6}$$

where

$$M^{2\times2} = \frac{R}{2r}\begin{bmatrix}m(r+\overline{x}) & m(r-\overline{x})\\ I_0 + m(\overline{x}^2 + r\overline{x} + \overline{y}^2) & -I_0 + m(-\overline{x}^2 + r\overline{x} - \overline{y}^2)\end{bmatrix} \tag{13.7}$$

After switching to actuator coordinate in (13.6), we will have

$$M_{mech}^{2\times2}\begin{bmatrix}\ddot{\theta}_1\\ \ddot{\theta}_2\end{bmatrix} + \begin{bmatrix}F_{r1}\\ F_{r2}\end{bmatrix} = \begin{bmatrix}F_1\\ F_2\end{bmatrix} \tag{13.8}$$

where

$$M_{mech}^{2\times2} = \frac{R}{4r^2}\left(I_0\begin{bmatrix}1 & -1\\ -1 & 1\end{bmatrix} + m\begin{bmatrix}(r+\overline{x})^2 + \overline{y}^2 & r^2 - (\overline{x}^2 + \overline{y}^2)\\ r^2 - (\overline{x}^2 + \overline{y}^2) & (r-\overline{x})^2 + \overline{y}^2\end{bmatrix}\right) \tag{13.9}$$

Based on the Amontons' laws of friction, the resistive forces F_{rk}, $(k=1,2)$ can be approximated as

$$F_{rk} = \mu N_k \operatorname{sign}(\dot{\theta}_k) \tag{13.10}$$

where μ is the constant friction coefficient, N_1 and N_2 are the weights supported by the right and left wheel, respectively. The friction coefficient can be further decomposed:

$$\mu = c + R_{ax}C_{ax} \tag{13.11}$$

where c is the wheel rolling resistance coefficient c, C_{ax} is the coefficient of axle friction, and R_{ax} is wheel axle radius.

A wheelchair, as a four-wheeled vehicle, is a statically indeterminate system which means that for accurate calculations, the deformations of the wheels and the ground need to be accounted for. Here we employ a simplified approach used in Ref. [17]. It rests on two assumptions: all wheels have the same elastic properties, normal force linearly depends on the position of COG:

$$N_1 = \frac{mg}{2}\left(1+\frac{\overline{x}}{r}\right)\left(1+\frac{\overline{y}}{L}\right) \tag{13.12}$$

$$N_2 = \frac{mg}{2}\left(1-\frac{\overline{x}}{r}\right)\left(1+\frac{\overline{y}}{L}\right) \tag{13.13}$$

where L is the distance between the front and the rear wheels (see Fig. 13.5) and g is the gravity acceleration. Now, we can write for the resistive forces:

$$\begin{bmatrix} F_{r1} \\ F_{r2} \end{bmatrix} = T_{mech}^{2\times2}\begin{bmatrix} \text{sign}\dot{\theta}_1 \\ \text{sign}\dot{\theta}_2 \end{bmatrix} \tag{13.14}$$

where

$$T_{mech}^{2\times2} = \frac{\mu mg(L+\overline{y})}{2rL}\begin{bmatrix} r+\overline{x} & 0 \\ 0 & r-\overline{x} \end{bmatrix} \tag{13.15}$$

The dynamic equations of the wheelchair model can, thus, be formulated as

$$M_{mech}^{2\times2}\begin{bmatrix} \ddot{\theta}_1 \\ \ddot{\theta}_2 \end{bmatrix} + T_{mech}^{2\times2}\begin{bmatrix} \text{sign}\dot{\theta}_1 \\ \text{sign}\dot{\theta}_2 \end{bmatrix} = \begin{bmatrix} F_1 \\ F_2 \end{bmatrix} \tag{13.16}$$

The mechanical model (13.16) can further be extended to include motor dynamics. For each force F_{rk}, $(k=1,2)$, exerted by the motors, we will use the standard linear DC-motor model:

$$F_k = \frac{G}{R}(K_T i_k - J_m\ddot{\theta}_k - \beta_m\dot{\theta}_k), k=(1,2) \tag{13.17}$$

where i_k are the motor currents, G is the gear reduction factor, K_T is the motor torque constant, J_m is the motor shaft moment of inertia, and β_m is the motor shaft damping. The dynamic equations with the motor current as the control signal becomes

$$M_{curr}^{2\times2}\begin{bmatrix} \ddot{\theta}_1 \\ \ddot{\theta}_2 \end{bmatrix} + C_{curr}^{2\times2}\begin{bmatrix} \dot{\theta}_1 \\ \dot{\theta}_2 \end{bmatrix} + T_{mech}^{2\times2}\begin{bmatrix} \text{sign}\dot{\theta}_1 \\ \text{sign}\dot{\theta}_2 \end{bmatrix} = \begin{bmatrix} i_1 \\ i_2 \end{bmatrix} \tag{13.18}$$

where

$$M_{curr}^{2\times2} = \frac{J_m}{K_T}\begin{bmatrix} 1 & 0 \\ 0 & 1 \end{bmatrix} + \frac{R^2}{4r^2GK_T}\left(I_0\begin{bmatrix} 1 & -1 \\ -1 & 1 \end{bmatrix} + m\begin{bmatrix} (r+\overline{x})^2+\overline{y}^2 & r^2-(\overline{x}^2+\overline{y}^2) \\ r^2-(\overline{x}^2+\overline{y}^2) & (r-\overline{x})^2+\overline{y}^2 \end{bmatrix}\right) \tag{13.19}$$

$$C_{curr}^{2\times2} = \frac{\beta_m}{K_T}\begin{bmatrix} 1 & 0 \\ 0 & 1 \end{bmatrix} \tag{13.20}$$

$$T_{curr}^{2\times2} = \frac{\mu mgR(L+\overline{y})}{2rLGK_T}\begin{bmatrix} r+\overline{x} & 0 \\ 0 & r-\overline{x} \end{bmatrix} \tag{13.21}$$

Eqs. (13.18)–(13.21) comprise a complete dynamic model of an electric wheelchair with DC motors.

13.3.2 Advanced Wheelchair Motion Control

Most commercially available wheelchairs come with PID (proportional-integral-derivative) velocity controllers. They usually have to be tuned for each individual operator by adjusting PID regulator coefficients, speed, and acceleration limits. The challenge is that wheelchairs are systems with changing parameters: rolling friction depends on the ground surface, the COG moves with the posture of the operator, and even the total mass of the system may vary. This affects the performance of PID controllers. An adaptive controller can solve this problem.

A number of adaptive motion control schemes have been proposed for generic differentially driven mobile robots. Some schemes, for example, [18, 19], consider only robot kinematics. With inertial robots, such as robotic wheelchairs, however, ignoring system dynamics leads to undesirable tracking errors. Other researchers have investigated various adaptive controllers based on robot dynamic models of various depth. An adaptive fuzzy logic-based controller [20] uses voltages as control signals of the system. Most of the researchers, however, when designing adaptive controllers for differentially driven robots, choose motor torques as control signals [21–25]. Among this group one of the most popular controller design techniques is backstepping from kinematics to dynamics [22–24, 26–28]. This method, while providing certain structure to control systems and simpler design, is not proved to be the most efficient one. Few researchers consider the problem in a more general formulation. In Ref. [27], the COG position is considered lying on the symmetry line of the robot which may not always be correct for a wheelchair. In Ref. [25], friction forces are ignored and the same simplified model is used for COG. An interesting approach was chosen in Ref. [28], where the inputs to the control system are wheelchair velocities, even though a dynamic model is considered. A common disadvantage of these implementations is the assumption that the COG of the system lies on the symmetry axis of the robot. With the wheelchair, however, the COG may be offset in both x and y directions. Moreover, the use case when the COG suddenly moves from left to right or any other direction is frequent in this application. It is especially interesting how an adaptive controller would react to such a change.

The research of adaptive control in its application specifically to robotic wheelchairs has been fairly limited. While certain results toward parameter self-adjustment and adaptive control for wheelchairs have been reported in Refs. [29, 30], most of the commercially available wheelchairs still employ standard PID controllers with fixed or manually configurable parameters. Ref. [31] is a notable work where the COG is not required to be on the robot symmetry axis. In that work, the output of the adaptive controller is a vector of linear and angular velocities assuming an additional low-level PD velocity controller.

The advantages of such approach when used on a wheelchair are somewhat questionable given that commercial wheelchairs are not typically equipped with velocity sensors, and thus do not implement velocity control. Moreover, the communication protocols for commercial wheelchair motor controllers are usually proprietary. These two factors usually require a replacement of the motor controller when any autonomy is desired. Once the controller is replaced, operating directly with motor voltages or currents seems a much more natural solution. In Ref. [32], another type of an adaptive controller with online parameter estimation is developed for a wheelchair using a dynamic model from Section 13.3.1. The controller uses motor currents as the control signals for the system. The rationale for that is that most commercial motor controllers communicate over a relatively slow interface such as RS-232 with a significant overhead of a text-based protocol. This may not allow to run a relatively fast closed-loop system with voltage being an input to the system. While the controller (Fig. 13.6) is demonstrated to be stable in simulation to some parameter changes, no real robot experiments are presented for it.

13.4 Wheelchairs With Low-Throughput HMIs

Quadriplegia and LIS pose additional challenges for wheelchair designers since standard joystick interfaces cannot be used by patients with these conditions. Over the last several decades a number of nonconventional HMIs have been developed to accommodate this audience [33]. Single-switch interfaces, tongue drive and eye-tracking systems, speech recognition and silent speech interfaces, various types of electromyography (EMG)-based interfaces, and brain-computer interfaces (BCI) have been successfully employed for wheelchair navigation [33, 34].

Due to the limited amount of information that these interfaces can pass through, they are known as low-throughput human-machine interfaces. Whenever human-machine aspect is understood from the context, they are simply called low-throughput interfaces (LTI).

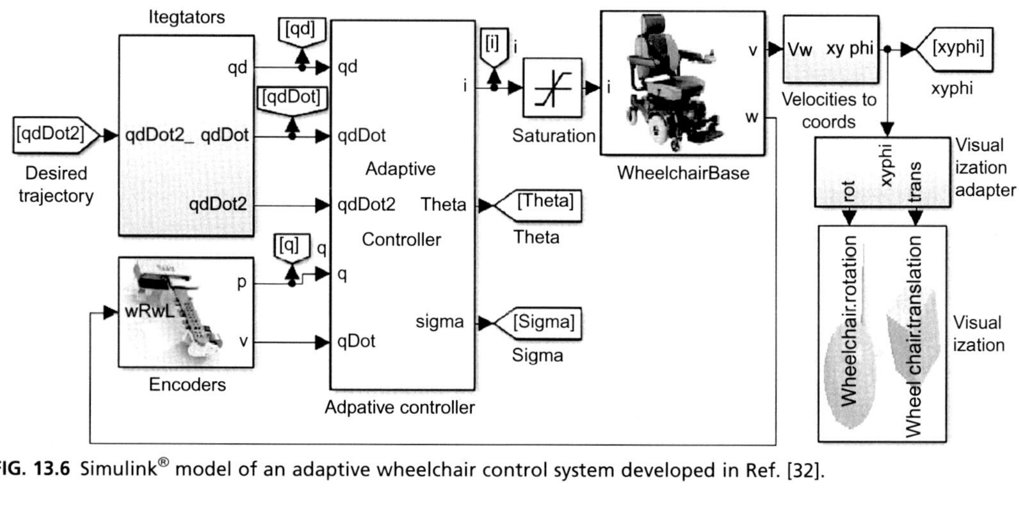

FIG. 13.6 Simulink® model of an adaptive wheelchair control system developed in Ref. [32].

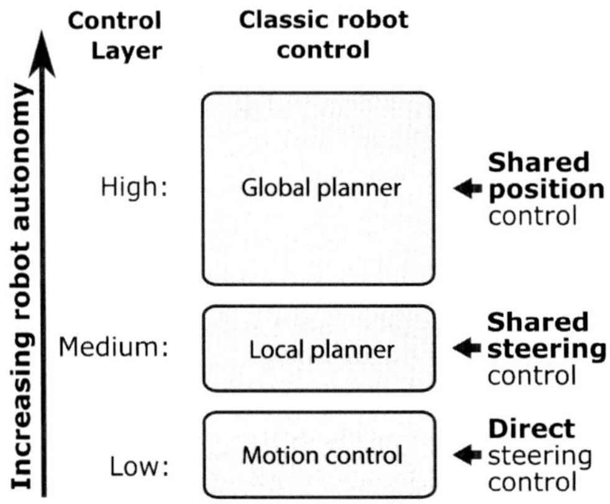

FIG. 13.7 Robot control levels.

Less information from the human means more autonomy of the robot. The diagram in Fig. 13.7 illustrates how various methods of navigation interact with the robot control system on different levels. Direct steering is a direct translation of human commands into the robot velocities, shared steering incorporates some robot intelligence to enable obstacle avoidance and/or direction prediction, and shared position control implements a fully autonomous navigation while the destination is inferred through an LTI.

13.4.1 Modeling Low-Throughput Human-Machine Interfaces

HMIs can be classified by temporal properties into synchronous (commands are selected at a certain constant frequency) and asynchronous (commands can be initiated by operator at any moment). Any asynchronous interface can be converted into synchronous by enforcing periodic inquiries to the user. The scanning interface is a classic example of turning a normally asynchronous button interface into a synchronous interface. The input device in this case is a single button, but multiple command options are displayed one after another. Whenever the desired command is presented, the operator has to press the button. Similarly, BCIs based on steady state visually evoked potential (SSVEP) often operate at a constant measurement period.

HMIs typically have limited resolutions, allowing the user to choose one option (command/letter) out of a finite number r of possible options (commands/letters). By selecting a specific option, the user conveys their intent to the machine. The set of all r permitted options is known as alphabet. For a regular joystick interface, depending on the joystick sensors, r can be so large that the notion of alphabet becomes less practical. For low-throughput HMIs, however, a typical value of r is relatively low. For example, SSVEP BCIs are usually designed for a selection of one pattern out of 2–4 stimuli displayed on the screen. Similarly, with scanning interfaces, only a limited number of options r can be presented to user.

HMIs can also be divided into deterministic and probabilistic based on the accuracy of intent detection. For example, joystick interface in a healthy user hand or a single-switch

Table 13.2 Interface Matrix

		Detected			
		L_1	L_2	...	L_r
Intended	L_1	$P(D_1\|I_1)$	$P(D_2\|I_1)$	...	$P(D_r\|I_1)$
	L_2	$P(D_1\|I_2)$	$P(D_2\|I_2)$	...	$P(D_r\|I_2)$
	...	...	...	...	...
	L_r	$P(D_1\|I_r)$	$P(D_2\|I_r)$	...	$P(D_r\|I_r)$

interface are normally deterministic (user commands are always detected correctly), whereas BCIs are naturally probabilistic (user-selected commands are detected correctly with a certain probability). In the latter case, the relation between the detected (D) and intended commands (I) can be described with a conditional probability matrix $P(D|I)$ (known as the interface matrix) of the following structure (Table 13.2):

Here $L_1 \ldots L_r$ are permitted commands (letters of the alphabet), $P(D_i|I_j)$ is the conditional probability that commands L_i is detected given that command L_j was intended. The interface matrix can be obtained by repetitive experiments which is, however, not a trivial task, since it may evolve over time due to the user fatigue, weariness of electrodes, and other factors. We can observe that deterministic HMIs are a special case of probabilistic HMIs where $P(D|I)$ is an identity matrix.

Resolution, accuracy, and frequency are mutually conflicting factors (a higher value of r can lead to lower accuracy or frequency, and so on). For a given HMI/operator, these parameters can be configured to maximize the information throughput. With this probabilistic model, a discrete low-throughput human-machine interface can be modeled as a noisy discrete information channel between a human intent and a robot control system (Fig. 13.8).

13.4.2 Direct Steering Control

Direct steering control is the most straightforward way to connect an LTI to a mobile robot. While this method has clear practical disadvantages (tendency for collisions with

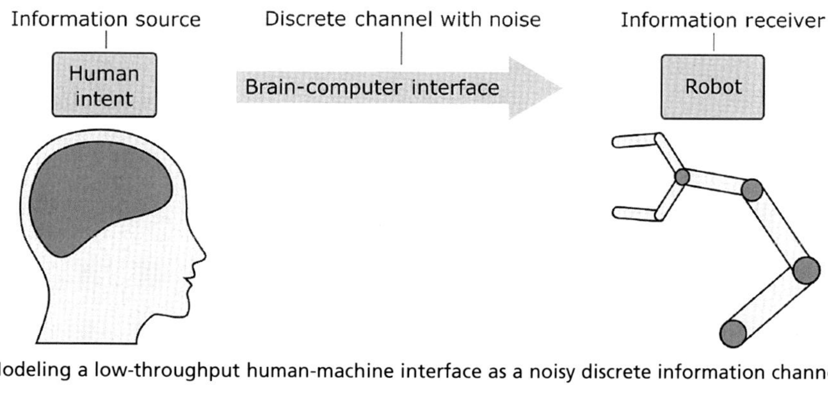

FIG. 13.8 Modeling a low-throughput human-machine interface as a noisy discrete information channel.

obstacles, slow or inaccurate navigation, early user fatigue), it is still popular among researchers, primarily as a method to demonstrate the advantages of a certain LTI.

In Ref. [35], three SSVEP stimuli are mapped to "forward," "left," and "right" commands that are passed to a small differentially driven robot. In Ref. [36], 10 elderly subjects steered a wheelchair in a virtual environment using standard directional commands ("forward," "left," "right," and "backward") detected with a P300-based BCI with tactile stimuli (four stimulators placed on left and right thighs, abdomen, and lower neck). After training, reported accuracy was 92.56%, while the information transfer rate (ITR) was 4.98 bits/min. Such an experiment would be impossible with a real wheelchair, however, due to inevitable collisions which are reported in this chapter. In Ref. [37], the wheelchair is steered with three commands: "forward," "left," "right." First, with an EMG armband from Thalmic Labs that is used to differentiate between three gestures: making a fist, hand release, and waving the hand right, and then with a consumer-grade Emotiv EPOC EEG-based BCI, where the commands are mapped to mental tasks: relaxing, math problem solving, and text reading. While authors report obstacle detection with a depth camera, the experimental results are only present for obstacle-free paths.

13.4.3 Shared Steering Control

Shared steering control combines robot intelligence with human intelligence in the velocity domain. Figs. 13.9 and 13.10 demonstrate a typical steering control interface. This method in the literature is associated with semi-autonomous navigation or assistive navigation. The shared control unit receives directional commands from the HMI and calculates the

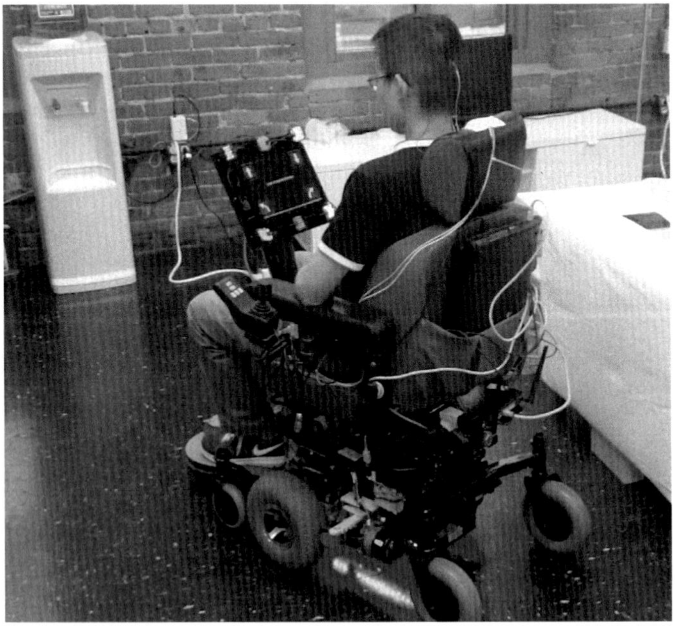

FIG. 13.9 SSVEP-based shared steering control in a home-like environment [38].

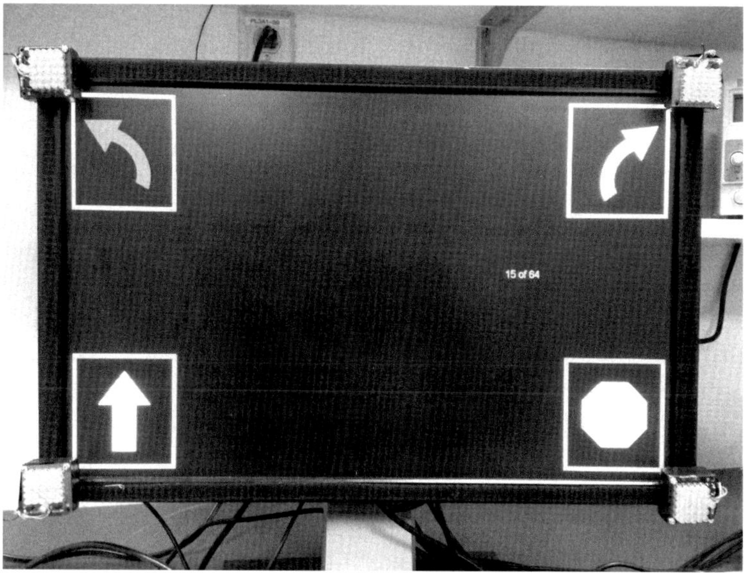

FIG. 13.10 A typical SSVEP-based steering control interface [38].

desired robot velocities that are passed to the motion control level. These velocities are calculated based on several additional factors. These typically are the information about obstacles around the robot (environmental knowledge), and/or, the human behavior model.

Obstacle configuration is usually obtained with robot sensors, such as laser-range scanners (LIDARs), depth cameras, vision, ultrasonic, and infrared sensors, while the behavioral model can be learned over time or manually configured in advance. By combining this data with the HMI commands, shared control can improve steering. Due to the decreasing cost of the sensors, a constantly improving performance of computing machines, and the advances in autonomous navigation algorithms, obstacle avoidance has become practical and affordable. Shared steering control methods can be divided into two categories: nonprobabilistic and probabilistic methods.

Nonprobabilistic methods are more common, though less efficient, since they do not utilize the maximum of HMI bandwidth, thus wasting some of the sparse information from the human. One of the simplest shared steering controls is implemented in Ref. [39] where obstacles are detected only in front of and behind the wheelchair with six ultrasonic sensors. If an obstacle is found in the direction of the human command, the command is not executed. In Ref. [40], the wheelchair is configured to normally move forward with constant velocity while avoiding obstacles. They used an interesting obstacle avoidance mechanism. Ten zones were defined around the robot. Whenever obstacles were present in a given zone, corrective velocities were calculated by modeling repulsive forces from the obstacles. Parameters for the calculation of corrective velocities for each zone were chosen empirically according to the wheelchair dynamics and sensor reliability. The operator could send "turn left/right" commands with a Motor imagery-based BCI. Whenever such a command is received, an attractor is placed in front of the wheelchair at an angle of 45°. It remains there until robot rotates 45° or a new BCI command is received. Ten close-range sonars and two off-the-shelf USB webcams are used to detect obstacles.

In Ref. [41], wheelchair steering control was implemented using the common four directional commands ("forward," "turn left," "turn right," and "stop") which were conveyed by the operator using an SSVEP-based BCI. An artificial potential field has been added to repel the wheelchair from obstacles detected by LIDAR. This enabled safe collision-free navigation.

Environmental knowledge obtained by robot sensors enables another steering model, namely opening selection which suits well Yes/No type of HMIs, such as P300 BCI. For example, in Ref. [42], virtual and real UAVs are navigated in an unknown two-dimensional (2D) space at constant altitude by six subjects using Yes/No commands mapped from the right- and left-hand motor imagery. The UAV stops at every crossroad (intersection) and determines feasible directions (openings) using range sensors. A scanning-like interface iterates through all openings and asks the operator if the given opening is the desired direction to go. The operator can answer with Yes or No. This system supports a manual control as well. When the operator discards all possible directions (openings), the left- and right-hand motor imagery commands are interpreted as turn left and turn right, while resting implies going forward. This model of selecting openings was initially proposed in Ref. [43].

In Ref. [44], a universal navigation framework incorporating voice control (CMU Sphinx and Google Glass) and facial expression control (Emotiv EPOC) with assistive navigation (obstacle avoidance) to enable shared steering control is presented. They implement tentacle-based obstacle avoidance (see below).

One of the earliest works on probabilistic shared steering control is found in Ref. [45]. The researchers mapped three EEG-based BCI mental tasks (left-hand motor imagery, relaxation, word task) to steering actions "forward," "left," and "right," but instead of simply accepting the command with the highest probability, the entire vector P_{EEG} of the three probabilities is kept. Then, another vector, P_{env} is constructed based on the obstacle configuration around the robot. This vector again assigns a probability to each of the three actions. The vectors are then fused using a simple element-wise product operation, and the action with the highest probability is chosen. By using the velocity increment steering model, they achieved a smooth context-aware wheelchair motion in a virtual environment. In Ref. [46], they report a real wheelchair navigation using this shared steering control method.

13.4.3.1 Shared Steering Control With Tentacle-Based Obstacle Avoidance

The general paradigm of tentacle-based obstacle avoidance [47] can be summarized as follows. First, a set of possible seeding velocities is identified based on the kinematic constrains, current velocity, acceleration, and velocity limits. Then for every seeding velocity, a trajectory is precalculated assuming that the velocity does not change over a configurable simulation time period. Then, every simulated trajectory is assigned a cost based on its proximity to obstacles and the difference between the desired user velocity and the given seed velocity. Finally, the seeding velocity that corresponds to the trajectory with the lowest cost is selected. The specifics of cost calculation vary. For example, the cost of the kth trajectory $C[k]$ can be calculated as a linear combination of obstacle cost $C_{obst}[k]$ and the velocity cost $C_{vel}[k]$:

$$C[k] = C_{vel}[k] + k_{obst}C_{obst}[k] \tag{13.22}$$

where k_{obst} is configured to reflect the desired relative importance of the two factors.

There are various methods to define the velocity cost which reflects the difference between the seed velocity $(V_{seed}, \omega_{seed})$ and the desired user velocity $(V_{user}, \omega_{user})$. As an example, it can calculated as a square of the distance between the velocities in a scaled velocity domain [44]:

$$C_{vel}[k] = (V_{user} - V_{seed}[k])^2 + \left(k_{ang}(\omega_{user} - \omega_{seed}[k])\right)^2 \tag{13.23}$$

where k_{ang} represents a configurable relative importance of the linear velocity compared to the angular velocity. There are various methods to calculate the obstacle cost as well. We discuss one approach that is implemented in the robotic operation system (ROS) [48] navigation stack. For every simulated trajectory, a set of poses along the trajectory are selected. The total cost of the k-th trajectory $C_{obst}[k]$ is calculated as a sum of the costs of each pose:

$$C_{obst}[k] = \sum_{i=1}^{n_k} C_{pose}[i][k] \tag{13.24}$$

Now, the cost of ith pose along the kth trajectory $C_{pose}[i][k]$ is calculated using a grid model of the environment. The data from various obstacle sensors are fused into the same grid map, such that every cell (x, y) on the map is assigned a cost $C(x, y)$. The cost is set "infinitely" large for cells that are identified as nontraversable and is assigned a high value in the nearby cells. Such map is known as a costmap. The cost of a pose $C_{pose}[i][k]$ is then identified as the maximum cost of a cell located within the robot support polygon (the projection of the robot on the horizontal 2D plane):

$$C_{pose}[i][k] = \max\left\{C(xy) : (xy) \in F_i^k\right\} \tag{13.25}$$

where F_i^k is the set of costmap cells that are covered by the robot support polygon if the robot is placed into the ith pose along the kth trajectory.

13.4.4 Shared Position Control

As its name implies, shared position control is implemented in the (x,y) domain rather than in the velocity domain. Destination selection is the simplest position control that works for multiple predefined locations. When the number of predefined destinations is not very high, they can be mapped directly to the HMI outputs. For example, multiple

SSVEP stimuli can be presented to the user, each corresponding to an individual destination [49]. When the number of destinations is higher than the number of options supported by the HMI, the destinations are repeatedly grouped into subsets. Among recent works, Ref. [50] is an example of such approach. They implement both shared steering and position control for a BCI-based wheelchair control. The position control is implemented for a Yes/No motor imagery interface and for a P300-based BCI. In the first case, a nonprobabilistic dichotomy is used to select the destination. For a P300 BCI, the destination IDs are arranged in a 10×4 matrix, and the selection process is designed to mimic P300 typing. Similar systems for destination selections are reported in Refs. [51–53].

More intelligent systems model user habits by assigning prior probabilities to each destination based on how often it is visited by the operator. This can enable a more efficient division into subsets that maximizes the amount of information received from human through the LTI. Overall, modeling user habits can improve the shared control experience [54]. At the border between steering and position control there is a notable work [55] where steering commands are used to infer one of the predefined destination poses (goals). The system is designed to control a mobile robot using a continuous or discrete HMI and provides a shared control that can recognize the user plan based on the steering commands. In this system, the navigation and inference processes are modeled with a partially observable Markov decision process (POMDP) in which the user intended plan (trajectory) is taken as the POMDP state, user commands (nine of them) are POMDP observations, and the wheelchair actions (nine of them) are POMDP actions. The system was demonstrated to be able to infer and execute complex maneuvers, such as parking next to a wall. One of the strong sides of that approach is an ability to model user capabilities in using the HMI. The main problem with this approach is that it limits user navigation to a specific set of goals. The demonstrated results are shown only in relatively simple environment with the maximum of only 21 user plans (hypothesis) maintained. Scaling this solution to a scenario with a significantly higher number of destinations will be a challenging problem.

A method for topological navigation is demonstrated in Ref. [56], where the robot calculates the most probable action and the operator can accept or discard it using an ErrP-based BCI. The environment is analyzed to determine places of interest (such as crossings). Overall, the systems with predefined lists of destinations limit the user in his freedom to move anywhere he wants.

One reason for steering navigation being popular in navigation with LTIs is that it naturally parallelizes the inference and navigation processes, thus often making it faster to navigate to the desired goal, as compared to first choosing the destination and then autonomously driving toward it. Such parallelization also gives the operator the feeling that he has a constant control over the moving robot.

There is group of share position control methods, however, known as local goal selection, that also provide such functionality. For example, in Ref. [57], a three-dimensional (3D) model of the local environment is constructed based on the sensor data. It is presented two the user with a set of possible local goals placed in the vertices of a polar grid centered at the current robot pose. The operator can then select one of the goals with a P300-based BCI, and the wheelchair would autonomously drive to the selected pose. Additionally, the interface allows the operator to select 90-degree left and right turns to choose local goals that are on the sides. A similar approach is developed in Refs. [58, 59], but in addition to local goals placed at fixed positions relative to the wheelchair, potential points of interest (such as tables) are identified based on depth sensor (Kinect) data and added to possible goals (Figs. 13.11 and 13.12). This enabled easy docking. Similarly, in Ref. [60] door ways are automatically detected to be quickly selected as destinations.

The shared position control method developed in Refs. [34, 61] uses a probability distribution over an entire known map to model the human intent estimate. To infer the destination, the map is repeatedly segmented into colored regions, each color corresponding to a specific LTI command (Fig. 13.13). The method parallelizes the inference and navigation processes by calculating probabilistically optimal waypoint locations based on the probabilistic human intent estimate. It allows the operator to navigate to any vertex of the grid map, maximizes the information throughput, and can model user habits in terms of popular areas, not just destinations.

13.5 Stair-Climbing Wheelchairs

Despite the increasing number of accessible public buildings, most private houses and even public infrastructure around the world, especially in rural areas, are not adapted for wheelchair users. One of the main challenges in such environments is ascending and descending the stairs in buildings that are not equipped with elevators.

Since the 1990s, a number of stair-climbing wheelchairs have been developed with a various degree of commercial success. The critical component of such systems is the

FIG. 13.11 Local goal selection using semantic goals: operator's view [58].

FIG. 13.12 Local goal selection using semantic goals: navigation [58].

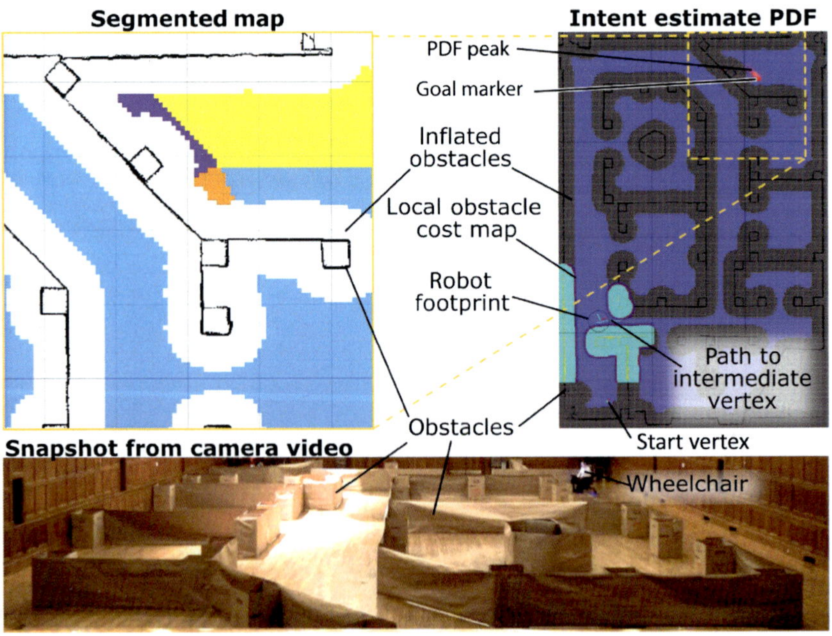

FIG. 13.13 NoVeLTI shared position control that uses colored map segmentation to gradually infer the destination [34, 61].

stair-climbing mechanism. It directly affects the cost, stability, control complexity, and energy consumption of the wheelchair [62]. There are five main types of such mechanisms.

- Track-based mechanisms utilize caterpillar tracks and their outer teeth to maintain a reliable contact with stair steps.
- Wheel cluster-based mechanisms use a set of 2–4 wheels evenly distributed around a central axis, known as a wheel cluster.
- Leg-based mechanisms are inspired by biological walking. They employ two or more robotic legs to propel the platform along the stairs.
- Hybrid mechanisms combine wheels and bar linkages.
- Nonconventional wheel geometry mechanisms propel the wheelchair using wheels with special geometric features that can provide the necessary traction on the stairs.

Only the first three categories have been successfully commercialized for electric wheelchairs and at least one commercial product exist for manual wheelchairs with nonconventional wheel geometry.

13.5.1 Track-Based Mechanisms

Track-based mechanisms are the most popular among the commercialized stair-climbing wheelchairs. The first successful machines in this category were developed at Nagasaki University. They designed several models with single and dual section tracks. The technology was patented in Japan in 1996 [62].

Over the past two decades, the mechanism evolved and now most of the commercially successful track-based stair-climbing wheelchairs are single track and are equipped with both: tracks for stair climbing and wheels for driving on a level ground. When mode switching is invoked by the operator, the wheels are raised up or the tracks are pushed down to engage the tracks with the stairs. Typical examples are TGR Explorer from TGR Instrument of Freedom company in Italy [63] and TopChair-S from TopChair Company in France [64]. Other notable examples are CaterWil GTS3 EXPORT (Fig. 13.14) from Katervil company in Russia [65] that was demonstrated to reliably drive even over slippery snow-covered stairs during the winter, and a stylish Scewo (Fig. 13.15) developed by five master's students from the Swiss Federal Institute of Technology and Zurich University of the Arts [66].

Most of the commercial track-based wheelchairs are relatively easy to control, do not require an assistant to operate, can climb stairs of various shape, including spiral stairs. The models with the dual wheel-track configuration have relatively low energy consumption and manufacturing cost since the same motors are used for the wheels and tracks. These properties make track-based systems very competitive on the market.

13.5.2 Wheel Cluster-Based Mechanisms

Wheel cluster-based mechanisms are also relatively cheap and compact. The iBOT (Fig. 13.16), one of the famous stair-climbing wheelchairs, was invented by Dean Kamen [67] and marketed through Johnson & Johnson. Back in 2004, it was greeted by FDA,

FIG. 13.14 CaterWil GTS3 EXPORT [65].

doctors, and patients as a potentially revolutionary device for the freedom of movement. The iBOT's wheels rotate up and over one another to go up and down stairs (or street curb). It uses gyroscopes to sense and help adjust a person's COG. The iBOT could also help a person reach the top shelf in the kitchen by taking the upright position.

In 2006, Medicare decided that lifting users and climbing stairs are not functions necessary for at-home care. Medicare would only cover the basic electric wheelchair price. The iBOT's price tag was $22,000, but Medicare only deemed it worth roughly $6000. Moreover, only a small fraction of paralyzed victims were eligible. For these reasons, Johnson & Johnson sold the last iBOTs, the first stair-climbing wheelchair in the United States, in the spring of 2009. In May 2016, Toyota partnered with Dean Kamen's DEKA company to facilitate the development and launch the next generation iBOT motorized wheelchair [68].

FIG. 13.15 Scewo [66].

On the other side of the price spectrum, in Ref. [69], a low-cost wheel cluster wheelchair prototype has been developed (Fig. 13.17) with the cost of additional components of 20% of the price an economical electric wheelchair.

Stair climbing with a wheel cluster is not as smooth as with track-based mechanism since it moves one step at a time. The resulting speed fluctuations make the motion less comfortable for the operator [62]. The fluctuation are lower with dual clusters systems, when the front wheels are also designed as a wheel cluster, for example in Ref. [69]. Additionally, wheel cluster mechanisms require more complex control than track-based system when climbing stairs. This affects both: the wheelchair control system and the operator skills. Most wheelchairs in this category require assistance in the climbing mode. This is the case with Scoiattolo 2000 [70] and iBOT. In Ref. [69], the wheelchair is shown to be independently driven over straight stairs by the operator, but stairs of more complicated shape, such as spiral, again, might not be that easily traversable with the proposed mechanism without an assistant. At the same time, driving over an even ground with wheel clusters is relatively energy efficient and switching between the modes is faster compared to tracks + wheels wheelchairs, because there is no need to shift the locomotion systems.

FIG. 13.16 Bill Clinton and Dean Kamen inventor of iBOT.

13.5.3 Leg-Based Mechanisms

Leg-based stair climbers can be divided into two subcategories: dual-locomotion systems and walking chairs. The mechanisms in the first category combine legs for stair claiming with wheels for driving on an even ground. C-Max U1 (Fig. 13.18) is one of several similar products from a UK-based company AAT GB Ltd. that can be classified as a dual-locomotion legged stair-climber [71]. The system, however, is not an independently driven wheelchair, but rather a special purpose device designed to help transporting people over stairs. It is equipped with actuated legs (pushers) that lift the platform one step at a time, while keeping the complex task of steering the chair with a human assistant. It can also be used as a pushing wheelchair due to a pair of passive wheels. C-Max U1 can operate even on narrow and winding staircases with steps up to 21 cm high.

A more sophisticated dual-locomotion mechanism (Fig. 13.19) is proposed and implemented at Seoul National University, Korea [72]. It consists of an actuated wheeled platform, two 3-DOF legs and a 1-DOF base. Due to a direction correction system, the

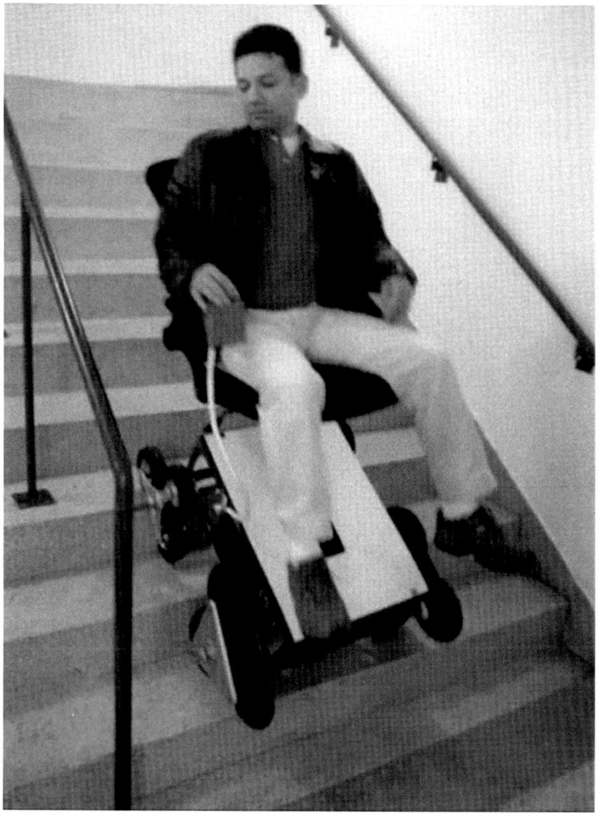

FIG. 13.17 Low-cost stair-climbing wheelchair (LCSCWC) [69].

wheelchair can climb without an assistant's help even slightly curved stairs with the slope angle up to 45°. It also supports standing and tilting modes that allow the operator to access objects at high positions and help reducing bedsores. This platform, however, is still a lab prototype.

Walking chairs are even more advanced systems and belong more to the class of humanoid biped robots than wheelchairs. They typically have complex leg mechanisms with 6–8 DOFs per leg and require sophisticated dynamic control algorithms. Due to a higher dimensionality of the configuration space, walking robots can execute advanced maneuvers, such as walking sideways and climbing obstacles of complicated shapes. At the same time, however, the cost and energy consumption of such mechanisms is much higher, while the speed of motion on a flat terrain is still relatively low. None of such system has reached the market, but several prototypes have been developed. A few examples are the WL-16 (Fig. 13.20) series of robots designed at Waseda and Kanagawa Universities in Japan [74, 75], HUBO FX-1 (Fig. 13.21) developed at KAIST, South Korea [76] and Toyota I-Foot [77].

FIG. 13.18 C-Max U1 push wheelchair [71].

13.5.4 Hybrid Mechanisms

Hybrid mechanisms use a combination of linkages and wheels to climb stairs. Due to the complexity of such systems, they are not as popular as other categories.

In Ref. [78], a hybrid stair-climbing wheelchair is proposed based on front and rear wheel clusters connected to the chair via powered linkages (Fig. 13.22). This design is motivated by two main requirements: always maintain at least four points of contact with

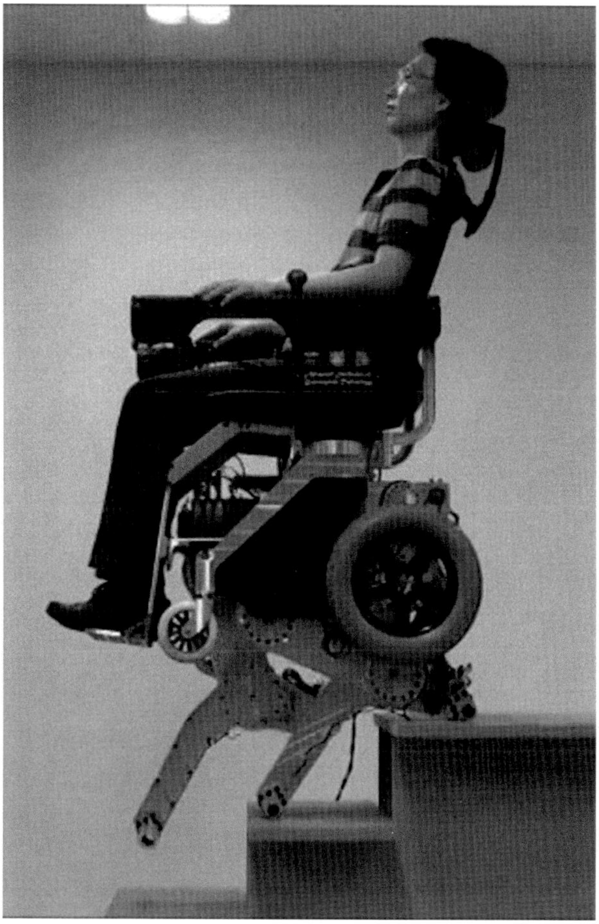

FIG. 13.19 Compact 7-DOF stair-climbing wheelchair [72].

the ground to ensure stability (safety) and autonomously climb steps as high as 75 cm (e.g., to load itself into a van). They demonstrated a numerical simulation of the system and a 1:6.25 scale model with a basic control system successfully executing the expected maneuvers. No full scale prototype, however, was presented.

Another hybrid wheelchair TBW-1 was developed with the purpose of reducing the pressure on stairs and particularly eliminating a concentrated load on the front edges of steps which is typically associated with track-based stair climbers. Additionally, the platform was expected to always maintain static stability for safety [79]. The wheelchair is designed as a “walking parallelogram” with a total of eight wheels attached to the vertices of two 4-bar linkages. A prototype of the vehicle was developed, but only stair descending was demonstrated.

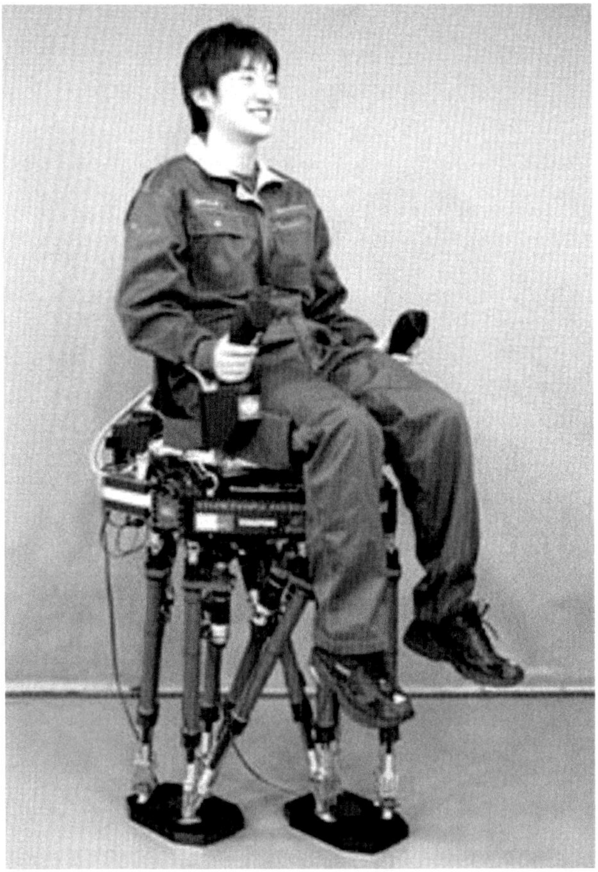

FIG. 13.20 WL-16RIV [73].

The primary requirement for the hybrid wheelchair designed in Refs. [80, 81] was to guarantee the comfort of the passenger. To achieve that, the researchers wanted to keep a near null inclination of the seat at all times, thus allowing the platform to move the center of mass along arbitrary spatial trajectories. The solution was a vehicle shown in Fig. 13.23 where the front and rear axles are attached to the base via 4-bar linkages. In addition to the motors driving the four wheels, the system has two pairs of linear actuators. One pair is used to independently move the linkages, and other two actuators serve as lifts or "ladders" for front and rear axles, respectively. The system can successfully climb stairs up and down and also supports a standing mode.

Overall, hybrid wheelchairs, may provide a good combination of high energy efficiency on an even ground due to the wheels and very versatile climbing capabilities, such as reaching very high steps, due to the linkage assemblies, but the complexity of the mechanical design and control systems of these mechanisms seem to have prevented them from leaving laboratories to reach the end user.

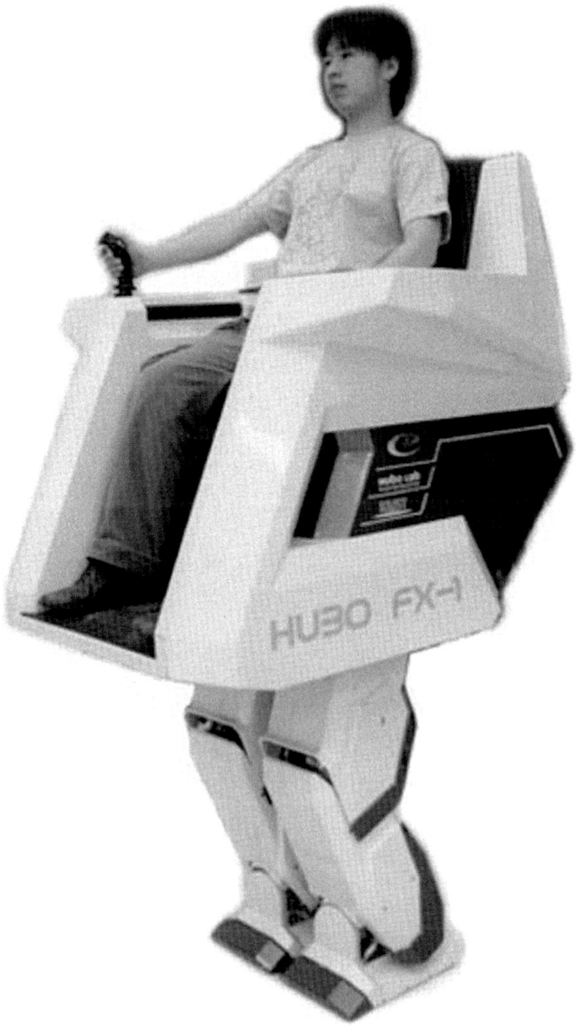

FIG. 13.21 HUBO FX-1 [76].

13.5.5 Nonconventional Wheel Geometry Mechanisms

Nonconventional wheel geometry mechanisms historically were the first type to be implemented in working prototypes. In 1964 Meredith Thring and Brian Shayer at Queen Mary College, London University developed two proof-of-concept stair-climbing wheelchairs [82]. Each of the four wheels of the "miniclimber" (Fig. 13.24) had an additional hub with four spring-loaded "blades" (Fig. 13.25). They passively protruded out of the hub to propel the wheelchair over the stairs and were pushed into the hub by the weight of the system to

FIG. 13.22 A 1:6.25 scale model of a wheelchair with a high single-step capability [78].

enable smooth driving on a flat terrain. Another prototype (Fig. 13.26) by Meredith Thring had four rimless wheels with the rubber-covered spikes enabling the climbing. Driving that platform, however, produced noticeable oscillations on a level ground at 3 mph. While these early systems had many challenges to address, such as poor stability and the absence of a basic steering mechanism, those were not the fundamental problems of the approach, but rather solvable issues of early prototypes. Nevertheless, the concept of using special geometry wheels (possibly wheels with changing geometry that could adapt to the shape of the steps) seems to be unfairly underrepresented among the modern stair-climbing wheelchair systems.

A notable example is a dual-locomotion system Gradi-Standart [84] patented in 2009 by a Russia-based company Gradi-Proizvodstvo. It is based on a conventional manual wheelchair that is equipped with extra four cross-shaped “wheels”. When hand levers are pushed forward, the wheelchair lifts on the crosses which are then engaged with the main wheels through a chain drive. In this mode, the operator can move up and down the stairs on the crosses by turning the main wheels. This rather minimalistic design,

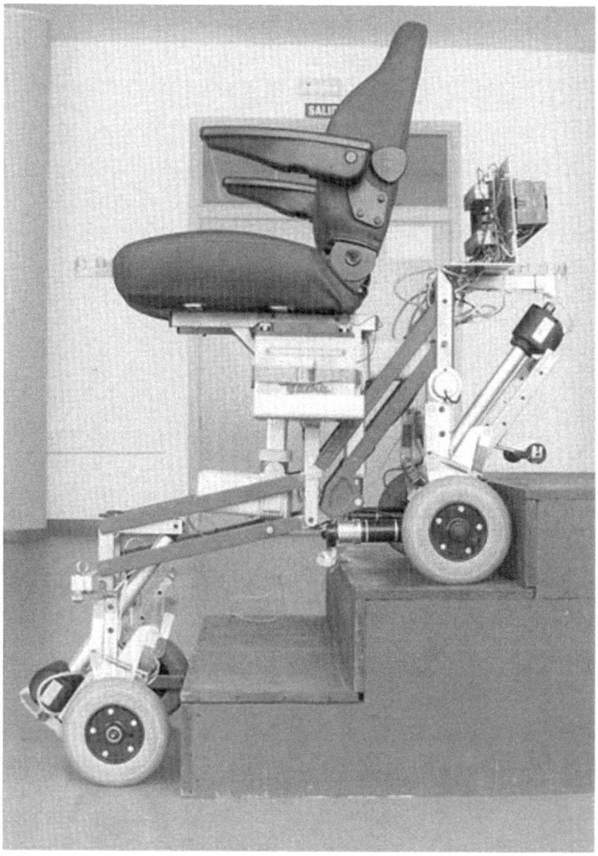

FIG. 13.23 Hybrid stair-climbing wheelchair by Morales et al. [80].

effectively achieves the main goal: it can climb obstacles as high as 20-cm under full control of the wheelchair operator.

In the past several decades, a number of stair-climbing wheelchairs have been developed. Today we finally see many of the systems that used to be lab prototypes on the market. Track-based and wheel cluster-based systems seem to be most commercially successful. There are multiple aspects that must be considered when stair-climbing wheelchairs are designed or compared. In addition to the typical criteria that are applicable to any wheelchair, such as the cost of manufacturing and operator's learning curve, for stair-climbers, we have to take into account multiple other factors such as the need for an assistant, the maximum height of the step that can be climbed, the supported shape of the stairs (straight-only, slightly curved or spiral), damaging effect on the steps, time/complexity of switching from the horizontal locomotion into stair-climbing mode and back, stability (static and dynamic), energy consumption, and operator's comfort in both of the modes. Additionally, some of the systems may provide additional

FIG. 13.24 "Miniclimber" by Thring and Shayer [82].

FIG. 13.25 "Miniclimber" wheel by Thring and Shayer [82].

FIG. 13.26 Sketch of another stair-climber by prof. Thring; for more details, see Ref. [83].

functionalities: standing mode, moving while standing, and tilting. A successful strategy on the way to reach a wider audience is to maximally reuse the same components in different modes of operation, for example tracks are usually driven by the same motors as wheels, while wheel cluster mechanism uses the same wheels in all modes of operation. This approach reduces the cost of the system and the learning curve for the operators.

13.6 Assisted Walking

According to the 2017 Disability Statistics Annual Report, 6.6% of Americans have difficulty walking [1]. People need assistance in walking when their lower limbs have limited or no functionality. This can be due to old age, nerve damage, missing limbs, muscle fatigue, atrophy, and a variety of injuries. Additionally, someone who has a compromised sense of balance due to blindness or inner ear functionality can benefit from these devices. And even in healthy individuals, augmenting normal walking to handle higher loads, improve balance, etc. has obvious benefits in situations such as hiking.

13.6.1 Types of Walking Aids

Devices designed to aid walking can be either passive or active, used to assist everyday walking or used for rehabilitation/training. These devices come in three general categories: orthotics, prosthetics, and immobile rehabilitation aids.

Many of these devices are covered in other chapters so the focus here is on the more traditional walking aids (Fig. 13.27), such as crutches, canes, walkers, and brace crutch

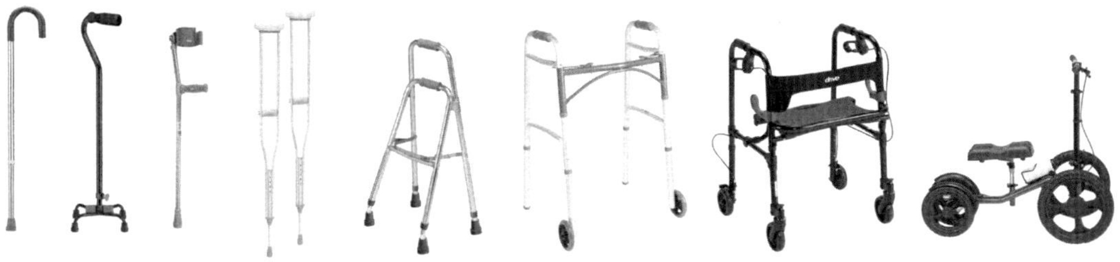

FIG. 13.27 Assisted walking devices by Medical Depot, Inc. [85]. *From left to right: round handle cane, quad cane with small base, bariatric steel forearm crutch, aluminum crutches, side walker, Deluxe folding walker, Clever-Lite walker, all-terrain knee walker.*

hybrids, such as the common Lofstrand crutches. These fall into a subset of orthotic devices. These devices allow the upper body to aid the lower body in walking by increasing their reach. With the arms acting as additional legs the stability improves through a larger footprint and shared loads. In addition, the arms' contact with the ground through a simple mechanical extension transmits some tactile information of the environment.

13.6.2 Device Selection

The selection of a proper device is based on several factors including the amount of load support needed, the environment to be used in, and the physical ability of the user. Someone who has only small balance deficiencies would likely use a cane, where as someone with more severe balance deficiencies would resort to a walker which has a larger support polygon and allows for more weight bearing than a cane. However, because all these devices require a certain level of physical ability to use, especially in the upper body, these would not be the correct option for many of the more severely disabled individuals.

A cane is not designed to be used as a load-bearing device, however some studies have shown that it reduces the peak load on knees and hips during ambulation [86, 87]. As seen in Fig. 13.28, it is instead primarily used to increase the support polygon in order to improve balance [88, 89]. Stability of canes has been improved by the addition of multiple feet connected to the cane by as ball joint increasing the quality of contact with the ground while allowing the upper portion of the cane to rotate freely. For those who are blind it also serves as an extension of their senses. Smart canes have been developed that can alert users to objects and even detect falls [90, 91]. It is the simplest of these devices to use and does not generally impede the ability to navigate a confined or uneven environment. As they are not used to bear a substantial load they do not cause any significant increase in fatigue of the upper body.

Crutches can be used to support loads, but improper or prolonged use can cause discomfort and secondary injuries [92–94]. Additionally, crutches are difficult and even dangerous to use on stairs and other extremely uneven terrain. They are often used when only one limb is limited and can be used to replace or share the load that goes to that leg while the healthy limb is used in the phase when the crutches are not supporting weight.

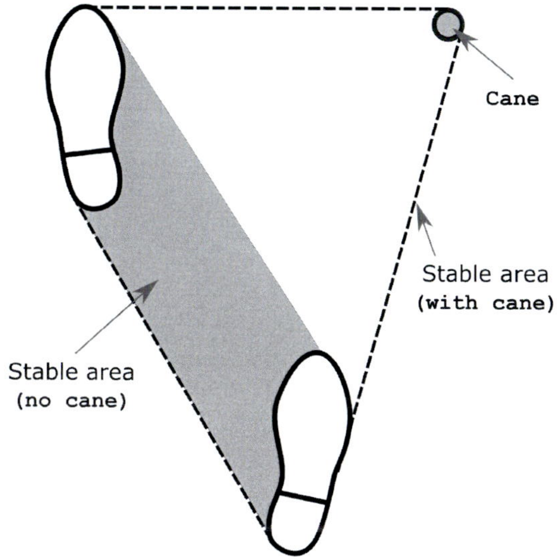

FIG. 13.28 Enlargement of stable footprint using an ambulation aid [88].

Brace crutch hybrids transfer the load to the arm in a way that reduces torque and permits greater loads. This improves comfort while decreasing the chance of injury. This in turn leads to longer and more varied use. Disabled athletes using the Lofstrand crutches as seen in Fig. 13.29 attach at the forearm near the elbow, are capable of dynamic athletic movements.

FIG. 13.29 Lofstrand crutches used in athletic competition.

Walkers are common among the elderly as they provide a large support base for those that need maximum stability. However, the need of two wheeled and no wheeled walkers to be at least partially lifted while walking is difficult as those that need a walker generally have diminished upper body strength as well. Four wheeled walkers have been used as well, but the ability to roll freely can significantly reduce the stability, particularly in the direction of movement. Brakes and motors, with easy to use push buttons, have been implemented into walkers to move with the human when necessary and remain stationary when not [95]. Ergonomic gait analysis allows for proper walker speed to maintain proper posture [96].

13.6.3 Problems Associated With Walking Aids

Several problems with walking aids remain unsolved. Improper selection and fit can increase the risk of falls, injuries due to prolonged use, and the abandonment of use all together. In 1987, over 20,000 visits to the emergency room were from accidents involving crutches, canes, or walkers [92]. Soon after receiving their aid, 30%–50% stop using them [97]. Even with the proper walking device, fall correction can be impacted negatively due to changes in balance and attention [98]. Additionally, stairs, objects, and uneven terrain can make these devices a tripping hazard [99]. The divergence from a normal walking gait caused by use of these can cause many secondary injuries especially with prolonged use [92, 98, 100].

13.7 The Challenge of Innovation in (Semi-)Autonomous Wheelchair Design

The current state of the art in the field demonstrates wheelchair systems capable of stair climbing, obstacle avoidance, simultaneous localization and mapping, path planning, and motion control with a shared autonomy framework. Semi-autonomous wheelchairs will dramatically enhance the quality of life for elderly as well as for individuals with physical and cognitive disabilities. The research on (semi-)autonomous wheelchairs has been on-going for >20 years; however, there are no commercially available systems in the market. It seems necessary to investigate the personalization of smart wheelchairs in three aspects: interfaces for interaction, controllers for action (top level, middle level, and low level), and feedback in interaction. Authors of this chapter (TP and DS) performed more than 120 interviews with powered wheelchair users, manufacturers, therapists, policy makers, and nonprofit organization staff. Findings based on that study revealed that the acceptability of fully autonomous systems by the users is still challenging and highly dependent on the severity of the disability. Furthermore, the cost, ease-of-use, and personalization are the most important factors in commercializing smart wheelchair technologies.

The summary of findings based on the above mentioned discovery interviews can be structured as following:

I DON'T WANT MORE DRAMA IN MY LIFE. Users see their wheelchairs as part of their bodies. They are challenged daily to carry out ADLs, and live ordinary lives. Any technology solutions should be seamless to integrate, and easy to use. Users who are accustomed to their wheelchairs have little to no tolerance to failures of "new features." As a result, full autonomy or "more autonomy than user desires" in a wheelchair will pose acceptability challenges.

THEY WANT IT ALL. The open-mindedness toward a self-driving wheelchair is dependent on the severity of the individual's disability. For example, as a person with ALS (pALS) progressively loses his or her ability to control their environment, they are open to using new technologies that will provide them with independent mobility. We identify individuals with progressive deterioration in motor skills as the early evangelists of the semi-autonomous wheelchair technologies. There are many well-established not-for-profit organizations in support of pALS, and working with these organizations is an essential component to achieve transition to practice.

USER CO-CREATION IS ESSENTIAL. Not all disabilities are the same, not one individual with a physical disability is the same from day to day. Therefore, one solution cannot fit all. Co-designing assistive technology with the users is essential in its acceptability and usability. The challenge is to maintain the technology in a personalized manner as the user's abilities often change over time. Modular and reconfigurable design principles must be adopted.

HOW MUCH IS EXPENSIVE? Interview responses indicated that it takes 8–16 weeks, and costs approximately $25,000 to train Labrador retrievers as companions for individuals in wheelchairs. Most of the costs are covered by nonprofits. Still further, interview responses strongly indicate that can provide safety and more situational awareness to a wheelchair user, for example while backing up into a transit van, will be affordable for most if the cost does not exceed the price of a high-end personal computer ($2000).

SHORT-TERM, SHORT-DISTANCE AUTONOMY IS FINE. Most users will be open to the use of technology to cross doors, traverse narrow hallways, and get on and off of vans using short-term autonomous behaviors. The term autonomy is still very disturbing to most, but short-term, short-distance-assisted control features will be acceptable.

ASSISTIVE TECHNOLOGY, BETTER LET ME KNOW WHAT IT IS UP TO. A personalized channel of feedback from the semi-autonomous wheelchair to the user is an important feature for the technology to be acceptable to users. Individuals do not want to be surprised by the actions of their robotic wheelchair. Since each person's ability is different, this aspect needs to be accounted for at the co-design process.

HOW TO INNOVATE IN THE CURRENT HEALTH-CARE SYSTEM? It is a well-known fact that it is very difficult to introduce new hardware systems within the current health-care system. Hospitals, for example, can be the perfect playground for autonomous wheelchairs, yet the liability concerns are forbidding. On the other hand, assisted living places, and user-centered nonprofit organizations are open to new ideas to improve the quality of life for individuals. Therefore, they are the part of the ecosystem for introducing new innovative technologies. Interview responses reflects a fact that the supply chain for users simply has too many voices. The design and deployment of wheelchair systems to users include input from manufacturers, distributors, installers, and of course users. In addition, most of these sources are not decision makers in terms of what hardware will be funded. Unfortunately, it appears that the wheelchair users are least likely to affect change in this ecosystem. This difficulty will prove to be a significant challenge for product adoption of any durable medical device.

REALIZATION OF PERSONALIZED SMART WHEELCHAIRS. Here, personalization is defined at three levels: interfaces for interaction, controllers for action, and feedback in interaction. Interfaces for interaction include various forms of detecting human intent including but not limited to voice, facial gestures, body gestures, muscle motions, and natural movements. Controllers for action can be viewed as a library of robot control methods structured in a hierarchical manner such as adaptive control with online parameter estimation scheme, fuzzy-logic based controllers, constrained optimal control, autonomous navigation, and assisted control algorithms. Feedback in interaction is composed of a set of cues so that the wheelchair can provide the human with feedback on its actions. The cues can be visual (LED lights, visualizations on tablets) or physical (haptic feedback). Interview responses are indicative that individuals with disabilities would easily abandon their assistive technology solutions if they feel insecure and/or if they cannot easily comprehend the actions taken by the wheelchair. That is why meaningful techniques for personalized feedback in human-robot interaction need to be investigated.

Even if the reliable autonomous wheelchair technology is available, are the users ready for it? The short answer is it depends. It depends on the severity of the disability, and on the individual's overall morale and attitude toward his or her condition. It also depends on how quickly and completely one can put support systems, trained technicians, and services in place. In conclusion, there is a substantial group of early evangelists who are ready to invest in smart wheelchair technologies to improve their mobility and as a result productivity. More work is required to bring this technology to them.

References

[1] L. Kraus, 2016 Disability Statistics Annual Report, University of New Hampshire, Durham, NH, 2017.

[2] Spinal Cord Injury Facts & Statistics (Online), Available from: http://www.sci-info-pages.com/facts.html (accessed 26.03.17).

[3] E. Smith, M. Delargy, Locked-in syndrome, BMJ 330 (7488) (2005) 406–409.

[4] Orphanet: Locked in Syndrome, Orphanet (Online), Available from: http://www.orpha.net/consor/cgi-bin/OC_Exp.php?Lng=GB&Expert=2406 (accessed 27.03.17).

[5] gocognitive, niels birbaumer Prevalence of Locked in Syndrome (Online), Available from: https://www.youtube.com/watch?time_continue=83&v=JL5e2bu2eyw, 2011 (accessed 25.04.17).

[6] F.E. Pisa, E. Biasutti, D. Drigo, F. Barbone, The prevalence of vegetative and minimally conscious states: a systematic review and methodological appraisal, J. Head Trauma Rehabil. 29 (4) (2014) E23–E30.

[7] N. Watson, B. Woods, History of the Wheelchair (Online), Encyclopædia Britannica, 2015. Available from: https://www.britannica.com/topic/history-of-the-wheelchair-1971423 (accessed 16.03.18).

[8] BBC—A History of the World—Object: Bath Chair (Online), Available from: http://www.bbc.co.uk/ahistoryoftheworld/objects/qcI7cMgiR0qmLnD_QPyIGQ (accessed 16.03.18).

[9] C. DiGiovine, T. Berner, K. Betz, T. Roesler, M. Schmeler, RESNA position on the application of ultra-light manual wheelchairs, Rehabil. Eng. Assist. Technol. Soc. North Am. (2012) 576–582.

[10] World Health Organization and Others, Guidelines on the Provision of Manual Wheelchairs in Less Resourced Settings, (2008).

[11] G. Wu, et al., ISB recommendation on definitions of joint coordinate systems of various joints for the reporting of human joint motion—Part II: shoulder, elbow, wrist and hand, J. Biomech. 38 (5) (2005) 981–992.

[12] M.M.B. Morrow, W.J. Hurd, K.R. Kaufman, K.-N. An, Upper extremity joint kinetics expression during wheelchair propulsion, J. Rehabil. Res. Dev. 46 (7) (2009) 939–944.

[13] Out-Front: The SmartWheel Clinical Tool (Online), Available from: http://www.out-front.com/smartwheel_overview.php (accessed 23.09.18).

[14] R.A. Cooper, SMARTWheel: from concept to clinical practice, Prosthetics Orthot. Int. 33 (3) (2009) 198–209.

[15] M.M. Rodgers, G.W. Gayle, S.F. Figoni, M. Kobayashi, J. Lieh, R.M. Glaser, Biomechanics of wheelchair propulsion during fatigue, Arch. Phys. Med. Rehabil. 75 (1) (1994) 85–93.

[16] M.B. Sabick, K.D. Zhao, K.N. An, A comparison of methods to compute the point of force application in handrim wheelchair propulsion: a technical note, J. Rehabil. Res. Dev. 38 (1) (2001) 57–68.

[17] B.W. Johnson, J.H. Aylor, Dynamic modeling of an electric wheelchair, IEEE Trans. Ind. Appl. IA-21 (5) (1985) 1284–1293.

[18] Z.-P. Jiang, H. Nijmeijer, Tracking control of mobile robots: a case study in backstepping, Automatica 33 (7) (1997) 1393–1399.

[19] R. Carelli, E. Oliveira Freire, Corridor navigation and wall-following stable control for sonar-based mobile robots, Robot. Auton. Syst. 45 (3–4) (2003) 235–247.

[20] T. Das, I.N. Kar, Design and implementation of an adaptive fuzzy logic-based controller for wheeled mobile robots, IEEE Trans. Control Syst. Technol. 14 (3) (2006) 501–510.

[21] R. Fierro, F.L. Lewis, Control of a nonholonomic mobile robot: backstepping kinematics into dynamics, in: Proceedings of the 34th IEEE Conference on Decision and Control, vol. 4, 1995, pp. 3805–3810.

[22] T. Fukao, H. Nakagawa, N. Adachi, Adaptive tracking control of a nonholonomic mobile robot, IEEE Trans. Robot. Autom. 16 (5) (2000) 609–615.

[23] D. Soetanto, L. Lapierre, A. Pascoal, Adaptive, non-singular path-following control of dynamic wheeled robots, in: 42nd IEEE Conference on Decision and Control, 2003. Proceedings, vol. 2, 2003, pp. 1765–1770.

[24] W. Dong, K.-D. Kuhnert, Robust adaptive control of nonholonomic mobile robot with parameter and nonparameter uncertainties, IEEE Trans. Robot. 21 (2) (2005) 261–266.

[25] P. Petrov, Modeling and adaptive path control of a differential drive mobile robot, in: Proceedings of the 12th WSEAS International Conference on Automatic Control, Modelling & Simulation, 2010, pp. 403–408.

[26] R. Fierro, F.L. Lewis, Control of a nonholonomic mobile robot using neural networks, IEEE Trans. Neural Netw. 9 (4) (1998) 589–600.

[27] F. Künhe, J. Gomes, W. Fetter, Mobile robot trajectory tracking using model predictive control, in: II IEEE Latin-American Robotics Symposium, 2005.

[28] F.N. Martins, W.C. Celeste, R. Carelli, M. Sarcinelli-Filho, T.F. Bastos-Filho, An adaptive dynamic controller for autonomous mobile robot trajectory tracking, Control. Eng. Pract. 16 (11) (2008) 1354–1363.

[29] S. Katsura, K. Ohnishi, Advanced motion control for wheelchair in unknown environment, in: IEEE International Conference on Systems, Man and Cybernetics, 2006. SMC'06, vol. 6, 2006, pp. 4926–4931.

[30] Z. Tian, W. Xu, Electric wheelchair controller based on parameter self-adjusting fuzzy PID, in: International Conference on Computational Intelligence and Natural Computing, 2009. CINC'09, vol. 1, 2009, pp. 358–361.

[31] C. De La Cruz, T.F. Bastos, R. Carelli, Adaptive motion control law of a robotic wheelchair, Control. Eng. Pract. 19 (2) (2011) 113–125.

[32] D.A. Sinyukov, T. Padir, Adaptive motion control for a differentially driven semi-autonomous wheelchair platform, in: 2015 International Conference on Advanced Robotics (ICAR), 2015, pp. 288–294.

[33] C.G. Pinheiro, E.L. Naves, P. Pino, E. Losson, A.O. Andrade, G. Bourhis, Alternative communication systems for people with severe motor disabilities: a survey, Biomed. Eng. Online 10 (2011) 31.

[34] D.A. Sinyukov, Semi-Autonomous Robotic Wheelchair Controlled with Low Throughput Human-Machine Interfaces, Worcester Polytechnic Institute, 2017.

[35] P.L. Lee, H.C. Chang, T.Y. Hsieh, H.T. Deng, C.W. Sun, A brain-wave-actuated small robot Car using ensemble empirical mode decomposition-based approach, IEEE Trans. Syst. Man Cybern. – Part Syst. Hum. 42 (5) (2012) 1053–1064.

[36] A. Herweg, J. Gutzeit, S. Kleih, A. Kübler, Wheelchair control by elderly participants in a virtual environment with a brain-computer interface (BCI) and tactile stimulation, Biol. Psychol. 121 (Part A) (2016) 117–124.

[37] G. Kucukyildiz, H. Ocak, S. Karakaya, O. Sayli, Design and implementation of a multi sensor based brain computer interface for a robotic wheelchair, J. Intell. Robot. Syst. (2017) 1–17.

[38] F. Quivira et al., BCI-based semi-autonomous wheelchair control using a human-in-the-loop cyber physical system approach, Proc. Sixth Int. Brain-Comput. Interface Meet. BCI Past Present Future.

[39] F. Achic, J. Montero, C. Penaloza, F. Cuellar, Hybrid BCI system to operate an electric wheelchair and a robotic arm for navigation and manipulation tasks, in: 2016 IEEE Workshop on Advanced Robotics and its Social Impacts (ARSO), 2016, pp. 249–254.

[40] T. Carlson, J.D.R. Millan, Brain-controlled wheelchairs: a robotic architecture, IEEE Robot. Autom. Mag. 20 (1) (2013) 65–73.

[41] Y.T. Lin, C.H. Kuo, Development of SSVEP-based intelligent wheelchair brain computer interface assisted by reactive obstacle avoidance, in: 2016 IEEE International Conference on Industrial Technology (ICIT), 2016, pp. 1572–1577.

[42] T. Shi, H. Wang, C. Zhang, Brain computer Interface system based on indoor semi-autonomous navigation and motor imagery for unmanned aerial vehicle control, Expert Syst. Appl. 42 (9) (2015) 4196–4206.

[43] X. Perrin, R. Chavarriaga, F. Colas, R. Siegwart, J.D.R. Millán, Brain-coupled interaction for semi-autonomous navigation of an assistive robot, Robot. Auton. Syst. 58 (12) (2010) 1246–1255.

[44] D.A. Sinyukov, R. Li, N.W. Otero, R. Gao, and T. Padir, "Augmenting a voice and facial expression control of a robotic wheelchair with assistive navigation," Presented at the 2014 IEEE International Conference on Systems, Man and Cybernetics (SMC), San Diego.

[45] G. Vanacker, et al., Context-based filtering for assisted brain-actuated wheelchair driving, Comput. Intell. Neurosci. 2007 (2007), e25130.

[46] F. Galán, et al., A brain-actuated wheelchair: asynchronous and non-invasive brain–computer interfaces for continuous control of robots, Clin. Neurophysiol. 119 (9) (2008) 2159–2169.

[47] F. von Hundelshausen, M. Himmelsbach, F. Hecker, A. Mueller, H.-J. Wuensche, Driving with tentacles: integral structures for sensing and motion, J. Field Robot. 25 (9) (2008) 640–673.

[48] M. Quigley, et al., ROS: an open-source robot operating system, in: ICRA Workshop on Open Source Software, vol. 3, 2009.

[49] H. Nezamfar, D. Sinyukov, U. Orhan, D. Erdogmus, T. Padir, Brain interface to control a tele-operated robot, in: Presented at the Fifth International Brain-Computer Interface Meeting, 2013.

[50] R. Zhang, et al., Control of a wheelchair in an indoor environment based on a brain-computer Interface and automated navigation, IEEE Trans. Neural Syst. Rehabil. Eng. 24 (1) (2016) 128–139.

[51] B. Rebsamen, et al., A brain controlled wheelchair to navigate in familiar environments, IEEE Trans. Neural Syst. Rehabil. Eng. 18 (6) (2010) 590–598.

[52] L. Bi, X.A. Fan, N. Luo, K. Jie, Y. Li, Y. Liu, A head-up display-based P300 brain-computer interface for destination selection, IEEE Trans. Intell. Transp. Syst. 14 (4) (2013) 1996–2001.

[53] X.A. Fan, L. Bi, T. Teng, H. Ding, Y. Liu, A brain-computer interface-based vehicle destination selection system using P300 and SSVEP signals, IEEE Trans. Intell. Transp. Syst. 16 (1) (2015) 274–283.

[54] X. Perrin, F. Colas, C. Pradalier, R. Siegwart, R. Chavarriaga, J.D.R. Millan, Learning user habits for semi-autonomous navigation using low throughput interfaces, in: 2011 IEEE International Conference on Systems, Man, and Cybernetics (SMC), 2011, pp. 1–6.

[55] E. Demeester, A. Hüntemann, D. Vanhooydonck, G. Vanacker, H. Van Brussel, M. Nuttin, User-adapted plan recognition and user-adapted shared control: a Bayesian approach to semi-autonomous wheelchair driving, Auton. Robots 24 (2) (2007) 193–211.

[56] X. Perrin, Semi-Autonomous Navigation of an Assistive Robot Using Low Throughput Interfaces, (2009).

[57] I. Iturrate, J.M. Antelis, A. Kubler, J. Minguez, A noninvasive brain-actuated wheelchair based on a P300 neurophysiological protocol and automated navigation, IEEE Trans. Robot. 25 (3) (2009) 614–627.

[58] Z. Wei, W. Chen, J. Wang, H. Wang, K. Li, Semantic mapping for safe and comfortable navigation of a brain-controlled wheelchair, in: Intelligent Robotics and Applications, 2013, pp. 307–317.

[59] Z. Wei, W. Chen, J. Wang, 3D semantic map-based shared control for smart wheelchair, in: Intelligent Robotics and Applications, 2012, pp. 41–51.

[60] A. Goil, M. Derry, B.D. Argall, Using machine learning to blend human and robot controls for assisted wheelchair navigation, in: 2013 IEEE 13th International Conference on Rehabilitation Robotics (ICORR), 2013, pp. 1–6.

[61] D. Sinyukov, T. Padir, A novel shared position control method for robot navigation via low throughput human-machine interfaces, in: 2018 IEEE/RSJ International Conference on Intelligent Robots and Systems (IROS 2018), Madrid, Spain, 2018.

[62] W. Tao, J. Xu, T. Liu, Electric-powered wheelchair with stair-climbing ability, Int. J. Adv. Robot. Syst. 14 (4) (2017). 1729881417721436.

[63] TGR, Explorer (Online), Available from: http://portale.siva.it/files/doc/product/au5185_2_Depliant%20EXPLORER.pdf (accessed 10.03.18).

[64] The TopChair-S electric wheelchair (Online), Available from: http://www.topchair.fr/en/ (accessed 05.03.18).

[65] CaterWil GTS3 EXPORT | Caterwil (Online), Available from: http://caterwil.com/caterwil-gts3-export/ (accessed 05.03.18).

[66] SCEWO – Stair Climbing Wheelchair (Online), Available from: https://scewo.ch/ (accessed 05.03.18).

[67] D.L. Kamen, R.R. Ambrogi, J.D. Heinzmann, R.K. Heinzmann, D. Herr, J.B. Morrell, Control of a Balancing Personal Vehicle, US6443250B1, 2002.

[68] iBOT Poised for Comeback | Toyota USA Newsroom (Online), Available from: http://pressroom.toyota.com/releases/toyota-deka-research-partnership-may21.htm, 2016 (accessed 05.03.18).

[69] L.A.M. Riascos, A low cost stair climbing wheelchair, in: 2015 IEEE 24th International Symposium on Industrial Electronics (ISIE), 2015, pp. 627–632.

[70] SCOIATTOLO L | TGR – Strumenti di libertà (Online), Available from: http://tgr.it/prodotti/scoiattolo-largo/ (accessed 05.03.18).

[71] C-Max U1 by AATGB, the Push Wheelchair With Stair Climbing Function (Online), Available from: https://www.aatgb.com/c-max-u1.html (accessed 05.03.18).

[72] C.-H. Lee, K.M. Lee, J.H. Yoo, I. Kim, Y.-B. Bang, A compact stair-climbing wheelchair with two 3-DOF legs and a 1-DOF base, Ind. Robot. 43 (2016) 181–192.

[73] K. Hashimoto, Y. Sugahara, H. Lim, A. Takanishi, Optimization design of a stewart platform type leg mechanism for biped walking vehicle, in: Robotics Research, 2011, pp. 169–178.

[74] Y. Sugahara, et al., Development of a biped locomotor with the double stage linear actuator, in: Proceedings 2007 IEEE International Conference on Robotics and Automation, 2007, pp. 1850–1855.

[75] K. Hashimoto, et al., Avoidance behavior from external forces for biped vehicle, in: 2010 IEEE International Conference on Robotics and Automation, 2010, pp. 4715–4720.

[76] J.-H. Kim, J.-Y. Kim, J.-H. Oh, Adaptive walking pattern generation and balance control of the passenger-carrying biped robot, HUBO FX-1, for variable passenger weights, Auton. Robots 30 (4) (2011) 427.

[77] Toyota Motor Corporation, Toyota Global Site | Robot Project: Mountable Model 'i-Foot (Online), Toyota Motor Corporation Global. Available from: http://www.toyota-global.com/innovation/partner_robot/aichi_expo_2005/index03.html (accessed 09.03.18).

[78] M.J. Lawn, T. Ishimatsu, Modeling of a stair-climbing wheelchair mechanism with high single-step capability, IEEE Trans. Neural Syst. Rehabil. Eng. 11 (3) (2003) 323–332.

[79] Y. Sugahara, N. Yonezawa, K. Kosuge, A novel stair-climbing wheelchair with transformable wheeled four-bar linkages, in: 2010 IEEE/RSJ International Conference on Intelligent Robots and Systems, 2010, pp. 3333–3339.

[80] R. Morales, J. Chocoteco, V. Feliu, H. Sira-Ramírez, Obstacle surpassing and posture control of a stair-climbing robotic mechanism, Control. Eng. Pract. 21 (5) (2013) 604–621.

[81] R. Morales, V. Feliu, A. González, P. Pintado, Kinematic model of a new staircase climbing wheelchair and its experimental validation, Int. J. Robot. Res. 25 (9) (2006) 825–841.

[82] Popular mechanics, Pop. Mech., 1967.

[83] M.W. Thring, Robots and Telechirs: Manipulators With Memory, Remote Manipulators, Machine Limbs for the Handicapped, Halsted Press, Chichester [West Sussex]/New York/Ellis Horwood, 1983.

[84] Кресло-коляска - Кресла-коляски компании ООО «ГРАДИ» (Online), Available from: http://www.gradi.pro/wheelchair.html (accessed 12.03.18).

[85] Home page | Drive Medical (Online), Available from: http://www.drivemedical.com/index.php/ (accessed 23.09.18).

[86] M. Okuno, U. Wyss, P. Costigan, K. Deluzio, R. Chew, Knee joint forces during contra-lateral cane-assisted gait, Proc. Int. Soc. Biomech. 15 (1995) 694–695.

[87] R.A. Brand, R.D. Crowninshield, The effect of cane use on hip contact force, Clin. Orthop. 147 (1980) 181–184.

[88] L. Bennett, M.P. Murray, E.F. Murphy, T.T. Sowell, Locomotion assistance through cane impulse, Bull. Prosthet. Res. 10 (31) (1979) 38–47.

[89] J.A. Ashton-Miller, M.W. Yeh, J.K. Richardson, T. Galloway, A cane reduces loss of balance in patients with peripheral neuropathy: results from a challenging unipedal balance test, Arch. Phys. Med. Rehabil. 77 (5) (1996) 446–452.

[90] V. Singh, et al., 'Smart'Cane for the visually impaired: design and controlled field testing of an affordable obstacle detection system, in: TRANSED 2010: 12th International Conference on Mobility and Transport for Elderly and Disabled Persons, Hong Kong Society for Rehabilitation S K Yee Medical Foundation Transportation Research Board, 2010.

[91] M. Lan, A. Nahapetian, A. Vahdatpour, L. Au, W. Kaiser, M. Sarrafzadeh, SmartFall: an automatic fall detection system based on subsequence matching for the SmartCane, in: Proceedings of the Fourth International Conference on Body Area Networks, ICST, Brussels, Belgium, Belgium, 2009. pp. 8:1–8:8.

[92] B.M. Joyce, R.L. Kirby, Canes, crutches and walkers, Am. Fam. Physician 43 (2) (1991) 535–542.

[93] I.S. Moon, J.K. Hwang, J.I. Kim, Recurrent upper extremity embolism due to a crutch-induced arterial injury: a different cause of upper extremity embolism, Ann. Vasc. Surg. 24 (4) (2010) 554.e7–554.e12.

[94] A.L. Brooks, S.B. Fowler, Axillary artery thrombosis after prolonged use of crutches, JBJS 46 (4) (1964) 863–864.

[95] E. Einbinder, T.A. Horrom, Smart walker: a tool for promoting mobility in elderly adults, J. Rehabil. Res. Dev. 47 (9) (2010) xiii–xv.

[96] W.L. Yeoh, P.Y. Loh, S. Muraki, Smart walkers from an ergonomics perspective: a review, Jpn. J. Ergon. 53 (Suppl. 2) (2017) S392–S395.

[97] D.B. Gray, L.A. Quatrano, M.L. Lieberman, Designing and Using Assistive Technology: The Human Perspective, Paul H, Brookes, 1998.

[98] H. Bateni, B.E. Maki, Assistive devices for balance and mobility: benefits, demands, and adverse consequences, Arch. Phys. Med. Rehabil. 86 (1) (2005) 134–145.

[99] J.A. Stevens, K. Thomas, L. Teh, A.I. Greenspan, Unintentional fall injuries associated with walkers and canes in older adults treated in U.S. emergency departments, J. Am. Geriatr. Soc. 57 (8) (2009) 1464–1469.

[100] S. Li, C.W. Armstrong, D. Cipriani, Three-point gait crutch walking: variability in ground reaction force during weight bearing, Arch. Phys. Med. Rehabil. 82 (1) (2001) 86–92.

14

Feeding Systems, Assistive Robotic Arms, Robotic Nurses, Robotic Massage

Marko B. Popovic

WORCESTER POLYTECHNIC INSTITUTE, WORCESTER, MA, UNITED STATES

CHAPTER OUTLINE

14.1 Feeding and Hygiene Assistance, Vocational Aid

People with disabilities and the elderly often require assistance with feeding tasks and with personal hygiene tasks such as bathing, brushing teeth, using restroom, combing hair, shaving, etc.

Many of these assistive tasks can be uncomfortable both to the people with reduced motor abilities and to their caregivers. It may take a lot of effort and skill, patience and compassion on the side of the caregiver to fully understand one's preferences and needs and to devise adequate strategies to successfully perform these tasks. In addition to accomplishing these tasks, it is also critical not to injure one's sense of dignity and privacy or to cause painful, unpleasant, or uncomforting sensations.

Depending on the level of need and type of assistance, having a personal caregiver may be quite expensive and hence it may cause a huge financial burden on care-receivers and their families.

Another observation deserving serious consideration is the world's growing population and increasing life expectancy. Some of the world's oldest nations in terms of median age of their population are Japan and Germany with a median age of 47.3 and 47.1 years, respectively, then Italy with 45.5, Greece 44.5, Austria 44.0, Croatia 43.0, Spain and Estonia both with 42.7, Serbia 42.6, Finland 42.5, Switzerland 42.4, etc. [1]. In many of these countries the number of elderly people requiring continuous close assistance for activities of daily life (ADL) may soon outmatch the number of potential caregivers.

Biomechatronics. https://doi.org/10.1016/B978-0-12-812939-5.00014-8

The US is a relatively young nation with a median age of 38.1 years in 2017 [1]. However, even in the US, about 11 million severely disabled people, that is, 3.4% of the total population need personal assistance for daily activities [2]. And nearly 18 million US older adults, that is, 5.6% of the total population, or nearly half of everyone 65 and older, report that they need some assistance with routine daily activities [3].

Hence, robotic assistance might be beneficial for a multitude of reasons. Introduction of robotic systems may not just resolve some of the above-mentioned issues, but it may also provide more self-confidence and sense of independence to people with disabilities and the elderly. Robots may also assist with vocational needs so that one can work and potentially generate income.

Historically, the first wheelchair-mounted robotic arm was the four degrees of freedom (DoF) "Mason" developed at the VA Prosthetic Center in New York, US, by Carl Mason in the 1970s [4, 5]. Although characterized with a very large workspace the robot was apparently too "springy" and the hook-shaped end-effector had not performed as well as expected. The control was quite basic, being operated on a joint-by-joint basis. However, users complained that it was not intuitive and that robot operation was tedious and time consuming.

The first assistive robot with a high-level command-type interface was the Seamone-Schmeisser Robot arm/worktable, shown in Fig. 14.1, developed in 1974 [5–7]. This robot did not require the utilization of low-level commands, that is, user control of a joint-by-joint motion. It was developed at the Johns Hopkins University Applied Physics Lab. The robotic arm had five coupled DoF as well as translational motion across the worktable. Motive power was provided by three actuators, two of which were located within the arm and one within the translational drive.

The goal was to assist spinal cord injury patients to eat, take care of personal hygiene, and perform common vocational operations. High-level commands were communicated with the chin manipulator and body-powered switches. With a high level of primarily feed-forward components, the robot was preprogramed, for example, to feed users. A clinical study performed at VA Medical Centers, showed the feasibility of self-feeding, operating a handset telephone, loading typewriter paper, turning pages in a book, and moving light-weight journals between different storage units. However, the robot was very slow, the food would frequently fall off from the utensil, there were numerous restrictions based on robot workspace, and the system with a chin controller was often inadequate for reclining users.

Other examples of robotic arm/worktable systems in the US were those developed as part of the desktop vocational assistant robots (DeVAR) project and its continuation, professional vocational assistant robots (ProVAR) project in 1990s [5, 8, 9] (Fig. 14.2, left). This research was performed at the Palo Alto VA Hospital and Stanford University. The systems designed mainly for people with quadriplegic disabilities were developed for office activities such as printing and other computer-related tasks. They utilized a PUMA-260 six DoF robotic arm and an Otto-Bock Greifer gripper with a maximum payload of 2.3 kg. The robotic arm can move on a rail attached to the ceiling. It was shown that this increased the robot's workspace and reduced potential obstructions. Various human-machine interfaces were explored as part of these projects including vocal recognition systems as the input interface.

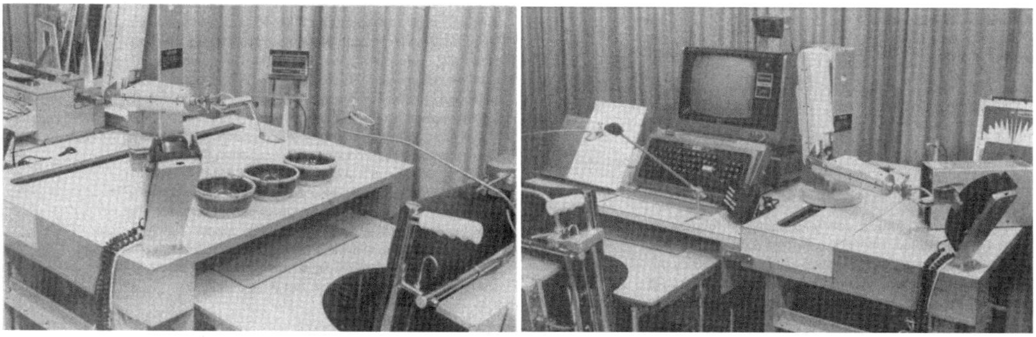

FIG. 14.1 Johns Hopkins Applied Physics Lab Robot arm/worktable from mid-1970s [6, 7].

Probably the most advanced systems developed in Europe in 1990s were RAID 2 and MASTER 2 (1993–96) as part of the technology initiative for disabled and elderly people (TIDE) program [5]. System shown in Fig. 14.2 on the right consisted of RAID workstation and robotic arm, RT200 (though testing has been also performed with the ABB IRb1000 industrial robot) and the human machine interface using controller and control language MASTER [10–12]. The evaluation of these systems was carried out in Sweden, France, and UK. The robot manages paper handling in the office environment and it had been commercialized by Oxford Intelligent Machines Ltd. in England [13]. About ten RAID robots have been sold to different rehabilitation centers in Sweden, France, and England. The RAID robot can handle books, A4 paper, and floppy disks on command from the user, which, using a keyboard or mouse set, controls the robot via a personal computer. The system had two end effectors, "book gripper" and "page turner." The page turner had a knife for turning multiple pages, suction cup for lifting a single page, clamp for grasping single pages, clamp for paper handling, and push button/switch for detecting page surface. The cycle time for turning one page was typically on the order of 15 s.

The Handy 1, Rehab Robotics Limited, UK, was developed in 1987 by Mike Topping from the Centre for Rehabilitation Robotics, Staffordshire University, to assist an 11-year-old boy with cerebral palsy to eat unaided [14, 15] (Fig. 14.3). The system was probably the most successful low-cost, commercially available robotic system during the 1990s

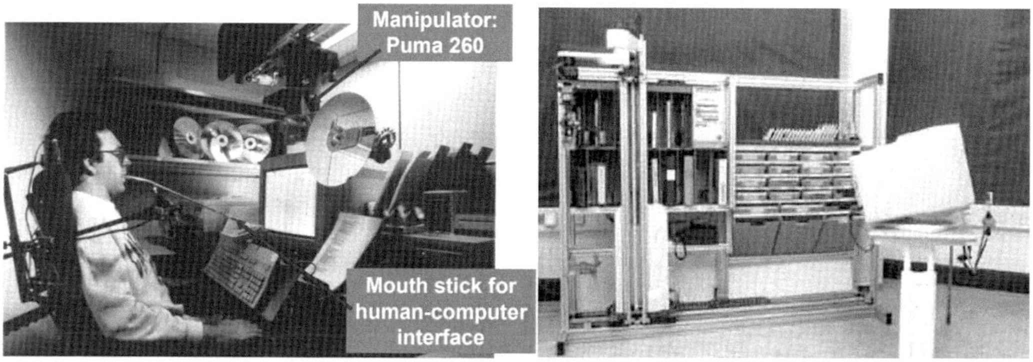

FIG. 14.2 DeVAR and ProVAR system, left, and RAID system, right.

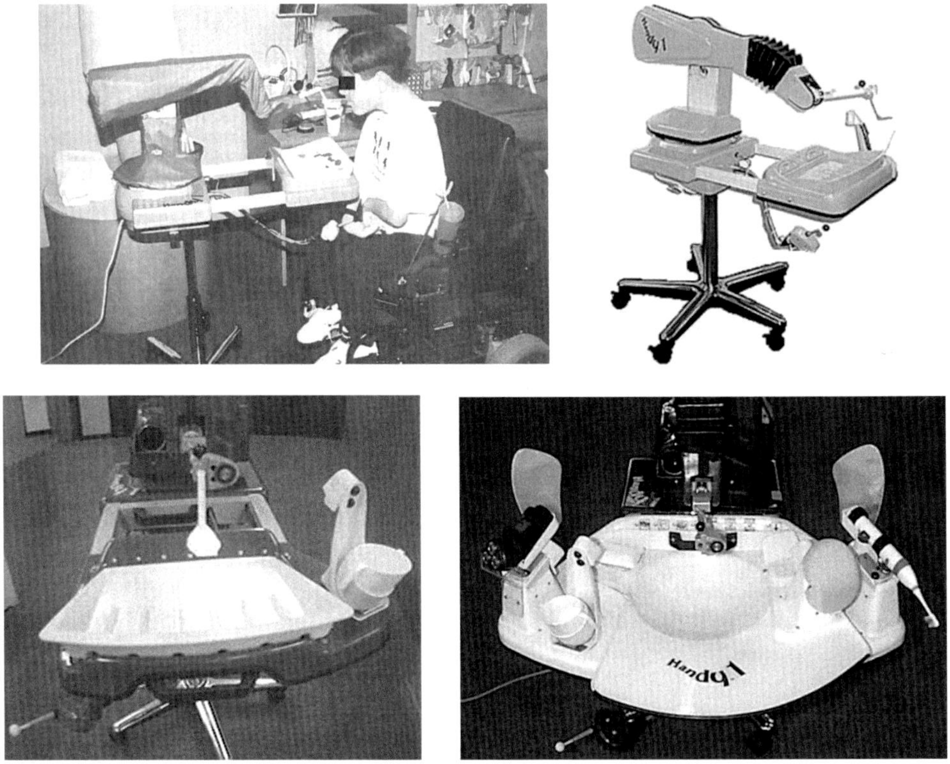

FIG. 14.3 First version of Handy 1, top left, more modern version, top right, food tray, bottom left, personal hygiene tray, bottom right.

and early 2000s, capable of assisting the most severely disabled with several everyday functions such as eating, drinking, washing, teeth cleaning, shaving, and make-up application.

A scanning system of lights allows the user to select food arranged in the walled columns of the food dish. The user waits for the light to scan behind the column of food that he/she wants to eat, and then presses the single switch that sets the Handy 1 in motion. The robot proceeds onto the selected section of the dish and scoops up a spoonful of the chosen food, presenting it at the user's mouth position. The onboard computer keeps track of where the food has been selected from the dish and automatically controls the scanning system to bypass empty areas. The use of walled dishes ensures that the food does not escape when the spoon scoops into it.

Another robotic arm mounted on an electrical wheelchair was introduced by the company Exact Dynamics in 1991. At that time, their first robotic manipulator, the Manus, was developed and manufactured [16–18]. The first prototype had six DoF with a symmetrical gripper at the end-effector with a maximum payload of about 1 kg. The first commercial version was controlled mainly with a keypad and a joystick (Fig. 14.4). The first successor of Manus was the assistive robotic manipulator (ARM) robotic appendage, which eventually evolved to the smaller, lighter and better intelligent ARM (iARM). As advertised by Exact Dynamics, the 9 kg iARM can even assist with drinking a beer with friends, eating

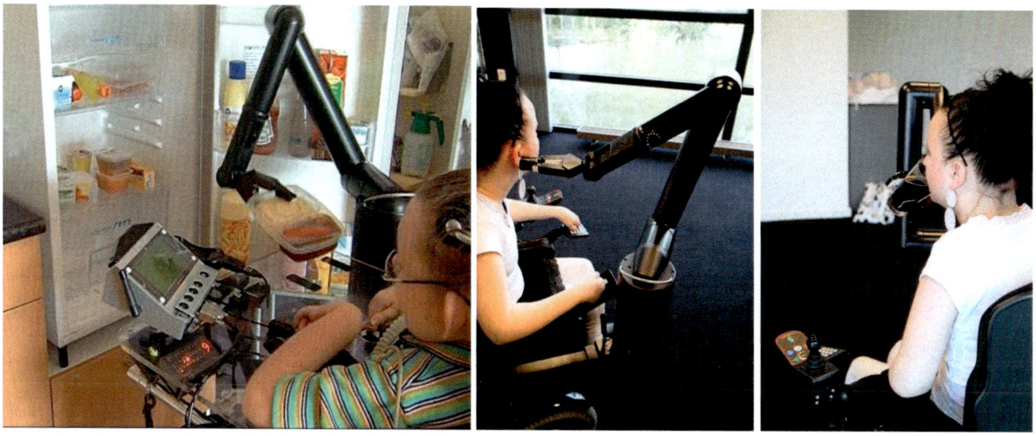

FIG. 14.4 Manus/ARM assistance with getting food from fridge, modern iARM assistance with putting earrings, middle, and drinking, right. *Reprinted with permission from Exact Dynamics.*

dinner in a restaurant, taking a walk with a dog, shopping in the supermarket and many other typical activities people would do on a daily basis.

In 1991, the Laboratory of Intelligent Systems in Japan began the development of another feeding robotic system named MySpoon [19, 20]. This five DoF robot became commercially available in 2002 from the SECOM Company in Japan. The user controls a utensil via a joystick or simply a switch. The robot may pick up a small food portion up to 50g. Unfortunately, one cannot eat soup with a spoon; liquid in the cup can be sipped only through a straw. And the food needs to be cut first into bite-size pieces to allow a spoon or fork to easily scoop it. The "stop" position of the fork/spoon is configured before each meal to a position just in front of the mouth. The spoon will not contact the face as long as the user does not lean in front of the "stop" position. If for any reason the spoon comes in contact with the face, the low power output motor ensures that the user is not injured.

MySpoon can operate in manual, semi-automatic, or automatic mode allowing different levels of low-level control to the user. The control's body appendages may be located on the chin or fingers, palm, or anywhere else. Shown in Fig. 14.5 is a reinforced joystick

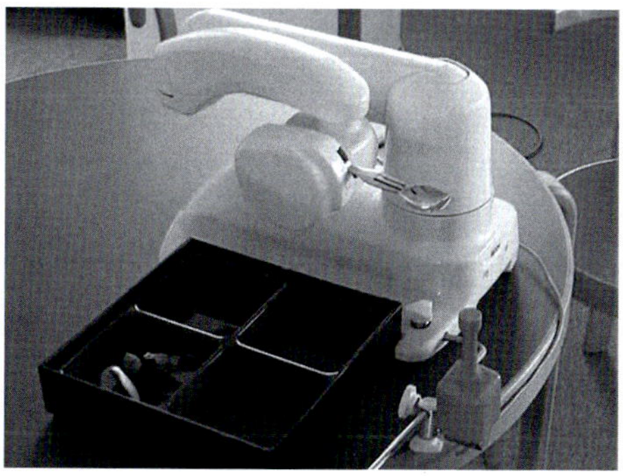

FIG. 14.5 MySpoon robotic feeding system, left, and reinforced joystick with large button, right.

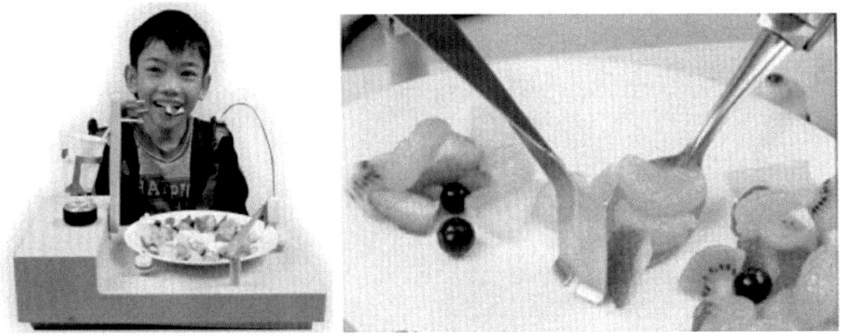

FIG. 14.6 The Winsford Feeder, left, "pusher" moves the food onto the spoon, right.

recommended for users who cannot accurately control a joystick due to trembling hands. With more resistance built into the joystick, it requires large, deliberate commands to operate. A large push-button is included for mode changes.

Other specialized feeding systems definitively deserving mention are the Winsford Feeder developed in 1999 [21, 22] and the Neater Eater developed in the mid-1990s [23].

The Winsford Feeder, shown in Fig. 14.6, commercially available through North Coast Medical, Inc. utilizes a powered arm that pivots to the preset position of the user's mouth. It uses a mechanical pushing device to fill a spoon on a pivoting arm and raises the spoon to the height of the user's mouth. The rotation of the plate places more food. The device can be controlled by either a chin switch or a rocker switch. If the switch is simply held down, the product will continuously feed the user.

The Neater Eater, commercially available through the UK-based company Neater Solutions, Inc. has a two DoF arm and one plate providing both manual and automatic operation (Fig. 14.7). The Neater Eater manually operated system was designed over 25 years ago to reduce essential tremor when eating, for people with ataxia. The viscous damping combined with spring loading absorbs shakes and uncontrolled movements, making it suitable for clients with athetoid cerebral palsy, Huntington's chorea, Friedrich's ataxia, Parkinson's disease, and intention tremors associated with multiple sclerosis. The user holds onto the arm of the eating machine, which acts as an interface to their food with a self-leveling spoon and rotating plate. Robotic version has a programmable mouth position, the device can be operated with a single switch, the spoon cycle can be paused with a light-touch switch and the plate turns-switch can be either user controlled or automatic for each spoon cycle.

The first product from the consumer robotics company Desin, Obi, the robotic dining companion, got a lot of attraction on the market in recent years (Fig. 14.8). The FDA approved system costs around $4500 USD. After a caregiver divides the meal into Obi's four separate bowls, users are able to feed themselves through a simple interface: one button moves the arm between the bowls, and another selects the food, dips the spoon in and brings it up to the diner's mouth. The user interface can be customized, depending on the specific needs and abilities of the user. They may include big bright "Buddy Buttons"

FIG. 14.7 The Neater Eater robotic version, left, and manually operated version, right.

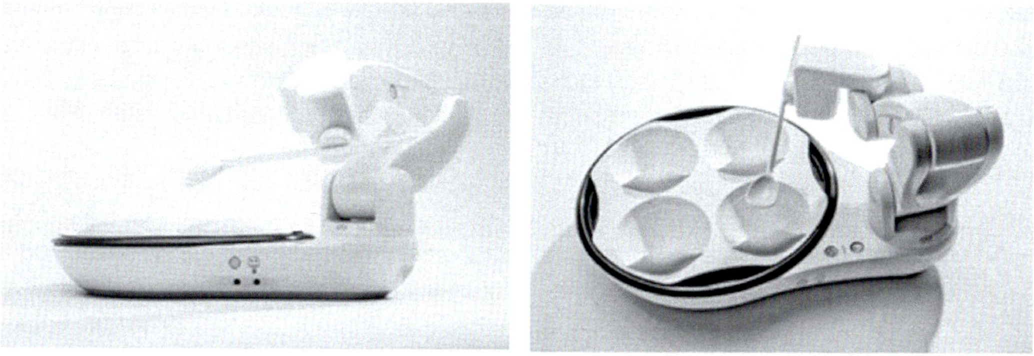

FIG. 14.8 Obi, the robotic dining companion; side view, left, and top view, right.

that can be placed on the table or on the floor, squeezing pillows that may be used for head- and cheek-based activation, and a small mouth piece switch that can be triggered through sip or puff actions. In the "Teach Mode" the robotic arm can be positioned manually to the desired location where the user mouth is located, the robot then stores this location and subsequently delivers food to the same location. The Obi also has a detection collision system that prevents a utensil from colliding with eye or hand of the user.

Like most other feeding assistance devices, Obi also requires the individual to have some level of controlled movements of the head and neck in order to be able to take in the food from the utensil when it is close to the mouth.

14.2 Robotic Nurses

As previously discussed due to increasing life expectancy in many countries the number of elderly and disabled people requiring continuous close assistance for ADL may soon outmatch the number of potential caregivers. In the US today roughly between 5% and 10% of the entire population require continuous close assistance. And many states including California have already proclaimed a shortage of nurses and professional caregivers [24]. Still further, depending on the level of need and type of assistance, having a personal caregiver may be quite expensive and hence it may cause a huge financial burden on care-receivers and their families.

One potential solution is robotic caregivers that could assist those in need at homes or retirement and nursing homes. Or pretty much everywhere else where regular nurses provide assistance such as hospitals, ambulatory care, clinics, outpatient surgery centers, nursing homes and extended care facilities, among other places. But this is all easier said than done.

Since the introduction of industrial robots in the early 1960s, most robots worldwide work at factories. And these robots typically do not physically interact with humans, they work alone. Moreover, if a factory safety system detects the presence of a human nearby, the alarm sounds and all robotic machinery immediately stops operating. The reason for this is that these robots are not really made to interact with humans. Their powerful limbs are continuously trying to attain in advance set desired position trajectories in a tightly or high gain controlled fashion. And they are not really aware of their surrounding including potential human visitor that could be easily smashed by an operating robot.

Many roboticists, especially during the past two decades, are closely working to address this challenging problem.

In terms of basic physical actions, a robotic caregiver must be able to perform locomotion (move from point A to point B in potentially dynamic environment, for example, with a cat running across the floor) and manipulate objects (e.g., pick and hold a tray with food plate and glass of water). A robotic caregiver will also need to sense and understand its environment (e.g., see a chair and know that this is a chair) and understand both verbal and nonverbal commands and cues such that it could act accordingly in a physically intelligent manner. For example, consider a trivial command: "Open the gallon of water." To address this command the robotic caregiver will need first to localize the gallon of water, then move towards the gallon of water, and finally perform a set of nontrivial actions to open that gallon of water in the same way that a human would do.

Moreover a robotic caregiver must be able to physically interact with humans; for example, help someone transition from a reclined to sitting position or assist a person walking by holding a person's arm to prevent a fall. Some of these physical interactions will clearly require larger, well-timed forces but that are still applied gently so as not to cause any harm to the patient. To do so, the robot will need to have advanced haptic and tactile senses and good understanding of human biology as well as the specific needs of the patient. Still further, the robotic caregiver will need to be able to learn, that is, to be taught by human health professional, for example, nurse, or by a patient, in order to optimize its performance. Each home environment and each patient's need and expectations are quite specific, quite individual and this is not some knowledge or routine that robot could possibly have preprogrammed.

This robot can be visualized as a super advanced, highly biomimetic, human-like robot. And indeed there is a good chance that this may be the case sometime in the not so distant future.

ROBEAR, shown in Fig. 14.9, developed by RIKEN, Japan, is a true co-robot (collaborative robot) that can help a person rise from a sofa and sit in a wheelchair. From Riken website: "it includes actuator units with a very low gear ratio, allowing the joints to move very quickly and precisely, and allowing backdrivability, meaning that the force encountered by the actuators as they perform their tasks can be quickly fed back into the system, allowing softer movement. It also incorporates three types of sensors, including torque sensors and Smart Rubber capacitance-type tactile sensors made entirely of rubber, which allow for gentle movements, ensuring that the robot can perform power-intensive tasks such as lifting patients without endangering them." [25]

ROBEAR is the second generation of advanced assistive co-robots developed by RIKEN. Its predecessor robot for interactive body assistance (RIBA) was the first robot able to lift or set down a human from a bed or wheelchair in a superbly gentle way. RIBA utilized tactile guidance based on sensory input from many tactile sensors. RIBA's arms had almost 500 sensors. And soft urethane foam was used for the robot's skin. The first RIBA prototype could lift up to 135 pounds (61 kg). The robot was able to respond to audio commands, visual input and touch, and was trained to recognize both faces and voices [2].

FIG. 14.9 ROBEAR developed by RIKEN, Japan and Sumitomo Riko Company Limited. For more information see Ref. [25].

14.3 Robotic Massage

Massage is used for the purpose of overall health and wellbeing especially to help blood circulation, to relax tense muscles, and to relieve stress.

Robotic massage chairs are popular in some regions of the world, typically in the world's most developed countries. For example, in Japan various estimates point toward roughly 20% of households with robotic massage chairs. Estimates for US suggest a more modest 1% of households. Hence, this Japan market size can be estimated at about 10 million and US market size at about 1.3 million robotic massage chairs typically priced in between \$1000 and \$10,000 USD.

Advanced robotic massage chairs can be considered a biomechatronics system. This system can understand the shape and weight distribution of the human user and it can also decide on the best ways to apply massage through action of its actuators. Some of these chairs, like the Inada Sogno Dreamwave Massage Chair introduced by Osaka, Japan-based company Inada in 2008 has been the research result of a \$50 million grant that involved seven universities. And many still consider it to be the top robotic chair in the world [26]. It obtained minor updates (software update) in 2012 and it is now called Inada Dreamwave [27].

Inada Sogno Dreamwave or Inada Dreamwave Massage Chair is one of most expensive chairs on the market probably with a good reason. The chair has 101 airbags which cushion, compress, and stretch the body from fingertip to toe, eight main massage programs or 16 preprogrammed massage sessions providing full body massage covering 1200 $in^2$of active area and based on the result of infrared surface scan, utilizing an optical sensor to trace body's shape, it provides customized massage for one of 106 stored body types [28, 29]. It has 30 actuators that move their multiple rollers along complex 3D trajectories with a goal to achieve the same effect as when the person is undergoing a professional shiatsu massage. These are referred to as 3D quad rollers with adjustable speed, sometimes also called 4D rollers. Heating function is provided for seat and lumbar area. The chair can also fully recline and achieve the zero-gravity state with most evenly distributed body-chair pressure. The chair operation is extremely quiet. Some reviews [27] suggest that only the iRobotics7 Medical Massage Chair [30] made by an Arlington, Texas-based company Luraco Technologies might be even quieter with its 100 air-cells and 12 powerful DC motors. Both massage chairs are featured in Fig. 14.10.

Some chairs perform a nonoptical body scan using sensors embedded in actuators that move rollers. Some chairs monitor the distribution of pressure in air cells to understand body orientation and weight distribution.

The Sanyo Zero Gravity Massage Chair includes sensors that monitor galvanic skin response, which relate to the user's pulse rate and perspiration. When the chair detects an increase in these metrics, it interprets that the area currently being massaged is particularly tense or stiff. The chair's microprocessor alters the normal pattern to spend more time on that area [31].

A few massage chairs even include voice-response software to interpret and respond to vocal commands from the user [31].

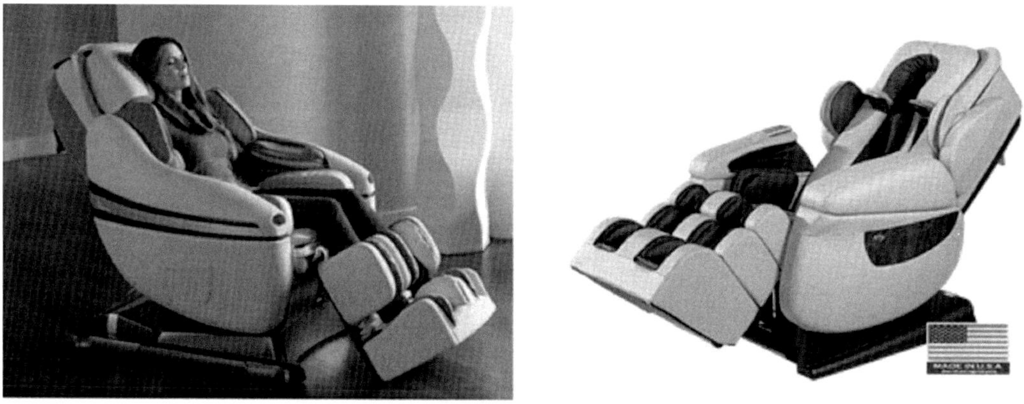

FIG. 14.10 Inada DreamWave Massage Chair [29] (left) and Luraco Technologies iRobotics7 Medical Massage Chair [30] (right).

References

[1] The World Factbook: Field Listing: Median Age, Central Intelligence Agency. https://www.cia.gov/library/publications/the-world-factbook/fields/2177.html (retrieved 11.03.18).

[2] M.B. Popovic, Biomechanics and Robotics, 351 pages, Copyright © 2014, Pan Stanford Publishing Pte. Ltd., Singapore, 2013doi:10.4032/9789814411387 ISBN 978-981-4411-37-0 (Hardcover), 978-981-4411-38-7 (eBook).

[3] H. Gleckman, Nearly Half of All Seniors Need Help With Daily Activities, Far More Than We Thought, Forbes, Personal Finance, https://www.forbes.com/sites/howardgleckman/2014/10/08/nearly-half-of-all-seniors-need-help-with-daily-activities-far-more-than-we-thought/, 2014 (retrieved 11.03.18).

[4] C.P. Mason, E. Peiser, Medical manipulator for quadriplegics, in: Proceedings of the International Conference on Telemanipulators for the Physically Handicapped, Rocquencourt, France, September 4–6, 1978, pp. 309–318.

[5] B. Abdulrazak, M. Mokhtari, Assistive robotics for independent living, in: A. Helal, M. Mokhtari, B. Abdulrazak (Eds.), The Engineering Handbook of Smart Technologies for Aging, Disability and Independence, John Wiley & Sons, Inc., Hoboken, NJ, 2008

[6] W. Seamone, G. Schmeisser, Early clinical evaluation of a robot arm/worktable system for spinalcord-injured persons, J. Rehab. Tes. Devel. (1985) 38–57.

[7] W. Seamone, G. Schmeisser, Evaluation of the JHU/APL robot arm workstation, in: R. Foulds (Ed.), Interactive Robotic Aids, World Rehabilitation Fund Monograph, New York, 1986.

[8] M. Van der Loos, VA/Stanford rehabilitation robotics research and development program: lessons learned in the application of robotics technology to the field of rehabilitation, IEEE Trans. Rehabil. Eng. 3 (1) (1995) 46–55.

[9] The ProVAR Homepage, Stanford University, 1998. http://web.stanford.edu/group/rrd/Projects/provar/provar_stanford.html (retrieved 11.03.18).

[10] J.L. Dallaway, R.D. Jackson, RAID—a vocational robotic workstation, in: ICORR 92—Conference Proceedings, Keele University, UK, 1991.

[11] H. Eftring, G. Bolmsjo, RAID—a robotic workstation for the disabled, in: Proceedings of the 2nd European Conference on the Advancement of Rehabilitation Technology (ECART 2), Stockholm, Sweden, 1993 http://www.arkiv.certec.lth.se/doc/raida/ (retrieved 11.03.18).

[12] M. Busnel, R. Cammoun, F. Coulon-Lauture, J.-M. Détriché, G. Le Claire, B. Lesigne, The robotized workstation "MASTER" for users with tetraplegia: description and evaluation, J. Rehabil. R&D 36 (3) (1999).

[13] T. Jones, RAID—toward greater independence in the office & home environment, in: ICORR '99: International Conference on Rehabilitation Robotics, Stanford, CA, 1999.

[14] M. Topping, J. Smith, The development of Handy 1, a rehabilitation robotic system to assist the severely disabled, Ind. Robot: Int. J. 25 (5) (1998) 316–320, doi:10.1108/01439919810232459.

[15] M. Topping, An overview of the development of handy 1, a rehabilitation robot to assist the severely disabled, J. Intell. Robot. Syst. 34 (2002) 253, doi:10.1023/A:1016355418817.

[16] H. Kwee, Spartacus and Manus: telethesis developments in France and Netherlands, in: Interactive Robotic Aids – One Option for Independent Living, World Rehabiltation Fund, 1986, pp. 7–17.

[17] G. Verburg, H. Kwee, A. Wisaksana, A. Cheetham, J. van Woerden, Manus: the evolution of an assistive technology, Technol. Disabil. (1996) 217–228.

[18] "About us" Exact Dynamics, http://www.exactdynamics.nl/site/?page=about (retrieved 12.03.18).

[19] R. Soyama, S. Ishii, A. Fukuse, The development of meal-assistance robot: my spoon, in: Proceedings of the 8th International Conference on Rehabilitation Robotics (ICORR 2003), 2003, pp. 88–91.

[20] Meal-assistance Robot My Spoon, SECOM, https://www.secom.co.jp/english/myspoon/faq.html (retrieved 12.03.18).

[21] R. Hermann, A. Phalangas, R. Mahoney, M. Alexander, Powered feeding devices: an evaluation of three models, Arch. Phys. Med. Rehabil. 80 (1999) 1237–1242.

[22] W.-K. Song, J. Kim, Novel assistive robot for self-feeding, in: Robotic Systems - Applications, Control and Programming, 2012. doi:10.1007/978-1-4614-4379-7_4.

[23] Manual Neater Eaters, Neater, http://www.neater.co.uk/neater-eater/ (retrieved 12.03.18).

[24] J. Dulaney, Nursing Students Being Turned Away Amid Faculty Shortage in Cal State System, Press Telegram, 2013. https://www.presstelegram.com/2013/11/30/nursing-students-being-turned-away-amid-faculty-shortage-in-cal-state-system/ (retrieved 10.03.18).

[25] The strong robot with the gentle touch, RIKEN Press Release, http://www.riken.jp/en/pr/press/2015/20150223_2/, 2015 (retrieved 10.03.18).

[26] Inada Sogno Dreamwave vs Osaki Massage Chair Review 2018, Chair Institute, https://chairinstitute.com/inada-sogno-dreamwave-vs-osaki/, 2018 (retrieved 10.03.18).

[27] J. Roberts, Best Massage Chair Reviews, Wellnes Geeky, 2018. https://www.wellnessgeeky.com/massage-chair/ (retrieved 10.03.18).

[28] Best Massage Chair Reviews 2018—(Comprehensive Guide), Massage Chair Land, https://www.massagechairland.com/ (retrieved 10.03.18).

[29] Sogno DreamWave Massage Chair – A revolution in robotic massage chair technology, Inada, https://www.dreamwave.com/videos/sogno-dreamwave-massage-chair-a-revolution-in-robotic-massage-chair-technology/ (retrieved 10.03.18).

[30] The World's #1 Medical Massage Chair®—The Power of American Engineering™, Luraco Technologies, http://www.luraco.com/products.php?v=6 (retrieved 10.03.18).

[31] J. Strickland How Massage Chairs Work, HowStuffWorks. https://electronics.howstuffworks.com/gadgets/home/massage-chair4.htm (retrieved 10.03.18).

Further Reading

[32] M. Singleton, Obi is a Robotic Dining Companion for People With Disabilities, The Verge, Tech, Circuit Breaker, 2016. https://www.theverge.com/circuitbreaker/2016/7/20/12242070/obi-is-a-robotic-dining-companion-for-people-with-disabilities (retrieved 12.03.18).

[33] Obi, https://meetobi.com/ (retrieved 12.03.18).

15

Robotic Surgery

Pinar Boyraz*,†, Ivo Dobrev‡, Gregory Fischer§, Marko B. Popovic§

*CHALMERS UNIVERSITY OF TECHNOLOGY, GOTHENBURG, SWEDEN †ISTANBUL TECHNICAL UNIVERSITY, ISTANBUL, TURKEY ‡UNIVERSITY HOSPITAL ZURICH, UNIVERSITY ZURICH, ZÜRICH, SWITZERLAND §WORCESTER POLYTECHNIC INSTITUTE, WORCESTER, MA, UNITED STATES

CHAPTER OUTLINE

15.1 Overview of Robotic Surgery

Robotic surgery is one of the fastest developing areas within the biomechatronics field, as it provides certain advantages to both surgeons and patients. From the perspective of surgeons the robotic systems may improve both safety and effectiveness of procedure, as well as reduce the physical and cognitive effort thus helping alleviate tiredness, especially in suture-intensive operations. From the perspective of the patients, robotic surgery usually means less postoperative pain, shorter hospital stays and less damage to tissues.

According to [1], the first robotic minimally invasive surgery (MIS) was performed in 1987 in the form of laparoscopy. Laparoscopy is a surgical procedure in which a fiber-optic instrument is inserted through the abdominal wall to view the organs in the abdomen or permit small-scale surgery. Since then, the robotic MIS has been accepted in many

Biomechatronics. https://doi.org/10.1016/B978-0-12-812939-5.00015-X

operations involving urology [2], brain and neural system, reproductive system [3], gastronomical, and cardiovascular tracts. Some of these operations do not even need an incision such as natural orifice transluminal endoscopy (NOTE) [4].

15.1.1 Introduction: Traditional Robotic Surgery

Surgical robots assist during surgical procedures. They have been used since the mid-1980s. Today, the majority of prostatectomies in the United States are robot-aided procedures as chances for successful operation are higher with robotic aid than without.

Robotic surgeries are typically *minimally invasive.* This feature has been around long before the introduction of robots. It is a broad concept that encompasses many common procedures, such as a laparoscopic cholecystectomy, or gall bladder excisions. The procedure refers to a method that avoids long cuts by operating on the body through small (usually 1 cm) entry incisions. Surgeons use long-handled instruments to operate on tissue within the body. Such operations are guided by viewing equipment called endoscopes. These are thin tubes with a camera attached to the end of it that allows the surgeon to view highly magnified real-time three-dimensional images of the operation site on a monitor.

The current benefits of robotic surgery include better accuracy, precision, dexterity, tremor corrections, scaled motion, and more recently haptic corrective feedback. These benefits result in more successful surgeries and smaller necessary incision cuts. Overall, the robotic systems have better accuracy and precision than unaided surgeons. Surgical robots are able to position the surgical tools closer to the "right spot" and deviate less from the "right trajectory." The robot's end-effectors can be much smaller and more dexterous than a human hand. They can record and filter out a surgeon's natural hand tremor and rescale movement to increase precision and reduce the chance for error. Lastly, the robot could restrain the surgeon's movement into undesired directions through haptic feedback.

Researchers are formulating new ways to address motion and tissue resistance. For example, the surgical robot could synchronically move with a beating heart such that their relative speed is close to zero. Another possible improvement is the ability for the robot to automatically adapt to the dynamical tissue resistance over time as the sensitivity scale of these processes goes beyond human capabilities.

Typically, robotic surgery can be classified as either (i) supervisory-controlled, (ii) telesurgical, or (iii) shared-controlled.

The supervisory-controlled approach is the most automated of the three methods. The RoboDoc from Integrated Surgical Systems Inc. is an example of a supervisory-controlled system used in orthopedic surgeries. After the surgeon positions the RoboDoc's bone-milling tool at the correct position inside the patient, the robot automatically cuts the bone to just the right size for the orthopedic implant.

Prior to the surgical procedure, the surgeon needs to prepare the operation through the planning and registration phase. In the planning phase, images of the patient's body are used to determine the right surgical approach. Common imaging methods include computer tomography (CT) scans, magnetic resonance imaging (MRI) scans, ultrasonography, fluoroscopy, and X-ray scans. Next, in the registration phase, the surgeon must locate the points on the patient's body that correspond to the images created during the planning phase. These points are matched to a 3D model, which can be updated by images seen through cameras or other real-time imaging techniques during surgery. After the robot finds the best fit between the model and reality, the surgical procedure is performed.

The tele-surgical approach allows the surgical robot to be tele-operated, that is, operated from a distance by a human surgeon. In practice, the robot and the surgeon are only a couple of meters apart. Tele-operation is also possible across larger distances. However, problems such as time delays (i.e., tele-surgical latencies) and the available bandwidth (i.e., the amount of information that can be transferred per unit time) need to be considered.

The tele-surgical approach is used by the da Vinci Surgical System, which was invented by Philip S. Green and developed by Intuitive Surgical Inc. This system currently dominates the surgical robot market. Initially dubbed Mona (after Leonardo's *Mona Lisa*), the system was rechristened the da Vinci Surgical Robot in 1999; according to Mr. Green "…in honour of the man who had invented the first robot." Although da Vinci never invented or built a real robot (credit for that goes to Tesla), he made many drawings of various mechanisms.

The da Vinci System consists of three primary components: (1) a viewing and control console that is used by surgeon, (2) a vision cart that holds the endoscopes and provides visual feedback and (3) a surgical robot's manipulator arm unit that includes three or four arms, depending on the model. The instruments that are attached to the arms are highly specialized. Functions for them include clamping, cutting, suturing, tissue manipulation, cauterizing, etc.

It takes some time for surgeons to get accustomed to the da Vinci System. According to a study, even with initial training program, provided by the Intuitive Surgical, it takes about 12–18 operations before surgeons feel comfortable performing the procedure. Often, during this period, surgeons complain on lack of tactile force feedback or ability to "feel" the tissue.

The shared-controlled approach refers to the method by which the robot is not just motion tele-operated as it can decide to resist the surgeons' intended movement if it deems that it would not be beneficial. Typically, the work space is split into several segments and the system behaves differently based on different localization according to *safe, close, boundary, or forbidden* classification. For example, if surgeon moves a cutting tool in the direction of tissue that should not be damaged, the robot will apply the force haptic feedback that will grow stronger as the cutting tool comes closer to the fragile tissue. In other words, here, surgeon again "feels" the virtual representation of tissue that may have preprogrammed specifications different from the real tissue as well as somewhat different localization in space.

15.1.2 Multiarm, Hyperredundant, Continuum, and Soft-Robotic Platforms for Robotic Endoscopy

Although the first widely used and FDA approved robotic platforms for surgery were based on industrial six degrees of freedom (DOF) serial and rigid robotic manipulator such as [5], the robotic platforms are currently often designed with more DoF and with some level of flexibility. The hyperredundant robots with more DoF and dexterity are proposed to address difficult passages. Theoretically, a continuum robotic platform has infinite number of DoF and hence it is anticipated to be able to improve surgical procedures still further. Besides, the tubular/telescopic continuum robotic platforms can be miniaturized easily due to their simple mechanic construction. Further improvement to surgical platforms can be obtained by employing the inherent compliance of the soft materials. This chapter addresses above mentioned aspects and reviews some of multiarm, hyperredundant, continuum, and soft-robotic platform examples from literature and case studies from selected research laboratories.

The forerunners of robotics surgical systems are endoscopic platforms since the inception of this field. An endoscopic process typically involves a small incision through which the rigid or flexible endoscopic devices is inserted to provided view on the tissue that is being operated. Because the insertion point/trocar limits the movement of the robotic arm in two directions, see for example Ref. [6], the remaining DOF for the endoscopic robotic platforms are totaled to four DOF (see Fig. 15.1). Although this usually limits the dexterity and manipulation of the end effector, endoscopic robots are still advantageous because they cause less pain, have a better cosmetic look after the operation, and the cost is much lower.

In Ref. [6], a comprehensive classification of endoscopes is given including the categories of laparoscopes with or without camera, the manipulator size, and if the sterilization is possible.

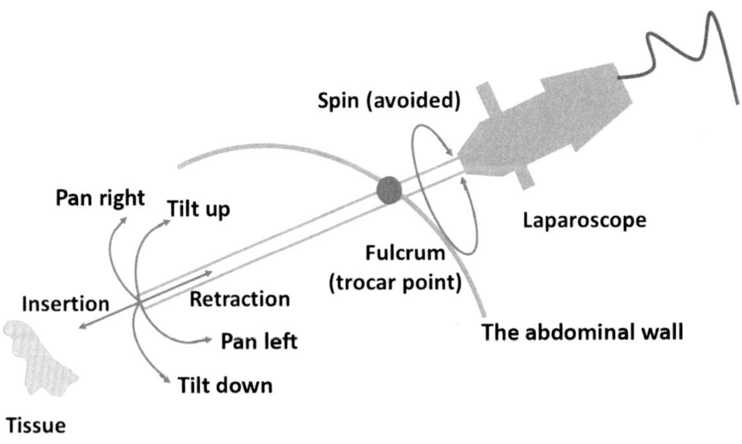

FIG. 15.1 Reduced DOF for endoscopic tool after insertion [6].

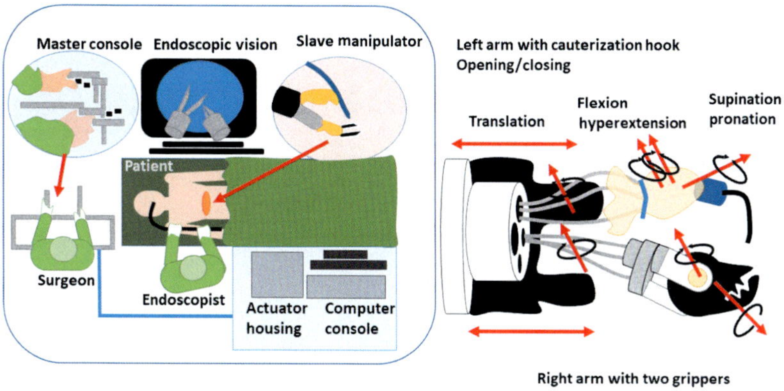

FIG. 15.2 (A) Overview of MASTER system and (B) structure of the slave manipulator. *Original illustration by P. Boyraz.*

A consensus by European Association of Endoscopic Surgeons (EAES) was reached in 2014 [7] defining the types of operations that can be performed using robotic endoscopy and the related requirements for safe and efficient execution. Ref. [8] summarizes robotic flexible endoscopy platforms for diagnosis and treatment of gastrointestinal diseases according to articles accessible through PubMed published between 2010 and 2015. According to that review Neoguide (Intuitive Surgical, United States) is FDA approved while commercially available robotic flexible endoscopic platforms include: Invendoscope (Invendo Medical Gmbh, Germany), Aer-O-scope (GI View Ltd, Israel), Endotics (ERA Endoscopy SRL, Italy), Endomina (Endo Tools Therapeutics, Belgium), and Viacath (Hansen Medical, United States). The other notable system mentioned in this review is MASTER (EndoMASTER Pte, Singapore) system, presented in Ref. [9] in adequate detail (see Fig. 15.2). This robotic system has a master and a slave manipulator driven by tendon-sheath actuation. The system also supports force-feedback for the surgeon.

Clinical study involving 42 patients that undertook endonasal trans-sphenoidal surgery has been reported in Ref. [10]. There, a rigid endoscope is used with iArmS robotic platform as a support robot. More advanced applications of robots in endoscopy involve flexible robotic platforms [11] and endoscopic capsules (ECs) [12].

As discussed in Ref. [11], the ideal robotic endoscope has not been realized yet as there are numerous challenges that need to be overcome especially for flexible endoscopes. These challenges are mainly concerning the use of the robots for manipulation of the endoscopes themselves for the remote insertion and incorporation of such robots not only for NOTE type of operations but also for endoscopic submucosal dissection (ESD). The setup for flexible endoscopic application development is shown in Fig. 15.3.

As a frontier in robotic endoscopy, thc ECs [12] can open a new solution window to the problems with tethered and rigid systems. The first applications of EC were aimed at the drug delivery and diagnosis; however, recently they are being developed for the operational purposes. The operative capabilities of ECs are still under development and the movement of the device is usually provided by an external magnet.

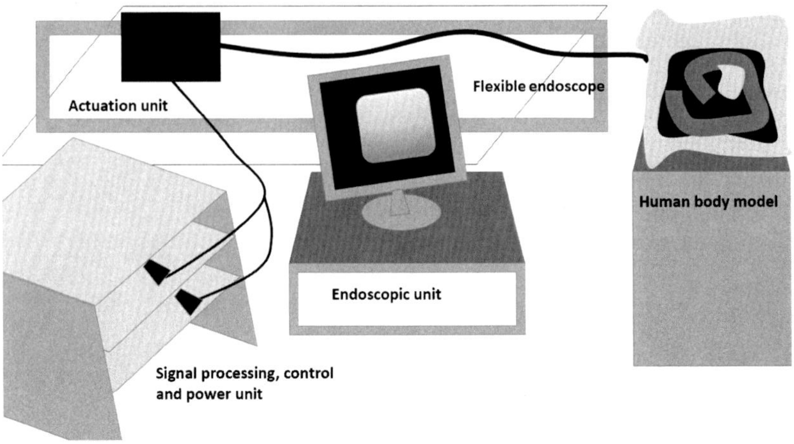

FIG. 15.3 The setup for flexible endoscopic application development [11]. *Original illustration by P. Boyraz.*

15.2 Platform-Based Classification of Robotic Surgery

In this section, the robotic platforms used as autonomous, semiautonomous or assistive systems in the operations are classified based on the structure. The robotics platforms may increase the number of DoF either by addition of robotic arms to form multiarm structure, discussed in Section 15.2.1, or by having hyperredundant backbones, discussed in Section 15.2.2. In terms of minimizing the dimensions and having almost limitless DOF, the continuum robotics is considered in Section 15.2.3. Recent research focused on soft robotics in the context of robotics surgery is addressed in Section 15.2.4. Several examples of hybrid systems that may eventually lead to general robotic platform benefiting from increased DoF, compact structure and inherent compliance altogether are reviewed in Section 15.2.5.

15.2.1 Multiarm Robotic Platforms for MIS

Over the last several decades there was a steady progress that finally led to sophistication of current gold standard in robotics surgery field in the form of the da Vinci Surgical System (Intuitive Surgical, Sunnyvale, CA, USA); detailed overview of these early historical developments is provided in Ref. [13].

In fact, the surgeons in the operation room need assistance from the other surgeons or staff nurse to hold some of the gripper carrying manipulators, *endo*-luminal devices or guidance catheters while they operate on the tissue with minimal stress and maximum safety. Therefore, the obvious and very first use of the robotics in the surgery has been the adaptations of the robots as third arm [14] or use multiple-arms of robotic platforms for dexterous manipulation [15]. There are even some cases in which the multiarm robot is used for a very specific surgical operation requiring three dimensional (3D) navigation and involving bone grafting [16]. Although being beneficial and safer, the multiarm robotic platforms have some disadvantages considering the need and cost of the optimization for the configuration space according to the patient's body and the operable area [15]. Because of such disadvantages and costly installation/optimization process, other alternatives such as hyperredundant robot-arms and continuum robotic platforms are proposed. These two specific design ideas will be detailed in the next sections.

15.2.2 Hyperredundant Robotic Platforms

Robotic platforms of the hyperredundant or piece-wise continuum design mostly use rigid or semirigid backbones. Recently proposed hyperredundant modular structures can be found in Refs. [17–19]. In Ref. [17], high dexterity and high-stiffness requirements for actual operation are provided although the design has limited compliance. The design in Ref. [17] also has intricate mechanical structure which can be difficult to miniaturize for medical applications required to work in limited operation space. More advanced prototypes can have control over the stiffness [18], however, they may have poorer module-based controllability. Even though it is possible to have applications with better control [19] featuring continuum elastic backbone while segmenting the structure into modules with coupling plates, the compactness cannot be guaranteed. Most hyperredundant platforms have cable driven structures because they are lightweight and compact. However, there is an inherent limitation of such mechanisms due to cable friction and interdependency between the sections of the modules.

Early examples of hyper-redundant manipulators with full solution on kinematics and dynamics are presented in Refs. [20–22]. Although the kinematics and dynamics are widely studied, the relevant control algorithms suggested for such mechanisms are not mature. For example, a modular control scheme is proposed in Ref. [23] with some promising results, however, more research efforts are needed in control aspect of hyperredundant manipulators. In that respect, the work in Ref. [24] presents a hybrid motion/force control application. Although the actual algorithm is developed for multibackbone continuum robots, it can be modified for hyperredundant robotic platforms using simplified kinematics. What is more exciting in hyperredundant platforms' research is that very recent developments are taking place providing controllable stiffness in robotic arms. In Ref. [25], the researchers proposed such a platform to be used in laparo-endoscopic single-site surgery. The structure can be switched between five DOF rigid status to seven DOF compliant status according to the task required in the operation (see Fig. 15.4).

15.2.3 Continuum Robots for MIS

A comprehensive survey on the continuum robots for medical applications [26] can guide the reader on which type of surgical interventions these type of robots should be preferred. A continuum robot has infinite DOF and can be driven by cables, hydraulic actuation or shape-memory-alloys based on the application and the level of required miniaturization.

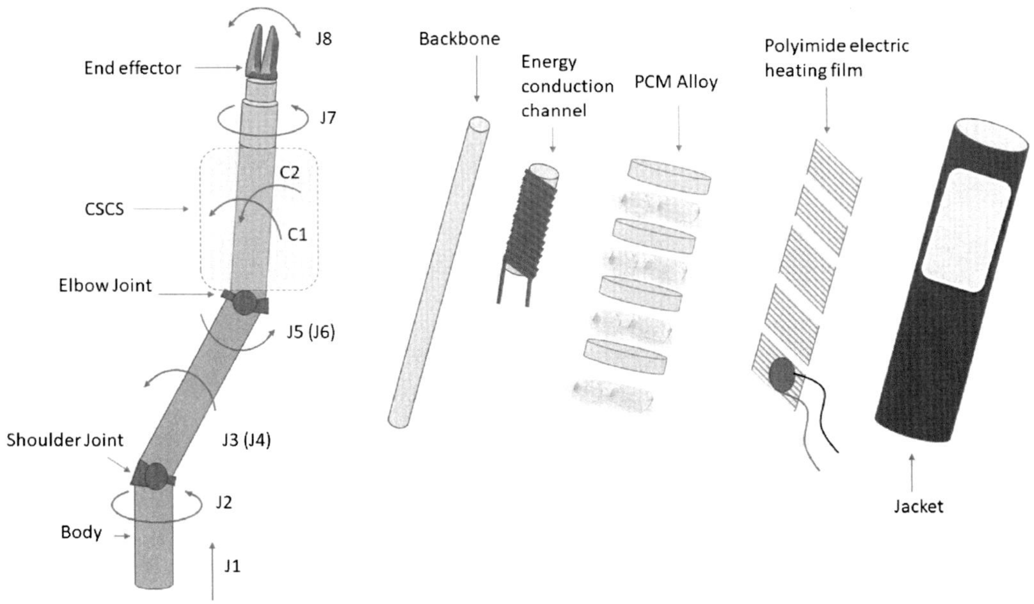

FIG. 15.4 The hyperredundant robotic arm with an adjustable section employing PCM alloy for the stiffness control via temperature change. *Original illustration by P. Boyraz prepared based on J. Wang, S. Wang, J. Li, X. Ren, R.M. Briggs, Development of a novel robotic platform with controllable stiffness manipulation arms for laparoendoscopic single-site surgery (LESS), J. Med. Robot. Comput. Assist. Surg. 14 (2018) 1–16.*

In Ref. [27] a co-centric tube type of robot is proposed using piezoelectric actuation. The co-centric tube type of continuum robots needs to be precurved according to their task and workspace in the operation. This might turn to be a disadvantage if the same medical device is desired to be used on different patients requiring different operational workspace. In addition, this type of continuum platforms may have limited capabilities of translational motion. A novel continuum manipulator design using serially connected double-layer planar springs [28] overcome these disadvantages to some degree (see Fig. 15.5). In addition, the proposed structure behaves like a helical spring, but its contraction and bending motion are decoupled, therefore easier to control.

15.2.4 Soft Robotics for Surgical Applications

Soft-material robotic platforms open up a new path for obtaining compliant and safe platforms. A good example of such an application can be seen in Ref. [29]. In this work, the researchers used flexible fluidic actuators allowing dexterous and safe navigation. It was possible to overcome the restrictions that are normally posed in a situation where the rigid manipulator has to work in a cavity with multiple organs in close proximity. Fig. 15.6 depicts the advantage of the soft robot in terms of dexterity in a confined region.

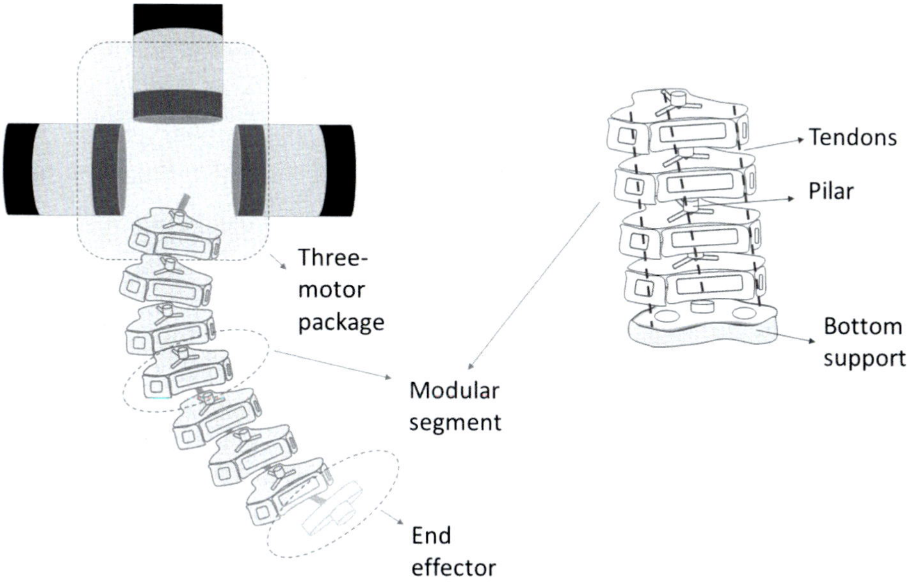

FIG. 15.5 A novel continuum robotic platform with multilayer deformable springs. *Original illustration by P. Boyraz prepared based on P. Qi, C. Qui, H. Liu, J.S. Dai L.D. Seneviratne, K. Althoefer, A novel continuum manipulator design using serially connected double-layer planar springs, IEEE/ASME Trans. Mechatr. 21(3) (2016) 1281–1292.*

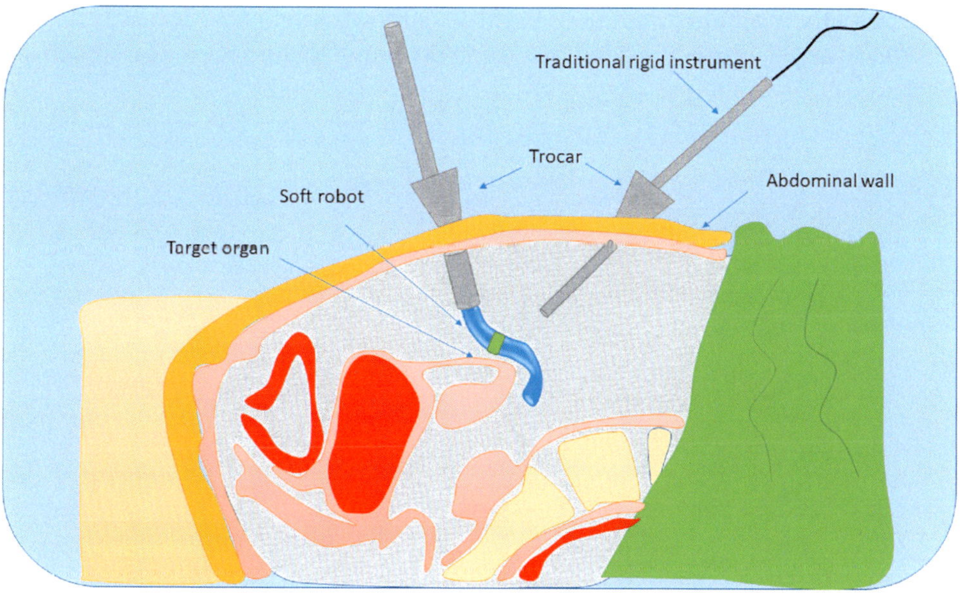

FIG. 15.6 An illustration of dexterous advantage of the soft robot compared to rigid robot. *Original illustration by P. Boyraz prepared based on H. Abidi, G. Gerboni, M. Brancadoro, J. Fras, A. Diodato, M. Cianchetti, H. Wurdemann, K. Althoefer, Highly dexterous 2-module soft robot for intra-organ navigation in minimally invasive surgery, Int. J. Med. Robot. Comput. Assist. Surg. 14 (2018) 1–9.*

Another work can be taken as a guiding example for planning the navigation or locomotion of such robots in tubular environments [30]. Although the application is generic, it can be easily modified for cardiovascular or urological interventions because it prevents full occlusion of the tubular channel while navigating in it. This is a big advantage compared to traditional rigid structures as they tend to block such channels and in extreme cases cause tissue damage.

15.2.5 Hybrid Robotic Platforms

The hybrid robotic platforms strive to combine the best properties of serial rigid robotic arms with continuum and/or soft robots to achieve superior qualities such as position accuracy and compliancy at the same time. For example, the system described in Ref. [31] comprises of a serial manipulator which is connected to a continuum platform. In this combined hybrid structure the robot was observed to have superb positioning and tracking accuracy. A conceptual design of the hybrid robotic platform can be seen in Fig. 15.7. Normally, when entering into human body through the incision, two DOF provided by the rigid arm are sacrificed. However, in this setup it can be regained, thanks to the flexible continuum robot at the end.

Herein, the word "hybrid" can also refer to the combination of serial and parallel links (i.e., open and closed kinematic chains) to form a better spatial capability for the end effector [32]. Another possibility of obtaining a hybrid platform is to combine the continuum structures with soft materials, as seen in Ref. [33], incorporating also reconfigurability. This hybrid platform (see Fig. 15.8), although not used in medical robotics yet could be easily adapted to many surgical applications, provided that it can be manufactured in smaller dimensions.

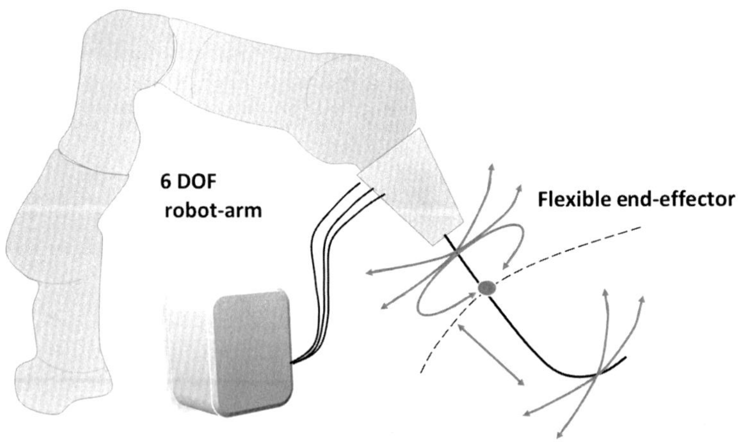

FIG. 15.7 Hybrid robotic system combining a flexible continuum robot and a rigid 6 DOF robotic arm [31]. *Original illustration by P. Boyraz.*

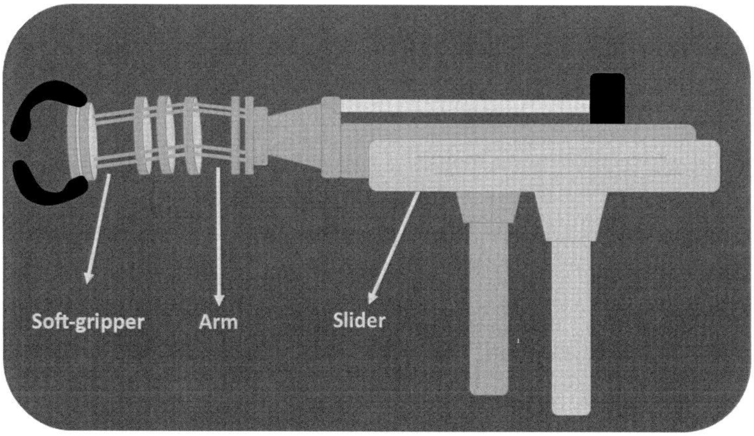

FIG. 15.8 Soft compliant manipulator with continuum segments, SIMBA [26]. *Original illustration by P. Boyraz.*

15.3 Human-Machine Interaction in Robotic Surgery

One of the grand challenges of the robotic surgery nowadays is in the area of designing efficient and ergonomic human-machine interaction (HMI) systems to improve the augmentation of the surgeon and the robot. Considering that not all the surgical platforms could be fully autonomous, HMI systems will always be needed in the operation rooms to facilitate the integration of robotic tools for surgery. A reasonable starting point of such technology involves the virtual-reality applications [34] to simulate the high-skill tasks for surgeons such as suturing. A more direct application of HMI systems could be teleoperated surgical systems. A recent and exciting example of such a system uses a haptic interface to manipulate and steer miniaturized untethered soft-magnetic grippers in Ref. [35]. In addition to new HMI system designs in robotic surgery, observing the human factors and ergonomics criteria is also very important. For example, in Ref. [36], human arm stiffness is examined during the virtual tele-operation of robotic manipulators. In that study, the researchers have proposed a nondisruptive method to study the stiffness of the human arm during different tasks within an operation. This type of performance analysis is required and must be performed to assess the stability and precision of the execution of the task, because the overall system is a combination of the human skills incorporated with the robotic system's dynamics. The collaborative robots could be considered as being on the very top of HMI research in the robotic surgery field. These systems are aimed at providing the surgeon more stable assistance, especially in effort-laden suture procedures. An example of such a system is detailed in Ref. [37]. This type of system must have the capability to recognize surgical gestures made by surgeon and should have an algorithm supporting the autonomous decision-making. The experimental setup for the proposed collaborative system is presented in Fig. 15.9.

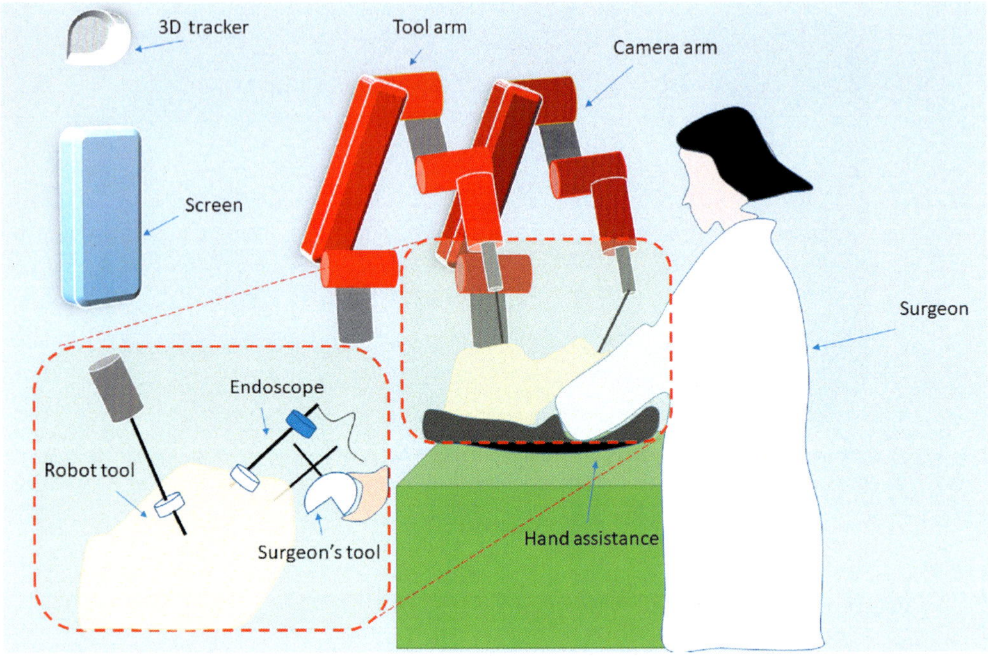

FIG. 15.9 Experimental setup for development of collaborative robot platform. *Original illustration by P. Boyraz prepared based on E. Bauzano, B. Estebanez, I. Garcia-Morales, V. Munoz, Collaborative human-robot system for HALS suture procedures, IEEE Syst. J. 10(3) (2016) 957–966.*

15.4 Autonomy Levels in Robotic Surgery

As with any system that is gradually evolving into autonomous operation modes, the surgical robots are also under development to make that transition in a safe manner. As it is also realized on time by the authorities, a new ISO standard has been compiled defining the autonomy levels in medical equipment [38]. In addition to standardization efforts by experts and authorities, the research community is also responding via well-funded research projects such as I-SUR (EU FP7 funded, Intelligent Surgical Robotics). In this project, the researchers developed innovative solutions [39] such as surgical planning and execution of movement of robot arms in contact with a deformable environment, an interface minimizing the cognitive load of the surgeon through supervision of the actions and intraoperative sensing together with reasoning to detect the transitions between the autonomous and tele-operated modes. In fact, in Ref. [40] this transition is detailed using a two-layered bilateral control, stabilizing the process of authority transition from robot to surgeon. More focused studies also exist, for example, integrating human surgeon in an automated suturing task as seen in Ref. [41]. In that study, the researchers were able to accelerate the suturing process by 79% using one-master and two-slave system with force sensing. The autonomous suturing process can be seen in Fig. 15.10.

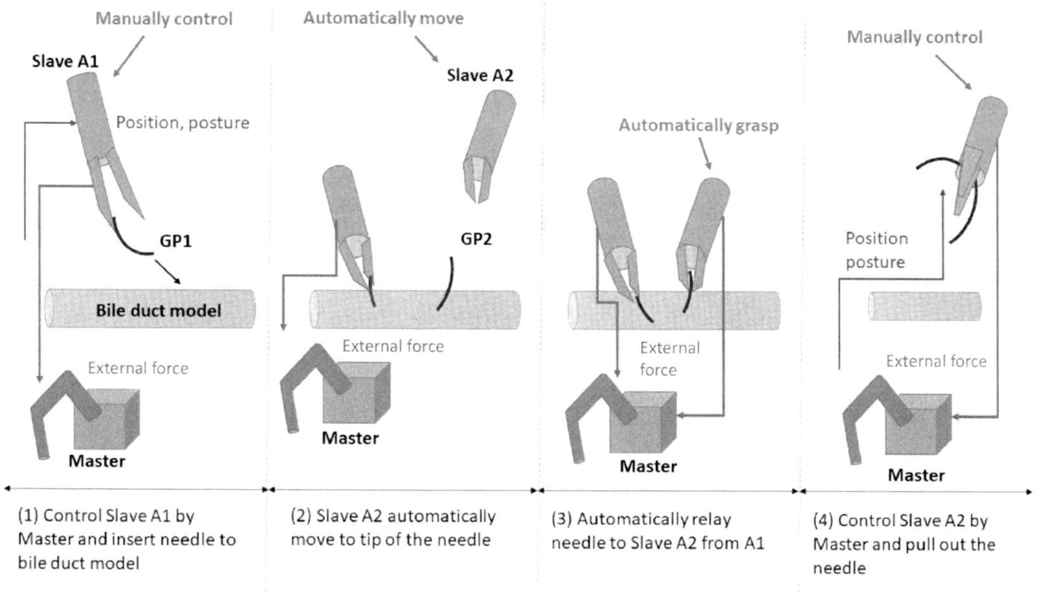

FIG. 15.10 Human-integrated autonomous suturing process. *Original illustration by P. Boyraz prepared based on K. Watanabe, T. Kanno, K. Ito, K. Kawashima, Human-integrated automation of suturing task with one-master two-slave system for laparoscopic surgery, IEEE Int. Conf. on Advanced Intelligent Mechatronics (AIM), Banf, Alberta, Canada, 2016.*

In order to execute a semiautonomous surgical process, task recognition is necessary for the surgical robot. In Ref. [42], a distance-based time series classification is used for task recognition. The recognition output is used to adjust the position of the camera upon recognizing the surgeon's movements in terms of kinematic time series. Using the (dis) similarity measures between the tasks, the robot is able to distinguish for example, the task of suturing from needle passing or knot tying movements.

15.5 Case Studies

15.5.1 Automated Ear Surgeries

Surgical manipulations of the ear, such as ones for implantable hearing aids, are able to help patients with hearing deficiencies. However, such operations require access to the middle and/or inner ear, which involves drilling a cavity into the mastoid (mastoidectomy), the skull bone behind the ear. Such manual procedures produce openings physically much larger (several centimeters) than optimally required for the insertion of the implantable hearing aid (a few millimeters), and this is done primarily to obtain visual access to underlaying key landmark features, used to help locate the target location, while navigating away from critical anatomical structures that should not be harmed. Even if the incision opening would be much smaller, the tool trajectory still needs to pass through a window of a few millimeters, where all critical biological structures and tissues are located. Thus, a precision smaller than 0.5 mm is needed for planning, navigation, guidance during operation.

There are several existing solutions to this challenge that have been applied in clinical and in vitro conditions, however all share common crucial procedural steps. Below is a brief overview of some of the existing solutions with emphasis on the common as well as unique parts of their methodology.

One solution to this problem [43] is based on the concept of a stereotactic frame, custom build for each individual operation and patient, which aids the guidance of the drilling instrument during the surgical procedure, in order to precisely reach the target location while avoiding crucial anatomical structures. The procedure consists of five major steps. The procedure starts with attachment of fiducial markers, in the form of bone anchored percutaneous screws, to the patient's head, surrounding the mastoid (the surgical access area). The design of the screws is such that they are clearly visible in the X-ray domain, and have a well-defined point of contact, accessible for physical contact with a registration tool. During the second step, the patient's head is preoperatively scanned using a medical X-ray computed tomography (CT) scanner, with a voxel size of approximately 0.3 mm, depending on the scan conditions. In the third step, the volumetric scan data is then analyzed, crucial anatomical features and structures are identified, and the geometry of the anchor screws is registered relative to their CAD model. Based on all these features, a preoperative planning is made, where the drilling trajectory, defined by the target (final) and entry locations, for the surgical procedure is established such that the anatomical structures, including the facial nerve and major blood vessels, are safely avoided. The anatomical feature segmentation and identification as well as the trajectory planning, is performed automatically in a matter of minutes, and the resulting trajectory is verified by a surgeon [44]. In the fourth step, custom stereotactic frame is manufactured such that it provides attachment points for custom legs, which anchor the frame to the screws in the skull. The stereotactic frame also features a channel to guide a medical drill along the preoperatively defined surgical entry trajectory, virtually defined in the previous step. The frame could be manufactured through means of rapid prototyping [45], which, while appealing for its ease of use, proved to be time consuming (48-h lead time) as it doubles or triples the total procedural time. This delay has been overcome, by modifying the design of the stereotactic frame to allow it to be manufactured by more traditional manufacturing techniques (e.g., via a CNC milling machine) on site, which decreases the leadtime down to a few minutes. In the final step, the stereotactic frame is attached to the anchor screws at the patient's head. Once the frame is firmly secured, the drilling can commence. Tests so far indicate drill jitter of approximately 70 μm on average, with target and trajectory accuracies in the order of 0.4 mm, verified in vitro. Total time for steps 2–4, which are at the core of the automated surgical process, is approximately 30 min. During the drilling process, a typical drill size of 1 mm diameter is used, which leaves <1 mm (sometimes <0.5 mm) of clearance from crucial anatomical structures to be avoided, as

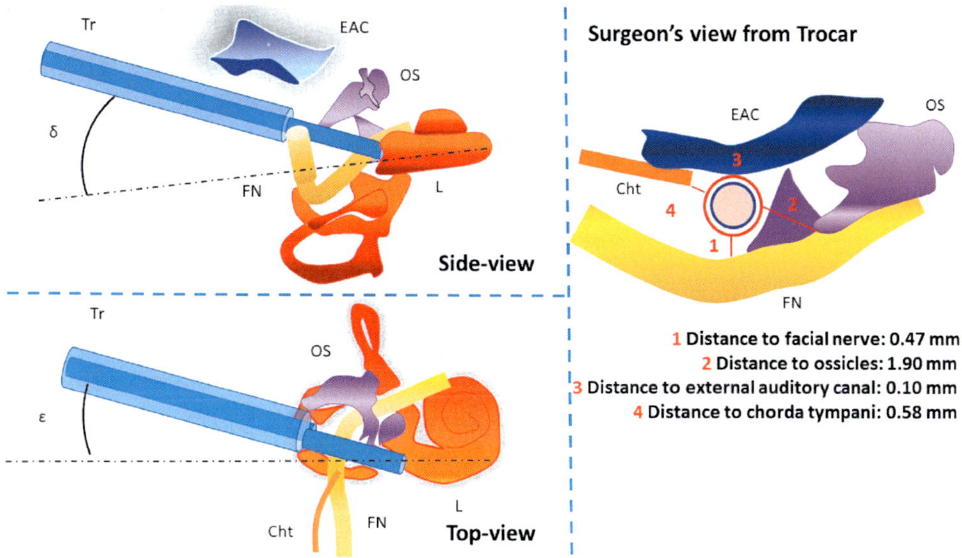

FIG. 15.11 Overview of a virtual trajectory definition with tolerance and minimal distances to crucial anatomical structures to be avoided during the drilling procedure. *Original illustration by P. Boyraz prepared based on N. Gerber, B. Bell, K. Gavaghan, C. Weisstanner, M. Caversaccio, S. Weber, Surgical planning tool for robotically assisted hearing aid implantation, Int. J. Comput. Assist. Radiol. Surg. 9(1) (2014) 11–20.*

shown in Fig. 15.11. The generality of such a procedure allows its application to a great variety of surgical procedures in the head, where the skull provides solid points for anchoring of the stereotactic frame.

While accurate, the above-mentioned approach to registration, between the physical space (patient and surgical tool) and virtual planning (CT scan data and trajectory), relies on semi-manual manipulation (assembly and adjustment of the stereotactic frame), which leaves room for an accidental error. This has been further improved in the approaches of two other independent research groups [46, 47], where the physical stereotactic frame has been substituted with an optical [46] or electromagnetic [40] tracking system, which provides precise (below 100 μm) location of the tool relative to the patient, via bone anchored fiducial screws. In addition, five or six DOF serial robotic manipulator is incorporated into the overall procedure to provide accurate and consistent way of holding and moving the surgical drilling tool intraoperatively. The combination of a robotic manipulator and an automatic guidance system allows for a fully automated execution of the registration with the patient's head and drilling via a predefined trajectory. In addition, one of the systems [47] includes a load cell at the tool holder interface to provide information on the exerted manipulation forces, which allows tele-operated haptic control mode, during which a surgeon guides the robot via a force-feedback controller. While such mode of operation seems to add additional complexity to the overall setup and procedure, it is a crucial feature for the adoption of such technologies, allowing further development and optimization.

15.5.2 Reconfigurable and Hyperredundant Robotic Platforms

In this part, the design of hyperredundant and modular robotic structures is detailed by emphasizing their features such as independent module/segment controllability and variable-adjustable stiffness. The proposed designs [48] are aimed at improving both position and force control of such structures employing whole-body shape control and local stiffness control in the robotic catheter on module basis. Three different module designs for hyperredundant mechanisms are depicted in Fig. 15.12. The first mechanism is called 'hybrid' comprising a universal joint placed in between two parallel plates. This mechanism has three DOF per module, having pan and tilt for adjusting the heading angle while using the translational movement to shrink or elongate while adjusting its stiffness. The helical spring around the middle shaft provides inherent compliance for the module.

The second mechanism is 3 × SPS having spherical-prismatic-spherical joints in each strut can be considered as a reconfigurable parallel mechanism. It provides stiffness adjustments by relocating the connection points of the struts on the lower plate. The struts can also elongate and contract along their axis so that the hyperredundant platform can be adjusted when passing along difficult cavities. Finally, the last module is named after the seahorse tail since it is inspired by the cross section of the biological structure. This mechanism can radially change dimensions in a similar manner of seahorse tail structure, using the oblique muscles. The modules are connected by a spherical joint in the middle. Since the radial struts are spring-loaded, therefore, the radial stiffness can be adjusted.

The main advantage of reconfigurable hyperredundant mechanisms is that each segment can be controlled separately, and the multidegree-of-freedom makes it possible to control the whole-body shape of the manipulator to reduce the risk of harming the tissues during the navigation task. If the modules also have variable stiffness or reconfigurability as it is shown here, the versatility and the safety of the hyperredundant platforms increase. Since robotic platforms should accomplish tasks such as navigation, diagnosis, and operation they may have to support different levels of stiffness.

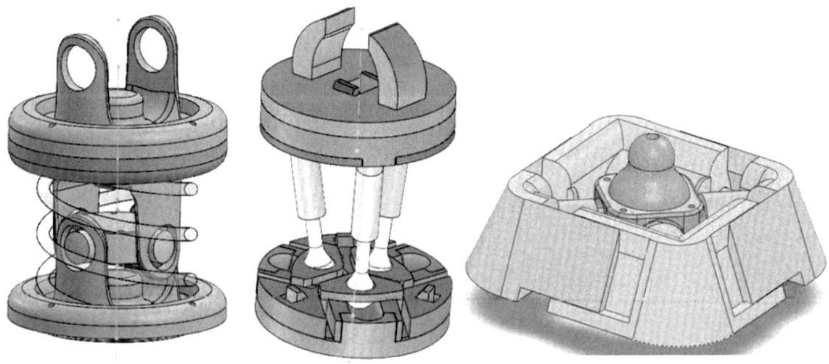

FIG. 15.12 Modules for hyperredundant backbone construction, from left to right: Hybrid-module, radially reconfigurable 3 × SPS parallel-kinematic mechanism, and seahorse tail section [48].

15.6 Conclusion and Future Trends

Robotic surgery has reached a certain maturity level in terms of application-specific solutions using multiple-arm robots and employing master-slave systems. However, several ongoing research directions involving for example soft-robotics, human-robot collaboration, and the transition to higher levels of autonomy, hold a promise to still further advance this field. Another trend for surgical robotics include micro-robots that could be delivered in gastrointestinal or cardiovascular tract to perform certain surgical operations, however, up to date research has only achieved a mesoscale prototype that could both navigate in torturous channels and perform an operation, which may require certain force levels. In summary, the field of surgical robotics is still an active area of research with the aim of obtaining minimal invasiveness while achieving autonomous navigation and operation.

The rapidly developing field of robotic surgery will also cause the rethinking of the role of surgeons in the operation room. While surgical robots could help in achieving more reliable, faster and standardized procedures, they will be still controlled (in fully manual or semiautonomous mode) by the surgeons, who will need to expand their field of expertise and knowledge, to use more efficiently the capabilities of such robotic systems. Surgeon will be still a key figure in the operating theater; however, surgeon's role will expand into augmenting existing surgical procedures or even designing completely new ones utilizing emerging novel systems. As the range of capabilities of surgical robotics expands, so will grow the importance of experienced surgeons to adopt new technologies and further advance the efficiency and success rate of surgical procedures.

References

[1] A.R. Lanfranco, A.E. Castellanos, J.P. Desai, W.C. Meyers, Robotic surgery: a current perspective, Ann. Surg. 239 (1) (2004) 14–21.

[2] J.S. Rassweiler, R. Autorino, J. Klein, A. Mottrie, A.S. Goezen, J.-U. Stolzenburg, et al., Future of robotic surgery in urology, BJU Int. 120 (2017) 822–841.

[3] J. Jayakumaran, S.D. Patel, B.K. Gangrade, D.M. Narasimhulu, S.R. Pandian, C. Silva, Robotic-assisted laparoscopy in reproductive surgery: a contemporary review, J. Robot. Surg. 11 (2017) 97–109.

[4] T. Arulampalam, S. Patterson-Brown, A.J. Morris, M.C. Parker, Natural orifice transluminal endoscopic surgery, Consensus Statement, Ann. R. Coll. Surg. Engl. 91 (2009) 456–459.

[5] M.E. Hagen, M.K. Jung, F. Ris, J. Fakhro, N.C. Buchs, L. Buehler, P. Morel, Early clinical experience with the da Vinci Xi surgical system in general surgery, J. Robot. Surg. 11 (2017) 347–353.

[6] K. Taniguchi, A. Nishikawa, M. Sekimoto, T. Kobayashi, et al., Classification, design and evaluation of endoscope robots, in: S.H. Baik (Ed.), Robot Surgery, 2010, p. 172. ISBN 978-953-7619-77-0.

[7] A. Szold, R. Bergamaschi, I. Broeders, J. Dankelman, et al., European association of endoscopic surgeons (EAES) consensus statement on the use of robotics in general surgery, in: Surgical Endoscopy, Springer, 2014, pp. 1–36. Consensus Statement.

[8] B.P.M. Yeung, P.W.Y. Chiu, Application of robotics in gastrointestinal endoscopy: a review, World Gastroenterol. 22 (5) (2016) 1811–1825.

[9] S.J. Phee, A.P. Kencana, V.A. Huynh, Z.L. Sun, S.C. Low, K. Yang, D. Lomanto, K.Y. Ho, Design of a master and slave transluminal endoscopic robot for natural orifice transluminal endoscopic surgery, Proc. IMechE Part C: J. Mech. Eng. Sci. 224 (2010) 1495–1503.

[10] T. Ogiwara, T. Goto, A. Nagm, K. Hongo, Endoscopic endonasal transsphenoidal surgery using the iArmS operation support robot: initial experience in 42 patients, Neurosurg. Focus. 42 (5) (2017) 1–5. E10.

[11] K. Kume, Flexible robotic endoscopy: current and original devices, Comput. Assist. Surg. 21 (1) (2016) 150–159.

[12] P.R. Slawinski, K.L. Obstein, P. Valdastri, Emerging issues and future developments in capsule endoscopy, Tech. Gastrointest. Endosc. 17 (2015) 40–46.

[13] M.E. Moran, Evolution of robotic arms, J. Robot. Surg. 1 (2007) 103–111.

[14] E. Abdi, M. Bouri, S. Himidan, E. Burdet, H. Bleuler, Third arm manipulation for surgical applications: an experimental study, in: New Trends in Medical and Service Robots, 2016, pp. 153–163.

[15] Z. Li, D. Milutinovic, J. Rosen, Design of a multi-arm surgical robotic system for dexterous manipulation, J. Mech. Robot. 8 (061017) (2016) 10.

[16] H. Zhao, J. Tian, D. Li, C. Ai, Design of a multi-arm robot for mandible reconstruction surgery, Appl. Mech. Mater. 654 (2014) 187–190.

[17] O. Salomon, A. Wolf, Inclined links hyper-redundant elephant-trunk-like robot, J. Mech. Robot. 4 (2012). 6 pp.

[18] Z. Li, H. Ren, P.W.Y. Chiu, R. Du, H. Yu, A novel constrained wire-driven flexible mechanism and its kinematic analysis, Mech. Mach. Theory 95 (2016) 59–75.

[19] M.M. Tonapi, I.S. Godage, A.M. Vijaykumar, I.D. Walker, A novel continuum robotic cable aimed at applications in space, Adv. Robot. 29 (6) (2015) 861–875.

[20] G.S. Chirikjian, J.W. Burdick, A modal approach to hyper-redundant manipulator kinematics, IEEE Trans. Robot. Autom. 10 (3) (1994) 343–354.

[21] G.S. Chrikjian, J.W. Burdick, Kinematically optimal hyper-redundant manipulator configurations, IEEE Trans. Robot. Autom. 11 (6) (1995) 794–806.

[22] G.S. Chrikjian, Hyper-redundant manipulator dynamics: a continuum approximation, Adv. Robot. 9 (3) (1994) 217–243.

[23] N.C. Cho, H. Jung, J. Son, K.G. Kim, A modular control scheme for hyper-redundant robots, Int. J. Adv. Robot. Syst. 12 (2015) 91.

[24] A. Bajo, N. Simaan, Hybrid motion/force control of multi-backbone continuum robots, Int. J. Robot. Res. 35 (4) (2016) 422–434.

[25] J. Wang, S. Wang, J. Li, X. Ren, R.M. Briggs, Development of a novel robotic platform with controllable stiffness manipulation arms for laparoendoscopic single-site surgery (LESS), J. Med. Robot. Comput. Assist. Surg. 14 (2018) 1–16.

[26] J. Burgner-Kahrs, C. Rucker, H. Choset, Continuum robots for medical applications: a survey, IEEE Trans. Robot. 31 (6) (2015).

[27] H. Su, G. Li, D.C. Rucker, R.J. Webster III, G. Fischer, A. Concentric Tube, Continuum robot with piezoelectric actuation for MRI-guided closed-loop targeting, Ann. Biomed. Eng. 44 (10) (2016) 2863–2873.

[28] P. Qi, C. Qui, H. Liu, J.S. Dai, L.D. Seneviratne, K. Althoefer, A novel continuum manipulator design using serially connected double-layer planar springs, IEEE/ASME Trans. Mechatr. 21 (3) (2016) 1281–1292.

[29] H. Abidi, G. Gerboni, M. Brancadoro, J. Fras, A. Diodato, M. Cianchetti, H. Wurdemann, K. Althoefer, Highly dexterous 2-module soft robot for intra-organ navigation in minimally invasive surgery‘, Int. J. Med. Robot. Comput. Assist. Surg. 14 (2018) 1–9.

[30] M.D. Gilbertson, G. McDonald, G. Korinek, J.D. Van de Ven, T.M. Kowalewski, Serially actuated locomotion for soft robots in tube-like environments, IEEE Robot. Autom. Lett. 2 (2) (2017) 1140–1147.

[31] G. Smoljkic, G. Borghesan, A. Devreker, A.V. Poorten, B. Rosa, H. De Praetere, J. De Schutter, D. Reynaerts, J.V. Sloten, Controlof a hybrid robotic system for computer-assisted interventions in dynamic environments, Int. J. CARS 11 (2016) 1371–1383.

[32] A. Singh, E. Singla, S. Soni, A. Singla, Kinematic modeling of a 7-degree of freedom spatial hybrid manipulator for medical surgery, Proc. IMechE Part H: J. Eng. Med. 232 (1) (2018) 12–23.

[33] A.K. Mishra, E. Del Dottore, A. Sadeghi, A. Mondini, B. Mazzolai, SIMBA: tendon-driven modular continuum arm with soft reconfigurable gripper, Front. Robot. AI 4 (4) (2017) 1–10.

[34] D.J. Kiely, W.H. Gotlieb, S. Lau, X. Zeng, V. Samouelian, A.V. Ramanakumar, et al., Virtual reality robotic surgery simulation curriculum to teach robotic suturing: a randomized controlled trial, J. Robot. Surg. 9 (2015) 179–186.

[35] C. Pacchierotti, F. Ongaro, F. Van den Brink, C. Yoon, D. Prattichizzo, et al., Steering and control of miniaturized untethered soft magnetic grippers with haptic assistance, IEEE Trans. Autom. Sci. Eng. Inst. Electr. Electron. Eng. 15 (1) (2018) 290–306.

[36] J. Buzzi, G. Ferrigno, J.M. Jansma, E. De Momi, On the value of estimating human arm stiffness during virtual teleoperation with robotic manipulators, Front. Neurosci. 11 (2017), 528.

[37] E. Bauzano, B. Estebanez, I. Garcia-Morales, V. Munoz, Collaborative human-robot system for HALS suture procedures, IEEE Syst. J. 10 (3) (2016) 957–966.

[38] ISO Standard, IEC/TR 60601-4-1:2017, Medical Electrical Equipment- Part 4-1: Guidance and Interpretation-Medical Electrical Equipment and Medical Electrical Systems Employing a Degree of Autonomy.

[39] R. Muradore, P. Fiorini, G. Akgun, D.E. Barkana, M. Bonfe, F. Boriero, A. Caprara, et al., Development of a cognitive robotic system for simple surgical tasks, Int. J. Adv. Robot. Syst. 12 (37) (2015) 1–20.

[40] F. Ferraguti, N. Preda, A. Manurung, M. Bonfe, O. Lambercy, R. Gassert, et al., En energy tank-based interactive control architecture for autonomous and teleoperated robotic surgery, IEEE Trans. Robot. 31 (5) (2015) 1073–1088.

[41] K. Watanabe, T. Kanno, K. Ito, K. Kawashima, Human-integrated automation of suturing task with one-master two-slave system for laparoscopic surgery, in: IEEE Int Conf. on Advanced Intelligent Mechatronics (AIM), Banf, Alberta, Canada, 2016.

[42] M. Fard, A.K. Pandya, R.B. Chinnam, M.D. Klein, R.D. Ellis, Distance-based time series classification approach for task recognition with application in surgical robot autonomy, Int. J. Med. Robot. Comput. Assist. Surg. 13 (e1766) (2017) 1–9.

[43] R.F. Labadie, R. Balachandran, J. Mitchell, J.H. Noble, O. Majdani, D. Haynes, M. Bennett, B.M. Dawant, J.M. Fitzpatrick, Clinical validation study of percutaneous cochlear access using patient customized micro-stereotactic frames, Otol. Neurotol.: Off. Publ. Am. Otol. Soc. Am. Neurotol. Soc. Eur. Acad. Otol. Neurotol. 31 (1) (2010) 94.

[44] J.H. Noble, F.M. Warren, R.F. Labadie, B.M. Dawant, J.M. Fitzpatrick, Determination of drill paths for percutaneous cochlear access accounting for target positioning error, Proc. SPIE 6509 (2007) 650925.1–650925.10.

[45] R.F. Labadie, J.H. Noble, B.M. Dawant, O. Majdani, R. Balachandran, J.M. Fitzpatrick, Clinical validation of percutaneous cochlear implant surgery: initial report, Laryngoscope 118 (2008) 1031–1039. 18401279.

[46] N. Gerber, B. Bell, K. Gavaghan, C. Weisstanner, M. Caversaccio, S. Weber, Surgical planning tool for robotically assisted hearing aid implantation, Int. J. Comput. Assist. Radiol. Surg. 9 (1) (2014) 11–20.

[47] R. Torres, G. Kazmitcheff, D. De Seta, E. Ferrary, O. Sterkers, Y. Nguyen, Improvement of the insertion axis for cochlear implantation with a robot-based system, Eur. Arch. Otorhinolaryngol. 274 (2) (2017) 715–721.

[48] P.R. Nasdar, P. Boyraz, T. Ortmaier, A. Raatz, Development of compliant hyper-redundant mechanisms for robotic catheters and analysis of controllability, Deutche Gesselschaft fuer Robotik (DGR) Days (2016) 29–30 Leipzig, Germany.

16

Biomechanics and Biomechatronics in Sports, Exercise, and Entertainment

Karen L. Troy*, Kimberly Tetreault[†], Adam D. Goodworth[‡], Songbai Ji*, Marko B. Popovic*

*WORCESTER POLYTECHNIC INSTITUTE, WORCESTER, MA, UNITED STATES [†]GAYLORD HOSPITAL, WALLINGFORD, CT, UNITED STATES [‡]UNIVERSITY OF HARTFORD, HARTFORD, CT, UNITED STATES

CHAPTER OUTLINE

16.1 Biomechanics Fundamentals

Biomechanics is the study of the structure and function of biological systems such as humans, animals, plants, organs, and cells by means of the methods of mechanics [1]. This chapter provides a comprehensive overview of biomechanics experimental procedures, data analysis, modeling, and simulation. This chapter also addresses advanced biomechatronics systems for sports, exercise, and entertainment.

Biomechatronics. https://doi.org/10.1016/B978-0-12-812939-5.00016-1

16.1.1 Development of Modern Age Biomechanics

Exercise motions are some of the earliest to be analyzed from a biomechanics perspective. In the 1860s, French biomechanicist, physiologist, and inventor Étienne-Jules Marey (1830–1904) [2] and English photographer Edward Muybridge (1824–1907) independently developed photographic methods to record whole body and joint motions during running, jumping, and other activities (Fig. 16.1).

In addition to his photographic work on movement of the body, Marey studied blood circulation, analyzed heart beats, respiration, and developed early electromyography (EMG) techniques. He studied medicine in Paris and became an assistant surgeon in 1855. He was a member of the French Academy of Medicine and the Academy of Sciences, of which he was made president in 1895. He held the chair of "Natural History of Organized Bodies" at the College de France from 1868 and published more than 150 scientific papers. In addition, he was the president of the French Photographic Society for many years (Fig. 16.2) [1].

FIG. 16.1 Photographic analysis of a polevault. Etienne Marey, 1890 [3].

FIG. 16.2 The chronophotographic gun invented by Étienne-Jules Marey [1].

To aid his studies, he developed many instruments for precise measurements. Marey's chronophotographic gun (1882) was capable of taking 12 consecutive frames a second recorded on the same picture (Fig. 16.3) [1].

Marey himself acknowledged that his innovations were the practical extension of the inventions of Janssen (1824–1907) and Muybridge (1830–1904) to meet the split-second accuracy that they sought. He was particularly inspired by the work of English photographer Edward Muybridge, who carried out "Photographic Investigation" in Palo Alto, California. In 1879, Muybridge published photographs of a galloping horse that proved that Marey was right when he wrote that during galloping, for a brief moment, horses have all four hooves off the ground (Fig. 16.4) [1].

These photographic techniques formed the foundation of modern motion capture and biomechanics analysis techniques. Although limited to two dimensions, photographic and video analyses allowed joint angles to be calculated by drawing lines between anatomic landmarks and manually measuring the angle. When the time between successive photographic images was known, joint angular velocities could also be calculated. Collectively, these kinematic measurements provided quantitative measures to distinguish one type of exercise from another.

Because locomotion is a common activity of daily living, walking and running are some of the earliest and most consistently studied exercises. Marey published many of his early ideas in a book titled, "Du movement dans les fonctions de la vie," or "Functional movements of life" (1868). In his book, he emphasizes the value of objective scientific

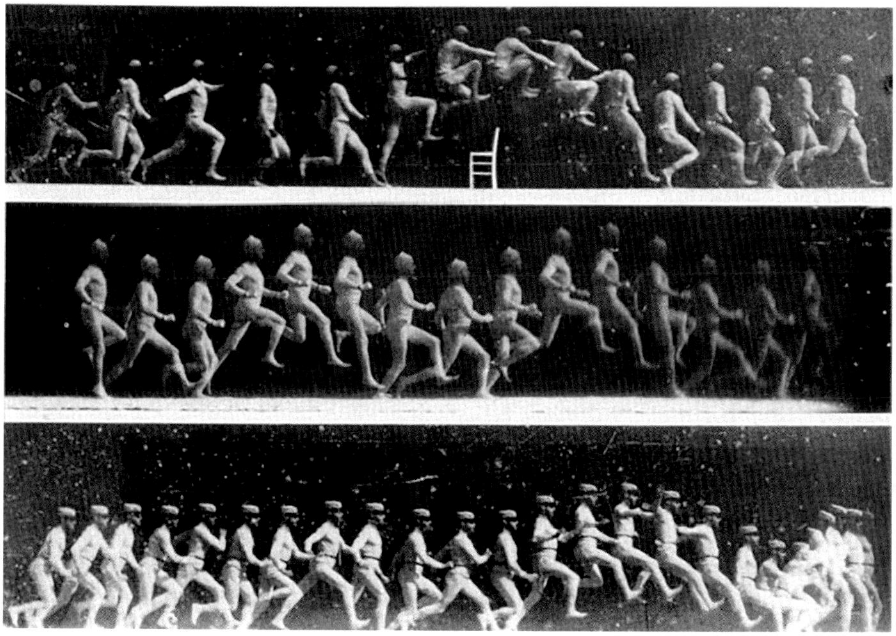

FIG. 16.3 Marey's "The human jumping" [1].

FIG. 16.4 Muybridge's "The horse in motion" and Scientific American, October 19, 1879 [1].

measurement. He also argues for applying physics and engineering concepts such as equilibrium and dynamics, to the analysis of living systems. His later work, "La method graphique dans les sciences experimentals et principalement en physiologie et en medicine" (1878; *Graphic methods for experimental sciences and principles of physiology and medicine*), expanded this to explain and present results from measurement techniques such as EMG. He invented the pneumatic switch (air chamber connected to Recorder), and in 1873, was the first researcher to record the duration of foot contact with the ground in level walking. These books, along with other writing from his extensive body of work, formed the basis for modern whole-body biomechanics analyses (Fig. 16.5).

While Marey and Muybridge pioneered photographic measurement techniques, Wallace O. Fenn (1893–1971), professor and chairman of physiology at the University of Rochester, introduced the first spring-based device for measuring ground-reaction forces in 1930. Just a few years later, the first modern force plates were developed in 1938 by American biomechanist and anatomist Herbert Oliver Elftman (1902–88) in collaboration with John T. Manter. Elftman was a professor of anatomy at the Columbia University, who researched foot and lower-extremity biomechanics (Fig. 16.6) [1].

Working with Elftman, J. T. Manter designed and built a spring-based platform in 1938, with a lever system to magnify small deflections as the platform was loaded by an animal's weight. Elftman and Manter visualized and measured the displacement of these levers in the same high-speed film images used to record the positions of the animal's limbs. Manter's elegant force-plate design stands out from others because it was the first to record forces in three axes. Elftman's study was on human walking and Manter's study was on cats [1].

16.1.2 Biomechanics Experimental Tools and Data Processing Techniques

Currently, there are many tools available to measure whole body kinematics and kinetics. Optical motion capture systems combined with force platforms remain the gold standard and are most common. Optical systems can include either passively reflecting, or actively

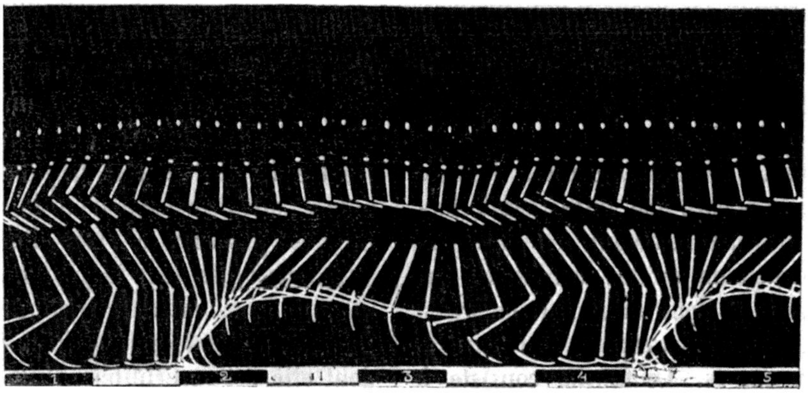

FIG. 16.5 Photographic sequence of running, taken by Etienne Marey, showing body segment angles [4].

FIG. 16.6 Wallace O. Fenn and Herbert O. Elftman [1].

emitting markers. In the first case, cameras that emit and detect light in the infrared spectrum are used to shine light onto reflective markers, and track the reflected points. After a volume in space has been calibrated, a point can be tracked in three-dimensional (3D) space, so long as at least two cameras are able to see it at any given time. Active marker systems use a similar principle, but rely on active marker emission of the light, rather than reflected light. Because each individual point being tracked has three degrees of freedom (position in the x-, y-, and z-directions), a minimum of three markers is required per body segment to completely describe the segment position and orientation in space. A primary advantage of these systems is the ability to measure the motion of multiple points with submillimeter resolution at high sampling rates. However, these systems are limited in that all measurements must be acquired in a laboratory setting and line-of-sight is required for optical tracking. To address these limitations, many alternative systems have been developed.

For example, electromagnetic tracking systems do not require line-of-sight and can be used in settings such as operating rooms for surgical navigation and other applications. These systems include an electromagnetic source, which generates electromagnetic dipole fields in three perpendicular directions and acts as a transmitter. A series of sensors detect not only their position in space relative to the source, but also their orientation with respect to the dipole field. Thus, each individual sensor has six degrees of freedom (position and orientation about x-, y-, and z-axes) and only a single sensor is required to completely describe a body segment position and orientation in space. The primary disadvantage of this type of system is potential interference and distortion of the electromagnetic field by metallic objects. The use of worn sensors to obtain out-of-laboratory measurements is also a rapidly growing trend. Recently, many investigators have used a combination of inertial measurement units, goniometers, force-sensing shoe insoles, and other tools for "mobile gait analysis."

Regardless of the details by which the data are obtained, this type of biomechanical study remains fundamentally an inverse problem in which kinematics (positions, velocities, and accelerations of body segments) are measured, but the muscle forces that generated them are unknown. Thus, *inverse dynamics* techniques are used to estimate the net joint reaction forces and moments, based on the kinematics, combined with a biomechanical model.

In contrast to a robot, where segment degrees of freedom, masses, and moments of inertia are known with some certainty, human segment properties must be estimated based on a combination of measurements and published average values. Key properties such as segment mass, center of mass location, and moment of inertia, have been calculated based on cadaveric measurements, for example, [5] and more recently, medical imaging, for example, [6]. Typically, these parameters are expressed as ratios, relative to segment length or mass, to allow for body-size scaling. This way, if the dimensions of an individual person are known, the body segment center of mass and inertial properties can be estimated. However, while proportions of body segments generally scale with body mass and height, these data are limited in that they do not account for individual variations in body type. Indeed, specific subpopulations such as children [7], obese individuals [8], or those with spinal cord injury [9] have been shown to have slightly different body proportions than the relationships most frequently used for biomechanical models.

In addition to assumptions that must be made about body segment mechanical properties, biomechanical models include assumptions about the degrees of freedom at a particular joint. These assumptions must be carefully examined, because they may limit the application of the model. For example, many human models treat the knee as a hinge joint that allows flexion/extension only. This has been shown to be reasonable in many instances such as normal gait, when the motion of the knee is not the primary target of investigation. However, this assumption would not be appropriate to answer research questions about how knee motion differs between athletes with and without a history of knee injury.

The end result of an inverse dynamics calculation is information regarding the net external forces and moments acting at each joint, over the time period in question. These net joint reaction forces represent the summed total of all forces crossing the joint that must have been present to produce the measured motion. Similarly, the net joint moment includes the summed total of all moments about that particular joint, which must have been present to produce the measured motion.

It is important to understand that the net joint reaction force and moment represent the sum total of the forces and torques at the joint, but do not provide any detail about the source of these forces and torques. Forces are assumed to arise from one joint surface pushing against another. However, the 3D shape of joint surfaces and the addition of ligaments and other passive tissues add uncertainty to calculations. Torques are assumed to arise from the action of muscles that cross a joint, or, in some types of robots, from a linear actuator that acts at a distance from the joint center of rotation. The torque or moment that is produced by the muscle or actuator is a result of the cross product of the force and muscle (or actuator) moment arm. Human joints have many redundant muscles that cross each joint, each of which can contract to produce a joint torque. The net joint moment that is determined from inverse dynamics is the sum total of all of these muscle contractions, plus any passive contributions to joint moment made by the joint surfaces and soft tissue constraints (usually assumed to be zero). Because of the number of degrees of freedom in living organisms, it is not possible to determine a single, unique solution of muscle contributions to a given net joint moment. To estimate individual muscle forces, several approaches have been developed. These include using EMG signals to verify when certain muscles are active, assuming the motor control system is attempting to optimize something like metabolic efficiency, or using muscle synergies to simplify and understand the motor control system. Each of these approaches has challenges and requires careful experimental validation. Improving and verifying the accuracy of these calculations remains an active area of research.

Muscles often work at a mechanical disadvantage, in that the muscle moment arm is usually quite small (on the order of centimeters) compared to the limb length. In humans, most daily activities involve external forces applied to the end of a limb. For example, during walking, the ground reaction force (GRF) is applied to the feet. During manipulation tasks, objects exert a force on the hands. If one imagines that a joint is like a pulley with a radius r, the moment that is provided by a muscle will be equal to its force, times the radius r. In contrast, the moment about that joint due to an external force such as a GRF or weight held by hand is equal to external force times the appropriate moment arm, which can be length of the limb, $R \gg r$ as shown in Fig. 16.7. To balance each other, the muscle force $= mg\frac{R}{r} \gg mg$. Here, it is apparent that a longer limb length and a smaller joint radius will both increase the amount of muscle force required to counteract external force. As this is a consequence of physics or mechanics of levers, the same principle holds true for linear force actuators used in robotic systems.

In addition to generating a moment about the joint, muscle forces cause compression between the two joint surfaces, resulting in internal joint contact forces that are many

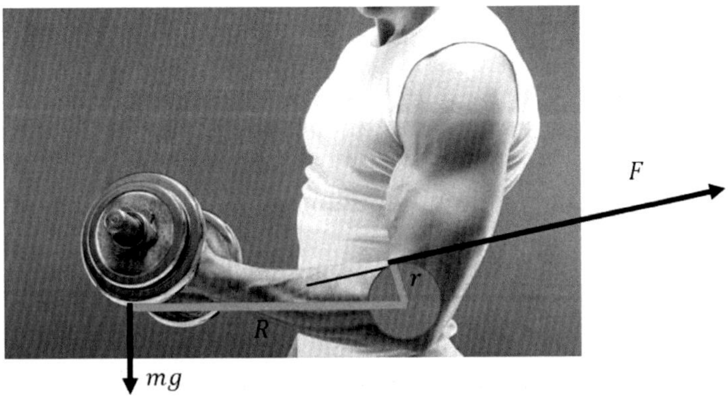

FIG. 16.7 Muscle force often needs to be much larger than external force in order to balance external moments.

times higher than the net joint reaction force. For example, during running, ankle joint contact force has been estimated to be about nine bodyweights, although the measured GRF is only 2–2.5 bodyweights [10].

16.1.3 Running, Jumping, and Landing Biomechanics

Sports activities such as running, jumping, and landing, are highly dynamic, involving large forces generated over short periods of time. Sports performance is often dictated by one's ability to generate power.

When considering sports and exercise biomechanics, it is often helpful to classify specific types of activities that occur in each specific sport. For example, basketball includes running, cutting (quick changes of direction), jumping, etc. Jumping can be further divided into one- or two-footed jumps. These could be further classified as those occurring during running, versus those occurring from a stationary position. Indeed, other sports also include cutting maneuvers that do not differ substantially from the cutting performed during basketball. This type of movement classification has allowed common mechanisms of injury between sports to be identified. For example, anterior cruciate ligament (ACL) [11] injuries frequently occur in basketball, soccer, and volleyball players. Each of these sports includes a high frequency of cutting and landing maneuvers, which have been implicated in ACL injury.

16.1.3.1 Running

Aside from walking, running is the most common form of locomotion. The primary distinction between walking and running is that running involves a *flight phase*. That is, during the running gait cycle, there is a period of time during which neither foot is touching the ground, and the GRF is zero. Because running is a fundamentally dynamic exercise, body segment inertial contributions to joint moment can be quite large, especially compared to walking gait. As a result, many aspects of running biomechanics depend on running speed. For an individual, running speed can be increased through either increased stride length, increased cadence, or (typically) a combination of the two. As a result, faster running speeds require greater torques at the hip and ankle, but not the knee (Fig. 16.8) [12].

Recently, there has been significant interest in running footstrike kinematics and footwear. Most adults running with shoes will adapt a rearfoot strike pattern, which results in a GRF pattern that includes a vertical impact peak, followed by a slower loading peak. This impact peak, which occurs over a 20–30 ms time period, causes rapid bone strain. Greater vertical impact peak along with higher average and vertical instantaneous load rates have been linked to tibial stress fractures, plantar fasciitis, and patellofemoral pain [13]. It has been hypothesized that reducing instantaneous loading rates may reduce the risk of these conditions [13]. Other evidence contradicts this notion and suggests that impact rates associated with heel-toe running have little influence on bone mechanical fatigue [14]. The relationship between running kinematics, kinetics, and the development of injury continues to be an active area of research.

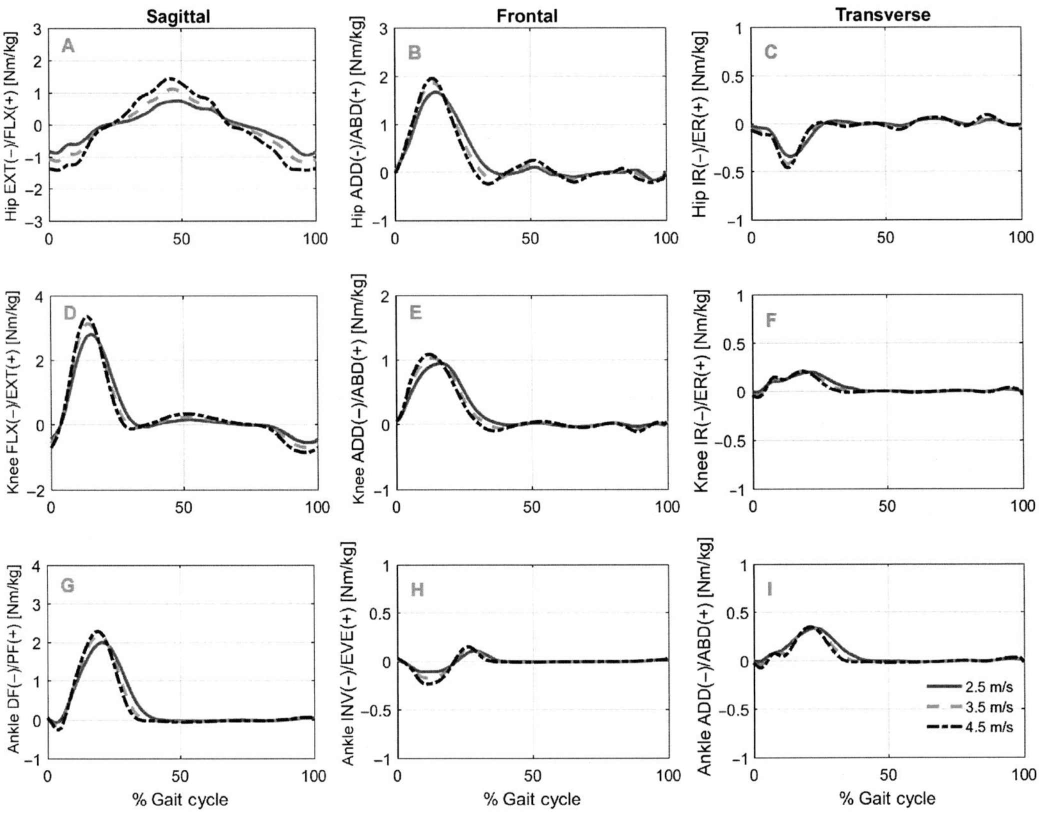

FIG. 16.8 Body mass normalized hip, knee, and ankle moments during running at three different speeds [12].

16.1.3.2 Jumping and Landing

Jumping and landing are common sports activities, the performance of which is primarily determined by power. Jump height is a common measure of jump performance, and is determined by the ability to accelerate the body vertically.

One can write a simple equation of motion for a standing vertical jump

$$\sum \vec{F} = \vec{F}_{GR} + m\vec{g} = m\vec{a} \tag{16.1}$$

Here, the forces on the left-hand side of the equation include the vertical GRF and the body weight. Since the propulsive GRF is generated over a finite period time, it is often useful to think about jump height in terms of momentum and impulse.

The momentum is expressed as

$$\vec{p} = m\vec{v} \tag{16.2}$$

and the time derivative of momentum is

$$\frac{d\vec{p}}{dt}=m\frac{d\vec{v}}{dt}=m\vec{a} \tag{16.3}$$

Hence impulse defined as change of momentum can be also expressed as

$$\vec{J}=\int\sum\vec{F}\,dt \tag{16.4}$$

Thus, the equation can be rewritten as

$$\vec{J}=m\triangle\vec{v} \tag{16.5}$$

with $\triangle\vec{v}$ being the total change in velocity of the person jumping, during the time period of interest.

Using this set of relationships, exit velocity, the velocity at which the body leaves the ground surface, can be calculated if the impulse and body mass are known. Similarly, jump height can be calculated based on projectile motion equations. Here, jump height can be obtained as

$$h=\frac{v_{exit}^2}{2g} \tag{16.6}$$

assuming no losses due to air friction.

Fig. 16.9 shows a typical GRF versus time curve for a countermovement jump. The total propulsive impulse is measured from the instant that the GRF deviates from 1 bodyweight. Initially, the GRF is less than 1 bodyweight, because the body center of mass is accelerating downwards as the person bends his knees. The total impulse includes both the negative impulse from this downward acceleration, as well as the positive impulse, which includes the braking and propulsive impulse.

While jump height is governed by the impulse of GRF that is generated during the jump, there are many biomechanical contributors to this impulse. The most important contributors are ankle plantarflexion moment and shoulder moment (from arm swing). Collectively, while jumping involves positive work production by muscles, landing is primarily negative work. Positive work is characterized by concentric (shortening) muscle contractions, which generates mechanical energy. In contrast, negative work is characterized by eccentric (lengthening) muscle contractions, which absorb and dissipate mechanical energy [15].

16.2 Modeling and Simulation: Simplified, Intermediate, and Detailed Models

16.2.1 Simplified Models

The human or animal body can be modeled with different levels of detail. Sometimes even the simplified, set in advance models can be very useful. These models may help the

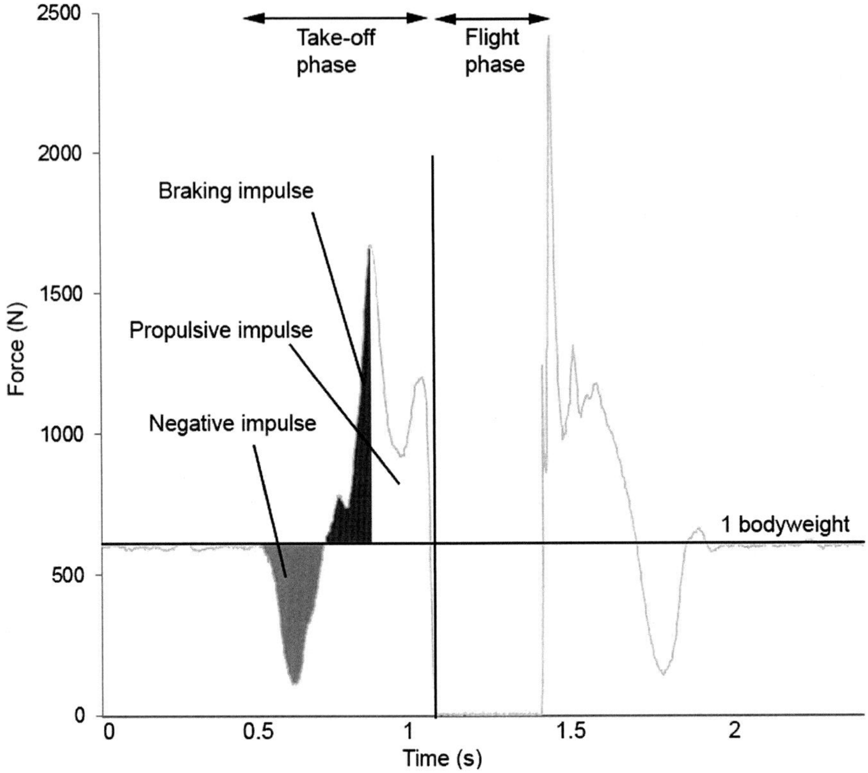

FIG. 16.9 The negative and positive impulses needed to initiate and brake the countermovement.

researchers to discover fundamental features of dynamics within experimental data for a specific class of movements. Alternatively, by using lower-dimensional reduction statistic methods [like principal components analysis (PCA)] on an entire data set, one could discover lower-dimensional subspaces explaining most of variance in data. This lower-dimensional fundamental dynamics can then be used as a diagnostics tool or performance metric in studies of various movement impairments including ambulation with prostheses or braces, or it can be used for motion planning and generation of joint-level trajectories (Fig. 16.10).

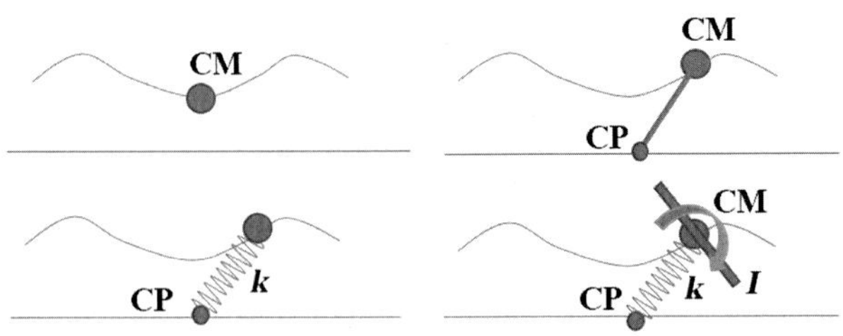

FIG. 16.10 Examples of simplified template models.

16.2.1.1 Body Translations

For example, one could analyze center of mass (CM) trajectory. To obtain the CM, one needs experimental motion caption (MoCap) data, intermediated skeletal model with well-defined joint degrees of freedom, and good intermediate model of mass distribution of human or animal body. Based on the placement of MoCap system's markers computer software can deduce joint angles. And, if body segments' CMs are known, one could then compute overall body's CM (Fig. 16.11).

Intermediate skeletal models typically have many degrees of freedom within a quasirealistic representation of the human body skeletal system. Their main purpose is in precise localization of various body parts as defined by the position of reflective markers placed at numerous bonny landmarks. However, they do not provide a realistic skeletal model (bones are represented by cylinders and joints are highly simplified) (Fig. 16.12).

The MIT mass inertial model [16–18] by Popovic and Herr is an example of an intermediate human mass distribution model; segments were modeled by truncated cones with uniform density in order to match segments' CM locations and segments' moment of inertia.

Clearly there are many potential sources of error; the "average" skeletal model and "average" segment mass distribution may not be an exact representative of the actual experimental subject and MoCap data can be a bit inaccurate. Hence, it is recommended that CM location is verified by several independent methods. For example, for static poses, one may verify if CM is positioned just vertically above the center of pressure (CP). Or for aerial phase if air resistance is ignored, one may verify if CM motion is expressible with projectile motion with zero moment about the CM. From these additional tests, one could estimate the CM error.

If CM trajectory is accurate enough, one could take the first derivative in respect to time and obtain CM velocity. By multiplying CM velocity with body mass, one could obtain body linear momentum. Still further one could take the second derivative and obtain CM acceleration. And by multiplying with body mass, one could obtain net force acting on the body. If net force is sum of weight and GRF, one could then deduce GRF. Alternatively, one could obtain overall body velocity and acceleration as sum of segments velocities and acceleration weighted by segments' relative masses. Finally, one could obtain GRF by direct measurement using force plates, which is probably the most reliable way.

16.2.1.2 Body Rotations

Similarly, one could compute angular momentum about the CM as a sum of individual segments' angular momenta about the CM. And, if angular momentum about CM is accurate enough, one could take the first derivative in respect to time and obtain moment about the CM. Alternatively, one could obtain moment about the CM as sum of rate of change of individual segments' angular momenta about the CM.

If only weight and GRF act on body, then moment about the CM is equal to moment generated by GRF. Hence, given the CM location, GRF, and moment about the CM, one could compute the point on a ground from which net GRF should act. This is

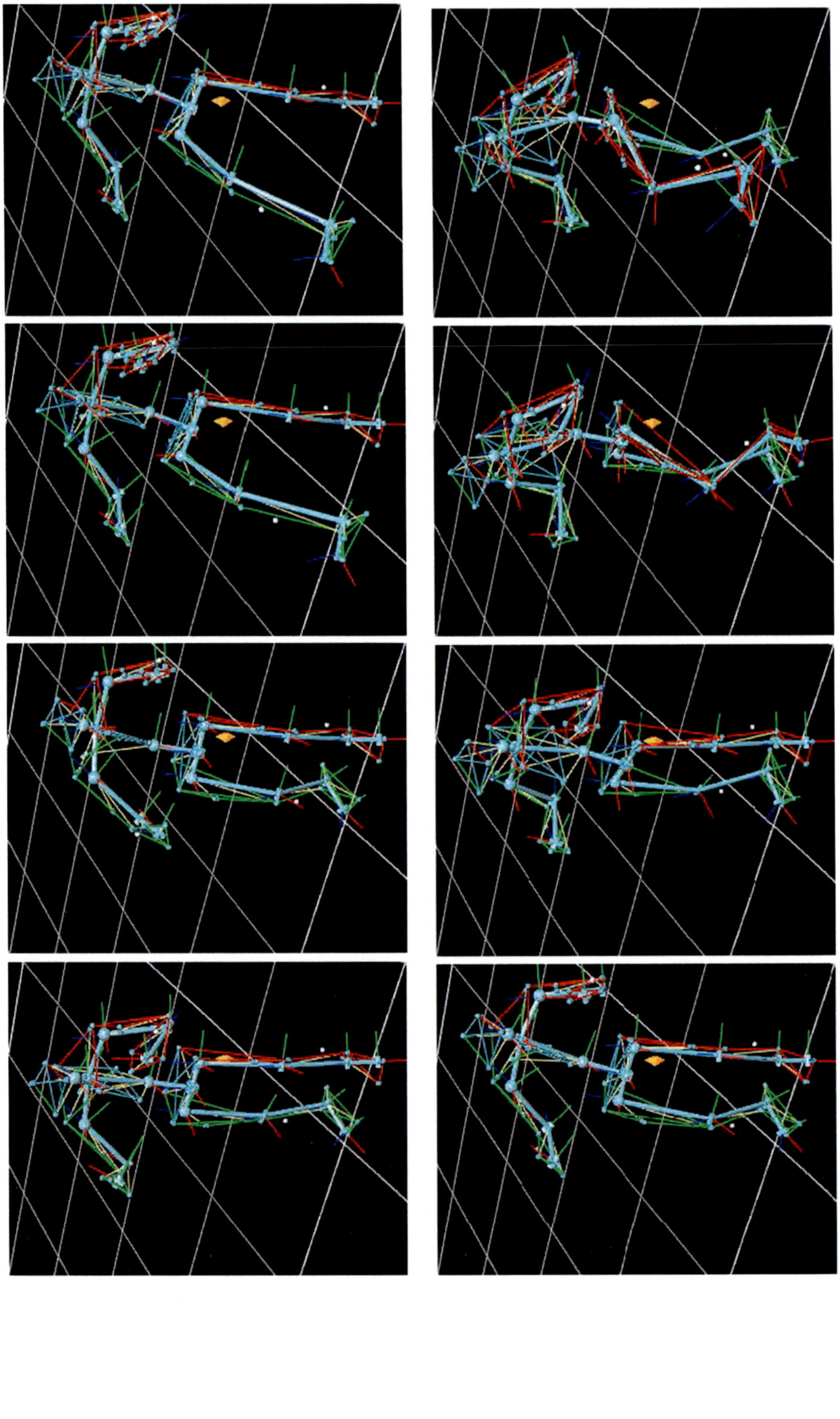

FIG. 16.11 The VICON intermediate skeletal model.

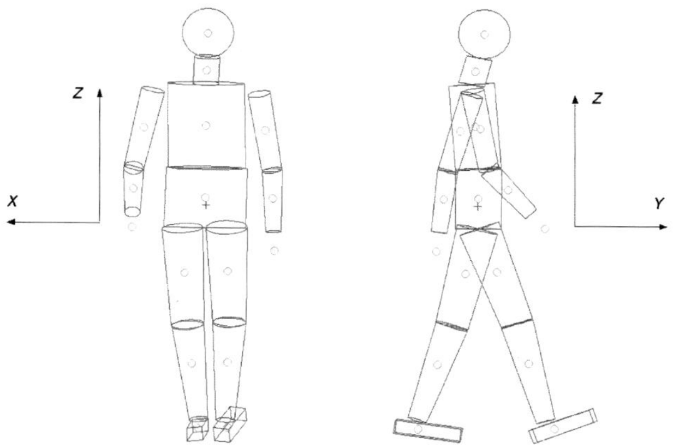

FIG. 16.12 The MIT mass inertial model [16–18] with segments represented with truncated cones.

zero-moment point (ZMP) which is identical to CP for flat surfaces [17–20]. Clearly, CP (or ZMP) must be located within the support polygon defined by body ground contact.

The modern force plates, see Chapter 4, provide net force and torque vectors as well as CP (or ZMP) location. The accurate CM model and accurate MoCap and force plate data can be then utilized to verify the moment model used to compute rotational dynamics of human or animal body.

16.2.2 Simplified Dynamic Template Models

Given these high level, integral quantities like CM, CP=ZMP, and GRF, one could address simplified template dynamic models.

For some classes of movements, for example, the simplified template of an inverted mathematical pendulum may be utilized to illustrate dynamics of upright body posture for humans (Fig. 16.13).

The CM can be visualized as a heavy mass on top of the pendulum while the bottom is located in the ZMP or CP location. The upright body posture may be thought of in the context of an inherently unstable equilibrium of the inverted mathematical pendulum.

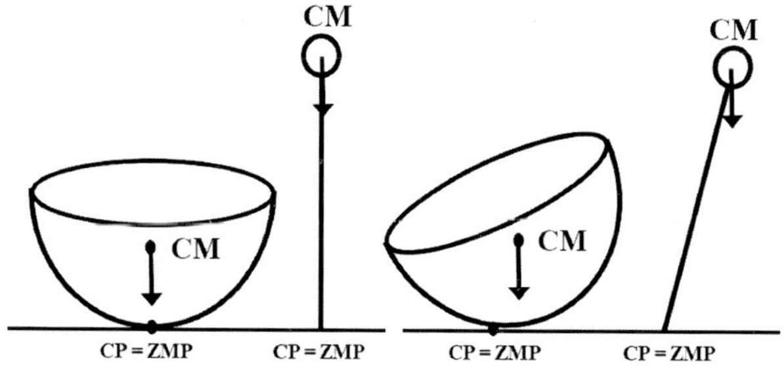

FIG. 16.13 Example of stable equilibrium (half sphere) and unstable equilibrium (inverted pendulum).

The equilibrium is unstable because if there is no active action performed by the human represented with the inverted pendulum, the human may fall due to external perturbation. This is very different situation from, for example, a half sphere whose mass distribution provides conditions for stable equilibrium.

Hence, the stability or ability to balance and maintain desired body posture is mainly the consequence of active actions performed by the human and not just the consequence of passive dynamics.

As introduced by Popovic and Herr [18], two main active strategies are used for balancing: (1) ZMP=CP repositioning and (2) generation of CM torque (moment).

Inverted pendulum strategy belongs to zero-moment balance strategy or ZMP=CP repositioning strategy. Similar to balancing an inverted pendulum on the tip of your finger, humans can balance by repositioning ZMP. For example, if CM and CP are vertically aligned but CM velocity is to the right, the CP should be repositioned to the right in order to create a restoring horizontal force. For larger perturbations or for purpose of performing activity (e.g., walking), the CP=ZMP repositioning may involve the change of support base in the form of stepping.

16.2.2.1 Zero-Moment Walking Class of Templates

To which extent can walking be expressed as zero-moment dynamics?

To answer this question, we need to address both rotational and translational dynamics (Fig. 16.14).

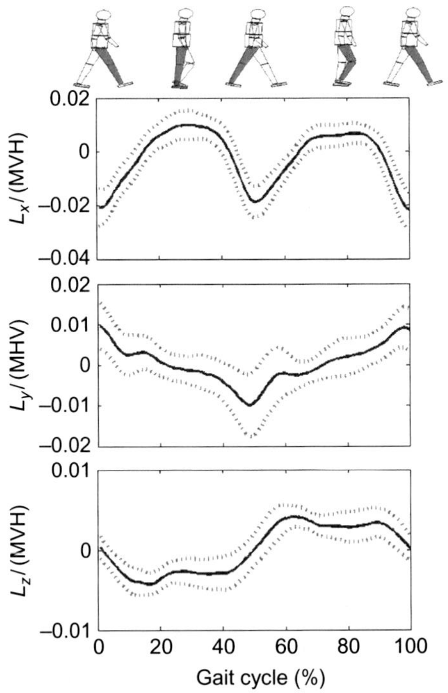

FIG. 16.14 The normalized (by CM average height, mass, and gait speed) angular momentum in walking [16, 18].

Angular momentum is a conserved physical quantity for isolated systems where no external moments act about a body's center of mass (CM). However, in the case of legged locomotion, where the body interacts with the environment (GRFs), there is no requirement for this relationship to hold.

Popovic and Herr [16, 18] found that angular momentum is highly regulated throughout the gait cycle about all three spatial directions ($|\vec{L}(t)| \approx 0$). Thus, horizontal GRFs and the center of pressure (CP) trajectory can be explained predominantly through an analysis that assumes zero net moment about the body's CM (Fig. 16.15).

Using a 16-segment human model and kinetic and kinematic gait data for 10 study participants, they found that calculated zero-moment forces closely match experimental values ($R_x^2 = 0.94$; $R_y^2 = 0.91$), and the zero-moment CP never leaves the ground support base, highlighting how closely the body regulates angular momentum in steady-state walking.

Hence, more than 90% of walking dynamics can be expressed by a zero-moment template.

Specific template of walking dynamics has been studied by Kajita et al. in the form of linear inverted pendulum model (LIPM) [21]. In difference to the inverted pendulum model, the LIPM model assumes constant CM height such that distance between ZMP = CP and CM varies throughout the gait cycle. Again, inverted pendulum and LIPM model both belong to the zero-moment class of templates.

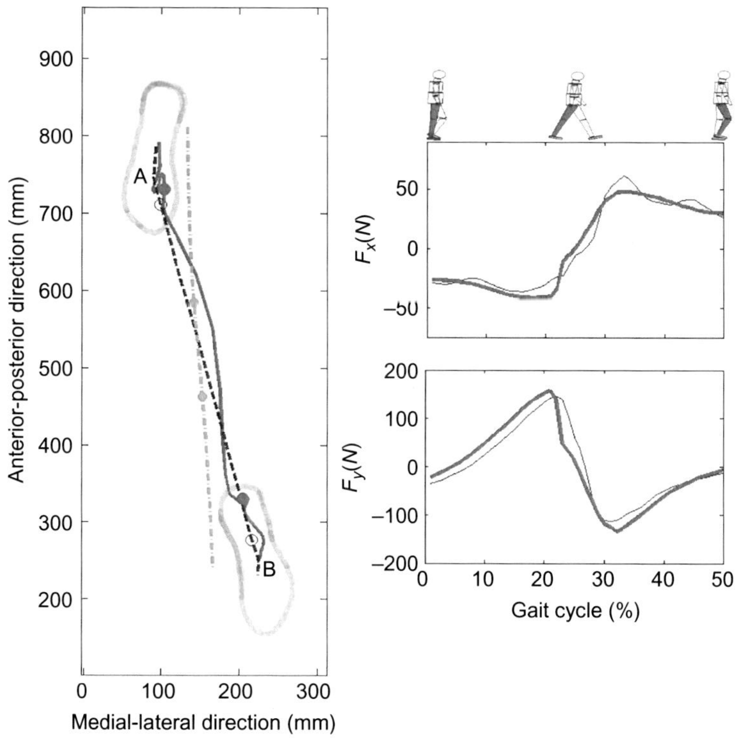

FIG. 16.15 CM trajectory ground projection (dashed-dotted line), ZMP=CP trajectory (dashed line), and theoretical zero-moment predicted ZMP=CP trajectory (solid line) in walking (left). Theoretical zero-moment forces (tick lines) and experimentally measured forces (thin lines) in walking for medial-lateral (top right) and anterior-posterior (bottom right) direction [16, 18].

Variant of LIPM model has been utilized to address control strategy through capture point concept introduced by Jerry Pratt et al. [22]. Based on velocity of CM when CM is just above the CP, the next foot placement is decided such that by assuming completely passive LIPM dynamics, the CM velocity is zero when CM is again above the CP.

However, CM height varies substantially through gait and hence neither LIPM nor capture point concepts are biologically inspired.

In fact as already discussed for human walking, the CM forces and CM torques closely resemble dynamics of a simplified inverted pendulum system. The CM trajectory is roughly sinusoidal along the vertical direction. The CM gravitational potential and CM kinetic energy are almost 180° out of phase. For a person with $m = 100\,\text{kg}$, an energy variation of 40 J in the CM gravitational potential energy during gait cycle at a self-selected speed corresponds to about 4-cm variation in the CM height. The CM potential energy, the kinetic energy of the CM, and the sum of segments' kinetic energy in the CM frame (i.e., comoving nonrotating frame whose origin coincides with CM) are shown in Fig. 16.16 [1].

If energy conservation holds and if one assumes that there is no energy loss to heat, impacts, friction, etc., shouldn't the sum of these three curves be a constant?

Apparently not. Energy can also be stored and released in the form of elastic energy of bones, muscles/tendons, and other tissues.

By summing CM kinetic energy and body parts' kinetic energy in the CM frame, we can obtain the whole body kinetic energy. In Fig. 16.17, the variation in potential energy is

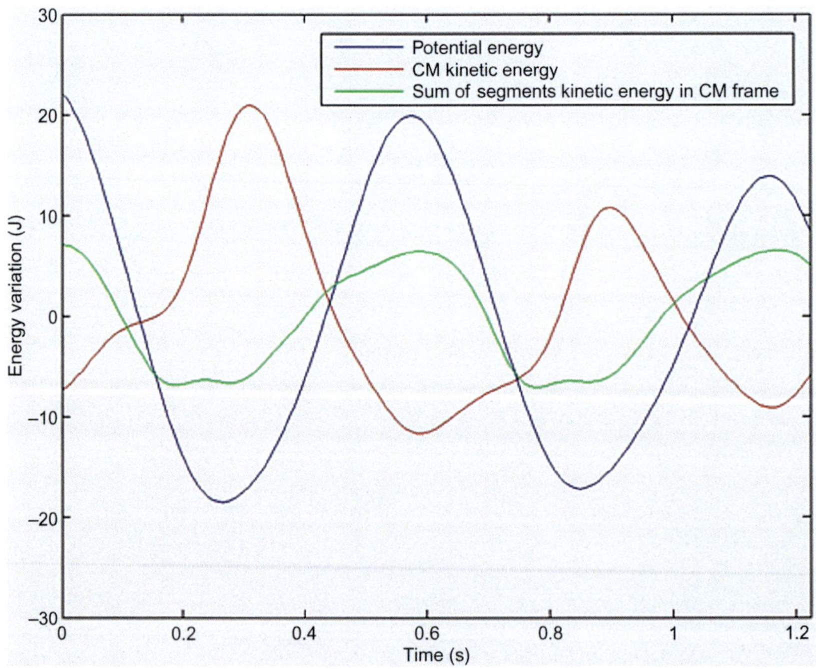

FIG. 16.16 Variations in energy: CM potential, CM kinetic, and segments kinetic energy in CM frame [1].

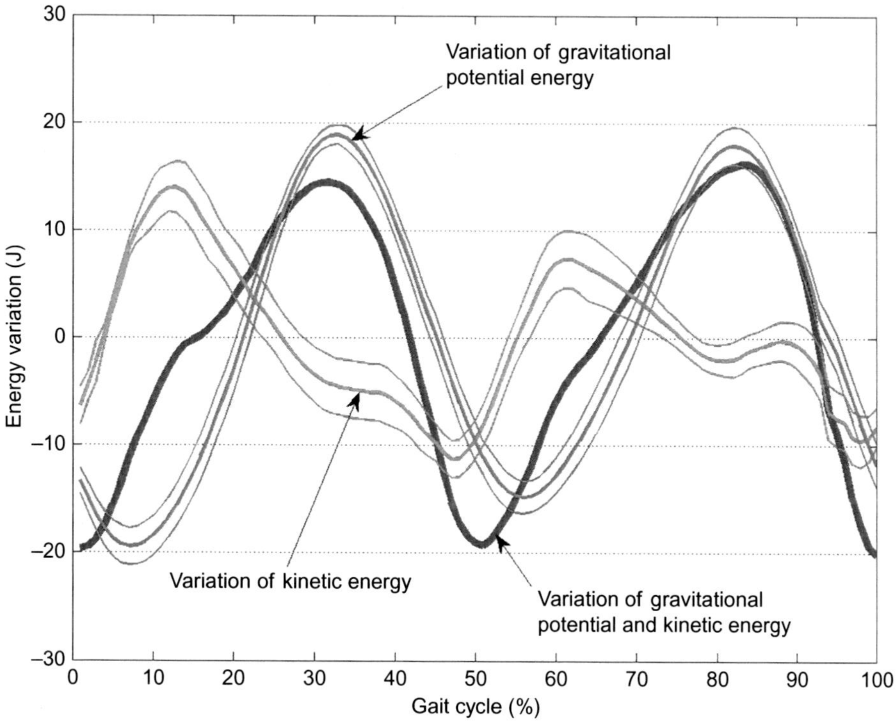

FIG. 16.17 Variation of kinetic energy (sum of segments' kinetic energies) and CM potential energy variation in walking [1].

plotted together with variation in kinetic energy. Not surprisingly, the extreme points of one curve and zeros of another curve are now grouped more closely together than in the previous case. The sum of kinetic energy and potential energy variations is close to sinusoidal. Part of that energy is continuously lost and actively regenerated by the muscles' work and part of the energy is exchanged with the body's elastic elements [1].

Inspired by elastic regeneration of body energy, numerous researchers, for example, Seyfart et al. [23] and Geyer et al. [24], has been studying the inverted pendulum model added with a spring.

16.2.2.2 Zero-Moment Running Template: Spring Mass Template

The mechanism of elastic recoil in running (in difference to pure inverted pendulum walking template discussed above) was first proposed in 1964 when Giovanni. A. Cavagna [25] and his collaborators noticed that the forward kinetic energy of the body's center of mass is in phase with fluctuations in gravitational potential energy. They hypothesized that humans and animals most likely store elastic strain energy in muscle, tendon, ligament, and perhaps even bone to reduce fluctuations in total mechanical energy (Fig. 16.18).

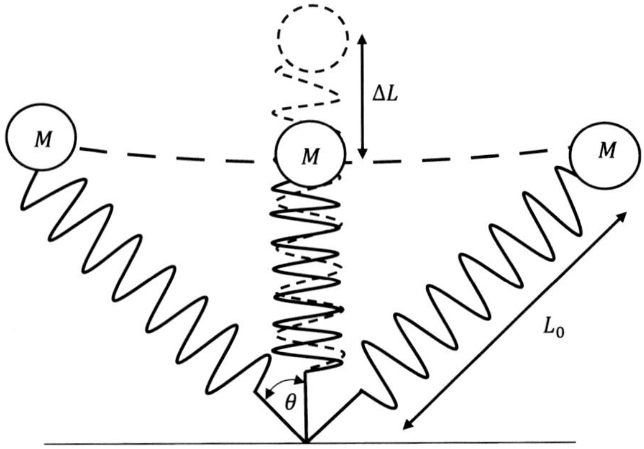

FIG. 16.18 The spring mass model.

Motivated by these findings, Blickhan in 1989 [26] and McMahon and Cheng in 1990 [27] proposed a simple model to describe the stance period of symmetric running gaits—a point mass attached to a massless, linear spring.

In the context of legged locomotion, there is an approximate invariant quantity [28] that relates k_{leg} and M for running. k_{leg} is experimentally obtained as the maximal spring force divided by maximal contraction (Fig. 16.19).

When experimentally obtained, k_{leg} and M are compared for equivalent running speeds (normalized by leg's length) among various species in the mass range of 0.1–140 kg and a very small variance is obtained for

$$\frac{k_{\text{leg}}}{M^{\frac{2}{3}}} = const. \tag{16.7}$$

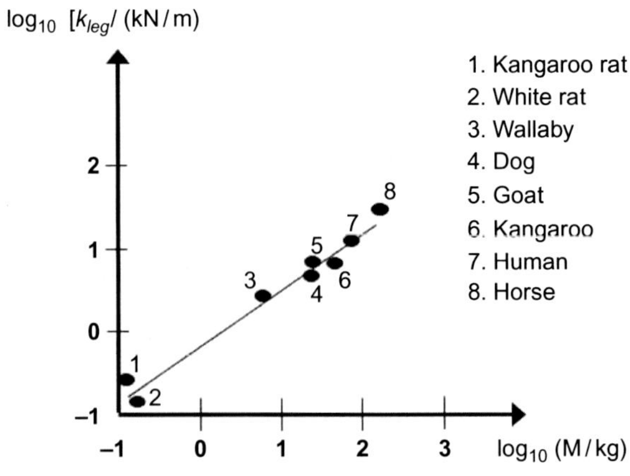

FIG. 16.19 The spring mass model across species: allometric invariance [1].

Here, if ZMP = CP is allowed to slide with constant speed across foot ground contact even more, then 95% of running dynamics can be expressed by a spring mass template.

16.2.2.3 Simplified Zero-Moment Models and Walk Run Transition

Based on an inverted pendulum model, the theoretical limit [1] for human walking speed is $v=\sqrt{gL}\cong 3.2\,\text{m/s}$, where L is approximately a leg length or more precisely a center of mass height. This is a speed for which the vertical GRF is zero corresponding to aerial phase (Fig. 16.20).

The transition typically happens at smaller speeds [29] described with a Froude's number, F, that is, $v=\sqrt{gLF}\cong 2.2\,\text{m/s}$, where $F\cong 0.5$.

The explanation why walk run transitional speed is less than that allowed by physics comes from energy considerations [1]. Apparently, it costs more metabolic energy to walk than to run for speeds larger than 2.2 m/s. However, analysis of musculoskeletal energetics requires a more detailed model of human body.

16.2.3 Detailed Musculoskeletal Models

In 1999, Pandy and Anderson [30–32] minimized energy needed to move from point A to point B for simulated musculoskeletal model with a large number of human leg skeletal muscles and obtained the resulting joint trajectories in good agreement with the biologically realistic human walking trajectories. This shows that energy considerations are clearly an important metric for understanding human locomotion strategies.

Detailed musculoskeletal model is also critical for understanding muscle functions in human test subjects. In fact, the highly detailed model is necessary to roughly quantify individual muscle performance.

In difference to simplified models in addition to MoCap and force plates data, one requires additional sensory input in the form of EMG signal and sometimes also oxygen

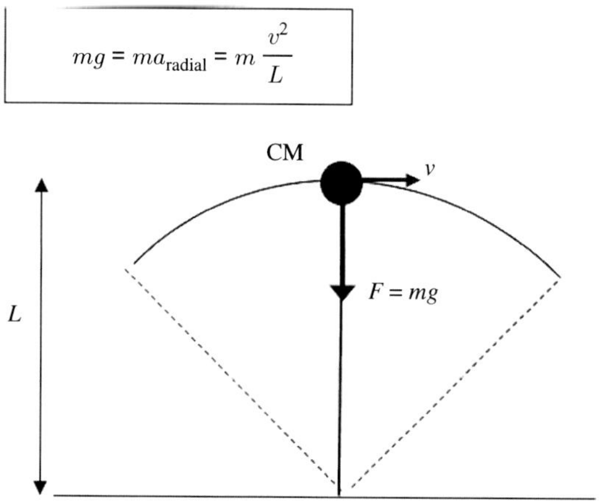

FIG. 16.20 The theoretical limit for human walking speed (ground force is zero) [1].

consumption, both addressed in Chapter 4. The reason for that is high redundancy of muscle actuation; namely combination of different muscle actuations may result in same joint dynamics. Hence, there is no unique solution for muscle active forces that can be obtained based just on joint states and joint torques. Still further this problem is even more difficult as "average" musculoskeletal model may not be exactly representative for a specific experimental subject. There is no experimental technique for direct measurement of muscle force in situ. The individual muscle force can be obtained indirectly, for example, by functional electrical stimulation in which case it is still difficult to explore all muscle recruitment possibilities and it may be hard to properly incorporate various effects of friction and viscoelastic properties of skin, tissues, ligaments, and tendons.

The EMG signal is also typically very noisy, as addressed in Chapter 4, and hence it is a highly nontrivial task to accurately correlate it with muscle force.

For all mentioned reasons deduced individual muscle performance from a detailed model is unlikely very accurate but it is still the best available tool to roughly estimate muscle function.

16.2.3.1 Examples of Detailed Human Body Models

The OpenSim platform contains [33] detailed full-body models that are used to create computer simulations of musculoskeletal structures and to simulate movements such as walking, cycling, running, and stair climbing.

The OpenSim "Gait2392" model of an average, adult male is shown in Fig. 16.21. It contains 92 muscle-tendon actuators representing 76 muscles in the lower extremities and torso. The upper extremity is simplified here to reduce computational time, but other models are available that contain additional degrees of freedom and slightly different sets of constraints.

The muscle-tendon actuators include the force-generating properties so that OpenSim can calculate the length, moment arm, force, and joint moment for any muscle in any body position.

The model was jointly developed by biomechanics research labs at several universities. All of the model parameters are from published sources and a bibliography is provided. In addition, the model is fully customizable.

Another example is the Motek Medical Human Body Model software [34] which calculates muscle forces and joint torques based on MoCap and force plate data in real time. Visualization includes changes in muscle form and color. In July 2014, the companies Motek Medical and Force Link merged. The new company, Motekforce Link, focuses on high-quality rehabilitation technologies and real-time feedback, using virtual reality techniques.

For a given muscle-tendon architecture, the joints' state defines the muscle-tendon lengths/speeds hence defining passive forces. However, there is an active force as well and there are more muscles than kinematic degrees of freedom. For example, there are 43 model muscles and only 7–8 joint DoF per leg.

The muscles' active forcing is said to be optimized to best match the joints dynamics. These results are said to be in good agreement with EMG predicted forces for healthy/average study participants (Figs. 16.22 and 16.23).

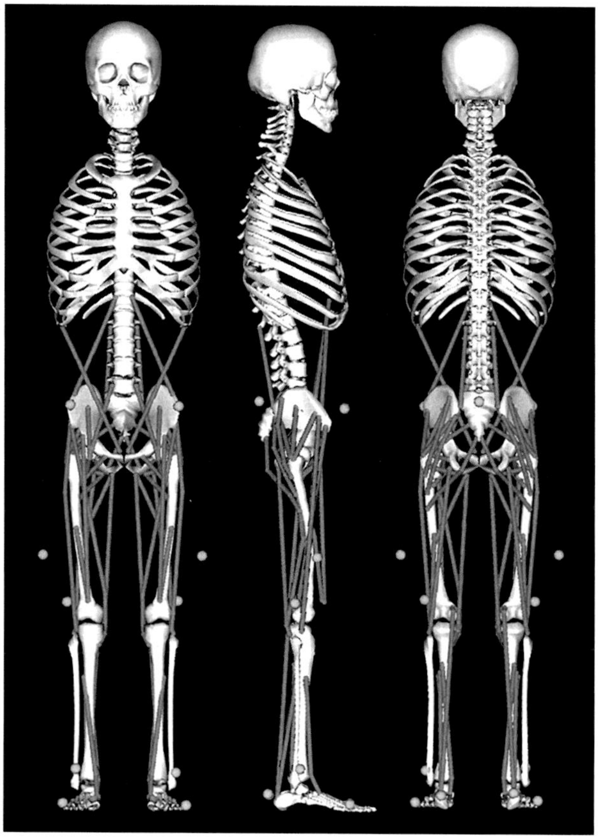

FIG. 16.21 The OpenSim Gait2392 model [33].

FIG. 16.22 Motek Medical Human Body Model visualization [34].

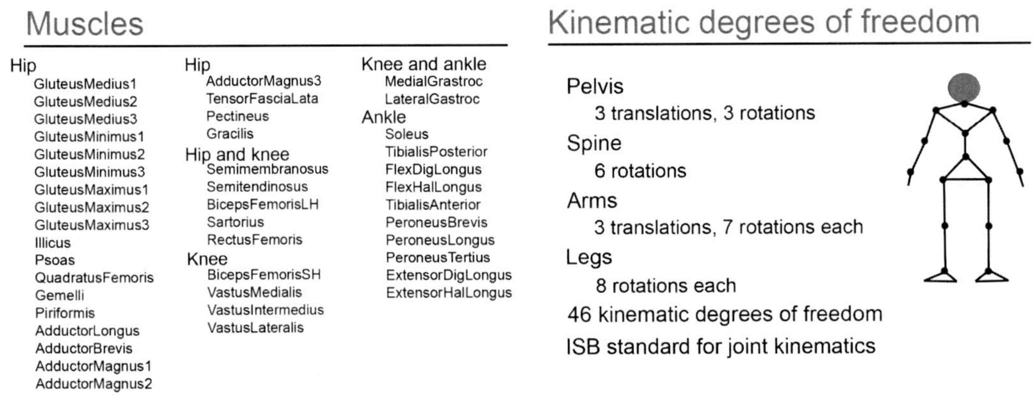

FIG. 16.23 Motek Medical Human Body Model leg muscles and joint DoF [34].

Another popular, very detailed model is AnyBody musculoskeletal model [35]. Redundancy of muscles actuations is technically dealt with assumption that muscles tend to share the work weighted proportional to their size. However, this sharing assumption is not fully supported by biomechanics. For example, the fatigued muscle is often less used then the rested muscle (Fig. 16.24).

16.2.4 Review of Exercise (and Games) Systems for Physical Therapy and Rehabilitation

Exercise plays a critical role in physical therapy and rehabilitation. To control movements and posture, humans (1) activate muscles in a coordinated fashion that (2) move body segments to their end goal within a specific environment while (3) integrating sensory

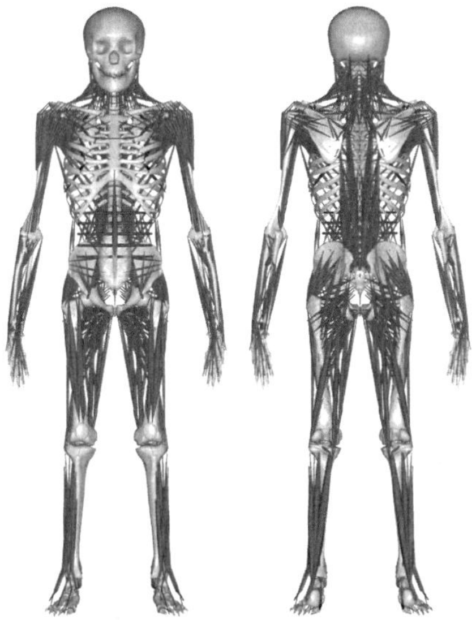

FIG. 16.24 AnyBody musculoskeletal model [35].

feedback to monitor the accuracy. Exercise can have a direct impact on each of these three aspects of motor control.

Muscle contractions are the primary way body segments are moved and positions maintained. Clearly, exercise can increase strength which is needed to successfully move. Exercise can also reduce adipose tissue which has the effects of reducing segment mass and moment of inertia. These reductions facilitate movement by requiring a smaller force to accelerate the body segment. Surrounding tissues only allow so much movement in each body segment. Examples of these tissues include skin, bone, muscles, tendon, ligaments, and joint capsules. Following an injury, the formation of scar tissue, muscle tightness, or contractures, can lead to limited available motion within body segments. Stretching serves as a key method to help to maintain and increase range of motion throughout the body each day or status post injury.

However, strength and range of motion by themselves are insufficient to account for the varied and necessary movements we do every day. Coordination is required. Coordination refers to the patterning of various body segments during a task. Typical mobility tasks include the movement of several muscles, which span across multiple joints. Coordination and motor control is highly complex and varied because the number of muscles activated with each movement exceeds the number of joints. Several theories exist to explain how coordinated movement is learned (and relearned after injury) through exercise and repetitive activities. Below is a summary of key concepts underlying role of exercise in learning coordinated movements.

Coordination is complex and requires exercise and repeated movements. The complexity of movement becomes evident when one considers the infinite ways that muscles could be activated to move body segments toward a goal. For example, the simple task such as touching one's finger to his/her nose becomes more complicated when considering the combination of motions that must occur across numerous joints: shoulder, elbow, wrist, and fingers. To add to the dimensions of this task, take into account the forces acting upon the body. When one segment accelerates relative to another, interaction torques are generated on adjacent segments. In addition, these movements are typically made in a gravitation field that exerts external forces. Consider the different impact of gravity when you touch your nose in a seated vs laying down position. One important mechanism used for controlling these variant factors is sensory feedback. Although we are rarely consciously aware of it, our brains are regularly receiving and integrating sensory feedback from receptors embedded in muscles, skin, joints, and tendon, along with receptors associated with vision and vestibular systems. Together, the feedback from these receptors is used to estimate body segment positions and velocities. When factoring in each of these key concepts contributing to coordination, simple tasks, such as touching one's finger to his/her nose, can transform into a more highly complex task in order to allow for timed, accurate, and smooth movements.

All of the above systems are honed and refined through practice. Thus, the role of exercise and repetition in learning coordinated movements is foundational. In physical therapy and rehabilitation, it is common to practice exercises to relearn a movement after injury. For example, following ACL surgery, a patient must regain strength, range of motion, and relearn aspects of their lower extremity coordination. Following spinal cord injury, it is common for a patient to relearn gait patterns and balance. A lower limb amputee will require similar learning of coordination in gait and balance because the biomechanics and sensory feedback inhere to a prosthetic device is quite different from an uninjured leg. A patient suffering from a stroke must relearn use of his/her neglected side through repetitive movements like reaching and grasping. In some cases, physical therapy is used to help people acquire basic movement patterns for the first time. For example, children with moderate-to-severe cerebral palsy require intentional exercise to help with head and trunk control to assist with their posture and balance in sitting.

Underlying the ability of one to learn coordinated movements is the notion of neuroplasticity. That is, with practice, neural connections and neural pathways can change and become more optimal to complete certain tasks. As one becomes more proficient in a particular movement pattern, he/she is better able to interpret and anticipate sensory feedback. An expert is also better able to predict and even take advantage of factors such as gravity and interaction torques to lower energy expenditure.

One of the main goals of physical therapy is to practice specific movements that will carry over into functional improvements. To maximize learning and retention, variety in movements are needed. In addition, it has been shown the exercises that are engaging, motivating, and provide the patient with some autonomy improve the speed of motor learning. Consequently, games (video games and virtual reality) have been used in physical therapy. Games can increase motivation by change the users perceived difficulty and demonstrate improvements (e.g., scores). Games can also enhance autonomy (e.g., choosing a character or background). In the context of physical therapy, gaming has been used to increase strength, coordination, and reduce pain.

In rehab, the most commonly used system to improve coordination, balance, and strength is the Nintendo Wii, Wii Fit, and Wii Sports, as well as the Xbox Kinect system. Newer devices, such as the FitMi (Flint Rehabilitation Devices, Irvine, CA), include a combination of rehabilitation games and devices. Typical interventions consist of three sessions per week for an hour each session, lasting 4–12 weeks. Many different patient populations have been studied, including those with general balance problems, ataxia, stroke, Parkinson's disease, elderly with fall history, brain injury, and amputees. Overall, results with this type of gaming have been very promising even though most studies were on a fairly small population [36] (Fig. 16.25).

Other systems have also been used to enhance coordination and balance and gait. For example, computer games and a dance video game were used to improve physical function in older adults [38]. Custom virtual reality systems have been used to help with upper extremity function where the success of the game requires coordinated arm motions and also gait [39]. While most gaming and virtual reality systems target inexpensive at-home use, the high-end completely immersive environments also exist in custom research laboratories and in military settings. One common immersive system is the computer-assisted rehabilitation environment (CAREN) [40]. The CAREN system includes a treadmill mounted on a 6 degree of freedom Stewart Platform, visual scene that surrounds the user, and optical tracking and force plate. The optical tracking and force plate measures

FIG. 16.25 FitMi motion interface and rehabilitative exercise system [37]. *Photograph provided by Flint Rehabilitation.*

body motion and acquires real-time feedback of segments that influence the visual display and tilting motion of the platform. Participants can walk, lean, and interact with an immersive environment. While the price tag on this device is relatively high, it represents an exciting virtual reality environment for rehabilitation (Fig. 16.26).

Games and virtual reality have also been used as a power tool for distracting a patient from pain during treatments in patients with cancer and burn wounds. The Nintendo Wii is often used, as well as custom systems. For example, one system, called "SnowWorld," offers individuals the unique opportunity to immerse themselves into a 3D virtual environment. This world allows patients to temporarily escape from unpleasant medical procedures by augmenting one's sensory environment through computer-generated auditory and visual stimuli [41]. The icy world was strategically created to distract individuals from the heated, traumatic memories attached to their injuries. Functional magnetic resonance imaging (MRI) studies have found that these systems serve as an analgesic mechanism [41]. Because of the analgesic effects, these systems have been used on patients with multiple blunt force trauma or burn injuries [42, 43]. In these studies, virtual reality was effective at reducing a patient's pain score during range of motion exercises when compared to no distraction intervention.

16.2.5 General Exercise and Game Systems

For anyone who has been around a teenager playing a video game for several hours, it is clear that gaming can be very engaging. Consequently, there have been many attempts to

FIG. 16.26 The Navel Health Research Centers CAREN system is expanded to include a high-performance treadmill, driving and laser riffle simulators, 3D projection, and realistic sounds and scents. Photograph provided by NHRC Public Affairs. The appearance of US Department of Defense (DoD) visual information does not imply or constitute DoD endorsement [40].

improve engagement in an activity through the use of video game technology and virtual reality. Thus, while most people imagine someone sitting down playing a video game with a handheld device, recent advances have altered the control of games by requiring specific movements across a range of body segments. With this specificity, along with the ability to customize difficulty, gaming combined with physical activity has increased in popularity.

It is common to describe gaming and virtual reality together, but it's more accurate to view them on a spectrum in terms of the degree to which the user is immersed. A virtual-real spectrum, ranging from "immersive virtual reality" to "physical reality," is used to identify to what extent a person is present in their virtual environment [44]. On this scale are three of the main categories of virtual reality therapy technologies. First, "immersive virtual reality" is considered the most powerful of the environments as it minimizes interactions with physical reality to the greatest degree. It aims to eliminate all physical stimuli through the use of specific equipment, such as headphones or head-mounted displays. Users are provided with additional equipment, like gloves with motion and position sensors, that allows them to directly interact with and manipulate the virtual environment. A less immersive category is called "augmented virtuality." In these systems, aspects from the physical world are built into the virtual world. A well-known example of this involves a video camera, which captures the image of the player's body and displays it within the computer-generated environment. The user is then able to perform specific body movements to manipulate virtual objects on the screen. "Mixed reality" is the least immersive of these three categories. It requires the individual to use physical controls, such as conventional joysticks or touch devices that affect the virtual environment.

Regardless of the degree of immersion, engineers and computer scientists leveraged technology to increase physical activity in various populations and settings, including typical physical exercises, maintenance of function in daily activities in individuals with long-term mobility impairments, and promote exercise in microgravity conditions. In all the cases, designers aim to seamlessly integrate the user with the virtual world.

16.2.5.1 Sensors and Feedback, Gamifying Physical Activity

To integrate gaming into exercise, designers must consider several questions. What movements are being encouraged? Who is the intended audience? How will movements be measured? How will the game be displayed and controlled?

Intended movement. One of the first exercise movements facilitated through gaming utilized the augmented stationary bike. Early versions of gaming included a very basic visual display (often dots on a screen) of speed and distance covered with a simulation of hills troughed varied resistance. Now, visual displays in stationary bikes may have vivid displays simulating different environments, including real-time competitions displayed on the screen. With modern technology, there are games that focus on nearly every movement used in exercises: dance and stepping games to facilitate lower extremity activity, climbing walls for upper extremity function, and resistance and power training to improve strength and cardiovascular health [45, 46]. It is important to note that the design of games (and success in the games) must be carefully matched to the intended movement (Fig. 16.27).

Audience/user. Another key design topic for games is to match the style, difficulty, and extent of motion to population of interest. Because games offer a powerful motivation, there is a chance that gaming can increase risk of injury. Some ways this could happen is if success in the game requires motions, force generation, or cardiovascular expenditures that are outside the safe range for a person. For example, a game designed for a 20-year-old athlete may be very frustrating or even dangerous for a geriatric population, if not adequately adjusted (Fig. 16.28).

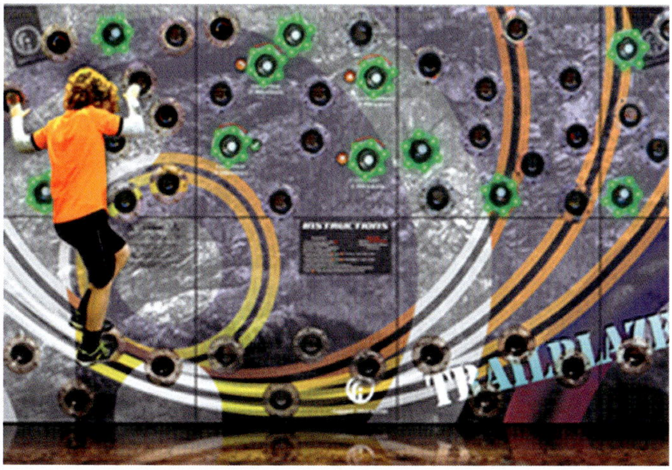

FIG. 16.27 Children climb around the wall scoring points by pressing the grips as they light up. Scores are built up as the players become fitter and stronger. TrailBlazer—provided by Exergame Fitness [45].

FIG. 16.28 With the CardioWall Pro-X, users are challenged with 14 different light pods positioned on the edge of the athlete's peripheral vision. Enhances sports training and rehab programs. CardioWall Pro-X—provided by Exergame Fitness [45].

Measuring movements/sensors. In all systems, the game is linked to the person's movements in some fashion. Thus, there is a critical need for implementation of appropriate sensors for gaming in exercise applications. Some systems, such as the Wii, measure center of pressure under someone's feet using a mat or measure motion of the arms through accelerometers and gyroscopes embedded into the controller. In stationary bikes, sensors are built into the tires to measure speed. In many applications, force sensors are used to measure contact with the body onto a surface (such as a punching game in the CardioWall Pro-X [45]). Other technologies take advantage of optical tracking of movements such as the Xbox Kinect and Sony EyeToy that uses a camera similar to a webcam to track motions using image processing techniques.

Display and control of games. Most often visual displays are paired with auditory stimuli. Both audio and visual modalities provide engaging and typically inexpensive ways to display a game. The game control refers to the linkage between measured body movements and display of the game. The control is the key component that targets intended movement.

Taken together, for a game to be successful, it must measure body motion and use this to update the audiovisual display in such a way that success in the game requires the intended movement that is safe, effective, and engaging for the designated participant. Given these requirements, how much success do games have in increasing energy expenditure? One recent study reviewed 27 research papers to address this question [47]. The 27 studies used various games and systems, including popular ones such as the Nintendo Wii, Xbox Kinect, Sony EyeToy, and Dance-Dance Revolution. Typical exergaming sessions were 8–25 min long. The main finding was that a strong correlation existed between exergaming and energy expenditure, as much as 300% above resting levels. Taken together, properly designed games have the potential to improve enjoyment and energy expenditure in exercise applications.

16.2.6 Future Directions: Virtual Reality and Video Games

We anticipate an increased use of virtual reality and video games in rehabilitation and exercise. With technology being common place in the United States and virtual reality integrated into mobile devices, it is reasonable to expect this technology to transfer into other aspects of life, such as rehabilitation and exercise. Moreover, there is growing evidence of effectiveness in reducing symptoms of pain and increasing engagement and energy output in rehabilitation exercises. However, the integration of games and virtual reality is a new field and requires even more research to optimize its use.

Major challenges include access, outcomes, and program specificity. At present, accessibility to off the shelf systems such as Kinect and Wii is very good. However, these systems typically lack the ability for rigorous outcome measures to track progress and lack specificity for subject-specific targeted movements. The majority of populations requiring specific movements and outcome measures do not have widespread access to gaming technology or virtual reality. Thus, a future direction will be for increased specificity to be implemented into existing systems and/or systems with specificity to become more affordable and widespread, which will result in the necessary training of clinicians on how to use these systems.

Finally, it is important to note that virtual reality and gaming will not pervade every aspect of physical therapy and exercise. Certain people are not interested in technology as an adjunct to therapy and may in fact become less motivated by these systems. Moreover, not all activities justify the use of virtual reality and gaming. Thus, clinicians will need to consider trade-offs across patient desires, cost of gaming systems, and reimbursable costs.

16.3 Several Examples of Biomechatronics Systems for Exercise, Rehabilitation, Games, and Sports

Example 1

There exist a number of commercial virtual reality physical simulators for different sports including golf, tennis, baseball, basketball, soccer, American football, rugby, cricket, hokey, etc. Common for all these systems is motion tracked trajectory of physical ball till its impact with a canvas also representing screen onto which virtual sport arena or field is projected. Based on this initial trajectory of the physical object, the virtual object trajectory is calculated and subsequently visualized on the screen. Limitation of these systems is that after initial throw or kick, there is no feedback from the system that would require human immediate attention or physical reaction beyond repetition of throw or kick (Fig. 16.29).

FIG. 16.29 Example of indoor multisport simulator [48].

Example 2

Company Athletec, a start-up in Manchester, UK, lunched at the 2016 Consumer Electronics Show in Las Vegas, introduced smart boxing tracker technology [49, 50] that originated from biomechatronics research at the Imperial College London in the lab of Dr. Ravi Vaidyanathan, a former boxer. Pair of wireless accelerometers is held by boxer's handwraps and advanced smart phone-based application generates insights into the performance of athlete. Dr. Ravi Vaidyanathan and another student are currently trying to expand the technology to martial arts (Fig. 16.30).

FIG. 16.30 Smart boxing tracker technology by Athletec. *Reprinted with permission by the Imperial College London. Nadia Barbu, Technology Developed in Biomechatronics LAB COULD CHANGE COMBAT SPORTS, 2017. https://www.imperial.ac.uk/news/177831/technology-developed-biomechatronics-lab-could-change/ (accessed 03.04.18); Corner The Smart Boxing Tracker. https://thecornerapp.com/#!/ (accessed 03.04.18).*

FIG. 16.31 SkyTechSport ski training simulator. *Image provided by Skytechsport. SkyTechSport Alpine-Simulator. http://www.skytechsport.com/alpine-simulator (accessed 03.04.18); Tim Newcomb, Experience G-force and Icy Slopes in a SkyTechSport Ski Training Simulator, Sports Illustrated, 2015. https://www.si.com/edge/2015/02/11/skytechsport-ski-training-simulator-us-ski-team-fis-alpine (accessed 03.04.18).*

Example 3

Developed in Munich, Germany, the Hollywood-based SkyTechSport ski training simulator [51, 52] provides immersive lifelike virtual reality experience for both novice and professional skiers. A machine that fully reproduces the sensation of going down the mountain. Dynamics simulator monitors user's movements to calculate resulting motion of the skies and subsequently provides big-screen visualization of chosen path within realistic representation of popular ski courses (e.g., Colorado's Beaver Creek). Four 2.5 horse power motors provide lateral forces that drive skis attached to parallel guides to reproduce sensation of inertial forces and recreate effective gravity due to virtual slope orientation. If user demands, the ski training simulator can replicate icy, hard-packed or spring snow by incorporating "powerful drives" with intense vibrations and compression effects (Fig. 16.31).

Example 4

The immersive lifelike virtual reality high-performance sail simulator [53, 54] has been developed in the lab of Dr. Ir. Jouke Verlinden at the TU Delft Sports Engineering Institute. Goal of the project is to allow competitive single crew sailing athletes (including Paralympic competitors) to engage in a competitive race in realistic conditions. The system includes the set of fans that could generate sensation of wind, reenacting both magnitude and direction, and realistic sail boat environment actuated by powerful Gough-Stewart like platform, that is, parallel robot with six prismatic actuators providing six actuated degrees of freedom.

This highly complex physical environment clearly poses very difficult task for dynamic simulator which needs to account for human movements and calculate in real-time interaction of wind with sail as well as interaction of boat with water streams and waves. And subsequently provide positional and force feedback supported by high-resolution large-screen visualization (Fig. 16.32).

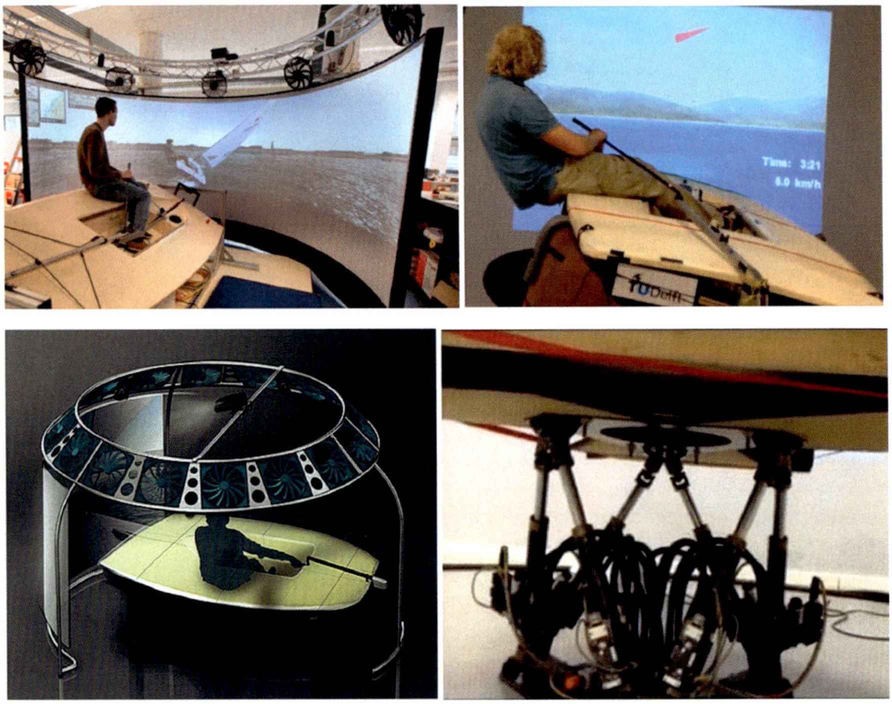

FIG. 16.32 TU Delft Sports Engineering Institute High-Performance Sail Simulator. *Courtesy of TU Delft, Faculty of Industrial Design Engineering, TU Delft Zeil Simulator. http://sportsengineering.tudelft.nl/portfolio_page/zeilsimulator/ (accessed 03.04.18); TU Delft High Performance Sail Simulator. https://www.tudelft.nl/en/ide/research/research-labs/applied-labs/high-performance-sail-simulator/ (accessed 03.04.18).*

Example 5

One of the authors (MP) designed and constructed two lively self-propelled robotic balls [55], namely robotic soccer ball and large beach ball in summer 2005 and then had some fun testing them on the MIT campus. The goal was to create an interactive toy that could enhance children play [56]. The omnidirectional soft balls weighted just a bit more than ordinary balls and they were almost as bouncy as their conventional counterparts. They were designed to be kicked and to robustly endure even fall from the multistore building (specifically the old MIT Media Lab). These initial prototypes could easily achieve 10 miles per hour (16 km per hour) and climb 30° slopes. The children at the MIT Day care center really liked these balls and they were open for interaction. The adults at the MIT campus typically ignored the presence of self-propelled ball, for example, moving at the same pace next to them. During one experiment one adult subject, unaware that beach ball is remotely controlled, engaged in nonverbal communication with the beach ball and assisted the ball to enter the elevator by holding its door. This spontaneous, accidental, and meaningful human robot interaction might be the first such case in history. No commercial product resulted yet from this research which was initially part of larger study

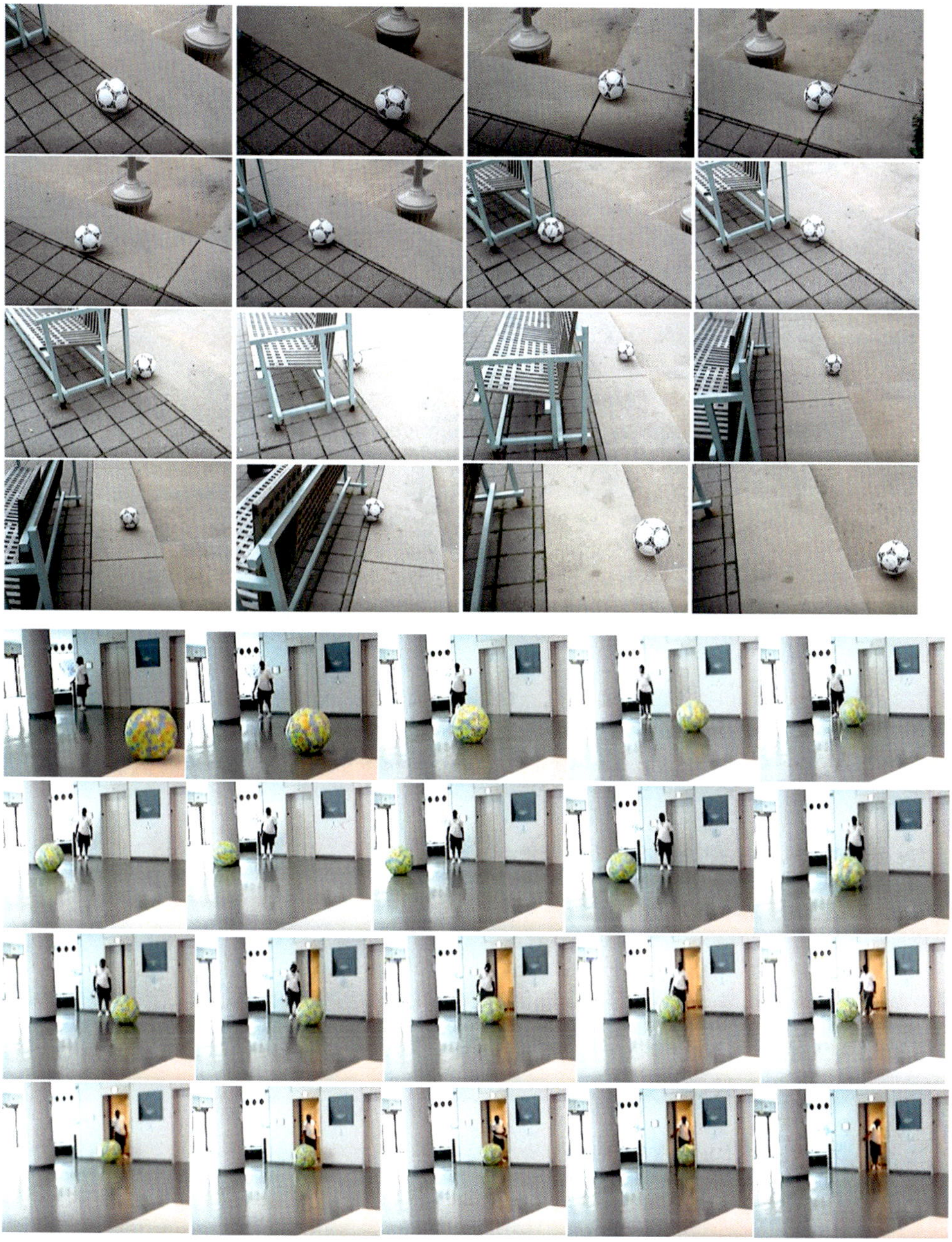

FIG. 16.33 Lively self-propelled robotic balls, robotic soccer ball (top) and large beach ball (bottom), MIT 2005 [55].

on different ways how modern technologies could be added to conventional playground for benefit of children (Fig. 16.33). It took ten years before another similar robot, that is BB-8, a droid character in the Star Wars franchise, appeared in both the 2015 Star Wars: The Force Awakens and 2017 Star Wars: The Last Jedi films.

Example 6

The Mobile Virtual Player (MVP) [57, 58], is a robotic human-size dummy originally designed for training exercise of American football to take jarring blow that, inflicted repeatedly, can injure the brains of human players. Estimated incidence of concussion during practice and games in youth, high school, and collegiate American football players was ~180,000 in 2015 [59]. The MVP is currently priced at $8250, weighs close to 200 lb., and can move with 20 miles per h (32 km per h) while being remotely controlled. Inspired by the 1970s Weeble Wobble toy, the MVP after being tackled can passively stand-up back on its own due to mass distribution and rounded shape. MVP was created in 2015 by two undergraduate mechanical engineering students at the Dartmouth University, Thayer School of Engineering, Quinn Connell and Elliot Kastner; football coach Buddy Teevens asked them to develop a dummy that could be controlled by a coach on the sidelines to limit tackling during practice. Initial approach was omnidirectional ball drive but due to technical challenges this approach was abandoned and exchanged with regular wheels (Fig. 16.34).

FIG. 16.34 The MVP tackle robot in action. Image provided by Mobile Virtual Player [57, 58].

Example 7

Ridell InSite [76] is a smart football helmet for impact monitoring introduced by company Ridell. It utilizes five zone sensor pad in the liner of the helmet to measure impact strength as function of time. The data can be then wirelessly transmitted to main unit alert monitor, for example, hand held by the team coach that could track simultaneously up to 150 players.

16.4 Head Injury Biomechanical Modeling

The biomechanics of brain injury has been an active research area for more than 70 years. Early work has focused on skull fracture and severe brain injury. Kinematic variables such as linear and rotational acceleration peak magnitudes have been widely used to assess the severity of head impact, and to predict the likelihood of head/brain injury. The most widely used injury metric, head injury criterion (HIC), was developed to assess the likelihood of skull fracture. It was established based on studies on the significance of linear acceleration on cadavers [60]. A critical limitation with HIC is that it does not incorporate head rotation that is known to be important to cause brain shear deformation responsible for concussive injury, diffuse axonal injury (DAI), and subdural hematoma [61]. Therefore, more recent injury metrics now explicitly include head rotational acceleration or velocity to assess the injury risk. Some of them are combinations of linear and rotational acceleration, and some are entirely based on head rotation [e.g., the rotational injury criterion (RIC) and power rotational head injury criterion (PRHIC, [62]), and the brain injury criterion (BrIC, [63])].

Regardless of the approach, an inherent limitation with kinematics-based injury metrics is that they treat the brain effectively as a single unit, but are unable to inform the location or distribution of brain deformation such as strain and strain rate thought to cause brain injury [61]. In large part, therefore, no consensus has been reached on an appropriate kinematic injury metric to date [64]. The ability to inform localized tissue deformation is likely important to predict concussion, a mild form of traumatic brain injury. This is because concussion is known to present widespread neuroimaging changes and a diverse spectrum of clinical signs and symptoms found in concussed athletes playing contact sports [65].

To address this challenge, computational models of the human head have been developed to estimate regional brain deformation using the measured head linear and rotational acceleration as input (see Fig. 16.35 for an illustrative example using the

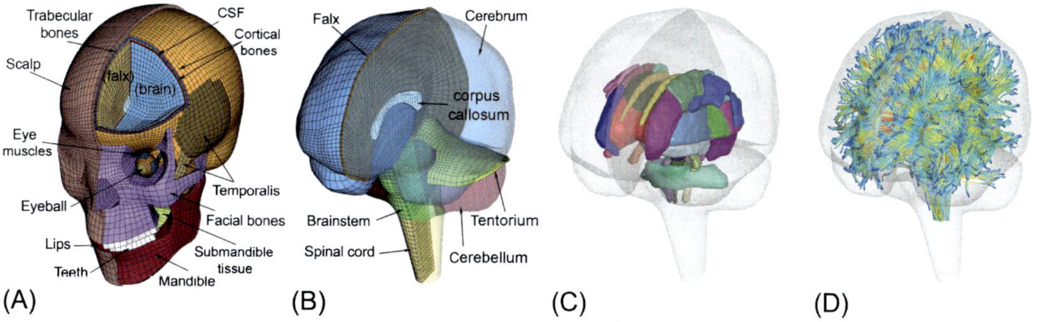

FIG. 16.35 The Worcester Head Injury Model (WHIM) showing the head exterior features (A), intracranial components (B), 50 deep white matter regions of interest (C), and tractography fibers color coded by their fractional anisotropy (FA) values (D) [66, 67].

Worcester Head Injury Model (WHIM) [66, 68]. Tissue response-based injury metrics have also been developed. They include the peak magnitude of maximum principal strain in the whole-brain or a given brain region such as corpus callosum, or the volume fraction of the brain above a given threshold (i.e., "cumulative strain damage measure" (CSDM) [63]). Studies have shown that they improve concussion prediction performance relative to kinematics-based injury metrics (e.g., [69, 70]).

Recent work in head injury model development involves further integration of advanced neuroimaging into biomechanical modeling. For example, neuroimages are used to identify specific white matter regions of interest in the deep brain [70]. This allows developing injury metrics in specific region of interests (ROIs) for more targeted investigation of brain injury. Additionally, with diffusion tensor imaging (DTI), structural anisotropy of the white matter has also been incorporated into the brain material properties [66, 69, 71]. This could improve the head injury model biofidelity [72]. Further, strain along white matter fiber bundles can also be obtained based on either DTI voxels or whole-brain tractography to develop improved injury criteria [66].

With the ever-growing model sophistication, the computational cost for impact simulation also increases. This could pose a practical barrier to deploying head injury models for real-world applications. A promising strategy to address this challenge is to precompute a large lookup table of impact simulations. For a given head impact, brain responses can then be "interpolated" without a direct simulation [73]. This approach may be especially effective for head impacts in contact sports because of their similar impact kinematic characteristics [74, 75].

16.5 Exercise Systems in Microgravity Conditions

Early astronauts and cosmonauts returned from orbit with severely diminished muscle and bone strength—a result of being in microgravity for an extended period of time. Because the bones and muscles adapt to their habitual mechanical loading environment, both tissues atrophy without the influence of gravitational loading. As a result, space agencies have long recognized the need for exercise while in orbit. Of course, without gravitational acceleration, this can be quite challenging and uncomfortable.

For many years, Russian cosmonauts wore "Pingvin" suits, which included elastic bands connecting the hands, feet, and torso. These were designed to provide resistance during movement to counteract the effects of microgravity. However, these suits were not comfortable or well liked. More recently, a "gravity loading countermeasure skinsuit" made with elastic fibers that pull the shoulders and the feet toward the waist to simulate the compressive loading experienced during gravity, have been designed [77]. While this appears to address some of the comfort issues, it has not been tested in microgravity.

Because worn clothing that provides elastic resistance have historically not been well accepted, most astronauts spend approximately 2 h each day exercising [78]. Exercises are performed on a variety of equipment adapted for microgravity. These include an adapted treadmill, stationary cycle, and weight machine. For example, the COLBERT treadmill uses elastic bungies connected to a shoulder harness, to provide a GRF during running [79] data collected on the International Space Station (ISS) from six astronauts showed that running on this treadmill produced similar GRFs to overground running, although the peaks were less than their 1G body weight [80]. Similar resistance methods have been used to adapt a stationary cycle.

Resistance training in microgravity poses a particular challenge, due to the lack of gravitational forces. The current system deployed in the ISS is the "advanced resistive exercise device (ARED)". This system uses adjustable resistance piston-driven vacuum cylinders along with a flywheel to system to provide loading. The variable resistance supplied by the flywheel assemblies is designed to mimic the inertial forces generated when lifting free weights on Earth (Figs. 16.36 and 16.37) [81].

Despite the rigorous exercise training program that most astronauts complete during spaceflight, they experience significant reductions in muscle mass and strength, as well as impairments to balance and physical function. This includes 8%–30% reductions in one-rep maximum (1RM) weight lifted, jump height, and other measures of physical capacity [82]. Although the number of individuals who require microgravity-adapted exercises is currently small, these devices may be more broadly applied to other ground-based scenarios. For example, individuals who experience prolonged bed rest, or have experienced a stroke or spinal cord injury, have similar changes to muscle and bone physiology due to lack of exercise.

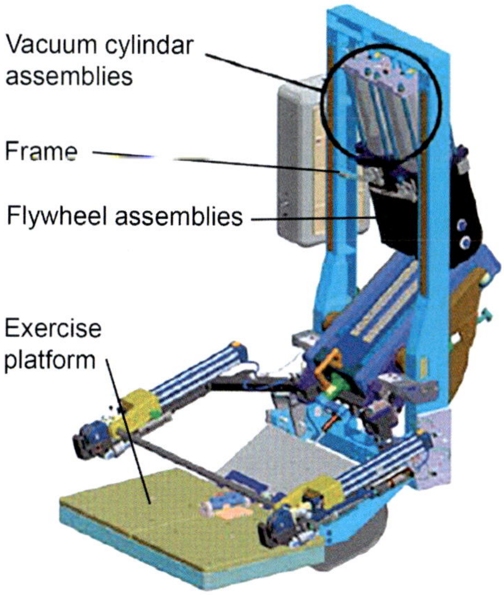

FIG. 16.36 ARED system. The resistance is provided by two vacuum cylinders and forces due to inertia that are experienced during FW exercises were simulated using a flywheel assembly [81].

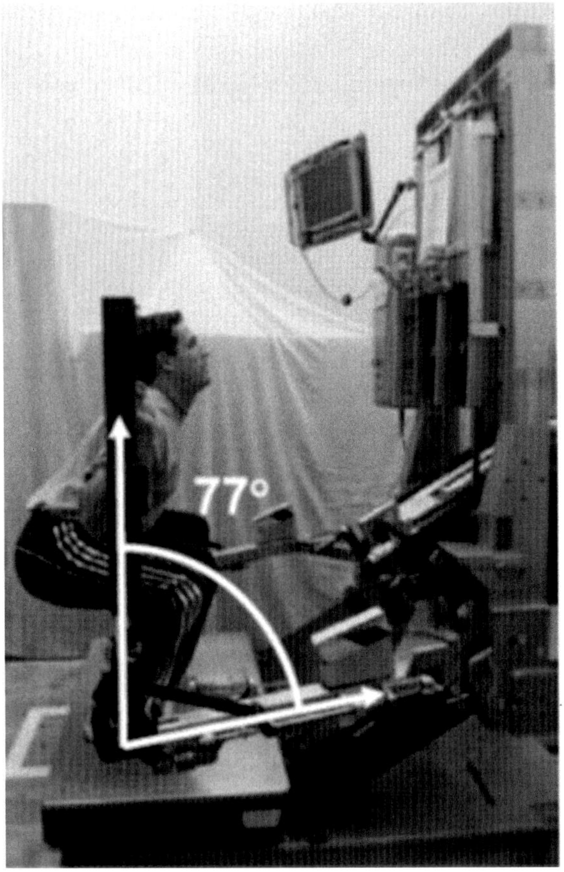

FIG. 16.37 NASA astronaut Steve Swanson, Expedition 39 flight engineer, works out on the advanced resistive exercise device (ARED) taken to the Earth-orbiting International Space Station on space shuttle mission STS-126 in November 2008 and put into service January 2009.

References

[1] Popovic, M.B. "Biomechanics and Robotics", 351 pages, Copyright © 2014 Pan Stanford Publishing Pte. Ltd., Singapore, ISBN 978-981-4411-37-0 (Hardcover), 978–981-4411-38-7 (eBook). 2013. https://doi.org/10.4032/9789814411387.

[2] About Biomechanics, The American Society of Biomechanics. http://www.asbweb.org/about-biomechanics/ (accessed 03.04.18).

[3] É.J. Marey, Du mouvement dans les fonctions de la vie: Leçons faites au Collège de France, Baillière, Paris, 1868. http://vlp.mpiwg-berlin.mpg.de/library/download2.html?litID=lit4270&pn=1. (retrieved from Max Planck Institute for the history of science, Berlin, accessed 03.04.18).

[4] E.J. Marey, La méthode graphique dans les sciences expérimentales, in: Physiologie Expérimentale. Travaux du laboratoire de M. Marey 2; retrieved from Max Planck Institute for the history of science, Berlin, 1876, pp. 133–219. http://vlp.mpiwg-berlin.mpg.de/library/data/lit29362 (accessed 03.04.18).

[5] Dempster, Space Requirements of the Seated Operator, WADC Technical Report, Wright-Patterson Air Force Base, Ohio, 1995, pp. 55–159.

[6] J. Wicke, G.A. Dumas, Estimating segment inertial parameters using fan-beam DXA, J. Appl. Biomech. 24 (2) (2008 May) 180–184.

[7] J.J. Bauer, M.J. Pavol, C.M. Snow, W.C. Hayes, MRI-derived body segment parameters of children differ from age-based estimates derived using photogrammetry, J. Biomech. 40 (13) (2007) 2904–2910 (Epub 2007 April 25).

[8] A.J. Chambers, A.L. Sukits, J.L. McCrory, R. Cham, The effect of obesity and gender on body segment parameters in older adults, Clin. Biomech. (Bristol, Avon) 25 (2) (2010) 131–136.

[9] Y. Fang, L.R. Morse, N. Nguyen, N.G. Tsantes, K.L. Troy, Anthropometric and biomechanical characteristics of body segments in persons with spinal cord injury, J. Biomech. 55 (2017) 11–17.

[10] S. Sasimontonkul, B.K. Bay, M.J. Pavol, Bone contact forces on the distal tibia during the stance phase of running, J. Biomech. 40 (15) (2007) 3503–3509.

[11] A.L. Gornitzky, A. Lott, J.L. Yellin, P.D. Fabricant, J.T. Lawrence, T.J. Ganley, Sport-specific yearly risk and incidence of anterior cruciate ligament tears in high school athletes: a systematic review and meta-analysis, Am. J. Sports Med. 44 (10) (2016) 2716–2723 (Epub 2015 December 11).

[12] R.K. Fukuchi, C.A. Fukuchi, M. Duarte, A public dataset of running biomechanics and the effects of running speed on lower extremity kinematics and kinetics, PeerJ 5 (2017), e3298, https://doi.org/10.7717/peerj.3298.eCollection2007.

[13] I.S. Davis, E. Futrell, Gait retraining: altering the fingerprint of gait, Phys. Med. Rehabil. Clin. N. Am. 27 (1) (2016) 339–355.

[14] L.L. Loundagin, T.A. Schmidt, W.B. Edwards, Mechanical fatigue of bovine cortical bone using ground reaction force waveforms in running, J. Biomech. Eng. 140 (3) (2018).

[15] P. DeVita, J. Helseth, T. Hortobagyin, Muscles do more positive than negative work in human locomotion, J. Exp. Biol. 210 (Pt 19) (2007) 3361–3373.

[16] M.B. Popovic, A. Hofmann, H. Herr, Angular momentum regulation during human walking: biomechanics and control, in: Proceedings of the IEEE International Conference on Robotics and Automation, New Orleans, Louisiana, USA, 2004, pp. 2405–2411.

[17] M.B. Popovic, A. Goswami, H. Herr, Ground reference points in legged locomotion: definitions, biological trajectories and control implications, Int. J. Robot. Res. 24 (12) (2005) 1013–1032.

[18] H. Herr, M.B. Popovic, Angular momentum in human walking, J. Exp. Biol. 211 (2008) 467–481.

[19] M. Vukobratovic, A.A. Frank, D. Juricic, On the stability of biped locomotion, IEEE Trans. Biomed. Eng., BME 17 (1) (1970) 25–36.

[20] M. Vukobratovic, H. Herr, B. Borovac, M. Rakovic, M.B. Popovic, A. Hofmann, V. Potkonjak, Biological principles of control selection for a humanoid robot's dynamic balance preservation, Int. J. Human. Robot. 5 (4) (2008) 639–678.

[21] S. Kajita, F. Kanehiro, K. Kaneko, K. Yokoi, H. Hirokawa, The 3D linear inverted pendulum mode: a simple modeling for a biped walking pattern generation, in: Proc. IEEE Int. Conf. Intell. Robots Syst, 2001, pp. 239–246.

[22] J. Pratt, J. Carff, S. Drakunov, A. Goswami, Capture point: a step toward humanoid push recovery, in: 6th IEEE-RAS International Conference on Humanoid Robots, 2006, IEEE, 2006, pp. 200–207.

[23] A. Seyfarth, H. Geyer, M. Günther, R. Blickhan, A movement criterion for running, J. Biomech. 35 (5) (2002) 649–655.

[24] H. Geyer, A. Seyfarth, R. Blickhan, Compliant leg behaviour explains basic dynamics of walking and running, Proc. R. Soc. Lond. B Biol. Sci. 273 (1603) (2006) 2861–2867.

[25] G.A. Cavagna, F.P. Saibene, R. Margaria, Mechanical work in running, J. Appl. Physiol. 19 (1964).

[26] R. Blickhan, The spring-mass model for running and hopping, J. Biomech. 22 (1989).

[27] T.A. McMahon, G.C. Cheng, The mechanics of running: how does stiffness couple with speed? J. Biomech. 23 (Suppl. 1) (1990).

[28] C.T. Farley, J. Glasheen, T.A. McMahon, Running springs: speed and animal size, J. Exp. Biol. 185 (1) (1993) 71–86.

[29] R.M. Alexander, Optimization and gaits in the locomotion of vertebrates, Physiol. Rev. 69 (4) (1989) 1199–1227.

[30] F.C. Anderson, A Dynamic Optimization Solution for a Complete Cycle of Normal Gait, (Ph.D. thesis), The University of Texas at Austin, Austin, Texas, 1999.

[31] F.C. Anderson, M.G. Pandy, A dynamic optimization solution for vertical jumping in three dimensions, Comput. Meth. Biomech. Biomed. Eng. 2 (1999).

[32] F.C. Anderson, M.G. Pandy, Dynamic optimization of human walking, J. Biomech. Eng. 123 (2001).

[33] https://simtk-confluence.stanford.edu/display/OpenSim/Gait+2392+and+2354+Models.

[34] MOTEK, MEDICAL Improving Human Performance, http://www.motekmedical.com/ (accessed 03.04.18).

[35] anybodytech.com, Frontpage, https://www.anybodytech.com/ (accessed 03.04.18).

[36] A. Staiano, R. Flynn, Therapeutic uses of active videogames: a systematic review, Games Health J. (2014) 351–365.

[37] FitMi Rehabilitation System, https://www.flintrehab.com.

[38] Molina, et al., Virtual reality using games for improving physical functioning in older adults: a systematic review, J. NeuroEng. Rehabil. 11 (2014) 156.

[39] Sviestrup, Motor rehabilitation using virtual reality. J. NeuroEng. Rehabil. 1 (2004) 1–8. https://doi.org/10.1186/1743-0003-1-10.

[40] Pinata Sessoms Enhancing Warfighter Readiness in a Virtual Environment, Naval Medical Research and Development, News Releases (accessed 15.08.16).

[41] Hoffman, et al., Using fMRI to study the neural corrleates of virtual reality analgesia, CNS Spectr. 11 (1) (2006) 45–51.

[42] Hoffman, et al., Virtual reality pain control during physical therapy range of motion exercises for a patient with multiple blunt force trauma injuries, CyberPsychol. Behav. 12 (1) (2009) 47–49.

[43] Hoffman, et al., Use of virtual reality for adjunctive treatment of adult burn pain during physical therapy: a controlled study, Clin. J. Pain 16 (2000) 244–250.

[44] W. Li, Development and Evaluation of a Virtual Reality Therapy System for Children with Hemiplegic Cerebral Palsy (Bachelor's thesis)University of Toronto, Toronto, 2007.

[45] Exergaming, https://www.exergamefitness.com/ (accessed 03.04.18).

[46] Resnick, Exergames: A New Step Toward Fitness?, Harvard Health Publishing, 2012 (accessed 20.11.17).

[47] Sween, et al., The role of exergaming in improving physical activity: a review, J. Phys. Act. Health 11 (4) (2014) 864–870.

[48] Multi Sports Simulators, http://www.sportsentertainmentspecialists.com/MultiSportSimulators/index.html (accessed 03.04.18).

[49] Nadia Barbu, Technology Developed in Biomechatronics LAB COULD CHANGE COMBAT SPORTS, https://www.imperial.ac.uk/news/177831/technology-developed-biomechatronics-lab-could-change/, 2017 (accessed 03.04.18).

[50] Corner The Smart Boxing Tracker, https://thecornerapp.com/#!/ (accessed 03.04.18).

[51] SkyTechSport Alpine-Simulator, http://www.skytechsport.com/alpine-simulator (accessed 03.04.18).

[52] Tim Newcomb, Experience G-force and icy slopes in a SkyTechSport ski training simulator, Sports Illustrated, 2015. https://www.si.com/edge/2015/02/11/skytechsport-ski-training-simulator-us-ski-team-fis-alpine (accessed 03.04.18).

[53] TU Delft Zeil Simulator, http://sportsengineering.tudelft.nl/portfolio_page/zeilsimulator/ (accessed 03.04.18).

[54] TU Delft High Performance Sail Simulator, https://www.tudelft.nl/en/ide/research/research-labs/applied-labs/high-performance-sail-simulator/ (accessed 03.04.18).

[55] WPI Robotics Balls, http://users.wpi.edu/~mpopovic/pages/balls.html (accessed 03.04.18).

[56] S. Seitinger, E. Sylvan, O. Zuckerman, M. Popovic, O. Zuckerman, A new playground experience: going digital? in: CHI'06 Extended Abstracts on Human Factors in Computing Systems, ACM, 2006, pp. 303–308.

[57] Elliot Kastner, How a Robot Football Player Will Prevent Concussions, IEEE Spectr. (2016). https://spectrum.ieee.org/robotics/humanoids/how-a-robot-football-player-will-prevent-concussions (accessed 03.04.18).

[58] Mobile Virtual Player, http://www.mobilevirtualplayer.com/the-store/ (accessed 03.04.18).

[59] Incidence of concussion during practice and games in youth, high school, and collegiate American football players, JAMA Pediatr. (2015).

[60] J. Versace, A review of the severity index, in: 15th Stapp Car Crash Conference. Coronado, CA, USA, 1971. SAE paper 710881.

[61] King AI, Yang KH, Zhang L, et al (2003) Is head injury caused by linear or angular acceleration? In: IRCOBI Conference, Lisbon, Portugal, pp 1–12.

[62] H. Kimpara, M. Iwamoto, Mild traumatic brain injury predictors based on angular accelerations during impacts, Ann. Biomed. Eng. 40 (2012) 114–126, https://doi.org/10.1007/s10439-011-0414-2.

[63] E.G.G. Takhounts, M.J.J. Craig, K. Moorhouse, J. McFadden, V. Hasija, Development of brain injury criteria (BrIC), Stapp Car Crash J. 57 (2013) 243–266.

[64] J.J.P. Mihalik, R.R.C. Lynall, E.E.B. Wasserman, et al., Evaluating the "threshold theory": can head impact indicators help? Med. Sci. Sports Exerc. 49 (2017) 247–253, https://doi.org/10.1249/MSS.0000000000001089.

[65] E.D. Bigler, Systems biology, neuroimaging, neuropsychology, neuroconnectivity and traumatic brain injury, Front. Syst. Neurosci. 10 (2016) 1–23, https://doi.org/10.3389/FNSYS.2016.00055.

[66] W. Zhao, S. Ji, White matter anisotropy for impact simulation and response sampling in traumatic brain injury. J. Neurotrauma (2018). https://doi.org/10.1089/neu.2018.5634.

[67] W. Zhao, J.C. Ford, L.A. Flashman, et al., White matter injury susceptibility via Fiber strain evaluation using whole-brain tractography, J. Neurotrauma 33 (2016) 1834–1847, https://doi.org/10.1089/neu.2015.4239.

[68] S. Ji, W. Zhao, J.C. Ford, et al., Group-wise evaluation and comparison of white matter fiber strain and maximum principal strain in sports-related concussion, J. Neurotrauma 32 (2015) 441–454, https://doi.org/10.1089/neu.2013.3268.

[69] C. Giordano, S. Kleiven, Connecting fractional anisotropy from medical images with mechanical anisotropy of a hyperviscoelastic fibre-reinforced constitutive model for brain tissue, J. R. Soc. Interface 11 (2014) 1–14, https://doi.org/10.1098/rsif.2013.0914.

[70] W. Zhao, Y. Cai, Z. Li, S. Ji, Injury prediction and vulnerability assessment using strain and susceptibility measures of the deep white matter, Biomech. Model. Mechanobiol. 16 (2017) 1709–1727, https://doi.org/10.1007/s10237-017-0915-5.

[71] S. Ganpule, N.P. Daphalapurkar, K.T. Ramesh, et al., A three-dimensional computational human head model that captures live human brain dynamics, J. Neurotrauma 34 (2017) 2154–2166, https://doi.org/10.1089/neu.2016.4744.

[72] R.W. Carlsen, N.P. Daphalapurkar, The importance of structural anisotropy in computational models of traumatic brain injury, Front. Neurol. 6 (2015) 1–6, https://doi.org/10.3389/fneur.2015.00028.

[73] S. Ji, W. Zhao, A pre-computed brain response atlas for instantaneous strain estimation in contact sports, Ann. Biomed. Eng. 43 (2015) 1877–1895, https://doi.org/10.1007/s10439-014-1193-3.

[74] S. Rowson, G. Brolinson, M. Goforth, et al., Linear and angular head acceleration measurements in collegiate football, J. Biomech. Eng. 131 (2009):061016, https://doi.org/10.1115/1.3130454.

[75] S.P. Broglio, B. Schnebel, J.J. Sosnoff, et al., Biomechanical properties of concussions in high school football, Med. Sci. Sports Exerc. 42 (2010) 2064–2071, https://doi.org/10.1249/MSS.0b013e3181dd9156.

[76] Ridell InSite Smart Football Helmet, http://www.riddell.com/InSite#howitworks (accessed 03.04.18).

[77] J.M. Waldie, D.J. Newman, A gravity loading countermeasure skinsuit, Acta Astronaut. 68 (7–8) (2011) 722–730.

[78] NASA, Exercising in Space, https://www.nasa.gov/audience/foreducators/stem-on-station/ditl_exercising (accessed 03.04.18).

[79] NASA Tumblr, Exercising in Space, https://nasa.tumblr.com/post/136706596374/exercising-in-space (accessed 03.04.18).

[80] J.K. De Witt, L.L. Ploutz-Snyder, Ground reaction forces during treadmill running in microgravity, J. Biomech. 47 (10) (2014) 2339–2347.

[81] J.A. Loehr, S.M.C. Lee, K.L. English, J. Sibonga, S.M. Smith, B.A. Spiering, R.D. Hagan, Musculoskeletal adaptations to training with the advanced resistive exercise device, Med. Sci. Sports Exerc. 43 (1) (2011) 146–156.

[82] N. Petersen, G. Lambrecht, J. Scott, N. Hirsch, M. Stokes, J. Mester, Postflight reconditioning for European astronauts—a case report of recovery after six months in space, Musculosk. Sci. Pract. 27 (2017) S23–S31.

17

Bioinspired Robotics

Krishnanand N. Kaipa*, Cagdas Onal†, Vukica Jovanovic*, Ana Djuric‡, Ming Luo§, Matthew P. Bowers†, Marko B. Popovic†

*OLD DOMINION UNIVERSITY, NORFOLK, VA, UNITED STATES †WORCESTER POLYTECHNIC INSTITUTE, WORCESTER, MA, UNITED STATES ‡WAYNE STATE UNIVERSITY, DETROIT, MI, UNITED STATES §STANFORD UNIVERSITY, STANFORD, CA, UNITED STATES

CHAPTER OUTLINE

17.1 Introduction: Bioinspiration

Animals and insects evolved distinct traits over millions of years of natural selection, which enable them to adapt to, and perform reliably in, a variety of unstructured environments. Nature abounds in examples of such natural creatures that exploit features like softness, compliance, and configurability to achieve effective interactions with their surroundings. Robotic replicas with similar adaptation capabilities can be created by copying some of these crucial elements into the robots' designs. Locomotion and manipulation are among the primary tasks for which these traits have been optimized over time. Therefore, we describe the representative mechanisms found in natural species from which bioinspired insights can be used to synthesize locomotion and manipulation capabilities in robotic systems.

The basic source of inspiration for a robot can be borrowed from any species belonging to the entire animal kingdom of vertebrates and invertebrates (Fig. 17.1). For example, a marine robot can be modeled after the swimming locomotion of a fish, while a robotic

Biomechatronics. https://doi.org/10.1016/B978-0-12-812939-5.00017-3

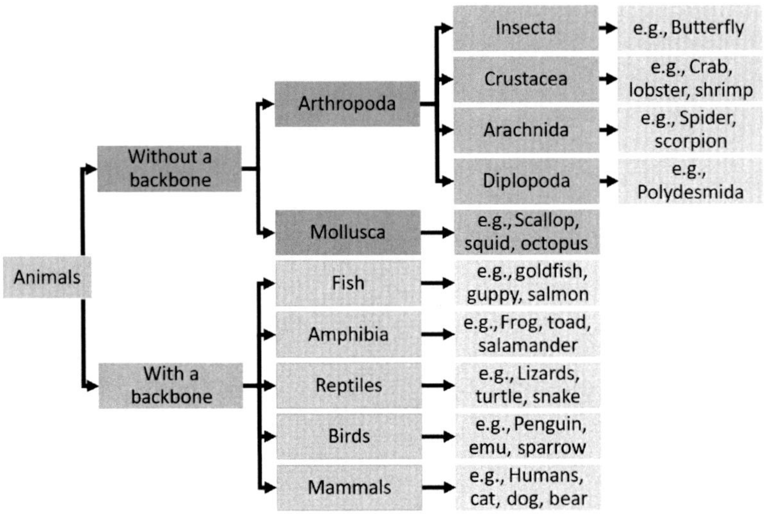

FIG. 17.1 The entire animal kingdom—vertebrates and invertebrates—offers sources of biological inspiration for robotic systems. *Courtesy: http://mapsandmemories.edublogs.org/files/2012/01/classification-animals-qybz56.gif.*

manipulator/gripper can be modeled after a human arm/hand. The principal traits of such a biological model of inspiration that can be mimicked in robot designs include *morphology, actuation, sensing,* and *control.*

Morphology is a branch of biology that deals with the form and structure of plant and animal, specifically with an organism or any of its parts. In this context, it refers to the overall body plan of the natural creature chosen as the inspiration source, which includes: (1) network of limbs and other body segments connected through specifically articulated joint degrees of freedom (DoF), (2) geometrical shapes and aspect ratios of dimensions of various body segments, and (3) compliance/stiffness properties of different body segments. The concepts of actuation, sensing, and control are discussed in more detail in Chapters 3–5, respectively.

The body structure of the animal impacts its form of interaction with the surrounding environment and its overall functionality in important ways. First, it determines whether the animal can operate in a particular environment or not. Second, it impacts how well the animal performs in a chosen environment. These two aspects can be explained using examples of locomotion patterns observed in various animals. While swimming, walking, and flying constitute three modes of locomotion, the body structures of most animals are best suited to only one locomotion mode. For example, purely aquatic creatures (e.g., fish, whales, dolphins, etc.) are suitable for swimming alone, while terrestrial animals (e.g., quadrupeds like cheetah, horse, etc.) are best suited for walking and running on land. Within quadrupedal mammals, the body plans of reptilian family (e.g., lizards) and cursorial mammals (e.g., cheetah) distinctly differ from each other, which reflect in their differing locomotion characteristics (Fig. 17.2). Lizards have their feet kept well out on either side of their body and move their legs sideways, allowing them to execute statically

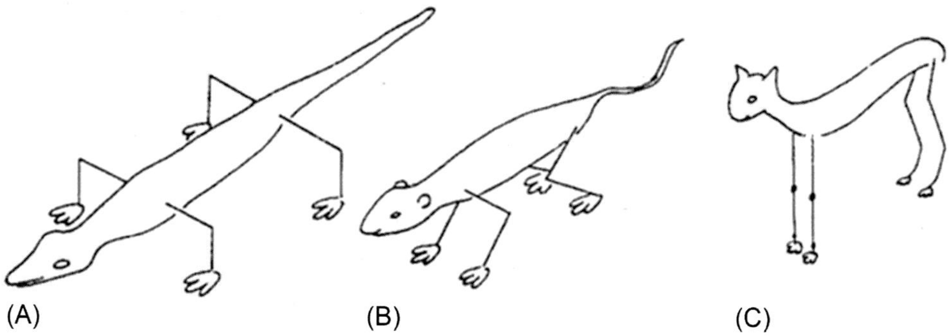

FIG. 17.2 Body plans of quadrupedal mammals afford different locomotion characteristic: (A) reptiles and newts have their feet kept well out on either side of the body, (B) small mammals like rats have their feet kept close under the body and run on bentlegs using an noncursorial posture, and (C) larger mammals stand and run on relatively straight legs using a cursorial posture. *Courtesy R. Alexander, The gaits of bipedal and quadrupedal animals, J. Robot. Res. 3(2) (1983) 49—59.*

stable crawling gaits (Figs. 17.3 and 17.4). However, larger mammals have their legs relatively straight, with feet under the body, and move them in a sagittal plane. This contrasting arrangement of body limbs results in losing static stability, but allows them to execute relatively much faster, dynamically stable, running gaits. Although, most birds are bipedal and are capable of walking slowly on land, their primary locomotion mode is flying.

An interesting evolutionary change can be observed in amphibians that can swim and walk as well. For example, salamanders are regarded as among the first amphibians that evolved morphological structures to transition from aquatic to terrestrial locomotion due

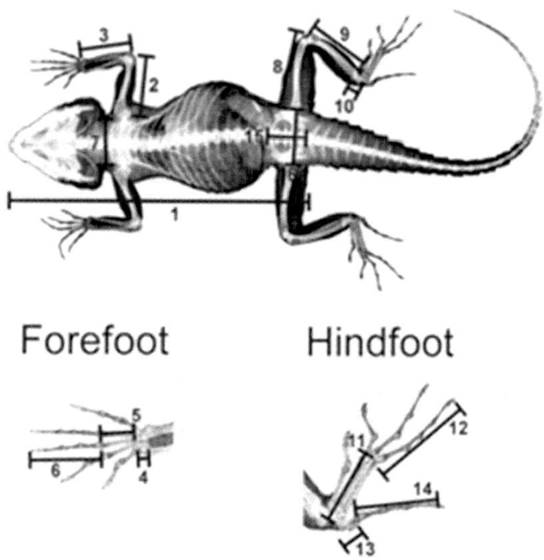

FIG. 17.3 Lizard morphology. *Courtesy R. Alexander, The gaits of bipedal and quadrupedal animals, J. Robot. Res. 3(2) (1983) 49—59.*

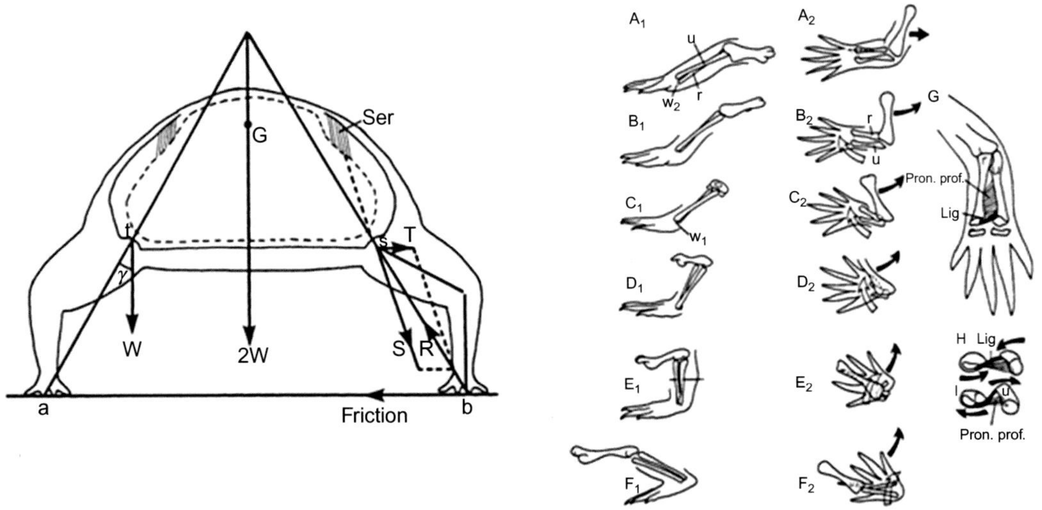

FIG. 17.4 Lizard biomechanics: (A) statics of sprawling posture and (B) integration of movements in shoulder, elbow, and wrist to maintain ground contact during stance phase. *Courtesy R. Alexander, The gaits of bipedal and quadrupedal animals, J. Robot. Res. 3(2) (1983) 49—59.*

to their close resemblance to the first vertebrates on land. Salamanders have articulated body segments similar to snakes that allow undulatory motion and legs similar to quadrupedal vertebrates. These body plans enable them to rapidly switch between swimming and crawling modes of locomotion. Another interesting example of amphibians is semi-aquatic freshwater turtles (e.g., Mauremys caspica). These turtles are reasonably good swimmers, yet are also agile crawlers, distinguishing them from terrestrial turtles that can walk on land efficiently but swim poorly. Freshwater turtles are able to morph their limbs between different postural configurations based on the locomotion mode. Features like posture, biomechanics, and gait patterns of the limbs during walking completely differ from those used during swimming. Moreover, their limbs transition through a continuum of configurations, enabling complex fluid-structure interactions (FSIs) that generate drag-based thrust during rowing strokes of a typical swimming gait. In all the above examples, it can be clearly seen that the animal's body plan primarily decides its mode of locomotion and how its mobility is affected in a particular environment.

17.2 Bioinspired Locomotion

Locomotion is the act of self-propulsion of the system in which entire, or most of, the system translates—that is, moves—in space. Natural and engineered systems use a variety of principles to locomote from one place to other. Traditional examples of engineered systems include terrestrial vehicles (e.g., automobiles) for locomotion on land, marine vehicles (e.g., ships, submarines, etc.) for locomotion in sea, and aerial vehicles (e.g., airplanes, helicopter, etc.) for locomotion in air. Till date, many mobile robots have been built by using this traditional inspiration—unmanned ground vehicles (e.g., robotic self-driving

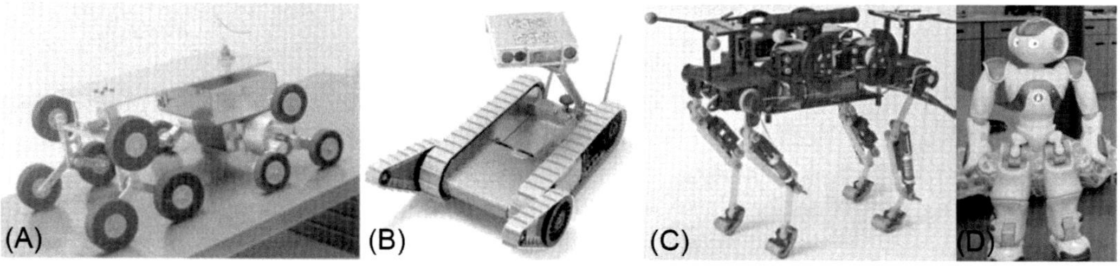

FIG. 17.5 (A, B) Examples of less biologically inspired solutions to locomotion—wheeled and tracked robots [1, 2]. (C, D) Examples of more bioinspired solutions to locomotion—legged robots (quadrupedal and bipedal humanoid) [3].

cars, tracked robots, etc.), unmanned underwater/surface vehicles (e.g., robotic boats/ submarines), and unmanned aerial vehicles (e.g., robotic helicopters, quadcopters, etc.). All these are less biologically inspired solutions to locomotion that depends on either wheels, rotors, and aerodynamic lift (heavier than air flight achieved by airfoil-based wings) for basic propulsion (Fig. 17.5A and B). However, more bioinspired solutions to robotic locomotion primarily involve mobility modeled after propulsion mechanisms found in nature (Fig. 17.5C and D). In the previous section, we looked at representative examples of how locomotion is achieved by various animals in the natural world. Now, we focus on several state-of-the-art bioinspired mobile robots.

17.2.1 Terrestrial Mobile Robots

Humans and many animals use legs to achieve locomotion. Accordingly, legged robotics is one of the most studied areas in bioinspired robotic locomotion. Walking/running robots can be broadly classified into multilegged robots that walk with more than two legs (hexapedal and quadrupedal) and bipedal robots that walk with two legs.

17.2.1.1 Multilegged Robots

The legged robotics research community has witnessed numerous examples of multilegged robots that walk using either eight, six, or four, or three legs. Here, we focus on hexapedal (six-legged) and quadrupedal (four-legged) robotic locomotion. Representative examples range from the adaptive suspension vehicle (six-legged) of the 1980s to recent quadrupedal robots like the Big Dog, the WildCat, LittleDog, LS3, Spot, and SpotMini.

Stability (maintaining balance and not falling over) is an important requirement during legged locomotion. Static stability refers to being stable at rest. Multilegged creatures achieve static stability by ensuring that the center of gravity is within a stability triangle formed by at least three legs that are in contact with the ground at any instant during locomotion. Dynamic stability refers to stability achieved by movement. Legged robots can be broadly classified into two types based on which form of stability is used during locomotion. For example, six- and eight-legged robots are almost always statically stable. Four-legged robots are either statically stable or dynamically stable, depending on the gait and speed of locomotion.

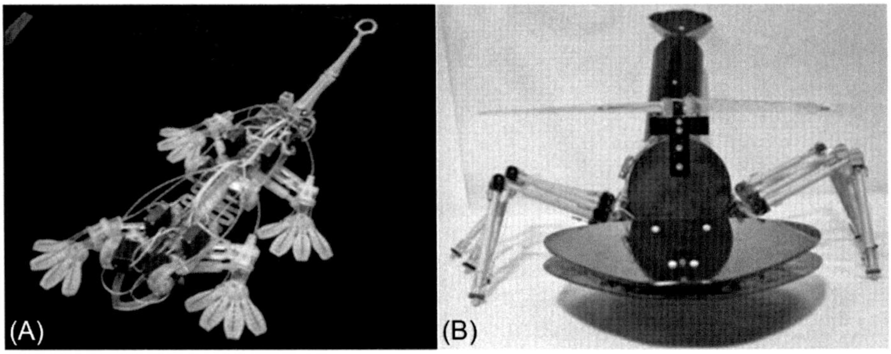

FIG. 17.6 Multilegged robots that borrow inspiration from insects in which the legs are spread out and move sideways: (A) stickybot, a gecko-inspired robot and (B) Lobster-inspired eight-legged robot [4].

Statically stable robots are suitable to tasks that involve transportation of heavy payloads, but are not optimized for speed. Representative examples include a stickybot (Fig. 17.6A), a gecko-inspired robot [4], stick-insect-inspired hexapedal robot [5], a lobster-inspired eight-legged robot (Fig. 17.6B) [6], a cockroach-inspired robot [7], and a mantis-inspired robot [8].

The hexapedal robot developed at the Technical University of Munich mimics the kinematics and gaits of a stick insect, *Carausius morosus* [5]. The lobster robot is an eight-legged bioinspired robot modeled after the biomechanical and neurobiological control principles observed in an American lobster. Ajax (Robot V) is a hexapedal robot that borrows inspiration from a death-head cockroach, Blaberus discoidalis. Ajax was built at the Bioinspired Robotics Laboratory, Case Western Reserve University (CWRU). The robot legs are driven by McKibben muscle-like actuators, which have a force-to-weight ratio relatively higher than that of traditional actuators. The robot has a total of 24 DoF: the two legs in the front have five joints each, the two legs in the middle have four joints each, and the two legs in the rear have three joints each. Although the entire range of motion achieved by the natural cockroach (seven DoF per leg) is not replicated by the robot, it still achieves agile walking capabilities. MantisBot is a mantis-inspired hexapedal robot with a total of 28 DoF, which was built at the same laboratory in CWU to study behaviors like prey tracking and striking. The robot uses a distributed reflexive posture controller that allows it to achieve postural balance through a series of reflexes observed in insects.

Quadrupedal robots are statically stable during slow walking (e.g., using an ambling gait), while they are dynamically stable during running (e.g., using trotting, cantering, and galloping gaits). Representative examples include the most popular lineup of quadrupedal robots like BigDog, WildCat, LittleDog, LS3, Spot, and SpotMini, which were built by Boston Dynamics in recent years.

BigDog is a quadrupedal robot that borrows inspiration from cursorial mammals like canine. It was built by Boston Dynamics in 2005. The target application of BigDog was to serve as a robotic pack mule, accompanying soldiers in rough terrains. The robot exploits the principle of dynamic stability to run up to 10 km/h. It is capable of carrying payloads

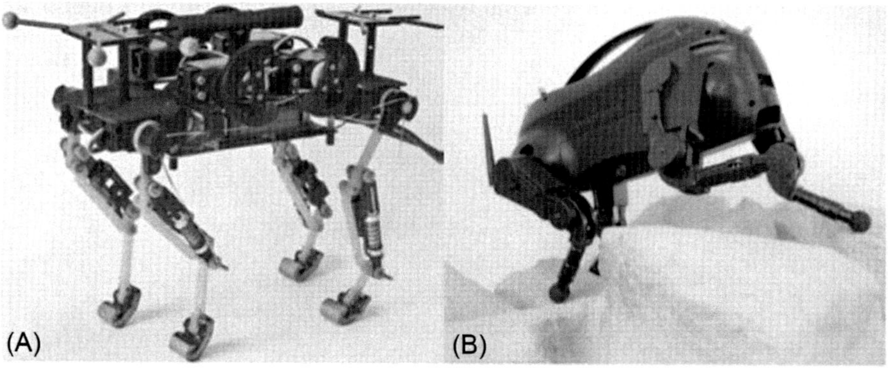

FIG. 17.7 Examples of quadrupedal robots: (A) Cheetah-cub [3] and (B) little Dog [9].

up to 150 kg, climbing inclinations up to 35°, and walking in a variety of terrains like rubble, muddy hiking trails, and snow. A combination of efficient sensing and control algorithms enable the robot to maintain balance even on slippery surfaces. WildCat sets the world record as the fastest quadrupedal robot by running at 32 km/h. WildCat uses gaits like trotting, bounding, and galloping to maintain balance while running. Cheetah-Cub [3] and LittleDog [9] are representative examples of miniature quadrupedal robots intended as test beds for research in robotic locomotion (Fig. 17.7). Legged Squad Support System (LS3) is a militarized version of BigDog, ruggedized with abilities to operate in hot, cold, wet, and other harsh environments. It can accompany marines and soldiers in rough terrains, while helping them in carrying loads up to 181 kg. Spot is another four-legged robot, with a payload capacity of 23 kg, designed for both indoor and outdoor operations. SpotMini is a small quadrupedal robot that inherits all of the mobility of Spot, with an additional ability to manipulate objects using a 5 DoF arm.

17.2.1.2 Humanoid Robots

Humanoid robots borrow inspiration from bipedal locomotion found in humans. Various developments of humanoid robots transpired in the previous century. The main term, "robot," led to early science fiction works that included the future of robots as human helpers, who look and act like their human counterparts. The word first appeared in Karel Čapek's 1921 [10, 11] book Rossum's Universal Robots (R.U.R.). Karel later explained that his brother, painter and writer Josef Čapek created the word, originating from the term "robota" that means "hard work" or "slave work" in many Slavic languages. The meaning of the word directly informs the main reasons why humans would even develop humanoid robots and what functions they would be assigned. These artificial helpers would be based on the electromechanical (mechatronics system), embedded with artificial intelligence that would work reliably and repetitively without injuring human beings. Having robots be "human-like" would make them more socially adaptable [12].

Biped Robots: Various biped robots have been created in the previous decades. Some of them were designed to showcase advanced capabilities of research departments from

FIG. 17.8 (A) Maschinenmensch "Robot" [13], (B) ASIMO at a Honda factory [14], (C) Sony Qrio Robot [15], and (D) NAO evolution [16].

companies like Honda or Sony (Fig. 17.8A–C). Others were developed as new platforms that would commercially become available for research and educational communities, such as the one shown in Fig. 17.8D [17].

The robot shown in Fig. 17.8A is a humanoid robot from the 1930s, designed by the engineer Alan Herbert Reffell and Captain W.H. Richards (shown in Fig. 17.8A with the robot). Captain and journalist William H. Richards was secretary of the Exhibition of the Society of Model Engineers [13]. The ASIMO humanoid robot (Fig. 17.8B) was designed and made by the Honda Motor Co. during their project the "Advanced Step in Innovative Mobility" [14], which was inducted into the Robot Hall of Fame, in Pittsburgh, PA in 2004, as the first humanoid droid capable of walking dynamically [14]. The NAO robot, in Fig. 17.8D, was developed by Aldebaran Robotics in 2006, later rebranded as SoftBank Robotics, and became one of the first humanoid robots placed on the mass market [17, 18]. It has "seven senses" integrated for natural interaction: moving, feeling, hearing and speaking, seeing, connecting, and "thinking" [18], and is a dynamically stable, bipedal robot that has controls based on the hybrid zero dynamics [19–22]. It is an autonomous robot with 25 DoF that can interact with the environment and adapt to events, such as falling to the ground, responding to someone talking to it, and alike. It has four directional microphones and loudspeakers, various sensors placed at its head, hands, and feet. It also has wireless connectivity with mobile devices [18]. The NAO robot at 2016 Robocup can be seen in Fig. 17.9. The NAO robot schematic is presented in Ref. [18, 23].

There are various applications of biped humanoid robots, such as in disaster response, which can be easily tele-operated by a human operator, as per the operator's decisions and because it is not completely autonomous [24]. These robots would be used for working on a disaster site, such as a nuclear plant accident. The DARPA Robotics Challenge (DRC), organized by US Defense Advanced Research Projects Agency (DARPA), focused on a nuclear plant accident and the use of robots for disaster response [24, 25].

FIG. 17.9 NAO robot at 2016 Robocup. *From Wikimedia, https://commons.wikimedia.org/wiki/File:Nao_Robot_(Robocup_2016).jpg.*

Various researchers have focused on developing an interface that includes mimicking human emotions through audio and visual effects, such as human speaking and gesturing [12], even by creating the face gestures [26]. Other researchers have focused on developing sensors that would detect the source location of sound to allow the robot to move with respect to the auditory feedback measured from the microphones [27]. Different safety measures are being designed so that robots can recover from falls and use methods that would assist the robots to avoid falling on a delicate object, or on a person, by direction-changing fall controls through intelligent stepping, as soon as the fall is predicted [28]. Similar to the human body, humanoid robots are designed so that they have various numbers of rotary joints. Depending on a number of these joints, robots will have a certain number of DoF, which can be controlled by one of the control procedures.

Education: As mobile robots have changed and improved due to recent technology developments, they are becoming more affordable. There are various autonomous robot platforms that are currently available on the market. Some of these platforms are humanoid robots, such as NAO and Pepper, and some are autonomous cars, such as SumoBots, BoeBots, and LEGO Mindstorms. They have been used in workshops held in science museums as educational instruction tools [29]. Robots have also become an integral component of teaching with technology at various educational levels [30]. They have been used for various courses, such as the teaching of a foreign language, programming, technology education, or as a device serving as an interface between a teacher and a student [23]. The NAO robot has been used to teach English to Iranian students in middle and high schools through interaction with students and playing games through questions, answers, and different robot responses [23]. Robots are also used in physiological and cognitive research in child development. For example, in working with autistic children and assisting in teaching of sign language [30, 31]. Another example includes the NAO robot as an instructional tool in teaching children how to write through better engagement with students who are less motivated in order to stimulate metacognition and empathy [32]. It has also been used in the preschool and kindergarten as a social assistive tool [33, 34].

Communal approach and expensive robots: In recent years, there have been numerous large-scale robotic competitions and several types of small-scale communal challenges that have proven very useful in spurring interest in robotics and advancing the state of both robotics hardware and software techniques. Some of these important large-scale initiatives utilizing real physical robots have been targeted more toward high school students, for example, first robotics competition (FRC) [35], and university students like the ones organized by American Society of Mechanical Engineers (ASME), or The Institute of Electrical and Electronics Engineers (IEEE) professional organizations. Some have also been more focused on the professional and cutting-edge roboticist community, for example, the DRC. The Hampton Roads Shipping Challenge, organized by IEEE, is an example of an undergraduate student robot competition (Fig. 17.10). The autonomous robot has

FIG. 17.10 An example of undergraduate student robot competition—Hampton Roads Shipping Challenge, organized by IEEE [36].

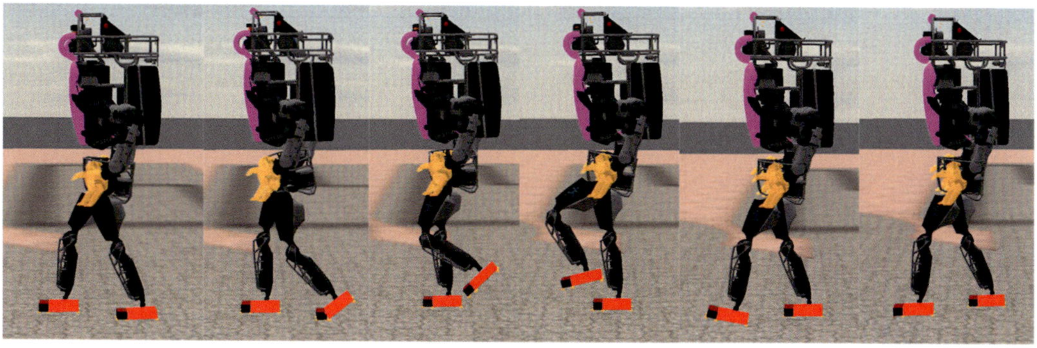

FIG. 17.11 Snapshots of advanced humanoid robot walking simulation; study of ATLAS robot emergent behavior due to novel toe-heel joint [38]. *Courtesy of Marko Popovic.*

90 s to complete a task related to identifying containers and their colors at the barge and then taking them to different zones: rail, ship, or a truck after robots pass the tunnel.

There have also been professional-level competitions that use very advanced real physical robots, for example, DARPA Robotics Challenge [37]. In the realm of the more professional, communal efforts, one expects the involvement of more advanced robots, for example, Petman or ATLAS by Boston Dynamics. A walking simulation of the ATLAS robot [38] is shown in Fig. 17.11. Unfortunately, the sophistication of robots like ATLAS (Fig. 17.12A) usually result in them being substantially more expensive and, hence, unattainable for most robotics researchers in the United States. To address that DARPA donated a handful of ATLAS robots to several carefully selected institutions in the United States.

The Willow Garage PR-2 robot (Fig. 17.12B), which has a mobile platform, an extensive sensor suite, and two multi-DoF arms, would have been a suitable candidate for use as a

FIG. 17.12 (From left to right) Atlas, PR2, and Valkyrie robots.

collaborative robot research platform; however, it is priced around $280,000, and as such, is another example of a robot that is unattainable for the majority of researchers. To address that Willow Garage donated some PR-2 robots to institutions in the United States (including WPI). To solve the problem of physical, advanced robots not being accessible to many institutions, NASA created a simulated virtual likeness of its R5 Valkyrie robot (Fig. 17.12C) [39] that can be experimented with in a purely software environment. In principle, the "robots" of this approach are more readily accessible to the larger community of professional roboticists; however, these robots exist as entirely model-based simulation, and are clearly very different from real physical robots situated in a real, physical environment.

The robots and formal competitions, in which they have participated, have incited impressive innovations in machine vision, path planning, and control. It can, therefore, be assumed that if advanced, physical robots were readily available to more researchers, the rate of this innovation could be increased. Unfortunately, it is still prohibitively expensive for an interested institution to have physical access to this advanced robot hardware, and as a result only a small percentage of research institutions in the United States can explore new concepts and perform practical experiments involving physical collaborative robots. One way to address this problem is to have advanced, physical robots accessible online to the professional robotics community. In other words, a potential solution is to create a collaborative, and perhaps open-source framework, which allows for smaller, less established, or otherwise more restricted institutions to experiment with robots like the ones mentioned above via internet tele-operation. In principle, this is an excellent approach, however the high-stiffness, high-gain nature of the standard research platforms (i.e., ATLAS, PR-2, etc.) necessitates very low-latency, high-frequency control loops with high bandwidths connections, beyond even what the rapidly increasing capabilities of the United States network infrastructure can achieve without highly specialized and cost-prohibitive hardware.

Most optimally, the universal co-robots will be physically very similar to humans. Physical similarity here refers to both appearance and function, for example, range of angles, velocities, force/torque, and passive viscoelastic properties. For example, a biomimetic arm/hand is often considered to be the most optimal physical realization of co-robot appendage because it can handle objects that are made for human use and its physical interaction with the human body can be more "natural" than with other types of robotic appendages.

The finite propagation speed of human neural signals causes significant latencies [40], that is, feedback delays that are multiple times larger than the current average network latencies in the United States. Similarly, the control frequency at which the human muscular system operates is likely not higher than 30 Hz (beta waves maximum frequency) [40]. This suggests that very low-latency communication is not necessary for controlling robots with biomimetic architecture and function. Finally, even if for some reason a network connection failed or another error were to occur, the truly biomimetic robot should be intrinsically safe since it would not be able to exert excessive forces onto its environment. Hence, it may be possible to build a system that is intrinsically safe which can be used as a tele-operable human-robot interaction (HRI) research tool for institutions that have historically not had access to them.

17.2.2 Bioinspired Aerial Robots

17.2.2.1 Aerial Locomotion, Flyers Versus Swimmers

Clearly, there is a huge difference between locomotion in water and air. The air mass density is about 800 times smaller than the water mass density ($\sim$1 g/cm^3) at 10°C. Both birds (0.6–1 g/cm^3) and fish ($\sim$1 g/cm^3) have comparable mass density [40]. Hence, buoyancy can be used efficiently only by fish and not by birds. Birds and insects need to use a lot of muscle work to flap their wings in order to lift their entire mass. This causes a huge difference in mass between swimmers and flyers. The heaviest flying animal today is the Great Bustard. A male is typically 90–105 cm (2 ft 10 in.–3 ft 5 in.) tall, with a length of around 115 cm (3 ft 9 in.) and has a 2.1–2.7 m (6 ft 10 in.–8 ft 10 in.) wingspan. The male can range in weight from 5.8 to 18 kg (13–40 lb), with an average of 13.5 kg (30 lb) [40].

The heaviest flying animal known that ever existed was Pterosaurs, a dinosaur-era reptile (existing between the Triassic and Cretaceous periods). This species is believed to have weighed up to 250 kg (550 lb), measured 7.9 m (26 ft) in total length [including a neck length of over 3 m (10 ft)] and measured up to 11 m (36 ft) across the wings [40]. The heaviest swimming animal that ever existed is the blue whale (*Balaenoptera musculus*) at 30 m (98 ft) in length and 180 metric tons or more in weight. And it still exists!

There are four main bird flying types: (1) parachuting—downward motion, angle with horizontal >45°, (2) gliding—downward motion, angle <45°, (3) flapping—upward with characteristic wings motion, and (4) soaring—upward with steering and energy harvesting. In gliding, the common wisdom is that the air moving over the top of the wing produces less pressure on the wing than the air moving under the wing. The result is same as for the airplane, that is, lift force is produced. By tilting its wings forward or backward, a bird can change its speed while gliding. Soaring animals may alternate glides with periods of soaring in rising air. Because soaring birds do not have to flap their wings very often, soaring flight takes very little energy.

How birds and insects generate lift forces by flapping wings?

17.2.2.2 Aerial Propulsion

Quite generally forces applied onto a streamlined body are modeled as pressure forces and drag forces. They are responsible for both lift forces opposing gravity and for trust forces. Typically in literature the trust force is either perpendicular to gravity or aligned with central body of flyer. And in most of the cases the lift force is perpendicular to the trust force.

Pressure forces are normal to the local surface. According to the classical kinetic theory of gases, pressure forces are equal to the rate of change of air particles' linear momentum normal to the stationary surface in air's co-moving inertial frame (i.e., frame in which air particles have zero group velocity). They are expressed in terms of the pressure, p, and the local surface area, A, as

$$F_{pressure} = pA. \tag{17.1}$$

Drag forces are present if air particles have nonzero group velocity in respect to the local surface co-moving inertial frame. Drag forces are split into pressure drag (or form drag) and skin drag (or friction drag).

Pressure drag is normal to the local surface. Pressure drag is equal to the rate of change of air particles' linear momentum normal to the local surface in a local surface's co-moving inertial frame minus pressure forces. They are

$$F_{drag\perp} = \frac{1}{2} C_D \rho A v_{\perp}^2 \tag{17.2}$$

with phenomenological drag coefficient, C_D, air density, ρ, and normal component of incoming air velocity relative to the local surface, $v_{\perp}$. If linear momentum, defined by air group speed in a local surface co-moving inertial frame, is annulled after collision, then $C_D = 2$ [40]. Skin drag is parallel to the local surface. If viscous forces are omitted, skin drag is equal to the rate of change of air molecules' linear momentum parallel to the local surface in a local surface co-moving inertial frame. They are

$$F_{drag} = \frac{1}{2} C_F \rho A v_{\parallel}^2 \tag{17.3}$$

with phenomenological friction coefficient, C_F, and tangential component of incoming air velocity relative to the local surface, $v_{\parallel}$. Here, the phenomenological friction coefficient is modeled according to the 1/7th power law with calibration such that

$$C_F = 0.0592 \times \mathrm{Re}_{\parallel}^{-\frac{1}{5}} \tag{17.4}$$

The Reynolds number,

$$\mathrm{Re}_{\parallel} = \frac{\rho L_{\parallel} v_{\parallel}}{\mu}, \tag{17.5}$$

is defined by the characteristic body length, $L_{\parallel}$, parallel to the tangential component of velocity, $v_{\parallel}$, and the dynamic viscosity, μ. There are also other analytic formulations of C_F; all depending on the Reynolds number and details of a particular system. Viscous forces are omitted here.

17.2.2.3 Two Simplified Hovering Models

The generation of lift force is illustrated here with two simplified models depicted in Fig. 17.13.

Let us consider benchmark model characterized by pair of infinitesimally thin, massless, and rectangular wings or plates with $L = 3\,\mathrm{m} \times W = 2\,\mathrm{m}$. Still further let us imagine that central body of flyer can be represented with point-like mass, $m = 100\,\mathrm{kg}$.

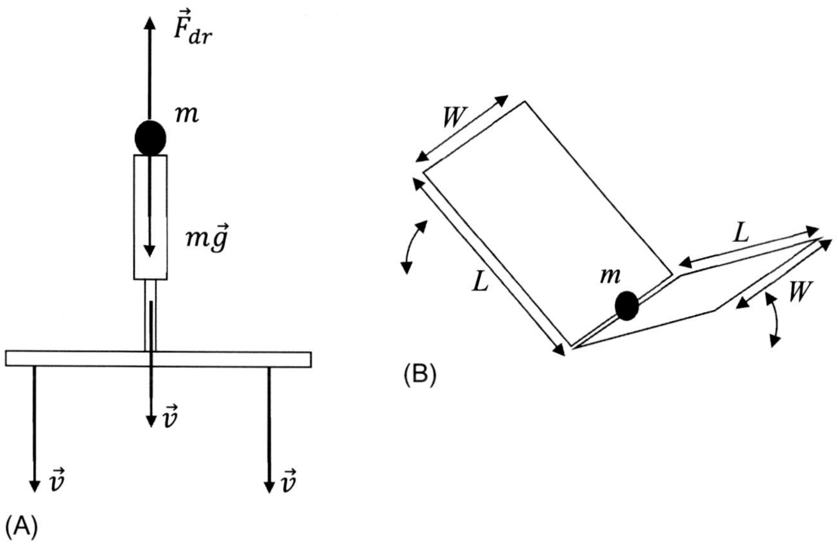

FIG. 17.13 Drag force created by a flat plate with telescoping mechanism (A) and pair of flapping wings (B).

The Reynolds number,

$$Re = \frac{\rho L v}{\mu}, \tag{17.6}$$

is expected to be in the range $2.3 \times 10^5 < Re < 6.7 \times 10^7$, based on the air density $\rho \sim 1.25\,\text{kg/m}^3$, the characteristic length equal to the chosen benchmark model wing or plate length $L = 3\,\text{m}$, the dynamic viscosity $\mu \cong 1.66 \times 10^{-5}\,\text{Pa s}$ at 0°C, and relative speed within the range $1\,\text{m/s} < v < 300\,\text{m/s}$. High Reynolds number typically indicates that the inertial forces are more significant than the viscous forces [40]. Hence, inviscid flow is used here, completely neglecting viscosity compared to inertial forces.

TELESCOPING PLATES

Consider a simple, idealized, nonconventional flying mechanism depicted in Fig. 17.13A. The aircraft's mass is concentrated in a single point and a telescoping mechanism pushes the plate in a direction perpendicular to the plate's surface. This motion generates the drag force equal to

$$F_{dr} = \frac{1}{2}\rho v^2 C_d A, \tag{17.7}$$

with air density, ρ, plate's speed, v, the drag coefficient, C_d, and the area of plate, A.

For N rectangular flat plates (length L and width W), the lift force balancing the weight is

$$F_{lift} = N\frac{1}{2}\rho v^2 C_d' WL = \text{mg}. \tag{17.8}$$

Thus, the required speed and power are

$$v=\sqrt{\frac{2\,\mathrm{mg}}{N\rho C_d WL}},P=Fv=mgv=\mathrm{mg}\sqrt{\frac{2\,\mathrm{mg}}{N\rho C_d WL}}=\sqrt{\frac{2(\mathrm{mg})^3}{N\rho C_d WL}} \tag{17.9}$$

For $\rho=1.25\,\mathrm{kg/m^3}$, $C_d\cong 2.00$, $W=2.00\,\mathrm{m}$, $L=3.00\,\mathrm{m}$, $N=2$, $\mathrm{m}=100\,\mathrm{kg}$ one obtains $v=8.09\,\mathrm{m/s}$, $P=7930\,W=10.63\,\mathrm{hp}$.

This provides an estimate of the minimal power required for any flying mechanism (of the same active surface area and mass) that mechanically pushes the air in order to hover. Here, minimal energy is lost on horizontal air motion, and thus, this is the least power requiring hovering mechanical model.

This is a very similar mechanism to the recoil aquatic locomotion; water, held in a body cavity, is forcibly expelled through an aperture, such as in jellyfish and chambered nautilus [40].

FLAPPING WINGS

Consider a simplified flapping wings system [40–42] with a pair of wings, each with the length $L=3\,\mathrm{m}$ and width $W=2\,\mathrm{m}$, depicted in Fig. 17.13B.

Imagine that this system can vertically takeoff and hover in place. Assume that during the downward stroke, wings swing with a constant angular speed ω between angles θ_0 and $-\theta_0$.

For an angle θ, assuming a negligible vertical velocity, the lift force equals

$$F(\theta)=2\int_0^L \cos\theta\frac{1}{2}\rho(\omega r)^2 C_d W dr=2\cos\theta\frac{1}{2}\rho\omega^2 C_d W\frac{L^3}{3}. \tag{17.10}$$

Given the drag coefficient of $C_d\cong 2$ and air density of $\rho\cong 1.25\,\mathrm{kg/m^3}$, the lift force is then

$$F(\theta)\cong\frac{5}{6}\mathrm{kg/m^3}\;WL^3\omega^2\cos\theta=45\,\omega^2\cos\theta\,\mathrm{kg\,m}. \tag{17.11}$$

Using the small angle approximation, $\cos\theta\cong 1$, the force is $F\cong 45\,\omega^2$ kgm. Therefore, to lift a 100 kg, this system requires $\omega\geq\sqrt{\frac{\mathrm{mg}}{45\,\mathrm{kg\,m}}}=\sqrt{\frac{981N}{45\,\mathrm{kg\,m}}}=4.7\,\mathrm{rad/s}$. If $\theta_0=\pi/6$, it would take approximately 0.23 s to complete a down stroke. The torque that each wing has to exert during the down stroke is

$$\tau=\int_0^L\frac{1}{2}\rho(\omega r)^2 C_d W r dr=\rho\omega^2 C_d W\frac{L^4}{8}\cong 1.1\,\mathrm{kNm} \tag{17.12}$$

which corresponds to the per wing power output of $\tau\omega=5.2\,\mathrm{kW}$, or about 7 hp (hp), which is $P=13.8\,\mathrm{hp}$ in total for both wings. This simplified analysis has not considered the effects of nonzero vertical velocity.

Hence to achieve hovering the two simplified models require power equal to 10.6 hp for pair of telescoping plates and 13.8 hp for pair of flapping wings. The current ordinary motorcycle engines can be rated up to 100 hp and the energy demanding takeoff phase may take only couple of seconds. After takeoff, the system can enter a less energy demanding gliding phase accompanied with less frequent flapping.

Hence, this intelligent flying system may prove to be energy efficient and a fast means of transportation with relatively small energy cost per distance traveled.

To optimize flying efficiency, wings can change their geometry or flap much slower during the upward stroke. System can be also added with "flappers" [40–42], Fig. 17.14A. Wing "flappers" passively open to allow air flow during the up stroke and passively close during the down stroke which partially eliminates unwanted drag forces during the up stroke while mimicking phenomena observed in nature. During the down stroke, feathers of many bird species stay together; whereas, during the up stroke, feathers bend and split to allow air passage [43, 44].

The overall mass of the system could be minimized by utilizing the one-to-many (OTM) principle [45–48] such that energy of a single motor is stored over the longer period of time, and then released to several mechanical DoF over the shorter period of time. A prototype [40–42], utilizing this principle was built and tested in a tethered condition, Fig. 17.14B and C. The motor does not directly engage the wing during the lift generating down stroke. Instead, the motor slowly builds up energy into springs during the slow up stroke and then disengage from the springs. The elastic energy is subsequently released quickly during the fast down stroke. Hence, the power output can be multiple times larger than the power input. With only 1/3 hp motor, the non-tethered, 29 kg ornithopter prototype was able to take off [40–42].

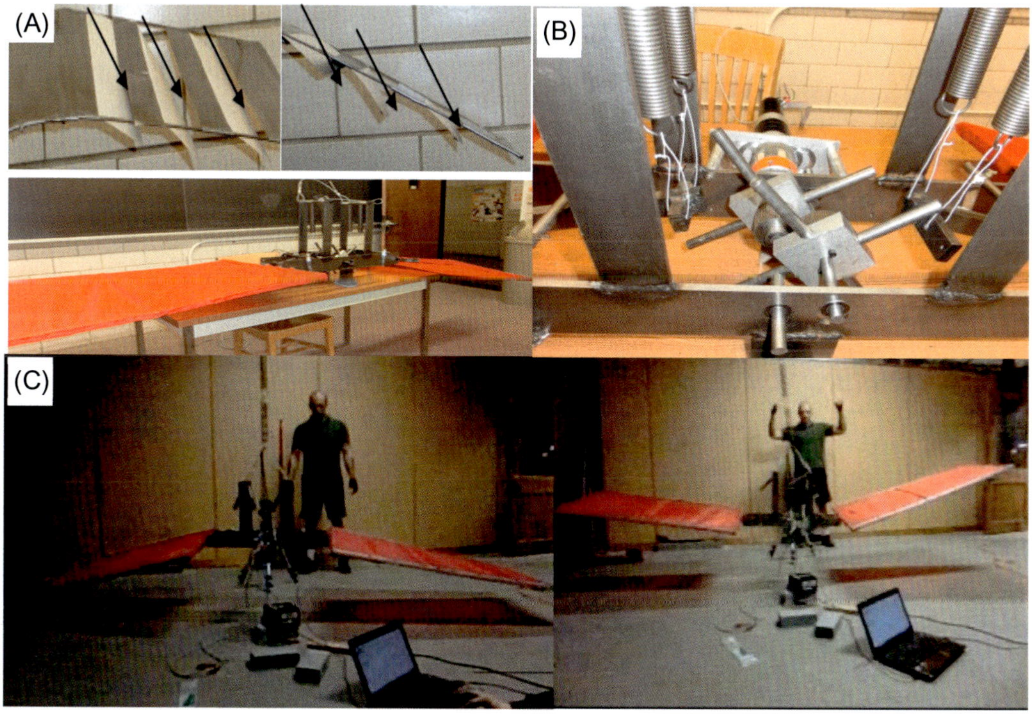

FIG. 17.14 WPI Popovic Lab's Ornithopters: small ornithopter's flappers (A), large ornithopter's elastic energy storing mechanism (B), and tethered tests of a large 29 kg ornithopter (C).

EFFECT OF MOVING AIR

The still air condition is assumed for both previous examples. However, that is clearly not true in the case of flapping wings. One of the authors (MP) is thankful to Russ Tedrake, MIT professor, who first pointed this interesting phenomena to him. As wings are moving upward the air is "sticking" onto wings, that is, there is no formation of "void" beneath the wings and hence air beneath the wings is also moving upward. At some point the wings suddenly change direction and start downward flap while air is still moving up. This simply means that relative wing-air speed is even larger than it is assumed by the previous simplified analysis. This also suggests that less power is required for the flapping wings to hover; though still more than it is required by the telescoping plates. However, details of this dynamics depend closely on fluid dynamics.

One approach to model this phenomenon and estimate its effect is to invoke Navier-Stokes equations and perform numerical simulation. Typically these numerical models have a few phenomenological parameters that can be estimated by comparing theoretical prediction and experimental result. Still further instead of Navier-Stokes-based model one could try to utilize simpler phenomenological aero-elastic type of model to address this problem potentially in real time.

17.2.2.4 Ornithopters

An ornithopter is a device that flies by flapping its wings. The word "ornithopter" (c. 1908) combines the ancient Greek words for "bird" and "wing".

One of the first ornithopters was constructed by Gustave Trouvé in 1870 (Fig. 17.15A) [40]. It used an unusual type of internal combustion engine; 12 gunpowder charges were fired successively into a bourdon tube to flap the wings. The Bourdon pressure gauge uses the principle that a flattened tube tends to change to be straightened or larger circular cross-section when pressurized. Eugene Bourdon patented his gauge in France in 1849. This ornithopter flew 70 m in a demonstration for the French Academy of Sciences.

Modern ornithopters, such as the ones constructed in 2007 by Robert Musters, use actively twisted wings made of foam (Fig. 17.15B) [40]. The appearance of these radio-controlled ornithopters is close to that of a real bird and they are being offered for use in bird control at airports.

Groups at Festo and MIT have also built nice examples of biologically inspired flapping bird robots.

FIG. 17.15 From left to right: Gustave Trouvé's ornithopter (1870), Robert Musters's modern ornithopter with realistic wings made of foam (2007), and Festo's ornithopter (2011).

In 2011, Festo AG announced a radio-controlled ultralight ornithopter with bending wings modeled after the herring gull. Festo's Smartbird (Fig. 17.15C) [49] has a wingspan of 2 m and weighs 450 g. The wings passively bend to exert thrust forces both during up strokes and down strokes. They are similar in function to the bending wings built by Erich von Holst in 1930, except that the wing twisting (not the bending action) is driven by a servo motor in each wing. This allows the amount of wing twist to be adjusted on the fly. SmartBird can take off, fly, and land by itself.

MIT's Phoenix can carry cargo up to 400 g and it was designed for mainly controls research [50, 51]. However, it requires some forward speed or appropriate head wind to fly.

Successful hovering at zero speed was achieved with small hummingbird robots (Fig. 17.16A) [52] and a new generation of insect-like micro aerial vehicles (MAVs) (Fig. 17.16B) [53, 55]. This class of flyers may have viable application in intelligence, surveillance, and reconnaissance missions [56]. Moreover, MAVs are also utilized to study the aerodynamics of biologically inspired flying [57].

In 2013 Festo has unveiled the BionicOpter (Fig. 17.16C), a fully functional robotic dragonfly that can fly forward, backward, hover, and even fly sideways. It can generate thrust in all directions. Flight is controlled by adjusting for wing-flapping speed, amplitude, and twist. The BionicOpter has 13 DoF including nine for the wings and four for head and tail movement. It is approximately 19 in. long with a wingspan of just over 27 in. and it weighs just 175 g.

17.3 Bioinspired Manipulation

17.3.1 Robotic Arms

As discussed in the Chapter 1, the first robot was created by Nikola Tesla, a Serbian-American inventor. The first patent for the industrial robotic arm was filed by George C. Devol in 1954 and awarded in 1961. Devol designed a mechanical arm that could be programmed to repeat precise tasks like grasping and lifting, and the first industrial robot based on it was created by Joseph F. Engelberger and George C. Devol 7 years later [58, 59]. Initially, they could not find an investor who trusted that their vision could indeed revolutionize the industrial world, but that soon changed.

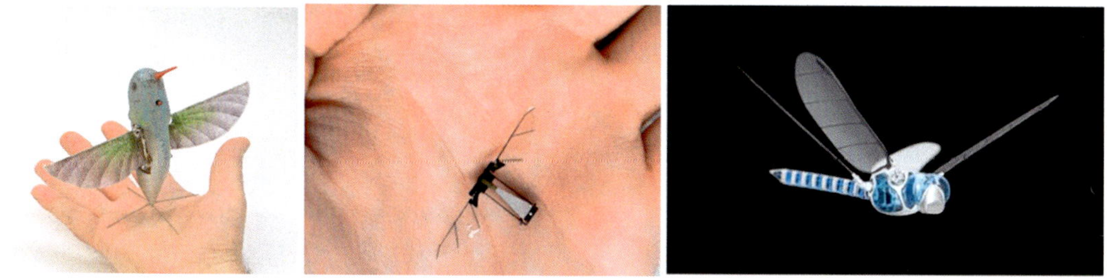

FIG. 17.16 From left to right: small hummingbird robots [52], Rob Wood's insect-like MAV [53], and Festo's dragonfly [54].

FIG. 17.17 Unimate industrial arm. *Photo credit: The Estate of George C. Devol.*

Engelberger formed a company, *Unimation Inc.*, in Danbury, Connecticut that was established for the Unimate industrial arm (Fig. 17.17). In 1961, *General Motors* put the first Unimate arm on an assembly line at the company's plant in Ewing Township, New Jersey, a suburb of Trenton. The device was used to lift and stack die-cast metal parts taken hot from their molds. *Chrysler* and *Ford* soon followed, along with many *Japanese companies*. The *Belgrade Hand* (Fig. 17.18) was developed by Rajko Tomovic and Miodrag Rakic at the University of Belgrade, between 1962 and 1964. It was the first robotic five-fingered hand prosthetic, and it was the first device with a sense of touch. It was anthropomorphic, and the size of a human hand. In 1963, *The Rancho Arm* was developed at Rancho Los Amigos Hospital in Downey, California. It required six joints to simulate human movements. Computer controls were integrated when acquired by a Stanford team in 1963.

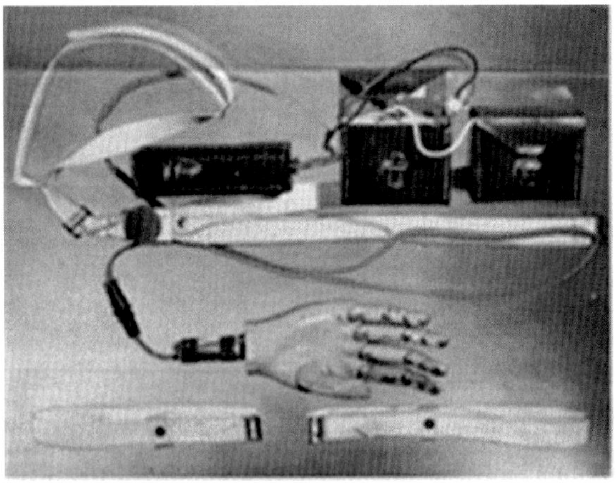

FIG. 17.18 The Belgrade Hand.

Robotics is the branch of Science, Technology and Art that deals with robots, that is, artificial "organisms" that convey life-like appearance by physical actions (movements, etc.) and perform tasks autonomously by using an active sensory-control system, or with guidance, typically by teleoperation, in which case, some of the functions of an active sensory-control system are taken over by the operator [40]. In industry, the term "industrial robot" is often used for reprogrammable machines that can do a variety of operations repeatedly and independently from humans [60]. Robotics manipulators can be found in many applications nowadays, such as in industry, environmental sensing, visual inspections, home and medical assistance, surgical systems, transportation systems, and safety applications (Fig. 17.19).

The automotive industry employs 39% of existing industrial robots, followed by the electronics industry (19%), metal products industry (9%), and, lastly, the plastic and chemicals industry (9%) [62]. As of 2014, China, Japan, the United States, South Korea, and Germany deployed 70% of total industrial robots [63]. Distribution of industrial robots across different US states is addressed in Ref. [62]. One million industrial robots currently in operation have been directly responsible for the creation of close to three million jobs [64].

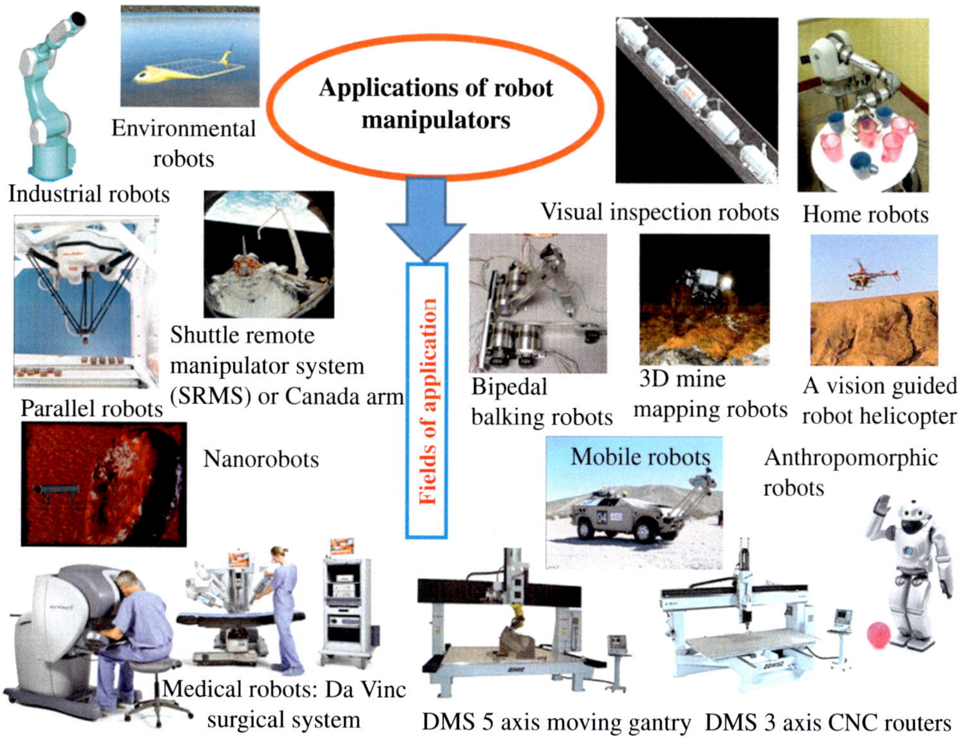

FIG. 17.19 Application of robot manipulators [61].

Industrial robots that are bioinspired manipulators are designed to perform strong, fast, and precise, and to achieve repetitive positions to support common manufacturing operations used in assembly lines [65]. Contemporary industrial robots are not suited to operate in unstructured environments, so have a limited range and complexity of performed tasks [66].

17.3.1.1 Collaborative Robots

Several definitions of cobot or co-robot are present in the literature [67–72]. A cobot or co-robot (from collaborative robot) is a robot intended to physically interact with humans in a shared workspace, aiding humans in performing their work [70]. This is in contrast to old fashioned 20th century industrial robots. Cobots are machines that were designed for direct physical interaction with a human operator; the same principle and idea as defined in Karel Čapek's 1921 book that introduced the use of the word "robot" as artificial machines that interact with humans. Modern collaborative robots used in industrial operations are capable of smooth, strong virtual surfaces, and other haptic effects within a shared human/cobot workspace. Collaborative robots had been applied in prosthetic and orthotic fields before robots were used in close proximity to humans on the manufacturing shop floor.

Assistive Collaborative Robots: The first collaborative robotics were used in assistive applications, with the main purpose to assist humans. Numerous research labs focused on autonomous systems and biomechatronics have been developing assistive and social collaborative robots for decades. Social robots have been used to assist autistic children to detect and recognize human emotions, as their interface is communication with the external world, and for various language building activities as a teaching aide. Other projects focused on using collaborative robots as assistants to the visually impaired, or as stable platforms providing care of the elderly or as assistance for nurses in medical applications.

One example of an assistive robot is the robot for interactive body assistance (RIBA) (Fig. 17.20); a nursing-care assistant robot developed by RIKEN-TRI Collaboration Center for Human-Interactive Robot Research (RTC) in a project led by Dr. Toshiharu Mukai. In 2011, Japanese nursing homes began testing the RIBA.

The RIBA is the first robot that can lift or set down a human from a bed or wheelchair in a superbly gentle way. The RIBA utilizes tactile guidance based on sensory input from many tactile sensors, as the RIBA's arms have almost 500 sensors. Soft urethane foam is used for the robot's skin. The first RIBA prototype could lift up to 135 pounds (61 kg). The robot responds to audio commands, visual input, and touch, and is trained to recognize both faces and voices. From Riken website: "it includes actuator units with a very low gear ratio, allowing the joints to move very quickly and precisely, and allowing backdrivability, meaning that the force encountered by the actuators as they perform their tasks can be quickly fed back into the system, allowing softer movement. The ROBEAR robot, shown in Fig. 17.21, is even more advanced version of RIBA robot.

Collaborative Industrial Robots: A collaborative industrial robot (CoBot) is a robot that can safely and effectively interact with human workers while performing simple industrial tasks [71, 72]. Each mechanical DoF in an industrial robot is typically controlled by a

FIG. 17.20 RIBA nursing-care assistant robot.

separate servo motor that is integrated in the industrial robot arm. These non-intelligent industrial robots are controlled by a high gain position. They have a limited level of joint torque sensing and some safety features. Still, they are yet to be upgraded with touch/ force-sensitive skin. The future of industrial collaborative robotics will focus on adding more safety features. Current commercial industrial collaborative robot examples are given in Fig. 17.22.

The integration of collaborative robots into the industrial setting requires strategies related to the system level, work-cell level, machine level, and worker level [71]. Collaborative industrial robotics are a relatively recent approach in advanced manufacturing, where a CoBot (often marketed by commercial robotics companies as cobot) is capable

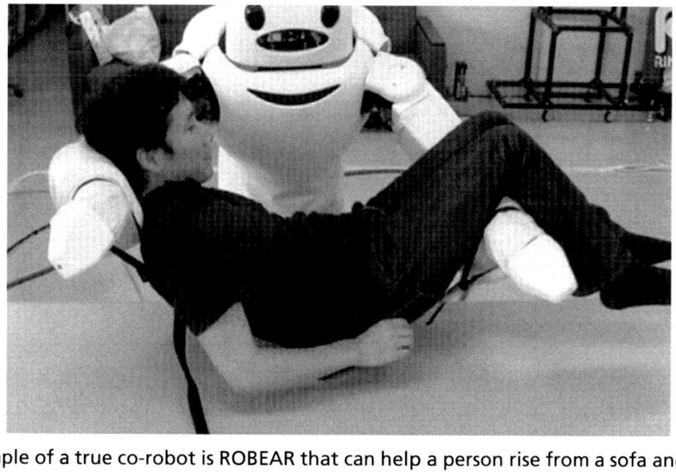

FIG. 17.21 The example of a true co-robot is ROBEAR that can help a person rise from a sofa and sit in a wheelchair.

FIG. 17.22 Examples of various industrial collaborative robots manufactured by different companies that are integrated into manufacturing operations [73–78]; from left to right: ABB Yumi, Fanuc, and Kuka.

of close proximity cooperation with the human worker in order to increase the manufacturing agility and flexibility [72]. Any integration of robots into the work cell poses safety concerns [79], since the main drawback from the introduction of the industrial arm in the 1960s was always workers' safety. These robots are mostly position controlled without much understanding of their environment and physical interaction with humans. Industrial collaborative robots are not 100% safe proof, and there are many areas where they can be made more intelligent and more human friendly.

Various industrial collaborative robot manufacturers are focusing on developing standards that would provide guidelines on safety, such as ISO/TS 15066 guidelines. Based on these guidelines, there are four different types of collaborative operations: (a) safety-rated monitored stop; (b) hand guiding; (c) speed and separation monitoring; and (d) power and force limiting. The safety-rated monitored stop is always on if the human is present in the workspace. The robot only works if the human is out of the work area; the robot and human do not work at the same time. Traditional safety features of regular industrial robots are required. Hand guiding operation is a CoBot that will move based on motion commands of the operator. Most of the industrial collaborative robots currently in use in the industrial setting are ones that are designed to limit power and force. If the robot detects a specific level of power or force, it stops so that it would not harm a human operator. However, this kind of functioning still requires in-depth risk assessment. The human worker and the robot would be able to move in the same workspace, but their movement is not unlimited and it has to adhere to the safe operating procedures, and along the analyzed pathways. The robot's speed and force are limited and cannot be used in applications that require higher payload and speed. These robots are used in conjunction with other safety devices, such as safety scanners. Current research is focused on using capacitive sensors that would measure the distance between a human and a robot, or on the development of more intelligent online motion planning and robots that would adapt to the movements of humans in the workplace [80].

Other suggestions in reducing possible hazards in the workplace are focusing on the use of multiple, smaller cobots, rather than large and heavy robots [79]. Other suggestions include using known robot joint angle values and accurate measurements of human positioning in the workspace coupled by the use of real-time measurements of separation distance, which is defined as a tightly coupled HRI [81]. Safety requirements related to the interaction of industrial robot and humans in a collaborative workspace are defined by the standard ISO 10218-½, published in 2011, and later expanded to ISO/TS 15066, which include additional guidance and more safety methods for HRI [82–84].

CoBots are currently being applied in various industrial settings. For instance, CoBots can be used in additive manufacturing platforms for positioning as a solution to teaching a 'median line'-based travel path in a bead deposition-based system. A designer can use a CoBot to generate path data that would be later imported into process planning software[85].

One example of a commercially available collaborative robot, which can be used in academic and industrial applications, is the Baxter humanoid robot manufactured by Rethink Robotics, a start-up company founded in 2008. Rethink Robotics, founded by Ann Whitaker and Rodney Brooks, launched pair of collaborative machines, Baxter and Sawyer, in 2011 and 2015, respectively; the company unfortunately shut down in October 2018. The Baxter robot is a dual arm, torso-based cobot on a movable pedestal. Each of its joints have series elastic actuators (SEAs), spring couplings between the motor and the link, with a built-in Hall effect sensor to measure deflection from desired trajectory; this deflection then serves as a prediction of torque when multiplied with the known stiffness constant. Baxter collaborative robots have two seven DoF arms, shown in Fig. 17.23, designed to provide kinematic redundancy, better manipulability, and increased safety [86]. Each arm has seven rotational joints and eight links coupled to an interchangeable gripper attached at the end of the arm. This robot has sensors in its hands and around its arms allowing it to detect and adapt to its surroundings. In this way, if a potential collision is detected, a robot would reduce its speed. The motor drives the spring, which drives the arms. The robot has other sensors and camera hands. Following the success of Baxter, Rethink Robotics recently added Sawyer to its collaborative robot line.

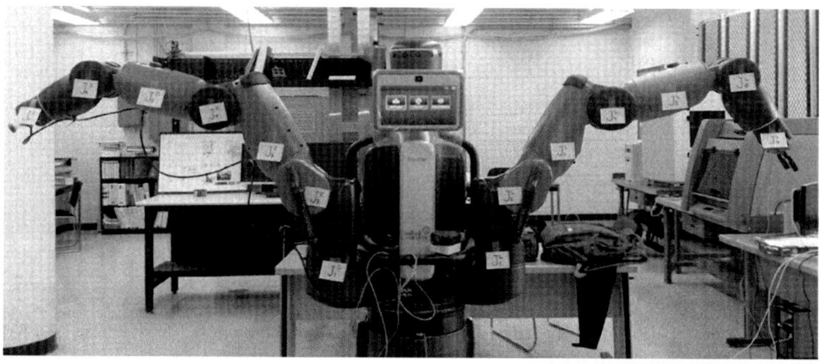

FIG. 17.23 Baxter cobot left and right joints (Rethink Robotics).

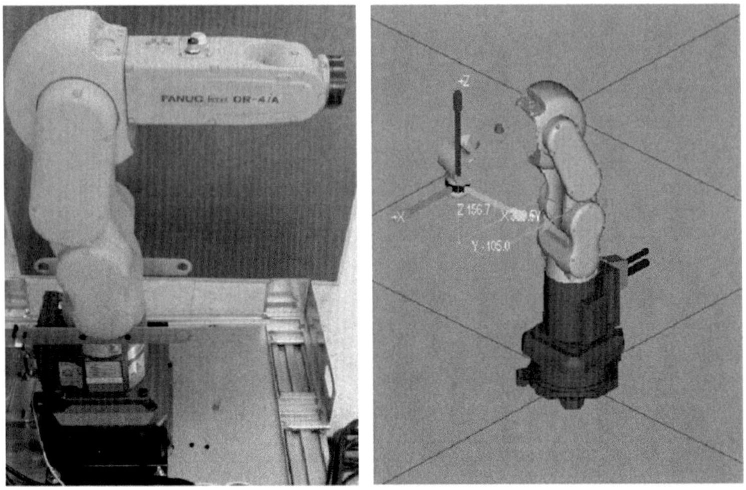

FIG. 17.24 Fanuc CR-4iA and its ROBOGUIDE model.

Another example is the Industrial Collaborative robot Fanuc CR-4iA. The robot, and its ROBOGUIDE model, is shown in Fig. 17.24.

The smallest Fanuc collaborative robot has six axes and a maximum payload of 4 kg. This robot is made for lightweight tasks that are tedious and highly manual. The FANUC Collaborative robots are made to work side by side with humans or to collaborate with them without the need for external safety devices. Six safety functions control the Collaborative robot safety:

(1) CONTACT STOP—FANUC Collaborative robots stop and retract after a collision with a fixed object or person. A safety system ensures that the robot stops at a maximum of 150 N. This force threshold can be reduced if required

(2) PASSIVE SOFT COVER—used for the larger Fanuc CR-35iA robot, which is fitted with a soft cover that eliminates pinch points and sharp edges and reduces contact pressure in case of impact

(3) SAFE STOP FUNCTIONS—Collaborative robots meet ISO 10218 and ISO 13849 Cat. 3 Pld

(4) PUSH BACK FUNCTION—made for the operators to push back the robot's two major axis to free themselves if the robot pose a risk

(5) ANTI-TRAP PROTECTION—for robots that are unable to create gaps of $<$120 mm for arms and 5 mm for fingers, anti-trap protection on the Collaborative robot meets ISO 13857; and

(6) DUAL CHECK SAFETY (DCS)—a safety system that monitors position and speed. It is a proven, and very well-known, technology used on many standard FANUC robots. Third-party safety equipment can be connected to DCS to provide an additional level of safety and can increase productivity.

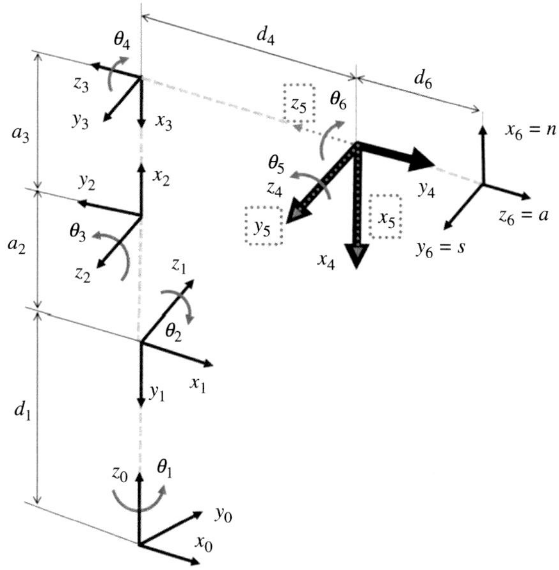

FIG. 17.25 Fanuc CR-4iA kinematic diagram.

Table 17.1 D-H Parameters for the Fanuc CR-4iA Cobot Arm

i	d_i	θ_i	a_i	α_i	Joint Type	Joint Variables
1	d_1	0°	0	−90°	Rotational	θ_1
2	0	−90°	a_2	180°	Rotational	θ_2
3	0	180°	a_3	90°	Rotational	θ_3
4	d_4	0°	0	−90°	Rotational	θ_4
5	0	0°	0	90°	Rotational	θ_5
6	d_6	0°	0	180°	Rotational	θ_6

These D-H parameters have been validated using direct kinematic equations and several Teach pendant pints.

The Fanuc CR-4iA kinematic models and D-H parameters are represented in Fig. 17.25 and Table 17.1.

17.3.2 Anthropomorphic Robotic Hands

As robotic manipulators are inspired by the human arm kinematics, their end effectors may be inspired by the human hand autonomy. Such anthropomorphic hands tend to be much more complicated than a simple two-finger gripper, in turn offering greater versatility for human-like interaction with the environment. In addition, such robotic hands find great use as prosthetic devices in recent years. Although lower limb amputations are more common, considering that currently there are over 500,000 Americans living with transradial (below-the-elbow) limb loss [87], there is a growing need for well-designed upper limb prosthetics to help amputees successfully complete tasks during activities of daily living.

The human hand is a complex structure made up of comprising bones, muscles, tendons, and ligaments. Carpals make up the compound wrist joint, metacarpals create the palm, and proximal, intermediate, and distal phalanges make up each finger. The thumb is the one exception, having only a proximal and distal phalange. Thirty-four muscles within the hand and forearm control the movement of this bony structure and create 22 DoF in total: each finger and the thumb have three flexion and one abduction and two more between the metacarpals of digits four and five to allow palm curling (Fig. 17.26) [88]. By controlling these muscles and finger movements, different prehension (or grasp) patterns are possible. Jaw Chuck, more commonly known as tripod grasp, tip grasps, and lateral grasps are the three most commonly used [89]. In cases when an object needs to be picked up tip grasps will occur 71% of the time and in cases of holding an object for longer-term use tripod grasp is used 88% of the time [89].

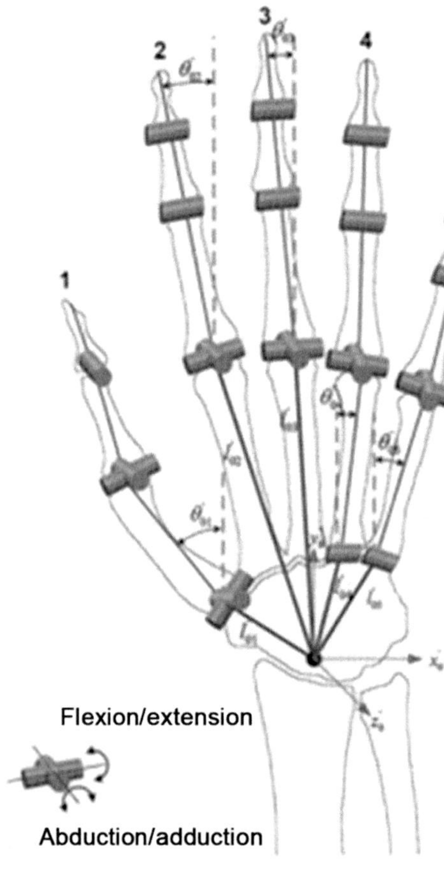

FIG. 17.26 Hand joint diagram and degrees of freedom [88].

Clearly, it is a significant challenge to accurately reproduce the complexity of the human hand. Over the years, there have been a number of noteworthy attempts by robotics researchers and the industry to approximate the hand functions with reduced active DoF.

The *Belgrade Hand* (Fig. 17.18) was developed by Rajko Tomovic (1919–2001), in collaboration with Miodrag Rakic at the University of Belgrade, between 1962 and 1964. It was the first robotic five-fingered hand prostheses and it was the first device with a sense of touch. It was anthropomorphic and it was the size of a human hand. Remarkably, the finger tips were equipped with pressure sensors, and when contact with an object was made, a single actuator began to close all five fingers until the pressure was equalized. This made the hand shape adaptable without effort from wearer.

Recent work in anthropomorphic robotic hands focused further in prosthetics. These devices are advantageous due to their increased grip force and grasping abilities and because they do not require the user to wear a harness or cable since they are battery powered and generally controlled by myoelectric signals, pressures, or strain gauges [90]. These features come at a higher price, are heavier, and make the prosthetic device dependent upon battery life. There are a variety of companies and prosthetic hand designs available. However, the i-limb ultra (Fig. 17.27), the Michelangelo Hand (Fig. 17.28), BeBionic, and Ada Hand are four of the leading prosthetic hands.

Among these, The Ada Hand by OpenBionics is a myoelectric hand, one that represents a more recent direction for prosthetic devices. The Ada Hand is completely open source and made largely from three-dimensional (3D) printed parts. This device can be bought online and assembled by anyone in under an hour using simple tools. Each finger in the hand is driven and movement is controlled by strings in tension that reach throughout the entire finger and connect to the motors in the palm. Five grips with varying weight limits are possible with the Ada Hand: fist grip—9.8 N, palm—49 N, tripod—6.9 N, pinch—3.9 N, and pointing. Although five DoF are possible with this device, compared to the other

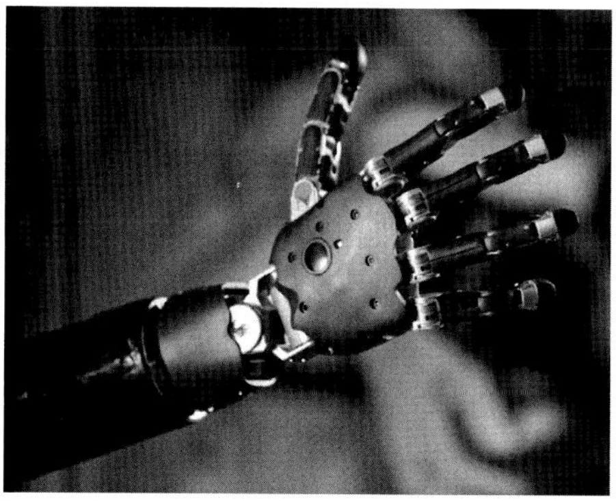

FIG. 17.27 Modern anthropomorphic prosthetic hand: i-limb ultra by Touch Bionics [91].

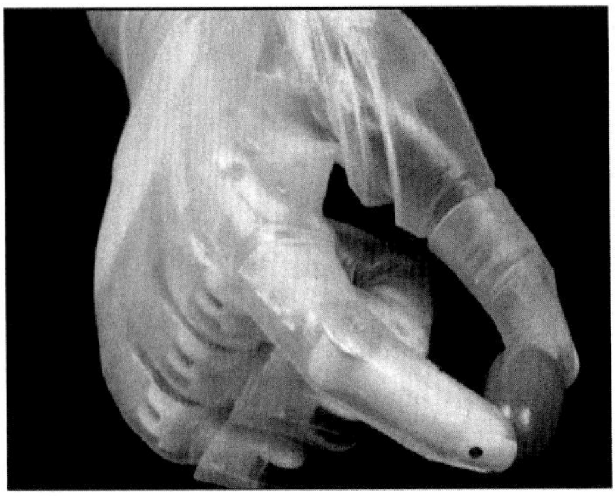

FIG. 17.28 Modern anthropomorphic prosthetic hand: Michelangelo Hand [92].

modern prosthetic hands discussed this simplistic hand lacks qualities and features of a high functioning hand [93].

As a recent contribution, a high-fidelity biomimetic hand, Fig. 17.29, actuated by nine stepper motors packaged within forearm casing was manufactured for <350 USD; it has 18 mechanical DoF, is 38 cm long, weighs 2.2 kg [94, 95]. The hand model has 3D printed replicas of human bones and laser cut tendons and ligaments. The user intent is deduced from EEG and EMG signals, obtained by Neurosky and Myoware commercial sensors, respectively. Three distinct EEG patterns trigger pinch, hook, and point actions. EMG signals are used for finer motor control, for example, strength of grip. A pilot test study on

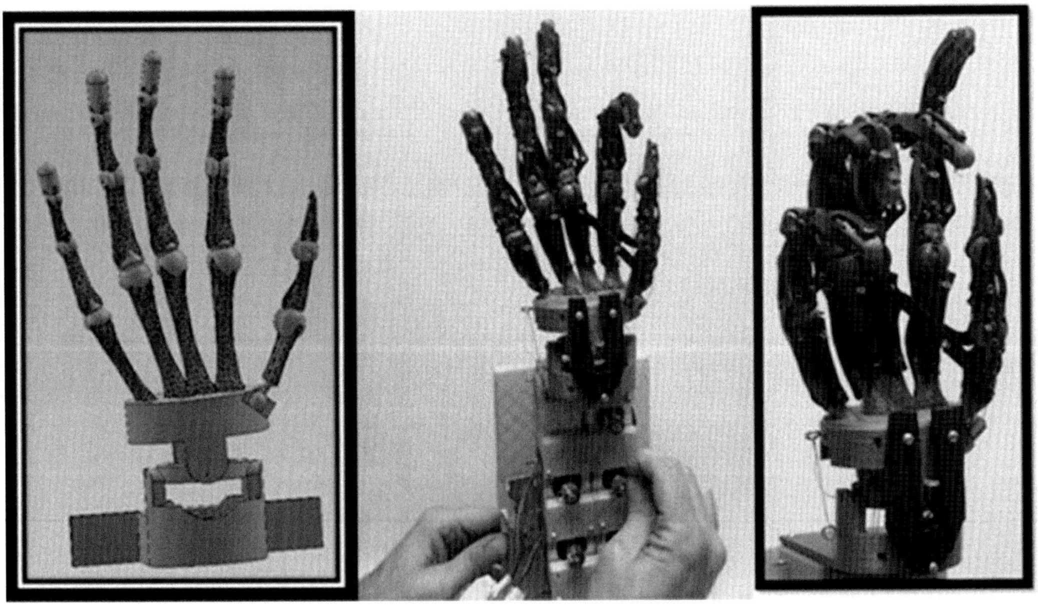

FIG. 17.29 WPI biologically accurate prosthetic hand is designed to follow the same kinematic structure as the human hand using stepper motor driven cable actuation representing the muscles and tendons.

three subjects showed that EMG can actuate the hand with an 80% success rate, while EEG allows for a 68% success rate. The system proved its robustness at the 2017 Cambridge Science Festival, using EEG signals alone. Out of approximately 30 visitors the majority could generate a "peace" sign after 1–2 min.

17.4 Bioinspired Soft-Robotic Systems

Generally, proprioceptive sensing and control are challenging for soft robots. The control hardware and the software are highly dependent on actuation methods utilized for soft robots [65, 96]. Soft-robotic systems typically utilize pneumatic or hydraulic actuation. The control system of most of these robots includes external fluid sources and a network of valves. This hardware tends to be bulky and heavy, many times larger than the robot itself. Therefore, they limit the mobility and portability of these robots. There are also some chemical fluid sources, but they are still in the very early stages of development. The most popular pneumatic actuator is the McKibben muscle discussed in Chapter 3, which is a compliant linear, soft actuator composed of elastomer tubes wrapped in fiber sleeves [97]. When pressurized, the McKibben actuator contracts, causing the motive force. In 1992, H. Tanaka's group developed a flexible microactuator (FMA) driven by an electropneumatic system which could grasp and twist tiny objects [98]. Walker's group developed soft, continuous backbone robot manipulators inspired by cephalopods [99] that are based on both electro-active polymers (EAP) and pneumatic McKibben muscles. In recent years, a new type of modular composition of pure fluidic elastomer actuators (FEA), addressed in Chapter 3, was developed. It is characterized by an extensible and adaptable, low-power soft actuator architecture [100]. G. M. Whiteside's and R. J. Wood's group developed a multigait, soft pneumatic quadruped capable of traversing difficult obstacles [101, 102]; a 3D printed, soft, jumping robot powered by the Combustible fuels [103]. The Octobot is the first robot to be made entirely from soft materials. Powered by a chemical reaction and controlled by a fluidic logic circuit, it heralds a generation of soft robots that might surpass conventional machines [104]. In D. Rus's group, an autonomous, soft-bodied, robotic fish (Fig. 17.30A), hydraulically actuated, and capable of sustained swimming in three dimensions [105, 106]. A soft-robotic manipulation system is capable of autonomous, dynamic, and safe interactions with humans and its environment

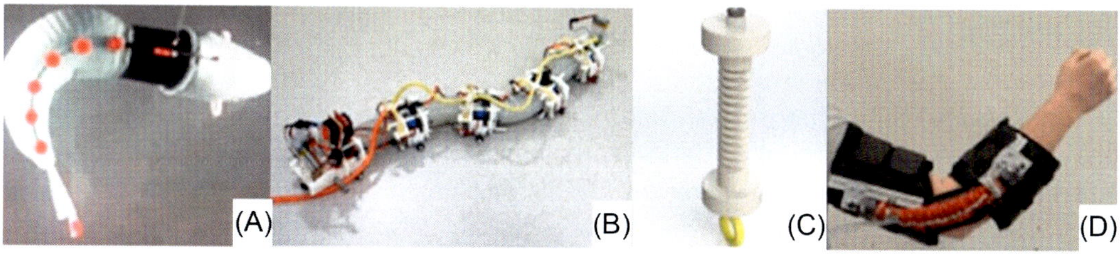

FIG. 17.30 Soft pneumatic or hydraulically robot: (A) soft pneumatic fish robot [105, 106], (B) soft-robotic snake [107–109], (C) reverse PAM [110, 111], and (D) hydro muscle [112].

[113, 114]. In Walsh's group, a benchtop cardiac simulator, and a Direct Cardiac Compression device, employing soft actuators in an elastomeric matrix can reduce the risk of heart failure treatment [115, 116]. A soft-robotic glove embedded with soft actuators is designed for restoring basic hand use for individuals who have peripheral nervous system conditions, including hereditary muscle disorders (e.g., muscular dystrophy), nerve diseases, and amyotrophic lateral sclerosis (ALS, Lou Gehrig's disease) [117, 118]. In Onal's group, a soft-robotic snake uses pneumatic pressure to actuate silicone rubber segments that make up the body of the snake, with a constant-curvature deformation along the length of each segment; resulting in a flexible, safe, and realistic motion, effectively mimicking the gait of a biological snake (Fig. 17.30B) [107–109]. The rPAM has also been invented, which is the combination of the McKibben actuator and the fluidic elastomeric actuators, and is controlled by the 1 Dof and 3 Dof robotic joints actuated by rPAM (Fig. 17.30C) [110, 111]. In Okamura's group, a class of soft pneumatic robot grows substantially in length from the tip while actively controlling its direction using onboard sensing of environmental stimuli [119]. In WPI Popovic Labs, Hydro muscle described in detail in Chapter 3 is currently a top contender for the best artificial muscle in terms of efficiency and strain, see for example, Ref. [112] (Fig. 17.30D) and reference within. It has already been used to actuate prostheses, braces, exoskeletons, exomusculatures, biologically inspired Robo-Dog and kangaroo robot, as well as humanoid robot legs. Hydro muscle generated a lot of practical interest with Lockheed Martin as well as independently with Toyota for an improved vehicle safety project. It generated the largest IP fee in WPI history. The fluid-actuated soft robots can achieve even smoother and more continuously shaped behavior than natural creatures. However, this approach introduces technical issues related to complex modeling of soft materials and challenging control of infinite DoF system that is also affected by various time delays. In addition, the size and weight of the primary fluid mover, e.g. pump, cannot be ignored in these systems, except in the Octobot, which is actuated by a self-contained chemical reaction, as mentioned above.

Soft-robotic systems also utilize mechanical tendons, like the variable length tendon cables pulled by the electrical actuators. The tendon driven soft robots are more compact than the soft pneumatic or hydraulic robots. There are many cable-driven continuum soft manipulators, which have been invented with the inspiration from an elephant's trunk and an octopus's arm because they could grasp any shape of objects by conforming their body shape to the object's surface, which the rigid robotic manipulator could not achieve. Laschi's group created a cable-driven soft arm with silicone rubber and tested it under water [120, 121]. Similarly, Pfeifer's group developed a cable-driven soft manipulator, to aid the Cardiac Ablation application [122].

Currently, the smart material actuation is a growing area of research in the soft-robotic field, because smart material actuation soft robots seem much more compact than other kinds of soft robots. The most representative is the shape memory alloy (SMA), addressed in Chapter 3, which can contract when heated beyond certain temperature and which can subsequently return to original state when cooled down. S. Kim's group developed a meshworm-like, soft mobile robot actuated by nickel titanium (NiTi) coil actuators

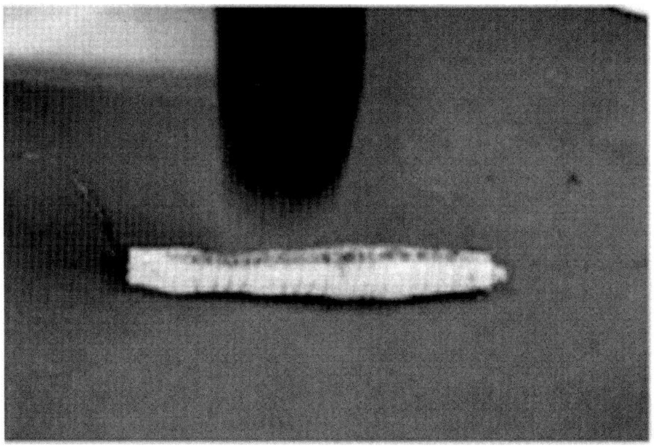

FIG. 17.31 Soft SMA-driven robot. Meshworm Robot [123].

(Fig. 17.31) [123]. In B. Trimmer's group, a soft-bodied robot called GoQBot closely mimics a caterpillar rolling, which was actuated by the SMA coil [124]. C. Laschi's group achieved local deformations of an octopus arm by attaching an SMA coil [125]. The disadvantages of SMA actuation: the cooling down process usually takes more time than the heating process, and the SMA response has hysteresis.

Dielectric elastomer actuators (DEAs), as a kind of electroactive polymer (EAP) actuators, stand out due to their many intriguing advantages: high energy density, fast response, lightweight, and low cost. DEAs have been widely employed for developing bioinspired/biomimetic soft robots [126]. J. Zhu's group developed the frog-inspired swimming robot and Jellyfish robot driven by DEAs [127, 128]. T. Li's group developed a DEA driven, soft fish robot capable of swimming at a speed of 0.4 cm/s (0.69 body length per second) [129]. G. Gu's group developed a DEA, three segment, annelid robot that can reach an average speed of 5.3 mm/s (1.871 body lengths min/3) (Fig. 17.32A) [130]. H.R. Choi's group developed a DEA annelid robot capable of providing up-down translational motion, and two rotational DoF motion (Fig. 17.32B) [131]. Cao et al. [132] is the first trial of a feedback-based controlled DE-based soft robot. Y. Hu's group developed a jumping robot, driven by electricity and sunlight, and based on the Rolled Carbon Nanotube Bilayer Composite.

In the soft-robotic field, sensor technology, which can be used to determine the kinematic and dynamic state of the robotics body, is challenging. The many DoF of the soft body, the stability of the dynamic response, the portability of the soft-robotic body, and the scalability are crucial factors for this technology. Three different methods to achieve this are addressed here, the first of which involves utilizing the changing properties of a material as it changes its shape. Examples of this include the commercially resistive flex sensor and the embedded liquid metals stretch sensor, which offer a simple and compact solution for embedded sensing in soft robotics [133, 134]. Nevertheless, these methods may suffer from dynamic artifacts, such as delayed response and drift [135]. The second method involves using optical fiber Bragg grating [136, 137]. This method can achieve accurate curvature measurements using a thin and flexible optical fiber. However, the

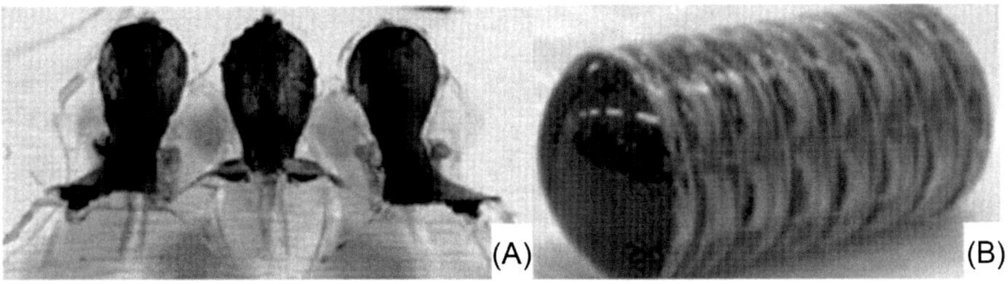

FIG. 17.32 Soft DEAs-driven robot: (A) soft three segments annelid robot [130] and (B) soft annelid robot with 3 DoF [131].

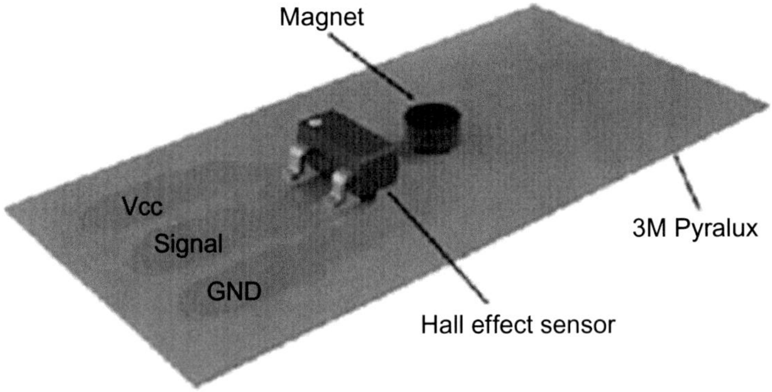

FIG. 17.33 Soft sensors. Magnet and Hall effect sensor [135].

accompanying hardware system, required to parse the raw data, is large and would occupy a significant portion of the soft-robotic body, especially for a small, tether-less mobile robot with many DoF. The last method involves the combination of a magnet and Hall effect sensor, e.g., addressed in Ref. [135] (Fig. 17.33), which measures the changes in magnetic field as body deformations cause the magnet to change position. This method can capture the high-frequency behavior of the soft actuator perfectly, as accurately as the optical method, while being significantly more compact overall. However, the space taken up by the magnet and sensor, both of which need to be inside the body being sensed, limits the miniaturization of this method.

17.5 Algorithmic Bioinspiration

17.5.1 Bioinspired Gaits

Animals and insects use different gaits while walking and running. Eight-legged organisms, such as scorpions and spiders, exhibit several interesting gait types. In the *tetrapod* gait of scorpion.

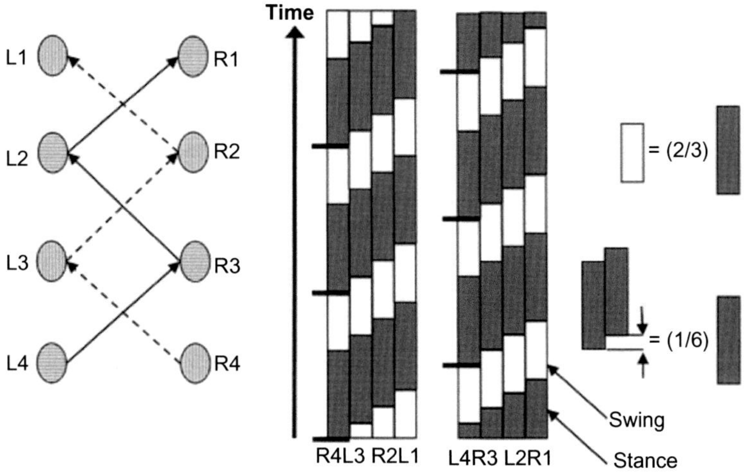

FIG. 17.34 The tetrapod gait of scorpion.

(Fig. 17.34), four legs work together as "tetrapods," but rather than all moving together the four legs in each tetrapod move in a wave-like succession—the two tetrapods working "approximately" 180° out of phase with each other.

The *Arachnid* gait of the hunting spider *Cupiennius salei* (Fig. 17.35) (large adult, 14 cm = 5.6 in.): when walking at 10 cm/s *Cupiennius* uses the 4-2-3-1 leg sequence about 68% of the time, but it also employs a 4-1-3-2 sequence about 15% of the time, and a 4-3-1-2 sequence about 13% of the time.

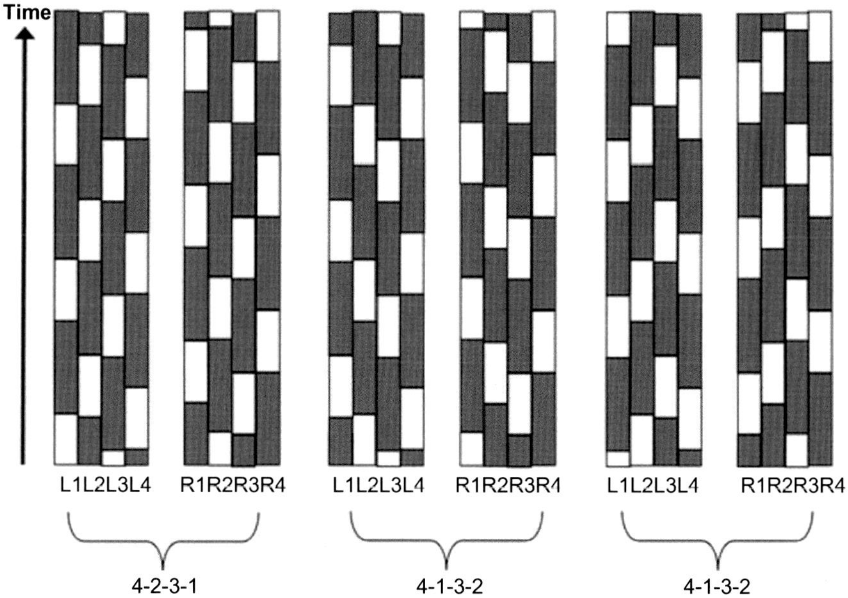

FIG. 17.35 The arachnid gait of the hunting spider.

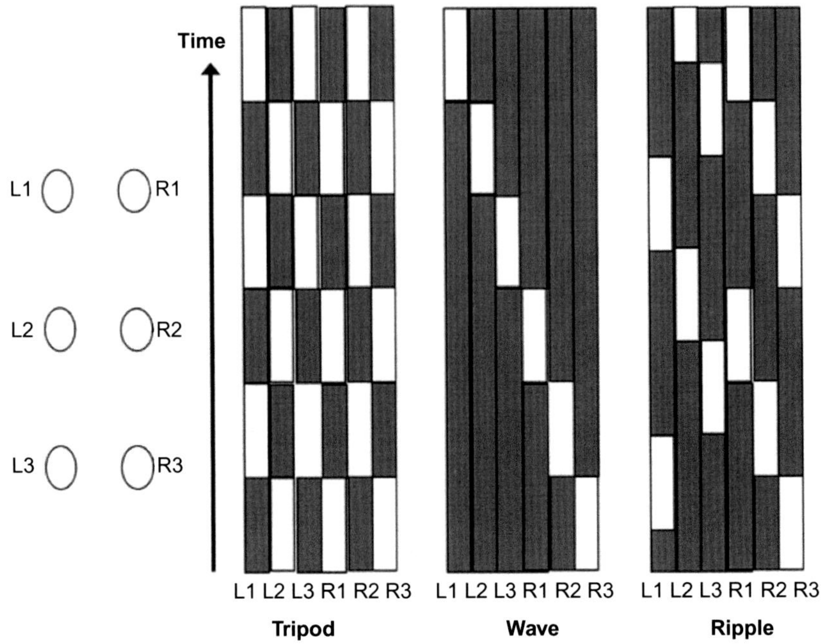

FIG. 17.36 The hexapod gaits: tripod, wave, and ripple.

Six-legged organisms, that is, insects, exhibit several types of Hexapod's gaits (Fig. 17.36). The *tripod gait* is the best-known hexapod gait. A tripod consists of the front-back legs on one side and the middle leg on the opposite side. For each tripod, the legs are lifted, lowered, and moved forwards and backwards in unison. During walking, a hexapod uses its two tripods not unlike a biped stepping from one foot to the other—the weight is simply shifted alternately from one tripod to the other. Since three legs are on the ground at all times, this gait is both "statically" and "dynamically" stable.

In the *wave gait*, all legs on one side are moved forward in succession, starting with the rear-most leg. This is then repeated on the other side. This is a relatively slow gait. In the *ripple gait*, on each side, a local wave comprising nonoverlapping lift phases is being performed, and that the two opposite side waves are exactly 180° out of phase with one another.

Four-legged animals, such as horses and dogs, have their species-specific quadrupedal gaits (Fig. 17.37). The *trot* is a two-beat diagonal gait (e.g., that of a horse) where the diagonal pairs of legs move forward at the same time. The speed of a regular working trot averages from 8 to 12 km/h (5–10 mph), up to 19 km/h (12 mph). Edward Muybridge was the first to prove, by photography in 1872, that there is a "moment of suspension" or "unsupported transit" during the trot gait.

The *canter* is a controlled, three-beat gait that is usually a bit faster than the average trot but slower than the gallop. The average speed of a canter is between 16 and 27 km/h (10 and 17 mph) depending on the length of the stride of the horse. The *gallop* is a fast four-

FIG. 17.37 Different gaits in a horse: (A) the trot, (B) the gallop, (C) the pace, and (D) the walk.

beat gait. It is the fastest gait of the horse, averaging from 25 to 30 mph (40–48 km/h). Like a canter, the horse will strike off with its nonleading hind foot, but the second stage of the canter becomes, in the gallop, the second and third stages, because the inside hind foot hits the ground a split second before the outside front foot. Then, both gaits end with the striking off of the leading leg followed by a moment of suspension when all four legs are off the ground. The *pace* is a lateral two-beat gait. In the pace, the two legs on the same side of the horse move forward together, unlike the trot, where the two legs diagonally opposite from each other move forward together. In both the pace and the trot, at least two legs are always off the ground. The *walk* is a four-beat gait that averages about 4 mph (6.4 km/h). When walking, a horse's legs follow this sequence: left hind leg, left front leg, right hind leg, right front leg, in a regular 1-2-3-4 beat. While walking, the horse will always have one leg raised and the other three legs on the ground.

The "*ambling*" gaits comprise of various footfall patterns and speed. The primary feature of the ambling gaits is that three of the four legs are on the ground at any time, reflected in the colloquial term, "singlefoot". In another four-legged species, specifically dogs, the *single suspension* gallop (Fig. 17.38) is a four-time gait. The dog supports its

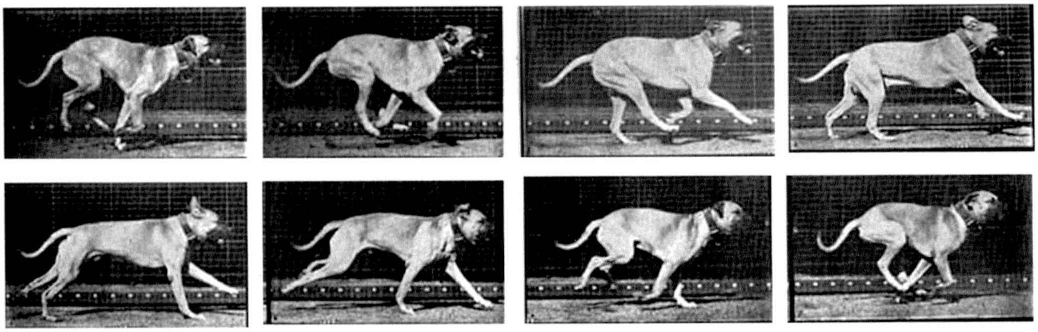

FIG. 17.38 The single suspension gallop.

weight with its feet in the unsymmetrical sequence: RF, LF, RH, and LH (it can happen that the two limbs LF and RH hit the ground simultaneously). Just after taking off from the front left foot, the dog achieves suspension.

The *double suspension* gallop (Fig. 17.39) is also a four-time gait. The dog's weight, however, is not supported by the feet in the sequence of the single suspension gallop. Just after taking off from the LF and just after taking off from the RF, suspension occurs. This is the only gait in which a dog is in full extension.

Elephants cannot trot, jump, or gallop. But they do have two gaits: a walk (at least two legs on the ground) and a faster gait that is similar to running (Fig. 17.40) (but at least one leg stays on the ground).

17.5.2 Central Pattern Generators

Central pattern generators (CPGs) (Fig. 17.41) are biological neural networks that produce rhythmic patterned outputs without sensory feedback. CPGs have been shown to produce rhythmic outputs resembling normal "rhythmic motor pattern production" even in isolation from motor and sensory feedback from limbs and other muscle targets. CPGs are also

FIG. 17.39 The double suspension gallop.

FIG. 17.40 The elephant gait similar to running. *Courtesy of John R. Hutchinson, the Royal Veterinary College, University of London.*

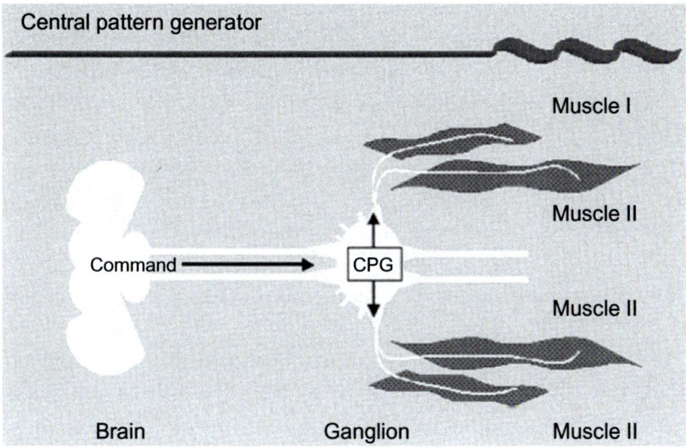

FIG. 17.41 Central pattern generator.

addressed in Chapter 5. To be classified as a rhythmic generator, a CPG requires (1) two or more processes that interact such that each process sequentially increases and decreases, and (2) that, as a result of this interaction, the system repeatedly returns to its starting condition.

The emergent whole body coordinated behavior due to CPG action is often referred to as the motor or motion primitives. Popovic et al. [138] performed one of the first studies of motor primitives that first extracted whole-body motor primitives utilizing the biomechanical analysis of human whole body movements and then used that knowledge for path planning and control of simulated robot. In that study, principal component analysis (PCA) revealed that only three angular momentum primitives explained 99% of the walking data for sagittal plane body rotations resulting in a huge dimensional reduction (original space was 16 dimensional). In addition, analysis showed that the angular momentum primitives are invariant with walking speed. With only the model's walking speed and stride length as an input, their control system searches for joint reference trajectories that minimize the error between the model's angular momentum distribution and the biologically determined distribution. Resulting model joint kinematics are in qualitative agreement with human gait data, suggesting that exploiting invariant angular momentum primitives in humanoid control may prove critical to achieving biological realism in legged robots and prostheses. The angular momentum primitive frameworks can substantially simplify the process of gait synthesis and enable the operator of a humanoid robot or powered leg prosthesis to easily change stride length and/or walking speed.

References

[1] G. Quaglia, D. Maffiodo, W. Franco, S. Appendino, R. Oderio, The Epi.q-1 hybrid mobile robot, Int. J. Robot. Res. 29 (1) (2010) 81–91.

[2] S. Ososky, A. William Evans III, F. Jentsch, From the lab to the field: observations from unmanned system field research and comparisons to laboratory counterparts, J. Manag. Inf. Syst. 7 (3) (2007) 1427–1431.

[3] A. Spröwitz, A. Tuleu, M. Vespignani, M. Ajallooeian, E. Badri, A.J. Ijspeert, Towards dynamic trot gait locomotion: design, control, and experiments with cheetah-cub, a compliant quadruped robot, Int. J. Robot. Res. 32 (8) (2013) 932–950.

[4] N.F. Lepora, P. Verschure, T.J. Prescott, The state of the art in biomimetics, Bioinspir. Biomimet. 8 (1) (2013) 013001.

[5] H.-J. Weidemann, F. Pfeiffer, The six-legged TUM walking robot, in: V. Graefe (Ed.), Intelligent Robots and Systems, Elsevier Science B.V., Amsterdam, 1995, pp. 549–557

[6] J. Ayers, J. Witting, Biomimetic approaches to the control of underwater walking machines, Philos. Trans. R. Soc. Lond. A: Math. Phys. Eng. Sci. 365 (1850) (2007) 273–295.

[7] D.A. Kingsley, R.D. Quinn, R.E. Ritzmann, A cockroach inspired robot with artificial muscles, in: IEEE/RSJ International Conference on Intelligent Robots and Systems, Beijing, 2006, 2006, , pp. 1837–1842.

[8] N.S. Szczecinski, et al., MantisBot: a platform for investigating mantis behavior via real-time neural control, in: S. Wilson, P. Verschure, A. Mura, T. Prescott (Eds.), Biomimetic and Biohybrid Systems. Living Machines 2015. Lecture Notes in Computer Science, vol. 9222, Springer, Cham, 2015.

[9] M.P. Murphy, A. Saunders, C. Moreira, A.A. Rizzi, M. Raibert, The LittleDog robot, Int. J. Robot. Res. 30 (2) (2011) 145–149.

[10] K. Čapek, RUR (Rossum's Universal Robots), Penguin, 2004.

[11] D. Zunt, Who Did Actually Invent the Word "Robot" and What Does it Mean, The Karel Čapek website, 2002.

[12] T. Kanda, H. Ishiguro, M. Imai, T. Ono, Development and evaluation of interactive humanoid robots, Proc. IEEE 92 (11) (2004) 1839–1850.

[13] Reffell, L. H., & Richards, W. H. (1932). Künstlicher Maschinenmensch, Roboter. In M. R. J. Bundesarchiv Bild Ed.), 102-13018, George the humanoid robot from the 1930s was constructed by motor engineer Alan Herbert Reffell and Captain W. H. Richards. Captain and journalist William H. Richards was secretary of the Exhibition of the Society of Model Engineers. The picture shows W. H. Richards and the Robot: Unbekannt, Aktuelle-Bilder-Centrale, Georg Pahl.

[14] ASIMO. (2011). Categories: ASIMO (2nd generation) Robot Hall of Fame, in: A. 4.28.11.jpg (Ed.), (pp. Asimo at a Honda factory. This humanoid robot was created by Honda Motor Co. as a part of the research project Advanced Step in Innovative MObility and was inducted into the Robot Hall of Fame in Pittsburgh in 2004 as the first humanoid droid able to walk dynamically): Wikimedia Commons—User: Vanillase.

[15] Reinecke, D. (2004). Sony Qrio Robot, in S.Q. Robot.jpg (Ed.), (pp. Presentation of the Sony Qrio Robot at the RoboCup 2004): Wikimedia Commons, the free media repository.

[16] SoftbankRobotics, NAO evolution, In: N. E. .jpg, (Ed), *Wikimedia Commons, the Free Media Repository*, 2017

[17] S. Shamsuddin, L.I. Ismail, H. Yussof, N.I. Zahari, S. Bahari, H. Hashim, A. Jaffar, Humanoid robot NAO: Review of control and motion exploration, in: Paper presented at the IEEE International Conference on Control System, Computing and Engineering (ICCSCE), 2011, 2011.

[18] SoftBank, Who Is NAO?, (2016). Available from:https://www.ald.softbankrobotics.com/en/cool-robots/nao (retrieved 14.10.16).

[19] A.D. Ames, E.A. Cousineau, M.J. Powell, Dynamically Stable Bipedal Robotic Walking with NAO Via Human-Inspired Hybrid Zero Dynamics, in: Paper Presented at the Proceedings of the 15th ACM International Conference on Hybrid Systems: Computation and Control, 2012.

[20] N. Kofinas, Forward and Inverse Kinematics for the NAO Humanoid Robot, Technical University of Crete, Greece, 2012.

[21] M. Vukobratović, B. Borovac, Zero-moment point—thirty five years of its life, Int. J. Human. Robot. 1 (01) (2004) 157–173.

[22] M. Vukobratovic, A. Rodic, Contribution to the integrated control of biped locomotion mechanisms, Int. J. Human. Robot. 4 (01) (2007) 49–96.

[23] M. Alemi, A. Meghdari, M. Ghazisaedy, Employing humanoid robots for teaching English language in Iranian junior high-schools. Int. J. Human. Robot. 11 (3) (2014) 1–25, https://doi.org/10.1142/S0219843614500224.

[24] S.i. Nakaoka, M. Morisawa, K. Kaneko, S. Kajita, F. Kanehiro, Development of an indirect-type tele-operation interface for biped humanoid robots, in: Paper Presented at the IEEE/SICE International Symposium on System Integration (SII), 2014, 2014.

[25] DARPA, DARPA Robotics Challenge, Available from:http://www.theroboticschallenge.org/, 2017.

[26] J. Monceaux, J. Becker, C. Boudier, A. Mazel, First steps in emotional expression of the humanoid robot NAO, in: Paper Presented at the Proceedings of the 2009 International Conference on Multimodal Interfaces, 2009.

[27] A. Magassouba, N. Bertin, F. Chaumette, Binaural auditory interaction without HRTF for humanoid robots: a sensor-based control approach, in: Paper Presented at the Workshop on Multimodal Sensor-based Control for HRI and Soft Manipulation, IROS'2016, 2016.

[28] A. Goswami, S.-k. Yun, U. Nagarajan, S.-H. Lee, K. Yin, S. Kalyanakrishnan, Direction-changing fall control of humanoid robots: theory and experiments, Auton. Robot. 36 (3) (2014) 199–223.

[29] A. Polishuk, I. Verner, Y. Klein, E. Inbar, R. Mir, I. Wertheim, The challenge of robotics education in science museums, in: The 4th Knowledge Cities World Summit, 2011, p. 319.

[30] O. Mubin, C.J. Stevens, S. Shahid, A. Al Mahmud, J.-J. Dong, A review of the applicability of robots in education, J. Technol. Educ. Learn. 1 (2013). 209-0015.

[31] H. Kose, R. Yorganci, E.H. Algan, D.S. Syrdal, Evaluation of the robot assisted sign language tutoring using video-based studies, Int. J. Soc. Robot. 4 (3) (2012) 273–283, https://doi.org/10.1007/s12369-012-0142-2.

[32] D. Hood, S. Lemaignan, P. Dillenbourg, When children teach a robot to write: An autonomous teachable humanoid which uses simulated handwriting, in: Paper Presented at the Proceedings of the Tenth Annual ACM/IEEE International Conference on Human-Robot Interaction, 2015.

[33] M. Fridin, Storytelling by a kindergarten social assistive robot: a tool for constructive learning in pre-school education, Comput. Educ. 70 (2014) 53–64.

[34] G. Keren, A. Ben-David, M. Fridin, Kindergarten assistive robotics (KAR) as a tool for spatial cognition development in pre-school education, in: Paper Presented at the 2012 IEEE/RSJ International Conference on Intelligent Robots and Systems, 2012.

[35] A. Melchior, F. Cohen, T. Cutter, T. Leavitt, N.H. Manchester, More Than Robots: An Evaluation of the First Robotics Competition Participant and Institutional Impacts, Heller School for Social Policy and Management, Brandeis University, 2005.

[36] V. Jovanovic, Hampton Roads Section Challenge IEEE SoutheastCon 2016, Norfolk, VA, 2016.

[37] G. Pratt, J. Manzo, The DARPA robotics challenge [competitions], IEEE Robot. Autom. Mag. 20 (2) (2013) 10–12.

[38] S. Agarwal, M. Popovic, Study of toe joints to enhance locomotion of humanoid robots, 2018 IEEE-RAS 18th International Conference on Humanoid Robots (Humanoids 2018), November 6–9, Beijing. China, 2018..

[39] N.A. Radford, P. Strawser, K. Hambuchen, J.S. Mehling, W.K. Verdeyen, A.S. Donnan, et al., Valkyrie: Nasa's first bipedal humanoid robot, J. Field Robot. 32 (3) (2015) 397–419.

[40] M.B. Popović, Biomechanics and Robotics, CRC Press, Pan Stanford Publishing Pte. Ltd., Singapore, 2013

[41] N. Deisadze, W.C. Jo, B.R. Seo, S. Nestinger, M.B. Popovic, Toward Biologically Inspired Human-Carrying Ornithopter Robot Capable of Hover, WPI, April 29, 2013. (MQP Report), http://www.wpi.edu/Pubs/E-project/Available/E-project-042613-114244/unrestricted/Wing_MQP_Report-2013-04-27_final.pdf, 2013 (retrieved 25.12.13).

[42] M.B. Popovic, Designing large ornithopter for human transport via ExoMusculature and sensitive robotics, in: Presented at Robotics Science and Systems (RSS) 2013, International Conference, Berlin, Germany, June 28, 2013.

[43] V. Tucker, Gliding birds—reduction of induced drag by wing tip slots between the primary feathers, J. Exp. Biol. 180 (1993) 285–310.

[44] H. Eder, W. Fiedler, X. Pascoe, Air-permeable hole-pattern and nose-droop control improve aerodynamic performance of primary feathers. J. Compar. Physiol. A: Neuroethol. Sens. Neural Behav. Physiol. 197 (1) (2011) 109–117, https://doi.org/10.1007/s00359-010-0592-7.

[45] Hunt, T.R., Berthelette, C.J., Iannacchione, G.S., Koehler, S., and Popovic, M.B., "Soft robotics variable stiffness exo-musculature, one-to-many concept, and advanced clutches", Invited Talk at the IEEE International Conference on Robotics and Automation 2012 WORKSHOP: Variable Stiffness Actuators Moving the Robots of Tomorrow, May 14, 2012. Presentation available at: http://www.ce.utwente.nl/car/ICRA2012/workshop.html (retrieved 25.12.13).

[46] T.R. Hunt, C.J. Berthelette, M.B. Popovic, Linear one-to-many (OTM) system: Many degrees of freedom independently actuated by one electric motor, in: Presented at the 5th Annual IEEE International Conference on Technologies for Practical Robot Applications (TePRA), Greater Boston Area, Massachusetts, USA, April 22–23, 2013.

[47] C.J. Berthelette, M. DiPinto, J.D. Sareault, Rotary "One-To-Many" (OTM) novel actuator, in: (Prof. M.B. Popovic advisor), The Winner of the 2013 WPI Mechanical Engineering Department Best MQP in Robotics Engineering Award. WPI, April 25, 2013, 2013 www.wpi.edu/Pubs/E-project/Available/E-project-042513-153455/unrestricted/Rotary (One-to-Many OTM Novel Actuator Final Report.pdf, retrieved 25.12.13).

[48] M. B. Popovic, C. J. Berthelette, M. DiPinto, J. D. Sareault, and T. R. Hunt "Actuation Systems and Methods" United States Patent and Trademark Office, Assignee Worcester Polytechnic Institute. Serial No.: 61/844,604. Filed: July 10, 2013. US Patent 20,160,138,690, 2016 (publication date 19.05.16)

[49] Send, W., Fischer, M., Jebens, K., Mugrauer, R., Nagarathinam, A., & Scharstein, F. (2012). Artificial hinged-wing bird with active torsion and partially linear kinematics. 28th International Congress of the Aeronautical Sciences, Göttingen, Germany 2012.

[50] R. Tedrake, Z. Jackowski, R. Cory, J.W. Roberts, W. Hoburg, Learning to fly like a bird, in: 14th International Symposium on Robotics Research. Lucerne, Switzerland, 2009.

[51] Subbaraman, N. (2009). Robotic Bird Flies Over MIT | Scope. Scope | The Publication of the Graduate Program in Science Writing at MIT. Available from: http://scopeweb.mit.edu/?p=671 (retrieved 28.03.13).

[52] L. Edwards, Robot hummingbird passes flight tests, in: Phys. Org., 2011 https://phys.org/news/2011-02-robot-hummingbird-flight-video.html (accessed 03.09.18).

[53] R.J. Wood, Liftoff of a 60mg flapping-wing MAV, IEEE/RSJ IROS, San Diego, CA, October, 2007.

[54] B. Yirka, Festo builds BionicOpter—fully functional robot dragonfly, Phys. Org. 1 2013. https://phys.org/news/2013-04-festo-bionicopterfully-functional-robot-dragonfly.html (accessed 03.09.18).

[55] W. Shyy, M. Berg, D. Ljungqvist, Flapping and flexible wings for biological and micro air vehicles. Prog. Aerosp. Sci. 35 (5) (1999) 455–505, https://doi.org/10.1016/S0376-0421(98)00016-5.

[56] S.S. Baek, F.L.G. Bermudez, R.S. Fearing, Flight control for target seeking by 13 gram ornithopter, in: IEEE/RSJ International Conference on Intelligent Robots and Systems (IROS), 2011, IEEE, 2011, pp. 2674–2681.

[57] A.A. Paranjape, M.R. Dorothy, S. Chung, K.D. Lee, A flight mechanics-centric review of bird-scale flapping flight, Int. J. Aeronaut. Space Sci. 13 (3) (2012) 267–281.

[58] Devol, G. (1954). Patent 2,988,237. Programmed Article Transfer. Filed December, 10.

[59] E. Mandfield, The diffusion of industrial robots in Japan and the United States, Res. Policy 18 (4) (1989) 183–192.

[60] L. Nocks, The Robot: The Life Story of a Technology, Greenwood Publishing Group, 2007.

[61] A. Djuric, V.M. Jovanovic, T. Goris, Preparing students for the advanced manufacturing environment through robotics, mechatronics, and automation training, in: Paper Presented at the 2015 ASEE Annual Conference & Exposition, Seattle, WA, 2015.

[62] D. Acemoğlu, P. Restrepo, Robots and Jobs: Evidence From the US, 2017.

[63] M. Hagele, Robots conquer the world [turning point], IEEE Robot. Autom. Mag. 23 (1) (2016) 118–120.

[64] P. Gorle, A. Clive, Positive Impact of Industrial Robots on Employment, International Federation of Robotics, METRA MARTECH Limited, 2013.

[65] S. Kim, C. Laschi, B. Trimmer, Soft robotics: a bioinspired evolution in robotics, Trends Biotechnol. 31 (5) (2013) 287–294.

[66] E. Mattar, A survey of bio-inspired robotics hands implementation: new directions in dexterous manipulation, Robot. Auton. Syst. 61 (5) (2013) 517–544.

[67] J.E. Colgate, W. Wannasuphoprasit, M.A. Peshkin, Cobots: robots for collaboration with human operators, in: Paper Presented at the Proceedings of the 1996 ASME International Mechanical Engineering Congress and Exposition, 1996.

[68] M.A. Peshkin, J.E. Colgate, W. Wannasuphoprasit, C.A. Moore, R.B. Gillespie, P. Akella, Cobot architecture, IEEE Trans. Robot. Autom. 17 (4) (2001) 377–390.

[69] J.E. Colgate, M.A. Peshkin, Cobots: Google Patents, 1999.

[70] M.J. Timms, Letting artificial intelligence in education out of the box: educational cobots and smart classrooms, Int. J. Artif. Intell. Educ. 26 (2) (2016) 701–712.

[71] A. Djuric, J. Urbanic, J.L. Rickli, A framework for collaborative robot (CoBot) integration in advanced manufacturing systems, SAE Int. J. Mater. Manuf. 9 (2016-01-0337) (2016) 457–464.

[72] A.R. Sadik, B. Urban, An ontology-based approach to enable knowledge representation and reasoning in worker–cobot agile manufacturing, Future Internet 9 (4) (2017) 90.

[73] ABB, YuMi®—Creating an Automated Future Together, You and Me, ABB Robotics, Available from: http://new.abb.com/products/robotics/industrial-robots/yumi, 2018 (retrieved 05.03.18).

[74] Bosch, APAS, Humans and Machines Hand in Hand, Robert Bosch Manufacturing Solutions GmbH, 2017. Available from:https://www.bosch-apas.com/en/home/ (retrieved 05.03.18).

[75] KUKA, Cobots in the Industry, KUKA Systems GmbH, Augsburg, Germany, 2018. Available from: https://www.kuka.com/en-us/technologies/industrie-4-0/industrie-4-0-cobots-in-industry (retrieved 05.03.18).

[76] Fanuc, Collaborative Robot, CR-35iA, Available from:https://www.fanucamerica.com/products/robots/productsbyseries/default.aspx?seriesId=25&robotseries=Collaborative%20Robot, 2018 (retrieved 05.03.18).

[77] Mabi, MABI Speedy 6 Cobot, Human-Robot Collaboration (HRC), Mabi-Robotics, 2018. Available from:http://mabi-robotic.com/en/products/mabi-speedy-6/ (retrieved 05.03.18).

[78] Universal Robots, Collaborative Robots From Universal Robots, Available from:https://www.universal-robots.com/products/, 2018 (retrieved 05.03.18).

[79] Å. Fast-Berglund, F. Palmkvist, P. Nyqvist, S. Ekered, M. Åkerman, Evaluating cobots for final assembly, Proc. CIRP 44 (2016) 175–180.

[80] S. Beak, Demystifying Cobot Safety: The Four Types of Collaborative Operation, Universal Robots, 2016. Available from:https://blog.universal-robots.com/demystifying-cobot-safety-the-four-types-of-collaborative-operation (retrieved 05.03.18).

[81] P.A. Lasota, G.F. Rossano, J.A. Shah, Toward safe close-proximity human-robot interaction with standard industrial robots, in: Paper Presented at the IEEE International Conference on Automation Science and Engineering (CASE), 2014, 2014.

[82] M. Bdiwi, M. Pfeifer, A. Sterzing, A new strategy for ensuring human safety during various levels of interaction with industrial robots, in: CIRP Annals-Manufacturing Technology, 2017.

[83] ISO, ISO 10218-1/2:2011—Robots and Robotic Devices—Safety Requirements for Industrial Robots—Part 1: Robots/Part 2: Robot Systems and Integration, 2011.

[84] ISO, ISO/TS 15066:2016—Robots and Robotic Devices—Collaborative Robots, 2016.

[85] A. Djuric, J. Urbanic, Using collaborative robots to assist with travel path development for material deposition based additive manufacturing processes, Comput.-Aided Design Appl. (2018) 1–14.

[86] Djuric, A., Rickli, J. L., Jovanovic, V. M., & Foster, D. (2017). Hands-on learning environment and educational curriculum on collaborative robotics Paper Presented at the 2017 ASEE Annual Conference & Exposition, Columbus, OH.

[87] K. Ziegler-Graham, E.J. MacKenzie, P.L. Ephriam, T.G. Travison, R. Brookmeyer, Estimating the prevalence of limb loss in the United States: 2005 to 2050, Arch. Phys. Med. Rehabil. 89 (3) (2008) 422–429, https://doi.org/10.1016/j.apmr.2007.11.005.

[88] S. Casley, T. Choopojcharoen, A. Jardin, D. Ozgoren, IRIS HAND: Smart Robotics Prosthesis, Worcester Polytechnic Institute, 2014, April.

[89] J. Billock, Upper limb prosthetic terminal devices: hand versus hooks, Clin. Prosth. Orthotics 10 (2) (1986) 57–65.

[90] R. Bowers, Prosthetic Devices for Upper-Extremity Amputees, Amputee Coalition of America, 2014.

[91] I-limb ultra Clinician Manual Issue 2. (2013). Available from: Touch Bionics http://www.touchbionics.com/sites/default/files/i-limb_ultra_clinician_manual_issue_2.pdf (retrieved 11.10.16).

[92] M. Brochure. Available from: ottobock http://www.ottobockus.com/media/local-media/prosthetics/upper-limb/michelangelo/files/michelangelo-brochure.pdf (retrieved 11.10.16).

[93] Ada V1.0-Datasheet. (2016). Available from OpenBionics www.openbionics.com (retrieved 11.10.16).

[94] E. Saint-Elme, M.A. Larrier Jr., C. Kracinovich, D. Renshaw, K. Troy, M. Popovic, Design of a biologically accurate prosthetic hand, Extended Abstracts Proceedings of the International Symposium on Wearable Robotics (WeRob 2017), Houston, Texas (USA), November 5–8, 2017.

[95] Brain Controlled Accurate Prosthetic Hand, Popovic Labs, Available from: http://users.wpi.edu/~mpopovic/pages/AccurateProstheticHand.html (accessed 17.03.18).

[96] D. Rus, M.T. Tolley, Design, fabrication and control of soft robots, Nature 521 (7553) (2015) 467.

[97] H.F. Schulte, The characteristics of the mckibben artificial muscle, in: The Application of External Power in Prosthetics and Orthotics, Washington, DC: Nat. Acad. Sci.-Nat. Res. Council, 1961.

[98] Suzumori, K., Iikura., S. & Tanaka, H. (1992). Applying a flexible microactuator to robotic mechanisms. IEEE Control. Syst., 12(1), pp. 21–27.

[99] I.D. Walker, D.M. Dawson, T. Flash, F.W. Grasso, R.T. Hanlon, B. Hochner, W.M. Kier, C.C. Pagano, C.D. Rahn, Q.M. Zhang, Continuum robot arms inspired by cephalopods, in: Unmanned Ground Vehicle Technology, vol. 5804, 2005, pp. 303–315 International Society for Optics and Photonics.

[100] C.D. Onal, System-level challenges in pressure-operated soft robotics, in: Micro- and Nanotechnology Sensors, Systems, and Applications VIII, vol. 9836, 2016, p. 983627. International Society for Optics and Photonics.

[101] R.F. Shepherd, F. Ilievski, W. Choi, S.A. Morin, A.A. Stokes, A.D. Mazzeo, X. Chen, M. Wang, G.M. Whitesides, Multigait soft robot, Proc. Natl. Acad. Sci. 108 (51) (2011) 20400–20403.

[102] S.A. Morin, R.F. Shepherd, S.W. Kwok, A.A. Stokes, A. Nemiroski, G.M. Whitesides, Camouflage and display for soft machines, Science 337 (6096) (2012) 828–832.

[103] N.W. Bartlett, M.T. Tolley, J.T. Overvelde, J.C. Weaver, B. Mosadegh, K. Bertoldi, G.M. Whitesides, R.J. Wood, A 3D-printed, functionally graded soft robot powered by combustion, Science 349 (6244) (2015) 161–165.

[104] M. Wehner, R.L. Truby, D.J. Fitzgerald, B. Mosadegh, G.M. Whitesides, J.A. Lewis, R.J. Wood, An integrated design and fabrication strategy for entirely soft, *autonomous robots*, Nature 536 (7617) (2016) 451.

[105] A.D. Marchese, C.D. Onal, D. Rus, Towards a self-contained soft robotic fish: on-board pressure generation and embedded electro-permanent magnet valves, in: Experimental Robotics, Springer, Heidelberg, 2013, pp. 41–54.

[106] A.D. Marchese, C.D. Onal, D. Rus, Autonomous soft robotic fish capable of escape maneuvers using fluidic elastomer actuators, Soft Robot. 1 (1) (2014) 75–87.

[107] C.D. Onal, D. Rus, Autonomous undulatory serpentine locomotion utilizing body dynamics of a fluidic soft robot, Bioinspir. Biomim. 8 (2) 2013, 026003.

[108] M. Luo, M. Agheli, C.D. Onal, Theoretical modeling and experimental analysis of a pressure-operated soft robotic snake, Soft Robot. 1 (2) (2014) 136–146.

[109] M. Luo, Y. Pan, E.H. Skorina, W. Tao, F. Chen, S. Ozel, C.D. Onal, Slithering towards autonomy: a self-contained soft robotic snake platform with integrated curvature sensing, Bioinspir. Biomimet. 10 (5) 2015, 055001.

[110] E.H. Skorina, M. Luo, S. Ozel, F. Chen, W. Tao, C.D. Onal, Feedforward augmented sliding mode motion control of antagonistic soft pneumatic actuators, in: IEEE International Conference on Robotics and Automation (ICRA), 2015, 2015, pp. 2544–2549.

[111] E.H. Skorina, W. Tao, F. Chen, M. Luo, C.D. Onal, Motion control of a soft-actuated modular manipulator, in: IEEE International Conference on Robotics and Automation (ICRA), 2016, 2016, pp. 4997–5002.

[112] S. Sridar, C.J. Majeika, P. Schaffer, M. Bowers, S. Ueda, A.J. Barth, J.L. Sorrells, J.T. Wu, T.R. Hunt, M. Popovic, Hydro Muscle—a novel soft fluidic actuator, in: IEEE International Conference on Robotics and Automation (ICRA), 2016, 2016, pp. 4014–4021.

[113] A.D. Marchese, R. Tedrake, D. Rus, Dynamics and trajectory optimization for a soft spatial fluidic elastomer manipulator, Int. J. Robot. Res. 35 (8) (2016) 1000–1019.

[114] A.D. Marchese, K. Komorowski, C.D. Onal, D. Rus, Design and control of a soft and continuously deformable 2d robotic manipulation system, in: IEEE International Conference on Robotics and Automation (ICRA), 2014, 2014, pp. 2189–2196.

[115] Roche, E.T., Horvath, M.A., Alazmani, A., Galloway, K.C., Vasilyev, N.V., Mooney, D.J., Pigula, F.A., & Walsh, C.J. (2015). Design and fabrication of a soft robotic direct cardiac compression device. In ASME 2015 International Design Engineering Technical Conferences and Computers and Information in Engineering Conference. American Society of Mechanical Engineers, pp. V05AT08A042–V05AT08A042.

[116] E.T. Roche, M.A. Horvath, I. Wamala, A. Alazmani, S.E. Song, W. Whyte, Z. Machaidze, C.J. Payne, J.C. Weaver, G. Fishbein, J. Kuebler, Soft robotic sleeve supports heart function, Sci. Transl. Med. 9 (373) (2017).

[117] P. Polygerinos, S. Lyne, Z. Wang, L.F. Nicolini, B. Mosadegh, G.M. Whitesides, C.J. Walsh, Towards a soft pneumatic glove for hand rehabilitation, in: IEEE/RSJ International Conference on Intelligent Robots and Systems (IROS), 2013, 2013, pp. 1512–1517.

[118] P. Polygerinos, K.C. Galloway, E. Savage, M. Herman, K. O'Donnell, C.J. Walsh, Soft robotic glove for hand rehabilitation and task specific training, in: IEEE International Conference on Robotics and Automation (ICRA), 2015, 2015, pp. 2913–2919.

[119] E.W. Hawkes, L.H. Blumenschein, J.D. Greer, A.M. Okamura, A soft robot that navigates its environment through growth, Sci. Robot. 2 (8) 2017, eaan3028.

[120] F. Renda, M. Giorelli, M. Calisti, M. Cianchetti, C. Laschi, Dynamic model of a multibending soft robot arm driven by cables, IEEE Trans. Robot. 30 (5) (2014) 1109–1122.

[121] M. Giorelli, F. Renda, G. Ferri, C. Laschi, A feed-forward neural network learning the inverse kinetics of a soft cable-driven manipulator moving in three-dimensional space, in: IEEE/RSJ International Conference on Intelligent Robots and Systems (IROS), 2013, 2013, pp. 5033–5039.

[122] T. Deng, H. Wang, W. Chen, X. Wang, R. Pfeifer, Development of a new cable-driven soft robot for cardiac ablation, in: IEEE International Conference on Robotics and Biomimetics (ROBIO), 2013, 2013, pp. 728–733.

[123] S. Seok, C.D. Onal, K.J. Cho, R.J. Wood, D. Rus, S. Kim, Meshworm: a peristaltic soft robot with antagonistic nickel titanium coil actuators, IEEE/ASME Trans. Mechatron. 18 (5) (2013) 1485–1497.

[124] H.T. Lin, B.A. Trimmer, Caterpillars use the substrate as their external skeleton: a behavior confirmation, Commun. Integr. Biol. 3 (5) (2010) 471–474.

[125] M. Cianchetti, M. Follador, B. Mazzolai, P. Dario, C. Laschi, Design and development of a soft robotic octopus arm exploiting embodied intelligence, in: IEEE International Conference on Robotics and Automation (ICRA), 2012, 2012, pp. 5271–5276.

[126] G.Y. Gu, J. Zhu, L.M. Zhu, X. Zhu, A survey on dielectric elastomer actuators for soft robots, Bioinspir. Biomimet. 12 (1) 2017, 011003.

[127] Y. Tang, L. Qin, X. Li, C.M. Chew, J. Zhu, A frog-inspired swimming robot based on dielectric elastomer actuators, in: IEEE/RSJ International Conference on Intelligent Robots and Systems (IROS), 2017, 2017.

[128] H. Godaba, J. Li, Y. Wang, J. Zhu, A soft jellyfish robot driven by a dielectric elastomer actuator, IEEE Robot. Autom. Lett. 1 (2) (2016) 624–631.

[129] T. Li, G. Li, Y. Liang, T. Cheng, J. Dai, X. Yang, B. Liu, Z. Zeng, Z. Huang, Y. Luo, T. Xie, Fast-moving soft electronic fish, Sci. Adv. 3 (4) 2017, e1602045.

[130] K. Jung, J.C. Koo, Y.K. Lee, H.R. Choi, Artificial annelid robot driven by soft actuators, Bioinspir. Biomimet. 2 (2) (2007) S42–S49.

[131] L. Xu, H.Q. Chen, J. Zou, W.T. Dong, G.Y. Gu, L.M. Zhu, X.Y. Zhu, Bio-inspired annelid robot: a dielectric elastomer actuated soft robot, Bioinspir. Biomimet. 12 (2) 2017, 025003.

[132] J. Cao, W. Liang, Q. Ren, et al., Modelling and control of a novel dielectric elastomer based soft crawling robot, in: IEEE International Conference on Robotics and Automation (ICRA), 2018, 2018.

[133] H.E. Holling, H.C. Boland, E. Russ, Investigation of arterial obstruction using a mercury-in-rubber strain gauge, Am. Heart J. 62 (2) (1961) 194–205.

[134] Kramer, R.K., Majidi, C., Sahai, R., & Wood, R.J. (2011). Soft curvature sensors for joint angle proprioception. In IEEE/RSJ International Conference on Intelligent Robots and Systems (IROS), 2011 (pp. 1919–1926).

[135] S. Ozel, E.H. Skorina, M. Luo, W. Tao, F. Chen, Y. Pan, C.D. Onal, A composite soft bending actuation module with integrated curvature sensing, in: IEEE International Conference on Robotics and Automation (ICRA), 2016, 2016, pp. 4963–4968.

[136] Y.L. Park, K. Chau, R.J. Black, M.R. Cutkosky, Force sensing robot fingers using embedded fiber Bragg grating sensors and shape deposition manufacturing, in: IEEE International Conference on Robotics and Automation, 2007, 2007, pp. 1510–1516.

[137] J. Yi, X. Zhu, L. Shen, B. Sun, L. Jiang, An orthogonal curvature fiber bragg grating sensor array for shape reconstruction, in: Life System Modeling and Intelligent Computing, Springer, Berlin, Heidelberg, 2010, pp. 25–31.

[138] M. Popovic, A. Englehart, H. Herr, Angular momentum primitives for human walking: biomechanics and control, in: IEEE/RSJ International Conference on Intelligent Robots and Systems (IROS 2004). Proceedings, vol. 2, IEEE, 2004, pp. 1685–1691.

Further Reading

[139] R. Alexander, The gaits of bipedal and quadrupedal animals, J. Robot. Res. 3 (2) (1983) 49–59.

[140] BeBionic3 Tech Manual. (2013). Retrieved October 11, 2016 from Steeper website: http://bebionic.com/distributor/documents/bebionic3_Tech_Manual_web.pdf.

[141] A. Rahman, A. Al-Jumaily, Design and development of a bilateral therapeutic hand device for stroke rehabilitation, Int. J. Adv. Robot. Syst. 10 (12) (2013).

[142] K. Jung, J.C. Koo, Y.K. Lee, and H.R. Choi. "Artificial annelid robot driven by soft actuators." Bioinspir. Biomimet., 2 (2), pp. S42–S49.

18

Biomechatronics: A New Dawn

Minas Liarokapis*, Kathleen A. Lamkin-Kennard†, Marko B. Popovic‡

*THE UNIVERSITY OF AUCKLAND, AUCKLAND, NEW ZEALAND †ROCHESTER INSTITUTE OF TECHNOLOGY, ROCHESTER, NY, UNITED STATES ‡WORCESTER POLYTECHNIC INSTITUTE, WORCESTER, MA, UNITED STATES

CHAPTER OUTLINE

18.1 Introduction

This chapter focuses on the new dawn of biomechatronics, by presenting and discussing biomechatronic technologies that could reshape our future and dramatically improve our quality of life. In order to put these promising and revolutionary technologies into context, some key aspects of biomechatronics are summarized, definitions are provided, and common misconceptions are clarified. More precisely, we define the terms biomechatronics,

Biomechatronics. https://doi.org/10.1016/B978-0-12-812939-5.00018-5

bionics, bioinspiration, and biomimicry and we discuss their similarities and differences. Then we describe some recent technological and scientific developments and discuss future directions of the field.

Born at the turn of the 21st century, word Biomechatronics refers to an interdisciplinary field that blends together biological and mechatronic systems. The biomechatronic field is quite wide, covering every system that closely relates to biological and mechatronic systems within a functionally unified entity. Biomechatronics overlaps with numerous fields including neuroscience and robotics. As discussed in Chapter 1, the boundaries of what devices should be characterized as biomechatronic are not always clear. For example, a bioinspired or biomimetic robot can be a completely autonomous system that does not rely on any type of interfacing with humans, thus, such a device cannot be considered a biomechatronic system. Typical examples of biomechatronic devices are prosthetic systems (e.g., upper limb and lower limb prostheses), wearable assistive devices (e.g., electrically actuated exoskeletons, exo-suits, etc.) and artificial organs and implants (e.g., cochlear implants, etc.). Most of these devices are used either to restore a human biological function or to enhance the human capabilities.

Regarding bioinspiration and biomimicry, bioinspiration draws inspiration from biological systems, identifying a key idea behind both the form and the function of a biological system and aiming to implement it technologically, while biomimicry is about copying a biological system as accurately as possible (morphology, biomechanics, sensing, control, etc.). Biomimetic systems can reproduce biological counterparts only to a certain extent. A fully biomimetic system would need to consist of exact copies of biological cells and this is not anticipated in the near future. Finally, the term bionics was coined by Jack E. Steele in 1958 but became quite popular due to its appearance in the US TV series "The Six Million Dollar Man" [1] and "The Bionic Woman" [2] that were based on the novel "Cyborg" of Martin Caidin [3]. The term bionics combines the Greek word "βίος" (bio-) which means life with the word electronics.

In the following sections, we describe how the field of biomechatronics can change our everyday life and have a tremendous impact on our future, rehabilitating patients, helping amputees regain their lost dexterity, enhancing human capabilities, and redefining the essence of being human. Although this book endeavored to cover as many topics in biomechatronics and in as many details as possible this effort proved to be very challenging because the entire field is advancing rapidly. In this chapter, several more topics are reviewed and more examples are provided. The final section of the chapter, focuses on the future of the "Biomechatronics age human" raising some philosophical and ethical questions, attempting to answer them and triggering a very interesting debate for the years to come. Thank you for joining us on this amazing journey into the new dawn of biomechatronics.

18.2 New Sensors and Actuators

In Chapters 3 and 4, we described the important role that the new materials, sensors, and actuators have in the development of biomechatronic technologies. More precisely, in

Chapter 3 we focused on different types and classes of actuators, presenting nonbiological actuators, synthetic muscles based on electroactive polymers, fluidic elastomer actuators, shaper memory alloy-based actuators, and variable stiffness actuators, while in Chapter 4 we focused on sensors, presenting the human nervous system, different types of receptors as well as artificial/synthetic sensors that range from chemical, electric, optical, and mechanical sensors to motion capture systems and sensors used in bioamplifiers [e.g., electromyography (EMG), electroencephalography (EEG), and electrocardiography sensors]. In this section, we present the latest developments in the fields of new materials, sensors, and actuators and we discuss some promising results that may revolutionize the field of biomechatronics devices.

The future of biomechatronic sensors and actuators will rely heavily on advances in materials science. Responsive materials, such as electroactive polymers, piezoelectric materials, or shape memory alloys, have already enabled the development of novel actuators and sensors that are being used in biomechatronic applications (see Chapters 3 and 4). Other materials, including phase change porous composites (Fig. 18.1), foams, edible, and fabrics are also being investigated. In this section, we discuss some recent developments in the field of biomechatronics, presenting new design paradigms of sensors and actuators and their applications (e.g., development of soft robots and multimodal artificial skins, for an example, see Fig. 18.2).

Some examples of these developments are as follows. Phase change materials can store and release large amounts of energy and have been successfully integrated into actuators to provide an alternative to traditional electromechanical actuators [5–7]. At the phase transition temperature of a material, a rapid expansion occurs, and significant mechanical force can be generated. Large, electrically driven deformations of soft elastomer membranes have been achieved using reversible, liquid-vapor-phase transitions [4, 8]. For example, a composite meta-material, comprising a polydimethyl-siloxane (PDMS)-based elastomer as the matrix and ethanol as the filler, was utilized to create a novel actuator capable of undergoing strains of up to 915% [4] (see Fig. 18.1). When the composite is heated to the liquid-vapor transition temperature, the ethanol boils, leading to a

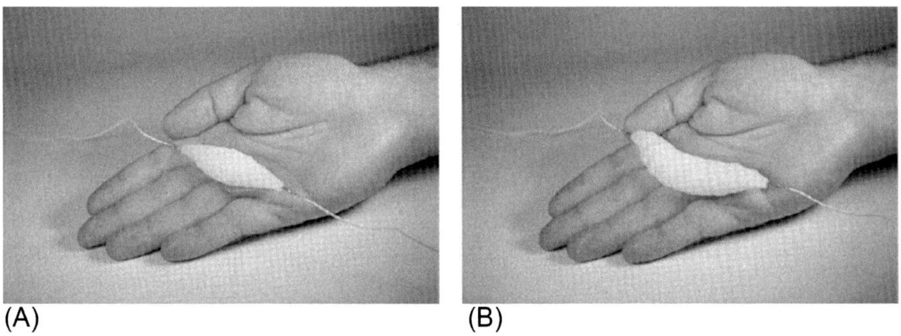

FIG. 18.1 Example of a soft artificial muscle that is composed of ethanol distributed throughout a solid silicone elastomer matrix: (A) the muscle in a rest position and (B) the muscle in a contracted state [4].

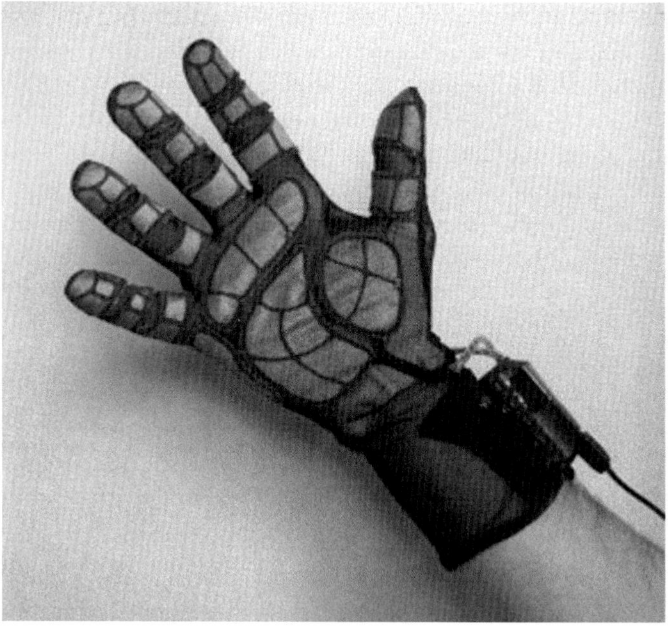

FIG. 18.2 A flexible, stretchable, durable, fabric-based tactile glove. *G.H. Büscher, et al., Flexible and stretchable fabric-based tactile sensor, Robot. Autonom. Syst. 63 (2015) 244–252.*

volumetric expansion of the elastomer. The expansion of the elastomer can be used to create an actuator capable of undergoing large deformations.

Open cell, elastomeric foams have been utilized to create novel actuators capable of generating large displacements and forces up to 20 N [9, 10]. A typical foam actuator consists of an open-celled elastomer foam, a strain-limiting layer, and a nonporous sealing layer [11]. Upon pressurization, the interconnected pore network causes the actuator to bend, extend, or twist [9]. The elastomeric foams can be customized by using different porogens and matrix materials to enable a range of actuator properties.

A biodegradable, biocompatible, edible actuator was synthesized using a combination of gelatin, glycerin, and water [12]. The actuator was molded to contain air chambers for pneumatic actuation. Gelatin-based or other edible actuators can be used as disposable robots for exploration or for medical applications. A multimodal fabric-based tactile sensing skin was synthesized in [13], using a combination of conductive, insulating, and resistive fabrics with embedded thermistors. The force and thermal sensing composite skin is capable of three sensing modalities, such as force, active thermal sensing, and passive thermal sensing. The development of composite materials that exhibit multifunctional sensing capabilities is likely to have a significant impact in the future on how robots interact with humans or their surroundings.

The aforementioned examples illustrate some of the possibilities for exploiting unique material properties and capabilities to develop novel actuators and sensors and integrate them in biomechatronic devices. The expansion of responsive material capabilities into the realm of multifunctional materials and composites is likely to have an even broader impact in the future. The addition of embedded computation would further enable the creation of a new class of robotic and mechatronic devices that integrates sensing, actuation, and computation to autonomously respond to the surroundings and closely mimic biological systems.

18.3 Brain Machine Interfaces

In Chapter 6, we described the most popular brain-machine interfaces (BMI) technologies that range from the noninvasive EMG and EEG to the invasive intracranial electroencephalography (iEEG) and electrocorticography (ECoG). Then we focused on direct neural interfaces presenting the theory of electrical recording, the structures and configurations of electrodes, the principles of electrical stimulation, the hardware of electrical stimulation, and finally various aspects of optical recording and stimulation. In this section, we focus on some open challenges in the BMI field, discussing the need for hybrid schemes that use combinations of biological signals and new sensor technologies. These promising research directions aim to revolutionize the way we interact and interface with biomechatronic devices, rendering the control of prosthetic, assistive, and human augmentation devices both intuitive and simple (see Fig. 18.3).

FIG. 18.3 A brain-controlled wheelchair that uses shared control (or context awareness) techniques.

18.3.1 Alternatives to the Surface Electromyography (EMG) Based Control

The simplest and most popular type of noninvasive BMI is an EMG-based interface that uses the myoelectric activations of the human muscles to control a technical device. The EMG-based interfaces were first used for the control of advanced prosthetic devices, 35 years ago [14]. During the last decades, the field has received increased attention as many applications have emerged. Some of these applications are: (i) EMG-based teleoperation of robot artifacts in remote or dangerous environments [15, 16], (ii) EMG-based control of advanced prosthetic limbs that help patients regain their lost dexterity [17], (iii) EMG-based control of exoskeletons that can be used for rehabilitation purposes [18], and (iv) muscle-computer interfaces, for human-computer interaction [19, 20]. Despite their possible applications and the advantages (e.g., easy to use, intuitive, can decode force and effort), EMG-based control schemes have also several drawbacks that have been identified and need to be addressed. Some of these problems are: (i) the high-dimensionality and complexity of the human musculoskeletal system that makes the identification of the correct muscles (for the execution of desired tasks) challenging, (ii) the nonstationarity of the EMG signals, (iii) the nonlinear relationship between the human myoelectric activity and the motion or force to be estimated, (iv) the noise created by the electrode perturbations, and (v) the effects of human fatigue and sweat in the EMG signals, etc.

Recent studies have tried to address the aforementioned problems of EMG interfaces by using alternative sensors, combinations of biological signals, or advanced control schemes. More precisely, in [21] the authors have used medical ultrasound imaging as a means of detecting in real-time the position and force configuration of the human hand and wrist (see Fig. 18.4). To do so, they used simple image processing techniques that were

(A)

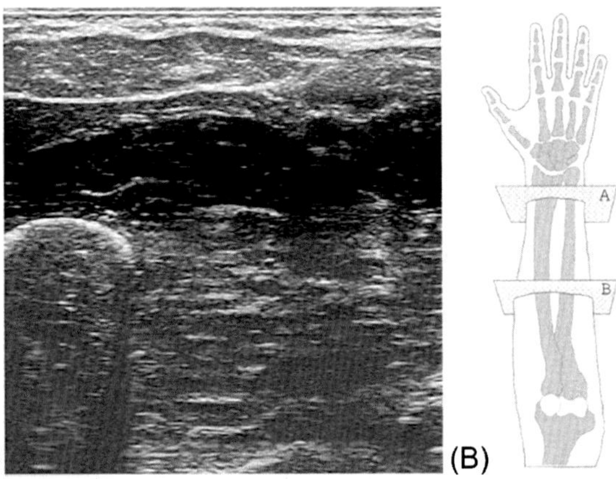

(B)

FIG. 18.4 (A) An ultrasound image that is used to predict the finger motions. (B) A graphical representation of the human hand and forearm. The planes A and B correspond to the sections from which the ultrasound images are collected. Standard image processing techniques can effectively derive from the ultrasound images the human finger motions [21].

applied to the collected ultrasound images providing significant estimation accuracy. Another approach uses topographic force maps that register the three-dimensional (3D) volume changes of the residuum as a dynamic map of muscle recruitment. Such maps can be created using pressure sensor arrays like in [22]. Similar to the ultrasound-based decoding of human motion, this approach is not dependent on precise anatomical placement of the sensors, and measures the soft tissue response to the neuronal activation. Regarding the combination of signals, in [23] the authors presented a hybrid EEG-EMG-based control approach that can be used in bio-robotic applications. Finally, a third option to improve EMG-based interfaces is by using advanced, semiautonomous, shared-control schemes [24] (more details are provided in the following sections). An overview of the different solutions that can address the problems of EMG-based interfaces can be found in [25].

18.3.2 Novel Neural Interfaces

Recently, a new type of neural interface was proposed, the neural dust platform [26] (see Fig. 18.5). The particular platform uses ultrasonic power and communication that facilitates an advanced interfacing with the nervous system using submillimeter-sized devices

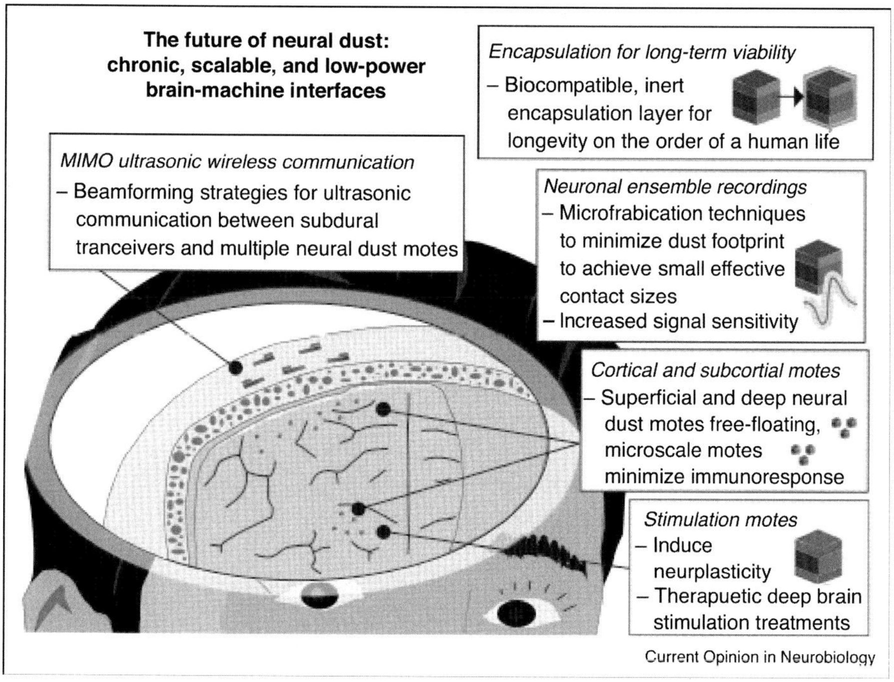

FIG. 18.5 Future characteristics of the neural dust platform. The goal is to realize clinically viable, wireless brain-machine interfaces. Challenges include: size reduction, noise reduction, implementation of electrical stimulation motes, ultrasonic multi-mode communication, biocompatible encapsulation, and implantation into various cortical regions [26].

that are implanted deep in the brain tissue. The platform offers a scalable, wireless, and battery-less interfacing and allows the transmission of high-fidelity electrophysiological data. The efficiency of the proposed platform was demonstrated in an in vivo environment. More precisely, a neural dust sensor called "mote" was developed using a piezoelectric crystal (acoustic domain antenna) and a custom integrated circuit with recording electrodes and epoxy encapsulation. The device was surgically attached to the sciatic nerve of a lab rat to measure both compound action potentials and electromyograms. Regarding future directions, the neural dust platform researchers plan to miniaturize it in the sub-mm scale as well as to deploy the platform in the human central nervous system.

18.4 Control Strategies, AI, and Machine Learning

In Chapter 5, we presented some fundamentals ideas and concepts of control theory (stability, feedback linearization, sliding control, adaptive control, etc.) and we attempted a brief introduction to the fields of artificial intelligence, data mining, general machine learning, and deep learning. In this section, we focus on advanced control topics and we discuss some current trends in the fields of artificial intelligence (AI) and machine learning and how they will affect the progress and applications of biomechatronics.

18.4.1 Shared Control of Biomechatronic Devices

In Section 18.3, we described some drawbacks of EMG-based interfaces and we presented some possible solutions. Some of these solutions are: (i) the formulation of real-time machine learning schemes where the prediction and control information is learned during the ongoing task execution (see Fig. 18.6 for details) and (ii) the formulation of semi-autonomous control schemes that have the human operator in the loop, assisting him/her in task execution by locally correcting/adjusting the behavior of the system [25]. The second solution leads to "shared control" schemes that provide minimal effort control to the

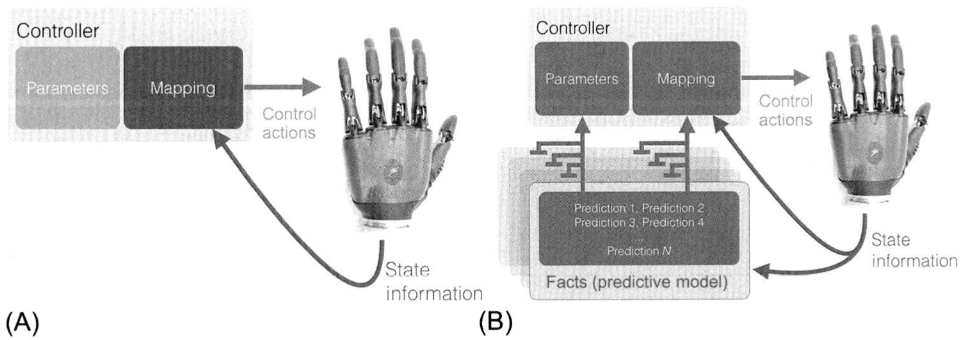

FIG. 18.6 An abstract representation of situational awareness/knowledge as a supplement in the formulation of myoelectric control schemes [25].

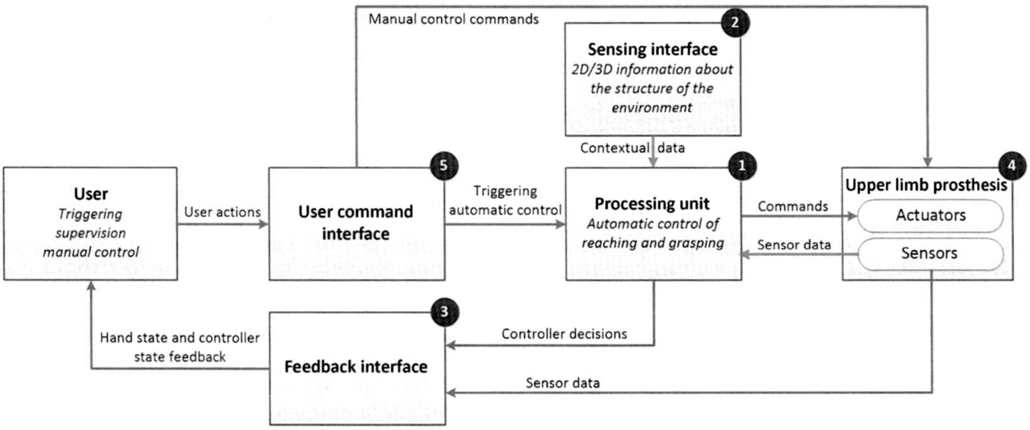

FIG. 18.7 Block diagram of a semiautonomous control scheme for prostheses [25].

user (see Fig. 18.7 for details). In Fig. 18.6 we present an abstract representation of the way that situational awareness (knowledge in the form of temporally extended predictions) can benefit the efficiency of a myoelectric control. We feel that shared control schemes can be of paramount importance for the efficient control of biomechatronic devices in the future, especially since the dexterity of these devices constantly increases and their control becomes more complicated. Shared control schemes can provide a significantly simpler and more intuitive control to the user [25].

18.4.2 Reinforcement and Deep Learning

Regarding advanced topics of machine learning, several methodologies based on incremental/reinforcement learning and deep learning have also been proposed for the control of biomechatronic devices (see Fig. 18.8). More precisely, in [27, 28] the authors present reinforcement learning approaches for the EMG-based control of prosthetic devices. In [27], the authors proposed a nonlinear incremental learning method that allows continual adaptation of the control parameters to the changes in the EMG signals. Incremental Ridge Regression and Random Fourier Features were combined to perform an EMG-based estimation of finger forces, both individually and in groups (i.e., grasping patterns). The method was used to execute online teleoperation tasks with a humanoid robot arm equipped with a prosthetic hand. The subject was able to reliably grasp, carry, and release the examined objects, irrespective of the EMG signal changes caused by the hand/arm movements and the wrist pronation/supination. In [28], the authors introduced a continuous actor-critic reinforcement learning method for the EMG control of dexterous prosthetic devices. The methodology requires only a sparse training signal, does not depend on a detailed description of the task, and provides online learning of new subject-specific motions using a single one-dimensional feedback signal provided by the user of the prosthesis.

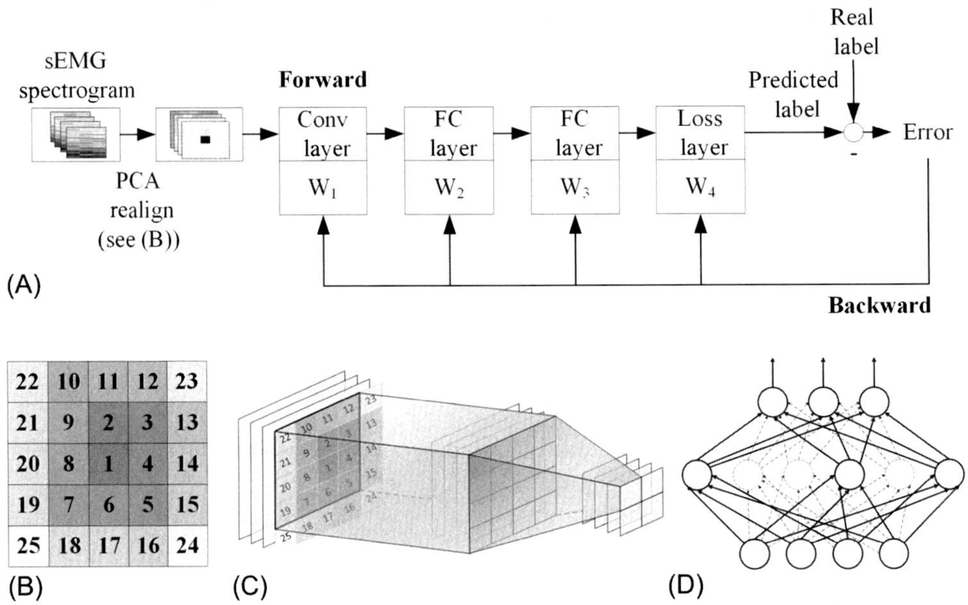

FIG. 18.8 Schematic of an EMG-based hand movement classification that uses convolutional neural networks. *X. Zhai, B. Jelfs, R.H.M. Chan, C. Tin, Self-recalibrating surface EMG pattern recognition for neuroprosthesis control based on convolutional neural network, Front. Neurosci. 11 (2017) 379. https://doi.org/10.3389/fnins.2017.00379.*

Deep learning was used in [29] to perform an EMG-based estimation of human motion. More precisely, the proposed methodology combines convolutional neural networks (CNNs) and recurrent neural networks (RNNs). The EMG signals are transformed to time-frequency frames that become the input to the model. The model is trained with the gradient descent and backpropagation methods. The efficiency of the model was tested for simultaneous and proportional estimation of limb movements in eight healthy subjects and was compared with the efficiency of support vector regression (SVR) and simple CNNs. The results were quite promising as the proposed model provided higher estimation accuracy than the other methods.

18.5 Bionic Tissue, Artificial Organs, and Implants

In Chapter 7, we presented a brief history of artificial organs and we focused on cardiovascular and respiratory devices (e.g., artificial hearts, pacemakers, respirators, etc.), metabolic and digestive devices (e.g., dialyzers, pancreatic islets), sensory devices (e.g., cochlear implants, artificial intraocular lenses, etc.), and orthopedic and reconstructive devices (e.g., artificial bones, joints, skin, etc.). In this section, we focus on recent developments in the field of bionic tissue and artificial organs and we present some promising future directions.

18.5.1 Multimodal Artificial Skin

A recent study [30] proposes a dynamic covalent, thermoset-based, rehealable, eco-friendly, and economical electronic skin (e-skin) that mimics the functionalities and mechanical properties of the human skin. The particular artificial skin is developed by doping the dynamic covalent thermoset with conductive silver nanoparticles. The result is a robust yet rehealable, fully recyclable, and malleable e-skin that has tactile, temperature, flow, and humidity sensing capabilities. Moreover, malleability enables the e-skin to permanently conform to the object geometry facilitating the execution of stable grasps. The truly remarkable fact is that the presented e-skin is not just a sensory system, it is an intelligent bionic tissue that incorporates a variety sensors and exhibits rehealing capabilities like the human skin, increasing the reliability and the robustness of the structure. All these properties guarantee that the proposed e-skin will be a revolutionary, key feature of future biomechatronics devices (e.g., bioinspired robots, advanced prostheses, assistive devices, etc.).

18.5.2 Future Directions: Bioprinting, Bioartificial Organs, and Biodegradable Electronics

Regarding future directions, one of the most promising fields is that of bioprinting of artificial organs (see Fig. 18.9). In [32], the authors used additive manufacturing of biological cells with structural and nanoparticle-derived electronic elements in order to create 3D printed bionic ears. In order to validate the efficiency of the proposed methodology they three dimensionally printed, in vitro, a bionic ear that exhibits enhanced auditory sensing for radio frequency reception. Other examples of futuristic bio-artificial organs are organs designed with computer-aided, jet-based 3D tissue engineering [33] and autologous, bio-artificial limbs created based on patient-derived cells [31]. The methodology presented in [31] is still under testing in animals. For example, in Fig. 18.9 we present a bio-artificial limb (rat forearm) created based on the "patient"-derived cells.

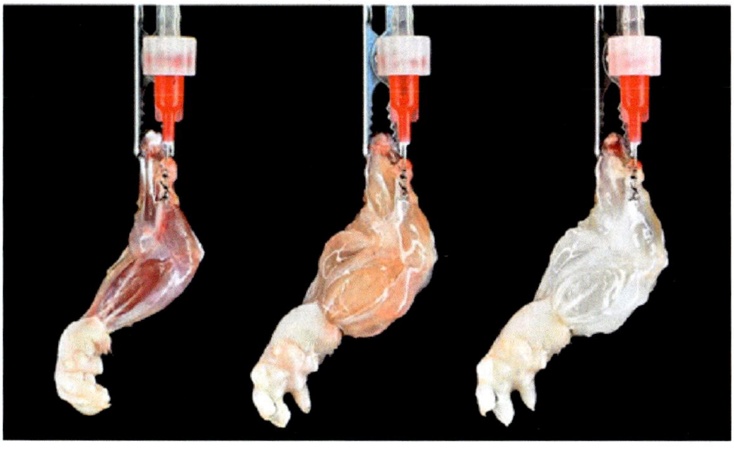

FIG. 18.9 Example of a bio-artificial limb (rat forearm) created based on the "patient"-derived cells [31].

18.6 Prosthetic, Assistive, and Human Augmentation Devices

In Chapter 9, we focused on upper limb and lower limb prostheses, presenting the different classes of prostheses, the major design considerations, the most common interfaces, the most popular control schemes as well as some social aspects pertaining to the use of prosthetic devices. Other topics that we covered in Chapter 9 are the kinematics, the materials, the actuators, and the sensors that are used in prosthetic devices, the design and fitting of sockets, the aesthetics and anthropomorphism of the devices and how they affect the prostheses acceptance. In Chapter 10, we focused on orthotic systems presenting different types of powered hip braces and waist support devices, powered knee braces, powered angle braces, shoulder braces, elbow and wrist braces, as well as hand and finger braces and robotic gloves. Most orthotic devices can also be described as wearable devices and typically combine rigid and soft materials with light-weight actuators and electronics, to increase the reliability, robustness, and the comfort and decrease the complexity, weight, and the cost. Finally, in Chapter 11, we focused on wheelchairs and other mobility assistance methods, presenting different types of robotic wheelchairs, different schemes of wheelchair motion control (including shared control), and the major challenges of the field. In this section, we focus on recent trends in the development of prosthetic, assistive, and human augmentation devices and we discuss some promising future directions.

18.6.1 Current Trends in Prosthetic Devices

As we have already discussed in Section 18.2, some recent developments in materials science and technology, promise to revolutionize the development of new, sophisticated sensors and actuators for biomechatronic applications. Still further, the new materials affect the mechanical design of various biomechatronic devices (e.g., prosthetic and assistive devices). For example, in [34] the authors proposed a new prosthetic device that combines 3D printed materials with elastomer structures used for the development of flexure joints. Different materials are blended together using appropriate molds and the concept of shape deposition manufacturing (SDM) is introduced [35]. The results of the experimental validation of the efficiency of the proposed prosthesis were quite promising, demonstrating an increased robustness, adaptability, and durability as well as a reliable performance. The developed prosthetic hand uses a single actuator for the eight degrees of freedom and can be either body-powered or externally powered. Other examples focus on composite materials that are used in modern orthopedic medicine and prosthetic devices that offer a superior strength to weight characteristic and excellent biocompatibility [36], as well as on materials that allow the online tuning of some of the mechanical properties of the biomechatronic devices (e.g., stiffness) [37]. Regarding variable stiffness, in [38] the authors proposed a quasi-passive, ankle-foot prosthesis that has the ability to quickly modulate the ankle stiffness between tasks.

Another trend in prosthetic devices is the development of highly biomimetic devices that try to replicate not only the functionalities but also the appearance of the human hand (highly anthropomorphic hands). In Chapter 9, we presented the WPI prosthetic hand, a high fidelity, low-cost, lightweight, under-actuated, biomimetic prosthesis that replicates the human form using 3D printed bones and laser cut tendons and ligaments [39]. The particular device is controlled using a hybrid EEG- and EMG-based interface that decodes human intention. Another example of a highly biomimetic prosthesis that can be easily replicated with off-the-shelf materials and desktop manufacturing techniques is the open-source, OpenBionics hand that is also based on flexure joints and which takes advantage of a selectively lockable differential mechanism that facilitates the execution of 144 grasping postures and gestures with a single actuator (see Fig. 18.10) [40]. More details regarding these two prosthetic devices can be found in Chapter 9.

18.6.2 Assistive and Human Augmentation Devices

Regarding "human augmentation," the particular term refers to the development of devices that can help humans increase some of their physical capabilities (e.g., strength, speed, stamina, precision, etc.). Prosthetic and assistive devices are intended to be used by amputees or patients who want to regain their lost dexterity. On the other hand, human augmentation devices are intended to be used by able body persons who desire to increase the capabilities of their bodies, by exerting bigger forces, running faster or for longer periods of time, jumping higher, or even becoming more robust to impacts with the

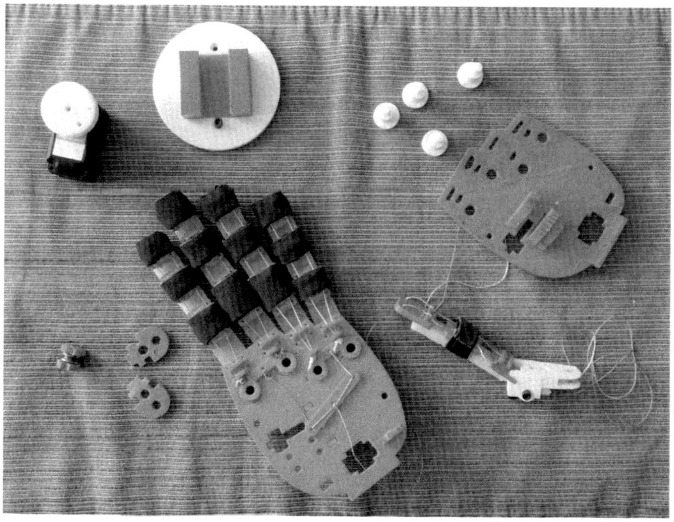

FIG. 18.10 Various components of the compliant, highly "under-actuated," anthropomorphic prosthetic hand developed by the OpenBionics initiative. The particular design is distributed in an open-source manner and can be replicated with off-the-self materials and a simple 3D printer [40].

environment (e.g., protective equipment). Wide variety of the potential applications of human augmentation devices range from devices that increase the task execution efficiency of people in workplace environments to hobby level devices that allow their users to have superhuman efficiency in desired tasks (e.g., devices that facilitate rock climbing or jumping) or new capabilities.

For example, one could argue that the wingsuits that are used in skydiving are actually human augmentation devices. The wingsuits modify the body area that is exposed to the wind to increase the amount of lift generated by the body with respect to the drag, during free fall facilitating gliding. Other examples of hobby level human augmentation devices are the ostrich-like bionic boots [41] and the air-trekker jumping stills [42] that allow humans to run faster and jump higher. Another example of a human augmentation device is a hand exoskeleton that allows the user to exert significantly greater forces with his/her fingertips. Such a device is useful not only in workplace environments (where demanding manual tasks may require a significant grip strength) but also in recreational activities such as rock climbing that requires a good grip.

Recently, researchers started working on extremely lightweight, fully compliant assistive devices that provide affordable and simplified dexterity. In [43], the authors proposed the ExoGlove Poly, a polymer-based soft, wearable, hand exoskeleton that enables individuals who have lost their hand function to regain their lost dexterity and stably grasp a variety of everyday life objects. The whole Exo-Glove Poly structure is made out of silicone, making the design essentially waterproof. The device supports three fingers, index, middle, and thumb using an antagonistic, tendon-driven actuation scheme. Index and middle fingers are flexed freely while the thumb is fixed in an opposed position, essential for grasping. The Exo-Glove Poly can be adapted to different hand sizes and shapes taking advantage of the silicone properties. A flexible silicone makes the glove stretchable and adaptable to different users. Thus, no further customization efforts are needed. The tendon-driven design facilitates an indirect transmission allowing the motor, the required circuit board, the battery, and other components of the device, to be positioned away from the glove reducing the weight and the size of the device and make it easier for the user to wear the device for longer periods of time.

One of the most impressive current directions in the field of assistive devices is the development of nonconventional robotic wheelchairs that can also assist the user to get off the bed, sit on a chair, stand up, avoid obstacles, or even climb stairs. Such wheelchairs act also as exoskeletons, thus the boundaries between the two types of devices have recently become less clear. In Fig. 18.11, the iBOT robotic wheelchair that was developed by a partnership between DEKA and Johnson and Johnson's Independence Technology division is depicted [44].

18.6.3 Techno-Biological, Bioartificial Devices

Regarding the future of prostheses and human augmentation devices, we feel that in the future biomechatronic devices will stop being purely artificial. A new class of

FIG. 18.11 The iBOT robotic wheelchair developed by a partnership between DEKA and Johnson and Johnson's Independence Technology division [44].

techno-biological, bio-artificial devices will emerge. This class may have technological parts integrated into bio-artificial limbs created with bioprinting methods and stem cells, or biological parts integrated on technical devices that use conventional actuators. Such technologies are expected to revolutionize the field of biomechatronic devices offering far more dexterous and efficient assistive, prosthetic, rehabilitation, and augmentation solutions.

18.7 Biomechatronic Technologies for Animals

Advances in human prosthetics and orthotics are now being applied to veterinary medicine as a means to assist in healing, rehabilitating injuries, or restoring gait of various animals. Restoration of mobility and optimal movements can play a significant role in the maintenance of an animal's overall physical health [45]. Rehabilitation is thus playing an increasingly important role in veterinary medicine and the areas of veterinary orthotics and prosthetics are likely to play an even greater role in the future. Development of animal orthotics and prosthetics represents a particularly unique challenge since there can be significant anatomical differences between species and attachment/fitting of the devices can be difficult.

18.7.1 Prosthetics

Missing limbs in animals may occur as a consequence of birth defects or amputation. Limb amputation is often indicated in cases of severe musculoskeletal trauma, severe congenital birth defects, cancer, infection, or for treatment of other life-threatening conditions [46]. Limb amputation can offer an economic alternative to euthanasia, particularly when treatment costs are prohibitive to the animal owner [47]. The classical approach to amputation in animals is proximal amputation, for example, at the level of the scapulohumeral joint or proximal femoral shaft, to avoid trauma at the stump during daily activities [47]. Three-legged animals are capable of ambulation, however, an increasing number of studies suggest that the biomechanical differences between tripedal and quadrupedal locomotion may have more significant pathological consequences than previously understood [48–51]. Short- and long-term consequences for tripedal animals can include limited endurance, increased metabolic demand, tissue breakdown, degenerative joint problems in remaining joints, weight gain, and chronic back or neck pain [46–53]. Over the past decade, there has been a significant increase in the use of animal prosthetics for restoring quadrupedal gait. However, the biomechanics of quadrupedal gait present unique challenges when designing prosthetic devices.

The two primary types of prosthetic limbs in use today for animals include socket-based devices and intraosseous transcutaneous amputation prosthesis (ITAP) devices [46]. Socket-based prosthetic devices consist of a socket that is placed over the residual limb stump and an extension that contacts the ground. The device is attached using straps or braces. The ITAP devices consist of an implanted endoprosthesis and attached exoprosthesis. The endoprosthesis, such as a titanium alloy screw, is implanted into the bone and is secured as a result of bone and skin growth around the implant [46]. The exoprosthesis attaches to the endoprosthesis and provides an extension to the ground. Socket-based prosthetic limbs are typically low cost, simpler to implement, and can be adapted to many levels of limb loss [46]. However, much like with human prosthetics, suspension and retention of the device on the residual limb is critical or soft tissue damage can occur. The advantage of ITAP devices is that there is direct skeletal integration so the exoprosthesis directly transmits forces to the skeleton via the endoprosthesis [46, 53]. The ITAP devices have been successfully used in dogs, cats, and horses to date [53–57]. However, ITAP devices are more expensive than socket-based devices and require invasive surgical intervention. Other novel prosthetic devices have also been developed for rehabilitation of wildlife. A computer molded, plastic beak was developed for an American bald eagle [56].

A rudder-like fin was developed for an Atlantic sea turtle missing three flippers [56, 58]. A rubber-silicon tail fin was also developed for a dolphin in Japan [56, 58]. Although still in its infancy, the field of veterinary prosthetics has significant potential to restore functionality to both domestic and wild animals.

18.7.2 Orthotics

Orthotic devices for animals are typically fabricated from thermoform plastics with a thermofoam lining for comfort [46]. Orthotic devices can be used to improve comfort, assist in healing, or rehabilitate some injuries [46]. Orthoses can also be used as preoperative, postoperative, or "no operative" solutions to restrict, block, enable, or guide range of motion [46]. Preoperative orthoses can be used to provide interim support during ambulation while postoperative orthoses can provide an alternative to traditional casting methods. For patients who are poor surgical candidates or for cases in which there are financial limitations, orthoses can offer viable "no-operative" solutions to maintain mobility. Prophylactic orthoses can prevent or reduce the severity of a joint injury [46, 55]. For example, a paw orthotic can be used to treat paw knuckling, in which the paw fails to flex properly during walking, due to arthritis or degenerative myelopathy. The orthotic keeps the paw in the correct position during load-bearing activities and can prevent sores from developing or other trauma [55].

Rehabilitative orthotics is used to support or stabilize joints while resisting range of motion following an injury or surgical procedure [55]. For example, the cranial cruciate ligament (CCL) stabilizes a dog's knee (stifle) joint similar to the anterior cruciate ligament in humans. The CCL tears are a common orthopedic injury and may require immobilization or surgical intervention [46]. Stifle orthotics can be used to provide support and stability in lieu of surgical intervention or during the surgical recovery period. A stifle orthotic device prevents external and internal rotation in the stifle joint and provides medial and lateral support [55]. Functional orthotics can be used to stabilize chronically unstable joints and limbs and provide long-term function and support [55]. Hip dysplasia, a potentially debilitating disease involving abnormal formation of the hip socket, can be managed using a bilateral hip orthosis. A hip orthosis controls the range of motion of the hip joint using mechanical joints and prevents unstable joints from dislocating [53]. The orthosis enables the animal to maintain a good stance position thereby enabling better leg movements and control during walking [55].

18.7.3 Additive Manufacturing of Veterinary Prosthetics and Orthotics

Additive manufacturing, commonly referred to as 3D printing, refers to processes in which 3D digital designs are fabricated layer by layer. The 3D printing is being increasingly utilized for the creation of custom veterinary orthotic and prosthetic devices [59]. Prosthetic design requirements can vary significantly between animal species due to extreme anatomical differences, however, 3D printing allows for rapid prototyping of designs customized to the specific anatomical needs of the species (see Fig. 18.12).

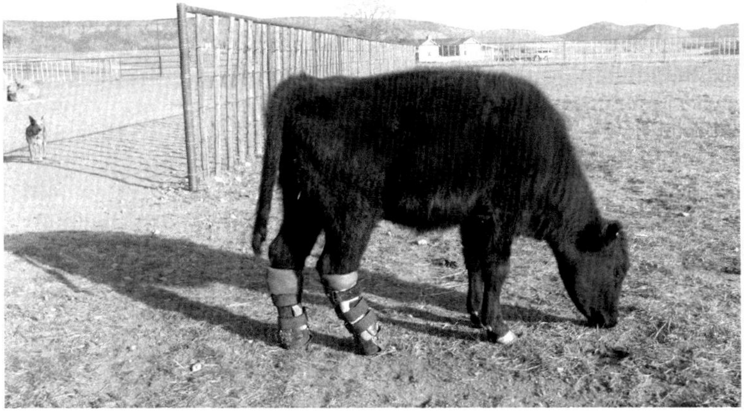

FIG. 18.12 A Black Angus calf that has two prosthetic legs. https://en.wikipedia.org/wiki/File:Meadow_(2).JPG.

The 3D printing has been used to date to create a broad range of novel designs for both domestic and wild animals. In 2014, a 3D printing company in Turkey printed a medical-grade titanium prosthetic jaw for a loggerhead sea turtle [59]. The 3D printing has also been used to successfully create silicone duck feet, titanium horse shoes, a sheep leg, hermit crab shells, and a toucan beak [59]. To date, 3D printing has been used successfully for creating a range of complex, custom animal orthotics and prosthetics although the functionality and durability of the 3D printed devices is still limited [59]. Despite current limitations with durability, 3D printing is likely to play a significant role in the design and development of veterinary orthotic and prosthetic devices in the future.

18.8 Other Human Oriented Applications

Regarding other human-oriented applications of biomechatronics, in Chapter 12 we focused on physical therapy and rehabilitation, presenting solutions for upper limb, lower limb, and balance therapy, as well as some critical rehabilitation assessment technologies. From the overview of the field it is quite evident that during the last decades, there has been a tremendous effort in developing new rehabilitation devices and that their progress follows the trends of prosthetic and assistive devices integrating new materials, sensors, and actuators (e.g., variable stiffness modules) and taking advantage of advanced control schemes and learning techniques. Other than the physical therapy and the rehabilitation devices, biomechatronics principles and technologies can be used for human-oriented applications in the fields of robotic surgery, medical care, sports science, and of course entertainment. Regarding other examples of medical care, in Chapter 14, we focused on feeding and hygiene assistance provided mainly by robotic platforms. In Chapter 15, we focused on robot surgery presenting: (i) a classification of the different robot surgery platforms based on the robotic structures they use (e.g., multi-arm robot platforms, hyper-redundant robot arms, soft robotics, etc.), (ii) the use of human-machine interaction or human-robot interaction in robotic surgery and how it can improve the overall surgeons experience, (iii) a variety of case studies, and (iv) some future trends of the field. Finally, Chapter 16, we focused on the remaining human-oriented applications of biomechatronics, presenting the sports and exercise biomechanics, game systems used for physical therapy and rehabilitation, modeling and simulation aspects, as well as games and entertainment applications. From the above it is quite evident that the field of biomechatronics has a plethora of human-oriented applications and the ability to dramatically improve the quality of life of millions of people around the world.

In the introduction of this book (Chapter 1), we used the car as an example of a mechatronic system that could easily become a biomechatronic system in the future. The use of advanced prostheses that integrate sophisticated computing systems and electronics, the use of implantable, biodegradable sensors and communication systems, and the development of simplified, compact, robust brain-machine interfaces will facilitate the interaction of a bionic human with all the mechatronic devices surrounding him/her. In the future, touch screens of electronic devices could be useless, as our connected brain could be able to interact with every device connected to the internet (including our car, office, and home appliances). Advanced implantable integrated circuits could be used to extend our memory and specialized learning modules could help us master a new hobby or a new skill in seconds. From the above it is evident that the possible applications are indeed countless and that the new dawn of biomechatronics is breaking.

18.9 The Future of the Biomechatronics Age Human

From the novel “Machine Man” by Max Barry [60] and the recent “Ghost in the Shell” [61] movie to the advanced neuroprostheses that can translate human intention to robot actions and recent developments in tissue level electronics and artificial organs, the idea of a biomechatronics age human is rapidly transforming from science fiction to reality. In science fiction, this exciting journey has been anticipated for a very long time (see Fig. 18.13). For example, four decades ago the phrase “Gentlemen, we can rebuild him. We have the technology” was introduced in the “Six Million Dollar Man” TV series [1]; the shattered body of the crash victim Steve Austin is rebuilt with bionic limbs and implants that he could control with his mind. Popular culture provided many examples of the biomechatronics age human as well as examples of future applications of biomechatronic devices. Artistic visions in the 16th century (Fig. 18.13) and science fiction in the 1970s evolved into nowadays reality, including mind-controlled hand prostheses [62], exomusculature soft braces [63, 64] and highly biomimetic walking leg prostheses [65] (Fig. 18.14).

The development of biomechatronics provokes many profound and perhaps controversial questions such as: “What does it mean to be human?,” “What are the boundaries of the human existence?,” or “Is the human biological body including brain just a fragile

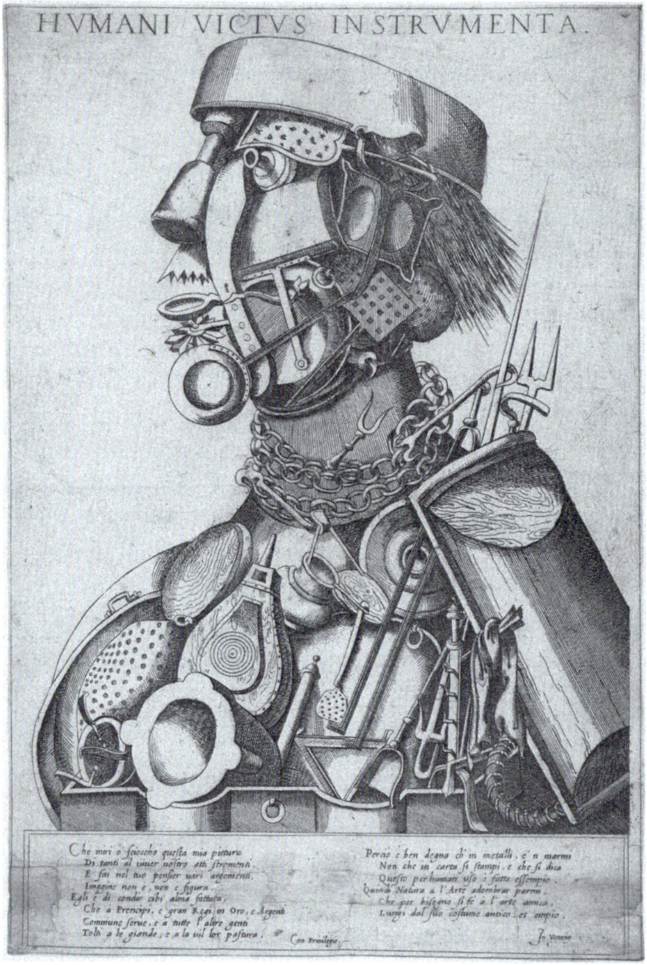

FIG. 18.13 Humani Victus Instrumenta: Ars Coquinaria (Italy, ~1569), Copper engraving, 379 × 277 mm, unknown master.

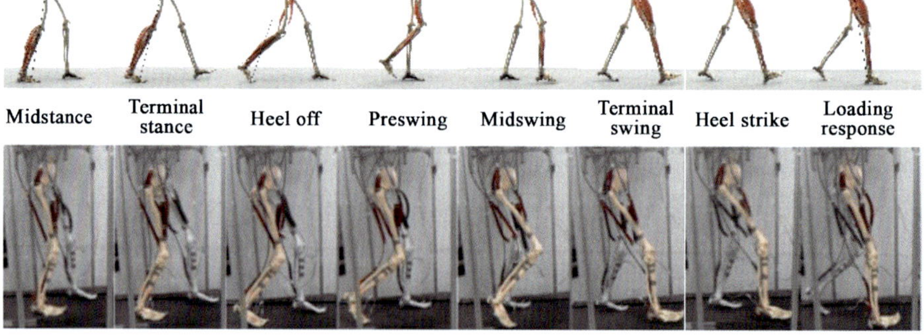

FIG. 18.14 WPI Popovic Labs' highly biomimetic legs consisting of bones, ligaments, tendons, and pneumatically actuated muscles (Hydro Muscles) are driven through biologically realistic joint trajectories; computer model (top) and physical prototype (bottom). *Courtesy of Marko Popovic.*

vessel that can be exchanged with something else?" The debate involving all these paramount questions has already started and it is also concerned with various ethical issues. Although many agree that humans should accelerate their evolution by taking advantage of the technological progress, many others still consider such scenarios as omens of a dystopian future.

In our opinion, the human body including brain is indeed a fragile "vessel" that can be enhanced using biomechatronic technologies. Nobody can predict what the extent of this enhancement will be. From advanced brain implants, for example, memory implants or neural prosthesis, and more conventional microprocessor-controlled limb prostheses to even surrogate humanoid robots that offer a new, improved "vessel," the possibilities appear endless.

The new biomechatronics age human will be shaped based on a unique trinity:

(1) Concerns and fears of "ominous" technological development and its effects on human race, requiring careful assessment and debate on the disadvantages and advantages.

(2) Celebration of technologies that can help us stay what we are, for example, with a brain implant that can eliminate Alzheimer disease, or help us enjoy life as everyone else, for example, with new advanced limb prosthesis that can let us walk again or that can help us regain lost dexterity.

(3) The pinnacle of human spirit that always wonders and tries hard to push boundaries—to explore beyond what is already known or to create beyond what has been already made. The human mind, love and spirit, is a very powerful engine that could create things that might improve the quality of life for all of us in ways that no one has yet imagined.

So, imagine...

References

[1] http://www.imdb.com/title/tt0071054/.

[2] http://www.imdb.com/title/tt0073965/.

[3] M. Caidin, Cyborg, Ballantine Books, 1978.

[4] A. Miriyev, K. Kenneth Stack, H. Lipson, Soft material for soft actuators, Nat. Commun. 8 (2017).

[5] S. Ogden, L. Klintberg, G. Thornell, K. Hjort, R. Bodén, Review on miniaturized paraffin phase change actuators, valves, and pumps, Microfluid. Nanofluid. 17 (2014) 53–71.

[6] E.T. Carlen, C.H. Mastrangelo, Electrothermally activated paraffin microactuators, J. Microelectromech. Syst. 11 (2002) 165–174.

[7] J.I. Lipton, S. Angle, R.E. Banai, E. Peretz, H. Lipson, Electrically actuated hydraulic solids, Adv. Eng. Mater. 18 (2016) 1710–1715.

[8] R. Altmüller, R. Schwödiauer, R. Kaltseis, S. Bauer, I.M. Graz, Large area expansion of a soft dielectric membrane triggered by a liquid gaseous phase change, Appl. Phys. A Mater. Sci. Process. 105 (2011) 1–3.

[9] B.C. MacMurray, X. An, S. Robinson, I. van Meerbeek, K. O'Brien, H. Zhao, R. Shepherd, Poroelastic foams for simple fabrication of complex soft robots, Adv. Mater. 27 (2015) 41.

[10] B. MacMurray, C. Futran, J.C. Lee, K. O'Brien, M. Amiri, B. Mosadegh, M. Silberstein, J. Min, R. Shepherd, Compliant buckled foam actuators and application in patient-specific direct cardiac compression, Soft Robot. (2017).

[11] https://softroboticstoolkit.com/foam-based-soft-actuators/design (02.04.18).

[12] https://spectrum.ieee.org/automaton/robotics/robotics-hardware/an-edible-actuator-for-ingestible-robots (02.04.18).

[13] J. Wade, T. Bhattacharjee, C. Kemp, Force and thermal sensing with a fabric-based skin, in: IROS Workshop on See, Touch, and Hear: 2nd Workshop on Multimodal Sensor-Based Robot Control for HRI and Soft Manipulation, 2016.

[14] D. Graupe, J. Salahi, K.H. Kohn, Multifunctional prosthesis and orthosis control via microcomputer identification of temporal pattern differences in single site myoelectric signals, J. Biomed. Eng. 4 (1) (1982) 17–22.

[15] P.K. Artemiadis, K.J. Kyriakopoulos, EMG-based control of a robot arm using low-dimensional embeddings, IEEE Trans. Robot. 26 (2) (2010) 393–398.

[16] J. Vogel, C. Castellini, P.P. van der Smagt, EMG-based teleoperation and manipulation with the DLR LWR-III, in: IEEE/RSJ International Conference on Intelligent Robots and Systems (IROS), 2011, pp. 672–678.

[17] C. Cipriani, F. Zaccone, S. Micera, M.C. Carrozza, On the shared control of an EMG-controlled prosthetic hand: analysis of user prosthesis interaction, IEEE Trans. Robot. 24 (1) (2008) 170–184.

[18] L. Lucas, M. DiCicco, Y. Matsuoka, An EMG-controlled hand exoskeleton for natural pinching, J. Robot. Mechatr. 16 (5) (2004) 482–488.

[19] E. Costanza, S.A. Inverso, R. Allen, P. Maes, Intimate interfaces in action: assessing the usability and subtlety of EMG-based motionless gestures, in: Proceedings of the SIGCHI conference on Human factors in computing systems, ACM, New York, NY, 2007, pp. 819–828.

[20] H. Benko, S. Saponas, D. Morris, D. Tan, Enhancing input on and above the interactive surface with muscle sensing, in: ITS '09 Proceedings of the ACM International Conference on Interactive Tabletops and Surfaces, 2009.

[21] D. Sierra González, C. Castellini, A realistic implementation of ultrasound imaging as a human–machine interface for upper-limb amputees, Front. Neurorobot. 7 (2013)17 https://doi.org/10.3389/fnbot.2013.00017.

[22] D. Yungher, M. Wininger, W. Baar, W. Craelius, A. Threlkeld, Surface muscle pressure as a means of active and passive behavior of muscles during gait, Med. Eng. Phys. 33 (2011) 464–471. https://doi.org/10.1016/j.medengphy.2010.11.012.

[23] T.D. Lalitharatne, K. Teramoto, Y. Hayashi, K. Kiguchi, Towards hybrid EEG-EMG-based control approaches to be used in bio-robotics applications: current status, challenges and future directions, Paladyn, J. Behav. Robot. 4 (2) (2013) 147–154.

[24] M. Marković, S. Dosen, C. Cipriani, D.B. Popović, D. Farina, Stereovision and augmented reality closed loop control of grasping in hand prostheses, J. Neural Eng. 11 (2013) 046001.

[25] C. Castellini, P. Artemiadis, M. Wininger, A. Ajoudani, M. Alimusaj, Bicchi, D. Farina, Proceedings of the first workshop on peripheral machine interfaces: going beyond traditional surface electromyography, Front. Neurorobot. 8 (2014) 22.

[26] R.M. Neely, D.K. Piech, S.R. Santacruz, M.M. Maharbiz, J.M. Carmena, Recent advances in neural dust: towards a neural interface platform, Curr. Opin. Neurobiol. 50 (2018) 64–71.

[27] A. Gijsberts, R. Bohra, D. Sierra González, A. Werner, M. Nowak, B. Caputo, C. Castellini, Stable myoelectric control of a hand prosthesis using non-linear incremental learning, Front. Neurorobot. 8 (8) (2014).

[28] P.M. Pilarski, M.R. Dawson, T. Degris, F. Fahimi, J.P. Carey, R.S. Sutton, Online human training of a myoelectric prosthesis controller via actor-critic reinforcement learning, in: IEEE International Conference on Rehabilitation Robotics (ICORR), 2011, pp. 1–7.

[29] P. Xia, H. Jie, P. Yinghong, EMG-based estimation of limb movement using deep learning with recurrent convolutional neural networks. Artif. Organs 42 (5) (2018) E67–E77, doi:10.1111/aor.13004.

[30] Z. Zou, C. Zhu, Y. Li, X. Lei, W. Zhang, J. Xiao, Rehealable, fully recyclable, and malleable electronic skin enabled by dynamic covalent thermoset nanocomposite, Sci. Adv. 4 (2) (2018).

[31] B.J. Jank, L. Xiong, P.T. Moser, J.P. Guyette, X. Ren, C.L. Cetrulo, D.A. Leonard, L. Fernandez, S. P. Fagan, H.C. Ott, Engineered composite tissue as a bioartificial limb graft, Biomaterials 61 (2015) 246–256.

[32] M.S. Mannoor, Z. Jiang, T. James, Y.L. Kong, K.A. Malatesta, W.O. Soboyejo, N. Verma, D.H. Gracias, M. C. McAlpine, 3D printed bionic ears, Nano Lett. 13 (6) (2013) 2634–2639.

[33] V. Mironov, T. Boland, T. Trusk, G. Forgacs, R.R. Markwald, Organ printing: computer-aided jet-based 3D tissue engineering, Trends Biotechnol. 21 (4) (2003) 157–161.

[34] A.M. Dollar, R.D. Howe, The SDM hand as a prosthetic terminal device: a feasibility study, in: IEEE 10th International Conference on Rehabilitation Robotics, 2007.

[35] A.M. Dollar, R.D. Howe, The highly adaptive SDM hand: design and performance evaluation, Int. J. Robot. Res. 29 (5) (2010) 585–597.

[36] M.-S. Scholz, J.P. Blanchfield, L.D. Bloom, B.H. Coburn, M. Elkington, J.D. Fuller, M.E. Gilbert, S. A. Muflahi, M.F. Pernice, S.I. Rae, J.A. Trevarthen, S.C. White, P.M. Weaver, I.P. Bond, The use of composite materials in modern orthopaedic medicine and prosthetic devices: a review, Compos. Sci. Technol. 71 (16) (2011) 1791–1803.

[37] F. Thorsteinsson, I. Gudmundsson, C. Lecomte, Prosthetic and Orthotic Devices Having Magnetorheological Elastomer Spring with Controllable Stiffness, https://patents.google.com/patent/US9724210B2/en.

[38] M.K. Shepherd, E.J. Rouse, The VSPA foot: a quasi-passive ankle-foot prosthesis with continuously variable stiffness, IEEE Trans. Neural Syst. Rehabil. Eng. 25 (12) (2017) 2375–2386.

[39] E. Saint-Elme, et al., Design of a biologically accurate prosthetic hand, in: 2017 International Symposium on Wearable Robotics and Rehabilitation (WeRob), IEEE, 2017.

[40] G.P. Kontoudis, M.V. Liarokapis, A.G. Zisimatos, C.I. Mavrogiannis, K.J. Kyriakopoulos, Open-source, anthropomorphic, underactuated robot hands with a selectively lockable differential mechanism: towards affordable prostheses, in: IEEE/RSJ International Conference on Intelligent Robots and Systems (IROS), Hamburg (Germany), 2015.

[41] http://www.bionicboot.com/.

[42] http://www.air-trekkers.com/.

[43] B.B. Kang, et al., Development of a polymer-based tendon-driven wearable robotic hand, in: IEEE International Conference on Robotics and Automation (ICRA), 2016.

[44] https://en.wikipedia.org/wiki/IBOT#/media/File:Clinton-kamen.jpg.

[45] P. Mich, The emerging role of veterinary orthotics and prosthetics (V-OP) in small animal rehabilitation and pain management, Topics Compan. Anim. Med. 29 (2014) 10–19.

[46] A. Desrochers, G. St. Jean, D. Anderson, Limb amputation and prosthetics, Vet. Clin. Food Anim. 30 (2014) 143–155.

[47] http://cal.vet.upenn.edu/projects/saortho/chapter_48/48mast.htm (02.04.18).

[48] B. Goldner, A. Fuchs, I. Nolte, N. Schilling, Kinematic adaptations to tripedal locomotion in dogs, Vet. J. 204 (2) (2015) 192–200.

[49] S. Jarvis, D. Worley, S. Hogy, A. Hill, K. Haussler, R. Reiser, Kinematic and kinetic analysis of dogs during trotting after amputation of a thoracic limb, Am. J. Vet. Res. 74 (9) (2013) 1155–1163.

[50] G. Cole, D. Millis, The effect of limb amputation on standing weight distribution in the remaining three limbs in dogs, Vet. Comp. Orthop. Traumatol. 30 (1) (2017) 59–61.

[51] W. Kano, S. Rahal, L. Mesquita, F. Agostinho, L. de Faria, Gait analysis in a cat with scapular luxation and contralateral forelimb amputation, Can. Vet. J. 54 (10) (2013) 990–991.

[52] M. Goldberg, J. Tomlinson, Physical Rehabilitation for Veterinary Technicians and Nurses, John Wiley & Sons Inc., 2018

[53] F. Diep, How 3-D printing made the perfect prosthetic legs for Derby the dog, Pop. Sci. (2014). https://www.popsci.com/how-3-d-printing-made-perfect-prosthetic-legs-derby-dog.

[54] http://www.k-9orthotics.com (02.04.18).

[55] http://advocacy.britannica.com/blog/advocacy/2010/02/animal-prosthetics-surviving-on-human-ingenuity-and-compassion/ (02.04.18).

[56] N. Fitzpatrick, T. Smith, C. Pendegrass, R. Yeadon, M. Ring, A. Goodship, G. Blunn, Intraosseous transcutaneous amputation prosthesis (ITAP) for limb salvage in 4 dogs. Vet. Surg. 40 (8) (2011) 909–925, doi:10.1111/j.1532-950X.2011.00891.x.

[57] http://www.horsetalk.co.nz/news/2011/02/026.shtml (02.04.18).

[58] https://3dprintingindustry.com/news/10-animals-who-got-a-2nd-chance-in-lifewith-3d-printing-51221 (02.04.18).

[59] K. Ten, G. Smit, P. Breedveld, 3D-printed upper limb prostheses: a review, Disabil. Rehabil. Assist. Technol. 12 (3) (2017) 300–314.

[60] M. Barry, Machine Man, Vintage, 2011.

[61] http://www.imdb.com/title/tt1219827/.

[62] E. Saint-Elme, M.A. Larrier Jr., C. Kracinovich, D. Renshaw, K. Troy, M. Popovic, Design of a biologically accurate prosthetic hand, in: IEEE RAS International Symposium on Wearable & Rehabilitation Robotics Houston, TX, November 5–8, 2017.

[63] S.B. Kesner, L. Jentoft, F.L. Hammond, R.D. Howe, M.B. Popovic, Design considerations for an active soft orthotic system for shoulder rehabilitation, in: 33rd Annual International IEEE EMBS Conference, August 30–September 02, Boston, USA, 2011.

[64] I. Galiana, F.L. Hammond, R.D. Howe, M.B. Popovic, Wearable soft robotic device for post-stroke shoulder rehabilitation: identifying misalignments, in: 2012 IEEE/RSJ International Conference on Intelligent Robots and Systems, October 7–12, Portugal, 2012.

[65] WPI Popovic Labs, Towards a Biomimetic Musculoskeletal Humanoid Robot with Human Movement Capabilities, 2018, http://users.wpi.edu/~mpopovic/pages/BiomimeticHumanoidRobot.html (Accessed on 2/12/2019).

19

Practice Problems

Marko B. Popovic[*], Kathleen A. Lamkin-Kennard[†], Hiroyuki Tashiro[‡], Philipp Beckerle[§,¶¶], Steffen Willwacher[¶], Minas Liarokapis[||], Michelle J. Johnson[#], Adam D. Goodworth[**], Pinar Boyraz[††,‡‡], Ivo Dobrev[§§]

[*]WORCESTER POLYTECHNIC INSTITUTE, WORCESTER, MA, UNITED STATES [†]ROCHESTER INSTITUTE OF TECHNOLOGY, ROCHESTER, NY, UNITED STATES [‡]KYUSHU UNIVERSITY, FUKUOKA, JAPAN [§]TECHNISCHE UNIVERSITÄT DORTMUND, DORTMUND, GERMANY [¶]GERMAN SPORT UNIVERSITY COLOGNE, COLOGNE, GERMANY [||]THE UNIVERSITY OF AUCKLAND, AUCKLAND, NEW ZEALAND [#]UNIVERSITY OF PENNSYLVANIA, PHILADELPHIA, PA, UNITED STATES [**]UNIVERSITY OF HARTFORD, HARTFORD, CT, UNITED STATES [††]CHALMERS UNIVERSITY OF TECHNOLOGY, GOTHENBURG, SWEDEN [‡‡]ISTANBUL TECHNICAL UNIVERSITY, ISTANBUL, TURKEY [§§]UNIVERSITY HOSPITAL ZURICH, UNIVERSITY ZURICH, ZÜRICH, SWITZERLAND [¶¶]TECHNISCHE UNIVERSITÄT DARMSTADT, DARMSTADT, GERMANY

Problem 1: Numerical Problem, Difficulty Level: *, Chapter 2

(Prepared by Marko Popovic)

(A) Use the virtual work method to estimate magnitude and direction of force acting on the end effector of the two-dimensional (2D) two-link system in horizontal plane depicted in Fig. 19.1. Torque in joint 1 is 2.5 Nm and torque in joint 2 is 3.0 Nm. Both torques are acting in positive z-direction (i.e., they tend to produce counterclockwise motion). System is static, that is, not moving due to that unknown force applied on the end effector. Here, $L_1 = 0.45$ m, $L_2 = 0.35$ m, $\theta_1 = 30^\circ$, $\theta_2 = 15^\circ$. Ignore friction.

(B) Consider a simple one-link system in vertical plane shown in Fig. 19.2. Ignore force F. Estimate required joint torque for time instances $t = 1, 3, 5$ s by using the Euler Lagrange method for the following motion $\theta_1 = 20^\circ + 10^\circ \cos(\omega t)$ with $\omega = 0.88$ rad/s. Here, $L_1 = 0.45$ m, $L_{CM1} = 0.30$, $m_1 = 1.2$ kg, $I_1 = 0.16$ kg m^2.

(C) Include external force $\vec{F} = 12\,\text{N}\,\hat{i}$ acting on the end effector and estimate joint torque. Everything else is as in part (B).

Biomechatronics. https://doi.org/10.1016/B978-0-12-812939-5.00019-7

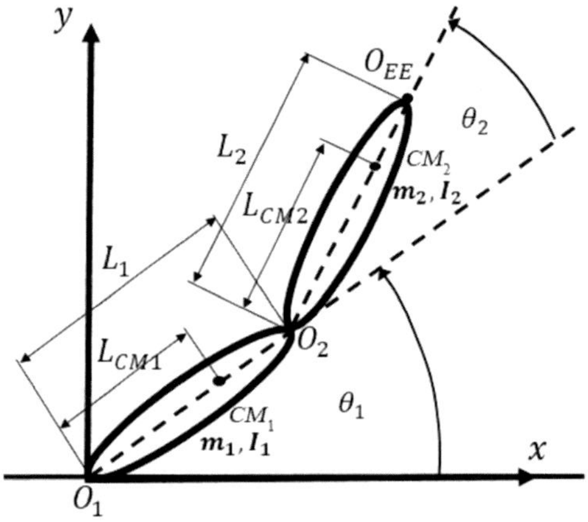

FIG. 19.1 A two-dimensional two-link system in horizontal plane.

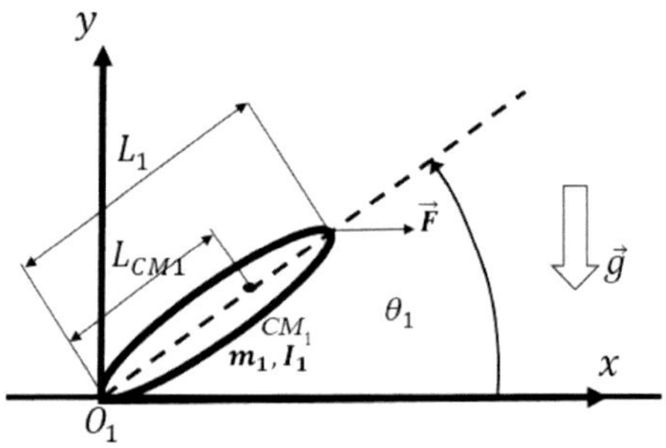

FIG. 19.2 A simple one-link system in vertical plane.

Problem 2: Numerical problem, difficulty level: *, Chapter 2

(Prepared by Marko Popovic)

In this problem we will study how to obtain joint torques from known joints motion. We will (maximally) simplify this problem and consider only upper arm and shoulder angle (ignore everything related to elbow angle and wrist position).

Trajectory of upper arm is defined as

$$\theta_1 = \frac{3}{8}\pi + \frac{1}{8}\pi\cos\left(\frac{\pi}{3\,\mathrm{s}}t + \frac{\pi}{16}\right) \tag{19.1}$$

with time t in seconds. The length and mass of the upper arm are $L_1 = 20$ cm and $m_1 = 3$ kg, respectively. Treat the segment's moment of inertia as the moment of inertia of rod. Assume that the gravity vector is pointing vertically downward within the arm plane.

Calculate shoulder joint torque as function of time and find maximal shoulder joint torque required to accomplish this task (Fig. 19.3).

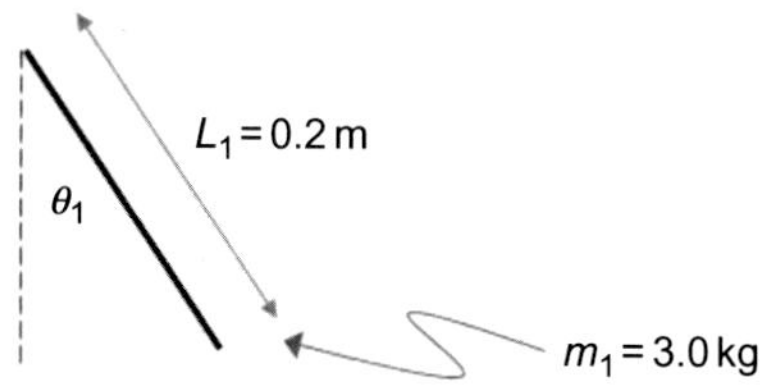

FIG. 19.3 Upper arm and shoulder angle.

Problem 3: Numerical problem, difficulty level: ***, Chapters 2, 12, 16, and 17

(Prepared by Marko Popovic)

After slipping on an icy surface, a 50 kg child is in contact with ground only with stance foot heel and body is tilted backward in respect to the vertical direction by some initial angle θ_0 (Fig. 19.4). You may assume that all initial velocities are zero. To stabilize body, swing leg and upper body can generate constant 200 Nm moment about body's center of mass for only 50 ms. What is the maximal angle θ_0 that can be still stabilized such that center of mass eventually ends up just above the heel of stance leg with zero final speed? Assume that distance between heel and center of mass is $L = 1.00$ m throughout the entire movement. Because balance is achieved through application of moment about the center of mass this is an example of *moment balance strategy.*

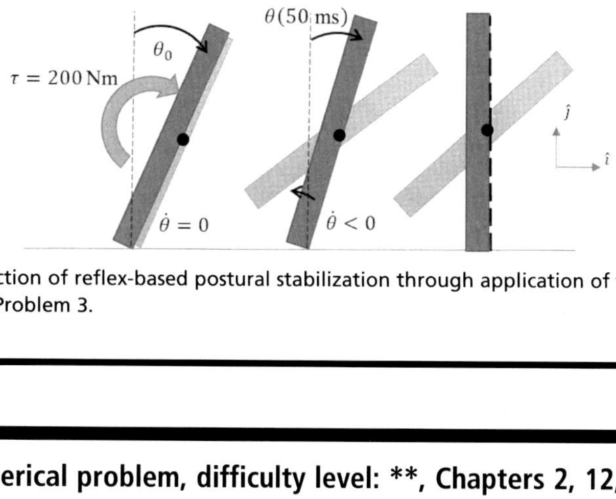

FIG. 19.4 Symbolic depiction of reflex-based postural stabilization through application of the moment balance strategy as specified in Problem 3.

Problem 4: Numerical problem, difficulty level: **, Chapters 2, 12, and 17

(Prepared by Marko Popovic)

Legged locomotion is under-actuated as foot is not bolted or glued to the ground.

To make things simpler, here we consider point-like foot located at the coordinate origin ($x=0, y=0$) as shown below (Fig. 19.5). Angle between ground and shank is deemed "ankle" angle θ_a.

Here, neither "ankle" θ_a nor "trunk" θ_t angles are directly controlled quantities; they are consequence of body dynamics expressed by "knee" θ_k and "hip" θ_h angles.

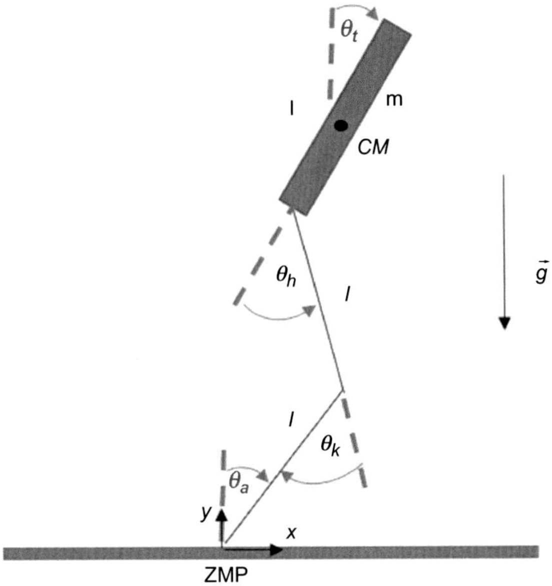

FIG. 19.5 Single leg and trunk balancing.

Imagine that you could extremely precisely control "knee" θ_k and "hip" θ_h angles and that entire mass is located within the trunk.

During single-support phase of walking, only one leg is in contact with the ground. Because swing leg is massless, dynamics of stance leg and trunk is completely independent from swing leg motion.

Goal here is to study how body posture (represented by stance leg and trunk) in respect to global frame is affected by "knee" θ_k and "hip" θ_h trajectories.

Leg is massless and trunk has mass m uniformly distributed along the rod shown in Fig. 19.5. Hence moment of inertia about its center of mass, also coinciding with CM of entire body, is

$$I = \frac{ml^2}{12}. \tag{19.2}$$

The torque, that is, moment of force is *into* the page for positive trunk angular acceleration and

$$\tau = I\ddot{\theta}_t = \frac{ml^2}{12}\frac{d^2\theta_t}{dt^2} = -\tau_z \tag{19.3}$$

(and with positive z-axis pointing *out of* the page).

Gravity gives zero contribution to torque and hence only ground reaction force contributes to CM moment generation.

The CM moment due to ground reaction force can be expressed as

$$\vec{\tau} = \left(\vec{r}_{ZMP} - \vec{r}_{CM}\right) \times \vec{F}_{GroundReaction} = -\vec{r}_{CM} \times \vec{F}_{GroundReaction} = -\vec{r}_{CM} \times \left(m\vec{a}_{CM} - m\vec{g}\right) \tag{19.4}$$

or

$$\tau_z = m\left[y_{CM}\ddot{x}_{CM} - x_{CM}\left(\ddot{y}_{CM} + g\right)\right] \tag{19.5}$$

with $g = 9.81\ \mathrm{m/s^2}$.

Note that we obtained two expressions for the same quantity, namely CM moment, and these two expressions must be equal, that is,

$$m\left[y_{CM}\ddot{x}_{CM} - x_{CM}\left(\ddot{y}_{CM} + g\right)\right] = -\frac{ml^2}{12}\frac{d^2\theta_t}{dt^2} \tag{19.6}$$

This is another way to state "zero-moment point (ZMP) realism" (i.e., that ZMP point must be within foot support polygon such that dynamics is *physically plausible*).

The center of mass is completely defined if "ankle," "knee," and "hip" angles are known and "trunk" angle in respect to the vertical direction is also known based on "ankle," "knee," and "hip" angles.

We can now use this condition to predict trajectory of "ankle" θ_a for given trajectories of "knee" θ_k and "hip" θ_h angles and for known initial condition for "ankle" angle and angular velocity.

Assume that $m = 30$ kg, $l = 0.50$ m, $\theta_a(t=0) = 0$, $\dot{\theta}_a(t=0) = 0$, and that

$$\theta_k(t) = \frac{\pi}{16}\left(1 - \cos\frac{\pi}{32}\frac{rad}{s}t\right) \tag{19.7}$$

$$\theta_h(t) = \frac{\pi}{8}\sin\frac{\pi}{32}\frac{rad}{s}t \tag{19.8}$$

Find trajectories of "ankle" θ_a and "trunk" θ_t angles.

If possible create a movie showing dynamics of body. How long it will take before it falls? Will it fall forward or backward? Is your result dependent on mass m and length l?

Problem 5: Numerical problem, difficulty level: ***, Chapters 2 and 17

(Prepared by Marko Popovic)

In this problem, we will study a simplified model of *terrestrial* locomotion in plane parallel to gravity, specifically single step of bipedal robot (with point-like feet).

You may consider legs to be massless and you may consider that the entire robot mass, $M = 100\,\mathrm{kg}$, is centered at hip joint location. The shank and shin are both $L = 0.65\,\mathrm{m}$ long and desired step length is $D = 0.75\,\mathrm{m}$. Initial velocity of the center of mass (i.e., hip) is $1.00\,\mathrm{m/s}$ in the horizontal direction and final velocity of the center of mass is same as initial velocity. The final state of the robot is the same as the initial state with left and right legs swapped (state means both joint angles and joint angular velocities). The boundary condition angles are labeled in Fig. 19.6 and joint angular velocities are unspecified. The right leg is drawn as a thick line (for both initial and final posture) in Fig. 19.6.

The joint angles are as labeled in Fig. 19.7. Hip joint angles are defined in respect to robot's trunk (i.e., angle between trunk and upper leg) and knee joint angles are defined as angles

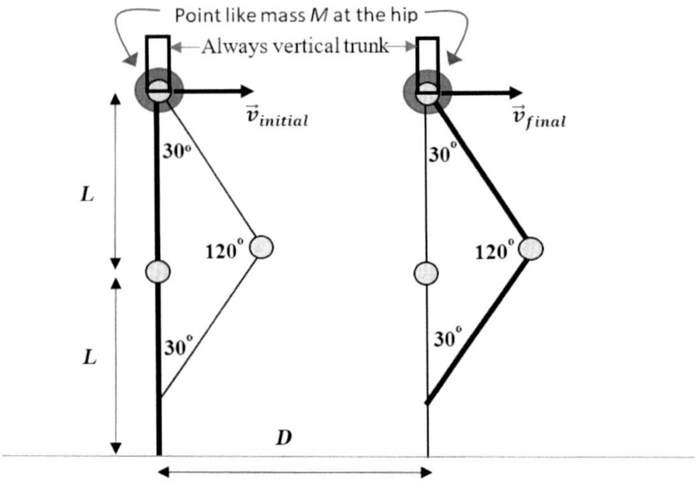

FIG. 19.6 A simplified bipedal model and steady-state walking boundary conditions.

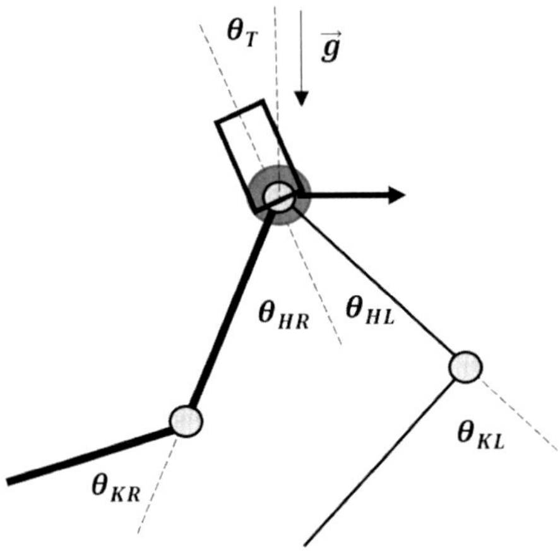

FIG. 19.7 A simplified bipedal model and joint angles convention.

between upper and lower leg segments. There is also the trunk angle defined as angle between trunk and vertical direction.

(A) Your task is to find joint angle trajectories for steady-state walking (with given initial and final conditions) that support desired step length and keep the trunk always vertical (i.e., parallel to gravity). Steady state assume left right legs symmetry; such that left leg final joint angles and joint angular velocities are same as right leg initial joint angles and joint angular velocities and that left leg initial joint angles and joint angular velocities are same as right leg final joint angles and angular velocities. Make sure that obtained trajectories are physically possible.

(Extra Credit)

(B) Make a stick figure movie with your walker walking (try to scale X and Y appropriately so it looks realistic and choose movie time step such that it indeed corresponds to physical time).

If possible, loop the movie (say 20, 30 times) so that one could see your walker in action walking steadily for at least 10–15 s.

The movie is another way that you could judge whether you successfully matched the proper boundary condition; if angles are fine and transition from end to start is smooth, you probably did a good job.

The transitions between single and double-support phase is also critical. Think what should be the good conditions for these transitions.

Is motion you obtained biologically realistic? How you could make it even more "biological"?

Problem 6: Numerical problem, difficulty level: ***, Chapters 2 and 17

(Prepared by Marko Popovic)

In this problem, we will study simplified model of *terrestrial* locomotion in plane parallel to gravity, specifically single step of bipedal robot (with point-like feet) while going upstairs.

Very similar to previous problem, here you may consider legs to be massless and you may consider that entire robot mass, $M = 100\,\text{kg}$, is centered at hip joint location. The shank and shin are both $L = 0.65\,\text{m}$ long. Desired step upstairs is shown in Fig. 19.8. The foot is placed in front of previous stair at distance $D_0 = 0.07\,\text{m}$ and behind the next stair at distance $D_1 = 0.21\,\text{m}$. The stair height is $H = 0.21\,\text{m}$. Initial velocity of the center of mass (i.e., hip) is $0.20\,\text{m/s}$ in the horizontal direction and final velocity of the center of mass is same as initial velocity. The final state of the robot is same as initial state with left and right legs swapped (state means both joint angles and joint angular velocities). The boundary condition angles are labeled in Fig. 19.8

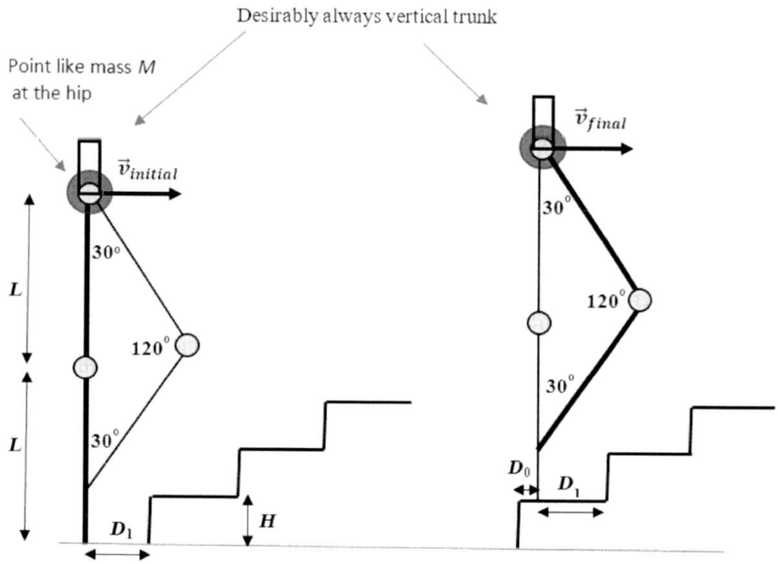

FIG. 19.8 A simplified bipedal model and steady-state walking upstairs boundary conditions.

and joint angular velocities are unspecified. The right leg is drawn as thick line (for both initial and final posture) in Fig. 19.8.

The joint angles are labeled as in previous problem.

(A) Your task is to find joint angle trajectories for steady-state walking upstairs (with given initial and final conditions) that support desired step length and keep the trunk always vertical (i.e., parallel to gravity). Steady state assume left right legs symmetry; such that left leg final joint angles and joint angular velocities are same as right leg initial joint angles and joint angular velocities and that left leg initial joint angles and joint angular velocities are same as right leg final joint angles and angular velocities.

(Extra Credit)

(B) Make a stick figure movie with your walker walking upstairs (try to scale X and Y appropriately so it looks realistic and choose movie time step such that it indeed corresponds to physical time). If possible, loop the movie (say 10 times) so that one could see your walker in action walking upstairs steadily for at least 10 s.

Problem 7: Numerical problem, difficulty level: *, Chapter 2

(Prepared by Marko Popovic)

A 1000 kg fixed wing plane with 12 m wing span has right triangle-shaped wing's cross section, see Fig. 19.9, with $a = 2.0$ m and $b = 0.75$ m. What is the speed and minimal thrust of the plane at the moment of takeoff? Assume air density 1.225 kg/m^3 and pressure drag coefficient $C_d \cong 2$. Ignore friction drag. Assume that the plane's central body can be represented by the point-like mass.

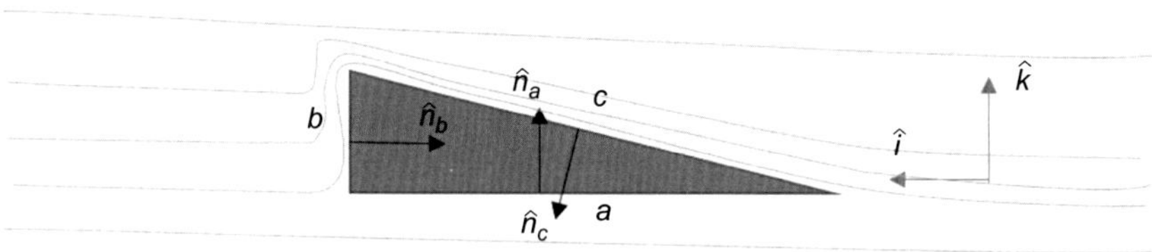

FIG. 19.9 A triangle-shaped wing's cross section and air streamlines.

Problem 8: Numerical problem and concept question, difficulty level: **, Chapters 2 and 17

(Prepared by Marko Popovic)

Estimate wing's frequency, that is, number of flapping cycles per second, for flyer shown in Fig. 19.10.

Easier (less realistic) version of problem: ignore vertical speed of flyer's central body.

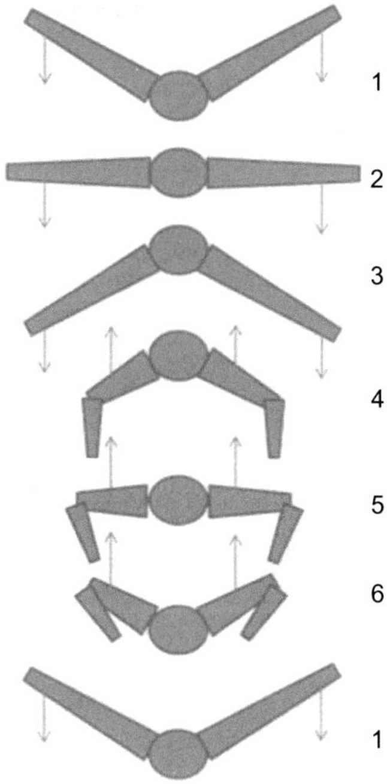

FIG. 19.10 Sequence of flapping wings flyer's poses.

Tougher (more realistic) version of problem: consider vertical speed of flyer's central body too.

Assume:

- constant angular speed of the massless wings,
- (−30, 30) degrees range of wing's angles with respect to horizontal,
- $R = 1\,\text{m}$ effective length of the wing during wings' downward motion,
- $R' = 0.5\,\text{m}$ effective length of the wing during wings' upward motion,
- $m = 3\,\text{kg}$ flyer's point-like mass and $g = 9.81\,\text{m/s}^2$,
- air density $\rho = 1.2250\ \text{kg·m}^{-3}$,
- coefficient of Newton's drag $C_D = 2$,
- Newton's drag on infinitesimal surface area vector, $d\vec{A} = \hat{n}_A dA = \hat{n}_A drdw$ with $0 \le r \le R$ $(or\ R')$ and $0 \le w \le W$ with $W = 1$ m can be expressed as $\frac{1}{2} C_D \rho v \left(\vec{v} \cdot d\vec{A} \right) (-\hat{n}_A)$ with $\vec{v}$ being the air surface relative velocity,
- the flyer keeps the steady average height,
- transition from poses 3 to 4 and 6 to 1 is instantaneous,
- the net vertical force is obtained after integrating vertical component of Newton's drag over both wings' lengths and subtracting weight, and
- the viscosity of air 1.78×10^{-5} kg/m s, or 17.8 μPa s, or 1.78×10^{-5} Pa s, at 15°C.

What is the value of Reynolds dimensionless number? Assume that linear dimension $D = 1\,\text{m}$ (i.e., fully extended wing's length), and that characteristic speed is equal to product of 1 m and angular speed.

Is the flow around wings laminar of turbulent?

Do you expect that substantial (measurable) "flyer's" energy will be lost on eddy currents and other turbulent phenomena?

Is assumption of Newton's drag appropriate?

Problem 9: Numerical problem, difficulty level: ***, Chapters 2 and 17

(Prepared by Marko Popovic)

In this problem, we study a simplified model of *aquatic* locomotion in a plane perpendicular to gravity, specifically axial locomotion with simplified motion of caudal fin.

From point A to point B, massless fin is moving with angular speed ω_1 and from point B to point C, fin moves with angular speed $\omega_2 < \omega_1$ (Fig. 19.11). From point C to point B, fin is again moving with angular speed ω_1 and finally from point B to point A, fin again moves with angular speed $\omega_2 < \omega_1$.

Assume that fin's length and width (in vertical plane) are $L = 1.50\,\text{m}$ and $W = 0.60\,\text{m}$, respectively (i.e., fin area is $0.90\,\text{m}^2$; ignore fin thickness). Fin is always completely submerged and perpendicular to water surface. Angle AOB and angle BOC are 15° each.

Assume that fish cross-sectional area is $0.04\,\text{m}^2$.

Goal of this problem is to estimate ω_1 for known $\omega_2 = 10.0\frac{\text{rad}}{\text{s}}$ such that fish is moving with an average forward velocity of $10.0\,\text{mph} = 4.47\,\text{m/s}$. Ignore effect of sideway (lateral) forces and consider only axial dynamics.

(If you need, you may use small angle approximation, for example, $\sin\theta \cong \theta$. Series expansion is also fine.)

(A) Is the Reynolds number small or large? Are drag forces proportional to speed or speed squared?

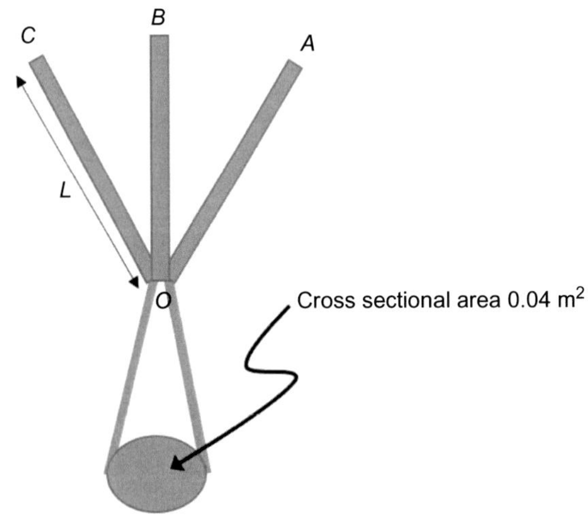

FIG. 19.11 A simplified model of a fish with maximal angular excursions of the caudal fin depicted.

Calculate slowing down drag on cross-sectional area $0.04\,m^2$.

Assume that frontal drag forces (slowing down fish) are equal to product of water density $1000\,kg/m^3$, (vertical) cross-sectional area, and function of speed $f(v)=v$ or v^2 (depending on your answer). For quadratic dependence, you may assume $C_d \cong 2$. Ignore friction drag.

(B) Assume that water is streaming from head to caudal fin with speed v in the fish rest frame and that fin is moving fast enough that it could still propel water backward and receive some thrust from drag forces (Fig. 19.12).

For example, consider small element of fin with area Wdr moving with speed $r\omega$. This element is colliding with water with speed equal to $r\omega - v\sin(\theta)$. Hence, drag force acting on that element is equal to $dF=\rho f(r\omega - v\sin(\theta))Wdr$ and force is perpendicular to fin. Therefore, the thrust component contribution of the same element is $\sin(\theta)\rho f(r\omega - v\sin(\theta))Wdr$.

In other words, for when fin is moving toward axial axis, the elements of fin more proximal to fish body will be slowing down the fish and more distal elements will tend to accelerate fish.

Integrate thrust component along fin to find net thrust. Note, when $\omega=\omega_1$, for r between $vsin(\theta)/\omega$ and L contribution is positive whereas drag forces slowing down fish are obtained for $0<r<v\sin(\theta)/\omega$. Note, when $\omega=\omega_2$, contribution of all elements will be to slow down fish.

At each instance of time, net fin force (positive or negative) will be function of v, $\omega=\omega_1$ *or* ω_2 and angle θ.

(C) Find net force acting on fish, that is, answer in (b) minus answer in (a). Ignore effect of sideway forces. When fin is moving toward axial axis, net force will be positive. When fin is moving away from axial axis, net force will be negative.

(D) Find acceleration of fish if fish mass is $m=30\,kg$ by using Newton's second law.

(E) Find condition for steady state. Note that at the end of a full cycle, the fish's speed should be equal as at the beginning. Hence, integral of acceleration over entire cycle should be zero.

(F) Determine desired ω_1. Is there a solution? Is it uniquely determined? What is the period of cyclic motion? How does it compare to corresponding period for fish of similar size moving with similar speed? [If you are citing some credible source include reference.]

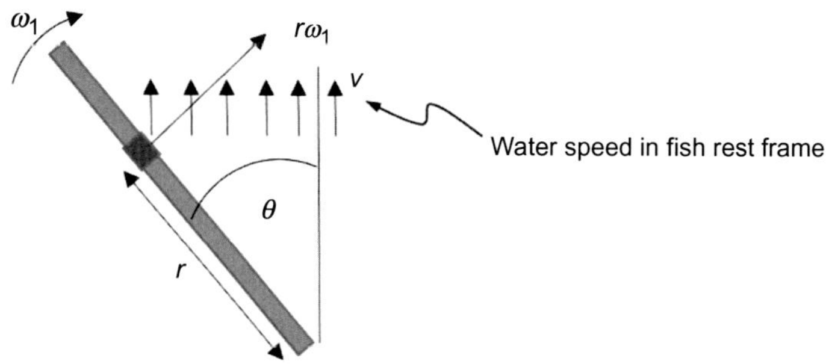

FIG. 19.12 Relative motion of caudal fin and water.

Problem 10: Numerical problem, difficulty level: *, Chapter 3

(Prepared by Marko Popovic)

Imagine a novel actuator consisting of two fixed solenoids with opposite direction of current and a single inner toroid-like solenoid that could rotate about the vertical axis. Outer radius of toroid-like solenoid is $R_2=1$ cm (assume that inner radius is zero) whereas radius of two fixed solenoids is $R_1=1.2$ cm; see Fig. 19.13. The total number of coil turns in toroid-like solenoid is $N_2=96$ and linear density of coil turns in ordinary solenoids is $n_1=10{,}000$ turns/m.

(A) Find the torque exerted by toroid-like solenoid generated by interaction between solenoids as function of currents I_1 and I_2. Assume that top and bottom bases of toroid-like solenoids are embedded within mumetal (75% nickel, 2% chromium, 5% copper, and 18% iron) which has relative permeability of 20,000. (Ordinary magnetic iron has relative permeability of 200.)

(B) If the motor shaft has radius $R_3=0.5$ cm, find the force in thin cable attached to the motor shaft for currents $I_1=I_2=1A$. (The motor force output can be increased by addition of several concentric toroid-like solenoids.)

(C) How would your results from (a) and (b) change if $I_1=I_2=10A$?

(D) Explain why your results do not depend on dimension (size) of two static solenoids.

(E) If motor rotates with 100 rpm (rotations per minute) for $I_1=I_2=10A$ while applying torque (moment of force) obtained in part (c) onto external load find motor efficiency if resistance of each static solenoid is 12Ω and resistance of toroidal solenoid is 15Ω. What is the input power? What is the output power? What causes losses of energy?

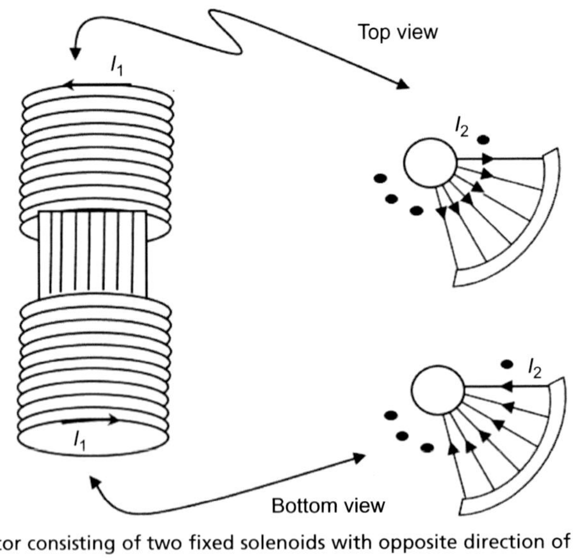

FIG. 19.13 A novel actuator consisting of two fixed solenoids with opposite direction of current and a single inner toroid-like solenoid that could rotate about the vertical axis.

(F) Imagine that this motor weighs only 0.5 kg. How does it compare (based on mass) with similar power output motors that are commercially available?

(G) Could this motor operate a small elevator with passengers with net mass of 450 kg at speed 0.75 m/s? What should be diameter of cable pulley attached onto motor shaft?

Problem 11: Numerical problem and concept question, difficulty level: *, Chapters 3 and 16

(Prepared by Marko Popovic)

Some believe that: "muscles are not optimized for force, they are optimized for motion." However, there is no scientific conclusive evidence to support this statement. In this problem, we will actually show that muscle might as well be optimized based on force considerations (in addition to motion considerations).

Consider the cable-driven system shown below (Fig. 19.14). The shoulder attachment point is located at $(x_1, y_1) = (7.95, 10.05)$ cm and the upper arm attachment point is located at (x_2, y_2). The coordinate frame origin is located at the arm center of rotation.

(A) Find the optimal attachment point (x_2, y_2) that minimize cable tension necessary to balance gravity in the range $(0, \pi/2)$ of the arm angles (angles between arm and gravity vector). For the 90° arm angle $y_2 = 5.55$ cm.

Hint: Maximal torque due to gravity corresponds to 90° arm angle. Minimal cable tension corresponds to 90 degrees angle at the shoulder attachment, that is, $\sphericalangle((0,0),(x_1,y_1),(x_2,y_2)) = \pi/2$.

(B) Find the optimal attachment point (x_2, y_2) that minimize component of cable tension that pulls the arm directly into joint center (and creates zero moment about shoulder joint).

Hint: This condition corresponds to 90° angle at the upper arm attachment, that is, $\sphericalangle((0,0),$-$(x_2,y_2),(x_1,y_1)) = \pi/2$.

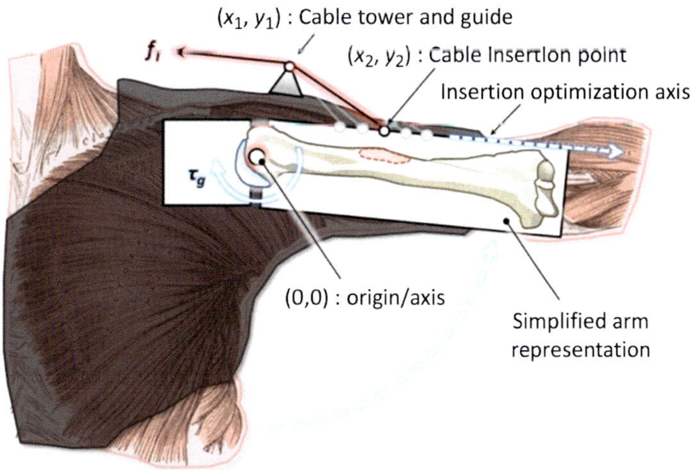

FIG. 19.14 Shoulder and cable (i.e., "muscle") attachment points.

(C) Compare the obtained result in (a) and (b) with experimental data; average deltoid muscle insertion into proximal humerus in study on 30 cadaver shoulders shows that the mean distance from the greater tubercle to the proximal and distal insertion is $x_{proximal}=0.061$ m and $x_{distal}=0.158$ m, respectively.

For students interested in this subject full reference for the above mentioned cadaver shoulders study is: S.J. Morgan, The deltoid muscle: an anatomic description of the deltoid insertion to the proximal humerus, J. Orthop. Trauma 20 (2006) 19–21.

(D) Biological muscle can only pull and not push. Hence, antagonist and agonist muscles often occur in pairs, such that they both pull from opposite sides. Comment on how this pairing can be used to modulate "stiffness" in joints. Define "joint stiffness." What is the relationship between "joint stiffness" and muscle tension and "muscle stiffness"? How is "muscle stiffness" modulated?

(E) How variable "muscle stiffness" affects muscle energetics? Comment on this based on ideal (no-loss) mechanical model and in the context of Hill's model. How do the two differ?

Problem 12: Numerical problem and concept question, difficulty level: **, Chapters 3 and 16

(Prepared by Marko Popovic)

Skeletal muscle has rather complex structure. For example, consider human biceps brachii. Total number of muscle fibers in human biceps brachii is between 150,000 and 450,000 (structures in parallel). Each muscle fiber contains 1000–2000 myofibrils (again structures in parallel). There are about 2000 myofilaments per myofibril (once again structures in parallel). Finally, there is about 100,000 sarcomers per myofilament (building blocks in series).

Make a rough estimate of (A) total number of sarcomeres in human biceps brachii and (B) average force per sarcomere for total 40 N biceps force output (roughly the force needed to hold arm straight while opposing gravity; here, the arm is perpendicular to gravity). Next, assume that required contraction is approximately 1 μm per sarcomere. Make a rough estimate of (C) work done by each sarcomere by assuming isotonic condition and force obtained in part (b) of this problem. Make a rough estimate of (D) number of adenosine triphosphate (ATP) molecules needed per each sarcomere for this condition and (E) number of ATP molecules per myosin head. Make a rough estimate of (F) number of ATP molecules needed for the same condition per single myosin head swivel.

Useful facts: Work = Force times displacement. 1 kcal = 4184 J. 1 mol = $6.02214179(30) \times 10^{23}$ or Avogadro's number of elementary entities.

The ATP is the major energy currency of the cell, providing the energy for most of the energy-consuming activities of the cell. When the third phosphate group (denoted by red wavy line) of ATP is removed by hydrolysis, a substantial amount of free energy is released. The exact amount depends on the conditions, but we shall use a value of 7.3 kcal/mol. The chemical process is ATP + $H_2O \rightarrow$ ADP + P_i where ADP is adenosine diphosphate and P_i is inorganic phosphate (Fig. 19.15).

(G) What is the "efficiency" of myosin head swivel? Estimate output work per single swivel and assume that integer number of ATP energies is needed as an input for this "biological

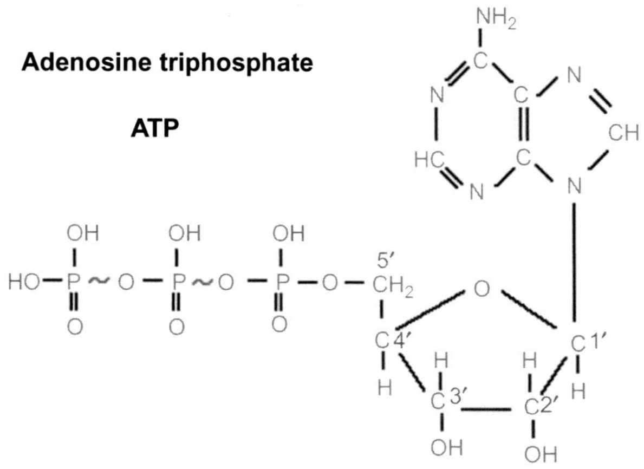

FIG. 19.15 Adenosine triphosphate (ATP).

machine." Is your result reasonable? What could cause small efficiency? List possible sources and comment on each.

(H) If you expect high efficiency for this process what would be your estimate of the number of muscle fibers in human biceps brachii and number of myofibrils per muscle fiber? Would you expect more or less of those?

(I) Assume 4 m/s contraction speed for $F = 40\,\text{N}$. If maximum isometric tension (or load) generated in the muscle is 200 N and speed for $F = 0$ is 8 m/s find coefficient of shortening heat, a, using the Hill's state equation (with $b = a\frac{v_0}{F_0}$). What is the relationship between velocity and tension near muscle's resting length? How is this relationship affected by coefficient of shortening heat? Would you expect higher efficiency with smaller coefficient of shortening heat or vice versa?

Problem 13: Concept question, difficulty level: regular, Chapter 4

(Prepared by Kathleen A Lamkin-Kennard)

What are some examples of synthetic sensors that could be used to impart a sense of "touch" to a robotic device?

Problem 14: Concept question, difficulty level: regular, Chapter 4

(Prepared by Kathleen A Lamkin-Kennard)

How are mechanoreceptors involved in audition?

Problem 15: Concept question, difficulty level: regular, Chapter 4

(Prepared by Kathleen A Lamkin-Kennard)

Provide examples of synthetic sensors that correspond to each of the following receptor types: chemoreceptors, mechanoreceptors, thermoreceptors, nociceptors, and electromagnetic receptors.

Problem 16: Concept question, difficulty level: regular, Chapter 4

(Prepared by Kathleen A Lamkin-Kennard)

What is the difference between sensor fusion and sensor integration?

Problem 17: Concept question, difficulty level: regular, Chapter 4

(Prepared by Kathleen A Lamkin-Kennard)

What types of motion capture systems would be most appropriate for motion capture of the following movements:

a. ski jumping,
b. running,
c. swimming,
d. sign language, and
e. head movements.

Problem 18: Numerical problem and concept question, difficulty level: *, Chapter 4

(Prepared by Marko Popovic)

In reality, sensory information is never ideal, in other words, there is some probability for distribution that relates, for example, actual position and position as reported by sensory system.

(A) Perform and report results of this simple experiment: lift your straight arm such that arm is perpendicular to gravity vector and parallel to sagittal plane and then while keeping upright posture face the wall such that your frontal (or coronal) plane is parallel to the wall and that your digitus medius manus (i.e., middle finger) is just barely touching the wall. Now estimate by "unaided visual inspection" how far the back of your foot is from the wall, record that estimate. Finally, perform more careful measurement of the same distance with ruler and compare the two results. Make sure that you don't move your feet (or that you marked point on the floor) while performing these two measurements. What is the absolute and relative error of your "unaided visual measurement" (here we assume that ruler-based measurement is the "ground truth")?

(B) If you think a little bit about it, you will realize that "unaided visual measurement" is not what you really did in this experiment as there are many nearby objects that you used as reference to help your estimate. Hence, imagine now that there are just two points (one is at the back of your feet and the other one is at the bottom of the wall, and line connecting them is perpendicular to the wall) at the same position in respect to your eyes and that there is nothing else in the entire universe. Having two eyes clearly helps to estimate three-dimensional (3D) location of points by triangulation. If you know that angular resolution of human eye is 0.02 degree, that distance in-between your eyes is roughly 10 cm and that height of your eyes in respect to floor is roughly 150 cm estimate the minimal error in locating two points and their distance on the floor by using triangulation.

(C) Discuss: Why is result in part (a) very different from the result in part (b)? What does that suggest about our vision system?

(D) Discuss: Can we estimate distance from point with single eye (assuming that there is nothing else in universe except us and that point)?

Problem 19: Numerical problem and concept question, difficulty level: *, Chapters 4 and 5

(Prepared by Marko Popovic)

Estimate the number of sensors in the human hand. Given that there is a large number of sensors of the same type how is their location differentiated within the nervous system; in other words, is their location encoded by "hardware," that is, by different innervation or by "software," that is, by differently encoded signal? Discuss general advantages and disadvantages of each of these approaches from the engineering perspective and also in the context of biological apparatus. This topic has not been discussed in great detail in the textbook. Look for answers elsewhere and reference the source.

Problem 20: Numerical problem, difficulty level: **, Chapters 4 and 5

(Prepared by Marko Popovic)

Imagine that you are using sensor "S" to measure some physical quantity "Y." This sensor directly measures another physical observable "X" which is then related to physical observable "Y." Sensor has threshold and gives $x=0$ output when actual $X \in (0, X_0)$. The $x=0$ output is then *translated* to $y=Y_0$.

(A) Discuss why it is sometime advantageous to have nonzero threshold in the sensory system.

In the reality, two physical observables are related as

$$Y = Y_{0r} + a_r X + b_r X^2 + c_r X^3 \tag{19.9}$$

but, whoever made this sensor had model that is slightly off and assumed

$$y = Y_0 + aX + bX^2 \tag{19.10}$$

with $Y_{0r} = Y_0$, $a_r = 1.02\ a$, $b_r = 0.97\ b$, $c_r = -0.01$.

(B) Assume that there is no error in measurement of X. What is the range of X and hence of Y for which relative error $\frac{\varepsilon(Y)}{Y} < 2\%$?

Now assume that there is an error in measurement of X (as well as abovementioned model error). Direct measurement of X reveals error of $\varepsilon(X) = \varepsilon_1$ for $X \in (X_0, X_1)$ and more complex $\varepsilon(X) = \varepsilon_1 + (X - X_1)\varepsilon_2$ for $X \in (X_1, X_2)$.

(C) What is the range of X and hence of Y for which relative error $\frac{\varepsilon(Y)}{Y} < 2\%$? You may consider only $X \in (0, X_2)$.

(D) How would the result in part (c) change if you had two identical sensors "S"? Describe the most optimal way to simultaneously use both sensors and improve accuracy of measurement? What is the range of X and hence of Y for which relative error $\frac{\varepsilon(Y)}{Y} < 2\%$? You may consider only $X \in (0, X_2)$.

Numerical values assumed in this problem are:

$$Y_0 = 10.0,\ a = 2.57,\ b = 0.89,\ X_0 = 1.00,\ X_1 = 30.0,\ X_2 = 100,$$
$$\varepsilon_1 = 2.20 \times 10^{-2},\ \varepsilon_2 = 1.34 \times 10^{-1} \tag{19.11}$$

Problem 21: Numerical problem, difficulty level: **, Chapter 5

(Prepared by Marko Popovic)

Desired trajectory of two-link manipulator shown in Fig. 19.16 is defined as

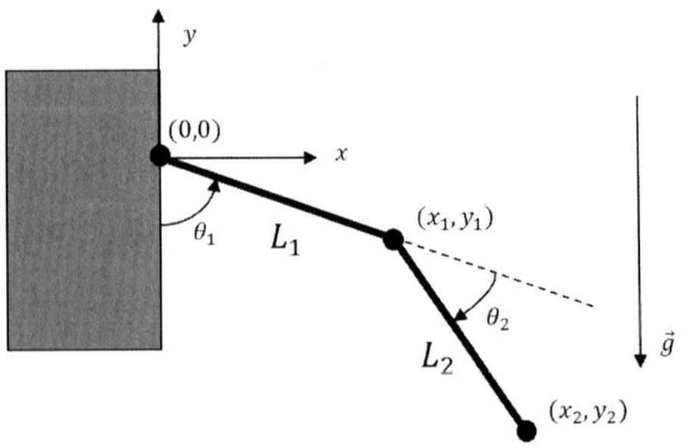

FIG. 19.16 Simplified arm model in the form of two DoF, two-link manipulator in vertical plane.

$$\theta_{1\,\text{tar}} = \frac{3}{8}\pi + \frac{1}{8}\pi\cos\left(\frac{\pi}{3\,\text{s}}t + \frac{\pi}{16}\right) \tag{19.12}$$

$$\theta_{2\,\text{tar}} = -\frac{5}{8}\pi - \frac{1}{8}\pi\cos\left(\frac{\pi}{2\,\text{s}}t - \frac{\pi}{16}\right) \tag{19.13}$$

With time t in seconds. This corresponds to desired trajectory of end effector

$$x_{2tar} = L_1 \sin\theta_{1\,\text{tar}} + L_2 \sin\left(\theta_{1\,\text{tar}} - \theta_{2\,\text{tar}}\right) \tag{19.14}$$

$$y_{2tar} = -L_1 \cos\theta_{1\,\text{tar}} - L_2 \cos\left(\theta_{1\,\text{tar}} - \theta_{2\,\text{tar}}\right) \tag{19.15}$$

The length and mass of the upper arm are $L_1=20$ cm and $m_1=3$ kg, respectively. The length and mass of the lower arm are $L_2=18$ cm and $m_2=2$ kg, respectively. Treat the segments' moment of inertia as moment of inertia of rods. Assume that the gravity vector is pointing vertically down within the arm plane.

Assume initial, =0, manipulator posture with $\theta_1 = \frac{\pi}{2}$ and $\theta_2=0$.

Choose one of the following metrics for successful tracking of desired trajectory during first 10 s:

(i) d_m maximal absolute distance between hand and target.
(ii) d_a average absolute distance between hand and target.
(iii) $d_m V_m$ with V_m being maximal magnitude of relative speed of hand in respect to target.
(iv) $(dV)_a$ average of product of absolute distance and relative speed.
(v) $(dV)_m$ maximal value of product of absolute distance and relative speed.
(vi) product of d_m and average jerk magnitude (jerk is third derivative of position in respect to time)

(A) Design PD control laws for tracking of joint angles and test it using chosen metric. Express PD laws in terms of angular accelerations, for example, for shoulder angle

$$\ddot{\theta}_1 = k_{p\,shoulder}\left(\theta_1 - \theta_{1\,tar}\right) + k_{d\,shoulder}\left(\dot{\theta}_1 - \dot{\theta}_{1\,tar}\right) \tag{19.16}$$

(B) Design PD control law for tracking of end effector Cartesian coordinates and test it using same metric.

Express PD laws in terms of end effector accelerations, for example, for y

$$\ddot{y} = k_{py}\left(y_2 - y_{2\,\text{tar}}\right) + k_{dy}\left(\dot{y}_2 - \dot{y}_{2\,tar}\right) \tag{19.17}$$

(C) Compare results in (a) and (b). Discuss difference and what the cause of it is.

(D) Calculate the joint torques for both (a) and (b) and determine maximal shoulder and elbow joint torques for each.

(E) Calculate joints positive work, that is,

$$\frac{1}{2}\int_0^{10s} \left\{\tau_1\dot{\theta}_1\left[1 + \text{sign}\left(\tau_1\dot{\theta}_1\right)\right] + \tau_2\dot{\theta}_2\left[1 + \text{sign}\left(\tau_2\dot{\theta}_2\right)\right]\right\}dt \tag{19.18}$$

and compare results in parts (a) and (b).

Problem 22: Numerical problem, difficulty level: **, Chapter 5

(Prepared by Marko Popovic)

You will design and test control strategy for cursor tracking target moving along x-axis. Target's trajectory in centimeters for every 100 ms is provided below, case I, second column, and case II, third column (Table 19.1). Cursor's initial position and velocity are both zero.

Table 19.1 Target Trajectory for Two Cases

t [100 ms]	*X* [cm] Case I	*X* [cm] Case II
0	0	1.1
1	0	0.2
2	0.2	0.4
3	0.6	0.5
4	1.7	0.1
5	2.6	0.3
6	3.1	0.7
7	4.6	1.9
8	3.9	3.7
9	3.5	5.2
10	2.4	7.9
11	2.9	9.8
12	1.4	6.2
13	1.7	4.8
14	4.5	3.0
15	5.7	2.0
16	6.5	1.3
17	7.8	0.5
18	8.2	0.3
19	8.4	0.4
20	8.5	0.6
21	6.1	0.8
22	5.4	1.8
23	4.9	3.5
24	4.1	5.1
25	4	7.8
26	3.8	9.4
27	5.2	9.6
28	5.6	9.7
29	5.5	8.9
30	2.3	1.1

Your control system could only modulate desired cursor's acceleration, $a_{C\ des}(t)$, in the range -20 to 20 cm/s^2. Based on 100 ms delayed sensory information on target's state (position and velocity) and instantaneous information on cursor's state the control system can affect actual cursor's acceleration with 100 ms delay. In other words acceleration of your system is based on 200 ms or older information on target's state and 100 ms or older information on cursor's state.

Still further, due to some technical issues (!) the actual cursor's acceleration, $a_C(t)$, is not equal to desired cursor's acceleration, $a_{C\ des}(t)$, but it varies in between 0.8 and 1.0 of desired cursor's acceleration with, as you may assume here, uniform probability distribution function (you may use function "rand" or similar if you are using MATLAB; you may use function "randbetween" if you are doing this problem in Excel). In other words, $a_C(t) \in [0.8 \times a_{C\ des}(t), 1.0 \times a_{C\ des}(t)]$.

Regarding physical simulator, you may assume that cursor's actual state relates to 100-ms old actual position, velocity, and acceleration as

$$v_C(t) = v_C(t - 100\,\text{ms}) + a_C(t - 100\,\text{ms}) \times 100\,\text{ms} \tag{19.19}$$

$$x_C(t) = x_C(t - 100\,\text{ms}) + v_C(t - 100\,\text{ms}) \times 100\,\text{ms} + 0.5 \times a_C(t - 100\,\text{ms}) \times (100\,\text{ms})^2 \tag{19.20}$$

Note that this is in agreement with

$$a_C(t - 100\,\text{ms}) = \frac{v_C(t) - v_C(t - 100\,\text{ms})}{100\,\text{ms}} \tag{19.21}$$

but

$$v_C(t - 100\,\text{ms}) \neq \frac{x_C(t) - x_C(t - 100\,\text{ms})}{100\,\text{ms}} \tag{19.22}$$

instead

$$\frac{v_C(t) + v_C(t - 100\,\text{ms})}{2} = \frac{x_C(t) - x_C(t - 100\,\text{ms})}{100\,\text{ms}} \tag{19.23}$$

You may assume similarly for target

$$v_T(t) = v_T(t - 100\,\text{ms}) + a_T(t - 100\,\text{ms}) \times 100\,\text{ms} \tag{19.24}$$

$$x_T(t) = x_T(t - 100\,\text{ms}) + v_T(t - 100\,\text{ms}) \times 100\,\text{ms} + 0.5 \times a_T(t - 100\,\text{ms}) \times (100\,\text{ms})^2 \tag{19.25}$$

TASKS:

(A) Describe your control strategy in detail.

(B) Test your control strategy: plot cursor and target trajectory, record maximal distance between cursor and target, $d_{CT\ max}$, calculate average distance between cursor and target $d_{CT\ ave}$.

(C) Repeat part (b) for new target's trajectory, see below, and comment on differences. Cursor's initial position and velocity are again both zero.

(D) EXTRA CREDIT Create video for parts (b) and (c). If possible slow it down by factor of, for example, 4 (so 3 s of actual motion is presented by 12 s video).

Problem 23: Numerical problem and concept question, difficulty level: *, Chapter 5

(Prepared by Marko Popovic)

Use table with conduction velocities (e.g., http://en.wikipedia.org/wiki/Axon) to roughly estimate range of reflex response times for reaction caused by

(A) pain sensation in the fingertip and

(B) hot sensation in the fingertip.

Assume that electromyography (EMG) signal precedes muscle peak force by 30–50 ms.

How would your results change for same sensation in foot if:

(C) only lower spinal cord circuitry is involved or.

(D) if cerebellum-based coordination mechanisms are involved too?

(E) How would your results (A)–(D) change if myelinated axons become 30%–50% demyelinated as consequence of aging or neurological disease?

(F) Given the reflex response times and feedback loop time periods what is your expectation on neural control frequency? Discuss it in relation to frequency of brain waves, especially those relevant for muscle contractions like betta waves.

You may want to read other references, for example:

S.N. Baker, Oscillatory interactions between sensorimotor cortex and the periphery, Curr. Opin. Neurobiol. 17(6) (2007) 649–655.

Problem 24: Concept question, difficulty level: regular, Chapter 6

(Prepared by Hiroyuki Tashiro)

Summarize the method of physical measurement used to record nerve activity.

Problem 25: Concept question, difficulty level: regular, Chapter 6

(Prepared by Hiroyuki Tashiro)

Describe the characteristics required of a recording electrode for DNI.

Problem 26: Concept question, difficulty level: regular, Chapter 6

(Prepared by Hiroyuki Tashiro)

Describe the characteristics required of a stimulation electrode for DNI.

Problem 27: Concept question, difficulty level: regular, Chapter 6

(Prepared by Hiroyuki Tashiro)

Describe side effects due to electrical nerve stimulation.

Problem 28: Concept question, difficulty level: regular, Chapter 6

(Prepared by Hiroyuki Tashiro)

Summarize the advantages/disadvantages of intracortical electrodes and electrocorticography (ECoG) electrodes.

Problem 29: Concept question, difficulty level: regular, Chapter 6

(Prepared by Hiroyuki Tashiro)

Summarize the noninvasive method for the measurement of brain activity.

Problem 30: Concept question, difficulty level: regular, Chapter 6

(Prepared by Hiroyuki Tashiro)

Consider the good and bad aspects expected from advances in BMI research in the future.

Problem 31: Concept question, difficulty level: regular, Chapter 7

(Prepared by Hiroyuki Tashiro)

Describe the advantages and disadvantages of treatment with artificial organs as compared to treatment by organ transplantation.

Problem 32: Concept question, difficulty level: regular, Chapter 7

(Prepared by Hiroyuki Tashiro)

Enumerate treatments by artificial organs and explain the principle briefly.

Problem 33: Concept question, difficulty level: regular, Chapter 7

(Prepared by Hiroyuki Tashiro)

Describe the characteristics required by biomaterials.

Problem 34: Concept question, difficulty level: regular, Chapter 7

(Prepared by Hiroyuki Tashiro)

Describe the principle behind the lost sensation being regenerated by a cochlear implant or artificial retina.

Problem 35: Concept question, difficulty level: regular, Chapter 7

(Prepared by Hiroyuki Tashiro)

Describe the advantages and disadvantages of regenerative medicine treatment as compared to artificial organ treatment.

Problem 36: Concept question, difficulty level: regular, Chapter 7

(Prepared by Hiroyuki Tashiro)

Consider the welfare and ethical issues arising from the advanced development of artificial organs.

Problem 37: Numerical problem and concept question, difficulty level: *, Chapter 8

(Prepared by Kathleen A Lamkin-Kennard)

In the 1966 movie "Fantastic Voyage," humans were miniaturized to fit in the microvasculature of the human body. How would the mass of a human change if they were miniaturized to 50 μm from a height of 175 cm? Assume that the human can be approximated as a cylinder with a radius of 12 cm and a density of 1 g/cm^3. Would you expect walking to become more difficult at this scale? Why or why not?

Problem 38: Numerical problem and concept question, difficulty level: *, Chapter 8

(Prepared by Kathleen A Lamkin-Kennard)

Show that electrostatic force for a parallel plate capacitor scales as l^2 if the voltage depends on the dielectric breakdown field, such that $E_{max} = \frac{V_0}{g_0}$. How would you expect the energy to scale if the energy, E, is calculated using $E = \frac{1}{2}CV_0{}^2$? Note: The capacitance, C, can be calculated using $= \frac{\varepsilon_0 \varepsilon_r}{g_0} wl$.

Problem 39: Numerical problem and concept question, difficulty level: *, Chapter 8

(Prepared by Kathleen A Lamkin-Kennard)

Derive the expected scaling with regards to length, *l*, for the following variables:

(A) mass,
(B) torque, and
(C) drag force.

Problem 40: Numerical problem and concept question, difficulty level: *, Chapters 8 and 17

(Prepared by Kathleen A Lamkin-Kennard)

Fig. 19.17 represents one leg of a six-legged water strider-like robot. The leg is assumed to be cylindrical in shape and is made of a nonwetting material. When the leg comes in contact with

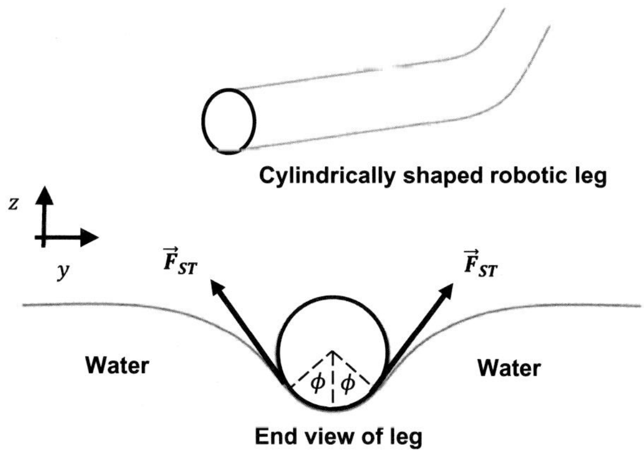

FIG. 19.17 The leg of a six-legged water strider-like robot.

the surface of water, it depresses the surface. The force on the leg due to surface tension (FST) reaches a maximum at a certain depression before decreasing.

(A) What is the expression for the resultant force in the z-direction due to surface tension as the leg depresses the surface?

(B) At what angle (φ) does the force reach a maximum?

(C) If the surface tension of the water is 0.07 N/m and the robot weighs 10 mg, what is the minimum contact length on the leg that must be maintained with the water for the robot to be able to stay on the surface of the water? Note: surface tension = force/contact length.

Problem 41: Concept question, difficulty level: regular, Chapter 8

(Prepared by Kathleen A Lamkin-Kennard)

Explain why microswimmers are not propelled forward if reciprocal propulsion mechanisms are used. What are some strategies utilized by microorganisms for propulsion that compensate for this effect?

Problem 42: Concept question, difficulty level: regular, Chapter 8

(Prepared by Kathleen A Lamkin-Kennard)

What are the advantages and disadvantages of using chemical, magnetic, acoustic, or optical methods for off-board propulsion of microrobots?

Problem 43: Concept question, difficulty level: regular, Chapter 8

(Prepared by Kathleen A Lamkin-Kennard)

Why cardiomyocytes or skeletal muscle cells might be used for propulsion of biohybrid robots instead of microorganisms? What are the disadvantages of using these types of cells for biohybrid robots?

Problem 44: Concept question, difficulty level: regular, Chapter 8

(Prepared by Kathleen A Lamkin-Kennard)

What are some critical barriers that must be overcome before micro- and nanorobots can be more widely used for treating cancer?

Problem 45: Concept question, difficulty level: regular, Chapter 9

(Prepared by Philipp Beckerle and Marko Popovic)

Amputation may be necessary due to various reasons. Please describe main causes of amputation and highlight differences for the upper and the lower limbs.

Problem 46: Concept question, difficulty level: regular, Chapter 9

(Prepared by Philipp Beckerle and Marko Popovic)

Amputation can have various levels, for example, partial hand amputation. Please describe the main levels of upper-limb and lower-limb amputations.

Problem 47: Concept question, difficulty level: regular, Chapter 9

(Prepared by Philipp Beckerle and Marko Popovic)

Prosthetic limbs can be distinguished as passive, semi-active, and active devices. Please give brief definitions and explain the differences.

Problem 48: Concept question, difficulty level: regular, Chapter 9

(Prepared by Philipp Beckerle and Marko Popovic)

Prosthetics is a very interdisciplinary field. Please list the disciplines involved and explain their interaction.

Problem 49: Concept question, difficulty level: regular, Chapter 9

(Prepared by Steffen Willwacher)

The transition between steps offers some biomechanical challenges to the users of lower-limb prosthetic systems. Please describe the main challenges and the potential benefits of using an actively powered lower prosthesis compared to a passive prosthetic system.

Problem 50: Concept question, difficulty level: regular, Chapters 9 and 16

(Prepared by Steffen Willwacher)

Passive running prostheses made of specific carbon fibers are frequently used by Paralympic athletes. Specify advantages and disadvantages of these prostheses for various locomotion tasks. Please justify your answer.

Problem 51: Concept question, difficulty level: regular, Chapter 9

(Prepared by Steffen Willwacher)

How does the use of a unilateral lower extremity prosthetic system affect the loading experienced by the affected and unaffected limb during everyday locomotion? Is there a difference between active and passive system?

Problem 52: Concept question, difficulty level: regular, Chapter 9

(Prepared by Philipp Beckerle and Marko Popovic)

Prosthetic design concerns technical, biomechanical, and other user-related factors. Please list five general requirements and describe their influence on device design.

Problem 53: Concept question, difficulty level: regular, Chapter 9

(Prepared by Philipp Beckerle and Marko Popovic)

Upper and lower-limb prosthetics do not receive equal acceptance by the users. Please describe the difference and discuss potential causes.

Problem 54: Concept question, difficulty level: regular, Chapter 9

(Prepared by Philipp Beckerle and Marko Popovic)

The Uncanny Valley is a concept historically introduced several decades ago to explain some pitfalls in human-like designs. Please draw the corresponding diagram, explain it, and describe how it might affect the design guidelines. Is did concept still given the same weight as several decades ago?

Problem 55: Concept question, difficulty level: regular, Chapter 9

(Prepared by Minas Liarokapis)

Upper-limb prostheses can be classified according to the actuation mechanisms used. Please describe the different classes and present their advantages and disadvantages.

Problem 56: Concept question, difficulty level: regular, Chapter 9

(Prepared by Minas Liarokapis)

What is the most common interface employed for the control of advanced prosthetic hands? What kind of biological signals does it use?

Problem 57: Concept question, difficulty level: regular, Chapter 9

(Prepared by Minas Liarokapis)

Upper-limb prostheses can be classified as adjustable or static and active or passive. Which classification is more general? Please discuss the differences between these classes.

Problem 58: Concept question, difficulty level: regular, Chapter 9

(Prepared by Minas Liarokapis)

The use of differential mechanisms is quite common in prosthetic hands. Please explain what their role is and how they affect the overall design.

Problem 59: Concept question, difficulty level: regular, Chapter 9

(Prepared by Philipp Beckerle)

Robotic lower-limb prostheses are required to operate autonomously. Please describe which challenges this imposes on actuator and control design.

Problem 60: Concept question, difficulty level: regular, Chapter 9

(Prepared by Philipp Beckerle)

Introducing elastic components to lower-limb prosthetic actuators can yield multiple benefits. Please list potential benefits and describe common solutions.

Problem 61: Concept question, difficulty level: regular, Chapter 9

(Prepared by Philipp Beckerle)

Powered lower-limb prostheses need controllers to command their actuators. Please explain the two fundamental control categories and discuss two exemplary methods per category.

Problem 62: Concept question, difficulty level: regular, Chapter 9

(Prepared by Philipp Beckerle and Marko Popovic)

Users' needs and experience are of paramount importance for prosthetic design. Please list and explain your reasoning, which data could be acquired from human participants to evaluate prosthetic design.

Problem 63: Concept question, difficulty level: regular, Chapter 9

(Prepared by Philipp Beckerle and Marko Popovic)

Sketch your personal concept of an upper and a lower-limb prosthetic device to be in the market in 2070. Please discuss why you chose certain materials or components and why you deem their integration realistic.

Problem 64: Concept question, difficulty level: regular, Chapter 12

(Prepared by Marko Popovic)

Use the facts below to answer the following:

What is the ratio of severely disabled people per PM&R physician in the United States? Is this appropriate? Assume 52-week year and 40 h/week for each PM&R physician. How many seconds could they allow for each severely disabled person in the United States?

Currently, there are between 8000 and 10,000 board-certified Physiatrists, that is, rehabilitation physicians in the United States. Physical therapists (PTs) held about 185,500 jobs in the United States in 2008. Approximately 54 million people in the United States have some type of disability, with physical, sensory, mental, and self-care effects ranging from mild to severe. About 35 million people of all ages are severely disabled.

Problem 65: Case study, difficulty level: *, Chapter 12

(Prepared by Michelle Johnson)

The patient (Millie) is a 75-year-old woman who emigrated from Jamaica (low-resource) to New York City (high-resource) 50 years ago. She is in relatively good health for her age—but has mild issues with mobility and right-side weakness, as well as mild, aging-related vision loss, hearing loss, and dementia. She lives independently and can complete basic self-care ADLs—but gets easily fatigued and struggles with some more tiresome activities, like grocery shopping or banking. She has two children and two grandchildren that live 2 h away from her and try to visit regularly to check up on her. At her most recent physician's visit, Millie's right elbow and right knee both scored 3/5 on the manual muscle testing (MMT). Millie's physician also found that she needs to drink more water (even though she claims to never be thirsty) and maintain a more active lifestyle. Millie currently uses a walker for mobility outside the home. Millie needs an AT system that (1) maintains her independence by helping her properly feed and hydrate herself and (2) strengthens her weak right-side limbs by increasing her physical activity.

Please use the human activity assistive technology (HAAT) model: areas of human (abilities/disabilities/needs), setting/context/environment of use, activities, and assistive technologies to propose and evaluate the devices that Millie can use.

- Potential users and stakeholders (e.g., competencies, abilities)
  - Physical skills, cognitive skills, insight versus awareness, emotional status
- Key activities:
  - The desired functional tasks that the user wants to do but cannot do without the device
  - Frequency
  - Analysis from user perspective
- Context: all environments in which the activities will be completed and the device will be used
  - Physical, social, cultural, institutional, micro, macro, *meso*
- Assistive technology: the desired tasks that the device needs to complete
  - Consideration for Interaction
  - Interface design
  - Caregiver
  - Environment
  - Independence/psychological factors

Problem 66: Case study, difficulty level: *, Chapter 12

(Prepared by Michelle Johnson)

Zinabu is a 13-year-old female who was diagnosed with hemiplegic cerebral palsy at age one. She lives is New York City in a two-bedroom apartment with her parents and an older brother. She goes to a school for children with disabilities and is in Grade 6. She has moderate functioning in terms of motor and cognition and is described as easily distracted. Zinabu is right-hand dominant. She uses a manual wheelchair that must be pushed by a caregiver. She is weak in both limbs. She and her caregiver desire her to strengthen her limbs to the point that she would be able to push herself. Based on the box and block tests, she is able to use her right arm to pick up 30 blocks and only 3 with her left arm.

(A) Use the HAAT model to determine the assistive technology key requirements?

(B) What should be the caregiver's retained role?

(C) Describe how the rehabilitation team would interact to provide support for Zinabu to integrate the device into her daily life?

Problem 67: Case study, difficulty level: *, Chapter 12

(Prepared by Michelle Johnson)

Edith is an 80-year-old female who now lives alone independently in a large old terrace house with steps. She has two supportive sons who live nearby. She has two cats. Edith is a retired pub owner and still has an active social life with many friends. She goes out nearly every day, either to meet up with friends, or go to the shops. Edith was experiencing symptoms of dizziness and headaches and feeling unsteady on her feet, which caused her to have a fear of falling. She exhibited some anxiety about going out of the house, for fear she would fall and lose her independence, becoming housebound. Her past medical history includes hypertension, type 2 diabetes (diagnosed in 1998), osteoarthritis, left total hip replacement (done in 2000), and phacoemulsification of the lens (cataract surgery). Both her hypertension and diabetes were poorly controlled, resulting in frequent hospital admissions. Edith's blood pressure was 190/95 mmHg. Edith reported feeling dizzy, particularly in the morning on getting out of bed. The dizziness had resulted in two falls (due to which she had a right shoulder subluxation) in 2012 and in 2015. On a recent examination (2016), she was found to have postural hypotension, and a medication review revealed that she was not taking her medicines as prescribed. She had stopped taking one of her antihypertensives. As Edith did not have any physical symptoms of diabetes, her perception of the possible consequences of the complications of the condition was low. She also reports tingling sensations in her right arm and difficulty with firmly gripping onto objects. As part of her medical care, Edith has been asked to monitor her blood sugar using *a glucose monitor (True2Go Blood Glucose Monitoring System)* and her blood pressure using *a blood pressure monitor (OMRON Automatic wrist blood pressure monitor)* three times a day. She has also been given *a reacher (Grip n' Grab)* to reach into shelves at higher or lower levels, enabling her to obtain materials she requires. Edith currently reports difficulty with using both the monitors as well as the reacher.

(A) You are attempting to understand Edith's use of the medical devices. For each of the three medical devices, please use the HAAT model: areas of human (abilities/disabilities/needs), setting/context/environment of use, activities, and assistive technologies to evaluate the devices proposed for Edith's use.

(B) In a Table device, evaluate each of the three devices in terms of the following to determine if they are usable by Edith.

A. Task requirements:

- **a.** selection, understanding, and learning to use,
- **b.** positioning self and device in preparation to use,
- **c.** approaches to and from device,
- **d.** interacting with device,
- **e.** communicating with device,
- **f.** positioning self and device during use,
- **g.** operating device,
- **h.** obtaining and using results, and
- **i.** maintaining device.

B. Device features:

- **a.** size,
- **b.** assembly and disassembly of parts,
- **c.** adjustable positioners and supports,
- **d.** controls (location, placement, and alternatives),
- **e.** displays (location, placement, and alternatives),
- **f.** compatibility with other systems,
- **g.** power sources,
- **h.** labeling, and
- **i.** alarms/warning signals/safety.

(C) You are now required to redesign the device to meet Edith's needs. Taking into consideration principles from HAAT design, design tensions, and human factors, propose redesign concepts for the ATs (1 per device). Please include sketches of concepts for each device—please label as needed. Pick your top choice and justify why it is appropriate for Edith and how it would improve her ability to use the AT.

Concept sketching tip: The goal is to produce hand-drawn representations of real objects that closely resemble the actual objects and that appear 3D. In this activity, you will focus on obtaining the correct shape and proportions of each object from a single "straight-on" or orthographic view and add shading to produce a more realistic 3D effect.

Problem 68: Case study, difficulty level: *, Chapter 12

(Prepared by Michelle Johnson)

A 45-year-old man, Mike, who lives in Philadelphia, USA, has a history of hypertension, diabetes mellitus, and hyperlipidemia. He was checked in to the emergency department (ED) with right-sided hemiparesis and mild aphasia. Imaging diagnosed an acute stroke.

Physical assessment found active range of motion (AROM—ability of subject to move limb on their own) on left upper extremity and lower extremity (LUE/LLE) within normal limits (WNL). AROM was decreased on right at end of range shoulder flexion and elbow extension. AROM was decreased with right ankle dorsiflexion. Active-assisted range of motion (AAROM) was WNL throughout. Strength was assessed by the MMT with the following results (5 being normal strength):

Left UE: 5/5 at shoulder, elbow, wrist, and grasp.

Right UE: 2/5 with shoulder flexion, abduction, 2/5 with elbow extension, 3/5 grasp.

Left LE: 5/5 at hip, knee, and ankle.

Right LE: 4/5 hip flexion, abduction/adduction, 3/5 knee extension/flexion and 2/5 ankle dorsiflexion, 3/5 plantarflexion.

Functional tasks: Patient required moderate assistance of one for rolling and bed mobility, minimal assistance of one for sit to stand transfers, and moderate assistance of one to ambulate 20 ft. with a rolling walker. During walking, the patient had with decreased swing on right due to foot drop and increased hip hiking to compensate on right during swing phase. His goal is to be able to take moderate walks with his grandkids again.

Cognition: The patient was oriented to place and person, but disoriented to time. Patient scored a 21/30 on the Mini-Mental State Examination (MMSE) demonstrating mild cognitive impairment.

Therapy: Mike was prescribed 8 weeks (three times per week) of gait training using BWS treadmill training, after which he had increased strength in both right LE knee 4/5 and 5/5hip flexion but still has impaired ankle gait.

(A) The subject, Mike is now disabled. Which disability would apply? Check all that apply.

(i) mobility/ambulatory,
(ii) cognitive,
(iii) independent living,
(iv) self-care,
(v) hearing, and
(vi) visual.

(B) Loss of abnormality in a body function or structure defines which international classification of function.

(i) impairment,
(ii) disability,
(iii) participation,
(iv) handicap, and
(v) activity.

(C) Because of his brain injury and resulting impairment, he would be limited in participation in which of the following daily activities. Check all that apply.

(i) bowling with family,
(ii) running for exercise,
(iii) cooking a meal,
(iv) watching a football game, and.
(v) walking slowly with kids.

(D) After therapy, what impairment was reduced and what disability was impacted?

(i) Spasticity—mobility
(ii) Muscle strength—mobility
(iii) Range of motion—mobility
(iv) Brain—cognitive
(v) Executive function—talking

(E) What other device from the list could be for training Mike during gait therapy (consider his level of abilities)?

(i) Cane
(ii) Lokomat
(iii) ESKO leg exoskeleton
(iv) Re-WALK leg device
(v) Nao humanoid robot
(vi) Treadmill
(vii) Exercise bike

(F) Considering Mike, cost, activity goal, age, and prior history and gait therapy outcome what gait assistive device(s) would you prescribe for Mike at the end of the therapy that he would use at home or in the community: check all that apply.

(i) Cane
(ii) Lokomat
(iii) ESKO leg exoskeleton
(iv) Re-WALK leg device
(v) Nao humanoid robot
(vi) Treadmill
(vii) Exercise bike

Problem 69: Numerical problem, difficulty level: *, Chapter 12

(Prepared by Adam Goodworth)

In some developing countries, individuals fill large containers of water from a river and walk several kilometers home with the container balanced on their head.

(A) If a person with a container on their head leans forward 9 degrees with respect to upright, what is the required corrective torque around the ankle joint to keep the body from falling over?

(B) How does this torque compare to the torque required without the water container?

(C) What is one biomechanical limitation that affects that ability to generate this toque?

Assume static conditions with the body represented as an inverted pendulum with mass = 65 kg, head to ankle height = 1.7 m, and center of mass COM height = 0.85 m. Assume the container is 20 L (Fig. 19.18).

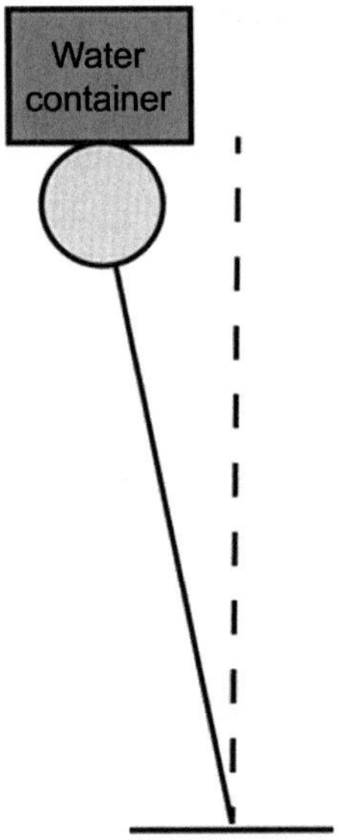

FIG. 19.18 Symbolic representation of person with the water container balanced on the person's head.

Problem 70: Numerical problem, difficulty level: *, Chapter 12

(Prepared by Adam Goodworth)

The following feedback diagram shown in Fig. 19.19 is a simple but well-documented description of how humans balance when standing on a tilting surface. Humans combine sensory feedback that orients the body vertical (visual + vestibular) with feedback that orients the body toward a tilted surface (proprioception). Each of these sources are independently weighted (Wv and Wp) and represent the relative reliance on each system, such that $Wv + Wp = 1$. The neural controller is modeled as a proportional-integral-derivative (PID) controller, s is the Laplace variable, J is the moment of inertia about the ankle joint, m is the mass, h is the center of mass height above the ankle joint, and the body is represented as a single-link inverted pendulum. The "body tilt response" is the angular tilt with respect to upright.

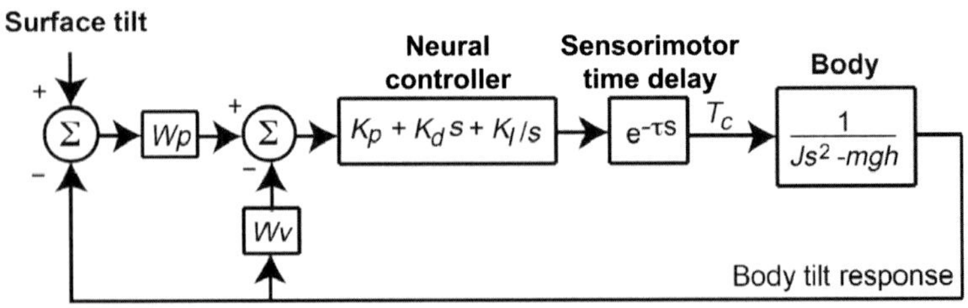

FIG. 19.19 Feedback diagram representing humans balancing when standing on a tilting surface.

(A) Beginning with a free body diagram (FBD), demonstrate that for small body tilts, the linearized transfer function of the "body" is

$$\frac{1}{Js^2 - mgh} \tag{19.26}$$

(B) Write out the transfer function between body tilt response and surface tilt.

Problem 71: Concept question, difficulty level: regular, Chapter 15

(Prepared by Pinar Boyraz)

Which robotic platform that can be used in the operating rooms was the first to be FDA approved?

Problem 72: Concept question, difficulty level: regular, Chapter 15

(Prepared by Pinar Boyraz)

How do you describe the autonomy levels for surgical robots?

Problem 73: Concept question, difficulty level: regular, Chapter 15

(Prepared by Pinar Boyraz)

What are the advantages of using a hyper-redundant mechanism for robotic surgery applications?

Problem 74: Concept question, difficulty level: regular, Chapter 15

(Prepared by Pinar Boyraz)

What are the different modes of human-machine interaction in robotic surgery?

Problem 75: Concept question, difficulty level: regular, Chapter 15

(Prepared by Pinar Boyraz and Ivo Dobrev)

How can we address the accuracy problem in navigation during automated ear surgery?

Problem 76: Numerical problem, difficulty level: regular, Chapter 16

(Prepared by Kathleen A Lamkin-Kennard)

If an athlete has a vertical jump of 98 cm on Earth, what would their vertical jump be on the moon? Assume that the athlete is able to generate the same impulse during takeoff on the Earth and on the moon and the gravity of the moon is $1.622\,\mathrm{m/s^2}$.

20

Solutions and Hints for Selected Problems

Marko B. Popovic*, Kathleen A. Lamkin-Kennard†, Michelle J. Johnson‡, Adam D. Goodworth§

*WORCESTER POLYTECHNIC INSTITUTE, WORCESTER, MA, UNITED STATES †ROCHESTER INSTITUTE OF TECHNOLOGY, ROCHESTER, NY, UNITED STATES ‡UNIVERSITY OF PENNSYLVANIA, PHILADELPHIA, PA, UNITED STATES §UNIVERSITY OF HARTFORD, HARTFORD, CT, UNITED STATES

Problem 1 Numerical Problem, Difficulty Level: *, Chapter 2

(Prepared by Marko Popovic)

Solution:

(A)

$$\tau_1 = -[L_1 \sin\theta_1 + L_2 \sin(\theta_1+\theta_2)]F_{x\,EE} + [L_1 \cos\theta_1 + L_2 \cos(\theta_1+\theta_2)]F_{y\,EE} \tag{20.1}$$

$$\tau_2 = -L_2 \sin(\theta_1+\theta_2)F_{x\,EE} + L_2 \cos(\theta_1+\theta_2)F_{y\,EE} \tag{20.2}$$

$$F_{y\,EE} = \frac{\tau_2 + L_2 \sin(\theta_1+\theta_2)F_{x\,EE}}{L_2 \cos(\theta_1+\theta_2)} \tag{20.3}$$

$$\tau_1 = -[L_1 \sin\theta_1 + L_2 \sin(\theta_1+\theta_2)]F_{x\,EE} + [L_1 \cos\theta_1 + L_2 \cos(\theta_1+\theta_2)]\frac{\tau_2 + L_2 \sin(\theta_1+\theta_2)F_{x\,EE}}{L_2 \cos(\theta_1+\theta_2)} \tag{20.4}$$

$$\tau_1 - \left(\frac{L_1 \cos\theta_1}{L_2 \cos(\theta_1+\theta_2)} + 1\right)$$

$$\tau_2 == \{-[L_1 \sin\theta_1 + L_2 \sin(\theta_1+\theta_2)] + [L_1 \cos\theta_1 + L_2 \cos(\theta_1+\theta_2)]\tan(\theta_1+\theta_2)\}F_{x\,EE} \tag{20.5}$$

$$F_{x\,EE} = \frac{\tau_1 - \left(\dfrac{L_1 \cos\theta_1}{L_2 \cos(\theta_1+\theta_2)} + 1\right)\tau_2}{-[L_1 \sin\theta_1 + L_2 \sin(\theta_1+\theta_2)] + [L_1 \cos\theta_1 + L_2 \cos(\theta_1+\theta_2)]\tan(\theta_1+\theta_2)} \tag{20.6}$$

$$F_{x\,EE} = \frac{2.5\,\text{Nm} - \left(\dfrac{0.45\,\text{m}\cos 30}{0.35\,\text{m}\cos 45} + 1\right)3.0\,\text{Nm}}{-(0.45\,\text{m}\sin 30 + 0.35\,\text{m}\sin 45) + (0.45\,\text{m}\cos 30 + 0.35\,\text{m}\cos 45)\tan 45} \tag{20.7}$$

Biomechatronics. https://doi.org/10.1016/B978-0-12-812939-5.00020-3

$$F_{x\,EE} = \frac{-5.224\,\mathrm{Nm}}{-0.4725\,\mathrm{m} + 0.6372\mathrm{m}} = -31.72\,\mathrm{N} \cong -32\,\mathrm{N} \rightarrow F_{x\,EE\,external} = 32\,\mathrm{N} \tag{20.8}$$

$$F_{y\,EE} = \frac{3.0\,\mathrm{Nm} + 0.35\,\mathrm{m}\sin(45)(-31.72\,\mathrm{N})}{0.35\,\mathrm{m}\cos(45)} = -19.59\,\mathrm{N} \cong -20\,\mathrm{N} \rightarrow F_{y\,EE\,external} = 20\,\mathrm{N} \tag{20.9}$$

(B)

$$\theta_1 = \frac{\pi}{9}\mathrm{rad} + \frac{\pi}{18}\mathrm{rad}\cos\left(0.88\frac{\mathrm{rad}}{\mathrm{s}}t\right) \tag{20.10}$$

$$\dot{\theta}_1 = -\frac{\pi}{18}0.88\frac{\mathrm{rad}}{\mathrm{s}}\sin\left(0.88\frac{\mathrm{rad}}{\mathrm{s}}t\right) \tag{20.11}$$

$$\ddot{\theta}_1 = -\frac{\pi}{18}\left(0.88\frac{1}{s}\right)^2\mathrm{rad}\cos\left(0.88\frac{\mathrm{rad}}{\mathrm{s}}t\right) \tag{20.12}$$

$$T = \frac{m_1 L_{CM1}^2 \dot{\theta}_1^2}{2} + \frac{I_1 \dot{\theta}_1^2}{2} = \frac{m_1 L_{CM1}^2 + I_1}{2}\dot{\theta}_1^2 \tag{20.13}$$

$$V = m_1 g L_{CM1}\sin\theta_1 \tag{20.14}$$

$$L = T - V = \frac{m_1 L_{CM1}^2 + I_1}{2}\dot{\theta}_1^2 - m_1 g L_{CM1}\sin\theta_1 \tag{20.15}$$

$$\frac{\partial L}{\partial \theta_1} = -m_1 g L_{CM1}\cos\theta_1 \tag{20.16}$$

$$\frac{\partial L}{\partial \dot{\theta}_1} = (m_1 L_{CM1}^2 + I_1)\dot{\theta}_1 \tag{20.17}$$

$$\frac{d}{dt}\left(\frac{\partial L}{\partial \dot{\theta}_1}\right) = (m_1 L_{CM1}^2 + I_1)\ddot{\theta}_1 \tag{20.18}$$

$$\frac{d}{dt}\left(\frac{\partial L}{\partial \dot{\theta}_1}\right) - \frac{\partial L}{\partial \theta_1} = (m_1 L_{CM1}^2 + I_1)\ddot{\theta}_1 + m_1 g L_{CM1}\cos\theta_1 = \tau_1 \tag{20.19}$$

$$\begin{aligned}\tau_1 &= (1.2\,\mathrm{kg}\,0.30^2\,\mathrm{m}^2 + 0.16\,\mathrm{kg\,m}^2)\ddot{\theta}_1 + 1.2\,\mathrm{kg}\,9.81\frac{\mathrm{m}}{\mathrm{s}^2}0.30\,\mathrm{m}\cos\theta_1 \\ &= 0.268\,\mathrm{kg\,m}^2\ddot{\theta}_1 + 3.532\,\mathrm{kg}\frac{\mathrm{m}^2}{\mathrm{s}^2}\cos\theta_1\end{aligned} \tag{20.20}$$

$$\ddot{\theta}_1(t = 1s) = -\frac{\pi}{18}\left(0.88\frac{\mathrm{rad}}{\mathrm{s}}\right)^2\cos\left(0.88\frac{\mathrm{rad}}{\mathrm{s}}1s\right) = -0.086\frac{\mathrm{rad}}{\mathrm{s}^2} \tag{20.21}$$

$$\theta_1(t = 1s) = \frac{\pi}{9}\mathrm{rad} + \frac{\pi}{18}\mathrm{rad}\cos\left(0.88\frac{\mathrm{rad}}{\mathrm{s}}1s\right) = 0.4500\,\mathrm{rad} \tag{20.22}$$

$$\cos\theta_1(t = 1s) = 0.8960 \tag{20.23}$$

$$\tau_1(t = 1s) = 3.142\,\mathrm{Nm} \cong \mathbf{3.1\,Nm} \tag{20.24}$$

$$\ddot{\theta}_1(t = 3s) = -\frac{\pi}{18}\left(0.88\frac{1}{s}\right)^2\mathrm{rad}\cos\left(0.88\frac{\mathrm{rad}}{\mathrm{s}}3s\right) = 0.1185\frac{\mathrm{rad}}{\mathrm{s}^2} \tag{20.25}$$

$$\theta_1(t=3s) = \frac{\pi}{9}\text{rad} + \frac{\pi}{18}\text{rad}\cos\left(0.88\frac{\text{rad}}{\text{s}}3s\right) = 0.1958\,\text{rad} \tag{20.26}$$

$$\cos\theta_1(t=3s) = 0.9809 \tag{20.27}$$

$$\tau_1(t=3s) = 3.496\,\text{Nm} \cong \mathbf{3.5\,Nm} \tag{20.28}$$

$$\ddot{\theta}_1(t=5s) = -\frac{\pi}{18}\left(0.88\frac{1}{s}\right)^2 \text{rad}\cos\left(0.88\frac{\text{rad}}{\text{s}}5s\right) = 0.041\frac{\text{rad}}{\text{s}^2} \tag{20.29}$$

$$\theta_1(t=5s) = \frac{\pi}{9}\text{rad} + \frac{\pi}{18}\text{rad}\cos\left(0.88\frac{\text{rad}}{\text{s}}5s\right) = 0.2956\,\text{rad} \tag{20.30}$$

$$\cos\theta_1(t=5s) = 0.9566 \tag{20.31}$$

$$\tau_1(t=5s) = 3.390\,\text{Nm} \cong \mathbf{3.4\,Nm} \tag{20.32}$$

(C)

$$\frac{d}{dt}\left(\frac{\partial L}{\partial\dot{\theta}_1}\right) - \frac{\partial L}{\partial\theta_1} = (m_1L_{CM1}^2 + I_1)\ddot{\theta}_1 + m_1gL_{CM1}\cos\theta_1 = \tau_1 - FL_1\sin\theta_1 \tag{20.33}$$

$$\sin\theta_1(t=1s) = 0.4352 \tag{20.34}$$

$$FL_1\sin\theta_1(t=1s) = 2.350\,\text{Nm} \cong 2.4\,\text{Nm} \tag{20.35}$$

$$\tau_1(t=1s) = \mathbf{5.5\,Nm} \tag{20.36}$$

$$\sin\theta_1(t=3s) = 0.1946 \tag{20.37}$$

$$FL_1\sin\theta_1(t=3s) = 1.051\,\text{Nm} \cong 1.1\,\text{Nm} \tag{20.38}$$

$$\tau_1(t=3s) = \mathbf{4.6\,Nm} \tag{20.39}$$

$$\sin\theta_1(t=5s) = 0.2913 \tag{20.40}$$

$$FL_1\sin\theta_1(t=5s) = 1.573\,\text{Nm} \cong 1.6\,\text{Nm} \tag{20.41}$$

$$\tau_1(t=5s) = \mathbf{5.0\,Nm} \tag{20.42}$$

Problem 2 Numerical Problem, Difficulty Level: *, Chapter 2

(Prepared by Marko Popovic)

Solution:

The joint angle trajectory is defined as

$$\theta_1(t) = \frac{3\pi}{8} + \frac{\pi}{8}\cos\left(\frac{\pi t}{3s} + \frac{\pi}{16}\right) \tag{20.43}$$

We need kinetic energy and potential energy so that we could obtain Lagrangian.

Kinetic energy is

$$T=\frac{1}{2}I(\dot{\theta}_1)^2=\frac{1}{2}\frac{1}{3}m_1L_1^2(\dot{\theta}_1)^2 \tag{20.44}$$

And potential energy is

$$V=-m_1g\frac{L_1}{2}\cos\theta_1 \tag{20.45}$$

Hence, Lagrangian is

$$L=T-V=\frac{1}{6}m_1L_1^2(\dot{\theta}_1)^2+m_1g\frac{L_1}{2}\cos\theta_1 \tag{20.46}$$

To express Euler Lagrange equations, we need

$$\frac{\partial L}{\partial\theta_1}=-m_1g\frac{L_1}{2}\sin\theta_1 \tag{20.47}$$

and also

$$\frac{\partial L}{\partial\dot{\theta}_1}=\frac{1}{3}m_1L_1^2\dot{\theta}_1 \tag{20.48}$$

Hence, we express Euler Lagrange equation as

$$\frac{d}{dt}\left(\frac{\partial L}{\partial\dot{\theta}_1}\right)-\frac{\partial L}{\partial\theta_1}=\frac{1}{3}m_1L_1^2\ddot{\theta}_1+m_1g\frac{L_1}{2}\sin\theta_1=\tau_{\theta_1} \tag{20.49}$$

We know that

$$\dot{\theta}_1=-\frac{\pi^2}{24s}\sin\left(\frac{\pi t}{3s}+\frac{\pi}{16}\right) \tag{20.50}$$

and that

$$\ddot{\theta}_1=-\frac{\pi^3}{72s^2}\cos\left(\frac{\pi t}{3s}+\frac{\pi}{16}\right) \tag{20.51}$$

and therefore

$$\frac{1}{3}m_1L_1^2\left[-\frac{\pi^3}{72s^2}\cos\left(\frac{\pi t}{3s}+\frac{\pi}{16}\right)\right]+m_1g\frac{L_1}{2}\sin\left[\frac{3\pi}{8}+\frac{\pi}{8}\cos\left(\frac{\pi t}{3s}+\frac{\pi}{16}\right)\right]=\tau_{\theta_1}(t) \tag{20.52}$$

From here one could, for example, find a maximal torque for a given period of time. Based on graph shown in Fig. 20.1 the maximal torque is 2.9 Nm.

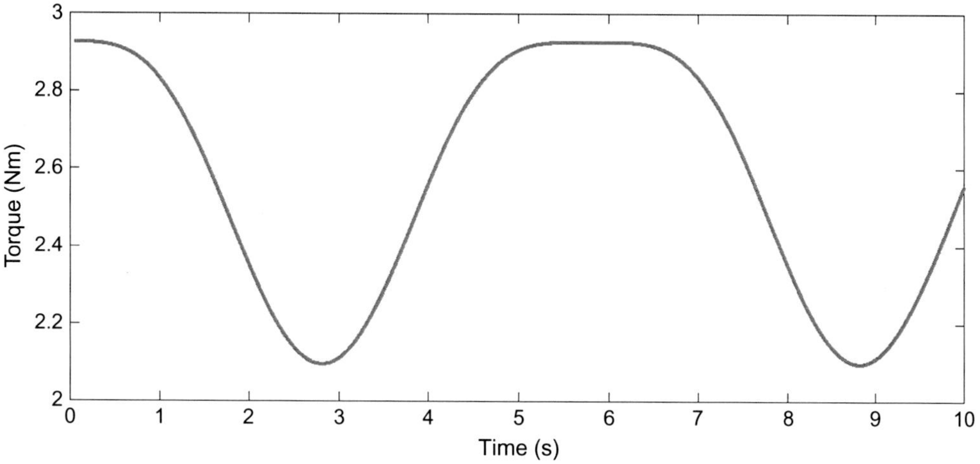

FIG. 20.1 Torque vs time.

Problem 3 Numerical Problem, Difficulty Level: ***, Chapters 2, 12, 16, and 17

(Prepared by Marko Popovic)

Solution:

After 50 ms line connecting heel and center of mass is tilted by some angle θ(50 ms). Condition that center of mass ends up just above the heel of stance leg with zero final speed can be expressed via energy conservation

$$mgL[1 - \cos\theta(50\,\text{ms})] = \frac{mL^2\dot{\theta}^2|_{50\,\text{ms}}}{2} \tag{20.53}$$

During the first 50 ms, the second Newton's law can be expressed in terms of radial and tangential components as

$$mL\left[-\dot{\theta}^2\left(\sin\theta\hat{i} + \cos\theta\hat{j}\right) + \ddot{\theta}\left(\cos\theta\hat{i} - \sin\theta\hat{j}\right)\right] = mg\left[-\cos\theta\hat{j}\left(\sin\theta\hat{i} + \cos\theta\hat{j}\right) + \sin\theta\left(\cos\theta\hat{i} - \sin\theta\hat{j}\right)\right]$$
$$+ F_R\left(\sin\theta\hat{i} + \cos\theta\hat{j}\right) + F_T\left(\cos\theta\hat{i} - \sin\theta\hat{j}\right) \tag{20.54}$$

with F_R and F_T being the radial and tangential component of ground reaction force, respectively.

By focusing on tangential component

$$mL\ddot{\theta} = mg\sin\theta - \frac{\tau}{L} \tag{20.55}$$

with $\tau = 200$ Nm and using

$$mL\frac{d\dot{\theta}}{dt}\frac{d\theta}{d\theta} = mL\dot{\theta}\frac{d\dot{\theta}}{d\theta} = \frac{mL}{2}\frac{d\dot{\theta}^2}{d\theta} \tag{20.56}$$

one obtains

$$\frac{mL}{2}d\dot{\theta}^2 = -mgd(\cos\theta) - \frac{\tau}{L}d\theta \tag{20.57}$$

or after integration

$$\frac{mL}{2}\dot{\theta}^2 = mg(\cos\theta_0 - \cos\theta) + \frac{\tau}{L}(\theta_0 - \theta) \tag{20.58}$$

Solving for $\dot{\theta}$ and performing one more integration gives

$$-\sqrt{\frac{L}{2g}}\int_{\theta_0}^{\theta(50\,\text{ms})} \frac{d\theta}{\sqrt{(\cos\theta_0 - \cos\theta) + \frac{\tau}{Lmg}(\theta_0 - \theta)}} = 50\,\text{ms} \tag{20.59}$$

For small angles, one could use expansion $\cos\varphi \cong 1 - \frac{\varphi^2}{2}$ such that

$$(\cos\theta_0 - \cos\theta) + \frac{\tau}{Lmg}(\theta_0 - \theta) \cong \left[-\frac{1}{2}(\theta_0 + \theta) + \frac{\tau}{Lmg}\right](\theta_0 - \theta) \cong \left(-\theta_0 + \frac{\tau}{Lmg}\right)(\theta_0 - \theta) \tag{20.60}$$

Note that we expressed $\theta_0 + \theta$ as $2\theta_0$ because main contribution to integral is obtained for $\theta \approx \theta_0$.

Hence, above integral can be approximated as

$$\sqrt{\frac{L}{2g}}\frac{1}{\sqrt{-\theta_0 + \frac{\tau}{Lmg}}}[\theta_0 - \theta(50\,\text{ms})] = 50\,\text{ms} \tag{20.61}$$

By combining other two equations one obtains

$$mgL[1 - \cos\theta(50\,\text{ms})] = mgL[\cos\theta_0 - \cos\theta(50\,\text{ms})] + \tau[\theta_0 - \theta(50\,\text{ms})] \tag{20.62}$$

which can be rewritten as

$$\cos\theta_0 = 1 - \frac{\tau}{Lmg}\sqrt{\frac{2g}{L}}\sqrt{-\theta_0 + \frac{\tau}{Lmg}}50\,\text{ms} = 1 - \frac{\tau^{\frac{3}{2}}50\,\text{ms}}{gm^{\frac{3}{2}}L^2}\sqrt{1 - \frac{\theta_0 Lmg}{\tau}} \tag{20.63}$$

One could now again use small-angle expansion of cosine function to estimate the angle θ_0

$$\theta_0 \cong \sqrt{2\frac{\tau^{\frac{3}{2}}0.05\,s}{gm^{\frac{3}{2}}L^2}\sqrt[4]{1 - \frac{\theta_0 Lmg}{\tau}}} \rightarrow \theta_0 \cong 0.23\,\text{rad} \cong 13^{\circ} \tag{20.64}$$

To check consistency of previous calculation, one could verify that $\theta(50\,\text{ms}) \cong 0.14$ rad.

Problem 4 Numerical Problem, Difficulty Level: **, Chapters 2, 12, and 17

(Prepared by Marko Popovic)

Hint:

Express CM position and velocity.

$$x_{CM} = l\left[\sin\theta_a - \sin(\theta_k - \theta_a) + \frac{1}{2}\sin(\theta_h - \theta_k + \theta_a)\right] \tag{20.65}$$

$$y_{CM} = l\left[\cos\theta_a + \cos(\theta_k - \theta_a) + \frac{1}{2}\cos(\theta_h - \theta_k + \theta_a)\right] \tag{20.66}$$

$$\dot{x}_{CM} = l\left[\dot{\theta}_a\cos\theta_a - (\dot{\theta}_k - \dot{\theta}_a)\cos(\theta_k - \theta_a) + \frac{1}{2}(\dot{\theta}_h - \dot{\theta}_k + \dot{\theta}_a)\cos(\theta_h - \theta_k + \theta_a)\right] \tag{20.67}$$

$$\dot{y}_{CM} = l\left[-\dot{\theta}_a\sin\theta_a - (\dot{\theta}_k - \dot{\theta}_a)\sin(\theta_k - \theta_a) - \frac{1}{2}(\dot{\theta}_h - \dot{\theta}_k + \dot{\theta}_a)\sin(\theta_h - \theta_k + \theta_a)\right] \tag{20.68}$$

If you are persistent, you could also express CM acceleration.

$$\ddot{x}_{CM} = l\left[\ddot{\theta}_a\cos\theta_a - (\ddot{\theta}_k - \ddot{\theta}_a)\cos(\theta_k - \theta_a) + \frac{1}{2}(\ddot{\theta}_h - \ddot{\theta}_k + \ddot{\theta}_a)\cos(\theta_h - \theta_k + \theta_a) - \dot{\theta}_a^{\,2}\sin\theta_a\right.$$
$$\left. + (\dot{\theta}_k - \dot{\theta}_a)^2\sin(\theta_k - \theta_a) - \frac{1}{2}(\dot{\theta}_h - \dot{\theta}_k + \dot{\theta}_a)^2\sin(\theta_h - \theta_k + \theta_a)\right] \tag{20.69}$$

$$\ddot{y}_{CM} = l\left[-\ddot{\theta}_a\sin\theta_a - (\ddot{\theta}_k - \ddot{\theta}_a)\sin(\theta_k - \theta_a) - \frac{1}{2}(\ddot{\theta}_h - \ddot{\theta}_k + \ddot{\theta}_a)\sin(\theta_h - \theta_k + \theta_a) - \dot{\theta}_a^{\,2}\cos\theta_a\right.$$
$$\left. - (\dot{\theta}_k - \dot{\theta}_a)^2\cos(\theta_k - \theta_a) - \frac{1}{2}(\dot{\theta}_h - \dot{\theta}_k + \dot{\theta}_a)^2\cos(\theta_h - \theta_k + \theta_a)\right] \tag{20.70}$$

Now, it may be suitable to express

$$\ddot{x}_{CM} = \ddot{x}_{CM1} + \ddot{\theta}_a l\left[\cos\theta_a + \cos(\theta_k - \theta_a) + \frac{1}{2}\cos(\theta_h - \theta_k + \theta_a)\right] = \ddot{x}_{CM1} + \ddot{\theta}_a y_{CM} \tag{20.71}$$

$$\ddot{y}_{CM} = \ddot{y}_{CM1} - \ddot{\theta}_a l\left[\sin\theta_a - \sin(\theta_k - \theta_a) + \frac{1}{2}\sin(\theta_h - \theta_k + \theta_a)\right] = \ddot{y}_{CM1} - \ddot{\theta}_a x_{CM} \tag{20.72}$$

With $\ddot{x}_{CM1}$ and $\ddot{y}_{CM1}$ not dependent on $\ddot{\theta}_a$, that is,

$$\ddot{x}_{CM1} = l\left[-\ddot{\theta}_k\cos(\theta_k - \theta_a) + \frac{1}{2}(\ddot{\theta}_h - \ddot{\theta}_k)\cos(\theta_h - \theta_k + \theta_a) - \dot{\theta}_a^{\,2}\sin\theta_a + (\dot{\theta}_k - \dot{\theta}_a)^2\sin(\theta_k - \theta_a)\right.$$
$$\left. - \frac{1}{2}(\dot{\theta}_h - \dot{\theta}_k + \dot{\theta}_a)^2\sin(\theta_h - \theta_k + \theta_a)\right] \tag{20.73}$$

$$\ddot{y}_{CM1} = l\left[-\ddot{\theta}_k\sin(\theta_k - \theta_a) - \frac{1}{2}(\ddot{\theta}_h - \ddot{\theta}_k)\sin(\theta_h - \theta_k + \theta_a) - \dot{\theta}_a^{\,2}\cos\theta_a - (\dot{\theta}_k - \dot{\theta}_a)^2\cos(\theta_k - \theta_a)\right.$$
$$\left. - \frac{1}{2}(\dot{\theta}_h - \dot{\theta}_k + \dot{\theta}_a)^2\cos(\theta_h - \theta_k + \theta_a)\right] \tag{20.74}$$

Now we are ready to express left-hand side of equation (19.6) as

$$m[y_{CM}\ddot{x}_{CM}-x_{CM}(\ddot{y}_{CM}+g)]=m[y_{CM}(\ddot{x}_{CM1}+\ddot{\theta}_a y_{CM})-x_{CM}(\ddot{y}_{CM1}-\ddot{\theta}_a x_{CM}+g)]$$
$$=m[y_{CM}\ddot{x}_{CM1}-x_{CM}(\ddot{y}_{CM1}-\ddot{\theta}_a x_{CM}+g)+\ddot{\theta}_a(y_{CM}^2+x_{CM}^2)] \quad (20.75)$$

as well as right-hand side as

$$-\frac{ml^2}{12}\frac{d^2\theta_t}{dt^2}=-\frac{ml^2}{12}(\ddot{\theta}_h-\ddot{\theta}_k+\ddot{\theta}_a) \quad (20.76)$$

One could now solve for ankle angular acceleration

$$\ddot{\theta}_a=\frac{x_{CM}(\ddot{y}_{CM1}-\ddot{\theta}_a x_{CM}+g)-y_{CM}\ddot{x}_{CM1}-\dfrac{l^2}{12}(\ddot{\theta}_h-\ddot{\theta}_k)}{y_{CM}^2+x_{CM}^2+\dfrac{l^2}{12}} \quad (20.77)$$

This second-order differential equation for θ_a is consequence of physics. (!) The ankle angular acceleration is completely determined by knee and hip angles, angular velocities, and angular accelerations and ankle angle and angular velocity. One straightforward way to solve it is using numerical methods.

Problem 5 Numerical Problem, Difficulty Level: ***, Chapters 2 and 17

(Prepared by Marko Popovic)

Hint:

There are many ways that you may solve this problem numerically. For example, you may start with a "good" guess on joint trajectories (say sin or cos functions or whatever you fancy) and then fine tune them such that motion is *physically possible* and that you also satisfy specified constraints.

For purpose of this homework, you may assume that time step of your numerical simulation is say 1/25 s, that is, 40 ms. As before you will define joint angle at those certain moments, indexed with ***i***, and then obtain velocity and acceleration as

$$\omega_i=\frac{\theta_i-\theta_{i-1}}{40\text{ ms}} \text{ and } \alpha_i=\frac{\omega_i-\omega_{i-1}}{40\text{ ms}} \quad (20.78)$$

If you believe that your approach may benefit from even smaller time step, please go ahead and use smaller time step.

Physical realism in this context means that zero-moment point (ZMP), or center of pressure (CoP), should coincide with stance point-like foot during single-support phase (one leg on the ground) or should be between the two point-like feet during the double-support phase (both legs on the ground). As legs are massless and entire robot's mass is located at the hip this means that there are no moments about the center of mass, and hence that the ZMP location is defined as intersection with the ground of line parallel to ground reaction force that also passes through center of mass. Ground reaction forces can be obtained from the second Newton law as

$$\vec{F}_{ground}=M(\vec{a}-\vec{g}) \quad (20.79)$$

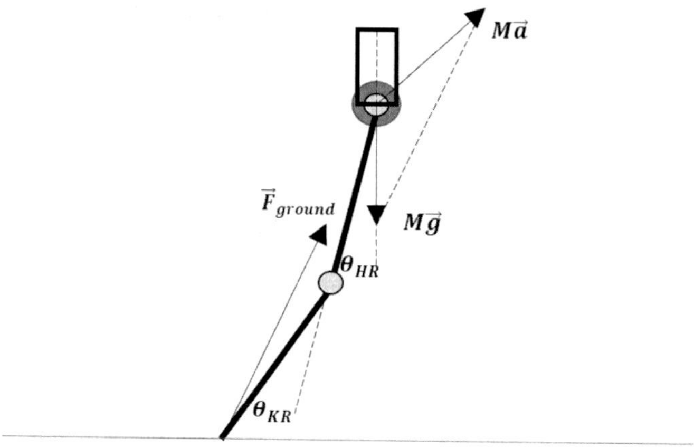

FIG. 20.2 Single support.

The acceleration $\boldsymbol{a}$ is the center of mass acceleration and can be obtained from joint-angle trajectories (Fig. 20.2).

Single-support phase

For example, during the single-support phase of say right leg the center of mass coordinates may be obtained relative to stance foot location as

$$x_{CoM} - x_{RFoot} = L(\sin(\theta_{KR} + \theta_{HR}) + \sin\theta_{HR}) \tag{20.80}$$

$$y_{CoM} = L(\cos(\theta_{KR} + \theta_{HR}) + \cos\theta_{HR}) \tag{20.81}$$

You could use this to obtain center of mass velocity by differentiating once in respect to time

$$v_{XCoM} = L((\omega_{KR} + \omega_{HR})\cos(\theta_{KR} + \theta_{HR}) + \omega_{HR}\cos\theta_{HR}) \tag{20.82}$$

$$v_{YCoM} = -L((\omega_{KR} + \omega_{HR})\sin(\theta_{KR} + \theta_{HR}) + \omega_{HR}\sin\theta_{HR}) \tag{20.83}$$

Or acceleration of center of mass by differentiating once again

$$a_{XCoM} = L\Big((\alpha_{KR} + \alpha_{HR})\cos(\theta_{KR} + \theta_{HR}) - (\omega_{KR} + \omega_{HR})^2\sin(\theta_{KR} + \theta_{HR}) + \alpha_{HR}\cos\theta_{HR} - \omega_{HR}{}^2\sin\theta_{HR}\Big) \tag{20.84}$$

$$a_{YCoM} = -L\Big((\alpha_{KR} + \alpha_{HR})\sin(\theta_{KR} + \theta_{HR}) + (\omega_{KR} + \omega_{HR})^2\cos(\theta_{KR} + \theta_{HR}) + \alpha_{HR}\sin\theta_{HR} + \omega_{HR}{}^2\cos\theta_{HR}\Big) \tag{20.85}$$

Hence the condition for physical realism that the ZMP coincides with the location of right foot leg during right leg single-support phase may be expressed in the Cartesian coordinates and their derivatives as

$$x_{Rfoot} = x_{CoM} - \frac{F_{groundX}}{F_{groundY}} y_{CoM} \tag{20.86}$$

or

$$\frac{y_{CoM}}{x_{CoM} - x_{RFoot}} = \frac{a_{YCoM} + g}{a_{XCoM}} \tag{20.87}$$

Or in terms of joint angles and their derivatives as

$$\frac{\cos(\theta_{KR}+\theta_{HR})+\cos\theta_{HR}}{\sin(\theta_{KR}+\theta_{HR})+\sin\theta_{HR}}=$$

$$\frac{-(\alpha_{KR}+\alpha_{HR})\sin(\theta_{KR}+\theta_{HR})-(\omega_{KR}+\omega_{HR})^2\cos(\theta_{KR}+\theta_{HR})-\alpha_{HR}\sin\theta_{HR}-\omega_{HR}{}^2\cos\theta_{HR}+\frac{g}{L}}{(\alpha_{KR}+\alpha_{HR})\cos(\theta_{KR}+\theta_{HR})-(\omega_{KR}+\omega_{HR})^2\sin(\theta_{KR}+\theta_{HR})+\alpha_{HR}\cos\theta_{HR}-\omega_{HR}{}^2\sin\theta_{HR}} \quad (20.88)$$

Another condition for physical realism is that ground reaction force is always pointing away from ground (i.e., foot is not glued or bolted to ground) so that

$$a_{YCoM}+g>0\rightarrow\frac{g}{L} > (\alpha_{KR}+\alpha_{HR})\sin(\theta_{KR}+\theta_{HR})+(\omega_{KR}+\omega_{HR})^2\cos(\theta_{KR}+\theta_{HR})+\alpha_{HR}\sin\theta_{HR}+\omega_{HR}{}^2\cos\theta_{HR} \quad (20.89)$$

For left leg single-support phase, you could now simply change index R to L in the two above formulas.

Double-Support Phase

Here, as shown in Fig. 20.3, you need to satisfy condition for kinematic chain, that is,

$$\cos(\theta_{KR}+\theta_{HR})+\cos\theta_{HR}=\cos(\theta_{KL}+\theta_{HL})+\cos\theta_{HL} \quad (20.90)$$

$$\sin(\theta_{KR}+\theta_{HR})+\sin\theta_{HR}-\sin(\theta_{KL}+\theta_{HL})-\sin\theta_{HL}=\frac{D}{L} \quad (20.91)$$

And condition that ZMP is in between the right and left foot can be expressed as

$$x_{Rfoot}<x_{CoM}-\frac{F_{groundX}}{F_{groundY}}y_{CoM}<x_{Lfoot} \quad (20.92)$$

or

$$-(x_{CoM}-x_{Rfoot})<-\frac{F_{groundX}}{F_{groundY}}y_{CoM}<-(x_{CoM}-x_{Lfoot}) \quad (20.93)$$

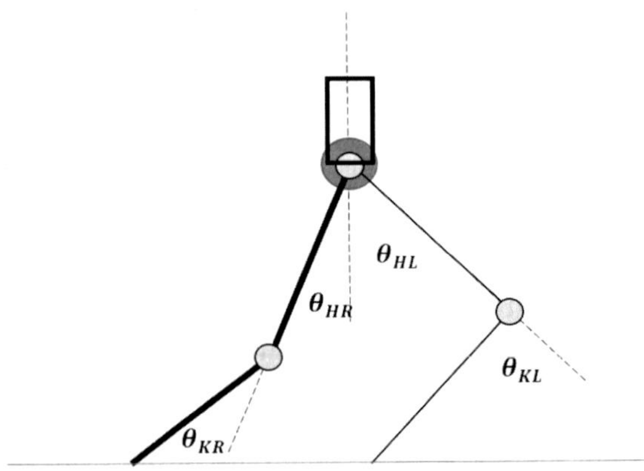

FIG. 20.3 Double support.

that is, for $F_{groundX}>0$ we need $x_{CoM}-x_{Rfoot}>\frac{F_{groundX}}{F_{groundY}}y_{CoM}$ or $\frac{y_{CoM}}{x_{CoM}-x_{RFoot}}<\frac{a_{YCoM}+g}{a_{XCoM}}$.

and for $F_{groundX}<0$ we need $x_{CoM}-x_{Lfoot}<\frac{F_{groundX}}{F_{groundY}}y_{CoM}$ or $\frac{y_{CoM}}{x_{CoM}-x_{LFoot}}>\frac{a_{YCoM}+g}{a_{XCoM}}$.

Even more hints

During single-support phase as legs are massless you can do whatever you wish with the joint angles of leg that is not in contact with the ground (i.e., swing leg), as that won't change system dynamics. Hence, if you know the complete trajectory of one joint of stance leg, you can solve for trajectory of other joint of stance leg by using equation for ZMP physical realism. In other words you will solve for angular acceleration of other joint and then automatically generate its angular position and velocity at next moment and so on. In case you run into numerical instabilities, you may approach this problem with parametrized splines or a similar method.

During double-support phase if you know joint angle trajectories for one leg, then you automatically could solve for both joints of other leg based on kinematics chain equations. Hence, again you could start with single joint trajectory during double-support phase, solve then for another joint of the same leg using ZMP physical realism, and finally deduce the joints of the other leg using kinematics chain equations.

Hence, if you are solving this problem numerically you just need single joint degree of freedom trajectory at any point of time and then you could try to match the boundary conditions.

There are infinitely many solutions for this problem, you have the freedom to choose the one that does the job for you.

Problem 6 Numerical Problem, Difficulty Level: ***, Chapters 2 and 17

(Prepared by Marko Popovic)

Hint:

Similar as for previous problem.

The main differences: (1) one needs to worry that legs do not collide with stairs and (2) double-support phase is a bit more complex.

Condition for physical realism during double-support phase can be now expressed as

$$x_{Rfoot}<x_{CoM}-\frac{F_{groundX}}{F_{groundY}}y_{CoM(R)} \tag{20.94}$$

$$x_{CoM}-\frac{F_{groundX}}{F_{groundY}}y_{CoM(L)}<x_{Lfoot} \tag{20.95}$$

with $y_{CoM(R)}$ being a height of center of mass in respect to stair at which right leg is in contact with and with $y_{CoM(L)}$ being a height of center of mass in respect to stair at which left leg is in contact with, that is,

$$y_{CoM(R)}=L(\cos(\theta_{KR}+\theta_{HR})+\cos\theta_{HR}) \tag{20.96}$$

$$y_{CoM(L)}=L(\cos(\theta_{KL}+\theta_{HL})+\cos\theta_{HL}) \tag{20.97}$$

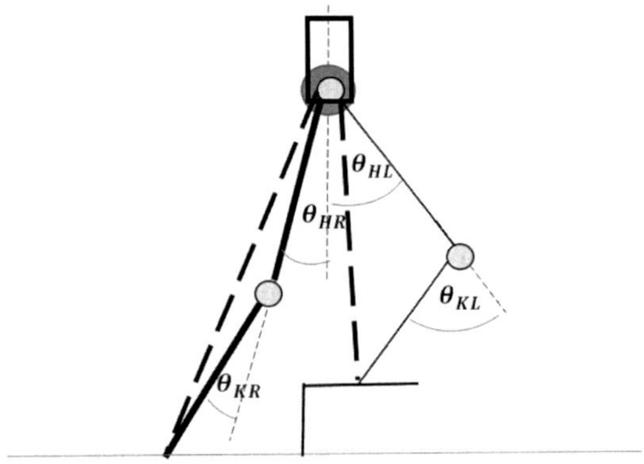

FIG. 20.4 Double support while walking upstairs.

during double-support phase.

Also, the condition for kinematic chain, see Fig. 20.4, can be now expressed as

$$\cos(\theta_{KR}+\theta_{HR})+\cos\theta_{HR}-\cos(\theta_{KL}+\theta_{HL})-\cos\theta_{HL}=\frac{H}{L} \tag{20.98}$$

$$\sin(\theta_{KR}+\theta_{HR})+\sin\theta_{HR}-\sin(\theta_{KL}+\theta_{HL})-\sin\theta_{HL}=\frac{D_0+D_1}{L} \tag{20.99}$$

The movie is another way that you could judge whether you successfully matched the proper boundary condition; if angles are fine and transition from end to start is smooth you probably did a good job.

The transitions between single and double-support phase is also critical. Think what should be the good conditions for these transitions.

Is motion you obtained biologically realistic? How you could make it even more "biological"?

Problem 7 Numerical Problem, Difficulty Level: *, Chapter 2

(Prepared by Marko Popovic)

Solution:

It should take the same amount of time Δt in average for air streaming above the wing to reach the right end of the wing as it should take air streaming beneath the wing to reach the same point. Hence,

$$\Delta t=\frac{a}{v_B}=\frac{\frac{b}{2}+c}{v_T} \tag{20.100}$$

with v_T being the speed on top of the wing and v_B being the speed beneath the wing. Therefore,

$$v_T = \frac{\frac{b}{2} + c}{a} v_B \tag{20.101}$$

One can express

$$v_T^2 - v_B^2 = \frac{\left(\frac{b}{2} + c\right)^2 - a^2}{a^2} v_B^2 = \frac{\frac{b^2}{4} + bc + a^2 + b^2 - a^2}{a^2} v_B^2 = \frac{b}{a^2}\left(\frac{5}{4}b + c\right) v_B^2 \tag{20.102}$$

According to Bernoulli's equation $p + \rho g h + \rho \frac{v^2}{2} = const$. Hence,

$$\Delta p = p_B - p_T = \frac{\rho}{2}\left(v_T^2 - v_B^2\right) \tag{20.103}$$

or

$$p_T = p_B - \frac{\rho}{2} \frac{b}{a^2}\left(\frac{5}{4}b + c\right) v_B^2 \tag{20.104}$$

One can now estimate force due to pressure on each side of the $l = 6$ m long wing. Force on side "a" is

$$\vec{F}_a = p_B a l \hat{k} \tag{20.105}$$

Force on side "b" due to pressure is

$$\vec{F}_b = -p_T b l \hat{i} \tag{20.106}$$

And force on side "c" due to pressure is

$$\vec{F}_c = p_T c l \frac{b\hat{i} - a\hat{k}}{c} \tag{20.107}$$

Hence, total force due to pressure per wing is

$$\vec{F}_{pressure} = (p_B - p_T) a l \hat{k} = \frac{\rho}{2} \frac{b}{a^2}\left(\frac{5}{4}b + c\right) v_B^2 a l \hat{k} = \frac{\rho}{2} \frac{b}{a^2}\left(\frac{5}{4}b + \sqrt{a^2 + b^2}\right) v_B^2 a l \hat{k} \tag{20.108}$$

For two wings, total lift due to pressure is obtained as

$$\vec{F}_{lift} = \rho v_B^2 \frac{b}{a}\left(\frac{5}{4}b + \sqrt{a^2 + b^2}\right) l \hat{k} \tag{20.109}$$

In the moment of takeoff the lift force is equal in magnitude to weight, that is, $F_{lift} = mg$ and hence

$$v_B = \sqrt{\frac{mg}{\rho \frac{b}{a}\left(\frac{5}{4}b + \sqrt{a^2 + b^2}\right) l}} = \sqrt{\frac{1000 \cdot 9.81}{1.225 \cdot \frac{0.75}{2}\left(\frac{5}{4} \cdot 0.75 + \sqrt{4.5625}\right) 6}} = 83.36 \frac{m}{s} = 186.5\, mph \tag{20.110}$$

Drag force acts on side "b"; for both wings it is equal to

$$\vec{F}_{drag} = -2\frac{C_d}{2}\rho v_B^2 lb\hat{i} \tag{20.111}$$

At the moment of takeoff thrust may be larger than drag force, that is, plane could be still accelerating in the $\hat{i}$ direction, hence

$$F_{trust} \geq F_{drag} = 2 \times 1.225 \times 83.36^2 \times 6 \times 0.75\,\mathrm{N} = 76{,}610\,\mathrm{N} \tag{20.112}$$

representing minimal thrust at the moment of takeoff.

Problem 8 Numerical Problem And Concept Question, Difficulty Level: **, Chapters 2 and 17

(Prepared by Marko Popovic)

Hint:

For easier (less realistic) version of problem: ignore vertical speed of flyer's central body. Total lift force, see Fig. 20.5, from both wings is

$$|F_V(\theta)| = 2\cos(\theta) W \int_0^R \frac{1}{2} C_D \rho r^2 \omega^2 dr = 2\cos(\theta) W \frac{C_D}{2} \rho \omega^2 \frac{R^3}{3} \tag{20.113}$$

In the time domain if we set $t=0$ for pose 1, then one has

$$F_V(t) = 2\cos\left(\frac{\pi}{6} - \omega t\right) W \frac{C_D}{2} \rho \omega^2 \frac{R^3}{3} \tag{20.114}$$

for $0 < t < \frac{\pi}{3\omega}$ (for downward wings motion) and

$$F_V(t) = -2\cos\left(-\frac{\pi}{6} + \omega\left(t - \frac{\pi}{3\omega}\right)\right) W \frac{C_D}{2} \rho \omega^2 \frac{Rr^3}{3} \tag{20.115}$$

for $\frac{\pi}{3\omega} < t < \frac{2\pi}{3\omega}$ (for upward wings motion).

Condition that vertical velocity do not change over entire cycle (duration T) can be expressed as

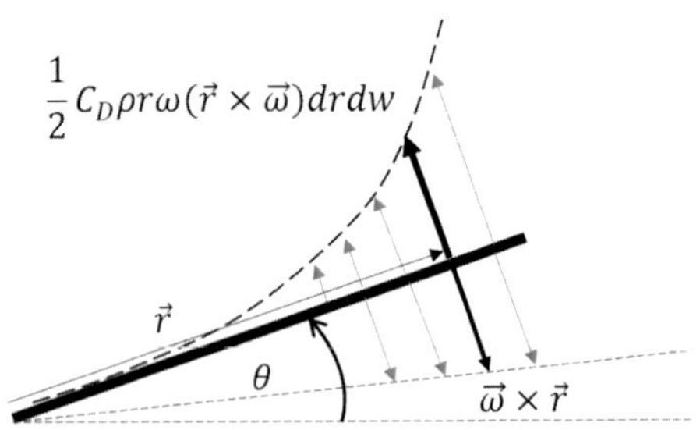

FIG. 20.5 Distribution of relative speed and force.

$$\Phi(T) = \int_0^T [F_V(t) - mg]dt = 0 \tag{20.116}$$

Condition that height do not change over entire cycle is

$$\int_0^T \Phi(t)dt = \int_0^T \int_0^t [F_V(t') - mg]dt'dt = 0 \tag{20.117}$$

(Although you probably won't need both of these conditions for easier version of the problem; only one equation should be enough to obtain flapping frequency.)

Frequency is $1/T$ in Hz and angular speed is $\omega = 2\pi/3T$.

For tougher (more realistic) version of problem: consider vertical velocity v of flyer's central body too.

At the beginning of downward motion of the wings, the flyer's central body has largest speed downward.

For $v<0$ and downward motion of the wings

$$F_V(\theta) = 2\cos(\theta)W\int_0^R \frac{1}{2}C_D\rho(\omega r - v\cos\theta)^2 dr = 2\cos(\theta)W\frac{C_D}{2}\rho\frac{1}{\omega}\frac{1}{3}\left[(\omega R - v\cos\theta)^3 - (v\cos\theta)^3\right] \tag{20.118}$$

For $v>0$ and downward motion of the wings

$$F_V(\theta) = -2\cos(\theta)W\int_0^{\frac{v\cos\theta}{\omega}} \frac{1}{2}C_D\rho(\omega r - v\cos\theta)^2 dr + 2\cos(\theta)W\int_{\frac{v\cos\theta}{\omega}}^R \frac{1}{2}C_D\rho(\omega r - v\cos\theta)^2 dr$$
$$= 2\cos(\theta)W\int_0^R \text{sign}(\omega r - v\cos\theta)\frac{1}{2}C_D\rho(\omega r - v\cos\theta)^2 dr \tag{20.119}$$

For $v>0$ and upward motion of the wings

$$F_V(\theta) = -2\cos(\theta)W\int_0^{R'} \frac{1}{2}C_D\rho(\omega r + v\cos\theta)^2 dr = -2\cos(\theta)W\frac{C_D}{2}\rho\frac{1}{\omega}\frac{1}{3}\left[(\omega R' + v\cos\theta)^3 - (v\cos\theta)^3\right] \tag{20.120}$$

For $v<0$ and upward motion of the wings

$$F_V(\theta) = -2\cos(\theta)W\int_0^{R'} \text{sign}(\omega r + v\cos\theta)\frac{1}{2}C_D\rho(\omega r + v\cos\theta)^2 dr \tag{20.121}$$

Here

$$F_V - mg = m\frac{dv}{dt} \tag{20.122}$$

Downward and upward wings motion corresponds to $0<t<\frac{\pi}{3\omega}$ and $\frac{\pi}{3\omega}<t<\frac{2\pi}{3\omega}$, respectively. In difference to previous case, both conditions are needed here, that is,

$$\int_0^T [F_V(t) - mg]dt = m[v(T) - v(0)] = 0 \tag{20.123}$$

$$\int_0^T v dt = z(T) - z(0) = 0 \tag{20.124}$$

to solve simultaneously for T and $v(0)$. One straightforward way to solve it is using numerical methods.

Problem 9 Numerical Problem, Difficulty Level: ***, Chapters 2 and 17

(Prepared by Marko Popovic; author acknowledge assistance from his student Shannon Moffat)

Solution:

(a) Reynolds number

$$R_e = \frac{\rho v L}{\mu} \sim \frac{1000\frac{kg}{m^3}4.47\frac{m}{s}1.5m}{1.002\,mPa\cdot s} = 6,705,000 \tag{20.125}$$

is quite large hence quadratic drag is more appropriate.

Slowing down drag due to frontal area is equal to

$$F_{SD} = \frac{C_D}{2}\rho A v^2 = 1000\cdot 0.04\cdot 4.47^2 N = 799\,N \tag{20.126}$$

(b) Net thrust

A to B ($\theta = 15^o \rightarrow \theta = 0^o$)

$$\begin{aligned}
F_{thrust} &= \sin\theta \int_0^L \frac{C_D}{2}\rho W (r\omega_1 - v\sin\theta)^2 \,\mathrm{sign}(r\omega_1 - v\sin\theta)dr \\
&= \rho W \sin\theta \left[-\int_0^{\frac{v\sin\theta}{\omega_1}} (r\omega_1 - v\sin\theta)^2 dr + \int_{\frac{v\sin\theta}{\omega_1}}^L (r\omega_1 - v\sin\theta)^2 dr \right] \\
&= \frac{\rho W}{3\omega_1}\sin\theta \left[-(r\omega_1 - v\sin\theta)^3 \Big|_0^{\frac{v\sin\theta}{\omega_1}} + (r\omega_1 - v\sin\theta)^3 \Big|_{\frac{v\sin\theta}{\omega_1}}^L \right] = \frac{\rho W}{3\omega_1}\sin\theta\left[-(v\sin\theta)^3 + (L\omega_1 - v\sin\theta)^3\right] \\
&= \frac{\rho W}{3\omega_1}\sin\theta (v\sin\theta)^3 \left[-1 + \left(\frac{L\omega_1}{v\sin\theta} - 1\right)^3 \right] = \frac{\rho W}{3\omega_1} v^3 \sin^4\theta \left[-1 + \left(\frac{L\omega_1}{v\sin\theta} - 1\right)^3 \right]
\end{aligned} \tag{20.127}$$

B to C ($\theta = 0^o \rightarrow \theta = -15^o$)

$$F_{thrust} = \sin\theta \int_0^L \frac{C_D}{2}\rho W (r\omega_2 - v\sin\theta)^2 \,\mathrm{sign}(r\omega_2 - v\sin\theta) dr = \rho W \sin\theta \int_0^L (r\omega_2 - v\sin\theta)^2 dr$$

$$= \frac{\rho W}{3\omega_2}\sin\theta (r\omega_2 - v\sin\theta)^3 |_0^L = \frac{\rho W}{3\omega_2}\sin\theta\left[(L\omega_2 - v\sin\theta)^3 + (v\sin\theta)^3\right]$$

$$= \frac{\rho W}{3\omega_2}\sin\theta (v\sin\theta)^3\left[\left(\frac{L\omega_2}{v\sin\theta} - 1\right)^3 + 1\right] = \frac{\rho W}{3\omega_2} v^3 \sin^4\theta \left[\left(\frac{L\omega_2}{v\sin\theta} - 1\right)^3 + 1\right] \quad (20.128)$$

C to B ($\theta = -15^o \rightarrow \theta = 0^o$)

$$F_{thrust} = \frac{\rho W}{3\omega_1} v^3 \sin^4\theta \left[-1 + \left(-\frac{L\omega_1}{v\sin\theta} - 1\right)^3\right] = \frac{\rho W}{3\omega_1} v^3 \sin^4\theta \left[-1 - \left(\frac{L\omega_1}{v\sin\theta} + 1\right)^3\right] \quad (20.129)$$

B to A ($\theta = 0^o \rightarrow \theta = 15^o$)

$$F_{thrust} = \frac{\rho W}{3\omega_2} v^3 \sin^4\theta \left[\left(-\frac{L\omega_2}{v\sin\theta} - 1\right)^3 + 1\right] = -\frac{\rho W}{3\omega_2} v^3 \sin^4\theta \left[\left(\frac{L\omega_2}{v\sin\theta} + 1\right)^3 - 1\right] \quad (20.130)$$

(c) Net force

$$F = F_{thrust} - F_{SD} \quad (20.131)$$

(d) Acceleration

$$a = \frac{F}{m} \quad (20.132)$$

(e) Condition for steady state

$$0 = \Delta v = \int_0^T a dt = \int_0^T a \frac{d\theta}{d\theta} dt = \int_0^T \frac{a}{\omega} d\theta = \int_{15^o}^{0^o} \frac{a}{\omega_1} d\theta + \int_{0^o}^{-15^o} \frac{a}{\omega_2} d\theta + \int_{-15^o}^{0^o} \frac{a}{\omega_1} d\theta + \int_{0^o}^{15^o} \frac{a}{\omega_2} d\theta \quad (20.133)$$

A to C is symmetrical to C to A so we can integrate just one of these, for example C to A

$$\int_{-15^o}^{0^o} \frac{a}{\omega_1} d\theta + \int_{0^o}^{15^o} \frac{a}{\omega_2} d\theta = \frac{\rho W}{3m} v^3 \left\{ \frac{1}{\omega_1} \int_{-\frac{\pi}{12}}^{0} \sin^4\theta \left[-1 - \left(\frac{L\omega_1}{v\sin\theta} + 1\right)^3\right] d\theta - \frac{1}{\omega_2}\int_0^{\frac{\pi}{12}} \sin^4\theta \right.$$

$$\left. \left[-1 + \left(\frac{L\omega_2}{v\sin\theta} + 1\right)^3\right] d\theta \right\} - \frac{799N}{m}\frac{\pi}{12}\left(\frac{1}{\omega_1} + \frac{1}{\omega_2}\right) = 0 \quad (20.134)$$

Here,

$$\int_{-\frac{\pi}{12}}^{0} \sin^4\theta \left[-1 - \left(\frac{L\omega_1}{v\sin\theta} + 1\right)^3\right] d\theta = -\frac{1}{16}\left[24\frac{L^2\omega_1^2}{v^2}\theta - 12\frac{L^2\omega_1^2}{v^2}\sin 2\theta - 4\left(4\frac{L^2\omega_1^2}{v^2} + 9\right)\frac{L\omega_1}{v}\cos\theta\right.$$

$$\left. + 4\frac{L\omega_1}{v}\cos 3\theta + 12\theta - 8\sin 2\theta + \sin 4\theta\right]_{-\frac{\pi}{12}}^{0} \quad (20.135)$$

and

$$\int_0^{\frac{\pi}{12}} \sin^4\theta\left[-1+\left(\frac{L\omega_2}{v\sin\theta}+1\right)^3\right]d\theta = -\frac{1}{4}\frac{L\omega_2}{v}\left[\left(4\frac{L^2{\omega_2}^2}{v^2}+9\right)\cos\theta + 3\frac{L\omega_2}{v}(\sin 2\theta - 2\theta) - \cos 3\theta\right]_0^{\frac{\pi}{12}} \tag{20.136}$$

You can also solve this graphically, for example, plot steady state condition as function of ω_1 and find intersection with zero.

(f) Solution is unique

$$\omega_1 = 14.7\,\text{rad/s} \tag{20.137}$$

Period can be obtained as

$$\frac{\pi}{6}\left(\frac{1}{\omega_1}+\frac{1}{\omega_2}\right) = 87.9\,\text{ms} \tag{20.138}$$

which is a few times shorter than the one (~300 ms) observed in nature for similar dimension fish.

Problem 10 Numerical Problem, Difficulty Level: *, Chapter 3

(Prepared by Marko Popovic)

Solution:

Magnetic field inside the mumetal inside the fixed solenoid is equal to $\mu_0\mu_{\text{eff}}n_1I_1$. This magnetic field will produce a torque on the horizontal segments of toroid-like wire. This torque per segment can be estimated by integration $\tau = \int_0^{R_2} rI_2B\,dr = \frac{1}{2}R_2^2I_2B = \frac{1}{2}R_2^2I_2\mu_0\mu_{\text{eff}}n_1I_1$. The total torque on toroid-like solenoid will be equal to $\tau_{\text{total}} = 2N_2\tau = \mu_0\mu_{\text{eff}}n_1N_2R_2^2I_1I_2$.

(a)

$$\tau = \mu_0\mu_{eff}n_1N_2R_2^2I_1I_2 = 2.41\frac{Nm}{A^2}I_1I_2 \tag{20.139}$$

(b)

$$\tau = 2.41\,Nm \rightarrow F = \frac{2.41\,Nm}{0.5\times10^{-2}m} = 4.82\times10^2N \tag{20.140}$$

(c)

Result for part (a) is same. Result for part (b) is 100 times larger.

(d)

B field depends on n_1 and it is independent of R_1.

(e)

$$\omega = 100\,\text{rpm} = 10.47\frac{\text{rad}}{\text{s}}, \tau = 241\,\text{Nm} \rightarrow P_{out} = 25.23\,\text{W}\times100 = 2.52\times10^3\,\text{W} \tag{20.141}$$

$$P_{in} = 2 \times 10A \times 12\Omega \times 12\Omega + 10A \times 15\Omega \times 15\Omega = 5.13 \times 10^3\,\text{W} \tag{20.142}$$

Efficiency is 49.12%. Energy losses are due to friction, heat and radiation.

(f)

Outstanding power for such small mass.

(g)

$$P_{required} = 450\,\text{kg}\,9.81\,\frac{\text{m}}{\text{s}^2}0.75\,\frac{\text{m}}{\text{s}} = 3.31 \times 10^3\,\text{W} > P_{out} \tag{20.143}$$

Hence, answer is no.

Problem 11 Numerical Problem And Concept Question, Difficulty Level: *, Chapters 3 and 16

(Prepared by Marko Popovic)

Solution:

(a)

$$l_{01} = \sqrt{7.95^2 + 10.05^2}\,\text{cm} = \sqrt{164.2}\,\text{cm} \tag{20.144}$$

$$l_{02} = \sqrt{5.55^2 + x_2^2}\,\text{cm} = \sqrt{x_2^2 + 30.8}\,\text{cm} \tag{20.145}$$

$$l_{12} = \sqrt{(x_2 - 7.95)^2 + (10.05 - 5.55)^2}\,\text{cm} = \sqrt{(x_2 - 7.95)^2 + 20.25} \tag{20.146}$$

Condition for 90 degrees angle

$$164.2 + (x_2 - 7.95)^2 + 20.25 = x_2^2 + 30.8 \tag{20.147}$$

$$153.65 - 2\,x_2 7.95 + 63.20 = 0 \tag{20.148}$$

Optimal attachment point is obtained for

$$x_2 = 13.64\,\text{cm} \tag{20.149}$$

(b)

$$x_2^2 + 30.8 + (x_2 - 7.95)^2 + 20.25 = 164.2 \tag{20.150}$$

$$x_2^2 - x_2 7.95 - 24.98 = 0 \tag{20.151}$$

Hence, optimal attachment point is obtained for

$$x_2 = \frac{7.95 + \sqrt{63.2 + 99.92}}{2}\,\text{cm} = 10.36\,\text{cm} \tag{20.152}$$

(c)

Above results are located just in the central region of what is observed in biological system

$$6.1\,\text{cm} < 10.4\,\text{cm} < 13.6\,\text{cm} < 15.8\,\text{cm} \tag{20.153}$$

(d)

"Joint stiffness" is the ratio of joint torque and angular excursion from equilibrium position. In the context of antagonistic muscle pair type of equilibrium, the torques generated by two muscles will be exactly balanced in the joint equilibrium state. Small joint angular excursion away from this equilibrium means that one muscle is more elongated and the other is less elongated compared to their joint equilibrium state. Hence, for given recruitment defining "muscle stiffness" and attachment points in the context of joint anatomy (defining moment arms, etc.), one could estimate "joint stiffness." Muscles can modulate tension, stiffness, and torque via recruitment.

(e)

In an ideal no-loss mechanical model, variable stiffness does not affect muscle energetics. In the context of Hill's model to increase tension for the same state, one expects larger coefficient of shortening heat meaning more losses and less efficient movements.

Problem 12 Numerical Problem And Concept Question, Difficulty Level: **, Chapters 3 and 16

(Prepared by Marko Popovic)

Solution:

(a) *Total number of sarcomeres in human biceps brachii*

Between $1.5\times10^5\times1\times10^3\times2\times10^3\times1\times10^5$ and $4.5\times10^5\times1\times10^3\times2\times10^3\times1\times10^5$, that is, **between 3.0×10^{16} and 1.8×10^{17}** sarcomeres.

(b) *Average force per sarcomere for total 40 N biceps force output*

Total 40 N is shared by elements in parallel, hence average force per sarcomere is between $\frac{40\,N}{3\times10^{11}}$ and $\frac{40\,N}{1.8\times10^{12}}$, that is, **between 1.33×10^{-10} N and 2.22×10^{-11} N.**

(c) *Work done by each sarcomere by assuming isotonic condition and force obtained in part (b) as well as by assuming that required contraction is approximately 1 μm per sarcomere*

Work is equal to product of force and displacement hence the required work per sarcomere is **between 1.33×10^{-16} J and 2.22×10^{-17} J.**

(d) *Number of ATP (adenosine triphosphate) molecules needed per each sarcomere for this condition*

Energy released per ATP molecule is $\frac{7.3\times4.184\,\mathrm{J}}{6.022\times10^{23}}=5.08\times10^{-20}\mathrm{J}$ (in agreement with 10^{-19} J estimate in D. Bray, Cell Movements, Garland, New York, 1992, p. 6). Hence, number of ATPs needed per sarcomere for this condition is between $\frac{1.33\times10^{-16}\mathrm{J}}{5.08\times10^{-20}\mathrm{J}}$ and $\frac{2.22\times10^{-17}\mathrm{J}}{5.08\times10^{-20}\mathrm{J}}$, that is, **between 2.62×10^3 and 4.37×10^2.**

(e) *Number of ATP molecules per myosin head*

There are about 300 molecules of myosin with two heads each. Hence, the number of ATPs needed for this condition per myosin head is between $\frac{2.62\times10^3}{6\times10^2}$ and $\frac{4.37\times10^2}{6\times10^2}$, that is, **between 4.37 and 7.28×10^{-1}.**

(This estimate is based on energy considerations and it is assuming 100% efficiency (!).)

(f) *Number of ATP molecules needed for the same condition per single myosin head swivel*

There is about 50 swivel cycles. Therefore, the number of ATPs per swivel is **between 8.74×10^{-2} and 1.46×10^{-2}.**

(Again, this estimate is based on energy considerations and it is assuming 100% efficiency (!).)

(g) *What is the "efficiency" of myosin head swivel?*

Efficiency is **between 8.7% and 1.5%**. Clearly energy is lost as energy needed for other chemical processes and energy released as heat. Both numbers are reasonable (however, 8.7% is a bit likelier). Though this result should be only considered as a rough estimate of an order of magnitude given many assumptions used in this calculation.

(h) *If you expect high efficiency for this process what would be your estimate of number of muscle fibers in human biceps brachii and number of myofibrils per muscle fiber?*

Higher efficiency is obtained for smaller number of muscle fibers in human biceps brachii and smaller number of myofibrils per muscle fiber in general.

(i) Assume 4 m/s contraction speed for $F = 40\,\text{N}$. If maximum isometric tension (or load) generated in the muscle is 200 N and speed for $F = 0$ m/s is 8 m/s find coefficient of shortening heat, a, using the Hill's state equation (with $b = a\frac{v_0}{F_0}$). What is the relationship between velocity and tension near muscle's resting length? How is this relationship affected by coefficient of shortening heat? Would you expect higher efficiency with smaller coefficient of shortening heat or vice versa?

Consider Hill's equation

$$(v+b)(F+a) = b(F_0+a) \tag{20.154}$$

With $b = a\frac{v_0}{F_0}$, one obtains

$$\left(v + a\frac{v_0}{F_0}\right)(F+a) = a\frac{v_0}{F_0}(F_0+a) \tag{20.155}$$

$$a = \frac{vF}{v_0\left(1-\dfrac{F}{F_0}\right) - v} \tag{20.156}$$

Here, $v = 4\frac{\text{m}}{\text{s}}, v_0 = 8\frac{\text{m}}{\text{s}}, F = 40\,\text{N}, \text{and}\ F_0 = 200\,\text{N}$.

Hence, **coefficient of shortening heat is $a = 66.6$ N.**

Hill's equation demonstrates hyperbolic relationship between velocity and force near resting length.

Larger coefficient of shortening heat indicates more energy losses and hence smaller efficiency.

Problem 13 Concept Question, Difficulty Level: regular, Chapter 4

(Prepared by Kathleen A Lamkin-Kennard)

Solution: capacitive sensors, piezoresistive sensors, etc.

Problem 14 Concept question, difficulty level: regular, Chapter 4

(Prepared by Kathleen A Lamkin-Kennard)

Solution: mechanical vibrations act on hair cells in the cochlea.

Problem 19 Numerical Problem And Concept Question, Difficulty Level: *, Chapters 4 and 5

(Prepared by Marko Popovic)

Solution:

There are about 100 touch receptors in each fingertip (within area on the order of 1 cm^2). Hand area is on the order of 100 cm^2, hence, there is about order of 10,000 touch receptors in the human hand. The sensor ID number is typically hardware encoded and recognized within the spinal cord and brain by its anatomical location. Similar is true for actuators. That's the reason why there is a very nice mapping of human body parts within motor cortex. And that's the reason why electrical stimulations of frog's spinal cord could successfully finely tune frog response to stimuli.

Problem 20 Numerical Problem, Difficulty Level: **, Chapters 4 and 5

(Prepared by Marko Popovic)

Solution:

(A) For many systems and tasks, it is advantageous to have a nonzero threshold for sensory system as it is not necessary for system to react to very small errors. Often it is passive structure that deals with small errors and no active system response is needed. Or better accuracy than the one defined with threshold is simply not needed (could you think of any tasks that would require positioning tip of your finger with precision better that part in millimeter or 10^{-5} m?). Hence, by hardwired sensory threshold, the system performs even more efficiently both in terms of information flow that is processed by control system and also in terms of efficiency of actual physical system.

(B) In reality, $Y=Y_{0r}+a_rX+b_rX^2+c_rX^3$ but sensory output is $y=Y_0+aX+bX^2$ with $Y_{0r}=Y_0$, $a_r=1.02\,a$, $b_r=0.97\,b$, $c_r=-0.01$. Also, for $X\in(0,X_0)$, $y=Y_0$. Numerical values assumed in this problem are: $Y_0=10.0$, $a=2.57$, $b=0.89$, $X_0=1.00$, $X_1=30.0$, $X_2=100$, $\varepsilon_1=2.20\cdot10^{-2}$, $\varepsilon_2=1.34\times10^{-1}$ Assuming that there is no error in measurement of X our goal is to find a range of X and hence of Y for which relative error $\frac{\varepsilon(Y)}{Y}<2\%$. Using the brute force numerical method to find $\frac{\varepsilon(Y)}{Y}=\frac{|y-Y|}{Y}$ we find that range of X is 1.1–3.5 and range of Y is 13.91–29.32.

(C) If direct measurement of X has error of $\varepsilon(X)=\varepsilon_1$ for $X\in(X_0,X_1)$ and more complex ε-$(X)=\varepsilon_1+(X-X_1)\varepsilon_2$ for $X\in(X_1,X_2)$ then again by the brute force numerical method one could find range of x 1.1–2.9 and range of Y 13.91–24.62.

(D) With two identical sensors, their distribution multiplies and we have $ae^{-\frac{(x+l-\mu)^2}{2\sigma^2}}ae^{-\frac{(x-l-\mu)^2}{2\sigma^2}}\to ae^{-\frac{(x-\mu)^2}{2(\sigma/\sqrt{2})^2}}$ or joint measurement error is $\frac{1}{\sqrt{2}}$ of single sensor measurement error. Again brute force approach gives range of x 1.1–3.1 and range of Y 13.91–26.12.

Problem 23 Numerical Problem And Concept Question, Difficulty Level: *, Chapter 5

(Prepared by Marko Popovic)

Solution:

(a) Assume 1 m for arm length. Assume that pain sensation travels at 3–30 m/s speed. Therefore, it will take between 30 and 300 ms for sensory input. Motor signal travels much faster with 80–120 m/s. Therefore, it will take about only 8–12 ms for command signal. It also takes some time before the actual force is indeed generated in the muscle $\sim\frac{1}{2}(30\,\text{ms} \rightarrow 50\,\text{ms}) \Rightarrow 15\,\text{ms} \rightarrow 25\,\text{ms}$. Hence, $\Delta t_{\min} \approx 50$ ms for pain feedback loop.

(b) Heat sensation travels even slower with 0.2–2 m/s speed. Therefore, it will take between 500 and 2000 ms for sensory input. Hence, $\Delta t_{\min} \approx 500$ ms for pain feedback loop.

(c) Assume 1 m for leg length. Hence, similar results: $\Delta t_{\min} \approx 50$ ms for part (a) and $\Delta t_{\min} \approx 500$ ms for part (b).

(d) Assume twice the length, that is, 2 m. Hence, $\Delta t_{\min} \approx 100$ ms for part (a) and $\Delta t_{\min} \approx 1$ s for part (b).

(e) Nonmyelinated conduction speed is maximally 25 m/s hence: $\Delta t_{\min} \approx 50$–85 ms for (a), $\Delta t_{\min} \approx 500$–550 ms for (b) and (c), and: $\Delta t_{\min} \approx 1.0$–1.1 s for part (d).

Problem 38 Numerical Problem And Concept Question, Difficulty Level: *, Chapter 8

(Prepared by Kathleen A Lamkin-Kennard)

Solution:

Use Eq. (8.14), $F = \frac{\varepsilon_0 \varepsilon_r w l}{2 g_0^2} V_0^2$. To show that the force scales as l^2, substitute $V_0 = g_0 E_{\max}$ into Eq. (8.14), such that $F = \frac{\varepsilon_0 \varepsilon_r w l}{2 g_0^2}(g_0 E_{\max})^2$. Canceling terms yields $F = \frac{\varepsilon_0 \varepsilon_r {E_{\max}}^2}{2} wl$. The w and l terms have units of $[l]$, so the dimensions of F scale as l^2.

Applying the same methodology for energy yields, $E = \left(\frac{1}{2}\frac{\varepsilon_0 \varepsilon_r}{g_0} wl\right)(E_{\max} g_0)^2$. Canceling terms yield $E = \left(\frac{\varepsilon_0 \varepsilon_r E_{\max}^2}{2}\right)(wlg_0)$. The g_0, w, and l terms have units of $[l]$, so the dimensions of E scale as l^3.

Problem 39 Numerical Problem And Concept Question, Difficulty Level: *, Chapter 8

(Prepared by Kathleen A Lamkin-Kennard)

Solution:

Mass scales with l^3, torque scales with l^3, and drag force scales with l^2.

Problem 40 Numerical Problem And Concept Question, Difficulty Level: *, Chapters 8 and 17

(Prepared by Kathleen A Lamkin-Kennard)

Solution:

(a) Using trigonometry, the force (F_{Tot}) on a single leg is $F_{Tot} = 2F_{st} \sin \varphi$. The 2 occurs since there are tension forces on either side of the leg.

(b) The maximum force, $2F_{st}$, occurs at 90°.

(c) To calculate the minimum contact length, equate the maximum surface tension force and the weight of the robot. Calculate weight using $w = mg$, where w = weight, m = mass of the robot, and g is the force due to gravity. Set the expression for weight equal to the surface tension, such that $2F_{st}L = mg$. Solving for L yields the minimum contact length, such that: $L = \text{mg}/2F_{st} = (0.00001\,\text{kg})(9.8\,\text{m/s}^2)/((2)(0.07\,\text{N/m})) = 0.0007\,\text{m} = 0.7\,\text{mm}$. Note: since there are six legs, the minimum contact length per leg is the total contact length divided by six so each leg needs a contact length of 0.117 mm.

Problem 41 Concept Question, Difficulty Level: regular, Chapter 8

(Prepared by Kathleen A Lamkin-Kennard)

Solution:

At low Reynolds numbers, flows are fully reversible and time independent. The net propulsive effects from the reciprocal motions will cancel and the microswimmer will not move forward. Some bacteria utilize helical tail motions while others rely on nonreciprocal motions.

Problem 43 Concept Question, Difficulty Level: regular, Chapter 8

(Prepared by Kathleen A Lamkin-Kennard)

Solution:

Advantage: higher force output, actuation direction can determined by joint configuration; disadvantages: cell sources, require more complex joint design

Problem 44 Concept Question, Difficulty Level: regular, Chapter 8

(Prepared by Kathleen A Lamkin-Kennard)

Solution:

Propulsion, controls, sterilization, more predictable methods for controlled drug delivery

Problem 64 Concept Question, Difficulty Level: regular, Chapter 12

(Prepared by Marko Popovic)

Solution:

Theoretically

$$\cong \frac{9\,000 \times 52 \times 40 \times 3\,600}{35\,000\,000} = 1925s \tag{20.157}$$

However, in reality in practice, the best case scenario is probably about one-third of the above result.

Problem 68 Case Study, Difficulty Level: *, Chapter 12

(Prepared by Michelle Johnson)

Solution:

(A) i–iv; (B) i; (C) all except iv; (D) ii; (E) ii–vi; (F) iii and iv

Problem 69 Numerical Problem, Difficulty Level: *, Chapter 12

(Prepared by Adam Goodworth)

Solution:

(A) A FBD would show two downward forces: force 1 located at the center of mass and force 2 located at the top of the head. These forces each generate a destabilizing torque that moves the body away from upright.

Torque 1 = body mass × COM height × g × 9° * PI/180 = 85.1 N m. Since the body tilt is <10°, the sine of body sway is approximated with body sway in radians.

Torque 2 = mass of water × head to ankle height × g × 9° * PI/180 = 52.4 N m. The mass of water equals 20 L × 1 kg/L = 20 kg.

Torque 1 + *Torque 2* must be offset by a corrective torque around the ankle = 137.5 N m in the opposite direction.

(B) Torque 2 from the water is substantial, about 60% of Torque 1, thus the required corrective torque with the water on the head is about 1.6 times the torque required without the water.

c. One biomechanical limitation is the length of the foot (base of support). During a static body tilt, the vertical projection of the COM must fall within the feet. Without the water, the COM projection is COM height × 9*PI/180 = 0.13 m. This is less than most adult's feet length anterior to their ankle, but it should be noted that the CoP extend to the tips of one's toes and remain stable. Thus, a 9° tilt is on the marge of stability, depending on one's muscle strength and limits of stability.

With the water container, the new COM projection is a weighted sum of the body and the water container, equal to (mass container × projection of COM container + mass body × projection of COM body) / (mass body + mass water). This equals (20 kg × 0.267 m + 65 kg × 0.13 m)/ (70 kg + 20 kg) = 0.156 m. The COM projection now extends past many adult's base of support and therefore, a static 9 degrees tilt would not be possible for most adults with the full water container on the head.

Problem 70 Numerical Problem, Difficulty Level: *, Chapter 12

(Prepared by Adam Goodworth)

Solution:

(A)

The FBD should have J × angular body tilt acceleration = mgh × angular body tilt—Tc. Note, the torque due to gravity and the corrective torque (Tc) are in the opposite directions and the difference between these torques is equal the body tilt acceleration. Also, the "linear" transfer function implies that tilt angles are small and the sine of body tilt can be approximated as body tilt. Finally, substituting s^2 into angular body tilt acceleration, one can solve for body tilt/Tc.

(B)

$$\frac{\text{body tilt response}}{\text{surface tilt}} = \frac{Wp\,(K_p + K_d s + K_I/s)e^{-\tau s}}{Js^2 - mgh + (K_p + K_d s + K_I/s)e^{-\tau s}} \tag{20.158}$$

Note, the transfer function is frequency dependent such that body tilt response (gain and phase) vary across frequencies of surface tilts. However, the reliance on sensory feedback (Wp and Wv) impacts only the magnitude of the response. If vestibular + visual feedback is large, Wp is small, the numerator is small, and a small body tilt response would result from a surface tilt. In contrast, if the reliance on proprioception feedback is large, the numerator is large indicating a large response to surface tilt.

Problem 76 Numerical Problem, Difficulty Level: regular, Chapter 16

(Prepared by Kathleen A Lamkin-Kennard)

Solution:

Calculate the height of the jump as $h_{jump} = \frac{1}{2}\frac{v(0)^2}{a}$

On the moon, the height is $h_{moon} = \frac{1}{2}\frac{v(0)^2}{1.622}$

Solve for v(0) then substitute to get $h_{moon} = h_{earth} * \frac{9.81\ \text{m/s}^2}{1.622\ \text{m/s}^2}$

Finally, $h_{moon} = 98\ \text{cm} \times 6.05 = 592.9\ \text{cm}$.

Index

Note: Page numbers followed by *f* indicate figures, *t* indicate tables, and *b* indicate boxes.

B

C

D

P

Printed in the United States
By Bookmasters